D1591451

The SOUTHEAST *in* EARLY MAPS

THE

FRED W. MORRISON

SERIES IN

SOUTHERN STUDIES

THE SOUTHEAST *in* EARLY MAPS

WILLIAM P. CUMMING

THIRD EDITION, REVISED & ENLARGED BY LOUIS DE VORSEY, JR.

THE UNIVERSITY OF NORTH CAROLINA PRESS

CHAPEL HILL & LONDON

First edition published by Princeton University Press, 1958
Second edition published by the University of North Carolina Press, 1962

Designed by Richard Hendel
Set in Monotype Garamond and Trajan types by Eric M. Brooks
Manufactured in the United States of America
The paper in this book meets the guidelines for permanence
and durability of the Committee on Production Guidelines for
Book Longevity of the Council on Library Resources.

Library of Congress Cataloging-in-Publication Data
Cumming, William Patterson, 1900–
The Southeast in early maps / by William P. Cumming.
— 3rd ed. / rev. and enl. by Louis De Vorsey, Jr.
p. cm.
Includes bibliographical references and index.
ISBN 0-8078-2371-6 (cloth : alk. paper)
1. Cartography—Southern States.
2. Southern States—Maps.
I. De Vorsey, Louis.
II. Title.
GA405.C8 1998
912.75—dc21
97-46332
CIP

02 01 00 99 98 5 4 3 2 1

TO MY WIFE

Contents

Preface to the Third Edition

In his advance commentary on the manuscript of "The Southeast in Early Maps" Lawrence C. Wroth, at the time the doyen of historical bibliographers, wrote that it was:

> a significant contribution and a definitive one. The historian of this general area, or of any considerable division of it, must hereafter be familiar with this book if he is to interpret properly original plans of colonization and later territorial development. Its value to the historian of geographical development and the historian of cartography is great and is unique in character.

Book reviewers proved him eminently correct, employing in their printed critiques terms such as "monumental," "indispensable," "infinitely useful," "a basic reference work," "a major contribution," "a primary reference tool," and "written in a notably lucid and urbane prose" to describe the first edition of *The Southeast in Early Maps* when it was issued by the Princeton University Press some forty years ago. Needless to say, William Patterson Cumming's richly illustrated opus was quickly snapped up by cartophiles, map curators, librarians, and antiquarian map dealers around the world. "Cumming number __," "as described in Cumming," or, to add cachet to a particular map variant, "not in Cumming" soon became familiar phrases in museum exhibit catalogs and dealers' sales brochures alike.

Can anyone be surprised that such a well-heralded and useful book was out of print within a year of its publication? Or that then-director of the University of North Carolina Press Lambert Davis was eager to have his press replace Princeton as its publisher and bring out a corrected second edition in 1962? Fortunately for me, this second edition appeared at the affordable price of $12.50 while I was immersed in my Ph.D. research with the early maps and documents in the British Museum, Public Record Office, and Royal Geographic Society. What a godsend *The Southeast in Early Maps* proved to be as I worked in those venerable and ofttimes intimidating repositories! When I left London to take a teaching position at East Carolina College (now East Carolina University) in Greenville, North Carolina, high on my list of things to do was meet the author of this invaluable book.

My heavy teaching load and continuing field and archival research, coupled with the fact that the Old North State is mighty broad between Greenville and Davidson—where Professor Cumming taught at Davidson College—forestalled that meeting until June, when I headed west, with my wife, Rosalyn, and our new baby, to teach a summer session at the University of Victoria in British Columbia. In our self-modified camping van we drove through the rain to a campground near Davidson. The next morning, after making ourselves presentable, we found the Cumming home on a tree-shaded street just off the Davidson campus. Bill and his wife, Betty, were warm and gracious in their welcome of three rain-spattered strangers. I can't recall what early map questions Bill clarified for me that rainy spring morning, but I treasure the inscription he penned on the title page of my already much thumbed copy of the UNC Press edition of *The Southeast in Early Maps*. What my wife and I do remember vividly is the cultured and easy informality of the hospitality we enjoyed in the Cumming home. They did not know it, nor did we at the time, but Bill and Betty were role models for us as we entered the lifestyle of the world of scholarship and teaching that they epitomized.

Bill and I began a continuous professional correspondence that ended only with his death in 1989. After joining the geography department at the University of North Carolina in Chapel Hill, I had an opportunity to work more closely with Bill in connection with the annual meeting of the Southeastern Division of the Association of American Geographers which was to be hosted by the department. A prominent place on the program of that meeting was regularly reserved for a session of papers dedicated to research on the Southeast. Bill Cumming was the first of a trio of early map scholar-experts who agreed to take part in such a session. Once I had Bill's agreement, it was not hard to persuade the chief of the Library of Congress Geography and Map Division, Walter Ristow, and Herman Friis, founder of the cartographic division of the National Archives, to join us. The positive reception received by that session on the early cartography of the Southeast encouraged the editor of the division's journal, *The Southeastern Geographer*, to invite us to revise our papers as manuscripts to form the first single-theme issue in the journal's history. That issue of *The Southeastern Geographer* appeared in 1966 and featured Cumming's essay, "Mapping of the Southeast: The First Two Centuries," Ristow's "State Maps of the Southeast to 1833," Friis's

"Highlights of the Geographical and Cartographical Activities of the Federal Government in the Southeastern United States, 1776–1865," and my own "The Colonial Southeast on 'An Accurate General Map.'"

On a personal level our families enjoyed sporadic contact at conferences and scholarly meetings and when Bill and Betty would stay with us en route to Florida after I had joined the faculty at the University of Georgia in Athens. They became "honorary grandparents," as Bill put it, to our three children, thanks to a wonderfully relaxed week the Cumming and De Vorsey families spent on the beach at Hilton Head Island as guests of the island's pioneer developer and map collector, Charles Fraser. In the morning Bill and I would study Charles's rare maps. The rest of a long day was typically spent swimming in the surf, exploring alligator pools, or sailing on Fraser's yacht, *Compass Rose*. One of our most prized home movie reels shows Bill Cumming an intrepid skipper at the wheel of a steeply heeled *Compass Rose* skimming the waters of Dawfuskee Sound as a large gallery of very young children clung to the rail with feet in the wash. In an effort to show our appreciation, Bill and I collaborated on a review report and set of recommendations to guide Charles Fraser in the maintenance and future development of his impressive map collection—Bill as master cartobibliographer and tutor and I as eager apprentice.

Our most fascinating interaction came in the 1980s when Bill and I were called to serve as expert witnesses by opposing attorneys general in an original action tried before the Supreme Court of the United States—Bill to testify for South Carolina and I for the state of Georgia. At issue was the location of the boundary between the states in the lower Savannah River. Hundreds of early maps were entered as evidence and our expert opinions, along with that of world-renowned cartographer Arthur Robinson, who joined Bill in the South Carolina ranks, were elicited and meticulously examined by the lawyers of both states. Evenings following long days of courtroom testimony and lawerly argument were made enjoyable when Bill, Betty, and I would find a quiet corner table and have a pleasant meal and conversation away from the all-too-often intensely trial-occupied attorneys. I should note that convenience to the courthouse and bureaucratic budget constraints resulted in the litigation teams for each state choosing to stay in the same hotel. While there was no written rule prohibiting our conversations, we couldn't help but feel slightly conspiratorial during these dinner table chats.

When well into his eighties, Bill began to experience a decline in physical strength. In 1988, when the Association of American Geographers invited him to their annual meeting to accept their Honor Award for his scholarly contributions, his doctor advised against undertaking the long trip to the meeting site in Phoenix. At Bill's request, I stood in for him and accepted the honor on his behalf at the meeting. My wife and I were delighted to carry the award to Davidson, where it was presented to Bill by the president of Davidson College at a special convocation.

While his extraordinary intellectual powers remained, Bill's physical health continued to decline, and, in the early summer of 1989, I felt some unease when Betty phoned to say that Bill would like me to come to Davidson. I lost no time in booking a flight to Charlotte, where Bill and Betty's son, Robert, met me. On the drive to Davidson Robert explained that his father's condition was terminal and that he wished to discuss his ongoing work on a third edition of *The Southeast in Early Maps*. I knew that for several years Bill had been urging the University of North Carolina Press to undertake the publication of a new edition, and from time to time I would send him suggestions that grew out of my use of the early maps of the region. The soaring price of the out-of-print original editions, coupled with frequent inquiries from collectors, scholars, and dealers, had ultimately convinced the Press of the desirability of a new edition, and Bill had begun working with them toward that end. Seen in retrospect, Bill's request that I take over the work that he was no longer able to accomplish on the new edition may not seem surprising. At that moment, however, I was overwhelmed! This book, more than any other scholarly effort, was William P. Cumming's lifework, and—as the reviewers quoted above make clear—it was a monumental accomplishment. I masked my trepidation and told Bill I would see his work clear to the publication of the third edition of *The Southeast in Early Maps*, and we discussed how that would be best accomplished through what was for him a very tiring afternoon in July 1989. We were in agreement that whatever changes might be made, every care possible would be taken to maintain the viability and usefulness of the two earlier editions. This book is the product of my effort to carry out the trust William P. Cumming placed in me not long before he died.

The reader familiar with the earlier editions will immediately note the fact that the text of the book has been typeset rather than produced directly from typewriter copy, as it was some forty years ago. The second most obvious change from the earlier editions is the inclusion of a gallery of color reproductions of maps. All the original maps from which these color photographs were made are in the collection of early maps, documents, and publications that Professor Cumming presented to Davidson College and now make up the Cumming Collection housed in the E. H. Little Library of that institution. The late Dr. Helen Wells, a world-recognized expert in the history of cartography, termed the Cumming Collection "the finest collection of American maps of the Southeast in private hands." She emphasized further that "the donation of this collection to the E. H. Little Library has made Davidson College one of the rare-map centers of North America." The UNC Press and I are grateful for Davidson College's cooperation in permitting the photographing of twenty-four of those maps for use in this edition.

Readers including myself found some aspects of the organization of the earlier editions confusing. After a great deal of consultation and analysis, I decided to make a few structural changes to remedy them. In the earlier editions, at pages 93–102, there is a section, headed "Reproductions of Maps," comprised of a short description and discussion of each of the sixty-seven map reproductions that follow. Potential for confusion arose from the fact that not all of the maps reproduced were included in the cartobibliographically most important part of the book—the section titled "List of Maps." To eliminate this potential in the present edition, descriptive text for all of the reproduced maps is included in the List of Maps. The reader is directed to these discussions by the Map List numbers found at the end of each reproduction plate caption. With all the reproduced maps included in the List of Maps, the Reproductions of Maps section is unnecessary and has been removed.

Thanks to the cooperation I received from the University of North Carolina Press, a number of additional black-and-white map reproductions appear in this edition. In some cases they are photographs of maps listed in earlier editions, while in other cases they reflect maps that were not known to Cumming when he compiled his original works in the late 1950s and early 1960s. In an effort to enhance the value of this edition as a true research tool, I have reproduced some of the Southeast's most crucial maps at a much more legible scale by presenting them in four quadrants. As these additional photographs are introduced, they are clearly indicated through the use of letters (A, B, C, etc.) appended to the numbers employed in earlier editions (e.g., 40A). This is done in an effort to ensure that curators, dealers, and collectors will remain able to use the Cumming map numbers as they have in the past with a minimum of confusion. A similar system is employed in the List of Maps, where new maps have been included chronologically and indicated by letter as in the case of new reproductions.

In some cases additional research has resulted in the change of the date for a particular map. Where this has occurred, no change has been made in the map's number or placement in the List of Maps, but the corrected data is given in the entry for the map within the list. Thus, it should be kept in mind that the organization of the List of Maps is chronological in most, but not all, instances. Referring to the Chronological Title List of Maps, following the List of Maps, provides a convenient way for readers to check the correct date of a particular map. Professor Cumming was particularly concerned that we not radically depart from his original scheme of map number identification, which had proven so useful since 1958.

Two of the appendixes found in the earlier editions—Appendix B: Indian Tribes and Settlements, and Appendix C: Political Divisions, Boundary Lines, and Roads—proved virtually unused by all readers surveyed, so it was decided to omit them in the present book. Professor Cumming had been careful to point out that "the list of maps given in Appendices B and C is not exhaustive; the special investigator should examine other maps in his period." We feel certain, therefore, that he would have approved of their omission in this corrected and expanded volume. The works listed in Appendix D: Bibliographies Which List Maps of the Southeast Made after 1775, have been included in the selected bibliography at the beginning of the List of Maps, so this appendix has also been omitted here. The fourth appendix from previous editions (Appendix A: Chief Type Maps in the Cartography of Southeastern North America) remains but has been moved to the end of Cumming's introductory essay.

I have added an original essay titled "American Indians and the Early Mapping of the Southeast." It grew from a research theme I began while working for my Ph.D. degree almost four decades ago and continue to find fascinating. While I never had the opportunity of discussing such an addition with Dr. Cumming, I am confident that he would welcome it as a complement to his seminal introductory essay, "The Early Maps of Southeastern North America."

The index has been expanded to include topics in my new essay as well as Cumming's original introductory essay. Additional information and corrections found in the supplement to the second edition have been incorporated as appropriate in the text of this edition. Finally, credits for the map photographs have been collected in a single list, which appears at the end of the book.

Literally hundreds of scholars, editors, librarians, archivists, curators, map collectors, dealers, and friends have assisted in the preparation of the revised and enlarged edition of *The Southeast in Early Maps*, and I am deeply in their debt. Their help, suggestions, and support have been profoundly appreciated. My regret is that I cannot thank them all personally on this page. In 1958, Professor Cumming dedicated the original edition of this book to his wife. I, in turn, dedicate this revised edition to my wife, Rosalyn.

Louis De Vorsey, Jr.
Cambridge, New Brunswick
August 29, 1997

Preface to the First Edition

This is a study of the historical cartography of the southeastern region of the North American continent before the American Revolution. It attempts to analyze the manuscript and printed maps of that area, showing the expansion of geographical knowledge through the periods of discovery and colonization, and at times relates these maps to other primary documents of the period. Through these cartographical records one can trace the origin and development of the fascinating misconceptions of the continent in the minds of the early explorers; they show vividly the expansion of the frontier and the shifting location of Indian tribes; they throw light on the complex history of the imperialistic struggles of France, Spain, and England during the period; and they delineate—often erroneously—shifting political divisions and boundary surveys. The very misconceptions and errors concerning the New World, graphically portrayed on these maps, are as important to the student of history in understanding the deeds and actions of our ancestors as are their correction and improvement on later maps.

A cartographical study of this kind requires the setting of limitations, often arbitrary, to prevent unwieldy growth. The List of Maps attempts to make an exhaustive check of all regional maps of the Southeast and of local maps south of Virginia and north of the Florida peninsula. The inclusion of all local maps of Florida and Virginia would have more than doubled the size of the volume and greatly increased the ancillary cartographical problems; their inclusion would also have largely duplicated other map lists of those regions already made or now in progress. The List of Maps does not attempt to include world maps or maps of the North American continent (with a few notable exceptions), although the Introductory Essay does contain a survey of sixteenth-century world maps important in their delineation of the Southeast.

The data for this study has been gathered during the academic vacations of over twenty years; the maps analyzed and tabulated have been examined in libraries scattered from Massachusetts to California and in two European countries. The correlation of maps and of the information on them, collected over such an expanse of time and space, will undoubtedly result in errors of omission and commission, no matter how carefully checked.

Two persons to whom I owe a special debt of gratitude are Dr. Lawrence C. Wroth of the John Carter Brown Library and Mrs. Clara Egli Le Gear, Bibliographer, Division of Maps of the Library of Congress. From my first visit there in 1935 I have never entered the John Carter Brown Library without anticipation nor left it without profit, both from its rich resources in early Americana and from the equally rich knowledge and scholarship of its librarian. I have repeatedly called upon Mrs. Le Gear's unexcelled knowledge of American cartography and her excellent judgment in bibliographical matters. To Dr. Wroth and Mrs. Le Gear are due many of the good qualities of this work but none of its faults.

I acknowledge a special indebtedness also, reaching back to the mid-thirties, to Professor Louis C. Karpinski, then professor at the University of Michigan, and to Mr. A. S. Salley, then Secretary of the South Carolina Historical Commission, who encouraged and guided my beginnings in this study.

Among the many who have helped me with their particular knowledge I want to express my especial appreciation to the following: Dr. John R. Swanton, formerly of the Bureau of American Ethnology, Smithsonian Institution; Dr. C. C. Crittenden and Mr. David L. Corbitt of the North Carolina Department of Archives and History; Mr. Charles E. Rush, former Librarian of the University of North Carolina; Miss Mary Thornton of the North Carolina Room, Library of the University of North Carolina; Mr. Colton Storm, formerly of the William L. Clements Library; Mr. Coolie Verner of Florida State University; Dr. Carl H. Mapes, Chief Geographer, Rand McNally and Company; Mr. J. H. Easterby, Director of the South Carolina Archives Department; Mr. Lloyd A. Brown, Director of the Chicago Historical Society; Dr. Walter W. Ristow, Assistant Chief, Division of Maps, Library of Congress; Mr. Lyle H. Wright, Head of the Reference Department, Henry E. Huntington Library; and Mr. R. A. Skelton, Map Room, British Museum. I am also indebted to the late S. Whittemore Boggs, Geographer of the Department of State.

I have acknowledged under the List of Reproductions the cooperation given by many individuals and libraries, here and abroad, in permitting reproduction of their maps. Especially I wish to thank Mr. Alexander O. Vietor, Curator of Maps of the Yale University Library, for his helpfulness in making available, for direct collotype reproduction, maps from the rich resources of the Yale Map Collection. Dr. Arch C. Gerlach, Chief, Map Division of the Library of Congress, the John Carter Brown Library, and Mr. Thomas W. Streeter of Morristown, N.J., have also been particularly generous in their cooperation.

Mr. Henry P. Kendall of Boston made available to me for several days of study his magnificent collection of Caroliniana and has shown a continual interest in this work. He supervised personally the photography of the Moseley map of North Carolina, of which he possesses the only known undamaged copy. He has thus made available in this volume the first accurate reproduction of this historically important map.

The reproductions reflect the skill of the Meriden Gravure Company and of its manager, Mr. Harold Hugo. Mr. Hugo's knowledgeable interest in cartographical subjects and his unusual care in obtaining, whenever possible, direct negatives from the originals for reproduction have resulted in the exceptional clarity of the plates used for this volume.

I am indebted to Mr. Charles Lloyd and Mr. E. C. Cumming for reading the manuscript. I thank Mr. J. B. Black, Mrs. Tom Daggy, Mr. and Mrs. Dyer McCrory, and Mrs. James Johnson for their aid in typing at various stages of the manuscript.

The editors of the *American Historical Review*, the *Journal of Southern History*, and the *North Carolina Historical Review* have kindly given permission for the use of extracts from articles published in those journals.

I owe a special statement of thanks to the different foundations that have given me grants-in-aid for research at various times. The Social Science Research Council, the Carnegie Foundation, and the Southern Fellowships Fund have made possible the investigations at many different libraries. The administration of Davidson College has supported this study in many ways, especially by summer grant funds and by a half-year Sabbatical leave.

I appreciate particularly the generous gift toward publication of this volume of Mr. Andrew McNally III, President of Rand McNally and Company, the publisher of many fine books on maps.

My warm personal thanks are due to Mr. Archibald Craige of Winston-Salem, N.C., for his gift, which has enabled the Princeton University Press to include sixty-four reproductions of maps instead of the thirty-two originally planned, thus greatly increasing the value of this work to everyone who will use it.

William P. Cumming
Davidson, North Carolina
June 3, 1957

The SOUTHEAST *in* EARLY MAPS

The Early Maps of Southeastern North America

AN INTRODUCTORY ESSAY

The mapping of the Southeast before the American Revolution may be divided into three periods, primary, transitional or descriptive, and modern.[1] Within each of these periods there is a generally progressive improvement of the methods used and an increasing accuracy of detail. Within the early periods type maps or mother maps appeared which incorporated the results of new discoveries and explorations, followed by derivative maps published by imitative mapmakers. These derivative maps often degenerated in accuracy and detail; the influence of a mother map, however, might linger for a hundred years after new and better maps had corrected its errors. Thus one finds maps based upon those of an earlier type or period being made long after discoveries or settlements of a later generation were made known.

The first or primary period lasted from the earliest maps which showed the coastline of the continent until the final decade of the sixteenth century. The maps of this period were based upon the reports and charts of early official expeditions, supplemented by information received from individual pilots or adventurers. Few original charts have survived from this period; the extant maps usually are the production of closet geographers and mapmakers in Europe who attempted to synthesize the incomplete, vague, often erroneous and contradictory information which they had gathered. The data supplied were based on crude observations of latitude, dead-reckoning longitude, and eye-sketches and lists of rivers, bays, and headlands sighted; the resultant maps, artistically designed, filled in with pictures of curious animals and geographical extravaganza, are often confused, contradictory, and inaccurate, valuable as they are for their evidence of increasing geographical knowledge. To attempt an interpretation of these maps and the identification of places and names by laying them alongside modern accurate maps is a fundamentally fallacious procedure. They have to be interpreted in the light of the limitations under which they were made, of the reports of the expeditions upon which a particular map is based when those are available, and of the cartographical conventions of the age. Even then, their value usually lies more in evaluation of the implied expansion of geographical knowledge than in the identification of specific locations.

Two maps of this period need special comment because of their importance in the cartographical evolution of the region. Mercator's world map of 1569 is the first map to show the Appalachians as a continuous mountain range stretching parallel to the coast in a southwest-northeasterly direction (Map 3A; Plate 7). The representation of the Appalachians remained one of the most confusing problems for geographers of this region for centuries, particularly in its southern reaches, where the general contour breaks up into smaller ranges extending in various directions. Mercator's delineation of the southern mountain region by a roughly inverted Y is in general not improved until the early part of the eighteenth century, in the so-called Barnwell map of ca. 1721, which shows the information gained by the Indian traders (Maps 184 and 184A; Plates 48, 48A–D). It was not until Guyot explored the southern Appalachians in the middle of the nineteenth century that an adequate survey of the region was shown on any map.[2] Ortelius's map of Florida (1584), which was based upon information derived from the Spanish royal cartographer Chaves, included the territory south of the present latitude of Virginia and west to New Mexico (Map 5; Plate 9). This map is not as accurate as Mercator's and its nomenclature is based upon early Spanish explorers; but it is the first printed regional map and remained the basis for the charts of many continental mapmakers for over a century. It is a mother map of the first importance, for its general geographical outline is found in many maps, in which the details were revised and corrected upon occasion as additions to geographical knowledge were acquired, until the beginning of the eighteenth century.

The transitional or descriptive period extends from the end of the sixteenth to the middle of the eighteenth century. Maps of this period were based upon actual though crude surveys; the details are much fuller and can usually be identified on modern maps; the place-name nomenclature is that of the early colonizers, Indians, or settlers; but the delineation is informative rather than accurate and becomes more distorted as the area shown extends away from the settlements and depends upon the impressions of explorers and Indian traders. The originals of the printed type maps of this period are frequently still preserved. White's map of Virginia (1590), which shows the North Carolina coastal region, is the first printed map of this degree of detail and accuracy for any part of the present area of the United States (Map 12; Plate 14). The possible original of this map, though not the immediate copy from which the engraver made his plate, is White's manuscript map of the same region, preserved in the British Library and probably made during White's first trip to Virginia in 1585 (Map 8; Plate 12). Le Moyne's map of Florida (1591) extends the coastline from North Carolina southward to the Cape of Florida (Map 14; Plate 15). Le Moyne's map is fairly accurate for the country around the French settlement on the banks of St. Johns River in 1564, where Le Moyne obtained the information for his map and possibly drew the original; for the interior and along the South Carolina coast the information is distorted and gave rise to numerous subsequent cartographical misconceptions. These two maps, both published by De Bry in his *Grands Voyages*, give the best coastal delineation of any like extent of the North American continent for nearly a generation. They were followed by important expansion of information in such maps as Ogilby's "A New Discription of Carolina" (ca. 1672), Gascoyne's "A New Map of the Country of Carolina" (1682), Thornton-Morden-Lea's "New Map of . . . South Carolina" (ca. 1695), Crisp's "A Compleat Description of the Province of Carolina" (1711), and the unsigned manuscript chart ascribed to Barnwell (ca. 1721) (Maps 70, 92, 118, 151, and 184). Barnwell's large manuscript map (Public Record Office, Colonial Office, North American Colonies. General. 7) was never printed but is a great advance over preceding maps in its general conception of the interior geography of the Southern District, and was used in the making of such later maps of North America as Popple's (1733) and Mitchell's (1755) (Maps 184, 217, and 293; Plates 48 [and 48A–D], 55, and 59).

The third or modern period begins with the first published work of De Brahm in 1757 (Map 309). The maps of this period are the product of professional surveyors using refined instruments and methods. They form a striking contrast in accuracy of detail to the maps which preceded them. De Brahm, who became His Majesty's Surveyor General of the Southern District of North America, followed his "Map of South Carolina and a Part of Georgia" (1757) with numerous maps for nearly thirty years, extending his surveys to the interior and southward to "The Ancient Tegesta, now Promentory of East Florida" (1772). Working with De Brahm and later independently was Bernard Romans, whose "Chart of the Coast of East and West Florida" (1774) is one of the finest and rarest cartographical items of that region.[3] Des Barres made use of the work of De Brahm and others in the admiralty surveys for the *Atlantic Neptune* (ca. 1774–81). The work of these men was added to but not surpassed until the Coast and Geodetic Survey maps and the topographical quadrangles of the United States Geological Survey began to appear.[4]

PRIMARY CARTOGRAPHY: THE DISCOVERY

From modern maps it is a far cry to the cartography of the years immediately following the discoveries of Columbus and his successors. The confusion which existed in the minds of the early explorers as to the identity, size, and shape of the newly discovered lands in the New World is reflected in the variety of names given by the mapmakers to the little-known territory to the north of the West Indies. Honest confessions of ignorance are shown in such legends as "Ulterius Terra Incognita,"[5] confusion of the northern continent with Asia or with Cuba,[6] and strange names such as "Zoanamela."[7] In October 1511 King Ferdinand granted permission to Juan de Agramonte, a native of Lérida in Catalonia, to explore and settle a "tierra nueva" reputedly rich in gold "according to two Indians he brought with him." Agramonte's projected voyage apparently never materialized. Is it possible that this "new land" may have some relation to the fabled isle of Bimini?[8] Bimini was the great land of which the Lucayan Indians of the Bahamas told the Spaniards; Peter Martyr had heard of it in Spain by 1511, when he wrote of the fountain of youth on an island north of Hispaniola and showed the island on an accompanying map. In 1512 Ponce de León was given license to discover and settle the island, with the title of adelantado, or governor.

Florida, however, was the land that Ponce de León discovered; landfall was on March 27, 1513, Easter Sunday (Pascua Florida).[9] Though he knew that the coast he named Florida was not Bimini, he apparently died without knowing it was not an island. A few early mapmakers also show it as an island, though the first known dated map (ca. 1520) to use the new name, that of Pineda,[10] was drawn to show the explorations of Francisco de Garay along the Gulf of Mexico, explorations which proved that Florida was not an island. The name was soon applied to the whole southeastern region of the continent, and though in some maps it is found only in the peninsula, especially when the names of new colonies and settlements are given to the north, Florida remained the common and accepted name for the whole area for 200 years, whatever other political subdivisions arose, both by European cartographers and American colonists. Thus Governor Archdale of Carolina, one of the lords proprietors of the province, states in his *New Description of that Fertile and Pleasant Province of Carolina* (London, 1707) that Carolina is only the northern part of Florida. "Florida," he writes, "which begins at Cape Florida, in the Latitude of about 25 . . . runs North East to 36½, and is indeed the very Center of the habitable part of the Northern Hemisphere."[11] Since the peninsula itself was left without a specific name, some writers and mapmakers of the seventeenth century, beginning in about 1630 with De Laet's map of "Florida et Regiones Vicinae," called it Tegesta after the name of a tribe of Indians living along the southwestern coast of the peninsula who played a part in early French and Spanish settlements (Map 34; Plate 24). "Tegesta" is found on the Spanish, French, German, Dutch, English, and American maps of the eighteenth century. But with the permanent growth of the English colonies along the coast, "Florida" was more and more restricted to the territory occupied by the Spanish, and "Tegesta" fell into disuse.[12]

The landing of Ponce de León on the coast of Florida is the first European arrival on the southeastern shores of the North American continent of which there is an authenticated account. Examinations of the earliest charts, globes, and mappa mundi including the New World show that there were discoveries made along the coast before that time, even if there were not other contemporary written references to substantiate them. Westward voyages of exploration by Spaniards, authorized by Ferdinand and Isabella to explore the coasts of the New World, were being made in the first decade after Columbus's discovery. No individual record of any of these voyages remains, but the results must have become known to contemporaries. In addition, there were surreptitious navigations. The sight of riches and Indian slaves and the reports of unlimited wealth lured adventurers from other countries upon explorations not licensed by the Spanish throne. Even during the regular journeys of commerce, accidental landfalls of vessels which had been blown off their course by a south wind while sailing along the coast of Cuba may have afforded the first sight of the continent.[13] Certainly by 1500 the Juan de la Cosa map shows a long, vaguely formed line to the northwest of the Island of Cuba.[14] It is the earliest map to delineate any part of the North American coastline, and its unbroken shore is curiously suggestive of the actual northeast-southwest trend of the land. To the north the elongated east-west line of the south coast of Newfoundland, with numerous names, reports without question the discoveries of John Cabot. It has no nomenclature for the southern part, and the indentations on the coast are unidentifiable. This does not mean that it is entirely imaginary; although the journal of Cabot's voyage has not come down to us, it is possible that Cabot went as far south as Carolina, or even Florida.[15] Cartographically, this problem is not important; no names appear on the coast south of Cape Breton Island, and any information that Cosa received from England about Cabot's voyage south of Nova Scotia was evidently general. Apparently both Cabot and Cosa thought that the coast explored was part of Asia. Any claim that Cosa's map delineates the southeast coast of North America as the result of actual exploration should be made with the gravest reservation.

The first map which shows a coastline with a Florida-like peninsula at the south is the Cantino map, probably made toward the end of 1502. The coast extends northward for approximately twenty-five degrees and then ends, an honest confession on the part of the cartographer that he knew not what was beyond. The definite configuration of this continental coast to the northwest of Cuba indicates that it is the report of some special voyage. An even earlier explorer than Cabot along this coast may have been Amerigo Vespucci.[16] In the account of his alleged first voyage of 1497–98 he mentions skirting 870 leagues of coast; on the assumption that his landfall occurred near the Gulf of Honduras, Florida and the Carolinas would come well within the distance of his discoveries.[17] Many authorities, however, deny that Vespucci made a voyage in 1497–98 or that he shared in the first discovery of the American continent.

The Cantino map gives a list of names along the coast which is found, with variations, misspellings, and additions, on a large number of maps made by the central European geographers for the next twenty years. Definite similarities of delineation and nomenclature put these maps together; the most famous of the group are the two great maps of Waldseemüller, the Cosmographia of 1507 and the Carta Marina of 1516. The frequency with which the Cantino type was copied is probably due to the Waldseemüller map of 1507, which was the first to use the name America and the first to show an unbroken coastline for the two American continents (Map 01; Plate 1). The only known copy of Waldseemüller's great map, printed from wood blocks and accompanied by a work entitled *Cosmographiae Introductio*, published at Saint Dié in the same year, was rediscovered by Professor Fischer in Wolfegg Castle, Württemberg, in 1901.[18] Waldseemüller states on the "Cosmographia" (1507) that the discoveries made by Vespucci are shown, and in the *Cosmographiae Introductio* he gives an account of the four voyages of Amerigo Vespucci and suggests that the newly discovered continent might appropriately be named after him.

An examination of the extant maps during the first quarter of the sixteenth century shows that the influence of the Cantino-type map and nomenclature remained dominant along the south Atlantic coast. Ponce de León's name, Florida, was being accepted toward the end of the period, together with an improved conception of the shape of the peninsula; but his explorations along the east coast did not extend northward beyond the peninsula and furnished little additional nomenclature for cartographers to use. By 1525 other explorations farther to the north were being made and reported; these later discoveries superseded the vaguer early geography and incorporated what was known of de León's discoveries with them.[19]

Ayllón is the first name given specifically by Europeans to what is now the Carolina coast.[20] Lucas Vásquez de Ayllón, a wealthy auditor of the island of St. Domingo, was a man of education and nobility of character, a licentiate in law (hence the name "Terra del Licenciado Ayllón"). He prepared an expedition to search out new lands which he put under Gordillo late in 1520; on June 30, 1521, Gordillo, together with Pedro de Quexós, a kinsman and leader of another caravel he had met on the way, made land at the mouth of a large river, at 33°30' N.L., which they named St. John the Baptist,[21] and formally took possession of the country. In 1525 Ayllón, who had obtained a royal cedula for settling the new land, sent Pedro de Quexós on a further voyage of exploration which covered 250 leagues of the coast. It was not until June 1526 that Ayllón, with a company of 600 persons in three large vessels, set sail and landed at the mouth of a large river, which they called the River Jordan after the name of one of Ayllón's captains. They estimated that it was in north latitude 33°40'.[22] Ayllón soon moved his settlement "40 or 45 leagues" farther down the coast and there died on October 18, 1526.[23] The settlement was abandoned, but the name of Ayllón was given to this part of the continent on maps for the next fifty years.

In connection with Ayllón's explorations a name appeared which was occasionally applied to the part of the coast between Charleston Harbor and Cape Fear River for the next 200 years. This was the land of Chicora, a name which the Spaniards accepted from the report of Francisco of Chicora, an Indian captured by Pedro de Quexós on his first expedition to the Rio de St. Juan Bautista. Francisco escaped from the Spaniards when Ayllón landed at the River Jordan in 1526, but the name Chicora, which Francisco's tales had made famous not only among Ayllón's followers but even in Spain, was long used and referred to by historians and on charts.[24]

The first known map to record the explorations of Ayllón and to establish a new type of nomenclature for the southeast coast is a holograph mappemonde by Juan Vespucci, nephew of Amerigo Vespucci, made in 1526, and now in the possession of the Hispanic Society of America (see Plate 2).[25] This map is apparently a draft or close copy of the official Spanish chart kept in the Hydrographic Department of the Foreign Office, the Casa de Contratación, in Seville. On this chart, known as the padrón real or (after 1526) the padrón general, were entered the corrections and new discoveries reported upon oath by pilots on their return to Spain. Juan was appointed pilot of the Casa de Contratación in May 1512; in 1515 he was a member of the junta that was brought together to improve and pass upon the existing charts; and he was a member of the Badajoz Commission of 1524. In the year that he made the map, he was appointed, with Miguel Garcia, to examine pilots in the place of the pilot major, Sebastian Cabot, who was then leading an expedition in Brazil. The map shows great care in giving only what Juan Vespucci felt was sufficiently certain to justify inclusion; while the Newfoundland coast is given, he omits the New England coastline entirely and starts again when he has reached the "trã nueua de ayllon," which he marks with a Spanish flag.

Besides "the new land of Ayllón," ten names are given along the south Atlantic coast. Since they do not include the "R. S. Juan" (Rio de San Juan Bautista), where Gordillo and Quexós landed in 1521, nor "aguada," which is usually found to the south on maps having "R. S. Juan," this map does not apparently report Ayllón's first expedition under Gordillo in 1521. Vespucci could hardly have included information derived from Ayllón's own attempt at settlement in 1526, for after Ayllón's death on October 18, 1526, the colonists endured part of the winter of 1526–27 before returning to their port of embarkation in San Domingo. Vespucci's map may therefore be taken to show the information gained by Pedro de Quexós on his voyage of exploration along the coast in 1525 and reported to the office of the pilot major in Seville.[26]

Two world charts appeared in 1529, designed and drawn by Diego Ribero, each of which has an extended legend concerning the Ayllón settlement and the Ayllón nomenclature along the coast.[27] One of these maps is in Weimar and the other in the Vatican (Map 04; Plate 4). Both maps have

"Tiera de Ayllon," followed by an account of the attempted settlement. The Vatican copy is larger and gives more detail. Where the Weimar map states that Ayllón discovered the country, went back to settle it, found it well equipped to furnish bread, wine, and all the things of France, and died there of a malady, the Vatican map states that Ayllón left from St. Domingo, or Puerto de Plata, with his men to settle the country, that they took few provisions, the Indians fled from fear, many died from hunger, they gave up the settlement, and returned to Hispaniola.[28] The Ayllón-type map had a wide and rapid effect upon the cartography of the southeast coast. The nomenclature of the southeast coast is basically that of the Spanish cartographers for the rest of the century, though there are reversions to names found on earlier maps. Indeed, it would be hard to find many maps with exactly the same nomenclature, though by 1584, the year in which the Chaves-type map of Florida was published in Ortelius's *Theatrum Orbis Terrarum* (Antwerp, 1584), the names along the coast were fairly well established (Map 5; Plate 9). Thereafter the Spanish-type maps often copied Ortelius, though in the seventeenth century the great Dutch mapmakers used the French and English coastal names.

A voyage which contributed to a new and fascinating though erroneous conception of the geography of the New World was made in 1524. So far, only Spanish and possibly Portuguese navigators and explorers contributed to the cartographical knowledge of the southeast coast, though possibly from the time of Cabot on unknown voyages of discovery had been made. The first non-Spanish navigator known certainly to have explored the Carolina coast is Giovanni da Verrazano, an Italian, who early in 1524 was sent by Francis I, king of France, to search for a passage to Asia. Verrazano sailed westward from Madeira until he sighted land in early March at about 34° N.L., probably near Cape Fear. After sailing southward along the coast for about 150 miles without finding satisfactory harborage, he turned back north, went ashore near his first landfall, then followed the shoreline eastward past "The Forest of Laurels" and "The Field of Cedars" to "Cape Annunciata," apparently Cape Lookout. From there on he "found an isthmus a mile in width and 200 long." Thus he describes the Carolina Outer Banks; apparently the shoal shore kept him too far out to sea to observe the inlets to Pamlico and Albemarle Sounds. Verrazano thought he had seen the "oriental sea . . . which is the one without doubt which goes about the extremity of India, China, and Cathay." He named the isthmus "Verrazanio" and sailed on past Cape Hatteras, "following always the shore, which turned somewhat to the north." Verrazano followed the coast up to the neighborhood of Nova Scotia and returned to Dieppe, where he wrote his report to the French king on July 8, 1524. This report is lost; but he wrote at least three other reports to Italy, differing somewhat in detail from each other.[29]

The first known maps to record Verrazano's discoveries in detail are Italian: Vesconte di Maggiolo's world map of 1527[30] and the sea chart of the world by Verrazano's brother, Gerolamo, in 1529 (Map 03; Plate 3).[31] These maps have two characteristics found in no other map before them. In the first place, the nomenclature is entirely new, at least for the southern coast. In many instances the two maps differ between themselves in the nomenclature, and comparatively few maps adopted the hybrid Franco-Italian names given by them. In the second place, they both show a deep indentation of the Pacific or South Sea which leaves only a narrow isthmus of a few miles between the Atlantic and Pacific at the latitude of the North Carolina Banks. This erroneous geography was a stimulating conception to those who were seeking a passage to Asia; "Verrazano's Sea" is found on many maps of the New World produced during the middle of the sixteenth century. Toward the end of the century in England, the possibilities opened up by the supposed isthmus stimulated the imagination of Sir Humphrey Gilbert, Richard Hakluyt, and others. It was one of the motivating factors behind the establishment of the Roanoke colony; Governor Lane and Thomas Harriot were eager and credulous in their inquiries of the natives concerning the South Sea; and one of John White's MS maps (1585) in the British Library shows an actual channel between the Atlantic and the South Sea farther down the coast (Map 7).[32]

Ayllón's attempt at colonization was unsuccessful; Verrazano's voyage was too transitory to leave a name; yet geographers evidently felt that that great stretch of continent lying behind the south Atlantic coast needed some designation. Before 1570 the name "Apalachee," spelled in many different ways, began to appear.[33] The Apalachees were an Indian tribe referred to in Cabeza de Vaca's narrative of the Narváez expedition (1528) in the Florida Peninsula[34] and often mentioned thereafter by the Spanish, French, and English.

The Apalachees were a division of the Muskhogean Indians with a habitat extending as far north as the lower ranges of the Alleghenies. During the last quarter of the sixteenth century this name became the most popular designation for the southeastern area; it continued to be used by continental geographers until well into the eighteenth century, when it appears as a kind of generic term for the Indian country back of the foreign settlements. It was usually written in the interior across the mountains, for which the name "Montes Apalatci" is found as early as in Le Moyne's map (1591) (Map 14; Plate 15). In this manner it came to be applied to the great mountain range which extends from northern Alabama to the Gaspé peninsula in Canada, the Appalachians.

The so-called de Soto map,[35] which is the only extant contemporary map to illustrate the extensive explorations of the expedition which Hernando de Soto and his followers made in 1539–43, is an unsigned, undated sketch in the Archivo General de Indias (Map 1; Plate 5). In 1539 de Soto's expedition landed in Tampa Bay and started northward and then westward on an amazing journey of nearly 3,000 miles. After leaving Florida de Soto led his men through Georgia and South Carolina to the Hiwassee River in North Carolina.[36] There he turned westward, striking the Tennessee River and following it southwest to McKee Island in present Alabama. The long laborious westward trek through the foothills of the Appalachians in South Carolina, North Carolina, Tennessee, and northern Alabama, reported in the various accounts of the expedition, was probably the original and main source of the conception of a range of mountains beginning in the region of the Carolina Piedmont and stretching directly west across most of the continent. This great east-west range remained a cartographical fixture until the end of the seventeenth century. Such a mountain barrier, if it had really existed, would have prohibited a great river basin such as the Mississippi. The fact, which has been mentioned repeatedly, that Sanson and the other seventeenth-century cartographers did not portray the Mississippi as a great river, when reports of its size were made by the early explorers who had seen it, may be explained by this difficulty. They could not believe that a great river could exist where a mountain barrier extended parallel with and close to the Gulf shore.

After leaving the Tennessee River de Soto followed the Coosa-Alabama Rivers far south to within fifty or seventy-five miles of Mobile Bay. Then, after a fierce battle with the Indians at Mabila, he turned northwest, reaching the Mississippi, below Helena and above the junction of the Arkansas River, on May 21, 1541. Exactly one year later de Soto died of a fever and was buried in the Mississippi. Before his death de Soto appointed Moscoso governor of the expedition; after further wanderings in Louisiana and Texas, the remnants of the army reached Mexico.

The entire area covered by de Soto's army is given on the anonymous manuscript map, which shows the interior as far north as present North Carolina and Tennessee and the coast from the vicinity of the Combahee River in South Carolina to the Panuco River in Mexico. The rivers along the south Atlantic coast flow correctly in a southeasterly direction and the Appalachian mountains appear for the first time. The range extends westward but, unlike later maps, with numerous breaks through one of which the Mississippi flows. Many of the 127 names and legends inscribed on the map are found in various accounts of the expedition. The spelling and location of Indian settlements differ in many cases and are apparently independent of the relations by members of the expedition, the Gentleman of Elvas, Ranjel, Biedma, and of the first-hand reports compiled by Garcilaso de la Vega. It is possible that the additional names on the map and the legend concerning the numerous herds of cattle (an early reference to bison) in southwest Texas came from the field notes of Ranjel or Biedma. The author of the map is probably the Spanish geographer Santa Cruz, since on the back of the map is a statement that it was "among the papers that they took from Seville belonging to Alonso de Santa Cruz." Since Santa Cruz died in 1567, that is the latest possible date for the making of the map. He was archi-cosmographer of Charles V when he made a map of the world in 1542; shortly before his death he made a map of the Caribbean area and the Florida coast which includes the latest coastal names but which has none of the details relating to the de Soto expedition. (See under Map List No. 1.) The most commonly given date, ca. 1544, is a reasonable one.

The de Soto map is interesting because it is probably the first map based on actual exploration to show the interior of any part of the present United States, with its delineation of the Appalachian mountains, of the correct direction of the flow of rivers to the Atlantic, and of the location of about a hundred villages showing the location of Indian tribes at that time. Upon this map or upon one like it, together with coastal information from other Spanish sources, is based the Ortelius-Chiaves printed map of 1584, one of the half-dozen most important mother maps of southeastern North America (Map 5; Plate 9). This map probably had more influence than any other map in establishing the subsequent conception of Florida as including that part of the present United States from the peninsula of Florida northward to about 40° N.L. and westward to or beyond the Mississippi. "Florida et Apalche" by Wytfliet (1597), Acosta (1598), and

Matal (1598) follow it closely; Sanson (1657), Duval (1660), Seller (1679), and other geographers until the eighteenth century used this map of Florida as a basis for similar maps, making changes in the nomenclature as their geographical knowledge increased (Maps 18, 20, 23, 53, 57, and 87; Plates 17, 31, and 33).

TRANSITIONAL OR DESCRIPTIVE MAPS: EARLY EXPLORATIONS AND SETTLEMENTS

Attempts at Settlement, 1562–1663

In the latter half of the sixteenth century attempts at settlement by the French and English were unsuccessful but resulted in two printed maps which together delineated nearly the whole coast from the Cape of Florida to Chesapeake Bay. These two maps influenced the cartography of the region for a century.

Attempts were made by the French to settle Florida and harass the Spaniards. Under Ribaut at Charlesfort on Parris Island in 1562 and under Laudonnière, at a fort called la Caroline or Carolina, on the River May in 1564, there were short-lived French settlements. With Laudonnière's expedition went Le Moyne, an artist who drew vivid pictures of Indian life and a map of the region. This map and the pictures were not printed until 1591, when they were published in the second volume of De Bry's *Grands Voyages*.[37] No special name was used for the French settlement, though about the middle of the seventeenth century the French geographers began to indicate on their maps the region from the St. Johns River to St. Helena's Sound as "Floride Françoise."[38]

Le Moyne's map of Florida (1591) extends along the coastline from North Carolina southward to the Cape of Florida (Map 14; Plate 15). It is fairly accurate for the country around the French settlement on the banks of St. Johns River in 1564, where Le Moyne obtained the information for his map and possibly drew the original; for the interior and along the South Carolina coast the information is distorted and gave rise to numerous subsequent cartographical misconceptions.

The French settlement at Port Royal was abandoned by the French themselves; the second under Laudonnière was wiped out by the Spaniards in 1565 under Pedro Menéndez de Avilés with pious savagery and with thorough understanding of the strategic danger to Spanish interests of a French colony in that location. In the same year, Menéndez established St. Augustine, then built the forts San Mateo on the St. Johns[39] and San Felipe at Santa Elena near the site of the French Charlesfort.[40] The following year he established posts on the Georgia islands. Although Spanish settlements along the coast continued and even expanded under the Franciscans to the interior with fluctuating success, the geographical and cartographical knowledge derived from the Spaniards for the next hundred years is negligible.[41]

The English had cast covetous eyes on the south Atlantic coast at least as early as Hawkins's visit to Laudonnière's Fort Caroline in 1565. In 1584 Amadas and Barlowe entered Pamlico Sound through one of the inlets and made explorations. Upon the return of the expedition to England and its report, this "new land" was named Virginia in honor of Queen Elizabeth. Attempts at colonization were made under Governor Ralph Lane and Governor John White. Lane, whose chief training and career had been as a soldier, did not return after going back to England with the first colony of 1585–86. Not much has been known of the background or personal life of John White, but recently found records indicate that he was a citizen of London with senior standing in the guild of painter-stainers of London and furnish details concerning other members of his family, including the birth of his daughter Elynor. These show convincing proof of the identity of the painter and the governor. (See a fuller account of the controversy about John White and its recent solution in note 42.)

In 1587 Governor White planted his ill-fated colony upon the island of Roanoke and returned to England for reinforcements and supplies. He never again saw his daughter or Virginia Dare, his granddaughter, the first child born of English parents in the New World. But the persevering courage shown in his repeated attempts to return despite difficulties at home and Spaniards at sea (he was driven back by the Spaniards on one of his attempts) have placed him high among Hakluyt's "men full of activity, stirrers abroad, and searchers of remote parts of the world." His rise from apparent obscurity to the governorship of the Virginia colony is one of the examples of meteoric ascent in that Elizabethan era of versatile and brilliant men of action. He accompanied the 1584 voyage of exploration under Amadas and Barlowe and the 1585 voyage of settlement under Lane as surveyor and painter; it was probably during this latter period, from May 1585 to June 1586, that he gathered most of the information for the drawings and maps from which his fame chiefly derives. These "Trve pictvres," according to De Bry, who published twenty-eight of White's drawings in 1590 as an appendix to his edition of Thomas Harriot's (Hariot, Harriott) *A Briefe and True Report of the New Found*

Land of Virginia, were "Diligentlye Collected and Draowne by Ihon White who was sent thiter speciallye and for the same purpose by the said Sir Walter Ralegh the year abouesaid 1585. and also the year 1588. now cutt in copper and first published by Theodore de Bry att his wone chardges."[42] Thomas Harriot, the philosopher, naturalist, and brilliant mathematician who also accompanied Lane on the 1585 voyage, must have helped White in gathering the data for the maps. Professor David B. Quinn, whose scholarly and carefully documented study of the Roanoke voyages[43] throws light on much that has been obscure in these expeditions, suggests that Harriot and White were employed to gather information according to a carefully prepared list of instructions similar to a set made for one of the expeditions planned by Gilbert or his associates in 1582.[44] In these 1582 instructions the surveyor and his assistant were to make a cartographical study from the southern tip of Florida northward on "cardes" or maps composed of four sheets of paper royal, marking them in a series of letters to avoid confusion. They were to keep rigidly to a uniform scale and include distances of capes, headlands, and hills, depth and breadth of inlets and rivers, with the variation in vegetation and land-use, location of springs, various sorts of trees, etc., both in a journal and on the map. Here we have, apparently, the background for the remarkable and intelligent collection of drawings by White and the notes which Harriot used for his *Briefe and True Report*.

The first cartographical report of the Roanoke ventures and the earliest known detailed drawing of the North Carolina Outer Banks and Sounds area is an anonymous undated sketch map (Plate 10) preserved in the Public Record Office in London (Map 6). It has recently been identified by Professor D. B. Quinn as a rough note of the first mapping done by White and Harriot in 1585; he thinks that Lane may have sent it to England on the *Roebuck* or *Elizabeth* together with his letter of September 1585 to Sir Francis Walsingham.[45] The sketch has over a dozen place-names and legends, many of which can be linked to expeditions and discoveries made in the summer of 1585 and reported in other accounts by the colonists. The northern part of the sketch has some similarities to White's pictorial map, "The Arriual of the Englishemen in Virginia," which is Plate 2 in De Bry's 1590 engravings. No map corresponding to either the sketch or the engraving is found in White's extant drawings. D. B. Quinn believes that White probably used a copy of the sketch as the basis for the pictorial map, now lost, which De Bry used for his engraving.[46]

In the volume of White's original drawings[47] in the British Library are two maps of the southeast coast. The first extends from the West Indies to Chesapeake Bay; the second gives a more detailed representation of the Outer Banks and the country explored by the colonists back of the Sounds. For the general map White had to rely chiefly on sources outside his own observation, except for the region around the Outer Banks. The most obvious of these sources is the work of Le Moyne, which furnished the topography and the French names for the rivers from Port Royal southward to the Cape of Florida. By 1582 Le Moyne was living in London, where he entered the service of Sir Walter Ralegh; White, the painter for Ralegh's 1585 venture, must have seen his drawings and his map or maps of Florida. Le Moyne refused to sell his paintings and narrative to De Bry in 1587, but De Bry bought them from Le Moyne's widow in 1588. De Bry's engraving of Le Moyne's map is stylized and Latinized; White's manuscript map for that area is not as detailed as De Bry's but has independent value and incorporates interesting differences from the engraving.[48] A second outside influence on White's general map is found in the strait leading from Port Royal to a large, vaguely drawn body of water in the upper left section of the map. This is a revival of Verrazano's Sea or Bay in which, as has been mentioned earlier, Hakluyt and other Elizabethan explorers and geographers were much interested at the time. In about 1580 Hakluyt had found a large map made by Verrazano and also a globe made by him which showed the South or Western Sea to be only a short distance from the Atlantic Ocean at some point along the southeastern coast.[49] In his *Divers Voyages*, published in 1582, Hakluyt included a map by Michael Lok which showed this sea.[50] Dr. John Dee's map of the Northern Hemisphere, made for Sir Humphrey Gilbert, shows a river or strait leading from the Atlantic to a lake which in turn has a long passage to the Gulf of Mexico; to the north of this lake is Verrazano's Sea (Map 4B; Plate 8). The map which is most similar to White's in detail at this point is the so-called Harleian World Map (ca. 1544), of the Dieppe type and sometimes attributed to Desceliers.[51] The strait on this manuscript map is just below "R: de S^{a} helene" and above "mer osto"; the configuration of Verrazano's Sea is similar to that part of it shown by White. A third source is a map, probably Spanish in origin, from which the Bahamas are derived. The 1585 colonists had a Spanish map with them, for Ralph Lane wrote to Sir Francis Walsingham in a letter dated August 12 that "Thys Porte [Wococon] in ye Carte is by ye Spanyardes called St Marryes baye."[52] In location of the Bahama Islands and in nomenclature White's map is similar in type to a manuscript map now in the Archivo Histórico Nacional, Madrid, attributed to Alonso Santa Cruz and drawn about 1567.[53] White's map appears to incorporate fairly contemporary Spanish cartographical information.

White's detailed manuscript map of the region surrounding the Carolina Sounds (Plate 12) must be accepted, writes D. B. Quinn,[54] "as the major contemporary authority on the configuration of the coastline in the later sixteenth century, on the nomenclature and spelling of place-names, and on the location of villages" (Map 8). Unlike White's general map of the southeastern coast, this map is based entirely on the discoveries of White and his fellow colonists. Although most of the data upon which it was based must have been gathered, with Harriot's expert help, during the 1585 stay, it may also incorporate information about topography and Indian villages gained in 1584 and possibly also in the voyages which White made in 1587 and later, as governor. De Bry's engraved map (Plate 14) represents almost the same area as White's manuscript map and, apart from conventions of engraving and some minor changes and additions, is the same map. It has, however, sixteen names not on the manuscript, extends slightly farther westward on the mainland, and increases the number of inlets along the coastal islands; evidently the engraver of the map worked from a drawing by White different from the extant manuscript and possibly modified by information gained in the 1587 voyage (Map 12).[55]

In 1590 De Bry published Harriot's work and the commentary with White's engravings almost simultaneously in Latin, German, French, and English editions as the first volume of his *Grands Voyages*; even before this magnificent publishing venture made White's cartographical work generally available, other maps which showed the English colony of Virginia in the New World were being made. Richard Hakluyt furnished Filips Galle with information about English explorations which appeared in Galle's Novvs Orbis (Plate 13), engraved in 1587 to accompany Peter Martyr's *Decades*; it has "Virginea 1584" near the fortieth parallel of latitude (Map 9). Boazio's "The Famouse West Indian voyadge" of 1588 shows Francis Drake's itinerary by a "pricked line" which touches the coast at "Virginia." Hogenberg's map of 1589 also shows "Virginie," which he pictures as a settlement off the mainland coast. None of these maps, however, adds any appreciable information to the knowledge of the coast except the name Virginia. De Bry's engravings of White's map in 1590 and Le Moyne's in 1591 gave to cartographers a remarkably detailed delineation of a part of the coast which they were not slow to utilize, although the fact that the two maps left an intervening stretch of coast between them caused confusion and, on some maps, an erroneous interpolation of Spanish cartographical information which lasted for many decades. In 1593 a large map of North America appeared which used both maps for the delineation of the southeastern coast of the continent in the notable atlas published by Cornelis de Jode.[56] In 1597 Wytfliet published two maps, one of Florida and Apalche using Le Moyne's 1591 and Ortelius's 1584 maps as a basis, and another of Norumbega and Virginia, in which the lower part is based on White's 1590 map. These two maps appeared in Wytfliet's *Descriptionis Ptolemaicae Augmentum*, a work which in spite of its title owes nothing to Ptolemy in substance or method. The first edition of this work is sometimes called the first atlas of the New World, since it contains no maps except those of the Americas. The work was highly regarded in the seventeenth century, and Wytfliet's division of North America below "Virginia" was followed by Acosta (1598), Matal (1600), Laët (1630), Sanson (1656), Duval (1669), and others (Maps 20, 23, 34, 49, and 61).

If in their division of the regions many later mapmakers followed Wytfliet, they did not follow his interpretation of Le Moyne's map. In 1606 Jodocus Hondius published an edition of Mercator's Atlas with a plate "A New Description of the American Provinces of Virginia and Florida" in which he corrected the grosser latitudinal errors of Jode and Wytfliet and put on one map much of the detail found in De Bry's engravings. However, instead of making the course of the River May run southward with an inverted V, as Le Moyne did, he straightened the bend in the river and thus placed the great lake with the invisible shore to the northwest of the mouth of the river. Through this unfortunate mistake, the lake was located among the Appalachian mountains, to the south of the lake fed by the great waterfall. Six years later (1612) Marc Lescarbot, who was considered by the French an authority on American history and geography, published a map probably influenced by the Mercator-Hondius map but showing independent study of his sources (Map 30). J. G. Kohl said that Lescarbot moved the mouth of the River May north half a degree and that this caused later geographers to identify it as the Savannah River instead of the St. Johns, which he confused with R. des Dauphins (St. Augustine Harbor).[57] But while Lescarbot's map undoubtedly influenced Sanson, Du Val, and other later

French cartographers, he did not originate these errors, which were already made in the Mercator-Hondius map. They evidently resulted from a confusion of Le Moyne's River May with the great river which runs southeast from the Apalache Mountains in the earlier Mercator and sixteenth-century Spanish maps and which is called "R. Sola" or "R. Seco." The Mercator-Hondius map, which combined an area of great contemporary political interest with exceptional artistic beauty, proved to be a popular one (Map 26; Plate 20). Jan Jansson, Laët, Blaeu, Montanus, and others closely imitated it, and by the second half of the century, the great lake with the invisible shores, the waterfall, the lake and island of Sarrope, the names of the Indian villages, the auriferous mountains of Apalatcy, and many other features were being embodied in the atlas maps of nearly every important European cartographer. At this time most of the great atlases were published by the Dutch school of cartography.

A general knowledge of the great seventeenth-century European mapmaking families and their methods is helpful in understanding the cartography of this period. Gerhard Mercator (1512–94) did not himself publish any large atlases, although he persuaded his friend, Ortelius, to publish a collection of maps in 1570, *Theatrum Orbis Terrarum*. Ortelius's first edition had three plates and listed eighty-seven geographers and mapmakers whom he had used as references. With the Additamenta which were occasionally published and incorporated in later editions, by 1587 the same atlas had 108 plates and listed 137 references. In 1595, the year after Mercator's death, his son Rumoldus published a collection of plates in the first Mercator atlas. In 1606 Jodocus Hondius (1563–1611), who had taken over the establishment, published a greatly augmented volume which is usually called the Mercator-Hondius atlas. Editions of both the Ortelius and Mercator-Hondius atlases were published with the text in Latin, French, Dutch, English, and other European languages, and this practice became customary among other Dutch atlas makers. Upon the death of Jodocus Hondius the great tradition of the firm was carried on by his son Henricus and his son-in-law, (Joannes) Jansson or Jan Janszoon (1596–1664). Jansson himself became one of the greatest Amsterdam mapmakers and had hundreds of new plates engraved for the many atlases published under his name. After 1664 his plates were purchased by Peter Schenk and used in later atlases. Jansson's greatest rival was Willem Janszoon Blaeu (1571–1638). He began making globes by 1599 and in 1606 established his famous printing and publishing business in Amsterdam. As "Cartographer to the Republic" he obtained the right to examine the maritime records of the Netherlands seamen; this, together with his association with such scholars as Tycho Brahe, enabled him to achieve a high quality of output. His early work was carried on under the name of Guglielmus Janssonius, which occasionally causes confusion with his rival Joannes Jansson. His business was carried on by his son Jan Blaeu (1596–1673), who achieved an even greater reputation and is thought to have surpassed the rival house of Jan Jansson. In 1662 Jan Blaeu published his *Atlas Major, sive Cosmographia Blaviana*, in eleven volumes, the culmination of what is probably the finest series of atlases ever produced; this same work was published in French in 1662 and again in 1667 in twelve volumes. It is exceeded in magnificence only by the 100-volume *La Galerie Agréable du Monde*, which Pieter Vander Aa published at Leyden in 1729. The Blaeu printing establishment was destroyed by a fire in 1672; the few plates that were salvaged were sold to Frederic de Wit, who used them subsequently in his atlases. The plate of the Virginia-Florida map, which had been used without change since 1640, was evidently lost, as that map does not appear in atlases after the fire.

The technique of cartographical expression achieved a level during the golden age of Dutch cartography in the seventeenth century which has never been surpassed, as is well illustrated by the maps of the Virginia-Florida region by Mercator (1606), Jansson (1636 and 1641), Blaeu (1640), and Montanus (1671) (Maps 26, 39, 41, 42, and 67; Plates 20, 25, and 26). Each map is a skillfully composed geographical unit, with artistically distributed lettering and decoration. Particularly characteristic are the "swash" lines of the lettering, with sweeping curves filling up the otherwise empty spaces, and the use of decorative cartouches for the title and for the scales embellished with animals and scenes appropriate to the region. Not always so successful were the Dutch mapmakers in the geographical accuracy of their presentation. After Mercator and Ortelius they tended to be commercial artists rather than scientific geographers. Too often they filled in the blanks of knowledge with geographical features or information which lacked discriminatory use of their sources. As the century wore on the Dutch school began to be superseded by the rise of the scientific school of French geographers.

The creator of the French school and the founder of one of the great European cartographical families was Nicolas Sanson of Abbeville (1600–1667). He showed a precocious ability in geography which earned him the favor of Richelieu and later of Louis XIII, who greatly admired him and took lessons from him. In 1638 he established a printing business in Paris, and after 1640 he was geographer-in-ordinary to the king. He was offered the position of secretary of state but did not accept it, fearing, it was said, that the office would draw his sons away from their interest in geography. His political interest can be seen in the maps of Mexico and Florida (1656) and of Florida (1657), in which the ill-fated attempts at settlements by Ribaut and Laudonnière nearly a hundred years before are used to lay claim to the region from St. Johns to Port Royal, which he calls "Floride Françoise" (Maps 49 and 53; Plate 31). Sanson's earliest effort, a map of Greece, was based upon a study of the methods of Mercator and Ortelius, and he always remained under the influence of the Dutch school. His work exhibited a more scientific attitude, however; he reduced the decorative element and increased the number of informative geographical notes. He was succeeded by his sons Adrien and Guillaume, who continued until the end of the century to put out a prolific stream of maps under their father's name, and more notably by his son-in-law, Pierre Duval. Duval's various maps of southeastern North America, however, are largely copies of Sanson's, except for one (La Floride Françoise ca. 1669) which is based entirely on Lescarbot's of 1612 (Map 61). Gilles Robert de Vaugondy was Sanson's grandson, and Didier Robert de Vaugondy his great-grandson. Toward the end of the seventeenth century Sanson's plates were bought by Alexis Hubert Jaillot, who published the maps in his atlases without removing Sanson's name.

For the first fifty years of the seventeenth century and longer, the Carolina coast was but little troubled by the activities of the European colonizing powers. While geographers and the makers of sea charts of the Atlantic coast multiplied their conceptions of the shoreline and drew upon Le Moyne's and White's maps with varying degrees of inaccuracy, little new information was available even to the most assiduous seeker for increased knowledge. Smith's map of Virginia and the Chesapeake region (1608) was an excellent example of "canoe-surveying" with the aid of a compass, and it was promptly copied by the atlas makers, as was Champlain's map of the St. Lawrence (1612). But to the south of Jamestown a few generalized manuscript sketches and engraved maps whose lack of authenticity is patent comprise the cartographical efforts of the earlier part of the century not dependent on sixteenth-century sources.[58] The most interesting of the printed maps were those of Smith (1624) and Farrer (1651), though neither was copied by later mapmakers (Maps 32 and 47; Plates 23 and 29).

In about the middle of the century, however, in the neighboring colony of Virginia and over the waters in England, interest in the "Ould Virginia" region (the area settled by Ralegh's colonists, now eastern North Carolina) was being revitalized.[59] There are indications of colonizing interest around 1650 in the region between the Chowan and Roanoke Rivers by persons from Virginia. In about 1648 Henry Plumpton, Thomas Tuke, and several others bought from Indians all the land on the Chowan between the mouth of the Roanoke River and the mouth of Weyanoke Creek.[60] Apparently no settlement was made at that time; at least, in 1708 Henry Plumpton stated that he had lived in Nansemond County, Virginia, since 1624. In 1653, upon the petition of Roger Green, a clergyman, and several inhabitants along the Nansemond River (possibly those who had purchased the land from the Indians in 1648), the General Assembly of Virginia granted between ten and eleven thousand acres to the hundred persons who should first seat themselves south of the Chowan and on the banks of the Roanoke River.[61]

Although there is no proof of permanent settlements before 1660 in the documents so far used by historians, evidence exists of a settlement at the mouth of the Roanoke River which dates at the latest from 1655–56. The New York Public Library possesses a vellum manuscript map by Nicholas Comberford, entitled "The Sovth Part of Virginia Now the North Part of Carolina," which is dated 1657 and embodies information gained from a fresh survey of the region (Map 50; Plate 32). This map deserves special attention because of its evident use of original contemporary source material, the excellence of its composition, and the historical importance of the information it gives. It shows a neatly drawn house near the sound, on the neck of land between the mouth of the Roanoke River and Salmon Creek, with the legend "Batts House" extending from it into the sound. That Nathaniell (or Nathanial) Batts, and probably others, as will be shown, had settled near the sound between the Roanoke and Chowan Rivers is not surprising in view of the earlier interest taken by Plumpton, Green, and their Nansemond neighbors in this region. "Batts House" was built as a "factory" or trading post for the Indians for Batts in 1655 or earlier by Robert Bodman, a carpenter under contract with Colonel Francis Yeardley, an important merchant of Princess Anne County, Virginia.

From references to the region a few years later one can

infer not only that it was settled but also that it was regarded as one of the chief centers of population in the early years of the proprietary rule. Robert R. Lawrence had a plantation a few miles north of the mouth of the Roanoke River from 1660–61 to about 1667–68.[62] One of the earliest maps of the Albemarle Sound region made after the grant to the lords proprietors shows one of the governor's plantations at the mouth of the Roanoke River and to the north.[63] The governor had also, according to this map, another plantation on the south shore of the sound between the Roanoke and the Mackay Creek or Bull Bay, which is about four miles from the mouth of the Roanoke. Another bit of evidence of continual if not continuous settlement is that in 1676 the lords proprietors stated that they had instructed the governor and Assembly of Albemarle to establish three port towns in the County of Albemarle, "To bee the onely places where the Shipps shall lade and unlaid,"[64] namely, on Roanoke Island, on Durant's Neck, and on the neck of land between Salmon Creek and the Roanoke River, the last being the location of Batts's house on the Comberford map. Since Durant's Neck at this time was one of the centers of population in the province, it is reasonable to suppose that the lords proprietors chose the other two localities because of their existing settlements. Further evidence is supplied by a map drawn by James Lancaster in 1679, now in the Blathwayt Atlas in the John Carter Brown Library, which shows the various settlements in the Albemarle region (Map 78).[65] This map may have been made from the information sent by Sir Peter Colleton in February of that year to Blathwayt, who was the collector of his majesty's customs and who had asked Colleton for information concerning the Albemarle region. The Lancaster map has houses at the mouth of the Roanoke to both the north and the south and on both sides of Mackay Creek.

The "Batts settlement" appears by all the evidence, therefore, to have continued with little or no interruption into the proprietary period. There is no way to establish the exact size of the population in these early settlements at any given time, but presumably there was no sudden migration of all the settlers from this region after Albemarle County began to grow. The strict orders of the Virginia council in 1653 had been that not less than a hundred able-bodied settlers would be allowed to settle this region. We may perhaps presume, therefore, that any settlement with as continuous a record as the one to the north of the Roanoke was probably, if not demonstrably, permanent.

Why did Comberford choose to depict only Batts's house on his map if, as seems probable, there were other settlers in the vicinity? It was apparently because Captain Nathaniell Batts was the leading man of that region. "Governor of Roan-oak" is the title given to him by George Fox, the founder of the Society of Friends, in his journal. In September 1672 Fox went with William Edmundson "down the creek in a canoe to Macocomocock River, and came to Hugh Smith's, where people of other professions came to see us (no Friends inhabiting that part of the country) and many of them received us gladly." Fox then states: "Among others came Nathaniell Batts, who had been Governor of Roan-oak. He went by the name of captain Batts, and had been a rude, desperate man. He asked me about a woman in Cumberland, who, he said, he was told, had been healed by our prayers and laying on of hands, after she had been long sick, and given over by the physicians: he desired to know the certainty of it. I told him, we did not glory in such things, but many such things had been done by the power of Christ."[66] At the end of his trip to Carolina, Fox again visited Batts: "& soe wee left our boate where wee had borrowed her (Edenton), & tooke our Cannoe & came to Captain Batts, and there lay most of us by the fire that night, & after we came half a mile to Hugh Smiths."[67] In the following year, 1673, from the Worcester prison, where he had been incarcerated for eight months, Fox wrote to Friends in Virginia, again mentioning Batts as governor: "If you go over again to Carolina, you may inquire of Capt. Batts, the Old Governor, with whom I left a Paper to be read to the Emperor, and his Thirty Kings under him of the Tusrowres, who were come to Treat for Peace with the People of Carolina: Whether he did read it to them or no, remember me to Major General Benett, and Col. Dew, and the rest of the Justices that were Friendly and Curteous to me, when I was there, and came to Meetings."[68]

Circumstantial evidence thus indicates that prior to the patent to the lords proprietors there was a sufficiently extensive settlement on Roanoke Sound to justify some kind of administrative organization. Whether Batts had been appointed by Governor Berkeley because of the grants and patents given by Virginia in the region, or whether, as is more probable, Batts was chosen by the settlers themselves, we do not know. There is direct evidence, as we have seen, that he was not without neighbors in the preproprietary period,[69] and contemporary references seem to connect him with persons in Nansemond County who were interested in exploration south of Virginia.

Batts had owned land in Nansemond, for he purchased

900 acres from Samuel Stephens which he sold later to Thomas Francis, according to an entry in a patent book in the Virginia State Land Office.[70] This is the Captain Thomas Francis who, with Colonel Thomas Dew and other unspecified gentlemen planters whom they might ask, was commissioned by the General Assembly of Virginia in December 1656 to make discoveries between Cape Hatteras and Cape Fear.[71] Both of these men lived in Nansemond County and were persons of prominence in the colony. Colonel Dew had been chosen speaker of the General Assembly in 1652 and appointed to the Virginia council in 1655. Captain Thomas Francis was a member of the General Assembly in 1657.[72] Captain Batts was probably interested in this expedition, either as a neighbor of Captain Francis in Virginia or as one who was deputized for the expedition because he had already moved south from Virginia and knew the region. Within six months after the petition for exploration was granted and while Captain Francis was a member of the General Assembly, which was then in session, Batts received special privileges as a reward for an expedition into the region mentioned in the commission for Dew and Francis.[73] Batts's undertaking may be connected not only with Dew and Francis but with the Nansemond County inhabitants who applied for a grant between the Roanoke and Chowan in 1653. Evidently Batts made explorations along Albemarle or Pamlico Sounds, either for himself or as agent in some colonizing movement; the Comberford map probably records the knowledge gained in these expeditions. In 1660 he purchased a large tract of land on the west bank of the Pasquotank River from its mouth to New Begin Creek; it is the earliest known recorded land grant in North Carolina.

Batts's "rude, desperate" nature apparently harmonized but little with the proprietary rule, for he seems to have taken no part in the later affairs of the province. After the entry in Fox's journal he drops from sight, unless he is connected with Batts Island, later called Batts Grave. If, as seems likely, the eponymous Batts was our Nathaniell Batts, he moved, probably after 1672, to the island and died there before 1679.[74] The territory to the north of the Roanoke River continued to develop without the presence of Batts, for by 1723 this region, containing the earliest continuous settlement for which there is documentary evidence, had become one of the most flourishing in North Carolina.[75]

Meanwhile, events in England were taking place which changed the ownership, the name,[76] and the forces working for colonization of the land south of Virginia.

The Establishment of Carolina, 1663–1700

In 1663 Charles II of England granted to eight men who had been of service to him in regaining the throne lost by his father the proprietorship of a province (which Charles II named in his own honor Carolina) with a latitude from 31° to 36° N and extending from the Atlantic to the Pacific.[77] This munificent gift cost him nothing, and two years later he extended the bounds south and north from 29° to 36½°. In spite of numerous claims and appearances to the contrary, the name Carolina apparently was not used for, or applied to, this region before 1663.

To examine the local use of the name and to clarify the problems involved, it will be necessary to go back to the French attempts at settlement on the south Atlantic coast during the middle of the sixteenth century. Historians of the Carolinas, such as Ashe,[78] Oldmixon,[79] Governor Glen of South Carolina,[80] Dr. Alexander Hewat,[81] eighteenth-century mapmakers like Delisle[82] and Homann,[83] and modern scholars like J. G. Kohl[84] and Professor F. Oger[85] have stated that the French called their colony Caroline or Carolina. These writers, however, have given no specific reference to documents for their statements, which can therefore be dismissed. A recent defense of the French priority in naming Carolina is that by Dr. St. Julien Ravenel Childs of Charleston, South Carolina, who quotes a French book published in 1616 and says that "it seems to prove that the French had begun to apply the name 'Carolina' to the country as a whole, a point that has been vigorously denied." His quotation follows: "Ribaut and Laudonnière,—having gone to Florida in a fine array by the authority of Charles IX, in the years 1564, 1565, and 1566, to cultivate the land, and there having extended Carolina to the 30th parallel of north latitude—took possession as far as the 38th and 39th parallels . . ."[86] But is it not the translation (by Thwaites) which is misleading, and not Biard? Biard writes: "& y ayant edifié la Caroline au 30. degré d'eleuation: ils prindrent possession iusques au 38. & 39. degré & par ainsi voila les anglois hors de leur Virginie, suiuãt leurs propres maximes."[87] Surely "edifié" means built, and the correct translation is also a true statement of the fact: "having built la Caroline at 30 degrees . . ." Biard is writing about the French fort la Caroline, which on Le Moyne's map is exactly on the thirtieth degree of latitude (Plate 15). In the numerous contemporary relations, reports, and maps published, two temporary settlements, Charlesfort and la Caroline (or Carolina), are mentioned; but the country itself is called "Florida" or "Floride Françoise."

In 1629 Charles I of England granted to Sir Robert

Heath a charter for the land extending from 30° to 36° of latitude (the present Georgia, South Carolina, and North Carolina to Albemarle Sound) and westward within those parallels to the Pacific, with the generosity characteristic of the kings of that period in granting land in the New World. In the charter printed in the Colonial Records of North Carolina,[88] transcribed from the originals in the British Public Record Office, both CarolAna[89] and CarolIna are used as the name of the province thus granted. But in the subsequent correspondence with Baron Sancé over the colonization of some French Huguenots under this grant, a correspondence starting in 1629 and lasting for some years, the briefs of the letters made from the originals in the Public Record Office, as published in the *Collections of the South Carolina Historical Society*,[90] use CarolIna only, and not CarolAna. An examination made of Patent Roll C. 66/2501/M26 (the Heath charter of 1629) and of the correspondence of Baron Sancé concerning the French Huguenots, shows that in both cases the only form in the Latin and English grants and in the Sancé letters is CarolAna.[91] Thus the incorrect form "Carolina," in the *Colonial Records of North Carolina* and in the *Collections of the South Carolina Historical Society*, is due to incorrect transcriptions or poor proofreading in these nineteenth-century publications.

In 1654 Francis Yeardley, son of the Governor of Virginia, wrote to John Ferrar, formerly deputy treasurer of Virginia, about discoveries to the south of Virginia. In this letter, as first printed in the *State Papers of John Thurloe* (London, 1742) and reprinted from Thurloe later by Peter Force in his *Tracts*, Washington, 1836, and by Mr. A. S. Salley in *Narratives of Early Carolina*, New York, 1911,[92] Yeardley uses only Carolina, and not Carolana. Yeardley's letter of 1654 is found in the state papers of John Thurloe, secretary to the Commonwealth under Oliver Cromwell, which are now preserved in the Bodleian Library MS Rawlinson A 14. An examination of the original manuscript shows that both within the letter and in the endorsement, only the form Carolana, not Caroline, is used.

In the New York Public Library is a manuscript map, dated 1657 and made in London by Nicholas Comberford, with the title "The Sovth Part of Virginia now the North Part of Carolina" (Plate 32).[93] But Mr. Victor Hugo Paltsits, former curator of manuscripts and chief of the Division of American History in the New York Public Library, has examined the map and states that the second line of the title, "now the North Part of Carolina," is added in darker ink in a later seventeenth-century hand and is not part of the original 1657 title. This conclusion has been substantiated by the recent discovery of another copy of this map, also dated 1657, now owned by the Greenwich Maritime Museum, which is identical except that it does not have this second line.

In a widely used map made in 1657 by Nicholas Sanson, geographer royal of France, the name Caroline appears halfway up along the coast of what is now South Carolina.[94] This map of "Florida" by Sanson, which has Carolina at Port Royal, does not apply it to the region but refers to a marked settlement at the mouth of the river entering Port Royal (Map 53). This is the location of Charlesfort and is evidently the result of a confusion of Laudonnière's fort "la Caroline" on the River May and Ribaut's Charlesfort at Port Royal.

Thus an examination of these documents before 1663 in which the name Carolina occurs shows that it was used only for the Laudonnière fort of 1564 and not for any region or settlement, and that the English grant to Sir Robert Heath was for Carolana, not Carolina. The name Carolina as applied to this area was first officially used in the fifth section of the First Charter of Charles II, where the king states: "and that the country . . . may be dignified by us . . . we of our grace . . . call it the Province of Carolina."[95]

Even before the lords proprietors obtained their charter from Charles II, renewed explorations along the coast were in progress. The Shapley manuscript map of 1662 is a report of a voyage made by William Hilton earlier the same year (Map 58). In 1663 Hilton made another and more extensive voyage in the ship *Adventure* in the interests of the Barbados adventurers, of whom Peter Colleton, later one of the lords proprietors, and Thomas Modyford were the agents and chief promoters. A map showing the information gained by Hilton was published by Horne in 1666 (Map 60). This is the only printed map to show the result of English explorations in the Carolina region from White (1585) to Lederer (1672), Blome (1672), and Ogilby (ca. 1672), with the possible exception of Smith's map (1624), whose new nomenclature for the Albemarle Sound region is of dubious significance and no observable influence. Horne's map is therefore of considerable cartographical importance.

Hilton sailed from Spikes (Speight's) Bay on August 10, 1663, spent some time in the vicinity of Hilton Head Island, which still bears his name, and St. Helena Sound, then sailed north along the coast to Charles River, which he explored a second time. He reported his explorations in a promotion tract, *A Relation of a Discovery lately made on the Coast of Florida (from Lat. 31° to 33°45' N.L.)* . . . , London, 1664.[96] The pamphlet has no map, and the map which illustrates Hilton's discoveries is published with Horne's *A Brief Description*, London, 1666, which nowhere specifically refers to Hilton (Map 60).[97] Above "C Fear" Horne's map has "Charles River," the earlier name given to the Cape Fear River by the brief English settlement in 1664. Near the mouth of "Charles River" and to the west is the name of the settlement, "Charles Town." When the geographer Richard Blome made a map of Carolina to accompany his often reprinted colonial promotion volume, *A Description of the Island of Jamaica: With the other Isles and Territories in America*, London, 1672, he used the Horne 1666 map as a basis (Map 69; Plate 34). He was misled, however, by "Charles Town" on Horne's map, which he evidently confused with the later 1670 settlement on the Ashley River; he therefore put "Ashly R" adjacent to and just to the west of Cape Fear or Charles River, to which he gives no name, and leaves "Charles Town" where it was on Horne's map. In this way "Charles Town" does duty for two different places and the coastal distance between the two is contracted to a few miles (see Blome's map, Plate 34). Blome's error apparently gave rise, in turn, to a settlement called "Carolina" at the same location on the Cape Fear River, found in a series of eighteenth-century maps by Visscher, Kocherthal, and Homann.[98]

A comparison of Horne's 1666 map with a manuscript map in the Blathwayt Atlas (No. 18; see Lancaster 1679? MS A) makes probable an earlier map common to both which derives eventually from the voyage made by Hilton in 1663. A number of names on the Charles River found on the Shapley map are omitted, and new names found on the Horne 1666 map are based on incidents reported in Hilton's *Relation* (1664): Turkey Quarters, Stagg Park, Rocky Point, Highland Point, Swampy Branch, Mount Skarie, Mount Bonny, and some other names are given by Hilton and his companions which are here recorded for the first time. The only names given by Hilton for places along the Cape Fear River which are still found on modern maps are Turkey Quarter, Rocky Point, and Stag Park.

The Horne map showed only the coastal region. In 1672 a small book was published in London which furnished geographers much new information (and misinformation) about the topography of the interior of the country. It confirmed the belief in a great inland lake of unknown size in the Southeast; it caused the addition to maps of such errors as a long savanna inundated with water half the year, extending along the base of the Virginia and Carolina Blue Ridge, and of a long narrow "Arenosa desert" covering roughly the region of the Carolina pine barrens. This book was *The Discoveries of John Lederer*, a report of three exploratory expeditions west and southwest of the Virginia settlements, made by a young German scholar and traveler, accompanied by "A Map of the Whole Territory Traversed by Iohn Lederer in His Three Marches" (Map 68; Plate 36).[99] The work was translated from the Latin account of Lederer, with additions from his personal conversations, by William Talbot, secretary of the Province of Maryland, who published it in London. John Lederer appeared on the American colonial scene in 1670, or shortly before, and returned to Europe in 1675.[100] He was the first European to explore the Virginia Piedmont and Blue Ridge Mountains and to leave a record of his discoveries. He helped to open the great Indian Trading Path toward the southeast for the fur traders of Virginia. He gave valuable commentary on the Indian tribes encountered during his travels. He was the friend and protégé of some of the leading personages then in the American colonies. Governor Berkeley of Virginia commissioned him to make explorations westward in the hope of finding an easy route to the South Sea. He became a citizen of Maryland, where he was befriended by Talbot, nephew of Lord Baltimore and secretary of the province, who returned to Ireland in the summer of 1671 as Sir William, baronet of Carton, and went on to a distinguished career in London. He went in 1674 to Connecticut, where he won the respect of Governor John Winthrop the Younger and other scholars and ministers in that colony. The correspondence by and about Lederer with Governor Winthrop, now in the Winthrop Papers of the Massachusetts Historical Society, gives a picture of a self-sacrificing and humanitarian student of medicine and a cultured scholar. Yet until recently his positive contributions have been largely disregarded; historically Lederer has been known primarily for the errors on his map and for the problems raised by the account of his journey into Carolina. His achievements have usually been underestimated or left unexamined; it is not his pioneer explorations of the Virginia Blue Ridge on his first and third trips, his courage and enterprise in opening the way to later trade routes, and his comments on Indian tribes which he encountered that writers have noted, but the misconceptions which he gives of the region traversed in his second journey. These have caused his work to be discounted and condemned.

It is his second journey and its influence on the cartography of the Southeast, however, that must here be examined.

On May 20, 1670, Lederer began a march with a party led by Major William Harris, composed of twenty colonists on horseback, and five Indians. They started from the falls of the James River, the present site of Richmond, and struck due west until they again encountered the James River in the extreme corner of present Buckingham County. Here Major Harris and his troop felt that the glories of discovery and its dangers did not equal the comforts of home and tried to force Lederer to return with them. He showed them his commission from Governor Berkeley for westward exploration, however, and they left him angrily; apparently they attempted later to protect their own reputation by spreading defamatory rumors about him. Lederer continued with one guide, a Susquehanna Indian named Jackzetavon, who accompanied him the rest of the way. He visited the Saponi Indians on Roanoke River below Long Island, the Akenatzy (Occaneechi, Occoneechee) Indians near the present Clarksville, Virginia, and from there went on to Oenock, an Indian settlement on the Eno River near the present town of Hillsborough, North Carolina. By June 21 he reached the Uwharrie Mountains, which he evidently thought, as indicated on his map and in his account, to be the foothills of the Appalachians. He saw only low hills after leaving the Uwharries and reported that the mountains "fall off due west" from there, as they do in the map found in the volume of José de Acosta to which he refers.[101] The Sara Indians were near the Indian Trading Ford. This famous ford on the Yadkin River was about a mile and a half below the present Greensboro-Charlotte highway and six miles northeast of Salisbury. From Sara, Lederer took a south-southwest course to Wisacky, traveling through a marshy terrain which took him three days to cross.[102] Wisacky, the Waxhaw village reached on June 25, was near the mouth of Sugar Creek, which joins the Catawba below the North Carolina–South Carolina boundary line.[103] Across this creek or across the Catawba River below Fort Mill, South Carolina, was Ushery, the town of the Catawba Indians.[104] The Esah villages on the Catawba probably mark the farthest limit of Lederer's actual journey; his description of what lay beyond the Catawba is geographically untenable. Here Lederer writes that he saw a great lake and tasted the brackish water of the lake, whose width he estimated at thirty miles because he could see the opposite shore. Its length he could not guess; the westerly end was invisible: "I judged it to be about ten leagues broad; for were not the other shore very high, it could not be discerned from Ushery. How far this lake tends westerly, or where it ends, I could neither learn nor guess."

After leaving the Usherys, he went east to avoid the marsh, but he escaped the Scylla of the "Wisachy marish" only to encounter the Charybdis of "a great desert" which took him twelve days to cross. Finally he reached the country of the Tuscarora Indians. Their chief took his gun and ammunition, and he hurriedly left before further demands were made, making his way back to Appomattox in Virginia.

As his later permit from the Maryland government shows, Lederer planned to open a fur trade along the path into Carolina which he had explored. He may not have been able to carry out his intentions, but the Virginia traders soon began to develop the route. The Needham and Arthur expedition sent out by Abraham Wood in 1673 is well known;[105] in 1684 Captain William Byrd wrote that five of his traders were killed by the Indians "in their return from the Westward, about 30 miles beyond Ochanechee,"[106] and in 1686 he noted that two of his traders had been killed "aboue 400 miles of."[107] In 1701 Lawson met among the Kadapaus (Catawbas) a Scots trader from Virginia who had traded with them for many years.[108] Col. William Byrd wrote of great trading caravans of a hundred horses that used to ply the Trading Path, each horse carrying 100 to 150 pounds of wares. He described in some detail the trade with the Catawbas, "called formerly by the general Name of the Usherees," although he said that by 1728 the caravans had diminished to less than half their former size, evidently because of the competition from Charles Town traders.[109]

The late David I. Bushnell of the Bureau of American Ethnology once said, "I would not say that Lederer did not think he saw anything that he said he saw." If one attempts to accept Lederer's good faith and supposes that his errors are based on a misinterpretation of the statements of Indians or on his own credulous exaggeration, how can one explain the savanna, the desert, and the lake? For each of these there is a basis that might explain Lederer's belief, although he grossly exaggerates that base. Concerning the savanna he wrote: "The valleys [of the Piedmont] feed numerous herds of deer and elks larger than oxen: these valleys they call Savanae, being marish grounds at the foot of the Apalataei, and yearly laid under water in the beginning of summer by flouds of melted snow falling down from the mountains." It is certainly probable that before the forest land was denuded and the topsoil washed away, the Piedmont had large marshy sections that have since largely disappeared. But his statement and the map, which included the whole Piedmont region of Virginia and North Carolina in the "Savanae," are clearly an exaggeration.

Lederer's description of the barren sandy desert, the "Arenosa desert" of his map, is also exaggerated. There are stretches of sandy pine barrens in the state of North Carolina, but none where one can go for twelve days in a northeasterly direction without crossing rivers. His return route did not cross the sandhill section of North Carolina during the heat of July; he may have been told that it extended for a considerable distance southward from where he crossed. Also in retrospect it would be easy to exaggerate the proportion of the return journey spent in that region and the extent of its aridity.

The great lake of Ushery seems to be based on the Mercator 1606 map (Plate 20). Lederer's reference to its invisible shore is too reminiscent of the phrase "Adeo magnus est hic lacus ut ex una ripa conspici altera non possit," which had been published on most maps of the region for the preceding half century, not to make clear its origin. He had gone in that direction; all the maps showed that the lake was there; he wished to convince his patron, Sir William Talbot, that his expedition had been of value. James Mooney, the ethnologist, has suggested that Lederer saw the Catawba in flood and thought it was a lake;[110] and Carrier's hypothesis is that log jams or beavers created a lake.[111] But where on the Catawba or when on the Catawba has the water been so wide that one could not judge that it was flowing? And Lederer, unfortunately, said that he tasted of it and found it brackish. At this point Lederer's critics appear right; his account seems, at the least, to be fictitious.[112] His substantiation of the myth of the great lake, however, became the strongest link in continuing the geographical misconceptions of the region begun in the Mercator map.[113]

The errors that Lederer made lived after him and became fruitful, for they were sown in fertile soil. The lords proprietors were anxious to spread information about their recently acquired (1663) Province of Carolina to prospective settlers, and the mapmakers were glad to get new material. In particular, John Ogilby, the royal cartographer to Charles II, wished to add a passage on Carolina to a free translation of Montanus's *De Nieuwe en Onbekende Weereld*, which he was publishing with the title *America*. He approached Peter Colleton, the lord proprietor who wrote the following letter to John Locke, at that time secretary to Lord Ashley, the earl of Shaftesbury, one of the most active of the lords proprietors:

> To my honoured frend Mr. John Lock
> Sr.
> Mr. Ogilby who is printing a relation of the West Indies hath been often with mee to gett a map of Carolina wherefore I humbly desire you to gett of my lord (Ashley) those mapps of Cape feare & Albemarle that he hath & I will draw them into one wth that of port Royall & waite vpon my lord for the nominations of the rivers, &c: & if you would do vs the favour to draw a discourse to be added to this map in ye nature of a description such as might invite people wth out seeming to come from vs it would very much conduce to the speed of settlemt. & bee a very great obligation to
> yr most faithful
> frend & servt
> P Colleton[114]

John Locke, the philosopher, was also the author of the "Fundamental Constitutions" of Carolina, a feudal and autocratic legal system which was accepted by the lords proprietors and which they attempted to enforce upon the settlers of their province. It showed little political sense of the liberal laws needed for an American colony settled by Englishmen and was unsuccessful in practice. It explains, however, how and why the nomenclature of the region should be referred to Lord Ashley, the lord chief justice of the province.

On the back of the letter that Colleton wrote, Locke wrote notes for the requested description of the new country. This description is evidently the rather obvious propaganda chapter which Ogilby printed between pages 204 and 212 of his *America*, published that year (1671). Among other notes by Locke on the back of Colleton's letter is a section listing the names of points along the Carolina coast.[115]

The map of Carolina referred to in Colleton's letter appeared in Ogilby's *America*, with the title "A New Map of Carolina, by Order of the Lords Proprietors" (Map 70; Plate 37). It has been known ever since as the First Lords Proprietors' Map and was undoubtedly sold as a separate to would-be settlers for the new province. In the same bundle of papers with Colleton's letter to John Locke is a map endorsed by Locke as "Map of Carolina [16]71" and with his notations on it (Plate 35).[116]

A comparison of the two maps shows that this manuscript map may have been used as an early rough draft for Ogilby's map. Though Ogilby used the printed Lederer 1672 map for numerous topographical details, Locke's map has many of the same names, given on no earlier map, which are found on Ogilby's. The printed map gives many more names than are found on Locke's draft; the names of the lords proprietors[117] are more frequently repeated. On Ogilby's official printed map, reading from south to north, are Craven River, Craven County, Colleton River, Berkeley County, Ashley River, Ashley Lake, Cooper River, Porte Carteret, Cape Carteret, Clarendon River, Clarendon County, Albemarle River, and Albemarle County. Some of these names, such as Cape Carteret, Ashley River, and Albemarle River, were already in use by 1670, but the names of the counties were evidently the result of some authoritative decision; and from the title's "By Order of the Lords Proprietors," the official approval of the new names is indicated. Thus, from Peter Colleton's letter to Locke's notes and rough chart, the manuscripts in the Public Record Office preserve the successive steps in the nomination of places along the Carolina coast, many of which are still in use.[118] Other place-names are given along the coast for the first time on any map; but the preponderance of the names of the lords proprietors shows the origin and authority for the nomenclature.

For the Albemarle and Pamlico Sound region, Ogilby used some manuscript not now known. The Clarendon (Cape Fear River) area is based on Horne's 1666 map (Map 60). South of Port Royal the influence is still Le Moyne's; nearly two score of Le Moyne's Indian Villages, "Sarropo" and "Edelano" Lakes, and the "R. May," into which Ashley Lake empties, show that the recently enlarged information of the English stopped abruptly at Hilton Head.

One name on Ogilby's map deserves special comment; below Charles Town is a large island named in honor of the secretary-philosopher "Locke Iland." But, alas, the island was a geographical error; between the Stono and the South Edisto are several islands, which the settlers discovered when they extended their explorations. Ten years later, when the "Second Lords Proprietors' Map" appeared in 1682, a number of unnamed islands are delineated where Locke Iland was (Map 92; Plate 39). The place was nonexistent; and on modern maps John Locke has not even a swampy inlet or a tidewater creek named in his honor, though he was once Landgrave in the titled nobility of the fair province of Carolina.

Other maps, based largely on Ogilby-Moxon's "Lords Proprietors' Map," followed in quick succession. A new edition of John Speed's *The Theatre of the Empire of Great Britaine*, probably the most famous seventeenth-century English atlas, appeared in 1676, with a map of Carolina based on Ogilby-Moxon and accompanied by a full description of John Lederer's exploration (Map 77). A smaller Speed atlas, *A Prospect of the Most Famous Parts of the World*, London, 1675, had been published the year before with the same material and proved so popular that it was reprinted in 1676. The marks of Lederer's influence are always clear and usually the same: the long narrow savanna in the Piedmont region, the great lake, and the long, narrow Arenosa desert. Usually the names of Indian villages are included, and in the earlier maps the path which he traveled is shown by a dotted line, as in the original map. The Daniel (1679) and Morden (1685) maps add this legend, apparently based on Lederer's description of a savanna quoted earlier in this essay: "This larg sauana lies as Nilus land from May to Sep^{t}. under Water & from Sep^{t}. to May perpetualy green Stock'd with dears Variety of beasts wild Turkies and other fowles Innumerable" (Maps 82 and 103). By the beginning of the eighteenth century continental mapmakers began to abandon the Mercator type of map, frequently using the Lederer topography in its place. Most of these maps show the influence of the Ogilby or the Speed map; but usually the great lake is unnamed. In the savanna region they usually add "Plaine couverte d'eau." One widely circulated map of Delisle, the great French mathematical geographer, has "Flame couverte d'eau!" At least two maps have the Lederer details with "Georgia" written below the lake, showing that the mapmakers were keeping abreast of political developments, if not of geographical discoveries.

It was ten years after Lederer and Ogilby before another map based on actual surveys and new explorations appeared. Gascoyne's map of 1682, known as the "Second Lords Proprietors' Map," marks a great advance over any previous printed map of Carolina in both detail and correctness, and it quickly became a type map for later cartographers who knew where to go for new information (Map 92; Plate 39). It discards entirely the erroneous conception of the interior given by Lederer and incorporated in Ogilby. In general, the rivers flow correctly from the "Apalatian Mountains" in a southeasterly direction, although the upper reaches are so regular and stylized that they are evidently based on no specific knowledge. The coastline is excellent and gives evidence of careful reports gained from the greatly increased traffic among the new settlements. For Albemarle County the nomenclature of many rivers and islands has been changed from Ogilby and brought up to date, though only two plantations are given by name, "Mo. Dimone" on "Yoopine R" and "Capt. Willobies Plan." on Blackwater in what is now Currituck County.[119] Soundings along the coast, entirely lacking in Ogilby's map, extend from the Albemarle Sound Inlet southward to what is now the Altamaha River in Georgia.[120] Detailed treatment of the country around Charles Town, which is shown at its new location on Oyster Point, gives evidence of the interest in that region and of recent reports by the surveyor general, Maurice Mathews. Numerous settlements are indicated by name and the locations of thirty-three others are numbered and listed in "A Table of the names"; in addition, the Ashley-Cooper Rivers inset gives the names of landowners in still greater detail. South of Port Royal the lack of exploration has caused the continuance of errors. As in Ogilby, St. Johns River is given twice, both as St. Matheo, the Spanish name, and R. May, the old French.

The R. May is confused with the Savannah River in its upper reaches. The "Combahee River" joins the May about 180 miles from the coast to form a great triangular island, apparently a relic of the similar juncture of two rivers of that region in Ortelius (1584) and many other sixteenth-century maps. The old Indian village names attributed to Le Moyne have gone and in their place are the contemporary locations of the "Westohs" on "R. May," the "Combahees" on the river of that name, the "Cofitaciqui" below the forks of the Santee, and the "Esaws" (Catawbas) on the upper Wateree. The information concerning the Indian tribes to the interior was probably originally furnished by Dr. Henry Woodward, who for fifteen years was the invaluable intermediary in the contacts of the proprietors and colonists with the native tribes.

The confusion on the map of the rivers on which the Indians lived indicates that Woodward's information was incorrectly transmitted. By September 1670 Governor William Sayle, who had himself explored the coast, praised highly Woodward's ability to give "a more exact account of the discovery of several places and rivers then ever we heard before"; and by 1682, as chief agent for the proprietors' Indian trade monopoly beyond the settlements, he probably knew more about the hinterland than any other individual at Charles Town. He had been left by Sandford at Port Royal in 1666 to learn the ways and language of the Indians. After escaping from the Spaniards, who captured and took him to St. Augustine, he went with Governor Sayle and the other colonists to Charles Town. During the summer of 1670 he had already made a two-week march to the northwest, where he found the emperor of "Chufytachyqj . . . a Country soe delitious, pleasant, and fruitfull, that were it cultivated doubtless it would prove a second Paradize."[121] In the summer of 1671 he made a courageous and remarkable trip overland to Virginia. There is some evidence that he went by the Esah (Ushery, Catawba) villages on the Catawba River near the mouth of Sugar Creek and from there followed the Trading Path route that Lederer had used the year before (see note 104). In 1674 he made a journey to the chief settlement of the fierce and trouble-making Westo Indians, the basis of a long descriptive report to Shaftesbury on the country through which he "voiaged."[122] If Dr. Henry Woodward's knowledge of the interior had been drawn on more heavily for the map, the location of Indian tribes of the Southeast during the seventeenth century, before the wars and migrations caused by the trade rivalries of the white man, would not be such an enigma now.

The chief excellence of the Gascoyne map of 1682 lay in the surveys of the region around Charles Town by Maurice Mathews. In about 1685 Mathews furnished much more detailed information for a large manuscript map of Carolina drawn by Gascoyne which is preserved in the British Library (Map 101).[123] While the general outlines remain unchanged from the printed map, the manuscript map shows the coastal islands and rivers from Charles Town south to Port Royal with more detail and accuracy, and the names of colonists and plantations located on the map, numbered and listed in the table, have increased from 33 to 250. The wealth of detail on this large map was too great for it to be reduced to folio size; but a printed map by John Thornton and Robert Morden, London mapmakers, appeared in about 1695 which used Mathews's map for a restricted area, giving in full the details and settlements from the "French Settlements" on the Santee River to New London on the Edisto (Map 118; Plate 42). It is the first map with the name "South Carolina."[124] A French edition of the Thornton-Morden map appeared in about 1696 under Sanson's name; the South Carolina area and the Charles Town inset of the Crisp 1711 map are based upon it or directly on Mathews; and Moll relied in turn upon Crisp for many of the details on his maps and insets of the South Carolina region.

Maurice Mathews is thus the author of two important types of maps. He was the best surveyor general that the province had, north or south, until Edward Moseley; it was nearly half a century before any part of North Carolina had a map approaching in quality those based on Mathews's work for the Charles Town settlement. The value of Mathews's achievement is in its detail of settlements and rivers, not in the professional accuracy of his surveying. He came to Carolina as a planter and became surveyor general because of the total incompetence or unfitness of those who preceded him.[125] He had sailed from England with his servants to the Barbados in 1669 and thence, in a dreadful six-month voyage in a sloop, to Kiawah (Charles Town Harbor). He had powerful friends in England; on his "Uncle's the Chalanors account" Lord Ashley made him his deputy on December 15, 1671. Mathews then became a member of the Council. He soon showed himself one of the most active and capable members of the new colony. In 1679 he was Lord Craven's deputy, and in 1680 he was referred to as "an ingenious gentleman and agent for Sir P. Colleton's Carolina affairs." He was made surveyor general on April 10, 1677. In 1682 he was appointed to set out Craven and Colleton Counties and made commissioner to the Indians. He was relieved by the proprietors of his offices as surveyor general and deputy from 1683 to 1685 because of his part in the Westo wars; but in 1686 they awarded him 1,000 acres for his negotiations in buying land from the Indians. He cleared the title by conferences with the Indian chiefs from St. Helena northward, in "one generall Deed from all the sd persons of all their said Lands from the Sea to the Appalatian Mountaines."[126] In 1690–91, with Captain (later governor) James Moore, he "journied over the Apalathean mount'ns for inland discovery and trade." They may have been looking for the gold mines of the Spaniards which Woodward had kept in mind.[127] It is also possible that Mathews and Moore were opening the way for the trans-Allegheny fur trade with the Indians which, according to a note on Barnwell's map of 1721, was in operation at a factory on the land at the forks of the Coosa and Tallapoosa Rivers in 1687 (Map 184; Plates 48, 48C). If so, Maurice Mathews, who died by 1694,[128] had initiated a program which was to have international complications in the following century. The chief accomplishment of the Mathews map is its repudiation of the various false conceptions of the interior found on the Mercator-Hondius, Lederer, and Ogilby maps and those of their imitators.

The English settlements in Carolina excited, and with justice, strong misgivings among the Spaniards in Florida. Don Juan Márquez Cabrera was governor of Florida when, incited by the English, the Indians of Santa Elena, San Simon, Santa Catalina, Sapala, and Gualquini rebelled against the Spanish rule. The situation was gravely aggravated when the cacique of the Yamasees in the province of Guale also defected to the English. A map of this period, sent by Cabrera to Spain on June 28, 1683, is of interest because it gives the location of the Spanish missions and the Spanish coastal nomenclature of the period. Except for a few local maps and plans of forts, very few new Spanish maps of Florida made during the seventeenth century are extant (Map 94). Therefore the Cabrera map is one of the most important Spanish maps of this region between 1600 and Arredondo's Florida of 1742.

EXPANSION OF THE FRONTIER AND THE SURVEYS

The Expansion of the Frontier in the Early Eighteenth Century

The struggle for possession of the old Southeast by the three great European powers, Spain, France, and England, which began with Menéndez's attack on the French Fort Caroline in 1565, became increasingly open and acute from the beginning of the eighteenth century. In this struggle Spain may have had greater legal rights and France more diplomatic skill, but the English had determination and advantages in trade with the Indians. During the early years of the century the English also possessed in a group of South Carolina traders, merchants, and landowners a remarkable set of aggressive and far-sighted men whose expansionist views were to play a major part in the struggle.[129]

From time to time during the sixteenth and early seventeenth century Spaniards like Pardo and Boyano (in 1566–67) and Pedro de Torres (in 1628) and Englishmen whose very names have been lost to history explored the Georgia and Carolina hinterland.[130] Even while the struggling settlement of Charles Town on the Ashley River was being established, the Indian Trading Path, followed from the Falls of the James River by Lederer and then by Arthur and Needham, and possibly by unnamed predecessors, opened the way to a thriving fur trade sponsored by Abraham Wood and William Byrd in Virginia.[131] In the last decade of the seventeenth century French coureurs de bois in the Mississippi Valley, like the renegade officer Jean Couture, found the way along the Tennessee River and across the mountain ridges down by the Savannah River or Combahee to Charles Town.[132]

One of the first maps[133] to indicate the trade routes to and from Charles Town west of the Alleghenies is by Delisle, who later became royal geographer to the king of France. Guillaume Delisle (de L'Isle), 1675–1726, was one of the foremost European cartographers of his time. Delisle began his apprenticeship under his father, Claude Delisle, a teacher of geography and mapmaker, and under Cassini, whose use of astronomical observations for determining longitude resulted in a map of the world (laid out in 1682, published in 1696) which was of great cartographical importance. Delisle's chief merit was in his use of scientific methods and the careful examination of original sources. Unlike the Amsterdam mapmakers, who made and sold their maps for profit and copied errors so often that they became accepted as truths, Delisle was continually revising and eliminating errors in light of new scientific information. Guillaume Delisle, working with his father, Claude, began a series of notable maps of North America with "L'Amerique Septentrionale" (1700), which improved longitudinal positions, correctly showed southern California as a peninsula, and revised the conception of the Mississippi River.[134] In 1699 Claude Delisle made a map and drew up the first of a series of careful instructions for observations and corrections which were presented to Iberville, then returning on his second voyage to oversee the new colony which he had founded near the mouth of the Mississippi. The information on Delisle's manuscript map (ca. 1701), based on data "given by Iberville in 1701," may have been derived from information presented after his return in the spring of 1700 or sent back during the third voyage, which extended from the fall of 1701 to the summer of 1702 (Map 131).[135] During the second voyage Tonty descended the Mississippi to see Iberville, and it may be then that he told of the defection of his lieutenant, Jean Couture, and other Frenchmen who rebelled against the strict French regulations on Indian trade. Delisle's manuscript map of the Mississippi region (cf. Map 131) has "Route que les François tiennent pour se rendre a la Carolinne" along the Tennessee River and another legend showing the English advance along the upper reaches of the Alabama and Tombigbee Rivers toward the Mississippi, "Chemin que tiennent les Anglois de la Caroline pour venir aux Chicachas." The later English maps of the period, beginning with Nairne's inset of the Southeast on Crisp's 1711 map, show the same English route, with extensions beyond the Chicachas.

Delisle's printed 1703 map of Mexico and Florida (Map 137; Plate 43) limited the English to the land east of the Alleghenies;[136] the Nairne inset on Crisp's map specifically extends the South Carolina possessions to and across the Mississippi River, except for a corner west of Mobile (Map 151; Plate 45). Thus began a battle for territory by the geographers which paralleled the struggle for empire waged by the traders, colonial representatives, and finally the European governments. This cartographical war reached its greatest intensity with the publication of Delisle's "Carte de la Louisiane" (1718), one of the most important mother maps of the North American continent (Map 170; Plate 47). In its detail of the Gulf-region continent from Mexico to Florida, in its mention of Texas for the first time on a map, in its improved detail concerning the lower reaches of the Mississippi and other rivers flowing into the Gulf (the upper reaches of the rivers and the interior mountain ranges are less detailed and often very misleading), and in its attempt to trace the course of de Soto's journey for the first time on a modern map, Delisle's work is of historical importance. It was soon copied widely and frequently by French, Netherlands, German, and even English mapmakers. Toward the south it delimited the English from Baye St. Matheo (St. Johns), as it was on the 1703 map, to a line on the Savannah River; even in the region to the north it emphasized the error about the naming of Carolina and infuriated the English by the misleading legend "Caroline ainsi nommez en l'honneur de Charles 9 par les François qui la decouvrirent en prirent possession et si etablirent lan 15. . . ." Governor Burnet of New York wrote in anger to the Lords of Trade on November 26, 1720: "I observe in the last mapps published at Paris with Privilege du Roy par M de Lisle in 1718 of Louisiana and part of Canada that they are making new encroachments on the King's territories from what they pretended to in a former Mapp publish[d] by the same author in 1703 particularly all Carolina is in this New Mapp taken into the French Country and in words there said to belong to them and about 50 leagues all along the edge of Pensilvania & this Province."[137]

When the French began their settlements in Louisiana in 1699, the English merchants and traders of Charles Town had already made practical use of the discovery that the way to overcome the barrier of the Appalachians was to avoid it by taking the southern route to the Mississippi Valley. The earliest English printed material to show the trade routes from Charles Town westward and to give the location of Indian tribes in the Southeast is the Nairne inset in the magnificent Crisp map of 1711 (Map 151; Plate 45). Crisp, who describes himself as "a Merchant trading to Carolina,"[138] apparently was little more than the sponsor of the map. The center is a large-scale map of the settlements from Santee to Savannah, giving the name and showing the location of colonists given on Mathews MS 1685 chart, with additions by John Love[139] showing many new settlements between "New London" on the Edisto River and "Beaufort Town."

The chief interest in Crisp's map lies in the inset of the Southeast by Captain Thomas Nairne, a leading figure in the Carolinas during the first fifteen years of the eighteenth century.[140] In 1702 he took part as captain of a company in Governor James Moore's unsuccessful attack on St. Augustine, and in the same year the Assembly employed him to regulate the Yamasee Indian trade among the planters. What the Carolina expedition against St. Augustine did not achieve, Nairne's Florida forays accomplished more destructively. He wrote in 1705: "We have these two . . . past years been intirely kniving all the Indian Towns in Florida which were subject to the Spaniards and have even accomplished it."[141] On the inset map of the Southeast, across lower Alabama and northern Florida is the significant legend: "no Inhabitants from hence to the Point of Florida,"[142] and at the headwaters of the St. Johns, "Here the Carolina Indians leave their Canoes when they go to War against the Florideans."[143] The road of the traders westward from Charles Town, with the divergent paths to various Indian tribes as far west as the Mississippi, and the number of villages and fighting men shows Nairne's interest in promoting the alliance with Indians for the purpose of forcing back Spanish and French influence. At or below Muscle Shoals on the Cussate (Tennessee River) is written: "A low Riff of Rocks a fitt place to Settle an English Factory."

In 1707 Nairne was a proponent of the Indian Act, which was passed by the South Carolina Assembly to regulate flagrant abuses in the Indian trade; he was made the first provincial Indian agent. At the same time he was the leader in a bitter and successful fight to lessen the monopolistic powers of the governor, Sir Nathaniel Johnson. The governor, angered by Nairne's effective attack on his special privileges, had him thrown into the Charles Town jail in June 1708 on a trumped-up charge of treason. From jail Nairne sent to the secretary of state in England a map, apparently no longer extant, which is probably the original of the Nairne inset in Crisp's map. This map accompanied his Memorial of July 10, 1708, "one of the most remarkable documents in the history of Anglo-American frontier imperialism," according to Professor Verner W. Crane.[144] In this document Nairne urged that South Carolina was the frontier against French and Spanish domination of the continent and that a strong, aggressive policy of frontier expansion, of alliances with the Indians, and of "factories" for the fur trade with the Indians on the Tennessee River were essential to the future of the British colonies. After five months Nairne was released, apparently without trial; in 1710 he went to England, won the favor of the lords proprietors, and was made judge advocate of South Carolina. On April 15, 1715, while he was engaged in talks with the aroused Yamasees at Pocotaligo Town, the great southern Indian War broke out. Nairne was burned at the stake after tortures which lasted, according to report, for several days. Tragically, he died in a revolt against the abuses of the Indian trade, for the reform of which he had labored since his appointment as Indian agent.[145]

Nearly twenty years after Crisp's map appeared, an interesting criticism of Nairne's inset was published in the London *Daily Journal*, October 14, 1730:

> Mr. Nairn's Map of Carolina is a very defective one; and indeed the whole of that Gentleman's Map (tho' a Commissioner for the Indian Trade) is full of Errors: The Coast is wrong laid down, the Rivers drawn at random, and particularly the Wattaree River is there made to run into the Sea, whereas it runs into the Santee above 100 Miles up the Country; in short, there is not any one Map, or any one Account that has been hitherto published of that Country, that can be depended upon, the Descriptions being taken by Hear-say, or by very ignorant Persons, who have hitherto made Journals of that Country.
>
> Near the Santee River, the General Atlas[146] makes Soto, in the year 1540, travel over the mountains as high as the Appalachean Mountains, whereas at that particular Spot of Ground, it is Pine Flat Land, the lowest in the Province.[147]

The self-assured tone of this criticism, which is accurate so far as it goes but does not indicate the new and illuminating information that Nairne's inset does give, stamps it as the comment of Sir Alexander Cuming, an eccentric and energetic Scot who had recently returned from Carolina. Extracts from Sir Alexander's journal that related his amazing 1,000-mile expedition from March 13 to April 13, 1730, through the Cherokee country of the southern Appalachians had been published in previous issues of *The Daily Journal* for September 30 and October 8, 1730. Posing as an envoy from the king of England, Cuming conciliated the disaffected Cherokee and persuaded six of them, including two chiefs, to return with him to England on the man-of-war *Fox*.[148] George Hunter, the surveyor who accompanied Cuming on part of his way, made a map listing the Cherokee towns visited by Cuming, giving detailed information of the Congaree Indian Trading Path and making notes on the settlements along the coast (Map 207).[149]

It was not Thomas Nairne but a wealthy Welsh adventurer, Price Hughes, who in 1713 became the active exponent of expansion and the real director of the plans for converting the southern Indians to English trade and alliance. He made a map of the Southeast which is unfortunately lost, though a rough draft made of it by Lieutenant Governor Spotswood of Virginia is preserved;[150] this draft gives the best conception of the northeast-southwest direction of the Appalachian range so far shown on a map. "Esq. Hughes," as he is called on the Barnwell 1721 and later maps, "Master You," as he became known in Louisiana, had come to South Carolina as a planter. Catching enthusiasm from his friend Nairne, however, he traveled among the Indians as a kind of self-appointed diplomat, more skillful and less erratic than Sir Alexander Cuming was in 1730. Soon he had official backing and letters; he advanced a project for a settlement of Welshmen on the Mississippi—unfortunately on Spotswood's copy the location, which Hughes refers to as on a map accompanying one of his letters, is not given[151]—to cut in two the French communications between the upper and lower Mississippi, expand the peltry trade, and to colonize the trans-Appalachian region. It was a country, he said, as superior to Carolina as the best parts of Wales were to the fens of Lincolnshire. By his enterprise, by his successful cultivation of Indian tribes whom he drew away from the French, Hughes aroused the latter's fears; Bienville had him captured, interviewed him in the stockade of Fort St. Louis, kept papers revealing his powers and designs, and let him go. On his way back to Charles Town, in the spring of 1715, he was caught and murdered by a party of Tohome Indians, bitter enemies of the English, not far from the mouth of the Alabama River.[152] Barnwell's map shows part of Hughes's route and marks the place of his assassination with a note that it was performed at the instigation of the French (Map 184; Plates 48, 48A–D). The other side of the picture is given by a French missionary, F. Le Maire, the author of a map in 1716 that Delisle used as a base for his famous 1718 map (Map 163). Le Maire, writing of the Carolina trading enterprise of 1713–15, wrote: "Dieu rompit ce coup et par le mort du ministre Yousse [Hughes], le chef de leur ambassade aux Indiens du Mississipi, et par la révolte des sauvages des environs de la Caroline."[153]

In England the most prolific designer and publisher of maps of Carolina in the first third of the eighteenth century was Herman Moll,[154] who was as open and effective a protagonist for British territorial claims in North America as was Delisle for the French. Moll was a Dutchman who was in England in about 1680 or soon after;[155] though he died in 1732, his maps and atlases continued to be published, with revisions, during most of the eighteenth century. Starting out as a mapmaker for others, he soon began publishing his own compilations and atlases. His cartographical style is distinctive; though not as artistic as the great seventeenth-century mapmakers nor as scientific a geographer as Delisle, he combines a rather blunt clarity of lettering and considerable detail without flourishes or extraneous design. He frequently scatters short explanatory legends over his map, especially in his later large folio sheets; as he frequently takes these verbatim from his sources, his indebtedness to other maps may often be traced.

Moll's "A New and Exact Map of the Dominions of the King of Great Britain on ye Continent of North America," dated 1715, has three insets which are taken from the Crisp 1711 map: Nairne's inset, Crisp's Charles Town, and that part of Crisp's main map which extends from St. Helena Sound to the Santee (Map 158). In 1720 Moll's counterblast to Delisle's 1718 map, "A New Map of the North Parts of America claimed by France," appeared.[156] Moll calls upon the English noblemen, gentlemen, and merchants interested in Carolina to note the "Incroachments" of the French map on their "Proprieties" and on the land of their Indian allies. This map presents details of the Southeast found in no other printed map. The chief source of this information is a large, unsigned, undated manuscript map of the Southeast in the Public Record Office,[157] from which Moll took much information on trading paths, Indian tribes, French, Spanish, and English forts and settlements, rivers, and other topographical data (Map 157). One of Moll's own printed maps is pasted in the bottom left-hand corner of the manuscript.

The careful enumeration of fighting men in the different Indian tribes and the number of French and Spanish at their forts show that the manuscript was prepared by someone promoting the Carolina expansionist movement. Moll writes on his 1720 map, "A Great part of this Map is taken from ye Original Draughts of Mr Blackmore, the Ingenious Mr Berisford now Residing in Carolina, Capt Nairn and others never before Publishe'd." Mr. Blackmore's contribution was for the Bay of Fundy region;[158] most of the detail for the Southeast, even including the details of Nairne's expedition to the Florida Everglades with its accompanying fourteen-line legend, follow the unsigned manuscript map, not the Nairne inset on Crisp's 1711 map. The supposition, therefore, that the unsigned map is the "Original Draught . . . of . . . the Ingenious Mr Berisford now Residing in Carolina" is strengthened when one learns that in 1716 Richard Berresford was in London with Joseph Boone as an agent for the

Assembly, pointing out to the Board of Trade and to the secretary of state the delinquencies of the Proprietors in their not protecting the colonies from the growing encirclement of the French.[159] In a memorial to the Board of Trade in December 1717 on "The Designs of the French to Extend their Settlements from Canada to the Mississippi behind the British Plantations," Berresford attacked the Crozat grant and the French subornation of the Iroquois alliance[160] in phrases remarkably similar to Moll's legend beneath his title on the 1720 map. Berresford was well equipped to give the information on the "Original Draughts"; as early as 1707 he had been sent by the South Carolina Assembly to reduce the Savannah Indians and on June 20 of that year received the Commons's thanks for his success with the Indians and his investigation of the abuses which had led to their unrest.[161] He was appointed one of the nine Indian commissioners by the Assembly; probably few persons in the colony except the Indian agent were in a position to know more about the back country than the commissioners, who met several times a month for the greater part of the year to hear complaints arising from the Indian trade. If Berresford, who died in 1721,[162] was the author of the Indian tribes map, he probably made it sometime between 1715 and 1717.[163] It was the best map of the Carolina back country until the great manuscript map attributed to Barnwell appeared.

Events in the Carolinas during the ten years after the publication of Crisp's map in 1711 moved rapidly. First, the disturbances and uprisings of the Indians, including the Tuscarora War of 1711–12 in North Carolina and the Yamasee War of 1715, were quelled by prompt and vigorous action by South Carolinians under James Moore, Jr., and John Barnwell.[164] The Indian outbreaks started with the death by torture of two makers of Carolina maps, John Lawson by the Tuscaroras in 1711 and Thomas Nairne by the Yamasees in 1715. Second, the efforts of the Crown to take over the colony from the lords proprietors, which began as early as the reign of James II, were assured of success by 1719, when South Carolina became a Royal Province. The lords proprietors formally surrendered their charter in 1729. Almost from the establishment of the lords commissioners for trade and plantations in 1696, the Board of Trade had been the object of petitions and appeals from Carolinians disgruntled with the ineptitudes of the lords proprietors. The Board of Trade, bumbling and awkward as it was during this period, with its heterogeneous and centripetal interests, proved to be a more than cooperative medium through which the anti-proprietary party achieved its purpose. Third, the struggle for the Southeast had intensified. During the first decade the direction of southern British colonial frontier expansion was primarily anti-Spanish. By 1717 the English, Spanish, and French were all attacking one another.[165] In that year the Spanish secured the allegiance of the Apalache and Creek Indians by sending seven chiefs to the viceroy in Mexico; and the French under Bienville, who had taken over the English factory at the forks of the Alabama River in 1715, built a fort there in 1717 called New Toulouse. These master counterstrokes of diplomacy by the French and Spanish did not long go unchallenged. The English in 1717 sent traders into the French territory who secured the trade even from the Indians around Fort Toulouse. In 1717 appeared the proposals of Sir Robert Montgomery for the Margravate of Azilia to be established between the Savannah and Altamaha Rivers in the territory that fifteen years later was established as the colony of Georgia. Boone and Barnwell, who were agents for the South Carolina Commons House in 1719–21, pressed for fortifications against the Spanish as far south as the Altamaha. The French also, it was pointed out, might push their sphere of influence toward the Carolina coast below Charles Town.[166] After 1721 the French and Spanish usually worked together to defeat their common rival. "Not Oswego in 1727, but Altamaha, in 1721, saw the inception of the British eighteenth-century scheme of frontier posts to counteract French expansion," according to Professor Crane.[167]

All these different conflicts are shown, directly or indirectly, on two almost identical maps: the Barnwell-Hammerton map, drawn in 1721 (Map 184A; Plates 48, 48A–D), and now in the Yale University Library, and the large, unsigned, undated Barnwell manuscript map, now preserved in the Public Record Office as Colonial Office, North American Colonies, General, 7 (Map 184). The route of expeditions and locations of battles against the Spanish and Indians in Florida in 1702, 1703, 1706, 1708, and 1709, against the Tuscarora in North Carolina in 1711 and 1712, and against the Yamasee in 1715 are given, with detailed listing of the location and number of men in the Indian towns, as far west as the Mississippi. The maps show the exploratory route of Capt. Welch to the Mississippi in 1698, the journey of Nairne in 1708, and that of Hughes in 1715 to the place where "hereabouts Esqr Hughs was murtherd by the Indians by order of the French." At Fort Toulouse is a legend stating that before its usurpation by the French in 1715 it had been "an English Factory for 28 Years without intermission till that Time"; at the mouth of the Apalachicola is a "Spanish Fort Built 1719"; in the forks at the mouth of the Flint River is the "Indian fort built in 1716 under Cherokeeleechee their Leader." Between the Altamaha and Savannah Rivers is "The Margravate of Azilia" and at the mouth of the Altamaha is Fort King George, built in 1721 by Barnwell. This fort was established as an outpost of the English forces soon after the arrival in Carolina of the royal governor, Nicholson; it symbolizes the taking over by the English government of the defenses and administration of Carolina from the lords proprietors.[168]

Though the Public Record Office map is undated and unsigned, reasons for ascribing its authorship to Colonel John Barnwell between 1721 and 1724 are strong (Map 184). In the University of Georgia Hargrett Library is a manuscript map, made about 1744, with so many detailed similarities in legends, topography, and area that it clearly derives from the P.R.O., C.O., N.A.C. General. 7. The title cartouche of the DeRenne map in the Hargrett Library reads, "The Original of this Map was drawn by Col. Barnevelt." The spelling Barnevelt (elsewhere on the map Barnwelt and Barnewel) is a scribal error for Barnwell; the maker of the Barnevelt map, or of the map from which he copied, made several similar mistakes showing no cognizance of local names, such as "Combetree R." for Combahee River (Map 255). The journal of the commissioners for trade and plantations refers to a map which Barnwell was making in 1720. On July 28, 1720, Mr. Ashley and Mr. Danson, two of the lords proprietors of Carolina, with their secretary Mr. Shelton, appeared before the Board of Trade. "As to forts or fortifications, they said, the only fort they had of strength was at Charles Town, though they had a slight fort with a garrison up the Savannah River. That they heard the French had deserted Pensacola in the Bay of Mexico. That most of the maps yet extant of those parts are erroneous, as Colonel Barnwell, who is reputed to be, among the English, the best acquainted with that country, reports, and who is now preparing a new map of it."[169] Barnwell, who was in London at the time, appeared before the commissioners later, on August 16, 1720, to plead that it was "immediately necessary for us to possess ourselves of the mouth of" the St. Johns River to prevent its prior occupation by the French, who laid claim to it.[170] Colonel Barnwell was appointed "Ingeneer" of the province later, and upon his return to Carolina with the royal governor—a triumph for Barnwell as a leader of the anti-proprietary party—he proceeded south to the Altamaha to build Fort King George. That Barnwell drew maps is further substantiated by his autographed map of the region near the mouth of the Altamaha and accompanying plan of a block house.[171] The *a quo* date of the large map, P.R.O., C.O., N.A.C. General. 7, is 1721, since Fort King George, built in that year on the Altamaha River, is shown on the map. The *ad quem* date, if Barnwell's authorship is accepted, is 1724, since Barnwell died in September of that year. It was certainly made by 1727, since in that year Henry Popple, clerk of the Board of Trade, used it for the large manuscript draft of the map which he published in 1733.[172] Since Barnwell was preparing his map in 1720, according to the statement of the lords proprietors, since C.O., N.A.C. General. 7 gives much evidence of detailed preparation, authority, and importance, and since a later manuscript copy of this map refers to Barnwell as the author, the map is referred to in this study as one made by Barnwell in about 1721.

Catesby used the Barnwell manuscript map in preparing the Carolina area of a map to accompany his *The Natural History of Carolina*, published in London in 1731, though he did not include the legends (Map 210).[173] The Bull 1738 MS map in the Public Record Office and the copy of the Bull map which is in the John Carter Brown Library in Verelst's 1739 manuscript volume are based upon Barnwell in topographical detail and in a score of legends written on the face of the map (Map 237). Both of these maps by Bull, however, agree as against Barnwell in other features; they omit some of the detail, give "The Colony of Georgia" instead of "The Margravate of Azilia,"[174] and they include the whole of the Florida peninsula instead of only the northern part. The Barnevelt ca. 1744 map in the University of Georgia Hargrett Library, which has already been referred to, is a close copy of Barnwell, although it adds a table of explanations about the Georgia Colony settlements and omits "Azilia" without adding "Georgia." Not until Mitchell's "A Map of the British and French Dominions in North America . . . 1755," however, did a printed map appear which made adequate use of the valuable data collected by Barnwell over thirty years before (Map 293; Plate 59).

Politically and historically, if not cartographically, the Mitchell map[175] is the most important in American history. Editions of it were used for the basis of negotiations between the English and American plenipotentiaries both in 1782 and in 1783; it has frequently been used as a basis in boundary disputes thereafter, as in the North Eastern Boundary Arbitration of 1843. It was by its origin as well as by its character that the Mitchell map achieved importance;

it was issued with the approval and at the request of the British government; it was dedicated to the Earl of Halifax, who was then president of the Board of Trade; it bears the endorsement of John Pownall, secretary of the lords commissioners for trade and plantations, dated February 13, 1755. From the text printed on the second edition of the map, moreover, it appears that Mitchell had access not only to the records of the Board of Trade, but also to those of the British Admiralty.

For the southern colonies, Mitchell's chief source of information was the Barnwell ca. 1721 map (Map 184 and 184A; Plates 48, 48A–D). Presumably he used the copy now in the Public Record Office, since he was largely compiling the map from the maps and reports sent by the governors of the royal colonies and provinces to the Board for Trade and Plantations. It was the first time Barnwell's map had been used intelligently or fully, and the result is that from the Appalachian region to the Mississippi, Mitchell's map marks a great advance in printed cartographical knowledge. For the Atlantic coastal settlements Mitchell makes use of the Barnwell map, but more sparingly, for he evidently had later and fuller sources of information. Details of creeks and their names, new settlements, and the position of roads for the "improved part" of Carolina are fuller than in any other preserved map of this date, and they posit a use of the written reports to the Board of Trade to which he had access.[176]

Mitchell does not go below 28° N.L. on the peninsula of Florida, though he is evidently puzzled by the maps that followed Nairne's 1711 map, which break up southern Florida into a kind of archipelago. This error is not given in Barnwell's map. Mitchell gives a river leading south in the middle of the peninsula and adds the legend: "River unknown to Geographers thus laid down by the People of Carolina, which they make to end in several Mouths on both sides of the Cape of Florida." This is one of the first appearances on a printed map, if not the first, of the Kissimmee River, which flows from the Tohopekaliga Lakes southward to Lake Okeechobee. The map has faults, but Mitchell's judgment is nearly always good based on information which he had at his disposal; the details are frequently documented with explanatory legend, and the improvement over previous maps is marked. The southern half of North Carolina still lacks detail; he uses Moseley (1733) (Map 218; Plates 50A, 51–54). For the northern half he uses Fry and Jefferson's map of Virginia (1751). Some details are found which show that he does not rely even here upon printed maps; "Brushy Mountains" is moved to the southwest slightly, and "Pilot Mountain," to the west of the Yadkin (and present Winston-Salem), is given. For the Indian Trading Path to Virginia from Wateree, South Carolina, he has "Reckoned about 400 m to Appomatox," and above it engraved "253 m survey'd." The mountains of southwestern North Carolina and Tennessee are unimproved; he uses the detailed but faulty conception of the Barnwell map, in which the Tennessee River does not flow far enough south.

The value of Mitchell's map was immediately recognized by other cartographers. Not only were foreign editions made of the map but also imitations without acknowledgement by European geographers appeared.[177] Regional maps, such as those of the southern district in the *London Magazine* in 1755 and in *The American Gazetteer* in 1762, used Mitchell as their source. The French and Indian War created a widespread desire for fuller knowledge which Mitchell's map opportunely supplied.

Thirty years had passed since Barnwell made his map as part of his successful plea to the Board of Trade to counterattack the French claims by frontier forts and an aggressive policy; much of the information on it which Mitchell used when he made "the first Drawing of this Map in 1750" was soon to be obsolete. The Seven Years' War, of which the French and Indian War was the American phase, ended in 1763 with the Treaty of Paris. By this treaty Great Britain won Florida from Spain and Louisiana east of the Mississippi from France, with the exception of New Orleans, which France had given to Spain along with Louisiana west of the Mississippi during the preliminary negotiations of 1762. As a result, the threat to the southern colonies from other European colonizing powers, which had existed in one form or another since the beginning of the Carolina settlements, was no longer a major problem. The westward expansion of the frontier after 1763 gave rise to complications in relation to the southern Indians which caused increasing concern to the British government. Its attempt to control western emigration by the Proclamation of 1763 and to supervise Indian trade and diplomacy through other administrative measures gave frontier problems in the Southern District new significance.

Although the Cherokee War terminated in a decisive English victory in 1761, the Indians continued to feel that they had serious grievances in regard to trade and land. In 1762 John Stuart was appointed Indian superintendent for the Southern District. For the next fifteen years Stuart played a major part in all the frontier controversies in the South. The first superintendent appointed to deal with the southern Indian tribes had been Edmund Atkin, who entered the duties of his office in 1756.[178] His services, hindered by personal faults and lack of official support, were not important. Upon his death in 1761, John Stuart was nominated for the position by Governor Thomas Boone of South Carolina and by others; his commission was dated in London, January 5, 1762, although he did not actually receive the document until February 1763.[179] Stuart, who was born in Scotland in 1718, arrived in Charles Town as a merchant in 1748. Although unsuccessful as a trader, he proved to be an able superintendent. As captain in the South Carolina provincials, he had taken part in the Cherokee War, had been captured during the massacre after the surrender at Fort Loudon in 1760, and had escaped through the aid of his friend Attakullakulla, the Cherokee chief. On his "Map of the Cherokee Country" Stuart drew "The Road by which Capt. Stuart escaped to Virginia" (Map 328).[180]

While superintendent of Indian affairs for the Southern District, Stuart directed the production of a notable series of maps of the southern region. Before or at the beginning of his superintendency he made the autographed map mentioned above. In 1764 came "A Map of the Southern Indian District" and in 1766 "A Map of the Indian Nations in the Southern Department" (Maps 341 and 352; Plate 61). Hillsborough, Dartmouth, and the Board of Trade wanted fuller and better maps of the Indian country than they possessed in order to deal with the westward drive of the pioneers and the pressure for large grants of land by speculators and promotion companies, which was resulting in encroachment on the territory reserved for the Indians and in shrewd bypassing of official procedures in purchasing land from the Indians. Stuart's position was of extreme delicacy and difficulty; he received pressures from all sides. His supervisors in England, the local colonial governors, the Indians whom it was his duty to see fairly treated and to keep as peaceful as possible, and the representatives of the land companies were perpetually presenting him with their contrary demands and claims. Stuart arranged for meetings with the Indians to negotiate new boundary lines; the resultant treaties usually required surveys. As the maps were usually made by surveyors appointed by the governors of the colonies and provinces involved, Stuart's name does not as a rule appear on them.[181]

To the earl of Hillsborough, secretary of state for the Southern Department, Stuart had written a letter on July 30, 1769, promising a large map. On February 11, 1771, Hillsborough wrote Stuart complaining that the promised map had not yet been sent; his successor, the Earl of Dartmouth, wrote Stuart on September 2 and again on September 27, 1772, that the promised map had not yet been sent.[182] Stuart's excuse was that the existing maps were too inaccurate to be relied upon. However, the pressure from London was sufficient to make some map necessary, and in the next three or four years several maps were drawn under the direction of Stuart. The surveys by Bernard Romans and David Taitt of West Florida and of the Choctaw country, by Taitt of the Creek country, by Gauld of the Gulf coast region, by Savery of the Georgia-Creek boundary, and other surveys were used in a large map once in the possession of General Gage and now in the W. L. Clements Library (Map 440). This map was made in about 1773.[183] Another large map, now in the Ayer Collection, Newberry Library, was made about the same time by Joseph Purcell in Stuart's office and under his direction.[184] In 1774 the draftsman of the lords commissioners for trade and plantations, Samuel Lewis, made a copy of one of Stuart's maps, the Stuart-Gage map or one very similar to it; this map by Lewis is now in the British Library; it is described under Map 446 in the Map List. Finally, in 1776 Stuart sent to the Earl of Dartmouth a very detailed chart of the Southern Indian District, which is now in the Public Record Office (Map 438A).[185] The delineation of the southern provinces on Stuart's maps follows printed maps: Fry-Jefferson for Virginia, Collet for North Carolina, Cook for South Carolina and for part of western North Carolina showing the 1722 survey.

The coastal areas of these maps from Stuart's office are, in fact, no longer merely descriptive, for they embody the careful surveys that had been in progress for more than twenty years.[186] For a study of the surveys along the Atlantic coast it will be necessary to retrace our steps and investigate the activities of a group of surveyors who began their work in about the middle of the eighteenth century.

The Surveys: The Third Quarter of the Eighteenth Century

The early official surveyors of the province were often unsatisfactory in training, character, and methods, as contemporary records abundantly testify. Their chief duty lay in defining specific grants of land to individual colonists, an activity for which they were paid fees; the appeals for general maps of the country by the lords proprietors and the Board of Trade usually went unheeded. Along the coast, however, need of accurate surveys to show soundings and good harbors for navigation resulted in a steady improvement of hydrographic charts.[187] In 1729 Captain John Gascoigne of H.M.S. *Alborough* made detailed surveys of Port

Royal Harbor and of the river and sound of D'Awfoskee; these were used by De Brahm in his map of South Carolina in 1757 (Map 310; Plates 59A–D).[188] In 1733 Edward Moseley published a large map of the North Carolina coast which was the result of more careful surveys than had before been made for that province (Map 218; Plates 50A, 51–54).[189] The rivers, sounds, and location of settlements and individual settlers are given from Winyah Bay, which North Carolina was at that time claiming as its southern boundary, to Virginia. The map gives evidence of considerable accuracy where settlements had developed. It has but little information for the soundings and channels along the coast and up the rivers. The insets of Ocracoke, Topsail Inlet, and the Cape Fear River, which remedy this lack on the general map, may have been supplied from surveys made by Governor Burrington, who wrote to the Board of Trade on January 1, 1733, that he had sent them drafts, made by his orders, of these three harbors, which he states optimistically "will admitt the largest merchants ships"; "it can hardly be imagined what pains I took in sounding the Inlets, Barrs and Rivers in this Province, which I performed no less than four times; I discovered, and made known the Channells of Cape Fear river and Port Beaufort or Topsail Inlett, before unused and unknown."[190] In 1738 James Wimble of Boston published a hydrographic chart of the North Carolina coast which supplements Moseley's map by more detailed soundings and surveys along the coast, in the sounds, and up the rivers to the main settlements (Map 241). During the 1750s Daniel Dunbibin, a North Carolina pilot, made extensive surveys along the Carolina coast.[191] The General Assembly paid his widow compensation for his expenses in 1764, but his chart does not appear to have been published until 1792. The maps of John White (1585), Comberford (1657), Lancaster (ca. 1679), Gascoyne (1682), the Thames School (1684), Wimble (1738), and Dunbibin (1792 for ca. 1756) probably afford the best illustrations available of the much disputed changes which have occurred in the cusps and inlets of the North Carolina Banks during 200 years of settlement before the Revolution (Maps 8, 50, 81, 92, 99, and 241).[192]

The provincial boundary surveys and the lines drawn as a result of treaties with the Indians came with greater frequency as the population expanded and problems arising from border disputes increased. Though these surveys were detailed in showing the location of roads, rivers, and other landmarks directly on the line, they increased the general topographical knowledge of the interior but little. One of these boundary surveys, however, appears to have resulted indirectly in an important contribution to cartographical knowledge. In 1749 William Churton and Daniel Weldon were appointed commissioners on the part of North Carolina to extend the Virginia–North Carolina boundary line. The commissioners for Virginia were Joshua Fry and Peter Jefferson. The line was run ninety miles farther west to Steep Rock Creek beyond the Blue Ridge. At this time, if not before, the Virginia commissioners must have become acquainted with Churton. As a surveyor in the land office of the Granville District, Churton was in a position to know what information had been gathered about the district.[193] In 1752 Bishop Spangenberg refers to Churton's work in laying out the purchases of the Moravians.[194] At this time, however, the bishop wrote that "there is neither a general surveyor's map of the Granville District nor of the individual Counties," although Francis Corbin, Lord Granville's agent, had told him that "he had been doing his best to have one made, and had given orders to the surveyor in each County to make a chart showing the land that had been taken up in his County."[195] If, as is probable, Churton furnished the Virginia surveyors with information for the Granville District, he contributed a good deal more after they drew the map in 1751, for the second edition in 1755 shows extensive increases in knowledge, particularly for the area which he surveyed for the Moravians and for new land grants to the west. No such acknowledgment, however, is made on the map or in Burwell's description of the sources of the map in his letter of transmittal to the Board of Trade.[196] In any case, Fry and Jefferson's map is the chief basis of the Mitchell 1755 map for the cartography of the northern Carolina area and was followed by most European mapmakers until Collet in 1770.

The inclusion of the Granville District must have been an afterthought, as the involved history of the Fry and Jefferson map shows. The need for a map of Virginia, long felt, became urgent when the contest with France for the Ohio Valley became an important international and political issue. A definite order for a chart or map of the boundaries of Virginia was sent to the council by the Board of Trade by the summer of 1750; Colonel Lewis Burwell, sworn in as acting governor on November 21, 1750, reported shortly afterward that he had "employ'd the most able Persons" to prepare the map. Both Joshua Fry and Peter Jefferson (the father of Thomas), the two men appointed by Burgess, were particularly well qualified for the project. The plan of such an undertaking had long been in Fry's mind;[197] he and Jefferson had made numerous surveys together, including the extension of the Virginia–North Carolina line in 1749 ninety miles beyond the Byrd-Moseley line of 1728.[198]

Their work did not stop with the drawing of the map in 1751;[199] Fry continued to gather material for a new edition, especially for the western region, for which both his wide interests and his activities afforded him unusual opportunities to get information. In 1752 Fry was one of four commissioners appointed to treat with the Six Nations, an undertaking which took him across the ranges of the Appalachians and resulted in the important treaty of Logstown on the forks of the Ohio. In company with Fry was Christopher Gist, a pioneer whose famous explorations on behalf of the Ohio Company must have given Fry a source of knowledge which contributed to the redrawing of the northwestern section of the map.[200] More significant in this study is the information that Gist could give him about the region around Gist's home on the banks of the Yadkin near Daniel Boone.[201]

In 1754, while stationed at Winchester in command of the Virginian Regiment under General Dinwiddie, Colonel Joshua Fry died and was succeeded in command by a young officer named George Washington. Under Fry's command, engaged in quartermaster's duties which took him all over Virginia, was a young Scots officer, John Dalrymple. Dalrymple returned to England in the summer of 1754, after Fry's death. Doubtless pursuant to some previous understanding with Fry, Dalrymple arranged for the publication in London of the 1755 edition of the Fry-Jefferson map (Map 281).[202] This "Dalrymple" edition is notable for the greatly increased detail in the trans-Allegheny region and in western North Carolina, showing in the latter case an increase of knowledge with which Dalrymple probably had little to do. The eastern part of North Carolina is unchanged, but the rapid and important expansion induced by Bishop Spangenberg's visit in 1752 and by the general westward migration shows in the western reaches.[203] "Unitas" (for the Unitas Fratrum) in "Wachaw" (from Zinzendorf's Wachau), "Anson County," "Gist, Junr, Mulberry Fields," "Cossart 4000 [acres]" at the "Yadkin Forks," "I Perkins" on the "Catawba R." beyond the "Brushy Mountains," and "Head of the Yadkin" near present Blowing Rock, North Carolina, show the rapid increase in knowledge and settlements between 1751 and 1754.

The most prolific mapmaker in the Southeast during the third quarter of the eighteenth century was John William Gerard von Brahm (later he signed his name William Gerard de Brahm), 1717–99, a German who had served as captain of the engineers under Emperor Charles VII.[204] In 1751 De Brahm came with 160 Salzburgers to Georgia, where he founded Bethany. The community's population and prosperity rapidly increased to an extent remarkable even among the thrifty Salzburgers. De Brahm was made one of the surveyors for Georgia in 1754 by the king's appointment. Three years later he had completed his extensive surveys of South Carolina and Georgia and published the result in his "Map of South Carolina and a Part of Georgia" (Map 310; Plates 59A–D).[205] This map, in its accuracy for the coastal area and its thoroughness for the region covered, was far superior to any cartographical work for the Southern District that had gone before. Before De Brahm's map, the size and shape of the coastal islands, the shape and direction of the rivers, and the location of settlements had for the most part depended upon the rough but very valuable maps of the Indian agents and boundary commissioners. With De Brahm we turn from the amateur to the professional, from the general outlines of the region to topographical accuracy. However much De Brahm must have owed to the other surveyors whom he mentions—Bull, Bryan, and especially Captain Gascoigne—he was the author, and to him should go the credit. The map did not attempt to give topographical details far back to the interior. The reports and surveys gathered by John Stuart, the Indian agent, were added to information derived from Cook's 1773 map for the 1780 edition of De Brahm's map, which gave greatly increased detail for the western part of South Carolina.

De Brahm's map of the two provinces neither stopped his other interests nor even wholly occupied his energies. In May of 1756 he joined Governor Glen's abortive expedition to the Cherokee Country to build a fort to protect the Indians from French raids.[206] He drew a plan of Fort Loudoun, which he constructed under orders from Glen's successor, Governor William Henry Lyttelton, in 1756;[207] in 1757 he fortified Savannah (Maps 308, 410, 414, 424, and 428). At Ebenezer he erected the fort and took lands on both sides of the river, for he was a speculator as well as an engineer (Maps 412 and 426). In about 1761 he built again Fort George on Cockspur Island (Maps 416 and 430). De Brahm was a military engineer; his knowledge of the most recent and effective architecture for military fortifications, acquired by his training in Europe, was urgently needed by the young provinces. His work was rewarded by the king; ten years after he had been appointed a surveyor for Georgia at a salary of £50 per annum, he was made, in 1764, surveyor general for the Southern District, at a salary of £150 a year, with a paid deputy, Bernard Romans. For the following ten years he conducted numerous and careful surveys of the Florida coast.[208] In 1770 the governor of East Florida, James Grant, suspended him for "incivilities" and other charges.[209] He continued his work and in 1773 finished his remarkable report as surveyor general to the Board of Trade. Not only does the report give maps and fortifications (see map list under the year 1773), it also includes a history of the provinces and notes on many germane subjects. Before this, in 1772, appeared his *The Atlantic Pilot*, printed in London. Though avowedly a manual of sailing directions, it is primarily a discussion of the supposed course and origin of the Gulf Stream and related currents, a pioneer appraisal of this important subject.[210] His fortunes fluctuated after leaving Florida for London; his son-in-law, Captain Mulcaster, was appointed surveyor in his place; De Brahm's wife died in 1774; he was recommended by the Lords of the Treasury for reinstatement as surveyor in 1775; and in 1776 he married Mary Fenwick, a Charleston widow. In 1777 he was banished by the Carolinians and "returned to England with great expenses" in the company of sixteen of the king's officers. He had been a prisoner at large in South Carolina for refusing to abandon his loyalty to the king, although that loyalty was later questioned when he was confused with his nephew, Ferdinand Joseph Sebastian De Brahm, who became a brevet lieutenant colonel in the Continental Army. In 1780 he wrote to Lord Shelburne from Topsham, near Exeter, that he had received orders to return to St. Augustine but was prevented by illness from going. He had not been reinstated in office nor had he received his annual salary; he requested his lordship's protection and aid.[211] He later returned to America, but apparently the Revolution left him with little source of income except for belated returns on his Florida land speculations. He lived on in Philadelphia until a year before the close of the century, writing on religious subjects. Plowden Weston, who edited De Brahm's "Philosophico-Historico-Hydrogeography of South Carolina" in 1856, says that he "lived in the memory of persons now alive, much addicted to alchemy and wearing a long beard."[212]

In North Carolina, William Churton of Edenton, whose probable contribution to the Granville District section of the Fry-Jefferson 1751 map has already been discussed, began to collect information for a map in about 1757, after some ten years of surveying for the settlements in the Granville District.[213] After ten more years of correction and improvement, he presented a draft of the map to Governor Tryon and to the Assembly, of which he had been an active member, and they allowed him a "handsome gratuity" of £155 to defray the costs of seeing the map through the press in England. Churton's draft was primarily an improvement of the Granville District shown in the Fry-Jefferson map on the basis of actual surveys; Governor Tryon was probably reinforcing the opinion of "Captain Gordon, chief engineer in America" when he said that "of this laborious work, I am inclined to think there is not so perfect a draft of so extensive an interior country in any other colony in America."[214] Governor Tryon told Churton that if he took pains to perfect the maritime and southern parts of the province, he would recommend him to the Board of Trade for his Majesty's benevolence in consideration of the extra expense and labor involved. When Churton, who had drawn the coastal region from old maps and other secondary sources, began to take actual surveys in 1767, he became so disgusted with the inaccuracies of his draft that he cut off that part of the map. While he was engaged in these maritime surveys, he wrote Governor Tryon that, if an accident should befall him, he left the map to the governor.[215] Shortly after this, in December 1767, Churton died; thus ended the efforts of a man who, though his name is on no extant published map as the author, probably did as much as anyone in the eighteenth century to extend the cartographical knowledge of the interior of North Carolina.

The results of Churton's labor were not to be lost; in the newly appointed commander of Fort Johnston, a young Swiss named John Abraham Collet, Governor Tryon found an able person to continue the work. Captain Dalrymple had died in 1766 in the midst of disturbances resulting from the Stamp Act so serious that he had, under orders, spiked the guns of the fort to prevent them from falling into the hands of the enraged citizens. Tryon had then appointed as commander Robert Howe, who was destined within ten years to be appointed one of the first brigadier generals by the Continental Congress.[216] The British government shrewdly wished someone in such a position to have fewer local connections than Howe had. Lord Shelburne sent the following letter to General Gage:

London August 2d 1767

Dear Sir:

I beg to recomment to your Favour and Protection the young man, who is the bearer of this, Mr. Collet, whom the King at my humble recommendation has appointed Governor of Johnson's Fort in North Carolina. He is a Swiss by birth, & appears to me to have had

> a very good Education, and to be very capable of recommending himself in the Profession to which he was bred that of an Officer and Engineer, having been some time in the French service. Mr. Henry Grenville at whose desire I first mentioned him to the King assured me that he was an extreme good character and of a very honest prudent disposition. At the same time I think it fit you should be acquainted with the particulars of his appointment, I shall be much obliged to you if you'll honour him with your Notice, and do him such service as from your observation of him you may think he merits. . . .
>
> I have the Honor to be with the most sincere Regard and Esteem.
>
> Sir Your most Obedient and most Humble Sr.
>
> Shelburne.[217]

Before the end of the year Collet had already reached North Carolina, for on December 13, 1767, Governor Tryon sent him to New York with the old Great Seal of North Carolina, which the royal warrant had required to be sent back to Whitehall upon receipt of the new seal. In December 1767 he also made a plan of Fort Johnston which he sent to Lord Shelburne.[218] In the next year Governor Tryon appointed Collet an aide-de-camp on his expedition against the Regulators at Hillsborough in September, a position which Collet filled with such gallantry and merit that Tryon recommended him for higher service and entrusted him with a report on the condition in the colony upon his return to London later in the year.[219]

Early in December 1768 Collet sailed for England with a large draft of a map of North Carolina, in three parts.[220] Tryon had lent Collet the map which Churton had prepared from actual surveys; Collet, with an assistant, made a large draft copy of Churton's work. Churton, it will be remembered, had in vexation over inaccuracies of the maritime part cut his map in two. Tryon, in his letter of October 28, 1768, to the earl of Hillsborough, which describes Collet's draft, notes that it is in three parts. This manuscript map is now in the Public Record Office. The first two parts, No. 1 chiefly of the Granville District and No. 2 of the coastal area, follow Churton's map with his ten years of laborious revision; No. 3, of "Mecklenburg County" west of New Bern and Bogue Inlet and south of New Bern, Cross Creek, and Catawba Town, is drawn on the same scale as Nos. 1 and 2 but does not fit together with them perfectly.[221] Governor Tryon referred repeatedly to this manuscript map in his letters to the authorities in England concerning the boundary line disputes with South Carolina and in his appeals to have surveys to establish lines so that bitter feeling over taxation and overlapping land grants could be settled. He requested that Collet be given funds and assistants to complete the survey. Not much evidence of new surveys is to be seen, however, on the Collet 1770 printed map, though in Mecklenburg County "Charlottesburgh" has been added to "Ct. House," the "Catawba Indian Tribe" rectangle surrounds "Catawba Town," and a few additional names of settlers dot the map. The maritime part of the chart gives details and soundings which are not on the 1768 manuscript; many of these apparently stem from Wimble 1738, although such changes as the closing of the Hatteras Inlet, substantiated by Dunbibin's surveys of the 1750s, show that Wimble's map is not the sole source of information.[222] Collet's is a handsome map, beautifully engraved by I. Bayly and is the basis for most subsequent maps of North Carolina until many years after the Revolution (Map 394; Plates 62A, 63–66). It is probably true that the chief credit for it should go to William Churton, whose fundamental contribution is acknowledged on neither the manuscript draft nor engraved work; it is certainly true that Collet's years in North Carolina ended in misfortune. The Cape Fear colonists resented his assiduous attempts to strengthen Fort Johnston, and rumors circulated against him. Apparently Collet was treated with undue severity by the Patriots, who in July 1775 burned the fort, destroyed his home, barn, and other property, driving him out of North Carolina with such serious accusations and in such anger that Governor Martin advised General Gage that his usefulness there was at an end.[223]

Probably the most finished set of plans made in the Carolinas before the Revolution was the result of surveys of the chief towns in North Carolina by Claude Joseph Sauthier, a Frenchman who had been brought to Carolina by Governor Tryon in 1767 as draftsman and surveyor. In October 1768 Sauthier surveyed Hillsborough; during the next year he surveyed Brunswick, Bath, New Bern, Edenton, Halifax, and Wilmington; and in 1770 he surveyed Cross Creek (Fayetteville), Salisbury, and Beaufort (Maps 363, 371, 372, 373, 374, 375, 376, 377, 378, 390, 391, and 392). In January 1771 Sauthier presented a set of plans of these towns to the North Carolina Assembly, together with the original survey of the province made by William Churton, to which was added a draft of the maritime region laid down by Sauthier himself from surveys given him by different persons. Governor Tryon wrote to the Assembly, recommending that Sauthier be recompensed for the "fatigue and expence of travelling," the "time in performing those survices," and "the integrity of the Gentleman." The Assembly evidently concurred in this instance with Governor Tryon, for they voted Sauthier £50, a sum validated by the council on January 22, 1771.[224] Later in the same year Sauthier drew several plans of the Battle of Alamance (Maps 395, 396, 397, and 398). None of these beautifully executed plans was published at the time, though some of them exist in two or more copies.

Sauthier was born in Strasbourg, Alsace, on November 10, 1736, the son of a saddler. The politically strategic position of Alsace necessitated careful military surveys, including highly technical cadastral mapping, in the 1750s, for which local draftsmen received training. The extent of Sauthier's participation is unknown, but numerous examples of his work, preserved in the library of the Grand Seminaire of Strasbourg, show high technical proficiency as well as artistic skill. The library also contains a treatise or textbook by Sauthier on landscape architecture and civil architecture, with beautiful colored designs, including drawings for a governor's mansion, made in 1763. It may be that Governor Tryon knew of this work when he employed Sauthier in 1767; at the time he was building a governor's mansion in New Bern, for which he had engaged a New York architect, John Hawks. Tryon expected it to be, when finished, the finest such edifice in the American colonies. Sauthier's plan of New Bern, surveyed and drawn in 1769, shows elaborate formal gardens between the mansion and the Neuse River. The garden may not have been completed; it has been suggested that Sauthier drew a plan that he had been told to lay out and execute.

In 1771, when Tryon was appointed governor of New York, he took Sauthier with him. During the next few years Sauthier made extensive surveys of the province and of the city of New York. In 1773 Tryon appointed him surveyor for the province to run the boundary line between New York and Quebec. At about this time he acquired from Governor Tryon 5,000 acres in the township of Norbury, present-day Vermont. He was not, however, destined to become a New England colonial. In 1774 he accompanied Tryon to England and returned with him in 1775 to the incipient turmoils of the Revolution. After General Howe landed his troops from Halifax on Staten Island in 1776, he ordered Sauthier as a military surveyor to map the island. The map, water-stained apparently from field use in the rain, is located at Alnwick Castle. When Percy made his attack on Fort Washington, Sauthier had surveyed the field and made a map, which is now in the Faden Collection of the Library of Congress. Percy evidently liked Sauthier and retained him on his staff when he made his headquarters in Rhode Island. In 1777, when Percy returned to England, Sauthier accompanied him and became his private secretary. Between 1785 and 1790 Earl Percy, by then the second Duke of Northumberland, had him make several estate maps. Sauthier returned to Strasbourg, where he lived at 14 de la Grand'rue, near his distinguished younger brother, Joseph Phillippe Sauthier (1751–1830), who at the time was professor of theology at the Grand Seminaire. He died on November 26, 1802, at age sixty-six.

In South Carolina the Commons House of Assembly had voted £18,000 in currency for a new survey of the entire province. The commissioners had appointed Tacitus Gaillard, a rich merchant, and James Cook, who in 1766 had published a draft of Port Royal Harbour, as the surveyors. On February 22, 1770, these two requested £4,000 for the expenses of engraving the map which they had completed.[225] On March 21, 1770, the map was presented to the Assembly,[226] and on the following day, although the commissioners had found "several little inaccuracies in the Map," it was voted to pay the balance of the fund assigned to Gaillard and Cook. On March 28, 1770, Cook appeared before the House and reported that the commissioners appointed by North and South Carolina had run the boundary line 11 miles south of the latitude it should have been run according to their instructions. Thereupon the House voted to take up the matter with the governor of North Carolina and, if no cooperation was given by him, with his Majesty.[227] On March 29, 1770, the House voted to pay Gaillard and Cook an additional £700 for adding the names of the owners of the houses which were marked upon the map. Cook also requested authority, without compensation asked, to survey an additional large body of land, lately added to the province, of 72 by 32 square miles, or 1,514,560 acres, "without a survey of which the map as now extant" would be incomplete.[228] In June 1772 James Cook was one of the surveyors appointed to run the boundary line between North and South Carolina.[229] Cook's 1773 map of South Carolina includes the new acquisition of 1772, the names of the owners of homes indicated on the map, and also gives evidence of surveys extending over the entire province as it was then defined, together with inset maps of the chief towns (Map 443; Plate 67). Very few copies of this fine map are extant.[230]

In 1775 appeared the large map of North and South Carolina by Henry Mouzon, Jr., which, because of its use by the American, British, and French forces, may be called the Revolutionary War map of North and South Carolina (Map 450). George Washington's copy of the map, folded and cloth-backed for saddlebag use, is in the American Geographical Society library. The copy of Lieutenant General J. B. D. de Vimeur Rochambeau, who with his French troops marched to Yorktown with Washington, is in the Library of Congress, as is a French 1777 edition of Mouzon's map, on which is marked in a contemporary hand with notations in red ink the March 1781 route of "Conouolis" (Cornwallis) to Guilford Courthouse in North Carolina. General Henry Clinton's British headquarters copy is in the William L. Clements Library in Ann Arbor, Michigan.

The identity of the author of the map is not certain. There were apparently two Henry Mouzons who were contemporaries, South Carolinians, and first cousins. Their grandfather, Lewis Mouzon, had two sons, Lewis, Jr., and Henry, Sr., both of whom had sons named Henry. The son of Henry, Sr., was Captain Henry Mouzon, Jr., of Kingstree, who served under General Francis Marion and was wounded in the Revolutionary War. He died on August 25, 1807.[231] There is no contemporary evidence that he was a surveyor or mapmaker. The son of Lewis Mouzon, Jr., Henry Mouzon of Craven County, who apparently also signed himself Junior after a custom not rare in the eighteenth century, was a surveyor. His will, signed October 19, 1775, leaves to his nephew "all my wearing apparel & surveyor's instruments."[232] The inventory of Henry Mouzon, Jr., deceased, made on April 17, 1777, lists "a parcel of surveyors instruments," "Sundry Maps & 2 copper plates."[233] A number of surveying plats and receipts signed by Henry Mouzon, Jr., are dated before the Revolution but none after. The signature of Captain Henry Mouzon for requisitions, audits, and the like that continue for many years after the Revolution appears to be in a different handwriting from that of the surveyor. The authorship of the maps of Henry Mouzon, Jr., has been accepted as that of the captain, son of Henry, by several previous writers.[234] On the basis of the documents mentioned above, however, the mapmaker is presumably Henry of St. Stephens's Parish, Craven County, son of Lewis, Jr. But the evidence is not completely conclusive, and the possibility that new and contrary evidence can still be found must be considered.[235]

For a generation, until Price-Strother's North Carolina map appeared in 1808, Mouzon's map, rather than Collet's 1770 map (on which Mouzon's is closely based), was the immediate source of most maps of North Carolina. Mouzon attempted, however, to bring his information up to date. Some new counties are given and they are distinguished more clearly than in Collet, although neither shows the boundary lines. Mouzon shows Tryon County, which though formed in 1768 is not on Collet. Mouzon also added Pelham County, in the northern province, above Brunswick. (Pelham was proposed as the name of a new county, but when it was created in 1784 the area was called Sampson.) Surry, Guilford, Chatham, and Wake Counties were formed in 1770 but are not on Mouzon's map, nor is Martin County, established in 1774; Beaufort and Hyde Counties, created in 1712, are on neither Collet nor Mouzon, though both are on Moseley 1733.[236] Some of the soundings along the coast differ from Collet and Moseley, indicating that Mouzon used some independent source. The topography west of the Catawba River is more detailed and accurate than on Collet and forms Mouzon's chief additional contribution to the North Carolina area; he showed increased knowledge of the course of the rivers west of the Catawba and adds roads, the names of smaller streams, and physical features like "White Oak or Tryon Mountains" and "Kings Mountain" (which he places in North Carolina). Mouzon made a special map of this area in 1772, showing the "New Acquisition" which South Carolina gained from the adjusted boundary line drawn in that year. As a compensation for the 1764 line, which was erroneously surveyed eleven miles below the thirty-fifth parallel, the governments of the two provinces agreed to run the extension of the line from the Catawba westward to Tryon Mountain in the White Oak Mountains eleven miles north of the thirty-fifth parallel.[237]

For South Carolina Mouzon based his map on two earlier maps of the province by Lodge-Cook (1771) and James Cook (1773), which in turn had been based on the great De Brahm map of South Carolina (1757), with its excellent surveys of the eastern precincts. Mouzon had himself made extensive surveys for the South Carolina legislature. He included on his map delineation of the rivers and Indian settlements west of the Cherokee Indian boundary lines run in 1766 and 1767.

In 1772 Mouzon made a map, mentioned earlier, of that part of Mecklenburg and Tryon Counties in North Carolina that had been allotted to South Carolina in the line run that year.[238] In 1773 he published a "Map of the Parish of St.

Stephen," engraved in England and printed in Charleston, which was the result of his surveys to find a suitable place for a canal between the Santee and Cooper Rivers (Map 439). The Revolutionary War interrupted the project; after the war Major Senf[239] surveyed and proposed a new plan and location for the canal, discounting and disregarding the three or four proposals of Mouzon. Senf's route was built across a ridge and navigation was often impossible because of a lack of water at the summit, which was sixty-nine feet above tidewater level. The subsequent unsuccessful history of the canal showed that Senf was wrong; in the selection of his route Mouzon's plan was completely vindicated and is approved by engineers to this day.[240]

In 1771 the governor of South Carolina, Lord Charles Greville Montagu, appointed Henry Mouzon, Jr., and Ephraim Mitchell "to run out and mark the Boundary Lines of the several Districts in this Colony."[241] On April 25, 1775, Mitchell and Mouzon, stating that their surveys had been completed, as they apprehended, to the satisfaction of the House, petitioned the Commons House of Assembly for the full amount of pay, only part of which had been approved. Perhaps the House's hatred of Governor Montagu's actions caused niggardly payment of any undertaking proposed by him, even to so loyal a patriot as Mouzon; at any rate, the House committee met on the next day and refused the petition, being of the opinion that the sum previously allowed was "fully sufficient for that service."

In the previous year, 1774, a long advertisement concerning his map, signed by Henry Mouzon, Jr., had appeared in the South Carolina *Gazette*.[242] Mouzon's advertisement cited the details in which his map proposed to differ from those of De Brahm and Cook. Actually, Mouzon's map differs from that of Cook very little, and a careful comparison of the two maps indicates that some of his claims are unjustified. The names of house owners in Cook tally with those in Mouzon, though the latter omits a few names. There are some boundary line improvements, and Mouzon's map extends beyond the Cherokee line, showing roads, villages, and forts that are not on Cook's map.

For North Carolina, Mouzon relies largely on Collet. He lacks a few items along the seaboard which are found in Collet, and the artistic detail and fine workmanship of Bayly, which make Collet such a beautiful example of cartography, are not approached by Sparrow, Mouzon's engraver. Mouzon's map adds information, however, not found on Collet's map. The soundings along the North Carolina coast are somewhat fuller. The chief difference between the two is found in the western reaches, from Charlottesburg (Charlotte) past King Mountain (shown by Mouzon as in North Carolina but not given by Collet) to the Tryon Mountains. This was the area surveyed for the new boundary line in June 1772, and the new location of the reservation for the "Catawba Nation" below Charlottesburg is given in Mouzon's MS map of that year. The rivers both north and south of the line are more correctly delineated than on earlier maps. Various roads or paths not shown on Collet's map are for the first time noted;[243] roads from north, east, south, and west meet at King Mountain; the road westward joins one from Virginia which continues south from Table Mountain. These two roads join slightly west of the present site of Rutherfordton, North Carolina, thence continuing southwest to Fort Prince George at Sugar Town on the Isundigaw (Keowee River) and so on to Tugeloo on the Savannah.

In general, Mouzon's map both used and improved upon the work of De Brahm and Cook in South Carolina and of Churton and Collet in North Carolina. For many years it was the basis on which maps of the two states were made. It was not until political boundary changes and the rapid western flow of population necessitated new surveys (which would be ordered by the state legislatures in the second or third decade of the nineteenth century) that mapmakers added significant detail to the delineation of the Carolinas shown on Mouzon's map.[244]

The exploration and charting of the Southeast was accomplished in a little over two and a half centuries after the first European voyagers sailed along the coast. In the first half of the sixteenth century maps of the period showed vague and often indeterminate outlines of the shore line. The abortive attempts at settlement in the second half of the century contributed, especially in the drawings of Le Moyne and White, a much clearer conception of the topography of the coastal areas. The development of actual colonization during the seventeenth century combined with the flowering of European cartographical art to produce a rapidly increasing number of regional maps which followed the expansion of geographical knowledge. In the first half of the eighteenth century the rivalry of English, Spanish, and French political powers and the fluctuating contests resulting from the Indian trade are reflected in illuminating detail by maps which add explanatory legends to the topographical description. During the third quarter of the eighteenth century De Brahm, Cook, Churton, Romans, and others made accurate surveys along the coast. The series of great maps drawn in the office of John Stuart just before the Revolutionary War were based on these scientific coastal surveys and recorded the chief topographical features of the interior, including the location of settlers and Indian tribes westward to the Mississippi. Thus the maps of the Southeast, taken as a whole, give a magnificent contemporary account of the historical development of geographical knowledge and the expansion of settlements and trade during the colonial period.

APPENDIX: CHIEF TYPE MAPS IN THE CARTOGRAPHY OF SOUTHEASTERN NORTH AMERICA

I. Ortelius type

- A. Ortelius-Chiaves 1584
- B. Laët 1630 A
- C. Sanson-Floride 1656 B
  - Delisle 1701 MS
- D. Delisle 1703
  [containing much new data from Iberville and other French explorers of the Mississippi Valley]
  - Le Maire 1716 MS A
- E. Delisle 1718
  [followed by Cóvens-Mortier, Homann, and other continental map makers for the rest of the century]

II. De Bry–Mercator type

[Le Moyne 1564–65: MS lost] White 1585 MS

- A. De Bry 1591
- B. De Bry 1590
- C. Mercator-Hondius 1606
  [followed by Blaeu 1640, Jansson 1641 Montanus 1671, Aa 1729, etc.]

III. First Lords Proprietors' Map type

Sources:

1. De Bry–Mercator
2. Drafts from Hilton's 1662, 1664 explorations; cf. Horne 1666
3. Lederer 1672

- A. Ogilby 1672 (The First Lords Proprietors' Map)
- B. Homann 1714

IV. Second Lords Proprietors' Map type

- A. Gascoyne 1682
- B. Thornton-Morden-Lea ca. 1685
- C. Thornton-Fisher 1689 (hydrographic chart)
- D. Sanson ca. 1696 A

V. Mathews type

[Maurice Mathews ca. 1682:hypothetical] → IV. A. Gascoyne 1682

- A. Mathews ca. 1685 MS → IV. B. Thornton-Morden-Lea ca. 1685
- B. Thornton-Morden ca. 1695
  - [lost Nairne ca. 1708 map; cf. inset in Crisp map]
- C. Crisp 1711
- D. Moll 1715

VI. Barnwell type

- A. Barnwell 1721 MS
- B. Popple 1733
- C. Mitchell 1755

VII. De Brahm type

- A. De Brahm 1757
- B. Cook 1773
- C. De Brahm–Stuart 1780

VIII. Collet type

- A. Fry-Jefferson 1751 [1753]
  [Churton:Collet 1768 MS]
  - Moseley 1733
  - Wimble 1738
- B. Collet 1770

IX. Mouzon type

- A. Mouzon 1775
  [The Mouzon map, derived largely from Cook 1773 and Collet 1770, was used as the basis of most maps of North and South Carolina until the nineteenth-century surveys]
- B. Stuart 1775 MS
  [The surveys of Florida and the Gulf Coast by Romans, Taitt, and Gault were combined with the Mouzon 1775 and Fry-Jefferson 1751 [1753] maps for the series of large MS Stuart maps which culminated in:]
- C. Romans 1776.

NOTES

1. Henry R. Wagner, in his *The Cartography of the Northwest Coast of America to the Year 1800*, Berkeley, Calif., 1937, calls the earliest period "imaginary geography." The terms imaginary or pictorial do not sufficiently recognize the serious and responsible attempts of contemporary cartographers to interpret and make use of the reports brought back by pilots and explorers, at least for the north Atlantic coast.

W. F. Ganong has divided the cartographical development along the Canadian Atlantic coast into early, transitional, modern, and precision periods in his "Crucial Maps in the Early Cartography and Place-Nomenclature of the Atlantic Coast of Canada," *Proceedings and Transactions of the Royal Society of Canada*, Third Series, XXIII (1929), 135–75; XXIV (1930), 135–87; XXV (1931), 169–203; XXVI (1932), 125–79; XXVII (1933), 149–95; XXVIII (1934), 149–294; XXIX (1935), 101–29; XXX (1936), 109–31. Ganong's articles were published as a book with the same title, edited with useful commentaries and map notes by Theodore E. Layng, by the University of Toronto Press and the Royal Society of Canada in 1964. T. E. Layng's edition has both volume pagination and the page numbers of the original *Transactions* articles.

2. Myron H. Avery and Kenneth S. Boardman, "Arnold Guyot's Notes on the Geography of the Mountain District of Western North Carolina," *North Carolina Historical Review*, XV (1938), 251–318.

3. For a description of the last two maps, consult P. L. Phillips, ed., *The Lowery Collection*, Washington, 1912, p. 366 (No. 556) and pp. 370–71 (No. 566).

4. The U.S. Coast Survey was organized in 1807 and began publishing charts soon after. It became the Coast and Geodetic Survey in 1878. The U.S. Geological Survey was organized in 1879; its first topographical quadrangles were issued in 1882. Surveying instruments and methods of printing greatly improved between the time that De Brahm and Romans made their surveys and the middle of the nineteenth century when geodetic and topographical surveys were undertaken on a large scale, but cartographic method did not change radically until aerial surveying became practical.

Following World War I the use of aerial photography in mapping became the subject of intensive study and development. During the interwar decades of the 1920s and 1930s, three more or less distinct developmental lines calculated to meet a particular need or solve a particular problem became apparent. The first of these was reconnaissance mapping of vast areas through the use of oblique photographs to produce small-scale maps of topographical patterns cheaply and quickly. The second was based on the use of stereoscopic plotting instruments and high-quality vertical photographs to produce large-scale, closely contoured maps at a high degree of accuracy. The third line of development was aimed at the production of medium-scale contoured maps of moderate accuracy by utilizing limited ground control and simple instruments.

With the Second World War came a phenomenal growth in the mapping sciences. Air travel and the rapid movement of forces over virtually the whole globe created undreamed of needs for maps of all sorts and scales. Not surprisingly all of the processes and techniques involved in the creation and reproduction of maps and charts underwent innovative rethinking and expansion to meet the challenge.

Following World War II the field of Remote Sensing was developed to make use of earth imagery and data that was being acquired first by rockets and high-flying aircraft and then by earth-orbiting vehicles and satellites. At present the cutting edge of research and advance in the mapping sciences involves the harnessing of the potentials of the computer and analytical photogrammetric instrumentation to process and order the almost limitless supply of current remotely sensed data. Thanks to the developing field known as Geographic Information Systems, or GIS, spatial questions of all sorts can now be asked and answered from computer-stored data bases without the need to array those data in conventional hard copy map form. The need for conventional maps as tools for decision-making may decline thanks to GIS, and increasingly printed maps are the products of computer cartography programs and printers. Without doubt the past couple of decades have seen the world of maps projected into what can only be termed the precision period of cartography.

For more comprehensive examinations of cartography, such as the history of mapmaking, projections, types, modern surveying methods, printing, etc. the following works are generally available: Leo Bagrow, *History of Cartography*, revised and enlarged by R. A. Skelton, Cambridge, Mass., 1964; Lloyd A. Brown, *The Story of Maps*, Boston, 1949; David Greenhood, *Mapping*, Chicago, 1964; Walter W. Ristow, *Guide to the History of Cartography*, Washington, 1973; Albert A. Stanley, "Sesquicentennial of Coastal Charting: 150 Years Service of the United States Coast and Geodetic Survey," *The Military Engineer* (Jan.–Feb. 1957) 1–11; Don W. Thomson, *Men and Meridians*, vol. 1, *Prior to 1967*, Ottawa, 1966; Norman J. W. Thrower, *Maps and Man*, Englewood Cliffs, N.J., 1972; John N. Wilford, *The Mapmakers*, New York, 1981; David Woodward, ed., *Five Centuries of Map Printing*, Chicago, 1975.

5. "Terra Vlteri[us] Incognita" of Waldseemüller, *Cosmographia* (1507), reproduced in this volume as Plate 1; "Vlterius Incognita Terra" on Schöner's 1515 globes, facsimile in Justin Winsor, *Narrative and Critical History of America*, Boston, 1886–89, II (1886), 118.

6. "Terra de Cvba, Asie Partis," in Waldseemüller's Carta Marina (1516), reprod. in I. N. P. Stokes, *Iconography of Manhattan Island, 1492–1909*, New York, 1905–28, II (1916), Plate 5; "Terra Cube," in Sylvanus's map (1511), facs. in Winsor, *America*, II, 122; "Terra de Cvba," on Schöner's globe (1520), facs. in Winsor, *America*, II, 119; "Terra de Cuba," in Grynaeus's map (1532), facs. in K. Kretschmer's *Die Entdeckung Amerika's Berlin*, 1892, Plate XIV, No. 3. H. Harrisse, in his *The Discovery of North America*, London, 1892, p. 95, calls attention to a curious reversion in the second quarter of the sixteenth century in Belgium, France, and Germany

to the merging of the coastline north of the equator with the Asiatic continent; this revival of a primary error marred for many years a whole family of maps and globes. The belief that the coast of the mainland was Asia was held by Columbus, Vespucci, and other early navigators; cf. Winsor, *America*, II, 121.

7. Found on a world map, "Typus Vniversalis Terre," in Gregorius Reisch's celebrated encyclopedia, *Margarita Philosophica*, Strassburg, 1515; facs. in Winsor, *America*, II, 114. A. E. Nordenskiöld, the nineteenth-century historical cartographer, suggested that this mysterious name for the North American continent was derived from John Cabot (called Messer Zoanna) and from Insula (island) corrupted to Mela. However, Franz von Wieser, in an article ("Zoana Mela," *Zeitschrift für Wissenschaftliche Geographie*, V (1884), 1–6), showed that the name apparently came from a statement by Peter Martyr that two islands discovered by Columbus were called Spagnola and Zoanna, the latter an Italian form of Iuana or Johanna. Nordenskiöld, in his *Facsimile Atlas*, p. 71, accepted this suggestion. Since nobody has offered a better solution than von Wieser's for the appearance of Zoana Mela, his conjecture is probably correct. Zoana Mela is a good example of the often inexplicable and evanescent names that appear along our coast on sixteenth-century maps and have contributed to the amusement and to the despair of cartographical scholars.

8. L. A. Vigneras, "A Projected Voyage of Juan de Agramonte to the Carolinas, 1511," *Terra Incognitae*, XI (1979), 67–70.

Probably Bimini is first found in Peter Martyr, *Decades* (1511), as "illa de beimeni parte," facs. in Winsor, *America*, II, 110; "Terra Bimene" on Reinel's map (ca. 1516), reprod. in Stokes, *Iconography*, II, Plate 4; "Tera bimini" on Maggiolo's map (ca. 1519), reprod. in C. O. Paullin, *Atlas of the Historical Geography of the United States*, Washington and New York, 1932, Plate 11A; "Tera Bimini" on the Weimar Military Library map of the Pacific (ca. 1518), facs. in Winsor, *America*, II, 217; "Terra Bimine" on the Munich-Portuguese map (ca. 1526), in the Bibl. Nat., Paris, facs. in Kretschmer, *Entdeckung*, Plate XII, No. 2; "Tera Bimini" in the Munich-Portuguese map (ca. 1520), reprod. in Stokes, *Iconography*, II, Plate 6. After the discovery of the mainland, Ponce de León sent a ship off to continue the search for Bimini under one of his captains, Perez Ortubia, who returned later to Puerto Rico with the news that his search had been successful: cf. Harrisse, *Discovery of America*, p. 135.

9. Thus the name may etymologically be the Land of Flowers, but the name was not given because of any floral profusion observed by its discoverers. Spanish explorers frequently named capes and rivers after the saint's day on which they were first sighted; this is so true that the arrival of a navigator at a certain place can sometimes be dated by the saint's name given to it.

A concise explanation of problems involved in the date of Ponce de León's discovery and landing and the reasons for dating it 1513 instead of 1512 is given in Frederick Davis, "The Record of Ponce de Leon's Discovery of Florida, 1513," *The Florida Historical Society Quarterly*, XI (1932), 14–15.

10. Pineda's map has the legend on the peninsula "la florida q̄ dezian bimjnj q̄ descubrio Jvanponce," facs. in Kretschmer, *Entdeckung*, Plate X, No. 3; and cf. bibliography and references in P. L. Phillips, ed., *The Lowery Collection*, pp. 18–19. See also David O. True, "Some Early Maps relating to Florida," *Imago Mundi*, XI (1954), 73–84. Florida is shown as an island in the so-called Leonardo da Vinci globe (1515–20), facs. in Winsor, *America*, II, 126, and in the Turin map (1523), reprod. in Paullin, *Atlas*, Plate 17B. "La Florida" (Ribero-Rome 1529), "Tera Florida" (Maggiolo 1527), "Terra Florida" (Verrazano 1529) are names found on many maps after 1530; by the middle of the century "Florida" is frequently found written across the whole southern part of the continent when not relegated to the peninsula by "Apalche" or "Ayllón"; and this wider use becomes common after Ortelius's "La Florida" (1584).

11. *Narratives of Early Carolina, 1650–1708*, edited by A. S. Salley, New York, 1911, p. 288.

12. De Laët's map, which has "Tegesta Prov.," first appears in his *Beschrijvinghe van West Indien*, Leyden, 1630, opp. p. 137, and in the Latin (1633) and French (1640) translations of the same work, which had a wide vogue and great influence. It also appears in Joannes Jansson, "America Septentrionalis," in his *Nuevo Atlas*, Amsterdam, 1653, II, No. 39; in several maps of N. Sanson d'Abbeville, the French royal geographer, such as "Tegesta Prov." in "Le Nouveau Mexique et Florida . . . 1656," *Cartes Générales*, Paris, 1658, No. 87, and in "La Floride . . . 1657," *L'Amerique en Plvsievrs Cartes*, Paris, 1657, No. 4; "Tegesta presqu'Isle" in "La Floride," P. Du Val, *La Geographie Vniverselle*, Paris, 1660, I, No. 8; "Tegests Peninsula" in Sir J. Moore, "Florida," *A New Geography*, London, 1681, No. 50; and "Peninsula of Tegesta" in "A New Map of North America," *A New Sett of Maps*, Oxford, 1700, No. 30. On the Southeast coast of the map which accompanies Arredondo's *Historical Proof*, ca. 1742, "Descripcion Geographica," copies in the Archivo General de Indias, Seville, and in the British Library (dated 1765) is note F, "Costa de Indias de la Nacion Tequesa que oy se dicen Indias Costas." Other continental mapmakers using the name are Lotter, Ottens, Seutter, Cóvens and Mortier, and Homann (see note by Lowery, referred to below). De Brahm uses it in his "The Ancient Tegesta, now Promentory [*sic*] of East Florida," *Atlantic Pilot*, London, 1772, No. 2. Bernard Romans at first ridiculed de Brahm's use of the term in his *A Concise History of East and West Florida*, New York, 1775, I, 296, 299, but later adopted it in his "A General Map of the Southern British Colonies in America," 1776. It is found on several other English and American maps during the next quarter century. Woodbury Lowery has a full bibliographical note on the origin of Tegesta and examples of its use in his *Spanish Settlements within the Present Limits of the United States: Florida, 1562–1574*, New York, 1905, pp. 440–43 (Appendix V, "Tegesta").

13. Harrisse, Henry, *The Discovery of North America*, London, 1892, Part I, Chapters VIII and IX, pp. 125–33.

14. The Cosa map is a Spanish map, showing to the north the discoveries of John Cabot along the south coast of Newfoundland. It is owned by the Spanish Marine Museum at Madrid. A recent study of the map is that by W. F. Ganong, "Crucial Maps in the Early Cartography and Place-Nomenclature of the Atlantic

Coast of Canada," *Proceedings and Transactions of the Royal Society of Canada*, XXIII (1929), pp. 140–75 (bibliography, pp. 141–42). This work has been published as W. F. Ganong, *Crucial Maps in the Early Cartography and Place-Name Nomenclature of the Atlantic Coast of Canada*, with an introduction, commentaries, and map notes by Theodore E. Layng, Toronto and Ottawa, 1964, 8–43, 469–73. Reproductions are in Stokes, *Iconography*, II, Plate 1, and in *Mapas Espanoles de América, Siglos XV–XVII*, Madrid, 1951, Plates 2 and 3, a magnificent color reproduction (with accompanying comments on pp. 11–19).

15. J. A. Williamson, *The Voyages of the Cabots*, London, 1929, pp. 195–96, places the southern extent of Cabot's first voyage near Penobscot, Maine, and his second to the mouth of the Delaware River. He thinks that Cosa, for diplomatic reasons, falsified his map and put Cuba at 35° N.L. in order to counter possible English claims; he holds also that Cabot placed his named coast west of Bristol, England, because his charter did not authorize him to explore south of it. Cosa, according to Williamson, had copies of Cabot's charts sent from England, and his map is the embodiment of the Spanish case to be presented to Henry VII. W. F. Ganong, "Crucial Maps," XXX (1929), 175 (*Crucial Maps*, 43), does not accept Williamson's interpretation of the Cabot charter as conclusive and disagrees with Williamson's identification of the Cosa-named coast; he places the coast farther to the north and supports his view with much new data.

The finding and publication of a letter by John Day by Dr. Louis-A. Vigneras in 1956 in the Archivo General de Simancas (Estado de Castilla, leg. e, fol. 6) throws new light on early English voyages across the Atlantic. John Day, an English merchant in Spain, wrote to the "Lord Grand Admiral" (Columbus?) in the winter of 1497–98 giving particulars of John Cabot's voyage of 1497 and referring to earlier northern trans-Atlantic discoveries known to men of Bristol. Vigneras published a translation of Day's letter in English with a brief commentary in "The Breton Landfall: 1494 or 1497? A Note on a Letter by John Day," *Canadian Historical Review*, XXXVIII (1957), 226–28. A fuller analysis of the letter and its implications are in J. A. Williamson's revised and enlarged edition of his work, *The Cabot Voyages and the Bristol Discovery*, with a useful essay by R. A. Skelton called "The Cartography of the Voyages," London, 1962, 29–31, 55–56, 211–14, 295–98. Another excellent treatment of the subject is D. B. Quinn, *England and the Discovery of America, 1481–1620*, New York, 1974, 11–21, 99–111.

16. The Cantino world map was made in Lisbon for the Duke of Ferrara by order of his envoy, Albert Cantino; it is now in the Biblioteca Estense, Modena, Italy. Maps following the general configuration and nomenclature of the Cantino map are called the Vespucci type by Ganong and the Lusitano-Germanic type by Harrisse. Since any voyage by Vespucci along the coast north of Florida is open to serious question and since this type of map was first made in Portugal and then widely copied by cartographers of central Europe during the first quarter of the sixteenth century, for our purpose the name Cantino or Lusitano-Germanic is more applicable.

A number of students of early American cartography hold that the land mass north of the Antilles is a cartographical error, since there is no proof that Vespucci or any other explorer sailed along the southeast coast of North America before 1502. George E. Nunn, in his *Geographical Conceptions of Columbus*, New York, 1924, states this view most fully. He follows the suggestion made by J. C. Brevoort in *Verrazano the Navigator*, New York, 1874, p. 72, that the entire land area north of "Isabella" represents Cuba. Professor Samuel Morison in *Portuguese Voyages to America*, Cambridge, Mass., 1940, pp. 138–40, and Charles E. Norwell in his review of *America la Bien Llamada, Hispanic American Historical Review*, XXX (1950), 139, among other recent scholars, accept Mr. Nunn's arguments. None of these arguments, however, is conclusive, and some of them are demonstrably wrong. Professor Morison does not know southern fauna when he writes (op. cit., p. 139) that C. delgato (cape of the cat) can apply neither to Cuba nor to Florida, there being no wildcats south of the Delaware; wildcats are native to the Southeast.

The arguments in favor of accepting the land mass to the north of the island of Isabella as the first clear representation of the Florida peninsula and Southeast coast overweigh those which have been advanced against it, in the opinion of the writer at the present time. The islands at the southern extremity strongly indicate a knowledge of the Florida keys and Tortugas; the general peninsula-like shape of the land mass at the southern extremity of most of the maps of this type and the general northerly direction of the Atlantic coast (the Florida coast actually trends northwest, not northeast; Savannah, Georgia, is almost a degree west of Miami, Florida) support the belief that the land mass represents the southeast part of the American continent. The mapmakers of this period, it must be remembered, were conscientious in their effort to give as exact and true a representation of the New World as they could. The names along the coast and the nearness to a correct delineation of the coastal topography are strong arguments in support of accepting the Lusitano-Germanic group of maps as the first to report some definite voyage of discovery along the coast.

In general, the biographers of Columbus have been the strongest attackers of Vespucci, from Las Casas to the present. The best recent defense of Vespucci's claims may be found in Roberto Levillier, *America la Bien Llamada*, Buenos Aires, 1948 (two volumes).

17. Cf. Dr. F. C. Wieder in Stokes, *Iconography*, II (1916), 7, suggests that "Vespucci-type" be applied to maps based upon the representation of the North American coast shown on the Cantino and Waldseemüller maps and those which followed them. He points out in further justification of this ascription that in 1508 Vespucci was created "piloto mayor," an office which included the instruction of navigators and the care of nautical maps of newly discovered countries. Dr. Wieder, at the time (1914) assistant librarian of the University of Amsterdam, compiled most of the information for the second volume of Stokes, *Iconography*.

18. This map was reproduced in full size with an accompanying volume by Hrsg. von J. Fischer und F. Von Wieser, called *Die*

älteste Karte mit dem Namen Amerika aus dem Jahre 1507 und Die Carta Marina aus dem Jahre 1516 des M. Waldseemüller (Ilacomilus), Innsbruck, 1903.

A late and degenerate form of the Waldseemüller map of 1507, found about forty-five years ago in the British Museum by Wouter Nijhoff, has this inscription for the east coast of North America: "Inue[n]ta p[er] rege[m] hispanie a 1497." This is an Amsterdam map, printed between 1534 and 1538; the inscription is not found on the Waldseemüller map. C. P. Burger, "Een Hollandsche Wereldkaart uit de eerste helft van de 16e eeuw," *Het Boek*, Den Haag, 1912, I, 291, and C. P. Burger, "Nog Lets over de 16de eeuwsche Hollandsche Wereldkaart," *Het Boek*, Den Haag, 1913, II, 197 (with facsimiles).

19. On Juan Vespucci's world map (1526) and Ribero's maps (1529) are found the Martyrs (Florida Keys) and Tortugas, which de Leon named, as well as the nomenclature given by de Leon on the west coast of Florida. But the places which he named on the east coast, Rio de la Crux (Matanzas Inlet?) and Cabo Corrientes (Cape Canaveral?), do not appear until later.

20. "Trã nueva de ayllon" in Vespucci's world map (1526) in the library of the Hispanic Society of America; "tierra del licenciado ayllon" in the Weimar map (1527), reprod. in Stokes, *Iconography*, II, Plate 9; "Tiera de Ayllon" in Ribero's Weimar and Rome maps (1529), reprod. in Stokes, *Iconography*, II, Plate 10; "tierra del licenciado ayllon," in Hernando Colon's map (1527), facs. in Woodbury Lowery, *The Spanish Settlements within the Present Limits of the United States: Florida, 1513–1561*, opp. p. 146 (Library of Congress, Kohl Collection, No. 38); "Licentiako Allon" in Ramusio's *Map in Indie Occidentali* (1534), drawing in De Costa, *The Verrazano Map*, New York, 1881, p. 53, but "Licentiato ailon" in a copy of Ramusio in the John Carter Brown Library; "Terra del Licencia dos Aulloh," in a British Library MS (No. 9814), ca. 1555, facs. in Winsor, *America*, II, 285; "Terra de lecenciados ailon" in the Battista Agnese portolan chart (ca. 1540) in the John Carter Brown Library and in the Agnese Atlas (1543) in the Library of Congress; "Tierra del Licenciado Aillon" in Alonso de Santa Cruz's map (1542), reprod. in Stokes, *Iconography*, II, Plate 18; "Tierra que Descvbrio Ellice[do] Aillõ" on Santa Cruz's Carta del Océano Atlántico septentrionale (1545) reprod. in *Mapas Españoles de América*, Plate 13; "Tierra del Licenciado Avlloh" in Gastaldi's world map (1546), reprod. in Paullin, *Atlas*, plate 12B; "Tierra del Licenciado Aulloh," in Pedro de Medina, *L'Art de Navigver*, Lyon, 1554 and 1569, opp. p. 48; "ailoh" in Callapoda, *Atlas*, facs. in Kretschmer, *Entdeckung*, Plate XXII; "Terra de Licenciad" (below "La Florida") in the globe (1570) of Franciscus Bassus Mediolenensis, Turin University Library, facs. in Kretschmer, *Entdeckung*, Plate XXIX; "Tierra del Licenciados Avlloh" in Gastaldi's world map (1560) in the British Library, reprod. in R. V. Tooley, *Maps and Map Makers*, London, 1952, Plate 18, opp. p. 22.

21. Winsor, *America*, II, 239 and 285, note 3; cf. also Lowery, *Spanish Settlements, 1513–1561*, p. 155. This was the Cape Fear River, North Carolina, or Winyah Bay, South Carolina: cf. John R. Swanton, *Early History of the Creek Indians*, Smithsonian Institution, Bureau of American Ethnology, Bulletin 73, Washington, 1922, pp. 35–36. Strangely, this name does not appear on the early maps showing Ayllón's discoveries, but begins to appear on later maps which give very few of the Ayllón names: "R. de S. Giovanni" in the Gastaldi world map (1542), reprod. in Paullin, *Atlas*, Plate 12B. The "C. S. Juan" which appears on the Ribero 1529 maps and on many later maps is placed north of Baya Santa Maria (Pamlico Sound or Chesapeake Bay). Chaves (ca. 1536) places it at 37° N.L. (Stokes, *Iconography*, II, 41), 38 leagues north of Cape Trafalgar; it is probably Cape Charles on the eastern shore of Virginia or Cape Henlopen, Delaware.

A careful study of sixteenth-century attempts at settlement on the South Carolina coast, based on a re-examination of the original documents, is found in Paul Quattlebaum, *The Land Called Chicora: The Carolinas under Spanish Rule with French Intrusions, 1520–1670*, Gainesville, Fla., 1956. He identifies St. John the Baptist with Winyah Bay (p. 9).

22. Shea, in Winsor, *America*, II, 240, note 4, and 285, and Swanton, *Creek Indians*, p. 35, state definitely that the Jordan is the Santee River. Ecija (1609) places its mouth at 33°11' N.L., which is almost correct for the Santee; cf. Swanton, *Creek Indians*, p. 35; Chaves, ca. 1536, places it at 33°30', four leagues below Cape Roman; cf. Stokes, *Iconography*, II, 41. Lowery, *Spanish Settlements, 1513–1561*, p. 155, note 2, identifies the Jordan as the Cape Fear River. A. S. Salley, in *Narratives of Early Carolina*, New York, 1911, p. 38, note 1, and p. 92, note 1, shows that Hilton and Sandford, English explorers, identified the Jordan as the Combahee; but this is at variance with the earlier Spanish statements and latitude, and Navarrete states that the English name for the Jordan is the Santee; cf. Swanton, *Creek Indians*, p. 35. Paul Quattlebrum, *The Land Called Chicora*, pp. 20–21, identifies it with the Cape Fear River.

23. The location of the settlement at St. Miguel de Gualdape is more uncertain than the Jordan, as not even the direction the Spaniards went from their first landing is clearly stated. Swanton, op. cit., pp. 38–39, is inclined to place it near the Savannah River in the later Spanish province of Guale; Lowery, op. cit., pp. 166, 447, 449, 451–52, on the basis of Oviedo, who writes that the Spaniards marched along the beach, places it at Winyah Bay (Pee Dee River). It would be impossible for the Spaniards to have made a march along the heavily indented coast south of the Santee; Swanton answers this objection by questioning the complete accuracy of Oviedo's report and by supposing that the ships which the colony had were used to transport the whole company instead of only the women and the sick. J. G. Johnson, "A Spanish Settlement in Carolina, 1526," *Georgia Historical Quarterly*, VII (December 1923), 344, suggests the Santee. Paul Quattlebaum, *The Land Called Chicora*, pp. 126–29 (Appendix B, "San Miguel de Gualdape," and Appendix C, "Waccamaw Neck"), places Ayllón's settlement on lower Waccamaw Neck, across Winyah Bay from Georgetown, South Carolina. With the appearance of Paul Hoffman's book, *A New Andalucia and a Way to the Orient*, Baton Rouge, 1990, new attention was directed toward the debate concerning the location of San Miguel de Gualdape. See J. Cook, ed., *Columbus and the Land of Ayllón: The Exploration and Settlement of the Southeast*, Darien, Georgia, 1992. Based on his analysis of a set of early

Spanish sailing directions by Alonso de Chaves, Hoffman forwarded an argument that places Ayllón's attempted colony on the coast of Georgia in the vicinity of Sapelo Sound.

24. Francisco's reports to Peter Martyr, Oviedo, and others contained so many wonderful tales about the Indians that the Spaniard Oviedo, as well as later historians, was inclined to discredit his statements; but many of these incredible stories were merely elements in the mythology of the southern Indians or misinterpreted condensations of his statements by Spanish writers. Francisco is not now considered to have been purposefully mendacious; cf. Swanton, op. cit., p. 36.

It is quite possible that Chicora is not an accurate phonetic transcription of the native Indian pronunciation; but this does not invalidate the fact that Chicora was long actually used for a province and is a name still found in South Carolina.

Chicora is often referred to in the early Spanish reports: it appears, with the neighboring province of Duhare, on Hakluyt's world map, published in his translation of Peter Martyr, *De Orbe Novo Decades Octo*, Paris, 1587, reprod. Plate 13; as "Chicora" in Pieter vander Aa, "De Vaste Kust Van Chicora Tussen Florida en Virginie," in his *Naaukerige Versameling*, Leiden, 1707, IX (1706), between pp. 88–89, and other works and atlases using Aa's plates; as "Provincia de Chicora" in Arredondo, "Descriptio Geographica" (1742), reprod. Plate 56, and in the British Library copy of Arredondo's map (Add. MS 17, 648a [1765]). "Chicola," a town on the River Jordan, is found in the same region on the maps of Florida by Blaeu (1640), reprod. Plate 26, Jansson (1641), Du Val (1669?), and Montanus (1671) in the seventeenth century. This name is probably derived from "Chiquola," an Indian tribe referred to by Laudonnière, *L'Histoire Notable de la Floride*, Paris, 1586, pp. 29–31; cf. Swanton, *Creek Indians*, p. 53.

See also the references to Chicora in Paul Quattlebaum, *The Land Called Chicora*, p. 12ff.

25. This mappemonde, a large roll measuring 8'3/8" × 2'9 3/8", has the identifying description on the North American continent: "Ju° Vespuchi piloto de sus mata. me fecit en seuj ilaño d. 1526." As it does not have the official seal of either the pilot major or the Casa de Contratación, it is probably not the padrón real itself. A brief analysis of Vespucci's chart, together with a facsimile of a part of the map showing the southern tip of Florida, is given in *Description of a Mappe Monde by Juan Vespucci . . .*, July 1914, pp. 3–8, published by Bernard Quaritch, who also describes it in his Catalogue, No. 332, July 1914. Stokes refers to it briefly in his *Iconography*, II, 10–11. It was purchased in 1914 by the Hispanic Society of America; see "Newly Discovered Map by Juan Vespucci," *The Geographical Journal*, XLV (January 1915), 83–84, and E. L. Stevenson, "The Geographical Activities of the Casa de Contratacion," *Annals of the Ass. of Am. Geographers*, XVII (June 1927), 39–59. In 1925 it was reproduced in a publication of the Hispanic Society: *Vespucci World Map MDXXVI*, New York, 1925. This work was reviewed unfavorably by the *Geographical Journal of the Royal Geographical Society*, LXVII (1926), 82–83, because the reproduction was made in 423 non-overlapping sections of 3 3/4" × 2 1/4" each.

26. In the land of Ayllón, reading from north to south, the following names are found:

Tr̃a nueua de ayllon. This name extends back of the coastal nomenclature in a west-east direction parallel with the coast.

C: de sã mã (Cabo de Santa Maria). See below.

baya de sã mã. Chesapeake Bay or the Carolina Sounds. Clifford M. Lewis, S.J., and Albert J. Loomie, S.J., *The Spanish Jesuit Mission in Virginia, 1570–1572*, Chapel Hill, 1953, pp. 7ff., identify Bahía de Santa María on Spanish maps as the Chesapeake Bay. Since Quexós was sent by Ayllón to explore the coast for 250 leagues, he might well have reached the Chesapeake. However, the wide, open indentation of the coast studded with islands, which is found on Spanish maps during most of the sixteenth century before Menéndez, is a more likely representation of the Carolina Sounds than of the Chesapeake. "Rio del Spiritu Santo" to the southeast of the bay is not a name given to the James River; Ribero-Weimar (1529) locates the bay at 34°–35°, too low for the Chesapeake; and the English with Governor Lane in 1585 evidently thought that Pamlico Sound was "The Port of Saynt Maris" (see Plate 10). L. D. Scisco, "The Voyage of Vicente Gonzales, 1588," *Maryland Historical Review*, XLII (1947), 99, thinks that Gonzales believed the Carolina Sounds and the Chesapeake were a continuous body of water, and this may have been true of earlier explorers. Father Lewis suggests that Pamlico Sound and the Chesapeake may have been confused (p. 11): certainly the conformation of the bay is represented on some maps as a long shallow indentation with a row of islands separating it from the Atlantic (clearly referring to the Sounds with the Outer Banks islands), and on other maps the bay is large and deep with a narrow entrance (more nearly an approach to the appearance of the Chesapeake). The subject is one that needs further investigation.

C: de Tr̃afalgar. Cape Lookout; 35°30' N.L., according to Chaves (ca. 1536); cf. Stokes, *Iconography*, II, 41.

rio da sa terazanas (boathouse?). Dr. L. A. Vigneras, "Is There a 'Verrazano River' on Juan Vespucci's 1526 Mappemonde?" *Terra Incognitae*, VII (1976), 65–68, states that the name is the Spanish *atarazana* (boathouse) and dismisses categorically the idea that it may refer to Verrazano's landfall at 34° N.L., at or near Cape Fear River, as suggested in earlier editions of this work and in Lawrence C. Wroth's *The Voyages of Giovanni da Verrazano*, New Haven, 1970, 168–70, and footnote 9, which gives a contemporary spelling of Terazano for Verrazano in a letter of December 24, 1527, from João Silveira to John III of Portugal. Dr. Vigneras's suggestion is possibly right; but he oversimplifies his interpretation. He fails to explain the form *terazanas* for *atarazanas* (with initial truncation of "a" and the change of "a" to "e" after the initial "t" in *atarazanas*). As the acting pilot major in charge of the padrón real, Vespucci probably knew of Verrazano; Wroth and others quote documents showing that the Spanish and Portuguese ambassadors in France kept their courts well informed of Verrazano's trans-Atlantic activities before and after his voyage in 1524.

r: de arezifes. Unidentified; Spanish arrecifes, reefs.

c: de sa njcolas. Also unknown.

R: Jordan. The Cape Fear River (or Santee). This name is found on maps for two hundred years.

R: de la X. (Rio de la Cruz). Cabo de Cruz in Chaves is 29°20' N.L., below St. Augustine; this name was later applied to Matanzas Inlet.

c: de sãta elena. Saint Helena Sound or probably Port Royal Sound; Chaves put it at 32°20' N.L.; cf. Stokes, op. cit., II, 41.

t͂ra floryda. (written across interior).

Canjto. (Cape of Florida).

For checking transcriptions of the names on the map the writer is indebted to Miss Ruth Kuykendall of the University of North Carolina Press.

27. An unsigned Weimar 1527 map, made by a Sevillan cosmographer usually identified as Ribero, has on it "tierra del licenciado ayllon" north of "La Florida." Harrisse suggests that this map has the same prototype as the Ribero 1529 maps (Harrisse, *Discovery of North America*, p. 572). But for the southeast there is no similarity except that both have "Ayllon." There are only two south Atlantic coastal names on the Weimar 1527 map: "R. de S. Juhan" and "aguada." These are not on the Ribero (1529) maps and probably are from the report of the Gordillo 1521 expedition. See L. A. Vigneras, "The Cartographer Diogo Ribeiro," *Imago Mundi*, XVI (1962), 73–83.

28. Harrisse, H., op. cit., pp. 570, 574. Good reproductions of the Ribero-type maps are in Stokes, *Iconography*, II, Plate 9 (Weimar 1527), Plate 10 (Weimar 1529 and Vatican 1529). The Ribero-Vatican (see Plate 4) has the following names along the coast, southward from the Pamlico Sound to the Chesapeake Bay area: b. de. s. mª, R. del espũ S., C. traffalgar, R. del principe, R. del sucio, R. de canãs, C. de Sa romã, R. Jordan, playa, S. elena, C. de S. elena, C. gruesso, r. solo, marbaxa baxa, La Florida (no names along the east-peninsula coast). The Ribero-Weimar 1529 map does not have R. de Sucio, R. de canãs, playa, R. solo; on the east Florida coast it adds C. roxo and R. Salado.

29. Only the Cellère Codex, found in the library of Count Giullo Macchi di Cellère of Rome in 1909, gives an account of the isthmus. It is now in the possession of J. P. Morgan and has been finely reproduced in Stokes, *Iconography*, II, Plates 60–81. A translation and notes are given in Edward H. Hall, *Fifteenth Annual Report of the American Scenic and Historic Preservation Society*, Albany, N.Y., 1910, pp. 134–226, and in L. C. Wroth's *The Voyages of Giovanni da Verrazzano*, New Haven, 1970, pp. 96–143.

A bibliography of earlier works on Verrazano is given in P. L. Phillips, ed., *The Lowery Collection*, pp. 35–40. A study of the nomenclature of the Verrazano type maps, with comparative tables of names, is in W. F. Ganong, "Crucial Maps," *Proceedings and Transactions of the Royal Society of Canada*, XXV (1931), 169–206, and in Ganong, *Crucial Maps*, ed. Theodore E. Layng, Toronto, 1964, 106ff. Professor E. H. Wilkins has made an interesting study of the influence of Verrazano's use of the place-name "Arcadia" in a work which was published under the auspices of the American Philosophical Society: *Arcadia in America*.

W. H. Hobbs, "Verrazano's Voyage along the North American Coast in 1524," *Isis*, XLI (December 1950), 268–77, attempts to prove that Verrazano's landfall was on the coast of Florida at about 27° N.L., since compass declination would have deflected him from the course he thought he was taking to that extent. Hobbs fails to prove his thesis, although it has been widely and uncritically accepted. He assumes that Verrazano used a magnetic compass without checking latitudes by the various other instruments that Verrazano expressly states that he used. He assumes that Verrazano, who elsewhere took latitudes which were accurate to within a fraction of a degree, landed on the coast but did not use astrolabe, quadrant, or cross staff to verify the latitude of his landing. He confuses the two Verrazano brothers. He is apparently ignorant of the discovery of the Cellère Codex and of Professor Alessandro Bacchiani's basic study of the text and the three variants; he bases part of his argument on an error of direction made in the text he was using, which is corrected in the Cellère Codex. Hobbs says that Verrazano must have sighted land in Florida, since Verrazano's letter reports that he sailed westward, and this could have been said only if he sailed along the Florida coast. The Cellère Codex, however, has "east," which is of course the direction he would have to sail if he coasted from Cape Fear to Cape Lookout. Verrazano's own statement of landfall near 34° N.L. and the traditional location of that landfall somewhere in the vicinity of Cape Fear has not been disproved by Hobbs's article.

30. Stokes, *Iconography*, II, Plate 12. A discussion of chief maps showing Verrazano's Sea, with excellent reproductions, is in L. C. Wroth, *The Voyages of Giovanni da Verrazzano*, New Haven, 1970, passim.

31. Plate 3.

32. G. B. Parks, *Richard Hakluyt and the English Voyages*, New York, 1928, p. 38ff; Lane's account in *Early English and French Voyages*, edited by H. S. Burrage, New York, 1930, pp. 252, 258; for maps showing the Verrazano Sea, cf. Stokes, *Iconography*, II, 40, and W. P. Cumming, "Geographical Misconceptions of the Southeast in the Cartography of the Sixteenth and Seventeenth Centuries," *Journal of Southern History*, IV (1938), 476, note 3. A discussion of chief maps showing Verrazano's Sea, with excellent reproductions, is in L. C. Wroth, *Voyages of Giovanni da Verrazzano*, New Haven, 1970, passim.

33. The name took many forms: Apalcen, Apalchen, Apalatci, Apalatcy, Palchen, Apalchem, Apalachee, etc. J. R. Swanton, in *Creek Indians*, p. 113, says that the form "Apalachen" seems to contain the Muskogean objective ending -n, which by a stranger would often be taken over as a necessary part of the word.

"Apalcen" in Mercator's world map (1569), reprod. Plate 7; "Apalchen" in Ortelius, "Americae," *Theatrvm Orbis Terrarvm*, Antwerp, 1570; "Apalchen" in Lok, "North America," in Hakluyt, *Divers Voyages*, London, 1582, facs. in Winsor, *America*, III, 40; "Apalchi" underneath "Virginia," in Jode's map of North America, *Specvlvm Orbis Terrae*, Antwerp, 1593, reprod. Plate 16; "Apalche" in "Florida and Apalche," Wytfliet, *Descriptionis Ptolemaicae Avgmentvm*, Louvain, 1597, reprod. Plate 17, and in maps with a similar title by Acosta (1598) and Matal (1600); "Palchen" in Blaeu's world map (1616), reprod. in Stokes, *Iconography*, VI, Plate 81A; "Costa di Apalchem" in "Mexicvm," in G. B. Nicolosi,

Dell' Hercole, Rome, 1660; cf. *Lowery Collection*, No. 151a; "Apalache" in "Amerique Septentrionale," in Sanson, *Atlas*, Paris, 1693, and in many other Sanson maps; "Apalache" in Delisle, "Carte du Mexique et de la Floride" in his *Atlas*, Paris, ca. 1703, reprod. Plate 43, and in other Delisle maps for half a century; "Apalaches" in Moll, "Carolina" in his *Atlas Minor*, London, 1729, reprod. Plate 50; "Pais des Apalaches" in "Carte de la Caroline et Georgie," in *Histoire Generale*, Paris, 1757; "Country of the Apalachees" in Georgia, in "A New and Correct Map of the Provinces of North and South Carolina, Georgia & Florida" in *American Gazetteer*, London, 1762.

34. J. R. Swanton, *Creek Indians*, p. 112. Cabeza de Vaca reached Mexico in 1536; his report was published as *La Relacion y Comentarios del Gouernador Aluar Nuñez Cabeça de Vaca*, Valladolid, 1555.

35. De Soto will be used, conforming to common American usage, although Soto is the more correct form.

36. The most exhaustive and careful monograph on the route followed by de Soto, which is here followed, is found in the Final Report of the United States De Soto Expedition Commission. Letter from the Chairman [John R. Swanton] United States De Soto Expedition Commission Transmitting the Final Report of the United States De Soto Expedition Commission. 76th Congress, 1st Session, House Document No. 71, Washington, 1939, pp. 1–400. More recently the question of the de Soto expedition's route through the Southeast has been addressed by anthropologist Charles Hudson. Employing archaeological findings not available to Swanton, Hudson and his associates argue for a number of significant changes in de Soto's route. See C. Hudson, "The Hernando de Soto Expedition, 1539–1543," *The Forgotten Centuries*, Athens, Ga., 1994, pp. 74–103.

Valuable recent studies of the de Soto map have been made by Barbara Boston, "The 'De Soto Map,'" *Mid-America*, XXIII (1941), 236–50, and by Jean Delanglez, *El Rio Espíritu Santo*, New York, 1945, pp. 61–64. Father Delanglez also has an interesting note (p. 71) on the dating of the Ortelius-Chiaves 1584 map; it must have been made prior to 1574, the year of Hierónymo Chaves's death, although it is not probable that Ortelius had the original in his possession much before the publication of the engraved map in 1584. For the suggestion that American Indian maps and geographical intelligence contributed to the so-called de Soto map see Louis De Vorsey, "Silent Witnesses: Native American Maps," *The Georgia Review* 47 (1992), pp. 709–26.

37. Jacques Le Moyne de Morgues, "Floridae Americae Provinciae Recens & Exactissima Descriptio," in his *Brevis Narratio*, Frankfurt, 1591, opp. p. 1.

The names on the southeast coast, found on Continental maps for nearly two hundred years, are those given by Ribaut in 1562 and recorded in a Spanish spy's tracing of the map by Barre, Ribaut's pilot. See W. P. Cumming, "The Parreus Map (1562) of French Florida," *Imago Mundi*, XVII (1963), 27–40.

38. Among the first maps with this name were N. Sanson's "Le Nouveau Mexique et La Floride" (1656), and Du Val's map "La Floride Françoise" in his *Diverses Cartes*, Paris, 1669?, Pt. II, No. 5; many subsequent French maps have "Floride Françoise." "Florida Gallica" is found on the maps of North America by Visscher (1684) and Ottens (1756?).

39. Most modern authorities identify Ribaut's River May and the Spanish Rio San Mateo with the present St. Johns River: cf. W. Lowery, *Spanish Possessions 1562–1574*, p. 391ff. and pp. 405–7 (Appendix G), J. R. Swanton, *Creek Indians*, p. 51, and Herbert M. Corse, "Names of the St. Johns River," *Florida Historical Quarterly*, XXI (October 1942), 127–34. Ribaut discovered the river on his voyage along the coast in 1562 and named it May because he reached it on May 1; Laudonnière founded Fort Caroline there June 1, 1564. Menéndez named the river San Mateo because he completed the capture of Fort Caroline on St. Matthew's Day (September 21, 1565). "Rio San Mateo" is found in most Spanish derroteros, and on most maps not using the French "R. May," throughout the seventeenth century. Rio de San Juan, however, is found on several maps of about the end of the sixteenth century: "R. de S. Juan," Thomas Hood 1592, reprod. in Stokes, *Iconography*, II, Plate 19; "R. de San Juan," Dutch planisphere ca. 1600, reprod. Ibid., Plate 20; "R. de S. Joã," Tattonus 1602, reprod., Ibid., Plate 20; "R de S Juan" between "R S Mathio" and "R de Guale" in Tattonus 1600, reprod. Plate 19. It appears as "Barra de S.ⁿ Ju.ⁿ" on a Spanish map almost a hundred years later, "Mapa De la Ysla de la Florida," ca. 1683, reprod. in V. E. Chatelain, *The Defenses of Spanish Florida, 1565 to 1763*, Washington, 1941, Map 7. On the Nairne inset of South Carolina in the Crisp 1711 map is "S. Juan"; subsequently the name became increasingly common.

A number of writers do not accept the identification given above. Albert H. Wright, "French and Spanish Settlements. Rio de May," *Our Georgia-Florida Frontier: The Okefinokee Swamp, Its History and Cartography, Studies in History No. 11*, Ithaca, N.Y., 1945, Part III, pp. 1–16, identifies the River May as the St. Marys, locates Fort Caroline on Reid's Bluff or Roses Bluff on that river, and argues that the great lake on the River May is the Okefenokee Swamp. Professor Wright's conclusions are based on a review of the original sources, examination of early maps, and personal trips along the coast and to the interior. The quotations from contemporary reports are not conclusive. Professor Wright's use of maps is elaborate but its value is unfortunately vitiated by a disregard of the comparative authority of the maps referred to. He does not seem to differentiate between the documentary value of the contemporary maps of Le Moyne and of John White (who knew Le Moyne), for example, and of eighteenth-century maps of European closet geographers. His references to the topography of scores of early maps are therefore often confusing. On the other hand, his first-hand knowledge of the terrain throws light on some of the early Spanish explorations. He is concerned primarily in finding which of the explorers could have reached, or did reach, the Great Swamp area. It is quite possible that the great lake with the opposite shores which are invisible, found on Mercator 1606 and many later maps, is a result of reports of the impassable stretches of Okefenokee Swamp. In the maps of Le Moyne (1591), Mercator (1606), and their imitators the mountain in the Apalatci Mountains to the north, from which gushes a great waterfall into a lake with the legend "the natives find grains of silver in the lake," refer, apparently, to the "Land of waterfalls" in western North Carolina.

40. San Felipe on Parris Island was built in 1565, near the site of Ribaut's Charlesfort, but abandoned by the Spaniards and burned in 1576 by the Indians, who had risen in revolt.

However, by the end of 1577, another fort, San Marcos, was built in the province of Santa Elena by Pedro Menéndez Marqués, the new Spanish governor of Florida and nephew of the great Adelantado, Pedro Menéndez de Avilés. A plan of this second fort is preserved in the Archives of the Indies at Seville (reprod. in J. T. Connor, *Colonial Records of Spanish Florida*, II, opp. p. 50), and a description of the settlement is given by Marqués in a letter to the King from Santa Elena dated March 25, 1580, *Archives of the Indies*, 54-5-9, 12, and translated in J. T. Connor, *Colonial Records of Spanish Florida*, Deland, Fla., 1930, II, p. 283. "This village [Santa Elena] is being very well built, and because of the method which is being followed, any of the houses appears fortified to the Indians, for they are all constructed of wood and mud, covered with lime inside and out, and with their flat roofs of lime. And as we have begun to make lime from oyster shells, we are building the houses in such manner that the Indians have lost their mettle. There are more than sixty houses here, whereof thirty are of the sort I am telling your Majesty." San Marcos was abandoned when the Spanish troops were withdrawn to St. Augustine after Drake's attack upon that town in 1586. In 1593 Mestas advised rebuilding the fort at Santa Elena and sent a proposed plan: see Mestas 1593 MS. Apparently no fort was built, though Spanish claims to the coast as far north as Port Royal continued until 1763: see V. E. Chatelain, *Defenses of Spanish Florida: 1565–1763*, Washington, 1941, p. 132, note 6. The Spanish governors at St. Augustine limited themselves to frequent coastal patrols and excursions to the interior like those of Pardo in 1566–67 and Pedro de Torres in 1628. The Franciscans were never absent long from the Guale Coast of Georgia and as late as 1655 had a mission at Santa Elena as well as at Chatuache, six leagues farther north (H. E. Bolton, ed., *Arredondo's Historical Proof of Spain's Title to Georgia*, Berkeley, Calif., 1925, pp. 20–21, 24–25).

William Hilton described the tabby ruins of San Marcos, which he mistook for Ribaut's Charlesfort, when he sailed along the coast for a possible place of English colonization in 1663 (William Hilton, *A True Relation*, London, 1644, reprinted in *Collections of the South Carolina Historical Society*, Charleston, 1897, V, 20–21).

Mary Ross, "French Intrusions and Indian Uprisings in Georgia and South Carolina (1577–1580)," *Georgia Historical Quarterly*, VII (September 1923), 251–84, gives an interesting account of relatively unknown French incursions on the coast in the sixteenth century.

41. A rude map of the coast appeared near the turn of the century (see Maymi ca. 1595 MS). The derroteros of Chaves (1536) and Gonzales (ca. 1609) are careful and detailed. No adequate modern study has been made of the early sixteenth-century Indian, French, and Spanish names along the coast and to the interior. However, a valuable series of appendices is found in Lowery, *Spanish Settlements, 1562–1574*, pp. 399–466 passim.

For identification of some of the Indian and Spanish names of the region during the sixteenth and seventeenth centuries, as well as accounts of the struggles of the Spaniards to hold onto their Florida possessions, see J. T. Connor, ed., *Colonial Records of Spanish Florida, 1570–1580*, Deland, Fla., 1925–30, 2 vols.; V. E. Chatelain, *Defenses of Spanish Florida, 1565–1763*, Washington, 1941 (with maps); Mary Ross, "French Intrusions and Indian Uprisings in Georgia and South Carolina (1577–1580)," *Georgia Historical Quarterly*, VII (December 1923), 251–81; J. G. Johnson, "The Yamassee Revolt of 1597 and the Destruction of the Georgia Missions," *Georgia Historical Quarterly*, VII (March 1923), 44–53; and "The Derrotero of Alvaro Mexia," translated with notes by C. D. Higgs and published as Appendix A in Irving Rouse, *A Survey of Indian River Archeology, Florida* [Yale University Publications in Anthropology, No. 44], New Haven, 1951, pp. 265–74.

42. The identity of John White has been the subject of much controversy only recently settled by newly found documents. P. L. Phillips, in *Virginia Cartography*, Washington, 1896, pp. 3–17, argues, from the preface to the English edition of De Bry's volume, that the governor John White and the "humble" painter John White are two entirely different persons. He points out that the name of the painter-cartographer is spelled "Ioanne With" on De Bry's engraved map and in all three of the non-English editions of his 1590 edition of Harriot's work and "With's" drawings. Woodbury Lowery, *Spanish Settlements, 1513–1561*, was convinced by "the lucid arguments" of Phillips. An attempt to summarize the issues and examine the evidence is found in W. P. Cumming, "The Identity of John White Governor of Roanoke and John White the Artist," *North Carolina Historical Review*, XV (1938), 197–203, with the conclusion that the two are almost certainly the same person, for reasons here briefly stated. (1) No more than one John White is mentioned on any voyage; Governor John White says that he was on five (all the known) voyages. (2) Governor White mentions his maps and pictures as among his damaged possessions found on the 1590 voyage. (3) De Bry's metatheses and omissions of letters are frequent: the/te, which/wich, Whit/With (in non-English versions). (4) The "With" of the non-English editions of De Bry, a major stumbling block to Phillips's acceptance of "White" as the cartographer's name, is "White" in the English edition, except on the engraved map. Possibly Phillips used only a Latin edition; no English edition is in the Library of Congress. (5) Thomas Moffett refers to the painter as "Candidus."

Professor D. B. Quinn, *Roanoke Voyages*, I, 43, 399, note 6, points out that De Bry himself, who certainly was in a position to know, considered the painter and the governor to be the same, since he refers to the painter as the John White who was sent back to the Roanoke colony by Ralegh in 1588 (see De Bry's statement in the text). In 1588 Governor John White made an unsuccessful attempt, as a passenger on a privateer, to reach the colony, as we know from his own account (Quinn, II, 564).

43. D. B. Quinn, ed., *The Roanoke Voyages 1584–1590: Documents to Illustrate the English Voyages to North America Under the Patent Granted to Sir Walter Raleigh in 1584*, Hakluyt Society, Second Series CIV and CV, London, 1955, 2 vols.

44. Quinn, I, 51, 55. A full commentary on these instructions, which include a wide range of observations to make and instruments to carry, is found in E. G. R. Taylor, "Instructions to a Colonial Surveyor in 1582," *The Mariner's Mirror*, XXXVII (1951), 48–62.

45. Quinn, I, 215–17; II, 846–47. See Map List No. 6 for a fuller description and transcription of the legends. Apparently it was placed, without justification, with a letter of 1618 from Captain John Smith to Sir Francis Bacon when the Colonial Papers were being assembled in the nineteenth century. In the opinion of the present writer Professor Quinn's identification is convincing; the places listed and legends given, which were along those trips of exploration made by the Roanoke colonists in 1585; the similarity to some of the information on De Bry's "The Arriual of the Englishemen in Virginia"; the identification of watermarks with paper of the period; the ignorance shown by Smith of the knowledge it contains in his map of Ould Virginia in 1624; etc. The authorship of the map is not identified; but Quinn states that the technique is not that of White nor the handwriting of the notes that of Harriot. The handwriting is not that of John Smith; a letter of August 4, 1953, from the Secretary of the British Public Record Office (P.R.O.) to the present writer, who at that time was studying the provenance of the map, states that the map (C.O. 1/1, 42(ii); now M.P.G. 584) is different from the "Secretary" hand of the letter apparently written to Bacon by Smith himself (C.O. 1/1, 42) and from the Italic hand of the accompanying list (C.O. 1/1, 42(i)) of the differences between Virginia and New England.

46. Quinn, I, 217; II, 846.

47. See White 1585 MS A for a fuller account of this volume. Quinn (I, 392–98) has made a study of the different groups of drawings which White made from his archetypal set and has found tentative evidence of six groups or sets. Of these sets and their derivatives the only group now known which is of cartographical interest is the portfolio, in White's own hand but without his signature (Brit. Mus. P.&D. 1906-5-9-1). Stefan Lorant, ed., *The New World: The First Pictures of America*, New York, 1946, reproduces in one volume the De Bry engravings of Le Moyne and White, the Clements Library copies of the British Museum White portfolio in color, and most of the text of De Bry's two volumes of 1590 and 1591.

48. Quinn (II, 460) points out that White's map shows evidences of another French version now lost but partly incorporated in Lescarbot's map of 1612 (Map List No. 30). White's bulging coastline below "R de May," the location of the inland lake flowing into the May River, and some coastal names differ from both Le Moyne and Lescarbot.

49. See Winsor, *America*, IV, 17, note 7, and Stokes, *Iconography*, II, 38; references are to Hakluyt's *Discourse concerning Westerne Planting*, Maine Historical Society, 2nd ser., 1877, II, 113, 114, and the *Dedication of Divers Voyages*, London, 1582.

50. Facsimile in Winsor, *America*, III, 40; IV, 44.

51. Brit. Mus. Add. MS 5413; reprod. in H. A. Biggar, *Voyages of Jacques Cartier*, Ottawa, 1924, opp. p. 128. Harrisse pointed out that Coote's dating of ca. 1536 was wrong, because of the map's representation of Cartier's discoveries. The general configuration of the coast and much of the nomenclature is Spanish. See also A. Anthiaume, *Cartes Marines*, Paris, 1916, I, 75–78, where it is dated "after 1542."

52. Quinn, I, 210, note 10. This interesting passage, in which the Spanish Bahia de Santa Maria is equated with "Wococon," is marked through in the letter; previous editors have not printed it.

53. Reprod. in *Mapas Españoles*, Madrid, 1951, Plate VIII; see under de Soto ca. 1544 MS (Map List No. 1) for a fuller description of this map, which does not, however, have Bahia de Santa Maria on the Carolina-Virginia coast. Quinn suggests Diogo Homem's map of 1568, reprod. in I. A. Wright, *English Voyages to the Caribbean*.

54. D. B. Quinn, *Roanoke Voyages*, II, 847.

55. It is unlikely that White contributed any details to De Bry's 1590 volume from his last voyage, since he did not reach Plymouth on his return until October 24, 1590.

In 1592 Emery Molyneux published a great terrestrial globe, engraved by Jodocus Hondius, which incorporated information from the De Bry–White map and apparently a few details directly from White or Hakluyt: see Quinn, II, 850–51 and other references given there.

56. See Jode 1593 for a more detailed account of this map. Both Jode and Wytfliet made the error of putting Virginia above C. de las Arenas (Cape May, Sandy Hook, or Cape Cod), which in turn placed the Chesapeake between 42° and 43° N.L., the latitude of Boston. This in turn continued the distortion of the New England coastline which Mercator made in his great 1569 world map.

57. Cf. P. L. Phillips, ed., *The Lowery Collection*, p. 118.

58. See, however, the comments on Zuñiga 1608 MS and Velasco 1611 MS in the Map List. Lowery refers to a printed map of Florida in Pedro de Medina's *Arte del Navigare*, Venice, 1609, which has not been identified by P. L. Phillips (*Lowery Collection*, p. 117) nor found in any of the volumes of Medina examined in this present study. William R. Shepherd (*Guide to Material for the History of the United States in Spanish Archives*, 1907, p. 80, No. 15) refers to a "Map of Florida, prepared by Andres Gonzalez, 1609." Gonzales made a derrotero of the coast, accessible in the Lowery Transcripts, Manuscript Division, Library of Congress.

59. W. L. Saunders, ed., *Colonial Records of North Carolina*, Raleigh, 1886, I, ix, x, 22, 355, 571, hereafter referred to as *CRNC*. In 1722 Dr. Daniel Coxe wrote that settlements had been made under the auspices of Lord Maltravers; but Coxe was not a disinterested witness, for he was the owner of the Heath patent and was contesting the validity of the charter of the lords proprietors, which expressly stated that the Heath grant (1629) was forfeit for lack of colonization. See D. Coxe, *Description of the English Province of Carolana*, London, 1722, p. 116. The whole matter is confusing because the patent given to Lord Maltravers in 1639 by the Virginia council included not the entire Heath grant of "Carolana" but only one degree of it (35°–36°) together with hundreds of square miles of Virginia territory (*CRNC*, I, 14–15). That the upper reaches of Nansemond River, included in the Maltravers

patent, were settled is very probable. See C. M. Andrews, *The Colonial Period of American History*, New Haven, 1934–38, III (1937), 186–91.

There is a tantalizing reference to the colonizing venture under the Maltravers grant in statements concerning a deputy governor, Captain William Hawley, who was in Virginia in 1640; see H. R. McIlwaine, *Minutes of the Council and General Courts of Colonial Virginia, 1622–1632, 1670–1672, with notes and excerpts from original Council and General Courts Records into 1683, now lost*, Richmond, 1924, p. 482; and Conway Robinson's "Notes from Council and General Courts Records," *Virginia Magazine of History and Biography*, XIV (July 1906), 195. In a London newspaper, *The Moderate Intelligencer*, April 26, 1649 (No. 215), pp. 2–3, the following notice appears: "There is a gentleman going over governour into Carolana in America, and many gentlemen of quality and their families with him . . . [a page of propaganda description of Carolana follows]. . . . If this that hath been said give incouragement to any, let them repair to Mr. Edmond Thorowgood, a Virginia merchant, living in White-Crosse-street, at the house that was Justice Fosters. He will inform you of the governour, from whom you will understand when and how to prepare themselves (not exceed August) and what conditions shall be given to Adventurers, Planters, and Servants; which shall be as good, if not better then have been given to other Plantations."

Dr. Coxe referred to a map made in the early years of Cromwell's Protectorate, the result of a survey on the coast of Florida (i.e., Carolana) "above Two Hundred Miles square" where there had been trading and land purchases from the Indians. "They," continues Dr. Coxe, referring to the promoters of a colony in the location, "named divers Places, expecially Rivers, Harbors, and Isles, by the Names of the Captains of ships, chief Traders, and other Circumstances relating to the English Nation. As by the said Map or Chart doth more fully appear." Daniel Coxe, op. cit., p. 117. This map has not been identified. In 1633 Henry Taverner gave Samuel Vassal of London a map and description of the Carolana coast, the result of an exploration Taverner made along the coast as far south as St. Helena Sound in the fall of 1632: see C. M. Andrews, op. cit., III, 190, note 3.

60. *CRNC*, I, 676.

61. *CRNC*, I, 17. To this minimum number of persons required for settlement was added a provision that they seat themselves advantageously for security and be sufficiently furnished with ammunition; the terrible massacres of the outlying settlers in preceding years by the Indians had forced the council to add this regulation in grants for new settlements. The grant of 1653 contains the phrase, "next to those persons who have had a former grant," which seems to imply the existence of a previous settlement. After the charter of 1663 to the lords proprietors, and even before, settlers who had received grants from the Indians were required to take out patents with Governor Berkeley of Virginia or the proper authorities (*CRNC*, I, 204, 355); but there is scant indication that there were Indian grants prior to 1660 for land which was settled. In short, though it is reasonable to suppose that settlements had been established before 1660, any date before then has so far been based on inference, and the number of settlers has probably been exaggerated.

62. On March 25, 1708 (1707?; the English year began on that date), Lawrence swore in a deposition that about forty-seven years before (1660–61) he seated a plantation three or four miles above the mouth of the Morattock (Roanoke) River and lived there about seven years (*CRNC*, I, 677). Among the oldest direct documentary proofs of settlement is the grant on March 1, 1662, to George Durant of some land bounded by the Sound and Perquimans River, in which the grantor, Kilcaconen, king of the Yeopim Indians, refers to an adjacent tract which "I formily sold to Saml Pricklove" (*CRNC*, I, 19, 20, 355).

63. British Museum Add. MS 5027.a.59. Photograph in A. B. Hulbert, *Crown Collection of American Maps*, Cleveland, 1908, Ser. I, Vol. V, Map 29. The map has been catalogued as "Drawn about 1660," but this is wrong, for "Albemarle River" is written on it, and the sound was called Albemarle after one of the lords proprietors around 1665. This map, however, is a valuable one with interesting details not elsewhere recorded. It was apparently made in the late 1660s.

64. *CRNC*, I, 229.

65. *CRNC*, I, 286. The Blathwayt Atlas is a collection of manuscript and printed maps in the John Carter Brown Library; the map referred to is No. 19. There are four manuscript maps of the Carolina region in the atlas, published as *The Bathwayt Atlas*, which includes Volume I, *The Maps*, Providence, R.I., 1970, and Volume II, by Jeannette D. Black, *Commentary*, Providence, R.I., 1975.

66. *CRNC*, I, 217.

67. *The Journal of George Fox*, Norman Penney, ed., Cambridge, 1911, II, 236.

68. George Fox, *A Collection of Many Select and Christian Epistles, Letters, and Testimonies*, London, 1698, Vol. II, Pt. I, p. 336.

69. See note 62. The following references may also be consulted: William P. Cumming, *North Carolina in Maps*, Raleigh, 1966, Plate IV, Comberford 1657, MS, and pp. 11–12, and Elizabeth G. McPherson, "Nathaniell Batts, Landowner on Pasquotank River, 1660," *North Carolina Historical Review*, XLII, 72–73.

70. "Elizabeth, Ann, & Susa. Francis. 900 Acres. To all, Etc., Whereas, etc., now Know ye that I, the said William Berkeley, Knight, Governor, etc., give and grant unto Elizabeth, Ann and Susanna Francis, the Orphans of Mr. Thomas Francis deceased, nine hundred acres of land in the County of Nancemond . . . The said Land being formerly granted unto Samuel Stephens by Patent dated the twentieth of July one thousand six hundred and thirty-nine . . . and by the Said Stephens Sold and Assigned unto Nathaniel Batts and by the same Batts sold unto Mr. Thomas Francis and now become due to the said Orphans . . . dated the twentieth of October one thousand six hundred and sixty five." *Patent Books*, V, 563–64. The early records of Nansemond County, which might have thrown further light on Nathaniel Batts, have been destroyed. See article by Elizabeth G. McPherson referred to in note 69.

71. W. W. Hening, *Statutes at Large*, New York, 1823, I, 442.

72. Ibid., I, 373, 379, 386, 408 (Dew); 430 (Francis).

73. In an abstract of the Minutes of Council and General

Records of Virginia for June 11, 1657, which is preserved in the manuscript notes of the historian Conway Robinson, made by him before the destruction of the minutes during the Civil War, the following note occurs: "p. 314 Privilege granted Nathaniel Batts for interest taken in the discovery of an inlet to the southward, p. 339, 392." The pages 339 and 392 refer to subsequent entries concerning Nathaniel Batts in the minutes, but of these Robinson made no further note, nor does he explain what the "privilege granted" was. Conway Robinson, "Notes from Council and General Court Records, 1641–1664," *Virginia Magazine of History and Biography*, VIII (October 1900), 165. The next entry in the Conway Robinson records is to page 317 in the original minutes. The three intervening pages (314–17) may have dealt with Batts's privilege or with unrelated matters. The privilege may have been merely for trade with the Indians or a grant to settle land. The Robinson notes were edited by the late Mr. William G. Stanard, editor of the *Virginia Magazine*, who indicated that he planned to write a footnote on this passage. The footnote was not printed, and his sister, Miss Stanard, after a careful search, could find no papers of her brother which contained it.

74. Under the date November 5, 1679, in the wills in the Secretary of State's office, Raleigh, North Carolina, it is recorded that Nathaniel Batts died intestate. His wife and administratrix married Joseph Chew. James Blount stood as surety. See *North Carolina Historical and Genealogical Register*, J. R. B. Hathaway, ed., Edenton, N.C., I (January 1900), 30.

Whatever the antecedents of Batts, he apparently left a strong imprint upon nomenclature along the coast. "Bats' Creek," a tributary to the Neuse on the south side, was the scene of a bloody massacre by the Indians in 1712 (*CRNC*, I, 864). On the Comberford Map "Battis Point," between the Pamlico River and Machapoungo River, may be another mark left by Batts in his "discovery to the southward." By 1672 the island in the sound at the mouth of Yeopim River had been changed to "Batts Island" from its earlier title of "Heriots Ile" (see the early sixteenth-century maps), for in that year George Fox writes: "and at the first house wee came to in corlina wee mett with an Indian Kinge a pretty sober man; The truth spreadeth, & as wee passed downe, wee passed by Batts Illand & by Kickwold youpen, & pekeque mines [Perquimans] River, where there is some friendly people" (*Journal*, II, 234). By the end of the century the island was called "Batts Grave," for in 1694 James Fewox brought suit against Benjamin Lakar for putting hogs "upon a certain piece of land called Batts grave or island" (*CRNC*, I, 414–15).

75. F. L. Hawks, *History of North Carolina*, Fayetteville, 1859, I, 61–67. The Moseley (1733), Wimble (1738), and Collet (1770) maps of North Carolina, based on new surveys, indicate several plantations in the region between the Roanoke River and Salmon Creek.

76. During the first half of the seventeenth century the usual cartographical nomenclature of the Carolina region, influenced by history and by the existing colonies, remained Florida for the southern half and Virginia for the northern. Many new local names are found for the region south of the Chesapeake. They disappeared with actual settlement in the fourth quarter of the century. Most maps followed John White, who named each area after the most important Indian town in it: Weopemeoc for the land between Albemarle Sound and the James River and Secotan for the land south of Albemarle Sound and the Roanoke River. Chawanook on both sides of the upper reaches of the Chowan River and Mongoack between the upper reaches of the Roanoke and Neuse Rivers are also found on the White map and some of the derivative maps. "Ould Virginia" on Smith's map (1624) and Farrer (1650 MS and 1651), "Rawliana" on Farrer 1651, "The South part of Virginia, the east part of Florida, and the intervening region" (in Latin) on Blaeu (1640), Jansson (1642), and Montanus (1671), are other designations on maps of this transition period.

77. An examination of the methods of granting charters and the economic factors back of the expansionist movement of the English people in the seventeenth century is not within the scope of this study, though a knowledge of the mercantile interests involved, such as is given in C. M. Andrews, *The Colonial Period of American History*, New Haven, 1934–38, I, pp. 40–43, IV, 370ff., "England's Commercial and Colonial Policy," aids the student of the period in understanding the powerful nonpolitical and nonreligious factors motivating the colonial backers in England such as Heath, Maltravers, and the lords proprietors.

The role of the English geographers in keeping their countrymen informed on the achievements of emigrants to the Americas demonstrates the intimate connection between the business forces in England that promoted expansion and the literary advocates who supported and justified the movement. In the beginning of the seventeenth century appeared the translations, compilations, and universal geographies of Hakluyt, Abbot, Heylyn, Carpenter, and Speed. Toward the middle of the century purely American works by Castell and Gardyner initiated a new and markedly promotional genre which reached a high level in Ogilby's *America*. Then came the handbooks of Moxon, Bloom, Crouch, Ligon, and Josselyn, which were devoted exclusively to the English settlements or to a single colony. Finally came the intensive promotion tracts, such as those which appeared in the year 1682 advertising Carolina; T. A[she], *Carolina; or a Description of the Present State of that Country*; R. F., *The Present State of Carolina with Advice to the Settlers; A True Description of Carolina*, an abbreviation of the previous pamphlet which was probably printed to accompany the second lords proprietors' map (Gascoyne 1682); and [Samuel Wilson], *An Account of the Province of Carolina in America*, London, 1682.

Mr. Mood considers the English geographical writers of this period more successful as exponents of English territorial designs than as geographers: Fulmer Mood, *The English Geographers and the Anglo-American Frontier in the Seventeenth Century*, [University of California Publications in Geography, Vol. VI, No. 9], Berkeley, Calif., 1944, pp. 1–33. For a discussion of Mr. Mood's work, see the book review by W. P. Cumming in the *Hispanic American Historical Review*, XXIV (November 1944), 668.

78. T. Ashe, *Carolina: or a Description . . .* , London, 1682. Reprinted in A. S. Salley, ed., *Narratives of Early Carolina*, New

York, 1911, p. 140. Mr. Salley pointed out the error in Ashe's statement in a note on page 140; later he corrected and greatly expanded this note, with copious illustrative quotations, in his *The Origin of Carolina, Bulletin of the Historical Commission of South Carolina*, No. 8, Columbia, S.C., 1926.

79. John Oldmixon, *The History of the British Empire*, London, 1708, I, 325–27. Reprinted in *Narratives of Early Carolina*, New York, 1911, p. 319.

80. James Glen, *A Description of South Carolina*, London, 1751.

81. Alexander Hewat, *An Historical Account of the Rise and Progress of the Colonies of South Carolina and Georgia*, London, 1779, I, 18. Reprinted in B. R. Carroll, ed., *Historical Collections of South Carolina*, New York, 1836, I, 23.

82. Guillaume Delisle, "Carte de la Louisiane et du Cours de Mississipi . . . ," dated June 1718 (Plate 47: Map List No. 170). Delisle, the French royal geographer and founder of modern scientific cartography, has the legend across the face of what is now North Carolina, "Caroline, ainsi nommez en l'honneur de Charles IX par les François qui decouvrirent en prirent possession et se etablirent lan 15. . . ." This map was designed for a political purpose, to invalidate the English claims in the imperial struggle for colonies between France and England; but the authority of Delisle and the general excellence of this map gave it widespread popularity with both the French and the English. Delisle's map continued to appear in atlases until after the American Revolution.

83. J. B. Homann, "Amplissimae Regionis Mississipi Sui Provinciae Ludovicianae . . ." about 1737, in many Homann atlases. This map is fundamentally a copy of Delisle's map, referred to above. Delisle's inscription concerning Carolina has been translated into Latin.

84. J. G. Kohl, *National Intelligencer*, July 22, 1864, "The French built on their Riviera May . . . a fort which they called Fort Caroline or Carolina. Some mapmakers and geographers applied this name, as an appellation of a country or territory, to the whole region. So we see, for instance, on a map of North America by Cornelis a Judaeis (1593), the whole French Florida called Carolina, in honor of Charles IX, King of France. [Kohl is incorrect; Jode's "Carolina" is written below a fort drawn on the R. Mayo; see Plate 16, Map List No. 16.] So we may say that we have three kings as godfathers to this province: Charles IX, of France, Charles I, and Charles II, of England." Kohl was the first great historical cartographer in this country.

85. "La Caroline fut ainsi nommé du roi de France Charles IX (en latin Carolus), par le navigateur Jean Ribaut, qui y fonda, en 1562, un établissement français, détruit par les Espagnols en 1565," in the article "Caroline du Sud," Charles L. Dezobry and J. L. Bachelet, *Dictionnaire Général de Biographie et d'Histoire*, 6th ed., Paris, 1873, I, 468. The article is signed by Oger, professor of history in the Collège Sainte-Barbe, Paris, who was the author of *Géographie de la France et Géographie Générale*, 6th ed., Paris, 1876.

86. St. Julien R. Childs, "French Origins of Carolina," *Transactions of the Huguenot Society of South Carolina*, L (1945), 29, quoting Pierre Biard, "Relation de la Novvelle France, de ses Terres, Natvrel du Pais, & de ses Habitans," Lyons, 1616, from the extract reprinted in *The Jesuit Relations and Allied Documents*, Reuben G. Thwaites, ed., Cleveland, 1896–1901, L (1897), 104–5.

87. *Jesuit Relations*, L, 104.

88. *CRNC*, I, 5–13.

89. Carolana: *Carolus*, Latin form of *Karl/ana*, Lat. suffix meaning of or belonging to. The use of the suffix -ina, -ine is more common for the formation of feminine titles; and for the -ana form, the euphonious variation -iana is more usual, as in Christiana. Possibly Charles I used the -ana to avoid any implied connection with the earlier French settlement at la Caroline.

90. Collections of the South Carolina Historical Society, V (1897), 200ff.

91. The English translation given in the Colonial Records of North Carolina, I, 5–13, is of an unrolled exemplication of August 4, Charles I (7), in the Shaftesbury Papers in the Public Record Office. The original document in the Shaftesbury Papers contains only the form "Carol*a*na."

92. Pp. 25–26. Mr. Salley, who noted the errors in the French claims many years ago (see note 78), made several helpful suggestions to the present writer when he began examining this particular problem in nomenclature in 1935.

93. See Comberford 1657 MS A (Plate 32, Map List No. 50).

94. Sanson's map continued to be used for various atlases until 1700 and influenced other maps. In about 1600 Duval also published a map with the same mistake; see note 98 also.

95. "First Charter granted by King Charles the Second to the Lords Proprietors of Carolina," reprinted in *CRNC*, I, 23.

96. Because of a storm, Hilton did not explore the coast between Port Royal and Cape Fear; the land between these two points is therefore given sketchily and incorrectly. Robert Sandford's *A Relation of a Voyage on the Coast of the Province of Carolina*, London, 1666, gives the fullest and most accurate account of the coast between Port Royal and Cape Fear written up to that time. Sandford's voyage, made under the auspices of the lords proprietors, had as its purpose the discovery of a place suitable for settlement to the southward. His account of a deep entrance into the Kiawah Country (Charleston Harbor) is especially detailed and probably led indirectly to the English settlement there four years later.

97. "Charles-River" is mentioned in the title of the pamphlet: *A Brief Description of The Province of Carolina On the Coast of Floreda. And More perticularly of a New-Plantation begun by the English at Cape-Feare, on that River now by them called Charles-River, the 29th of May. 1664 . . . Together with A most accurate Map of the whole Province.* London, Printed for Robert Horne . . . 1666. On page 2 of the pamphlet the new name of the river referred to as Charles-River in the title is given: "The entrance into the River, now called Cape-Feare River, the situation of the Cape, and the trending of the Land, is plainly laid down to the eye in the Map annexed."

98. See Kocherthal 1709 and Homann 1714 (Plate 46); probably the name "Carolina" at the Cape Fear shows the influence of the misplaced "Caroline" on the Sanson and Duval maps; see note 94.

99. See Map List No. 68 and Plate 36. An edition of Lederer's

work by the present writer was published by the University of Virginia in 1958; it includes unpublished biographical information about Lederer, the Lederer letters in the Winthrop Papers in the Massachusetts Historical Society, and an essay on the Indian tribes referred to in *The Discoveries of John Lederer* by Douglas L. Rights and W. P. Cumming. Some of the new information in this edition is used here by kind permission of Mr. John C. Wyllie, librarian of the University of Virginia, the general editor of the Tracy W. MacGregor Library series in which the edition is to be published. A list of the important maps influenced by Lederer's map is found in W. P. Cumming, "Geographical Misconceptions of the Southeast," *Journal of Southern History*, IV (1938), 489–92, Appendix B. The Lederer correspondence in the Winthrop Papers was first noted by Harold S. Jantz, "German Thought and Literature in New England, 1620–1820," *Journal of English and Germanic Philology*, XLI (1942), 17, note 43. Dieter Cunz, in *The Maryland Germans*, Princeton, 1948, pp. 30–39, and "John Lederer, Significance and Evaluation," *William and Mary Historical Magazine*, XXII (1942), 175–85, discusses Lederer's explorations and first called general attention to his years in Maryland after leaving Virginia. Lyman Carrier, "The Veracity of John Lederer," *William and Mary Quarterly*, IX (1939), 435–45, defends the integrity of Lederer ably, although neither of the specific routes proposed by Carrier or by Cunz can be accepted by the present editor or was accepted by the late Dr. Rights, whose careful study of the Indian Trading Path and the relation of Lederer's explorations to it continued through the years since his "The Trading Path to the Indians," *North Carolina Historical Review*, VIII (1931), 403–26. Clarence W. Alvord and Lee Bidgood, *The First Explorations of the Trans-Allegheny Region by the Virginians, 1650–1674*, Cleveland, 1912, pp. 63–99, 131–71, published the text of Lederer's work, pointed out that he was the first white man on record to look into the Valley of Virginia, and emphasized the historical and ethnological value of his report. Unfortunately they failed to understand the cartographical and commercial significance of Lederer's second expedition, misunderstood or failed to analyze some of the difficult passages, and dismissed too much of the account as fabrication.

100. The reference to March 1669 in the title and to March 9, 1669, as the date when Lederer began his first expedition, is our 1670; the year then began on March 25, not January 1, as it does now.

When and why Lederer came to Virginia is not known. He was born about 1644 and was apparently brought as an infant to Hamburg, Germany, by his father. The Lederers may have been refugees from the devastation of the Thirty Years' War. Young Johann matriculated in the Hamburg Academic Gymnasium on April 18, 1662: see Sillem, *Die Matrikel des Akademischen Gymnasiums in Hamburg, 1613–1883*, Hamburg, 1891, p. 46, No. 1032. Dr. K. D. Möller, Staatsarchivdirektor of Hamburg, has furnished the information concerning Lederer's Hamburg background. Although Lederer registered at the gymnasium as a Hamburger and in his application for Maryland citizenship in 1671 he calls himself a native of Hamburg, he does not seem actually to have been born there. At the time such a statement was customary if a person or his parents had resided long in or near a city. On August 18, 1646, Jürgen (i.e., George) Lederer, son of Johann Lederer senior, was baptized in the Evangelical-Lutheran Nicholas Church (cf. p. 185 of the Church registers). From this Dr. Möller infers that a younger brother of Johann, the matriculant of 1662 in the Academic Gymnasium, was born to Johann Lederer senior, who had come to Hamburg in the forties during the time of the Thirty Years' War, apparently after the birth of Johann the younger. The parent did not become a citizen (Bürger), for from January 24, 1646, to July 3, 1657, he paid a non-citizen's tax of one Taler (three marks): Hamburg, *Senatsakte*, CI. VII, Lit. Cc, No. 13, Vol. 1b., p. 149. In 1657 he was called to the tax office to confer about his taxes, with the evident implication, according to Dr. Möller, that he was prospering financially and that the tax was to be increased. Unfortunately the tax lists are missing thereafter for some time, and the result of the tax office order is not available. It is apparent that Johann Lederer senior was living in Hamburg from 1646 until at least 1657 and that by the latter date the tax authorities were not satisfied with the amount of tax paid. Since eighteen was the usual age for matriculating in the gymnasium, Johann junior must have been born about 1644 and was probably brought as a child by his parents to Hamburg. The data given above and John Lederer's own quotations from the New Testament in his letter of April 14, 1675, to Governor Winthrop make it equally improbable that he was the Lederer who was a Franciscan friar, the author of a dissertation on angels in 1661, or one of the Austrian Lederers of the Jewish faith. It is possible that he was the Johann Lederer who in 1691 published at Innsbruck a geographical and geological study of the Alps, *Descriptio Geographica et Geologica de Montibus Alpium*.

A little less than five years after his three expeditions in Virginia, Lederer sailed from Fairfield, Connecticut, for a return to his home in Germany on January 16, 1675; his last letter to Governor Winthrop is from the house of Judge Sharpe in the Barbados and is dated April 14. He reached home safely, for in 1681 the editor of the *Journal des Sçavans* (No. 6, March 3, 1681, pp. 75–76) published some extracts concerning the Indians of Virginia from the memoirs of John Lederer of Hamburg, who had returned from "that country after a ten years' stay." This statement may indicate that Lederer was in Virginia or in America five more years than is at present known, or merely that the memoirs had been written ten years before they came to the attention of the editor's correspondent, Dr. Jacob Spon. In this as in other earlier records about Lederer there is a tantalizing ambiguity.

101. See Map List No. 17: José de Acosta, *Geographische vnd Historische Beschreibung der . . . America*, Cölln, 1598, fol. 7, "Florida et Apalche." A considerable part of the confusion about Lederer's route lies in the difference between Lederer's conception of the terrain, based on contemporary maps, and the actual topography of the region which modern commentators know from accurate maps.

102. For more detailed accounts of the Trading Path from the Yadkin Ford southward see William Byrd's *Histories of the Dividing Line*, ed. Boyd, Raleigh, 1929, pp. 298–304, and Jethro Rumple, *History of Rowan County, North Carolina*, Salisbury, N.C., 1881,

pp. 6–7, 17–18. Traces of the old marshes in this area, such as the Ephraim Swamp in north Mecklenburg and swamps in the creek bottoms, still remain; but Lederer's description of a swamp extending from Salisbury to South Carolina in a piedmont terrain is an obvious exaggeration.

103. On Mouzon's 1775 map the Path bifurcates at Buffalo Creek near Concord, North Carolina, the westerly path going through Charlotte toward the Catawba, and the easterly branch through Indian Trail, Union County, North Carolina, toward Charleston, South Carolina. If Lederer actually got this far, as he apparently did, he probably took the easterly path.

It is possible that Arthur and Needham, trading with the Tomahitans in the mountains, took a prong of the Indian Trading Path that branched off at the Trading Ford and followed the Catawba River up to Swannanoa Gap near Old Fort, North Carolina: see Myer, "Indian Trails of the Southeast," *Forty-Second Annual Report*, Bureau of American Ethnology, Washington, 1928, p. 778 (Trail 39).

104. According to the great chart of Carolina made by Maurice Mathews about 1685, the Esah (Ushery) Indian villages were on the left (east) bank of the Catawba River just below Sugar Creek. Lederer said that he turned back at Ushery because of Indian reports that "to the southwest, a powerful nation of bearded men were seated, which I suppose to be the Spaniards." There were Spanish missions at the time to the southwest near the Gulf; it is possible, as Lederer evidently believed, that the Spanish were working mines in the lower Appalachian range (see also Glover's account below). Hawks's theory (II, 60) that the Indians were referring to one of the earlier English settlements on the Carolina coast is strained and unnecessary, although it has been followed by some commentators. It is possible that the Indians may have been reporting expeditions to the interior made from the new English settlement on the Ashley River. Lederer could hardly have known that in April 1670 the English had landed at Charles Town or that in this same June 1670 Dr. Henry Woodward made a fourteen-day trip to the "west and by north" from Charles Town. A year later, in July 1671, Dr. Woodward made a land trip to Virginia from Charles Town. He probably used the same Indian Trading Path that Lederer explored, for "the Emperor Cotachico . . . says that Woodward is got to Roanoke near Virginia" (*Collections of the South Carolina Historical Society*, V [1897], 186, 338 footnote 1, 388). "Cotuchike" is located on the Mathews ca. 1685 map just below the Esah (Ushery) villages on the Catawba River.

An independent account of Lederer's second journey, with references to Spanish articles at Ushery, was given to the Royal Society in London by the Virginia surgeon, Thomas Glover (*An Account of Virginia*, Oxford, 1904 [reprint], p. 10), in 1676: "Above five years since there was a German Chirurgeon, who obtained a Commission from Sr. Will Bartlet to travel to the South-west of Virginia, and to make discovery of those parts: He went along the foot of the Mountains as far as the Lake of Usherre, and discovered them to be passable in two places, and he gives a relation, that, while he was in an Indian town adjacent to the Mountains, there came four Indians on an Embassie to the King of that town, from a King that lived on the other side of the Mountains, who by the Commandment of the King on this side were all strangled, with which barbarous usage he was much abashed, fearing the like cruelty; but they proved more civil to him, permitting him to depart in safety.

"At his return he brought an Emerauld, and some Spanish mony, which he said he had of the Indians bordering on the Lake of Usherre, which caused some to think that some Spaniards are seated near upon the back of the Mountains."

105. Alvord and Bidgood, op. cit., pp. 210–26. For Lederer's Maryland citizenship and for his license to trade with the Indians encountered on his Carolina expedition, see *Archives of Maryland*, II (1884), 282–83, and V (1887), 84–85.

106. "Capt. Byrd's Letters," *The Virginia Historical Register*, I (1848).

107. "Letters of William Byrd, First," *Virginia Magazine of History and Biography*, XXV (1917), 51–52.

108. Lawson, *A History of North Carolina*, ed. Harriss, Richmond, 1937, p. 40.

109. William Byrd's *Histories of the Dividing Line*, ed. W. K. Boyd, Raleigh, 1929, pp. 298–300.

110. James Mooney, *The Siouan Tribes of the East*, Washington, 1894, p. 70.

111. Lyman Carrier, "The Veracity of John Lederer," pp. 435–45.

112. What part Sir William Talbot's own enthusiasm and Irish imagination played in preparing the text of *The Discoveries of John Lederer* for printing as a piece of promotion literature for Lord Ashley and the other lords proprietors can never be known. Talbot states on the title page that he prepared the work from his conversations with Lederer as well as from his translation of Lederer's report from the Latin. Talbot's uncle, Lord Baltimore, accused him vehemently of irresponsibility, and even his admirer, Governor Charles Calvert, states that he misrepresents facts (Calvert Papers, Baltimore, 1889, pp. 260, 261, 262). Lederer, on the other hand, appears to give most of his account with Germanic thoroughness and care; this same quality is present in the Winthrop letters. Lederer's narrative, one can guess, may have suffered from his "Discourse" with Talbot.

113. F. L. Hawks suggested that Lederer reached Lake Mattamuskeet or Lake Phelps in Bertie County. Hawks was so anxious for Lederer to reach a swampy country in Bertie County with a lake that he apparently forgot the mountains and the desert. This route not only supposes that Lederer, an experienced traveler, could go for three weeks in almost the opposite direction from that in which he thought he was going, but also is in direct variation with several later identifications of places on his route (F. L. Hawks, *History of North Carolina*, II, 50). S. A. Ashe, "Was Lederer in Bertie County," *North Carolina Booklet*, XV (1915), 33–38, shows several difficulties in Hawks's theory and suggests that Lederer may have gone to Sawratown on the Dan River in Stokes County; but Sawratown was not yet in existence. H. A. Rattermann suggested that Lederer reached Miccosukee Lake in northwestern Florida; but this is completely contradicted by Lederer's

own estimate of the mileage covered as well as by its inherent impossibility (H. A. Rattermann, "Der erste Erforscher des Allegheny-Gebirges—Johann Lederer," in *Der Deutsche Pioneer* [Cincinnati, 1869–87], VII [1877], 399–407). Cyrus Thomas attempted to examine Lederer's route, gave it up as a tangled mass of lies, and declared that he did not believe Lederer ever went into the Carolinas at all (Cyrus Thomas, "Was John Lederer in Either of the Carolinas?" *American Anthropologist*, V [1903], pp. 724–27; Cyrus Thomas and J. N. B. Hewitt, "Xuale and Guaxale," *Science*, XXI [1905], pp. 863–67).

Recent investigation has shown, however, that many of the difficulties of tracing the route had their origin in Lederer's inability to understand the statements or sign language of the Indians and also in the misinterpretation of Lederer's statements by modern commentators. The information given by Lederer, say Alvord and Bidgood, "seems to be remarkably correct and valuable . . . He gave occasion, moreover, for the production of a book of great historical and ethnological value . . . No material risk of inaccuracy is incurred in accepting his narrative where there is no external or internal evidence of its improbability" (Alvord and Bidgood, pp. 69, 64). The difficulty of dealing impartially with Lederer's account is shown by the following misconceptions of his narrative given by Alvord and Bidgood. They write (p. 68): "After he left the Saura village, no certainty can be evolved from the mass of palpable falsehood . . . It makes pleasant reading: Silver tomahawks, Amazonian Indian women, peacocks, lakes 'ten leagues broad,' and barren sandy deserts two weeks' journey in width, when located in the Carolina piedmont sound like the tales of Baron Münchhausen." But Lederer does not say that he saw silver tomahawks; he repeats what an Indian told him: "The men it seems should fight with silver-hatchets: for one of the Usheryes told me that they were of the same metal with the pomel of my sword" (p. 160). He says that the Indian women of the Oustack tribe, of whom he heard only through the Ushery Indians, "come into the field and shoot arrows over their husbands shoulders" and that the Ushery women delight in feathers of peacocks "because rare in those parts." He mentions one lake which he judged "to be about ten leagues broad: for were not the other shore very high, it could not be discerned from Ushery," and "a barren sandy desert." With the reduction of Alvord and Bidgood's plurals to the singular and a juster presentation of the other statements, the narrative becomes more credible.

114. British Public Record Office, Shaftesbury Papers, Section IX, Bundle 48, No. 82. Colleton's letter, with the notes on the back in Locke's hand, are given in *Collections of the South Carolina Historical Society*, V, 264–66.

115. The list is apparently based largely on a description of the coast given in a letter to the lords proprietors by Robert Sandford, who explored the coast south of Cape Fear in 1666; Sandford's *Relation of a Voyage on the Coast of the Province of Carolina* is reprinted in *CRNC*, I, 118–38.

Albemarle from 35½ to 36½. This may refer to Albemarle River (now Albemarle Sound, formerly called Roanoke River) or Albemarle County, both so named shortly after the proprietary was granted. Compare *CRNC*, I, 120, 153, 155. General George Monk, Duke of Albemarle, was preeminently the restorer of Charles II to the throne and was first Palatine of Carolina.

Ashley River. The present Ashley River; named by Robert Sandford during his voyage along the coast in 1666. Compare *CRNC*, I, 137. Anthony Ashley Cooper, Baron Ashley, later the earl of Shaftesbury, was chief justice of Carolina.

Berkeley. Berkeley Bay, between Cape Romain and Cape Fear, is the present Long Bay. Compare *CRNC*, I, 137. It was so named by Sandford in honor of Lord John Berkeley, chancellor of Carolina, and his brother, Sir William Berkeley, governor of Virginia. The name was never applied, apparently; it is not even given in Locke's MS map or Ogilby's first lords proprietors' map.

Carteret Cape Roman. The cape was renamed for Lord George Carteret, admiral of Carolina, by Sandford in 1666. Compare *CRNC*, I, 138. The name soon reverted to Cape Romain, one of the two oldest and best-known names along the entire coast. Cabo de Santa Elena and Cabo de Santo Roman were named during the Ayllón expedition, and Elena first appears on Juan Vespucci's World Map of 1526, now in the New York Hispanic Society Library.

Clarendon Cape Fear. Clarendon was the name of the county around the Cape Fear River (*CRNC*, I, 72), and on Ogilby's map (1672) it was also the name for the river. Cape Fear River is the Rio de Principe of the early continental mapmakers, the Cape Fair-River of Hilton's *Relation* (1664), the Charles River of Horne's *Brief Description* (1666) and the accompanying map, the "Charles River neer Cape Feare" of Robert Sandford's *Relation* and of the seventeenth-century MS map in the Blathwayt Atlas, John Carter Brown Library, Providence, R.I., the Clarendon River of Ogilby's map (1672), and the "Cape Fear R or Clarendon R" of the Gascoyne second lords proprietors' map (1682). Locke's name, Clarendon, after the earl of Clarendon, one of the lords proprietors, did not last long, for Cape Fear River continued to take its name from Cape Fear, first bestowed upon Cape Lookout in Sir Richard Grenville's narrative of 1585 and in John White's account of 1587. Compare also George Davis's "An Episode in Cape Fear History," *South Atlantic Magazine* (January 1879), reprinted as "Origin of the Name, Cape Fear," in James Sprunt, *Chronicles of the Cape Fear River*, Raleigh, 1916, pp. 1–7. John Hilton, who explored the Cape Fear River in 1662 and 1663, named about twenty branches, islands, and points on the river, many of them apparently after names of members of the expedition: Long Island (Anthony Long was on the ship with Hilton), Blower Ile (Pyam Blower was a master), Fabian River (Peter Fabian was a companion). Point Winslow, Goldsmith, Hory, Borges, and Brown suggest other unmentioned members of the expedition, as may Crane, Green, and Greenless. But these names, though they appeared on several manuscript and some printed maps for a few years, soon fell into disuse, as there was no permanent settlement made on the river for several decades. The only names given by Hilton for places along the Cape Fear River still in use are Turkey Quarter, Rocky Point, and Stag Park. Compare W. C. Ford, "Early Maps of Carolina," *Geographical Review*, XVI, No. 2 (April 1926), 264–73, and Sprunt, *Chronicles*, p. 28.

Colleton. "Ashpow als Colleton R" does not appear until Ogilby's map. The Ashepoo River, lying about halfway between the South Edisto Inlet and Combahee River, is an unnamed river referred to in Sandford's *Relation* (*CRNC*, I, 129). It is shown as Colleton's River on Gascoyne's map (1682), but soon after this the name was dropped for the Ashpow or Ashepoo. Sir John Colleton, a Barbados planter, was high steward of Carolina.

Craven North Side of port Royal. Locke may refer to Craven County or to Craven River (called Combahee except on Ogilby's map). The earl of Craven was of the English Privy Council and first high constable of Carolina.

Edisto 32 d's 30 m. S. C. "Orista was the Spanish interpretation of the name of the Indian tribe which the French called Audusta and the English subsequently Edisto." A. S. Salley, *Origin of Carolina*, Columbia, 1926, p. 20, note 1.

Port Perio 32 d. 25m. Sandford's name for St. Helena Sound; the name is found only here and in Sandford's *Relation* (*CRNC*, I, 129).

Kiwaha 32d 40m. The name given to the country of the Kiawah by Sandford (*CRNC*, I, 127); the name referred to the country around the Ashley River but was apparently not subsequently used by Locke or the early mapmakers.

116. See Map List No. 65; the eastern half of the map is reproduced as Plate 35. The map has been removed from the Shaftesbury Papers and is now identified as Public Record Office M.P. 1/11. On an original lightly penciled base, with some Lederer details and coastal information from the recent Hilton and Sayle explorations, are superimposed in heavy ink numerous additional details of Lederer's three explorations. It seems possible that the earlier information from various sources may have included data about Lederer's journey sent by Governor Berkeley from Virginia. To this were added details about Lederer derived from the printed map or directly from Sir William Talbot, who in the summer of 1671 had come from Maryland with a map in Lederer's own hand.

117. The original list of lords proprietors were Edward, earl of Clarendon; George, duke of Albemarle; William, earl of Craven; John, Lord Berkeley; Sir William Berkeley, governor of Virginia; Anthony Ashley Cooper, earl of Shaftesbury; Sir George Carteret; and Sir John Carteret.

118. The names of some of the places have been changed; Colleton River reverted to its earlier Indian name, Ashepoo; Porte Carteret is Charleston Harbor; Clarendon River is the Cape Fear; Cape Carteret reverted to its old Spanish name of Cape Romain; Ashley Lake is the non-existent Ushery Lake of John Lederer's map of 1672, which was incorporated with the rest of Lederer's imaginary geography and nomenclature on Locke's MS map and the Ogilby map and continued to be found on subsequent maps for eighty years and more.

119. In 1681 Samuel Wilson, secretary to the lords proprietors, wrote at their instruction to the governor of the north part of Carolina, Henry Wilkinson, "that you be sure as soon as you can to send home the mapp of the County mended by your owne or frds: experience." *CRNC*, I, 338.

120. Soundings along the coast are reported in Hilton's and Sandford's *Relations* (*Collections of the South Carolina Historical Society*, V, 19, 22, 61–62, 67, 71, 73) and on Shapley's 1662 MS map and the Horne 1666 map.

121. *Collections of the South Carolina Historical Society*, V, 186–87, 191.

122. Ibid., V, 188, note 1, 338, 456–62.

123. See Mathews ca. 1685.

124. See Thornton-Morden ca. 1695. About the same time the freebooter-mapmaker William Hack made a manuscript map of "North Carolina" giving the Sound and Banks region; its detail for the soundings and inlets of the North Carolina coast as they existed at that time is elaborate, but no settlements are given. It was not used for a printed map. See Hack ca. 1684 MS B (Map List No. 99; Plate 41).

125. Florence O'Sullivan, the first surveyor general, was so unfitted for his post that he was incapable of drawing a map of the place to send to the lords proprietors. He was an Irish soldier of fortune whose contentiousness, ill practices, and abuse of position became notorious. Sullivan's Island in Charleston Harbor was named after him. He had first to share and then entirely relinquish the surveyorship to John Culpeper. Culpeper was reasonably competent, as the two manuscript maps of Charles Town used for the inset of the first lords proprietors map by Ogilby show. However, he also was a malcontent, took part with O'Sullivan in raising disturbances against the local government, and fled to Albemarle to avoid the danger of being hanged. There he fathered Culpeper's Rebellion. *Collections of the South Carolina Historical Society*, V, 131, 248, 285, 381.

126. P.R.O., C.O. 5, Vol. 288, fol. 100 (microfilm, Library of Congress).

127. Crane, *Southern Frontier*, pp. 14, 119.

128. *Collections of the South Carolina Historical Society*, V, 332, note 2.

129. The following works, based upon study of the original documents, are among the most helpful examinations of various phases of this contest: Crane, *The Southern Frontier, 1670–1729*, Durham, N.C., 1928; John T. Lanning, *The Diplomatic History of Georgia: A Study of the Epoch of Jenkins' Ear*, Chapel Hill, N.C., 1936; Herbert E. Bolton, *Arredondo's Historical Proof of Spain's Title to Georgia*, Berkeley, Calif., 1925; John R. Alden, *John Stuart and the Southern Colonial Frontier: A Study of Indian Relations, War, Trade, and Land Problems in the Southern Wilderness, 1754–1775*, Ann Arbor, Mich., 1944; W. Stitt Robinson, *The Southern Colonial Frontier, 1607–1763*, Albuquerque, 1979.

130. Bolton, op. cit., pp. 24–25. More recently Juan Pardo's contribution's have been critically examined and made more accessible through the work of Charles Hudson, an anthropologist and expert on the de Soto expedition and its impact on the native peoples of the Southeast. See his *The Juan Pardo Expeditions: Exploration of the Carolinas and Tennessee, 1566–1568*, Washington, 1990. Hudson's book includes the documents relating to the Pardo expeditions in new transcriptions and translations by Paul E. Hoffman.

131. C. W. Alvord and L. Bidgood, eds., *First Explorations of the*

Trans-Allegheny Region, pp. 104–5; Douglas Summers Brown, *The Catawba Indians: The People of the River*, Columbia, S.C., 1966.

Mrs. Douglas Brown of Emporia, Virginia, has kindly referred the writer to several passages which indicate that Virginia trade with Indians as far as South Carolina may have existed before Lederer's journey. The Council of Virginia wrote to the lords commissioners of trade on October 19, 1708, protesting against high duties laid on Virginia traders by the government of South Carolina and stating that Virginia had "traded with those Indians before the name of Carolina was Known": *Executive Journals of the Council of Colonial Virginia*, Richmond, 1925–45, III (1928), 194. About 1733 William Byrd II wrote, "Of late the new colony of Georgia has made an act obliging us to go 400 miles to take out a license to traffic with these Cherokees, though many of their towns lie out of their bounds, and we had carried on this trade eighty years before that colony was thought of" (*A Journey to the Land of Eden*, ed. Mark van Doren, New York, 1928, p. 179). These are eighteenth-century statements made with the intent of emphasizing the priority of the Virginia trade; as such, the dates are suspect. How far south the Virginia traders themselves went is uncertain; Indians may have brought their furs north to Occaneechi or some other trading place. By 1670 the Westo Indians on the Savannah River, however, were already using guns and ammunition obtained by trade from Virginia (Verner Crane, *The Southern Frontier*, p. 12). The implication of far-ranging Virginia traders is strong.

132. Crane, op. cit., pp. 42–44. See also Verner Crane, "The Tennessee River as the Road to Carolina: The Beginnings of Exploration and Trade," *Mississippi Valley Historical Review*, III (June 1916), 3–18.

133. Both English and French manuscript maps before the turn of the century show that English traders had established contact with Indian tribes beyond the Appalachian range. In 1698 or 1699 an English expansionist in Virginia, Cadwallader Jones, made a map of the Mississippi valley region in which he has two rivers, the Ohio and the Tennessee, with the following legends between them to the west of "Pallatia" and "Appelatinian Mountains": "Cherico Indians, 30 Towns run to 30 degrees," and "S^{o} Carolina men trade with these Chericooes. Ugo river 35^{d}" This map, "Louisiana Pars Discovered by Father Louis Hennepin . . . Dedicated to W^{m} ye 3^{d} King of England 1698 in a .11. years search," accompanies "An Essay Louissiania and Virginia Improved by Coll^{d} Cadwallader Jones Esqr Dedicated to His Excellency ffrancis Nicholson Esqr His Maj^{ties} Leiv^{t} and Govern^{r} Gener^{ll} of Virginia—Qui timeat Nunquam Honorem habebit—Dated at York Town Jan^{ry} ye 17th 1698/9," and was sent by Governor Nicholson to the lords of trade (Public Record Office, Colonial Office Library, 5, 1310, p. 261ff.). The map, with the essay it accompanies, is reproduced in Fairfax Harrison, "Western Explorations in Virginia between Lederer and Spotswood," *Virginia Magazine of History and Biography*, XXX (1922), opp. p. 337. It is based on the one in Hennepin's *A New Discovery of a Vast Country in America*, London, 1698, a book that Jones saw while visiting Secretary Ralph Wormeley in England in 1698. Jones's map has numerous legends and comments.

Louvigny de la Porte's "Carte du fleuve Missisipi," ca. 1699, indicates the route from the Chickasaw Indians to Charles Town: from "5 v[illages] des sicaca" is the "chemin pour aller aux Anglois." A photocopy of this map, Bibl. Nat. 4040 C. 10, is in the Clements Library.

Although the tentacles of English trade reaching toward the Mississippi by the trading route south of the Appalachians thus began well before the end of the seventeenth century, exact names and dates are difficult to determine. Barnwell's map 1721 states that for 28 years prior to the erection of the French Fort Toulouse in 1715 at the forks of the Coosa and Tallapoosa Rivers an English factory had been operating "without interruption." According to this statement, by 1687 the English had established a trading post a few miles north of the present site of Montgomery, Alabama, well over halfway to the Mississippi from Charleston.

James Ramsey, *Annals of Tennessee*, Charleston, 1853, p. 63, says that as early as 1690 Doherty, a trader from Virginia, had visited the Cherokees; but this has no contemporary documentation. For the early trade from Virginia to the Cherokees in western Carolina, see W. Neill Franklin, "Virginia and the Cherokee Trade, 1673–1752," *East Tennessee Historical Society Publications*, IV (January 1932), 3–21.

Joshua Fry quotes from the journal of John Peter Salley, a member of Howard's expedition of 1742, who was captured by the French and later escaped from New Orleans and returned to Virginia: "We were told by some of the French, who first settled there, that about forty years ago, when the French first discovered the Place, and made an Attempt to settle therein there were pretty many English settled on both sides of the River Mississippi, and one Twenty Gun Ship lay in the River. What became of the Ship we did not hear, but we were informed, that the English Inhabitants were destroyed by the Natives by Instigation of the French." See Delf Norona, "Joshua Fry's Report on the Back Settlements of Virginia (May 8, 1751)," *The Virginia Magazine of History and Biography*, LVI (January 1948), 36.

The ship which Salley heard of was presumably the twelve-cannon English frigate of Captain Barr, bringing Dr. Coxe's intended settlers to "Carolana," which was met and turned back by Bienville on September 16, 1699. See J. Winsor, *America*, V, 20, and D. Coxe, *Carolana*, Preface. If Salley's rather vague and belated second-hand report can be accepted, soon after the exploratory trip of Captain Welch to the Mississippi in 1698, recorded on Barnwell's 1721 map, English traders were busy extending their trading posts as far as the Mississippi.

134. Phillips, *Lowery Collection*, pp. 219–21, with a bibliography of works on Delisle, p. 214. Claude Delisle had made three drafts of the map of North America as early as 1696. On the contributions of Claude Delisle to the maps signed by Guillaume Delisle, see Jean Delanglez, *El Rio Espíritu Santo*, New York, 1945, pp. 3, note 5; pp. 9, 136, 137, 143–44; and Jean Delanglez, "The Sources of the Delisle Map of America, 1703," *Mid-America*, XXV (1943), 275–98.

135. J. Winsor, *America*, V, 21. In 1702 Delisle made a manuscript map of the Mississippi River on five large sheets, showing so

much more detailed knowledge of the river that one is justified in assuming that the general map is earlier and the 1702 map is based on information from the third voyage. See Phillips, *Lowery Collection*, p. 220.

136. Delisle's printed map (1700) of North America also attempted to show the boundaries of various possessions in the New World: reproduction in C. O. Paullin, *Atlas of the Historical Geography of the United States*, Washington, 1932, Plate 22A.

137. *Documents Relative to the Colonial History of the State of New York*. Edited by J. R. Brodhead, Esq., Albany, 1855, V, 577. The importance of Delisle's map during the first half of the eighteenth century can be indicated by Mr. Calden's "Answers to Queries of the Lords of Trade," February 14, 1738, concerning nearness to French and Spanish settlements, reputed boundaries, and by whom made or decided upon: "These advantages [navigation by the Great Lakes] I am sensible cannot be sufficiently understood without a map of North America. The best I have seen, is Mr. De L'Isle's map of Louisiana publish'd in french in the year 1718, for this reason I frequently use the french names of places, that I may be better understood." Ibid., VI, 112.

138. Crisp so writes after his name in a petition to the commissioners for trade and plantations concerning the insurrection of the Indians; he is one of twenty-four signatories: cf. the original MS minute book of the Proprietors of Carolina, Vol. 7, N.C., preserved in the Public Record Office, London, under the date July 18, 1715. See Phillips, *Lowery Collection*, p. 263.

Crisp must have been a resident in South Carolina, since he is the author of the Crisp 1704 plan of Charles Town; he may be considered the author of the very similar Charles Town inset in the Crisp 1711 map. For thirty or forty years this inset was the basis of most of the numerous printed plans of Charles Town.

139. Very little is known about John Love, who is responsible for the valuable additions to the main map, based on Maurice Mathews's manuscript map of about 1685. The writer is indebted to Lawrence C. Wroth of the John Carter Brown Library for the suggestion that he may be the same individual as John Love, author of *Geodaesia: or the Art of Surveying*, London, 1688. On the title page of *Geodaesia* is a sentence or two saying that the work concerns itself to some extent with "how to lay out New Lands in America." In the preface is a reference to young men in America, "particularly in Carolina," finding it difficult to lay out lands for which no previous surveys existed. Later on in the book there is a short chapter on the subject of laying out new lands, especially in Carolina. John Love's *Geodaesia* was apparently a useful work; in addition to the first edition of 1688, the John Carter Brown Library has a 1720 edition and two later eighteenth-century editions. Local South Carolina records may be able to throw further light on relations between the author and the surveyor who contributed to the Crisp map.

140. The account of Nairne's career is scattered through the pages of V. W. Crane, *The Southern Frontier*. Unexploited Nairne MSS are in the H. E. Huntington Library. In the winter and spring of 1708 Nairne traveled west to the Mississippi River and south nearly to the Gulf of Mexico to draw the Indians to the English cause in the competition with France. His journal of that expedition and his memorial of July 10, 1708, a continuation of the journal, were edited and published by Alexander Moore under the title, *Nairne's Muskhogean Journals: The 1708 Expedition to the Mississippi River*, Jackson, Miss., 1988. Moore encountered the previously unpublished journal in the British Library's collection of manuscripts.

141. Crane, op. cit., p. 81.

142. The lower half of the peninsula is broken up into a number of islands. This conception of the Everglades part of Florida as an archipelago, found here for the first time on an English map, occasionally influenced cartographers for the next half century (see Scaciati-Pazzi 1663). On the first issue of the copy (from a new plate) of this map, which was used in various propaganda pamphlets for colonizing Georgia (see Georgia 1732), the legend about "no inhabitants" to the point of Florida is found; in later issues of the Georgia map, this and other inscriptions are erased from the plate. Spanish MS charts show the lower half of Florida as an archipelago by the 1690s.

143. On the Indian Tribes ca. 1715 map is a carefully annotated path leading down into southern Florida. Moll's "A New Map of the North Parts of America . . . 1720" shows this same route and adds a long explanatory legend, much of which corresponds with the annotations on the Indian Tribes map. Starting with "Explanation of an Expedition in Florida Neck, by Thirty-three Iamesee Indians accompany'd by Capt. T. Nairn," the printed legend lists under items A to N a detailed account of a foray into the Everglades. It begins "A. The Place where the Indians leave their Canoes to go a Slave-Hunting. it is 6 days Rowing from St. Whan's [St. Johns] Rivers Mouth," and ends after recording the capture of 35 slaves, "M. Here they took and Killed 33 men at 2 a Clock ye same day a very Numerous body of Indians came against them, they being but 33 Men, yet put them presently to Flight; they having no Armes but Harpoons, made of Iron and Fishbones; they were all Painted. N. Fresh water Creek."

144. Crane, *The Southern Frontier*, p. 93. The document is P.R.O., C.O., 5. 382/11. Nairne writes that he sent his map, based upon his own observations and travels, "to the End your noble Lordship may at one View perceive what part of the Continent we are now possest off, and what not, and procure the Articles of peace, to be formed in such a manner that the English American Empire may not be unreasonably Crampt up. Your Lordships may depend on the Inland topography to be as exact, as anything of that kind can well be. The numbers of the inhabitants I took with the greatest care." See *Great Britain. Public Record Office. Calendar of State Papers, Colonial Series, America and West Indies*. C. Headlam, ed. London, 1922, p. 421, No. 632.

145. Crane, op. cit., pp. 168–69.

146. See Senex 1721; this map appeared in John Senex's *A New General Atlas*, London, 1721, and is a copy of Delisle's 1718 map, in which the mountains referred to stretch along the east bank of the Santee River.

147. Quoted from an untitled contribution, signed W. K., *Magazine of American History*, IX (April 1883), 272; see also P. L.

Phillips, *Lowery Collection*, p. 235. In light of this scathing public criticism it is somewhat surprising to find James Edward Oglethorpe employing Nairne's map as the model for the first maps to show the new colony he and his associates founded a few years later. For an illustrated discussion of the role Nairne's map played see Louis De Vorsey, "Maps in Colonial Promotion: James Edward Oglethorpe's Use of Maps in 'Selling' the Georgia Scheme," *Imago Mundi* 38 (1986), pp. 35–39.

148. S. G. Drake, *Early History of Georgia, Embracing the Embassy of Sir Alexander Cuming*, Boston, 1872, and V. W. Crane, *The Southern Frontier*, pp. 277–98. The October 14, 1730, article in *The Daily Journal* also criticizes Nairne's enumeration of the Cherokee towns. There is a reference to Nairne's map in 1728 in the Public Record Office, C.O. 5, 360/C22; cf. Crane, op. cit., p. 93.

149. See Hunter 1730 MS. Hunter certified the map, which is in the Faden Collection, Library of Congress, on May 21, 1730, eight days after Cuming left for England on the Fox. In 1744 Hunter, then surveyor general, copied a map which John Herbert made in 1725; Herbert's original map, apparently no longer extant, was of the Southeast. According to a note by Hunter on his 1730 map, he used a map by Herbert, presumably the general map of 1725, as a basis for the restricted area shown on his much more detailed map.

150. See Hughes ca. 1720 MS and references thereunder.

151. Unpublished letter referred to in V. W. Crane, *The Southern Frontier*, p. 102; the account in Crane's work is the source of information on Hughes here used.

152. Crane, op. cit., p. 100.

153. F. Le Maire. Extrait d'un Mémoire sur la Louisiane . . . 1717. Ed. Gustave Devron. *Comptes-rendu de l'Athénée Louisianais*, September–October 1889, p. 13, note 105.

154. See Moll 1701 A, Moll 1701 B, Moll-Oldmixon 1708, Moll 1715, Moll 1728 A, Moll 1728 B, Moll 1729, Moll 1730.

155. *The Dictionary of National Biography*, XXVIII, 128, states that Moll came to London about 1698. R. V. Tooley, *Maps and Map-Makers*, London, 1952, p. 55, says that Moll came to London about 1680, at first engraving for other publishers and then setting up on his own account. He gives an incomplete list of Moll's atlases and the works in which Moll's maps appeared, beginning with an untitled general atlas published about 1700.

156. This two-sheet map is found in copies of Moll, *The World Described*, London, [1709–20], No. 9, where it follows Moll's 1715 "Dominions" map (see Moll 1715). The full title is "To Thomas Bromsall Esq: This Map of Louisiana, Mississipi &c. is most Humbly Dedicated by H. Moll Geographer [dedication cartouche, lower left] | A new Map of the North Parts of America claimed by France under ye Names of Louisiana, Mississipi, Canada and New France with ye Adjoyning Territories of England and Spain. A great part of this Map is taken from ye Original Draughts of Mr. Blackmore, the Ingenious Mr. Berisford now Residing in Carolina, Capt. Nairn and others never before Publishe'd, the South West Part of Louisiana is done after a French Map Published at Paris in 1718. and we give you here the Division or Bounds according to that Map, which Bounds begin 30 Miles S. West from Charles Town in Carolina and run on to ye Indian Fort Sasquesahanok 30 miles west of Philadelphia &c.
NB The French Divisions are inserted on purpose, that those Noblemen, Gentlemen, Merchants &c. who are interested in our Plantations in those Parts, may observe whether they agree with their Proprieties, or do not justly deserve ye name of Incroachments; and this is ye more to be observed, because they do thusly Comprehend within their Limits ye Cherakeys and Iroquois, by much ye most powerfull of all ye Neighboring Indian Nations, the old Friends and Allies of the English, who ever esteemed them to be the Bulwark and Security of all their Plantations in North America. The Projection of this Map is Call'd Mercator's, And it is laid Down according to the Newest and Most Exact Observations by H. Moll Geographer. 1720." [cartouche, bottom left] Size: 40 × 21½. Scale: 1" = ca. 95 miles. The map extends from Newfoundland to the Gulf of California and from the Yucatan Peninsula and the West Indies to "James Bay" (Hudson Bay), with insets and various explanatory legends.

157. London. P.R.O., C.O. 700 Carolina no. 3. See Indian Tribes ca. 1715 (Map 157).

158. Nathaniel Blackmore surveyed the Bay of Fundy and the surrounding region in 1711–12; Moll's reference is to his map of 1713. See W. F. Ganong, "Cartography of the Province of New Brunswick," *Proceedings and Transactions of the Royal Society of Canada*, second series, II, section II (1897), 366–67. Blackmore's chart of Nova Scotia is considered the first recorded chart to show seabed contour lines: W. P. Cumming, *British Maps of Colonial America*, Chicago, 1974, pp. 10, 86, plate 7; Arthur H. Robinson, "Nathaniel Blackmore's Plaine Chart of Nova Scotia," *Imago Mundi*, XXVIII (1976), 137–41; W. P. Cumming, "Nathaniel Blackmore, 1714/1715," *English Mapping of America, 1675–1715*, New York, Mercator Society, 1986, no. 20.

159. "Memorial from Mr Beresford, Representing the Present State of South Carolina (23rd June 1716)," *CRNC*, II, 229–34; ibid., p. 224.

160. P.R.O., C.O. 5/327.7.K116, quoted from Crane, op. cit., p. 209.

161. Crane, op. cit., p. 148.

162. E. McCrady, *History of South Carolina under the Royal Government 1719–1776*, New York, 1899, p. 482.

163. A fort at the forks of the Alabama River has the legend "A French fort since ye Warr." The French occupied the English factory there in 1715; it is doubtful that this legend refers to the new Fort Toulouse, built in 1717, though it would suppose a certain lapse of time after the Yamasee War of 1715. The important Indian fort built at the fork of the Flint River by the Indians in 1716 is not shown on the map.

164. E. McCrady, *History of South Carolina under the Royal Government 1719–1776*, p. 1. For the relinquishing of the proprietary charter, see C. C. Crittenden, "Surrender of the Charter of North Carolina," *North Carolina Historical Review*, I (October 1924), 338–402.

165. Crane, op. cit., p. 263.

166. This danger, presented to the Board of Trade by Barnwell

in 1720, is graphically shown by Moll in his "A New Map of the North Parts of America claimed by France. 1720," when French claims extend to the coast between Edisto River and the "Margravate of Azilia" along the Savannah.

167. Crane, op. cit., p. 234.

168. Crane, op. cit., pp. 233–34. More recently M. T. Hatley drew attention to the importance of Barnwell's maps of the Southeast in any serious interpretation of South Carolina–Cherokee relations in the eighteenth century. See his *The Dividing Paths: Cherokees and South Carolinians through the Era of the Revolution*, New York, 1993, which includes a photograph of a significant portion of the map as Figure 1 facing p. 147.

169. *Journal of the Commissioners for Trade and Plantations from November 1718 to December 1772*, London, 1925, p. 189.

170. Ibid., p. 198.

171. See Barnwell 1721 MS; P.R.O., C.O., Georgia. 2. Crane, op cit., p. 350, ascribes the first four Georgia maps in the Colonial Office to Barnwell, on the basis of similarity of handwriting.

172. British Museum Add. MS. 23615, fol. 72: cf. "Tocobbogga destroy'd 1709." This is not on the Herbert 1744 (copied from a non-extant 1725) map. Popple used the Barnwell map very sparingly on the 1733 key map; the large Popple 1733 has many details taken from Barnwell which show a fuller use of Barnwell than the 1727 draft or the small "key" Popple (Plate 55).

173. Catesby's use of Barnwell is shown by similarity of topographical details and by such phrases as "Quanasse, an English Factory," "Congaree, an English Factory," which appeared for the first time on a printed map.

174. John Barnwell was one of the early and enthusiastic backers of Montgomery's project for the Margravate of Azilia, though by 1721 he attacked it as a "bubble": cf. Crane, op. cit., p. 213.

175. A careful study of the sources and uses of the Mitchell map is in Edmund and Dorothy S. Berkeley, Dr. John Mitchell, *The Man Who Made the Map of North America*, Chapel Hill, 1974, pp. 175–213.

The best analysis of the various editions and states of the map is Richard W. Stephenson, "Table for Indentifying Variant Editions and Impressions of John Mitchell's Map," in Walter W. Ristow, *A la Carte*, Washington, 1972, pp. 109–13.

The bibliography of the Mitchell map is complicated by the numerous impressions, for it was published at various times and places. By 1775 four English editions of Mitchell's map had been published in London; of the first edition there were three impressions and of the third, two, or seven impressions in all. In addition, two Dutch impressions, published in Amsterdam, with English titles, are known, as well as at least eight French editions or impressions, some of them with titles in German as well as French, and two Italian piracies published in Venice.

Colonel Lawrence Martin, late chief of the Division of Maps of the Library of Congress, made a detailed study of the Mitchell map; brief abstracts and excerpts from his work have been published from time to time. A list of these articles, as well as an excellent resumé prepared by Col. Martin, is to be found in Hunter Miller, *Treaties and other International Acts of the United States*, Washington, 1933, II, 328–56. A list of the historic copies of the map, their owners and location, as well as a list of the different impressions, is given. A reproduction of the Steuben-Webster (Library of Congress) copy is folded at the back of the book. The reproduction of the most famous copy of this map, King George III's copy (British Museum), is found (rather illegibly reproduced) in Fite and Freeman, *A Book of Old Maps*, Cambridge, Mass., 1926, Plate 74, with an earlier impression in the same book, Plate 47. It has been thought unnecessary to tabulate all the different editions of Mitchell's map in the accompanying Map List. The Library of Congress has copies of all the different known impressions of the English editions with the exception of the first, of which it has a photostat taken from the Yale copy.

176. The southern trading Paths to the Indians differ from those given by Barnwell ca. 1721 or Hunter 1744 (1725); the small branches on the Upper Congaree River (Santee), Thirty-Mile Run, Reedy Creek, Turkey Creek, Mulberry Creek, 96, and others are listed; some of the branches of the Savannah and the headwaters of the Altamaha are shown. Most of these names agree with some MS map, such as the Hunter 1730 MS map in the Faden Collection in the Library of Congress; but they show more detail than the maps of that period now in the possession of the London Colonial Office.

177. G. Robert de Vaugondy's "Partie de l'Amerique Septentrionale . . . 1755" and J. B. d'Anville's "Canada Louisiane et Terres Angloises . . . Novembre MDCCLV," both dependent largely on Mitchell though without acknowledgment, appeared later in the same year.

178. John R. Alden, *John Stuart and the Southern Colonial Frontier*, Ann Arbor, 1944, p. 70. De Brahm claimed that he was responsible for Stuart's appointment: ibid., p. 136, note 44.

179. Alden, op. cit., p. 136.

180. Stuart, ca. 1761 MS.

181. Cf. Wyly 1764 MS (the Catawba lands between North and South Carolina); Pickins 1766 MS (South Carolina–Cherokee line); Savery 1769 MS A and Savery 1769 MS B (Georgia-Creek line); Stuart 1771 MS (Virginia-Cherokee line); Stuart 1772 MS (Georgia-Cherokee line). For an account of the various boundary lines negotiated by Stuart, see J. R. Alden, *John Stuart*, pp. 262–323 passim; for boundary-line maps and surveys of Virginia and Florida not listed here or in the following List of Maps, see Alden, op. cit., p. 366; A. B. Hulbert, *Crown Collection*, passim; Phillips, *Lowery Collection*, under De Brahm and Romans; and photocopies in the W. L. Clements Library of surveys in British archives. The fullest account of the Indian boundary line and pertinent maps is Louis De Vorsey, *The Indian Boundary in the Southern Colonies*, Chapel Hill, 1961.

182. Alden, op. cit., pp. 283, 287, with references there cited.

183. See Stuart-Gage ca. 1773 MS. Louis De Vorsey, "The Colonial Southeast on 'An Accurate Map,'" *Southeastern Geographer*, VI (1966), 20–32, is a study of the Stuart maps based on original research.

184. See Stuart-Purcell [1775] MS A.

185. See under Stuart-Purcell [1775] MS B.

186. The information on the Stuart maps is to be found in simplified form on the 1776 map by Bernard Romans, "A General Map of the Southern British Colonies, in America . . . by B. Romans, 1776. London. Printed for R. Sayer and J. Bennett." Size: 25⅛ × 19⅝. Scale: 1" = ca. 65 miles. See P. L. Phillips, *Lowery Collection*, p. 381, No. 585. It was reproduced on a separate sheet by the United States Constitution Sesquicentennial Commission, Washington, D.C., 1939.

The influence of French maps on British conceptions of the trans-Appalachian region in the eighteenth century, especially before the Stuart maps, was strong: see Susan M. Reed, "British Cartography of the Mississippi Valley in the Eighteenth Century," *Mississippi Valley Historical Review*, II (1915), 213–24.

187. For problems of communication and transportation, the importance of good harbors, and the condition of roads during this period, see C. C. Crittenden, "The Seacoast in North Carolina History," *North Carolina Historical Review*, VII (1930), 432–33; C. C. Crittenden, "Overland Travel and Transportation in North Carolina, 1763–1789," op. cit., VIII (1931), 239–57; and F. W. Clouts, "Travel and Transportation in Colonial North Carolina," op. cit., III (1926), 16–25.

188. See Gascoigne-Swaine 1729 MS. Captain Gascoigne's maps were published about 1776 by Jeffreys and Faden. Gascoigne's original MS of D'Awfoskee is apparently lost. Both of these surveys were made in 1729–33. For several years Gascoigne continued his surveys, for on April 3, 1731, Governor Rogers of the Bahamas wrote Mr. Alured Popple, secretary of the Board of Trade, that Captain Gascoigne, in command of three vessels, was now "sayleing hence for ye Bahamas in order to carry on his survey of them": Cecil Headlam and A. P. Newton, eds., *Calendar of State Papers, America and West Indies, 1731*, London, 1938, Vol. 38, p. 83, No. 123. For further information about John Gascoigne's surveys see W. P. Cumming, *British Maps of Colonial America*, Chicago, 1974, pp. 47–48; see also W. E. May, "The Surveying Commission of *Alborough*, 1728–1734," *American Neptune*, XXI (1961), 260–78. On February 28, 1733, Governor Johnson of South Carolina ordered the Surveyor General to lay out 48,000 acres to John Gascoigne, agent for John Roberts, who had purchased eight baronies of 12,000 acres each from Lord Carteret: Headlam and Newton, op. cit., 1733, Vol. 40, pp. xliii, 48, 49, Nos. 59–60.

189. Plates 50A, 51–54. Moseley had been at work on this map for some time, as he wrote the duke of Newcastle on May 22, 1731: "I am preparing a large Map of this Province for his Majesty's view, drawn from several observations I collected when I was Surveyor General of this Province and many helps I have received from several Gentlemen of this and neighboring Governments": *CRNC*, III, 137.

190. *CRNC*, III, 430 and 436. In this same letter Burrington condemns violently the surveying methods of Moseley and his deputies. He had transmitted the three drafts of harbors with his letter of November 14, 1732, at which time he wrote: "I had agreed to give ten guineas for a Map of the Country which was drawn for me but is sent as I am told to Coll. Bladen which is better then if I had pay'd for it being at this time very Poor": Ibid., III, 372. Mr. Burrington's map of Winyeau (Winyah) Bay is referred to by Alured Popple, who sent it to Mr. Carkesse of the London Customs House on behalf of Governor Johnson. See C. Headlam, ed., *Calendar of State Papers, America and West Indies, 1730*, London, 1937, under dates February 18 and 19, 1730. The map apparently was not returned or has been lost.

191. "Chart of the Coast of America from Cape Hatteras to Cape Roman From the Actual Surveys of Dr Dunbibin Esqr" [cartouche, upper left] Size: 32 × 20½. Scale: 2" = 10 miles. In John Norman's *The American Pilot*, Boston, 1792, No. 5 (and in the later 1794, 1798, and 1803 editions of *The American Pilot*). On May 19, 1757, Daniel Dunbibin, who had been a merchant in Newton (Wilmington) as early as 1740 and was a landowner in New Hanover, petitioned the North Carolina House of Assembly for a sum necessary to complete a survey of "a great part of the Coast of North and South Carolina," as the expense had been larger than he had expected and the subscriptions received insufficient. On February 11, 1764, the widow of Dunbibin petitioned the House that £50 be paid her "for services rendered this Province, by her deceased husband Daniel Dunbibin in surveying and Making a Map of the Sea Coast thereof." The House approved the petition on February 28 and forwarded it to the Council, which suggested the sum of £100 as being not unreasonable for "the care and time employed." The House turned down this proposal but reaffirmed its grant of £50, which was approved by the governor on March 9, 1764. See *CRNC*, IV, 504, 643; V, 847; VI, 1114–15, 1137–38, 1160, 1187–88, 1191, 1210, 1214.

Collier Cobb refers to a map of the North Carolina coast made by Dunbibin in 1764 in his "Some Changes in the North Carolina Coast Since 1585," *North Carolina Booklet*, IV, No. 9 (January 1905). This reference to a Dunbibin 1764 map has caused some confusion, for no printed map of that date has come to light. Miss Mary Cobb, his daughter, reports that no such map is to be found among Professor Cobb's papers, nor is such a map in the Carolina Collection or in the Geology Department Library at the University of North Carolina. It is probable, therefore, that Professor Cobb was referring to the map made by Dunbibin and mentioned in his widow's petition in 1764.

The following advertisement found in the *South Carolina Gazette*, September 23, 1756 (No. 1161), fol. 249, p. 3, col. 1: "Capt. Dunbibin having proceeded in his Survey of the Coasts on North and South Carolina, as far as Winyah: and now being come to compleat it as far as that of Port Royal, but the Subscriptions are not sufficient to enable him to do it. Those therefore that would not see so necessary an Undertaking dropped, or imperfectly executed, may promote it, by calling at the printer's, and subscribing to the Proposals in his Hands."

Dunbibin's hydrographic chart therefore presumably incorporates the results of surveys of the coast shortly before or during the year 1756.

192. For inlets recorded in the 1585–90 maps and reports, D. B. Quinn, *The Roanoke Voyages*, London, 1955, II, 852–72, remains the standard. Gary S. Dunbar, *Historical Geography of the North Carolina Banks*, Baton Rouge, 1958, has comments on various inlets

and cusps scattered through his notes, and in Appendix A ("Inlet Changes in Historic Times," pp. 215–17) he examines eleven inlets. David Stick, *The Outer Banks of North Carolina*, Chapel Hill, 1958, has a table of twenty-four inlets, with various names given to the same inlet and dates of opening and closing. The fullest study is Joseph Fisher, "Geomorphic Expression of Former Inlets along the Outer Banks of North Carolina," master's thesis, Chapel Hill, 1962 (copies in UNC library), Appendix 2, pp. 83–112.

193. *CRNC*, VII, 861; Governor Tryon said that he had held this position "near twenty years" before his death in December 1767.

194. Adelaide L. Fries, ed., *Records of the Moravians in North Carolina*, Publications of the North Carolina Historical Commission, Raleigh, 1922, I, 46. Except for exact adherence to Lord Granville's agent's orders Churton is, says Bishop Spangenberg, "a tractable Man."

195. Ibid., pp. 32–33.

196. Fairfax Harrison, *Landmarks of Old Prince William*, Richmond, 1924, II, 632; and Delf Norona, "Joshua Fry's Report on the Back Settlements of Virginia," *The Virginia Magazine of History and Biography*, LVI (January 1948), 30–41.

197. As early as December 15, 1738, the following entry was made in the Journal of the House of Burgesses: "A proposal of Joshua Fry, and Robert Brooke, to make an exact survey of this colony and publish a map thereof, in which shall be laid down the Bay, Navigable Rivers, with the soundings, the Counties, Parishes, Towns, or whatever else is remarkable: if this House think fit to encourage the same, was presented to the House and read." H. R. McIlwaine, ed., *Journals of the House of Burgesses in Virginia, 1727–1734, 1736–1740*, Richmond, Va., 1910, p. 379. The proposal was referred to the Assembly in 1740 and again in 1742. The need was obvious, and it was expected that the proposal would pass. But in 1744 it was rejected, possibly because of an earlier estimate in 1732 that the cost would be £5,000. Actually, for their work Fry and Jefferson were each voted £150 by the council on October 15, 1751: Wilmer L. Hall, *Executive Journals of the Council of Colonial Virginia*, Richmond, Va., 1945, V (1739–54), 354, 370; P.R.O., C.O. 5: 1327/415, and 1423/497, 519.

198. See Virginia–North Carolina Boundary 1749 MS (Map List No. 268). Interestingly enough, in 1749 the names of both Fry and Jefferson are found as members of a group who formed the Loyal Land Company, to whom the Council of Virginia granted 800,000 acres along the southern border of Virginia: see Delf Norona, "Joshua Fry's Report on the Back Settlements of Virginia," *The Virginia Magazine of History and Biography*, LVI (January 1948), 25.

199. Though the map was drawn in 1751 and that is the engraved date on the earliest issues, it was not actually published until 1753. Its appearance is noted and a brief description of it is given by M. Maty, *Journal Britannique*, XII (Septembre–Decembre 1753), 427–28; see Fry-Jefferson 1751 [1753] (Map List No. 281). For further comment on the dating of the map and on its preparation, see Delf Norona, "Joshua Fry's Report on the Back Settlements of Virginia," *The Virginia Magazine of History and Biography*, LVI (January 1948), 27–30; and Coolie Verner's "Checklist" in Dumas Malone, *The Fry & Jefferson Map*, Princeton, N.J., 1950, p. 13.

The fullest checklist of editions and impressions of the map is Coolie Verner, "The Fry and Jefferson Map," *Imago Mundi*, XXI (1967), 70–94.

200. Delf Norona, "Cartography of West Virginia," *West Virginia History*, IX (January 1948), 107.

201. On the 1755 edition of the Fry-Jefferson map is "Gist, Junr, Mulberry Fields," on the Yadkin in the extreme west of pioneer settlements in North Carolina; Gist was an explorer and surveyor. On October 31, 1750, the Ohio Company employed him to explore the Ohio River lands as far as the present Louisville, Kentucky. "His plans and surveys have been praised as 'models in mathematical exactness and precision'": see "Christopher Gist," *Dictionary of American Biography*, VII, 324; see also W. M. Darlington, ed., *Christopher Gist's Journals*, Pittsburgh, 1893.

202. Captain John Dalrymple returned to Virginia in 1755; Braddock appointed him commandant of Fort Johnston in North Carolina. He died in the fort on July 13, 1766: see Tryon's letter to Lord Barrington, *CRNC*, VII, 244.

203. In the 1751 [1753] edition only four names are given west of the coastal settlements. These are on or near the Ararat River close to the Virginia line: Mount, Loven on Rocky Creek, Easly on Johnsons Creek, and Peter King on Fishes River (Fisher River, near the present village of Round Peak in Surry County).

204. Wilbur H. Siebert, "De Brahm," *Dictionary of American Biography*, V, 182–83. Louis De Vorsey, ed., *De Brahm's Report of the General Survey in the Southern District of North America*, Columbia, S.C., 1971, pp. 7–59. De Vorsey has published several articles on De Brahm, including "William Gerard De Brahm: Eccentric Genius of Southeastern Geography," *Southeastern Geographer*, X (April 1970), 21–29.

205. See De Brahm 1757.

206. See De Brahm ca. 1755 MS and De Brahm 1757 MS. De Brahm commenced building the defenses of Charles Town in 1755; by May 1756 he had a number of bastions to the south and east and had damned out the sea water by the East Battery, the foundation of the sea wall which is still one of Charleston's greatest attractions. According to the plan the city was to be defended by walls, bastions, and a fortified canal to protect the town from enemies and from the encroachment of the sea. See *South Carolina Gazette*, May 6, 1756, for a description of the work at that time, and also E. McCrady, *History of South Carolina under Royal Government, 1719–1776*, New York, 1899, pp. 282–83. A manuscript volume in the South Carolina Historical Society, "Journal of the Commissioners of Fortifications, 1755[–1770] B," gives an account of the relation of De Brahm to the commissioners. The minutes of December 9 and 15, 1755, refer to a "profile or section of the Plan" requested by the commissioners and sent by De Brahm.

207. See De Brahm ca. 1757 MS; also De Brahm [1773] MS A.5 and 6. Fort Loudoun, designed as a protection against the French, was located a short distance above the junction of the Little Tennessee and Tellico Rivers in what is now Monroe County, Ten-

nessee. In May 1756, De Brahm accompanied Governor Glen, who was planning to build the Overhills fort which had been promised to the Cherokees. Glen was recalled by his successor, Governor William Henry Lyttelton, who arrived in Charles Town before Glen's expedition had reached its destination. Lyttelton appointed De Brahm to design and construct the fort. De Brahm quarreled with the officer in charge of the force sent to man and build the fort, Captain Raymond Demeré; Demeré was in favor of accepting the location for the fort proposed by John Pearson, a provincial deputy surveyor sent by Governor Glen early in 1756 to find a suitable place. Strategically De Brahm's choice was superior to Pearson's; its position atop a hill and the elaborateness of the construction, which caused Demeré to condemn De Brahm roundly in his report for extravagance, were vindicated in the long and bitter siege which the garrison had to endure from the Cherokees. The fort capitulated only because of lack of food. De Brahm's difficulties during the building of the fort increased to such an extent that on the evening of December 25, 1757, less than a week before the troops moved into the completed fort, he took unceremonious leave and returned to Charles Town, thus earning the sobriquet from the Indians of "The warrior who ran away in the night." See Alden, op. cit., pp. 49, 59–60, and Philip M. Hamer, "Fort Loudoun in the Cherokee War, 1758–76," *North Carolina Historical Review*, II (1925), 442–58.

208. See P. L. Phillips, Lowery Collection, pp. 340–41 and 366, for manuscript and printed maps of Florida by De Brahm.

209. Governor Grant's chief complaint was that De Brahm claimed to get his authority directly from the King's ministers and not from Grant. W. Knox, in a letter of December 18, 1771, dated from Whitehall, sent De Brahm a copy of the complaint against him made in a letter from Governor Grant dated April 23, 1771, to the earl of Hillsborough. Governor Grant lists a number of items under "Heads of Complaint," among them "That Mr. De Brahm instead of carrying on the General Survey himself & committg his office of Provincial Surveyor to Deputies, resided in St. Augustine for the sake of the proffits accruing from the private Surveys & trusted the General Survey to his Deputies who were very unequal to so important a trust": P.R.O., C.O. 5/72, p. 647 (transcript in Library of Congress).

210. Ralph H. Brown, "The De Brahm Charts of the Atlantic Ocean, 1772–1776," *Geographical Review*, XXVIII (January 1938), 124–32. In May 1775 De Brahm was provided by the Admiralty with an armed ship, the *Cherokee*, to continue his studies and investigations; he first announced his study of the Atlantic currents in a letter to the editor of the *Gentleman's Magazine*, Sylvanus Urban: *Gentleman's Magazine*, XLI (1771), 436.

De Brahm's equipment for his Florida coastal survey is listed in Louis De Vorsey, "Hydrography: A Note on the Equipage of Eighteenth-Century Survey Vessels," *Mariner's Mirror*, LVI (1972), 175–77.

De Brahm's Atlantic Pilot, London, 1772, was published in a facsimile reproduction, edited with introduction, notes, and index by Louis De Vorsey, in Gainesville, Florida, 1974.

A bibliographical summary of articles on the Gulf Stream is found in W. P. Cumming, "Early Maps of the Chesapeake Bay Area," in *Early Maryland in a Wider World*, ed. David B. Quinn, Detroit, 1982, pp. 296, 307–8, n. 50. While returning from London aboard the *Cherokee* to his duties in the Southeast, De Brahm conducted a detailed oceanographic survey which greatly refined his understanding of the Gulf Stream and the North Atlantic. Although the journal and map which De Brahm prepared during that voyage have not been published in full, his findings have been summarized. See Louis De Vorsey, "Pioneer Charting of the Gulf Stream: The Contributions of Benjamin Franklin and William Gerard De Brahm's 'Continuation of the Atlantic Pilot,' An Empirically Supported Eighteenth Century Model of North Atlantic Surface Circulation," *Oceanography: The Past*, New York and Heidelberg, 1980, ed. M. Sears and D. Merriman, pp. 718–33.

211. W. L. Clements Library, Shelburne Papers, Vol. 87, fol. 24: published in R. G. Adams, *British Headquarters Maps and Sketches*, Ann Arbor, 1928, pp. 109–10.

212. In P. C. Weston, ed., *Documents Connected with the History of South Carolina*, London, 1856. See also A. J. Morrison, "John G. De Brahm," *South Atlantic Quarterly*, No. 3 (July 1922), p. 252. Other facts relating to De Brahm are found in his autobiography in his Harvard [1772] MS and in letters by Professor C. M. Andrews which are laid in the same volume. The best life of De Brahm is De Vorsey, ed., *De Brahm's Report of the General Survey*, pp. 7–59.

213. *CRNC*, VII, 861. On November 6, 1765, the Moravians at Bethabara noted that he visited them to acquire some information for his map: A. L. Fries, ed., *Records of the Moravians in North Carolina*, Raleigh, 1922, I, 299. The chief facts of Churton's life are summarized in Mary C. Engstrom, "William Churton," *Dictionary of North Carolina Biography*, Chapel Hill, I (1979), pp. 370–71.

214. *CRNC*, VII, 861.

215. Ibid., VII, 861.

216. Ibid., VII, 181, 244, 445; X, 483.

217. W. L. Clements Library, Gage Papers, Box 45. Endorsed: "Lord Shelburne August 2d 1767. received 8 Jan.ry 1768. answd." A life of Collet is found in William P. Cumming, "John Abraham Collet," *Dictionary of North Carolina Biography*, Chapel Hill, I (1979), pp. 402–4.

218. See Collet 1767 MS. Among the Gage Papers, Vol. 85, No. 36, in the Clements Library is an undated report: "Remarques sur le Fort de Johnston, au Cap Fear avec son plan," signed "J. A. Collet, Commandant du dit Fort," which describes the strategic advantages of the fort and points out several advisable improvements.

Captain John Dalrymple made a plan of Fort Johnston which he sent to the Board of Trade; Governor Tryon, in a letter to the Board of Trade dated August 1, 1766, refers to this plan and describes the dilapidated condition of the fort, which was poorly designed and constructed of such miserable tabby that parts of the parapet crumbled whenever one of the guns fired a salute: see *CRNC*, VII, 246.

For an unsympathetic account of Dalrymple and of Collet, based largely on the documents printed in *CRNC*, see J. G. De-

Roulhac Hamilton, "The Site of Fort Johnston," in *James Sprunt, Chronicles of the Cape Fear River, 1660–1916*, Raleigh, 1916, pp. 52–55.

219. *CRNC*, VII, 829, 863.

220. See Collet 1768 MS. Justin Winsor, *A Narrative and Critical History of America*, VI, 537–38, note 8, refers to this map as in twelve sheets.

221. Governor Tryon's letter refers to the lines on No. 3 showing the boundary surveys between North Carolina and the Cherokee hunting grounds run in 1767, the survey between North and South Carolina westward to the Catawba Nation five and a half miles east of Catawba River, and the Granville Line. None of these is shown on the No. 3, though a boundary line is given running below Catawba Town at 35°, and the map extends west to the Cherokee lands; No. 1 shows the Granville Line, which stops at Coldwater Creek on Rocky River, about two miles east of the present town of Kannapolis on the north boundary line of what is now Cabarrus County.

222. Wimble, Collet, and Mouzon all show differences in soundings, as in the depth at the Middle Ground of the Cape Fear River: see C. C. Crittenden, "The Seacoast in North Carolina History," *North Carolina Historical Review*, VII (1930), 440–44.

223. Lord Shelburne nominated Captain Collet to the post of commander of Fort Johnston in the belief that it was one both of honor and of profit, whereas for Collet it turned out to be neither. Henry Howe, appointed by Tryon, apparently continued to collect the "perquisites" of five shillings for each incoming vessel until Governor Martin discharged him peremptorily, to Howe's anger. Collet, whose amiability and industry were repeatedly commended by his superiors until the events occurred which caused his departure, for some reason—possibly delicacy of feeling, according to Governor Martin—did not disillusion Lord Shelburne by informing him of disappointments. Hillsborough wrote Tryon, as Dartmouth wrote later to Martin, that they appreciated Collet's merits but saw nothing that they could do to help him. In 1771 or 1772 Collet returned to England in the hope that he might better his fortune; he returned to America in July 1772, with a letter of recommendation to General Gage: Clements Library, Gage Papers, Hillsborough to Gage, July 17, 1772, endorsed as "Received May 6, 1773 Pr. Mr. Collet." While in London before his return to New York in July 1772, Collet apparently made a contact which, however innocent, probably did his cause no good. A. M. Chamier of the British War Office wrote to General Barrington that the French ambassador was very solicitous of getting in touch with Collet. Barrington immediately wrote to General Gage, enclosing Chamier's letter.

> Private Pakett Oct. 5 1772
> Dear Sir:
>
> I know Capt. Collet and have always entertained a good opinion of him, and I do so still; but the french ambassadors inquiries look as if he was to be *tempted*. There can be no harm in puting [*sic*] your Excellency on your guard. Collet was strongly recommended to me by Mr. Henry Grenville & his Lady & I believe I have recommended him to you. He went to North America this summer, & I believe took passage in a ship bound for New York. I can do him no injury by writing this letter to a man who is Equity and Candour in perfection. I am ever Dear Sir Your most faithful obedient, humble Servt
>
> Barrington.

(Clements Library, Gage Papers, English Series 23.)

Governor Martin, who found strong grounds for disapproval of Henry Howe, Tryon's appointee to the command of Fort Johnston, dismissed Howe and made Collet the actual as well as the royally appointed governor of Fort Johnston. Collet began vigorously to repair the fort without previous allocation of funds for the purpose by the Assembly. Only after repeated pressure on them from the governor did the Assembly, on March 24, 1774, allocate £322/9/4 to Collet for expenses already incurred; they significantly pointed out that no further funds after the next Assembly were to be spent on the fort.

Collet disregarded this resolution and, according to Governor Martin, spent upwards of £1,500 for improvements. Probably not all of this was spent on the military improvements, for he built a small new dwelling and several other structures, including a carriage house, for himself near the fort. He purchased a schooner which, it was learned, had been sent to New York by Governor Martin's command to get a supply of ammunition. His zeal overreached prudence; he salvaged unjustly the contents of a vessel stranded near the fort, disregarding a writ thereon by the attorney general, provisioned the fort with corn and wine for which he did not pay, and reportedly concealed runaway slaves in the fort. The suspicion and anger of the Patriots were aroused by these and other reported acts, and on the evening of July 18, 1775, a large band of men, led by John Ashe, burned the fort. They returned the next day to complete the destruction, setting fire to the dwelling and houses of Collet, who had escaped to a transport off the shore with the four men who remained loyal to him. Ten days before, on July 8, 1775, he had written to General Gage describing the "forlorn situation I am in . . . reduced by desertion to 12 men and 2½ barrels of gunpowder" (Clements Library, Gage Papers, Vol. 131). With his four remaining men, the governor of Fort Johnston set sail for Boston and General Gage. Governor Martin, himself a fugitive at the same time, advised that Collet not be sent back until he "can make his peace with the people now to the last degree exasperated against him." Thus, unhonored and anathematized, Captain John Abraham Collet passed from the North Carolina scene; a month later he was requesting credit to pay fifty men whom he had enlisted in the past fortnight (Clements Library, Gage Papers, Vol. 134, letter of Collet dated August 21, 1775). Besides the references cited to the Gage Papers, see *Colonial Records of North Carolina*, VII, 829, 860, 863, 877; VIII, 65–66; IX, 944, 799, 993; X, 84, 102, 108, 140, 234.

224. *CRNC*, VIII, 369, 390, 452. Plate 62 (Plan of the Town of Salisbury) is an example of C. J. Sauthier's work; many of Sauthier's North Carolina town plans have been reproduced in F. B. Johnston and T. T. Waterman, *The Early Colonial Architecture of North Carolina*, Raleigh, 1952. Color photographs of Sauthier's town

plans, the originals of which are in the British Library's King George III's Topographical Collection CXXII, are in the North Carolina Division of Archives and History, Raleigh, N.C.

A cartobibliography of North American maps in Alnwick Castle from the Percy collection, including some by Sauthier, is in William P. Cumming, *British Maps of Colonial America*, Chicago, 1974, pp. 79–84; map reproductions from the same source are in William P. and Elizabeth C. Cumming, "The Treasures of Alnwick Castle," *American Heritage*, XX (August 1969), 22–33; and W. P. Cumming, "The Montresor-Ratzer-Sauthier Sequence of Maps of New York City, 1766–76," *Imago Mundi*, 31 (1979), 55–65, describes and reproduces Sauthier's contribution to the sequence.

The Cumming Map Collection in the Davidson College Library has microfilms of Sauthier's works in the Bibliothèque du Grand Séminaire, Strasbourg, MS 316, and a bound English translation of Sauthier's French work on landscape and civil architecture. It also has a cartobibliography of Sauthier's published maps by Faden and others.

The following maps by C. J. Sauthier are in the Library of Congress; no attempt is made here, however, to give an exhaustive list of Sauthier's work.

A. A Topographical Map of Hudson's River . . . William Faden, London, 1776.

B. [The same.] French edition by Le Rouge, found in the lower right-hand corner of Le Rouge's 1777 edition of De Brahm's Caroline Méridionale et Partie de la Géorgie.

C. A Plan of Fort George at the City of New York. MS; undated.

D. A Plan of the Operations of the King's Army under . . . Gen. Sir William Howe . . . Against . . . Gen. Washington . . . 1776 . . . on the White Plains. MS; undated.

E. A Map of the Province of New York . . . By Order of His Excellency William Tryon . . . C. J. Sauthier & B. Ratzer. London. Engraved by W. Faden, 1776.

F. [The same.] Carte des Troubles de l'Amérique. Le Rouge, 1778.

G. A Map of Part of New York Island, Showing a Plan of Fort Washington. MS; undated.

H. A Map of the Inhabited Part of Canada, From the French Surveys. London. Wm. Faden, 1777.

I. A topographical Map of the North[n] Part of New York Island, Exhibiting the Plan of Fort Washington. London. W. Faden, 1777.

J. A Chorographical Map of the Province of New York . . . Compiled from Actual Surveys Deposited in the Patent Office at New York, by Order of Major General William Tryon. London, W. Faden, 1779.

Fuller biographical accounts of C. J. Sauthier are to be found in Cumming's *British Maps of Colonial America*, pp. 72–74, 101, and in Cumming's "Sauthier," in the *Dictionary of North Carolina Biography*, ed. W. S. Powell, Chapel Hill, 1979. A cartobiographical list of Sauthier's manuscript and published works is in the Cumming Map Collection, Davidson College Library.

225. South Carolina Historical Commission, Columbia S.C., *Journal of the Commons House of Assembly*, Vol. 38, p. 276: "Thursday the 22[d] day of February 1770. Read a Petition of Tacitus Gaillard and James Cook; in the words following (viz[t]), That your petitioners being appointed to make a general Survey of this Province, have with indifaticable [*sic*] pains and labour compleated the same; a Map whereof being now finished in the best manner, according to agreement with the commissioners for the said Survey: and as your Petitioners are very desirous to have it executed on Copper plate, with dispatch, and not being able to advance such sums of Money as may be necessary for the Engraving the same,

Therefore humbly pray that this Honorable House will be pleased to advance to your Petitioners the sum of Four thousand Pounds Currency, part of the Eighteen thousand Pounds granted for the said Survey, to be lodged in some safe and proper persons hands to pay the expence of Engraving the said Survey on Copper Plate &c[a] and the Balance may be paid your Petitioners, to enable them to defray the expences that they have been at in the execution of the Survey."

226. The eastern or coastal half of this original draft, in rather dilapidated condition, is in the South Carolina Historical Society; see Gaillard-Cook 1770 MS.

The identity of James Cook has confused many because there were four contemporary James Cooks, all surveyors and members of the British admiralty. The "South Carolina" James Cook was master of the *Mars* and *Alarm* until cashiered for overzealousness in surveying against his captain's orders. His career is related in Jeannette D. Black with R. A. Skelton, "Too Many Cooks," *Map Collector* No. 34 (March 1986), 10–16, with an appendix of a list of 13 maps made by Cook between 1763 and 1775.

227. *Journal of the Commons House of Assembly*, Vol. 38, pp. 347, 358. In the previous year James Cook had called the error, which he had discovered in the surveys he was making, to the attention of the boundary committee of the Council. See A. S. Salley, *The Boundary Line between North and South Carolina*, Bulletin of the Historical Commission of South Carolina No. 10, Columbia, S.C., 1929, p. 25.

228. *Journal of the Commons House of Assembly*, Vol. 38, pp. 347, 358, 363, 541, 546–47. The payment of the £700 for the additional names was not made until February 1, 1775; ibid., Vol. 39, p. 210.

229. North Carolina–South Carolina Boundary 1772 MS.

230. Cook 1773; see Gaillard-Cook 1770 MS and Lodge-Cook 1771 MS for earlier drafts of Cook's map.

231. The primary facts concerning Captain Henry Mouzon, Jr., are found in a small book now in the possession of Mrs. W. Boddie, a descendant of Mouzon's sister-in-law. The pertinent information is in Henry Mouzon, Jr.'s own hand. The author is indebted to Mrs. R. W. Hutson of Charleston, S.C., for this and other information concerning Captain Henry Mouzon.

232. Records of the Secretary of the Province, Will Book VV (1776–84), 222–24, in South Carolina Archives Department (Columbia).

233. South Carolina Archives Department, Inventories, 1776–84, 39. The writer is indebted to Miss Wylma Wates of the South Carolina Archives Department, who has made a much

fuller study of the authorship of the map than is presented here and who has kindly put her notes and photographs of the holograph documents signed by the two Mouzons at the writer's disposal.

234. Cf. William W. Boddie, *History of Williamsburg*, Columbia: State Company, 1923, p. 123; Cumming, *Southeast in Early Maps* (1962 edition), pp. 59–60.

235. Further details concerning the life of Captain Henry Mouzon, Jr., may be found in McCrady, *South Carolina in the Revolution, 1775–1780*, pp. 577, 649, 748, 750; and in A. S. Salley, *Indents for Revolutionary Pay*, Lib. Y, Indent No. 3; and Book P, No. 553.

236. For detailed information about dates and boundaries of counties, see David L. Garbett, *The Formation of the North Carolina Counties, 1663–1943*, Raleigh, 1950.

237. This analysis of the authorship and features of the Mouzon map is taken from William P. Cumming, *North Carolina in Maps*, Raleigh, 1966 and 1986, pp. 21–22.

238. See Mouzon 1772 MS. James Cook and Ephraim Mitchell made the surveys and ran the boundary line.

239. Major Christian Senf, reportedly a Swede by birth, was captured with the Hessians at Saratoga and sent by Henry Laurens to be the state engineer of South Carolina. The canal was finished in 1800. See D. D. Wallace, *The History of South Carolina*, New York, 1934, II, p. 399, for further references.

240. S. G. Stoney, "Memoirs of Frederick Augustus Porcher," *South Carolina Historical and Genealogical Magazine*, XLIV (1943), 70–71.

241. This statement is made in the Petition of April 25, 1775: South Carolina Historical Commission, *Journal of the Commons House of Assembly*, Vol. 39, pp. 274–75. The House committee's disallowance of the petition is in the same volume, p. 275.

242. See under Mouzon 1775 (Map List No. 450), where the advertisement is reprinted.

243. Collier Cobb, "The Ramsgate Road," *The North Carolina Booklet*, XXIII (1926), 62, says that in spite of its high-sounding title, Mouzon's map was taken bodily from Collet's 1770 map, the only change in most cases being the omission of initials before the family names of freeholders along the roads. This is hardly just. While the upper half of Mouzon's map is based on Collet, Mouzon has attempted to bring the North Carolina section up to date. New counties are given and are distinguished more clearly than in Collet, though neither shows the boundary lines. Along Mouzon's additional counties in the northern province is Pelham County, above Brunswick; Pelham was proposed as the name of a new county, but when it was created the area was called Sampson County. The details west of the Catawba River are more numerous and more accurate, showing the increased knowledge of streams and adding roads and such topographical features as "King Mountain."

244. South Carolina expended $52,760 from 1816 to 1820 on a map of the state, under the direction of John Wilson: "Map of South Carolina, constructed and Drawn from the District Surveys, ordered by the Legislature by John Wilson, late civil and military Engineer of South Car^{o} Engraved by H. S. Tanner. Philadelphia . . . 10th Day of April, 1822." Four sheets, each 22½ × 29. The legislature spent $12,000 for the same purpose in 1825 to aid the production of R. Mills, *Atlas of the State of South Carolina*, Baltimore, [1825], with twenty-five maps. See E. McCrady, *The History of South Carolina under the Royal Government, 1719–1776*, New York, 1899, p. 115, note 2. Mills's atlas was reprinted by L. H. Bostick and F. H. Thornley, Columbia, S.C., 1938.

In North Carolina, in the early years of the nineteenth century, requests for financial aid in gathering information for a map of the state were refused by the General Assembly. A survey was made possible, however, through the support of David Stone and Peter Browne: see A. D. Murphey, "Memoir on the Internal Improvements Contemplated by the Legislature of North Carolina, Raleigh, 1819," *The Papers of Archibald D. Murphey*, edited by William H. Hoyt, Publications of the North Carolina Historical Commission, Raleigh, 1914, II, 105. A copy of the map is in the State Department of Archives and History in Raleigh: "To David Stone and Peter Browne Esq^{rs} This First Actual Survey of the State of North Carolina Taken by Subscribers is respectfully dedicated by their humble servants Jon^{a} Price John Strother 1808 Engraved by W. Harrison Philad^{a}." Size: 60 × 28½. Scale: 1" = 8 miles. Between 1816 and 1818 the North Carolina Sounds and several rivers were surveyed under the direction of Peter Browne, who had been appointed chairman of a board and commissioners charged with the surveys of the principal rivers: op. cit., II, 114–16. In 1819 Robert Brazier was appointed state surveyor at a salary of £300; in 1833 he produced a map published under the patronage of the legislature: "A New Map of the State of North Carolina. Constructed from Actual Surveys, authentic Public Documents and Private Contributions by Robt H. B. Brazier, Published under the Patronage of the Legislature . . . by John Macrae, Fayetteville, N.C. & N.C. & H. S. Tanner, Philadelphia . . . 1833." Size: 82⅞ × 34⅞. Scale: 1" = 5¼ miles. Brazier's map showed a sketchy knowledge of the country west of the Blue Ridge. In 1839 appeared Burr's map, which gave evidence of greatly expanded information for the western reaches of the state: "Map of North and South Carolina Exhibiting the Post Offices, Post Roads, Canals, Rail Roads, &c. By David H. Burr (Late Topographer to the Post Office). Geographer to the House of Representatives of the U.S." Size 48¼ × 35¾. Scale: 1" = 10 miles. In 1966 the North Carolina Department of Archives and History in Raleigh published a series of North Carolina and earlier regional maps, approximating the size of the originals, in a tubular container: White 1585 MS, White-DeBry 1590, Mercator-Hondius 1606, Comberford 1657 MS, Ogilby-Moxon ca. 1672, Moseley 1733, Collet 1770, Mouzon 1775, Price-Strother 1808, MacRae-Brazier 1833, Colton 1861, Bachman 1861, U.S. Coast Survey 1865, Kerr-Cain 1882, Post Route Map 1896. These were accompanied by a booklet, W. P. Cumming, *North Carolina in Maps*, with other maps; it was reprinted in 1985.

In Georgia not much improvement in maps occurred after B. Romans's "A General Map of the Southern British Colonies . . . 1776" until the Early-Sturges map of 1818: "Map of the State of Georgia Prepared from Actual Surveys and other Documents for

Eleazer Early. Daniel Sturges . . . 1818." Size: 60 × 44. Scale: 1" = 8 miles. Even in this map western Georgia has little detail beyond rivers, paths, and a few settlements.

Maps of Virginia are available in reproductions published in 1975 by the Virginia State Library, Richmond, in a portfolio, accompanied by a manual with the same title as the portfolio, *Description of the Country: Virginia's Cartographers and Their Maps 1607–1881*. Besides map illustrations in the booklets, the portfolio contains: Smith Virginia 1607, Senex Virginia 1719 (third issue of Herrman's Virginia 1673), Fry-Jefferson Virginia 1775 (third issue of second edition), Sartine Chesapeake Bay 1778 (French edition of Hoxton-Smith Chesapeake Bay 1776), Kitchen United States 1783, Boye Virginia 1859 (nine sheets), Crozet Internal Improvements of Virginia 1848, Mitchie Richmond 1865. The choice of maps and author of the manual is E. M. Sanchez-Saavedra, born and educated in Richmond and at the time of publication a librarian in the Virginia State Library.

An atlas of over a hundred maps is Richard W. Stephenson, *The Cartography of Northern Virginia: Facsimile Reproductions of Maps Dating from 1608 to 1915*, Fairfax County, Virginia, 1981. Stephenson writes in the preface, "The intent was to reproduce the relevant portions of the key or mother maps of Virginia, as well as the most significant maps of all or part of northern Virginia."

American Indians and the Early Mapping of the Southeast

BY LOUIS DE VORSEY, JR.

None of the Indians of the Southeast had anything approaching the system of writing possessed by the Europeans whom they first encountered during the early sixteenth century. But they did possess sophisticated systems of record keeping and communication that played vital roles in their societies. These were mainly oral systems augmented by mnemonic aids such as wampum belts, notched or carved sticks, inscribed or painted boards and trees, quipus of notted cord, scribed or painted bark or shell, or painted hides. In the last analysis, however, the Indians depended largely upon human memory and oral communication to ensure the transfer of knowledge, especially from one generation to the next.

In some Indian societies this transfer function was formalized into a special role, but generally it seems to have been the duty of the eldest or most experienced members of a community. In times of peace and stability these assisted oral record keeping systems appear to have served reasonably well, as attested to by a number of ethnological studies.[1] In times of great stress, however, living human repositories of group history, myth, and lore were often, by virtue of their advanced ages, at great risk. The death of one of these elders in the epidemics or violent conflicts that often followed early contact with Europeans was tantamount, in today's terms, to the loss of a national library or public archive. As a consequence, when the history of the Southeast's conflict-filled, sixteenth, seventeenth, and eighteenth centuries came to be written according to Europe's document-bound canons, the Indian participants were given faint, if any, voice. To help redress this now obvious gap in the region's historiography, a new generation of historians has been turning increasingly to the work of archaeologists and ethnohistorians in an effort to write a more balanced and satisfactory history of the Southeast. Evidence previously ignored or disregarded is undergoing reassessment, and heretofore silent witnesses to the history of Indian-white relationships in the Southeast are being heard and acknowledged.[2]

Early maps of the Southeast are important and formerly silent witnesses. A few have been clearly identified as being of Indian origin, but many others have long been regarded as European in origin. When some of these "European" maps are scrutinized closely, intriguing questions arise in regard to their authorship—questions that point toward important but heretofore unacknowledged historical roles played by the region's Indians.

From the time of Columbus's first landfall and safe passage through the reef-strewn Bahama archipelago to Cuba and Hispaniola, successful European explorers relied on the wayfinding and cartographic skills of the American Indians. Even a casual reading of Columbus's initial appraisal of the Indians he met on the beaches of Guanahani (now San Salvador) in October 1492 makes clear his recognition of those skills. In his famous "First Letter" to King Ferdinand and Queen Isabella, Columbus assured his patrons that the natives he called Indians were not "slow or unskilled but of excellent and acute understanding." Among them, he continued, "the men who have navigated this sea give an account of everything in an admirable manner; . . . As soon as I reached that sea, I seized by force several Indians on the first island, in order that they might learn from us, and in like manner tell us about those things in these lands of which they themselves had knowledge; and the plan succeeded."[3]

As he departed San Salvador, Columbus wrote that he "saw so many islands that I did not know how to decide which one I would go to first." His knowledgeable kidnapped guides, he continued, "told me by signs that they [the islands] were numberless." They also guided him along their well-traveled canoe routes to a landfall on Cuba, where he fully expected to find the Great Khan, fabled emperor of Cathay.[4] Later, in a royal encounter, Columbus's Indian pilots were called upon to demonstrate, in cartographic form, the precision and completeness of their valued geographic intelligence.

This opportunity came on the homeward voyage, when the storm-tossed *Niña*, with Columbus aboard, was forced to put into Lisbon harbor. A few days after his landing, Columbus was summoned to an interview with the Portuguese king, Don João, at a monastery outside Lisbon. During their conversation, Columbus assured the Portuguese ruler, as he had the monarchs of Castile and Aragon, that the Indians with him were intelligent and well informed as to the geography of the islands making up their homeland. As a test of their ability, as well as the truth of where Columbus reported those islands to lie, the king decided to interrogate the Indians personally. Although not a witness to what transpired, Bartolomé de Las Casas provided an account of the interrogation that has impressed scholars with its ring of authenticity. In his *History of the Indies,* Las Casas told of how Don João ordered a bowl of dried beans brought into the room. Scattering the beans on the table, he ordered one of the Indians to arrange them in the form of a map of his homeland. Deftly complying, the Taino quickly grouped clusters of beans to represent Cuba and Hispaniola while using single beans to represent smaller islands. "The king," Las Casas wrote, "observing this geographical game with a gloomy countenance, as if by inadvertance disarranged what the man had set forth, and commanded another Indian to play mapmaker with the scrambled beans."[5] The second Indian proved better than the first when he reassembled the bean chart "and added many more islands and lands, giving us an explanation of all that he had depicted and indicated in his own tongue."

It would stretch credulity to assume that Columbus, and others with access to his kidnapped Indian pilots, did not record their maps in more permanent media than a scattering of dry beans. Regrettably, however, no Indian maps transcribed by Columbus are known to have survived. A woodcut map that was published in Seville with the 1511 edition of Peter Martyr's *Oceani Decas* could, however, be based on such an Indian map.[6] The so-called Martyr map (Figure 1) has long intrigued scholars because it shows a named coastline north of Cuba where Florida is now known to lie. It should be recalled, however, that according to the historical record, Juan Ponce de León did not *discover* that coastline and name it "Florida" until 1513, or two years after the Peter Martyr map appeared. If it is accepted that the coastal area, named "illa de Beimeni parte," is indeed a portion of Florida, it could be taken as indirect evidence that the author of this map, be that Martyr or someone else, had access to cartographic intelligence or maps provided by Columbus's Taino pilots or some other Indian informant.[7] It should be added that Martyr was well acquainted with Columbus and entertained him and his Indian companions at his table in return for descriptions of discoveries in the West.

The significant role that Native American informants, guides, and mapmakers played for Columbus and his fellow "discoverers" grew in significance as others began to explore the new lands being made known to Europe. Figure 2 provides a useful framework or schema for a review of the Indians' role in the mapping of the Southeast, which is the subject of the rest of this essay.[8] This diagram should be recognized for what it is: a much simplified linking of a number of very complex processes and interrelationships, all of which contributed either positively or negatively to the creation of those graphic representations of the Southeast we recognize as maps. In the top box the totality of the things, conditions, processes, and events making up the Southeast in any period are embraced under a label of convenience, "The Geographic Environment." Through the contrasting value and belief systems underpinning their distinctive cultures, peoples of Amerindian and European backgrounds perceived and cognized differently the contents and qualities of the region comprising their shared "Geographic Environment."

The broad arrow drawn between the boxes labeled "Amerindian Perceptions" and "European Experience" is based on a still-to-be-tested hypothesis that argues that observed variation in Amerindian cartographic techniques and styles is a reflection of the adoption of both European cartographic conventions and expanded European spatial information fields. It is a hypothesis awaiting a corpus of Indian-drawn maps of known provenance sufficient to provide evidence for its testing. In no way, however, is the broad arrow meant to diminish the significance and endurance of uniquely Indian systems of cartographic communication.

On the left side of the diagram a heavy line leads directly from the "Amerindian Perceptions" box to the last box, "Southeast Area and Regional Maps." There are disturbingly few extant early maps of the Southeast, either drawn by Indians or copied directly from their drafts, that may be placed in this box.[9] It is anticipated, however, that growing awareness and interest in Indian cartography will lead to the discovery and identification of many more. It is entirely possible that Southeastern petroglyph and other artifactual motifs embody as-yet-unidentified cartographic elements.

Even a superficial review of this book's list of maps, with its descriptions of more than 450 early maps of the Southeast, will quickly reveal that the overwhelming majority of these maps are products of European or Euroamerican authors. A major purpose of Figure 2 is to suggest ways in which Indians entered into the processes contributing to the creation of many of these maps. It will be seen that Indians contributed most to what have been identified as the information-gathering and information-processing functions of Euroamerican map production.[10] Most often the Indians served as informants or guides for European sojourners and explorers as they surveyed the landscapes (Geographic Environment) of the Southeast. Invariably, as soon as European explorers left the tidewater, with its easi-

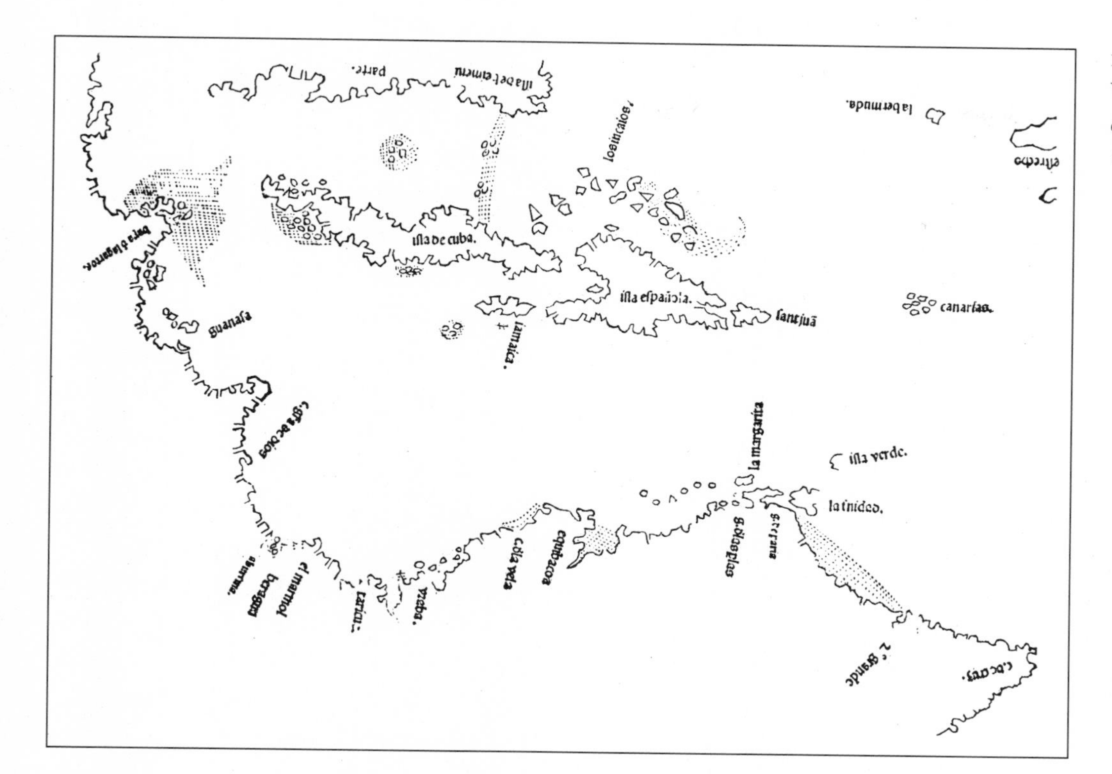

FIGURE 1.
Woodcut map from 1511 edition of Peter Martyr's *Oceani Decas*

ly followed estuaries, they moved along the Indian trails crisscrossing the region. As a consequence, their initial routes and ensuing perceptions and resource appraisals were strongly influenced by pre-existing aboriginal patterns of resource exploitation and systems of communication and movement.[11]

In case after case, Indian cartographers sketched out maps for the enlightenment of the Europeans. Unfortunately, most of these maps were ephemera chalked on a ship's deck or traced in the sand of a river bank or the ashes of a campfire. In some cases, however, they were drawn on paper or parchment in colonial towns or on plantations. Other maps were prepared in European chambers by Indians who had been transported there, not always voluntarily. As Figure 2 indicates, these Indian contributions were available for incorporation in the mapping efforts being undertaken by Europeans. But problems often intervened to complicate the smooth transfer of intelligence, geographic and otherwise, between members of such sharply contrasting cultures. These problems are graphically summarized in Figure 2 by means of the three shaded boxes labeled "Amerindian Hostilities," "Amerindian Misinformation," and "Misunderstanding of Amerindians." Notice that these "problem" boxes form potential interruptions or barriers in the line connecting "European Experience" to "European Perceptions" and the consequent creation of maps of the Southeast. They are junctures in the carto-information pathway at which the potential for distortion or outright error is at its greatest. In this essay a fuller awareness of the ways in which the Indians contributed both directly and indirectly to the early mapping of the Southeast will be developed through a discussion of their roles in the overall mapping process suggested by Figure 2.

THE SIXTEENTH CENTURY

Already a very wealthy man, in 1513 Juan Ponce de León received a royal license allowing him "to discover and settle the island of Bimini," which had provocatively been shown to the north of Cuba on the Peter Martyr map of 1511 (Figure 1).[12] The most indelible legacy of his first voyage to this new land is the name that Ponce de León gave it when he discovered it in 1513 on the day of the Feast of Flowers—Florida. Although nothing of exceptional promise was found and the Indians encountered turned out to be notably warlike and inhospitable, Ponce de León followed up his discovery with an attempt to found a settlement there in 1521. Probably encouraged by the news of Cortes's conquests and discovery of fabulous wealth in Mexico, he equipped and led a party of 200 colonists, complete with livestock, seed, and furniture, to a landing near the southwest tip of Florida. A fight broke out almost as soon as the first landing was made, in response to a party of Indians hailing from shore.

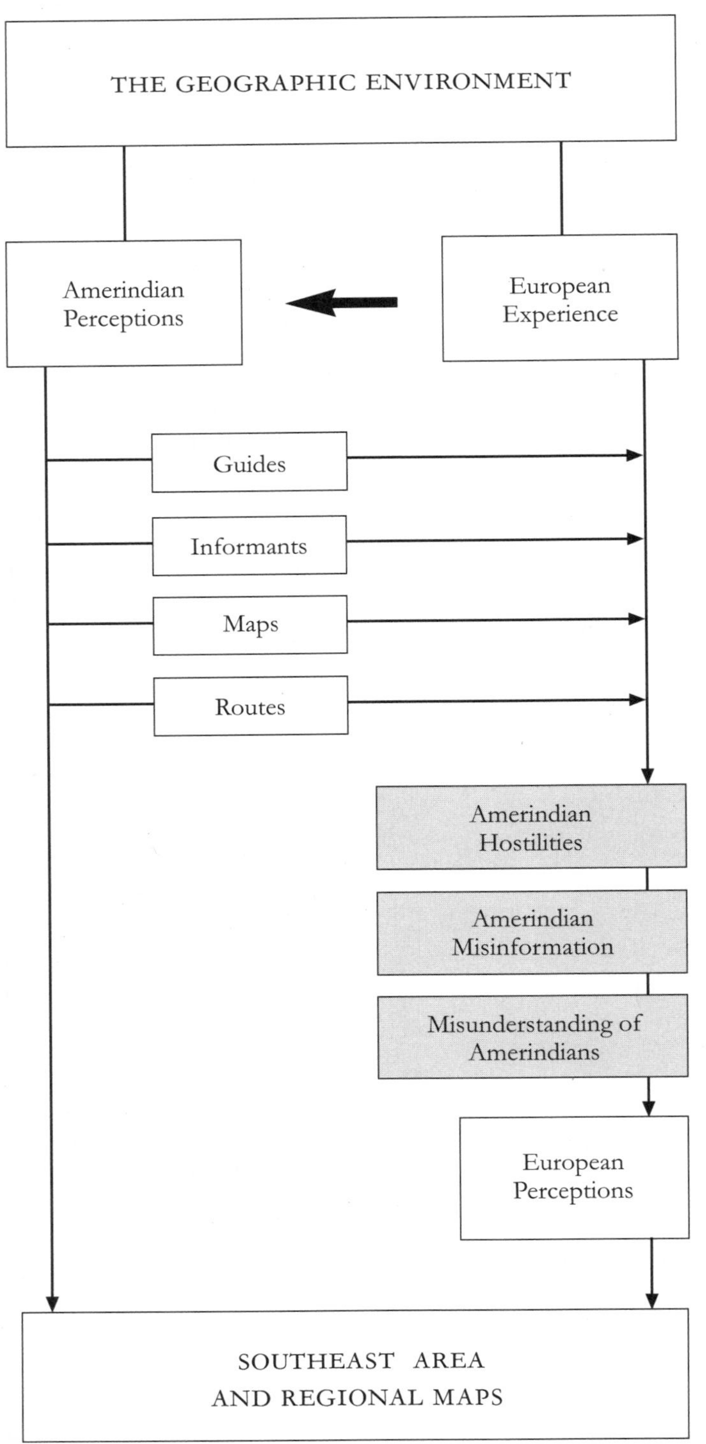

FIGURE 2. Schematic diagram of Indians' role in the early mapping of the Southeast

The Indians, armed with bows and arrows and spears tipped with "sharpened bones and fish spines," wounded two Spaniards while escaping any significant injuries to themselves.

In a second clash, four Indian women were taken captive to be utilized as informants and guides. They appear to have been queried as to the general nature of Florida, in response to which they described it as part of a landmass and not an island as Ponce de León and his fellows believed. The Spanish, however, retained their erroneous belief that Florida was an island in spite of the Indians' insistence that it was attached to the mainland and their ability to recite the names of its several "provinces" or chiefdoms. According to Ponce de León's Indian informants, the land now known as Florida was named "Cautio" because its people "covered certain parts of their body with palm leaves woven in the form of a plait."[13] Antonio de Herrera, historian of the Indies, later observed that it was not for some years after Ponce de León's explorations that the Spanish finally accepted the fact that Florida was not simply another island but instead part of a continental landmass. The first European map to employ the name "Florida" for the peninsula forming the eastern shore of the Gulf of Mexico is attributed to Alonzo Alvarez de Pineda, who commanded the expedition that explored the Gulf in 1519.

Pineda's sponsor, Francisco de Garay, was more interested in the Mexican shores of the Gulf, and he failed to make a successful bid for lands in the quarter originally sighted by Ponce de León. That bid came from another well-to-do Spanish official named Lucas Vásquez de Ayllón, who succeeded in winning a royal charter in 1523 granting him the right to settle lands located "in a 35°, 36°, 37° North-South line from the island Española, and . . . believed and considered sure to be very fertile and rich and apparently suitable for settling because in it there are many trees and plants like those of Spain, *and the people are of good intelligence and better suited for living in settlements than those of the island of Española*."[14]

Ayllón, like Ponce de León before him, had first learned of those new lands through his involvement in an unsuccessful Bahamian slaving expedition. In the words of the historian Gómera, the would-be slavers with whom Ayllón was leagued "failed to find men to barter or take by surprise so as to fetch them to their mines, their flocks or their farms."[15] Rather than returning empty handed from the now almost totally depopulated Bahama Islands, the slavers "resolved to go further north to look for a country where they might find them [slaves]." They pushed on to a landing on a coast that the pilot Pedro de Quexós determined to be at 33° 30' north latitude. Here, on June 30, 1521, "possession was taken on the bank of a river in this land; this land was named after St. John the Baptist; crosses . . . cut into the savin trees [arboles de sabinas]."[16] At the conclusion of about three weeks spent in exploring with the assistance of the hospitable local Indians, the Spaniards decided to return to Hispaniola. As they secretly prepared to weigh anchor, a large group of unsuspecting Indians was invited to the ships. Gómera wrote, "The Indians came onboard without any thought of treachery. Then the Spaniards raised anchor and set sail, and came with a good prize of Chicorans for Santo Domingo."[17]

In his colorful account of the Spanish landing in 1521 and the kidnapping of the Chicorans, Peter Martyr mentioned that the local Indian "sovereign" had "received them respectfully and cordially."[18] When the Spaniards "exhibited a wish to visit the neighbourhood," the local chieftain provided them with "guides and an escort." As he continued the story, Martyr revealed his disapproval of what transpired: "But what then? The Spaniards ended by violating this hospitality. For when they had finished their exploration, they enticed numerous natives by lies and tricks to visit their ships, and when the vessels were quickly crowded with men and women, they raised anchor, set sail and carried these unfortunates into slavery."[19] Martyr found the natives of Chicorana to have "a well browned skin, like our sunburned peasants," with black hair that both sexes wore to a great length.

One of Martyr's chief informants was "a native of Chicorana" whom Lucas Vásquez de Ayllón had brought to Spain as his servant while seeking his license from the king. "This man had been baptised under the Christian name of Francisco united to the surname of his native country, Chicorana," Martyr wrote.[20] While Ayllón was engaged in his lobbying efforts at court, Martyr "sometimes invited him and his servant Francisco Chicorana to my table."[21] It was doubtless here that the erudite priest Martyr determined that "this Chicorana is not devoid of intelligence." Martyr found that the Southeastern Indian informant "understands readily and has learned the Spanish tongue quite well."[22]

According to Martyr's information, the exploration of Chicorana's homeland by Quexós and the other Spaniards "occupied but a few days," during which they found that "it extends a great distance in the same direction as the land where the Spaniards anchored." Martyr goes on at some length to recount a fascinating collection of tall tales interspersed with bits of ethnological description concerning the Indians of Chicorana and their surrounding provinces or chiefdoms.[23] Exactly how much of this material was provided by the Indian informant, Francisco Chicorana, cannot be determined, but it is reasonable to assume that the cultural gulf separating Chicorana from Martyr and other courtiers resulted in a high degree of misunderstanding. Any attempt to sort out fact from fiction in Martyr's account would carry far beyond the scope of this essay. It should be noted, however, that the Spanish crown appeared to be convinced that the Indians of the Southeast were ruled "politically," to use the Aristotelian jargon of the day. That is to say, they lived according to "natural law" and, as a result, Ayllón's colonial enterprise was directed to proceed by incorporating the Indian polities into the Spanish Empire through peaceful rather than warlike means. Never one to suffer excesses of personal modesty, Peter Martyr wrote of how "thanks to us, the licenciate and royal counselor, Ayllón, succeeded in obtaining what he wanted," and "His Imperial Majesty accepted our advice, and we have sent him back to New Spain, authorising him to build a fleet to carry him to those countries where he will found a colony."[24]

In 1525 Ayllón's pilot, Pedro de Quexós, sailed with two caravels to select the landing location for the colonial effort Ayllón would personally lead the following year. Some of the results of this exploration are reflected in the map drawn by Juan Vespucci in 1526 (Map 02, Plate 2; note the prominent caption marking the land newly discovered by Ayllón). Based on his evaluation of Spanish legal documents, historian Paul E. Hoffman suggests that Quexós made a landfall at the mouth of the Savannah River on May 3, 1525.[25] Proceeding northward because his Chicoran Indian interpreter-guides could not understand the local Muskogean-speaking Indians, Quexós sailed on to the familiar Santee River–Winyah Bay–North Island complex, presumably where he had kidnapped his guides in 1521.[26] Continuing his explorations to the north, Quexós may have reached the Maryland coast and traversed in all, according to Hoffman, some 688 nautical miles along the coast of the Southeast before returning to Hispaniola in late July or early August 1525.

As Hoffman and others have detailed, Ayllón's attempted colony was even more ambitious than that of Ponce de León five years before. It is believed that the first landing was made on August 9, 1526, at the Santee-Winyah entrance on the coast of present-day South Carolina, where the flagship and its cargo of stores was lost. According to an account by the Spanish historian Gonzalo Fernández de Oviedo, Ayllón's Indian guides, presumably including his Christianized confidant Francisco Chicorana, "fled inland after a few days," leaving the Spaniards floundering in a sea of uncertainty.[27] Oviedo describes Ayllón's party as being unsure whether to believe anything they had been told by the Indians because, "along the whole coast, and as much of the interior as the Spaniards saw, they never managed to find any news of any province, port, river or settlement" that Chicorana or his fellows had described. As Oviedo concludes, Ayllón, who had effectively employed Chicorana's half-understood accounts and exaggerations concerning his homeland for political ends while in Spain, became their victim on the sands of South Carolina.[28]

The absence of Indian settlement and promising agricultural land in the country around the landing site induced Ayllón to send out scouting parties to seek a more propitious site for his colony. According to Oviedo, "They decided to go and settle further up the coast, towards the west coast, and they went to a large river (about forty or forty-five leagues distant) called Gualdape, and made their main camp on its banks: they set to building houses, for there were not any there." Oviedo described the countryside around the site of Ayllón's town of San Miguel de Gualdape as "all very flat and marshy," with the river "very full, and with lots of good fish."[29] As the autumn weather turned "very cold" and food supplies ran down, Oviedo reported that "many of [Ayllón's party] went ill, and many of those died." Ayllón himself sickened and died on October 18, 1526. After a period of Indian attacks and a mutiny that saw "some negro slaves independently" set fire to a house containing prisoners, San Miguel de Gualdape was abandoned. According to Oviedo's sources only 150 ill and starving survivors made it back to Hispaniola—considerably fewer than the 600 who so optimistically had been led to Chicorana by Ayllón only months before.

While the coastal outlines of the region were being revealed by explorers like Quexós, Ayllón, and Giovanni da Verrazano, the interior Southeast remained a tempting terra incognita to Europeans. By far the best known sixteenth-century incursion into that beckoning unknown land was led by Hernando de Soto in 1539–43.[30] When he began his expedition, de Soto was among the wealthiest of the veteran conquerors who had been in Peru with Francisco Pizzaro, and his enormous wealth made it possible for him to mount an expedition of the scale that he led from Spain to a landfall in Tampa Bay. Still a young man when he returned from the conquest of the Inca Empire, de Soto had married into Spain's aristocracy and was accepted in circles including the higher nobility. His energy and ambition, however, remained unquenched, and he petitioned the crown for the right to lead an entrada and establish a government to be known as the Province of Florida. In the process he was to enjoy the right "to conquer, pacify, and populate the lands . . . from the Province of the Rio de las Palmas to Florida," or all the unclaimed land bordering the Gulf of Mexico.[31]

An advance party was sent to reconnoiter a suitable landing place, and on their return to Cuba they brought with them four Indians who had been captured on the coast to serve as guides and interpreters.[32] The main force of some 600 fighting men and artisans sailed from Havana on May 18, 1539, and made a successful landing somewhere in Tampa Bay on Florida's west coast several days later. On one of their early reconnaissance forays into the surrounding countryside, de Soto's men made an amazing discovery—they encountered a Spanish survivor from an ill-fated expedition led by Pánfilo de Narváez a decade earlier.[33] His name was Juan Ortiz, and his knowledge of the local terrain and Indians was invaluable as the expedition made its way north to a winter encampment at present-day Tallahassee.[34]

In its rambling progress, de Soto's army, with its train of kidnapped Indian bearers, hundreds of forest-ranging hogs, pack of war dogs, and cavalry horses, trekked for more than 3,500 miles across parts of what are now Florida, Georgia, South and North Carolina, Tennessee, Alabama, Mississippi, Arkansas, Louisiana, Texas, and possibly Missouri. En route a legion of Indian guide-interpreters were thanklessly pressed into an often brutal servitude. Over most of his circuitous route de Soto followed the existing system of trails connecting one Indian village or chiefdom to another.

On at least one occasion de Soto had reason to seriously regret following the advice of an apparently well-qualified Indian informant. This was a teenage youth the Spaniards called Perico or Pedro. He and another youth appear to have been traveling with some Indian traders and claimed to know a great deal about the trails leading to the interior. Perico "stated that he did not belong to that country [around present Tallahassee], but was from a distant land in the direction of the sun's rising, from which he had been a long time absent visiting other lands; that its name was Yupaha, and it was governed by a woman, the town she lived in being of astonishing size, and many neighbouring lords her tributaries, some of whom gave her clothing, others gold in quantity."[35] One can imagine the rapt attention paid by de Soto's treasure-seekers when Perico "showed how gold was taken from the earth, melted, and refined exactly as though he had seen it done . . . so that they who knew aught of such matters declared it impossible that he could give that account without having been an eye-witness; and they who beheld the signs he made credited all that was understood as certain."[36]

With Perico showing the way, the army marched into southwestern Georgia, noticing as they went the distinctly different dwellings making up the Indian villages they encountered.[37] Perico was now out of his depth, and while at the Indian town called Patofa, he was either seized with, or pretended to have, a fit and "began to froth at the mouth, and threw himself on the ground as if he were possessed of the devil."[38] After an exorcism was said over him, Perico regained his senses and declared that "four days' journey from there, towards the sunrise was the province he spoke of."[39] The local Indians told the Spanish that they knew of no settlement in that direction, but that toward the northwest lay a well provisioned land of very large villages called Coosa. In spite of the obvious instability of his young guide, de Soto ill-advisedly decided to believe him and loaded a four-day supply of corn from Patofa's cribs for the anticipated journey.

From Patofa the expedition marched along a road that "gradually became less and on the sixth day . . . disappeared."[40] When Perico finally admitted that he had no idea where he was, de Soto threatened to throw him to the army's pack of ravenous dogs. His life was spared only because he was now the sole Indian left who spoke a language that Juan Ortiz understood.[41]

Limited space will not permit following the indefatigable de Soto across the Appalachian Mountains and the Mississippi River as he pressed into service one after another aboriginal guide along the network of Indian trails he found lacing the continent. When he met Plains dwellers, he sought to learn the route "to the other sea," as he sought an understanding of the vast land he vainly hoped would yield a second Inca or Aztec empire. Finally, in 1542, four years after landing in Cuba, he apparently decided to abandon his quest and return to the world he thought of as civilized. On the trail to the coast, Hernando de Soto, sick with fever from Indian-inflicted wounds, died and was committed to a watery burial in the Mississippi River. About 300 Spaniards in his force managed to survive their ordeal and reached the Mexican town of Panuco in the autumn of 1543. They were empty-handed and dressed in patches of hides taken from the Indians, and they had little beyond their memories to share with their curious fellows in Mexico, Cuba, and Spain.

Many of those memories and, incidentally, most of what is known concerning de Soto's exploration of the Southeast have come down to us as chronicles written or related by survivors of the ordeal. In addition to these chronicles, a little-understood map also appears to have been based on survivor memory. It is frequently referred to as "the de Soto map" and has been reproduced here as Plate 5 (and appears as Map 1 in the Map List). One of the reasons for considering the de Soto map a survivor's memory map is that it seems to show more detailed information from the later rather than the earlier phases of the long march across the Southeast. The map also demonstrates that its anonymous author owed a large debt to Indian informants and cartographers—as can be seen most clearly in its depiction of the region's major river systems.

One of the key diagnostic traits revealing Native American cartographic influence on European maps involves the manner in which the Indians treated river systems. These treatments were often copied by the Europeans, who otherwise remained silent about their debt to Indian informants. In nature (and at macro scale), river systems are dendritic or treelike: that is, small streams join as branches to form larger trunk streams that drain discrete basins and flow into the sea or other large bodies of water. On the de Soto map, however, some large rivers are shown to be anastomosed—connected to each other—a condition that almost never occurs in nature. Notice, too, that one large drainage system flows from the mountains to both the Atlantic Ocean and the Gulf of Mexico.

The reason for these "unnatural" depictions of rivers on the de Soto map is rooted in a basic difference in the ways Europeans and Native Americans treated networks in their mapping traditions. In European maps, asymmetrical or mixed networks retained their characteristics to the extent that a particular map would allow. Trails or portages were shown as different and distinct from the rivers they connected or paralleled in a region. By Indian cartographers, however, such systems were usually shown as combined or undifferentiated. To them the overall transport or communication network was of principal concern—not whether one segment was on land and had to be walked, while another was a waterway and required a canoe for passage. Thus, the European author of the de Soto map depicted some rivers as anastomosed and weirdly branching as a result of his misunderstanding of Indian route maps. As the box labeled "Misunderstanding of Amerindians" in Figure 2 suggests, such fundamental cultural differences should be kept in mind as potential sources of error and distortion in early European maps of the Southeast. For some other examples see Plates 9, 17, 19, 24, 31, 33, and 39.

In spite of the dismal reports from de Soto's survivors, interest in the exploration of the American mainland persisted, and further expeditions were sent out under Spanish missionaries such as Fray Luis Cancer (in 1549) and military commanders such as Tristan de Luna, Angel de Villafane, and Juan Pardo. These expeditions continued to press Southeastern Indians into service as indispensable guides and informants. While more often than not the documents are silent on the fates of these reluctant participants, it is clear that at least one survivor of the de Soto entrada returned to her Coosa homeland in north Georgia as an interpreter for the de Luna expedition that was in the area in 1559–61.[42] It is doubtful that many were as fortunate.

When the French, in the 1560s, and the English, in the 1580s, decided to challenge Spain's hegemony in the Southeast, the by-then-familiar pattern of reliance on local Indians was repeated time and again. The names of two present-day North Carolina communities, Wanchese and Manteo, are daily heralds of this fact. Manteo and Wanchese were two "very impressive Indians, said to be volunteers," who accompanied Ralegh's explorers, Philip Amadas and Arthur Barlowe, on their return from Roanoke in 1584.[43] They were invaluable to Ralegh's propaganda efforts in launching England's first American colony. Similarly, on their return to Roanoke, they continued to play crucial roles as informants and guides to the English. Particularly important was Manteo, who guided Thomas Hariot and traveled with both the explorer Ralph Lane and the colony's artist-cartographer, Governor John White.

THE SEVENTEENTH CENTURY

During the first decade of the seventeenth century Captain John Smith and his Virginia associates provided lucid accounts of their heavy reliance on Powhatan Indian guides and informants as they began exploring a corner of the Southeast not penetrated by de Soto's entrada. Soon after their first landing at Jamestown, in the spring of 1607, a party led by Captain Gabriel Archer undertook an exploration of the James River by boat. When they were about eighteen miles upstream, they encountered "8 salvages [Indians] in a canoa" and made friendly contact with them.[44]

Using sign language and pantomime, Captain Archer began to interrogate the Indians in an attempt to learn something of the river and country that lay before them. One of the Indians, in Archer's words, "seemed to understand our intention, and offered with his foote to Describe the river to us." Not a mere sign language description, but a *map* sketched in the sand with his foot! Seizing the opportunity to capture this vital intelligence in a more permanent medium, Archer "gave him a pen and paper (shewing first ye use) and he layd out the Whole River from the Chesseian bay to the end of it so farr as passadg[e] was for boats." The cartographically talented Indian went on to tell the Virginians "of two Iletts in the Ryver we should passe by . . . and then come to an overfall of water, beyond that of two kyngdomes which the Ryver Runes by, then a great Distance off, the Mountaines Quirank as he named them."[45] Archer ended this part of his account somewhat cryptically by noting that beyond the mountains "by his relation is that which we expected."

Later on in the exploration, Archer revealed that "that which we expected" to find beyond the mountains was copper. At the falls, now the site of Virginia's capital, Richmond, Archer related that "now setting upon the banck by the overfall beholding the same," the Indian mapmaker "began to tell us of the tedyous travell we should have if wee proceeded any further, that it was a Daye and a half journey to Monanacah and if we went [to] Quiranck, we should get no vittailes and be tyred, and sought by all means to Disswade our . . . going any further: Also he told us that the Monanacah was his enemye and that he came down at the fall of the leafe and invaded his countrye."[46] In apparent respect for this persuasive argument, the party decided to terminate the expedition and return to base at Jamestown.

The flow of geographic intelligence from Indians to Europeans continued, however. Archer noted:

> So farr as we could discerne the river above the overfall, it was full of huge rocks: About a mile off it makes a pretty big island: It runnes up between highe Hilles which increase in height one above another so far as we sawe. Now our kynde Consort's relayton sayth (which I dare well beleeve, in that I found not any one report false of the River so farr as we tryed or that he tolde us untruth in anything ells [else] whatsoever) that a Day's journey or more, this river Devydes it selfe into two branches which both came from the mountaynes Quirank. Here, he whispered to me that theer Caquasson [copper] was gott in the bites of Rocks and between Cliffes in certayne vaynes.[47]

Archer's endorsement of the Indian guide-mapmaker's veracity was well founded rather than simply a rhetorical flourish. Archer, like many early explorers, was careful to test and double-check information provided by Indian guides. An opportunity for such checking came when the Virginians were entertained by one of Powhatan's subchiefs, "kyng Arahatec," at his riverside town. Here, Archer wrote, "I caused now our kynde Consort that described the River to us, to draw it again before Kyng Arahatec, who in every thing consented to his draught, and it agreed with his first relayton." With some feeling, Archer concluded, "This we found a faythfull fellow, he was the one that was appointed guyde for us."

In the account of his intercourse with his "faythfull fellow" guide, Archer provides a revealing example of the sort of exchange symbolized by the broad, shaded arrow in Figure 2 discussed earlier. Archer, obviously wishing for a more durable statement of what he recognized as potentially important geographical intelligence, *taught* the Indian to record his spatial knowledge with pen and ink on paper when the Indian volunteered to sketch an ephemeral map in the sand. Further, he retained this "talking leaf," to use an Indian metaphor for European written messages, and tested it against a second rendition executed before an important and independent witness—chief or "kyng" Arahatec. The significance of all of this could not have escaped the Indians.

An equally colorful example of this sort of transfer from European to Indian is provided by Captain John Smith's account of his capture by a Powhatan hunting party led by Opechancanough. In writing of being presented to their chief by his captors, Smith told how he "gave him a round Ivory double compass dyall" and described its use.[48] "Much they marvailed at the playing of the Fly and Needle, which they could see so plainely, and yet not touch it, because of the glasse that covered them," he added. Smith, eager to keep his captors amused, went on to use the "Globe-like Jewell [compass]" to demonstrate for them "the roundnesse of the earth, and skies, the spheare of the Sunne, Moone, and Starres, and how the Sunne did chase the night round about the world continually." Nor did Smith neglect mother earth in his discourse; he told the Indians of "the greatnesse of the Land and Sea, the diversitie of Nations, varietie of complexions, and how we were to them Antipodes, and many other such like matters, they all stood as amazed with admiration." Rather than amazed, the Indians were more likely baffled by Smith's animated performance, for within an hour they had tied him to a tree and drawn their bows to kill him; but, as Smith wrote, "the King holding up the Compass in his hand" signaled that he should be spared. It is possible that Smith saved his life through this impromptu lecture-demonstration, which also contributed something to the Indians' deeper understanding of the aliens now among them.

At another point in his captivity Smith described how the Indian chief Powhatan himself "took great delight in understanding the manner of our ships, and sayling the seas, the earth and skies, and of our God: What he [Powhatan] knew of the dominions he spared not to acquaint me with, as of certaine men cloathed at a place called Ocanahonan, cloathed like me, the course of our river, and that within 4 or 5 daies journey of the falles was a great turning of salt water."[49]

Following this exchange and, as Smith termed it "after good deliberation," Powhatan

> began to describe mee the Countreys beyonde the Falles, with many of the rest, confirming what not onely Opechancanoyes, and an Indian which had beene prisoner to Powhatan had before tolde mee, but some called it five dayes, some sixe, some eight, where the sayde water dashed amongst many stones and rocke, each storme which caused oft tymes the heade of the River to bee brackish: . . . Hee described also upon the same Sea a mighty Nation called Pocoughtronack, a fierce Nation that did eate men, and warred with the people of Moyaoncer, and Pataromerke, Nations upon the toppe of the heade of the Bay [Chesapeake], under his territories, where the year before they had slain an hundred; he signified their crownes were shaven, long haire in the necke, tied on a knot, Swords like Pollaxes.
>
> Beyond them he described people with short Coates, and Sleeves to the Elbowes, that passed that way in Shippes like ours. Many Kingdomes hee described mee to the heade of the Bay, which seemed to bee a mightie River, issuing from mightie Mountaines betwixt the two Seas. The people cloathed at Ocanahonan he also confirmed, and the Southerly Countries also, as the rest that reported us to be within a day and a halfe of Mangoge, two dayes of Chawwonock, 6. from Roanoke, to the south part of the backe sea: he described a countrie called Anone, where they have an abundance of Brasse, and houses walled as ours. I requited his discourse, seeing what pride hee had in his great and spacious Dominions, seeing that all he Knewe were under his Territories.[50]

It is extremely doubtful that Smith, at this time, was able to comprehend all of what he was being told concerning the geography and geopolitical alignments existing in Powhatan's world. There can be little doubt, however, that Powhatan impressed the English with the sweep of his knowledge. Writing twenty years before Smith, Ralph Lane provided a similar assessment of another well-informed Indian who was his captive rather than captor. Lane's prisoner was the chief of Chawanoac, a town located on the Chowan River in eastern North Carolina. This account, along with many similar ones, suggests that Powhatan's performance was the rule rather than exception for members of Indian elites in the Southeast. Lane wrote:

> The King of the sayd Province is called Menatonon, a man impotent in his lims, but otherwise for a Savage, a very grave and wise man, and of very singular good discourse in matters concerning the state, not only of his owne Countrey, and disposition of his owne men, but also of his neighbours round about him as wel farre as neere, and of the commodities that eche [each] Countrey yeeldeth. When I had him prisoner with me, for two dayes that we were together, he gave me more understanding and light of the Countrey than I had received by all the searches and salvages [Indians] that before I or any of my companie had had conference with: it was in March last past 1586.[51]

An intriguing commentary respecting Powhatan's efforts to extend his information field is found in a letter sent by Spain's king to his ambassador in London in 1610. The document is endorsed, "July 1, 1610. Report on Virginia to the [Spanish] council of state. Report of What Francisco Maguel, an Irishman, learned/knew in the State of Virginia, during the eight months that he was there." Maguel was an Irish spy providing intelligence on the English sailing route to Virginia. In his report he mentioned Powhatan as follows:

> The Emperor of Virginia has sixteen Kings under his dominion; he and all his subjects deal peaceably with the English and attend a market which the English hold daily near the Fort and bring to them there the commodities of the country to exchange. . . . The natives of this country are a robust, well disposed race; and generally go about dressed in very well tanned deer skins as they understand very well how to prepare them. Their arms are bows and arrows. The Emperor sends every year some men by land to West India and to Newfoundland and other countries, to bring him news of what is going on there. And these messengers report that those who are in West India treat the Natives very badly and as slaves, and the English tell them that those people are Spaniards, who are very cruel and evil disposed. The English have some boys there among these people to learn their language, which they already know, at least some of them, perfectly. The Emperor sent one of his sons to England, where they treated him well and he returned him once more to his own country, from which the said Emperor and his people derived great contentment thro' the account which he gave of the kind reception and treatment he received in England. . . . This narrator returned to England in the same vessel with the said son of the Emperor.[52]

Powhatan's "son" was Namontack, who sailed for England with Captain Newport on April 10, 1608, and arrived there on May 21. He returned to Virginia with Newport, arriving in late September of the same year. Whether Powhatan's metaphorical "eyes" and "ears" directly reached as far as the Caribbean and Newfoundland is highly problematic. It is, however, entirely possible that he and other powerful Indian leaders shared intelligence and information along the same extensive network linkages that have most often been acknowledged as operating in the areas of trade and commerce. It is also possible that Maguel was embellishing his story in an effort to gain added recognition with his employers.

In another portion of his report, the spy Maguel discussed the assurances the Indians gave the English "that they can easily take them to the South-Sea by three routes."[53] One route was up the James River to its head and then a ten day's march, presumably to the west. The second route involved a long river flowing to the South Sea that was reported as being within a day and a half's march from the head of the James. The third route appears to have been up the Susquehanna River to its head and on to another "large river, which flows to the South Sea." If this Susquehanna construct is accepted, it is one more confirmation that the "South Sea" the Indians were describing was in fact one of the Great Lakes rather than the Pacific Ocean of the Englishmen's dreams.[54]

The vain hope of easily reaching the shores of the "Sea of China and the Indies" by traveling overland from Virginia lived on, however, well into the early eighteenth century. In his 1651 map caption placed under a medallion portrait of Sir Francis Drake in that sea behind "Virginia," John Farrer promised that New Albion on the Pacific could be reached "in ten dayes march with 50 foote and 30 horsmen from the head of Jeames River, over those hills and through the rich adjacent Vallyes beautyfied with as proffitable rivers, which necessarily must run into ye peacefull Indian Sea." Farrer's map is included here as Plate 29.

When later he encountered the "Sasquesahanock" (i.e., Susquehanna) Indians, Smith had reason to become more skeptical of the extent of Powhatan's knowledge. In his account of "the second voyage to discover the Bay," Smith told how the Susquehanna provided "many descriptions and discourses . . . of Atquanahucke, Massawomecke, and other people, signifying they inhabit the river of Cannida [Canada], and from the French do have their hatchets, and such like tools by trade." "These Indians," he concluded, "knowe no more of the territories of Powhatan than his name, and he as little of them."[55]

Much more could be written concerning Captain John Smith's interactions with Indian guides and informants in the early months of the founding of Virginia, but it will suffice to conclude with a final and exceedingly revealing episode in his ongoing relationship with Powhatan. In the autumn of 1608 Smith had been elected the colony's president and led a delegation to meet with the still-powerful Indian leader in his own country, Powhatan having refused to come to Jamestown. After his famous description of a greeting by nearly naked Indian dancing maidens, Smith continued:

> The next day came Powhatan; Smith delivered his message of the presents sent him, and redelivered him Namontack desiring him [Powhatan] to come to his Father Newport to accept those presents, and conclude their revenge against the Monacans, whereunto the subtile Salvage thus replied.
>
> If your king have sent me presents, I also am a king, and this my land, 8 daies I will stay to receave them, yor father [Newport] is to come to me, not I to him, nor yet to your fort, neither will I bite at such baite: as for the Monacans, I can revenge my owne injuries, and as for Atquanuchuck, where you say your brother was slain, it is a contrary way from those parts you suppose it. But for any salt water beyond the mountaines, the relations you have had from my people are false, whereupon *he began to draw plots upon the ground* according to his discourse of all those regions; many other discourses they had.[56]

When Smith's accounts of the interactions between Virginians and Indians are purged of their romantic hyperbole, heavy Eurocentric bias, and egocentric exaggeration, they reveal Powhatan to be a well-informed political leader holding sway over an expanding Indian polity with strong resemblances to the Mississippian chiefdoms encountered earlier by de Soto. In his dealings with the European intruders, Powhatan can be seen as following a perfectly rational and frequently intelligent strategy designed to cope with a group of poorly organized but militarily potent aliens who threatened to interfere in the drama unfolding as he maneuvered to consolidate and enlarge his chiefdom. Fortunately Smith was a careful, if not always fully understanding, chronicler of this drama, who realized the inestimable value of the Indians' geographical intelligence.

Even more fortunate is the fact that Smith synthesized this intelligence graphically in the form of a map (Figure 3) as well as verbally in his several published tracts. Experts in the history of cartography are laudatory in their evaluations of John Smith's 1612 Map of Virginia. Coolie Verner, for example, described it as "the most important map to appear in print during the period of early settlement and the one map of Virginia that has had the greatest influence upon map making for a longer period of time."[57] As Smith himself wrote, "This annexed mappe . . . will present to the eie, the way of the mountaines and current of the rivers, with their several turnings, bays, shoules, Isles, Inlets, and creekes, the breadth of the waters, the distances of places and such like." Unlike many European explorers, Smith is refreshingly free in his acknowledgment of the debt he owed to the Indians in the accomplishment of his map. On this score, in his next sentence he wrote, "In which Mappe observe this, that as far as you see the little Crosses on rivers, mountaines, or other places have been discovered; the rest was had by information of the Savages or set downe according to their instructions."[58]

Smith's no longer extant manuscript first map of Virginia was forwarded to London when Captain Newport sailed in late 1608. He described it as a "mappe of the Bay and Rivers" in the draft report he titled "Relation of the Countries and Nations that inhabit them." He doubtlessly prepared the map and report in response to instructions issued in 1606 by the London Council. The council had directed that considerable effort be expended in exploring the land included in its charter. Among their many instructions and admonishments, the colonists were advised to be certain that "your discoverers that pass over Land with hired Guides . . . look well to them that they slip not from them and for more Assurance let them take a Compass with them

FIGURE 3. John Smith's 1612 Map of Virginia, indicating the area depicted on the basis of Indian sources

and Write Down how far they Go upon Every point of the Compass for that Country having no way nor path if that Your Guides Run from you in the Great Woods or Deserts you Shall hardly Ever find a Passage back."[59]

Smith's adventure with his "round Ivory double compass dyall" is proof that he was equipped to conduct the compass traverses being advised by Virginia's planner-promoters. The primary motivation for a program of exploration and mapping on the part of the original Jamestown settlers was found, however, in a quite different instruction. It was one predicated on the then-widely-held belief that the Atlantic and Pacific Oceans were separated by only a narrow isthmus in the belt of latitude embraced by the Virginia charter. Smith and his fellows were to "observe, if you can, whether the river on which you plant doth spring out of mountains or out of lakes. If it be out of any lake, the passage to the other sea will be more easy, and is like enough, that out of the same lake you shall find some spring which runs the contrary way towards the East Indian Sea."[60]

No copy of Smith's manuscript map of 1608 is known to have survived, so his position in the annals of the history of cartography rests largely on the map engraved by William Hole for publication in 1612. It accompanied Smith's printed tract titled, "A Map of Virginia With A Description of the Country, the Commodities, People, Government and Religion."[61] Oriented with the west at its top, the map measures 16" x 12½" and has an elaborate "Scale of Leagues and halfe Leagues" located to the east of the Delmarva Peninsula, which lacks a distinctly drawn coastline on its Atlantic littoral. Two arresting illustrations flank the map itself. To the right is the figure of an Indian holding his bow in one hand and a cudgel in the other. The caption at the figure's foot reads, "The Sasquesahanougs are a Gyant people and thus atyred." In the upper left corner is an interior perspective view of a structure with a large group of Indians around a council fire. As the caption explains, it shows the "State & fashion" in which Powhatan received Smith when he "was delivered to him as a prisoner" in 1607.

There is no indication that Smith or anyone associated with the original Jamestown colony prepared sketches of the Indians or their houses. As a result, for his models engraver Hole was forced to adapt existing De Bry engravings of Indians first made for Thomas Hariot's *A briefe and true report of the new found land of Virginia* published in 1588. The "Gyantlike" Susquehanna Indian is closely modeled on De Bry's "A weroan or great Lorde of Virginia," and the Powhatan scene is an amalgam of De Bry's "Their manner of pra[y]inge with Rattels abowt t[h]e fyer," "The[i]r Idol Kiwasa," and "The Tombe of their Werowans or Cheiff Lordes."[62]

In spite of this plagerism, it is the Indian-related content of Smith's map that gives it its principal value. Writing in an era preceding the advent of modern archaeology, J. Hotchkiss emphasized this in his evaluation of Smith's map. Recognized in his day as Virginia's leading geographical and archaeological expert, Hotchkiss, in 1883, observed, "I am sorry to say that about the only information we have concerning the location of Indian tribes at the time of the settlement of Virginia is to be found on Smith's map; it is a marvel of results in representation of outline compared with the time occupied in procuring information. Smith had all the important features of our wonderfully developed [coastal plain] well shown."[63]

What Hotchkiss and other scholars consistently neglect to acknowledge is the role played by the Indians—through their service as guides and informants as well as mapmakers—in Smith's exploring and map compilation accomplishments. Simply put, the Indians were the reason Smith could produce such an invaluable and, for its day, extraordinarily accurate depiction in months rather than the years that might otherwise have been required. This failure to provide due acknowledgment cannot be blamed on Smith. In his accompanying tract, as noted above, as well as on the face of the 1612 map, he freely indicated his debt to the Indians. On the map is included a large number of small Maltese crosses that are explained in the map's legend as follows: "Signification of these markes, To the crosses hath bin discovered—what beyond is by relation" (see Figure 4). "By relation" meant compiled from verbal and cartographic descriptions provided by Indians since there were no other sources descriptive of the terrain beyond where Smith and his associates themselves had traveled. Figure 3 emphasizes those areas beyond the Maltese crosses on Smith's map. In a crude way it provides graphic testimony acknowledging what the Indians provided "by relation" as Smith compiled his much-lauded map. Recall that in his tract he had further clearly stated that his map's coverage beyond the crosses "was had by information of the Savages or set downe according to their instructions."[64]

It is interesting to note that the Maltese crosses on Smith's map are a case of the cartographer's art imitating nature. Smith told of how, during his exploration of Chesapeake Bay, "in all those places and the furthest we came up the rivers, we cut in the trees so many crosses as we would, and in many places made holes in trees, wherein we writ notes, and in some places crosses of brasse, to signifie to any, Englishmen had beene there."

That Smith's map can be considered a "silent" witness to Southeastern Indian geographical knowledge and cartographic capability is due only to the lack of scholarly acknowledgment that has been afforded Smith's explicit statements in the writing of later historians. The 1612 Smith Map of Virginia is long overdue for recognition as an eloquent and powerful memorial to the Powhatan Indians' contributions to the exploration and mapping of this quadrant of the Southeast.

A few decades later, in 1649, a party of Englishmen bound for Virginia were aboard a ship whose captain got lost after a long spell of bad weather that had made it impossible for him to determine their latitude. When they made a landfall on the Delmarva coast near Chincoteague Island, the party's leader, Henry Norwood, requested that they be allowed to land rather than continuing with the floundering ship. After a harrowing period of hardship the survivors reached the territory of the "king of Kickotank." This was probably the chief of the Awascecencas, an Eastern Shore band who lived in what is now Accomack County.

Norwood and the Indian leader found it impossible to understand one another's language until, as Norwood related: "The King attacked me again, with reiterated attempts to be understood, and I thought by these three or four days conversation, I had the air of his expression much more clear and intelligible than at first. His chief drift for the first essay seemed to be a desire to know which way we were bound, whether north or south; to which I pointed to the south."[65] Norwood continued by explaining that this revelation seemed to please the Indians, one of whom then "took up a stick, with which he made divers circles by the fireside,

FIGURE 4. Detail of John Smith's 1612 map, showing the legend explaining the area depicted from Indian sources

and then holding up his finger to procure my attention, he gave every hole a name; and it was not hard to conceive that the several holes were to supply the place of a sea-chart, shewing the situation of all the most noted Indian territories that lay to the southward of Kickotank." Where the spoken word and sign language had failed, cartography provided a communication bridge across the broad cultural gulf separating the English survivors from their Indian hosts.

Norwood went on to state: "That circle that was most southerly, he called *Achomack*, which tho' he pronounc'd with a different accent from us, I laid hold on that word with all demonstration of satisfaction I could express, giving them to understand, that was the place to which I had desire to be conducted." The friendly Indian chief was described as "in a strange transport of joy" to have finally achieved this communication breakthrough. The chief then dispatched "a lusty young man" to contact the English settlers in Northampton County, Norwood's destination. In due course an Englishman named Jenkins Price arrived to escort Norwood and his party to their destination.

At this point attention should be directed to the fact that Norwood mentioned that the Indian cartographer drew *circles* on his map by the fireside to represent the several Indian towns that lay to the southward. Analysis of this episode and several others discussed below suggests that Indian cartographers normally employed circular or ovoid bounded spaces when representing Indian groups or polities on their maps. This seems to have occured to the extent that it can be employed as a diagnostic trait of Indian cartographic convention. Conversely, some Native Americans appear to have employed rectangularly bounded areas to represent European settlements or colonies on their maps.

Norwood and his fellow survivors were far more fortunate in the treatment they received from Virginia's coastal Indians than was a group of Englishmen and their slaves that was cast upon the shore of Florida later in the century. Rather than vigorously trying to inform Jonathan Dickinson and his shipwrecked fellows correctly, the Florida Indians tried hard to misinform them. Dickinson wrote of how, once on shore and in captivity, the English "began to enquire after St. Augustine . . . and St. Lucie," which they knew lay to the north.[66] He continued, "but they cunningly would seem to persuade us that they both lay to the southward." The Indians' chief—or, as he was styled by Dickinson, "Casseekey"—was determined to bring the English to his town that lay to the south of where they had wrecked, and it took a great deal of persuading before he would allow them to begin their harrowing progress northward along the beaches of Florida from Jupiter Inlet to St. Augustine. Although Dickinson never admitted it, he and his fellow survivors might well have fared far better had they followed the Casseekey's urging to go south and sought rescue by Cuban fishermen who frequented the Florida Keys and adjacent shores.

It was not until explorations were undertaken by a German scholar-physician in the service of Virginia's governor in 1670 that anyone approached Captain John Smith for quality of insight and understanding of the Indians of the Southeast. The man in question was John Lederer, who had found his way from Hamburg to Virginia as a curious young adventurer. His book, *The Discoveries of John Lederer, In three several Marches from Virginia, To the West of Carolina, And other parts of the Continent*, was published in London in 1672 and ranks with Smith's writings as a revelatory source on the role of Indians in the early exploration and mapping of the Southeast.[67] The map Lederer prepared to accompany his book is included here as Plate 36. (See also the description of it in the Map List at number 68.)

After beginning his book with a short but enlightened exposition on the physiography of eastern North America, Lederer provides a section titled "Of the Manners and Customs of the Indians inhabiting the Western parts of Carolina and Virginia." It is here that one finds much of interest concerning Indian systems of record keeping and symbolic modes of communication:

> But before I treat of their ancient Manners and Customs, it is necessary I should shew by what means the knowledge of them hath been conveyed from former ages to posterity. Three ways they supply their want of letters: first by Counters, secondly by Emblemes or Hieroglyphicks, thirdly by Tradition delivered in long Tales from father to son, by which being children they are made to learn by rote.
>
> For Counters, they use either Pebbles, or short scantlings of straw or reeds. Where a Battel has been fought, or a colony seated, they raise a small Pyramid of these stones, consisting of the number slain or transplanted. Their reeds and straws serve them in Religious Ceremonies: for they lay them orderly in a Circle when they prepare for Devotion or Sacrifice; and that performed, the Circle remains still; for it is Sacreledge to disturb or to touch it: the disposition and sorting of the straws and reeds, shew what kind of Rites have there been celebrated, as Invocation, Sacrifice, Burial, etc.
>
> The faculties of the minde and body they commonly express by Emblems. By the figure of a Stag, they imply swiftness; by that of a Serpent, wrath; of a Lion, courage; of a dog, fidelity; by a Swan, they signifie the English, alluding to their complexion, and flight over the Sea.
>
> An account of Time, and other things, they keep on a string or leather thong tied in knots of several colours.[68]

Continuing his discourse with a consideration of the Indians' religious beliefs and creation myth, Lederer concluded:

> Though they want those means of improving Humane Reason, which the use of Letters affords us; let us not therefore conclude them wholly destitute of Learning and Sciences: for by these little helps which they have found, many of them advance their natural understanding to great knowledge in Physick, Rhetorick, and Policie of Government: for I have been present at several of their Consultations and Debates, and to my admiration have heard some of their Seniors deliver themselves with as much Judgement and Eloquence as I should have expected from men of civil education and Literature.[69]

It was while embarked on his second and longest "March," an expedition "From the Falls of Powhatan, alias James-River in Virginia, to Mahock in the Apalataean Mountains," that Lederer came to appreciate Indian geographic knowledge and wayfinding skills, as he related in the following manner:

> Here enquiring the way to the Mountains, an ancient Man described with a staffe two paths on the ground; one pointing to the Mahocks, and other to the Nahyssans; but my English Companions slighting the Indian's direction, shaped their course by the compass due West, and therefore it fell out with us, as it does with those Land-Crabs, that crawling backwards in a direct line, avoid not the Trees that stand in their way, but climbing over their very tops, come down again on the other side, and so after a day's labour gain not above two foot of ground. Thus we obstinately pursuing a due West course, rode over steep and craggy Cliffs, which beat our Horses quite off the hoof. In these Mountains we wandered from the Twenty fifth of May till the Third of June, finding very little sustenance for Man or Horse; for these place are destitute of Grain and Herbage.[70]

Lederer, like Smith before him, was convinced that access to the Pacific Ocean was within reach somewhere just beyond the mountains of western Virginia. It is revealing to observe how this strong preconception may have influenced his comprehension and interpretation of Indian geographical intelligence. In an episode that exemplifies the "Misunderstanding of Amerindians" box in Figure 2, he told of meeting "four stranger-Indians, whose Bodies were painted in various colours with figures of Animals whose likeness I had never seen." He continued, "by some discourse and signes which passed between us, I gathered that they were the only survivours of fifty, who set out in company from some great Island, as I conjecture, to the Northwest; for I understood that they crossed a great Water, in which most of their party perished by tempest, the rest dying in the Marshes and Mountains by famine and hard weather, after a two months travel by Land and Water in quest of this Island of Akenatzy."[71] The "Island of Akenatzy" is usually identified as the largest of the islands at the confluence of the Dan and Staunton Rivers near Clarksville, Virginia.

Lederer's strongly held geographical preconceptions led him to what he termed "the most reasonable conjecture that I can frame out of this Relation," namely, "that these Indians might come from the Island of new Albion or California, from whence we may imagine some great arm of the Indian ocean or Bay stretches into the Continent towards the Apalataean Mountains into the nature of a mid-land Sea, in which many of these Indians might have perished."[72] Obviously the many European maps showing the vestiges of Verrazano's Sea and California as an island had left their mark on Lederer.

With his well-established mental world map, it was easy for the German explorer to confuse Indian descriptions of the wavelike topography of the Ridge and Valley or other portions of the Appalachian system with imagined waves of an embayment of the Pacific Ocean lapping the continent somewhere on the flanks of the mountains he could see in the distance beyond the Shenandoah Valley. He describes how he arrived at such a conclusion: "I have heard several Indians testifie, that the Nation of Rickohockans, who dwell not far to the Westward of the Apalataean Mountains, are seated upon a Land, as they term it, of great Waves; by which I suppose they mean the Sea-shore."[73]

Heir as he was to the prevailing geographical beliefs of his culture and period, Lederer was unable to accept what the Indians were telling him in its literal sense, that is, that they lived in a *wavelike land*—a perfectly reasonable description of the Ridge and Valley physiographic province or other portions of the Appalachian system, likened by more than one observer to a congealed storm at sea.

To his credit, however, Lederer realized that access to the western or Pacific Ocean was not to be so easily attained as many of his contemporaries thought. On this score he wrote, "They are certainly in great errour, who imagine that the Continent of North-America is but eight or ten days journey over from the Atlantick to the Indian Ocean."[74] In an even more interesting line of argument he went further and cast serious doubt on the prevailing idea that navigable rivers rising on the western flank of the Appalachians could be followed to their outlets in the Pacific as those on the east were traced to the Atlantic.[75]

Lederer wrote of how the Indians had helped to shape his ideas on this controversial topic:

> Nevertheless, by what I gathered from the stranger Indians at Akenatzy of their Voyage by Sea to the very Mountains from a far distant Northwest Country, I am brought over to their opinion who think that the Indian Ocean does stretch an Arm or Bay from California into the Continent as far as the Apalataean Mountains answerable to the Gulfs of Florida and Mexico on this side. Yet I am far from believing with some, that such great and Navigable Rivers are to be found on the other side of the Apalataeans falling into the Indian Ocean, as those that run from them to the Eastward. My first reason is derived from the knowledge and experience we already have of South-America, whose Andes send the greatest Rivers in the world as the Amazones and Rio de la Plata, &c. into the Atlantick, but none at all into the Pacifique Sea. Another Argument is, that all our Waterfowl which delight in Lakes and Rivers, as Swans, Geese, Ducks, &c. come over the Mountains from the Lake of Canada, when it is frozen over every Winter, to our fresh Rivers; which they would never do, could they finde any on the other side of the Apalataeans.[76]

At about the same time that John Lederer was probing the Appalachians for the Virginians, farther to the north French eyes were also turned to the west. Writing in 1670 the expansionist intendant of New France, Jean Talon, stated, "I have dispatched persons of resolution, who promise to penetrate farther than has ever been done; the one to the West and Northwest of Canada, and the others to the Southwest and South. Those adventurers are to keep journals . . . they are to take possession, display the King's arms and to draw up proces verbaux to serve as titles."[77] As part of this French thrust, Louis Jolliet and missionary father Jacques Marquette were assigned to follow up on Indian reports of a great river flowing south from the country to the west of Lake Superior.

Needless to say, Indian informants played crucial roles in

the planning as well as the execution of their exploration. In *The Jesuit Relations* Marquette is described as one who "had long premeditated this Undertaking, influenced by a most ardent desire to extend the Kingdom of Jesus Christ." When it came to his preparations for so lofty an undertaking, Marquette wrote in a more practical tone, "We obtained all the information that we could from the savages who had frequented those regions; and we even traced out from their reports a Map of the whole of the New Country; on it we indicated the rivers which we were to navigate, the names of the peoples and the places through which we were to pass, the Course of the great River, and the direction we were to follow when we reached it."[78]

In the company of Miami Indian guides, he and Jolliet successfully navigated the Mississippi from the Wisconsin River downstream as far as the Arkansas River, where local Indian informants caused them to break off their progress. Here, Marquette wrote, "they [the Indians] assured us that we were no more than ten days' journey from the sea; that they bought cloth and all other goods from the Europeans who lived to the East, that these Europeans had rosaries and pictures; that they played upon instruments; that some of them looked like me, and had been received by these savages kindly."[79] Believing that they were dangerously close to Spanish settlements on the Gulf of Mexico, the small party decided to return to New France with a report of their discovery rather than risk capture. On the return trip the Frenchmen wisely accepted Indian advice and followed a route up the Illinois River to the Chicago River and Lake Michigan. This would be the route followed later by La Salle in his own descent of the Mississippi to the Gulf of Mexico.

Thanks to a deliberate policy of retrenchment, the French did not immediately follow up on the accomplishment of Marquette and Jolliet. Not until René-Robert Cavelier, Sieur de La Salle, received a royal grant of a trading monopoly in return for finding the mouth of the Mississippi did European exploration of the Indians' "great river" resume. With his royal privilege in hand, La Salle arrived in Canada in the autumn of 1678 and began to lay the foundation for his trading empire in the American heartland. During 1679–80 he built the ill-fated sailing ship *Griffon* on the Niagara River to provide the Great Lakes link in his grandiose scheme. Fort Crèvecoeur was built near the site of Peoria, and finally, in the winter of 1681–82, La Salle was ready to lead a small band of French and Indians down the Father of Waters to its mouth, wherever that might be.

Like his predecessors, Jolliet and Marquette, La Salle made frequent and sophisticated use of Indian informants and mapmakers to gain a detailed appreciation of what to anticipate en route to terra incognita. In one instance, for example, he had a young Illinois Indian warrior who had just returned from a raid downriver draw "a fairly exact map of it [the Mississippi] with charcoal." The same informant added to the value of his map when he "told us the names of the tribes living on its banks and of the rivers emptying into it."

Thanks to the recollections of one of La Salle's lieutenants, Father Louis Hennepin, the details of the process by which geographical intelligence was transmitted from Indian percipient to European recipient can be outlined here. Hennepin wrote:

> A few days later, fortunately we found means of ridding our men of the false impressions given them by the Illinois at the instigation of Monso, the Miami Chief. Indians of distant tribes arrived at the Illinois village. One of them assured us of the beauty of the great Colbert or Mississippi River. This report was confirmed by several of these Indians and by one Illinois who told us confidentially on our arrival that the river was navigable.

Even in light of these favorable reports many of the French remained apprehensive and under the sway of Monso's "false impressions" concerning the possibility of safely navigating the river downstream. Hennepin went on to say: "We wanted the Illinois themselves to confess the truth, although we had learned that they resolved in council to tell us always the same thing." The communication impasse was broken when, as he wrote, "A favorable occasion soon occured." It came in the following form:

> A young Illinois warrior who had taken some captives in the territory to the south was coming back ahead of his comrades and when he passed our stocks, was given corn to eat. Since he was returning from the lower part of the Colbert [Mississippi] River with which we pretended to be acquainted, the young man made us a fairly exact map of it with charcoal. He assured us he had followed its course in his pirogue and there were no falls or rapids all the way to the sea which the Indians call the great lake. Since the river became very broad, there were in places sand bars and mud which partially obstructed it. He also told us the names of the tribes living on its banks and of the rivers emptying into it. I wrote them down.[80]

Continuing his account, Hennepin revealed just how valuable this unexpected intelligence windfall was to the French:

> The next morning after our public prayers, we went to the village, where we found the Illinois assembled in the wigwam of one of the most important Indians. He was feasting them on bear meat, a food of which they are very fond. They made room for us in the center of a fine rush mat which they laid for us. We had one of our men who knew their language say that we wished to inform them that the Maker of all things, whom they call the great Master of Life, takes particular care of Frenchmen and had revealed to us the condition of the great river we named Colbert, concerning which we had been at a loss to know the truth ever since they had told us the river was not navigable. We then repeated to them what we had learned the preceding day.[81]

Obviously the local Indians had been deliberately misinforming the Frenchmen—probably in an effort to protect their lucrative intermediate position in the trading network connecting the lower Mississippi Valley and the Southeast with the Great Lakes basin. Armed with what they had learned from the young Illinois informant, the French made the locals believe that they "had learned all this by some extraordinary means. . . . They told us their only reason for hiding the truth was their desire to keep our chief [La Salle] and the gray-gowns or bare feet (as all the American Indians call Franciscan monks). They admitted everything that we had learned from the young warrior and since they have not denied it." While it can only be speculated upon, it seems unlikely that very many European explorers were as fortunate as La Salle in so successfully overcoming the obstacle of deliberate "Amerindian Misinformation," to use the terminology suggested in Figure 2 above.

The episode reported by Hennepin was no isolated incident for La Salle. Even as he was nearing his tragic death in the wilderness of eastern Texas, the French explorer continued to obtain geographical intelligence and maps from the Indians he encountered. The priest Anastasius Douay wrote of how, at the populous village of the Coenis Indians, "the sieur de la Salle made them draw on bark a map of their country, of that of their neighbors, and of the river Colbert or Mississippi, with which they are acquainted."[82]

When Pierre le Moyne, Sieur d'Iberville, rediscovered and correctly located the mouth of the Mississippi a dozen years after La Salle's murder, he too relied extensively on Indian guides and maps. These sources were crucial in his clearing away the confusion caused by La Salle's duplicity and the resulting erroneous depictions that showed the river finding its way to the sea on the coast of Texas. In his journal Iberville wrote that he was "convinced that so many different Indians, especially those who drew maps for me could not have lied. . . . They have presented abundant proof that Sieur de La Salle . . . had descended this river to the sea."[83]

In one instance Iberville reveals the thoroughness with which he tested his Indian mapmaker-informants. He told of taking a Taensa Indian aboard his longboat "to have him make a map of the country and see whether he will not talk differently when he is away from the others."[84] Once underway and on his own, the Indian assured Iberville that "the Malbanchya—that is the name of the Myssysypy—does not fork between here and the Acansa where he has been." Then, as Iberville related, "he made a map for me on which he shows me that on the third day of our journey we shall come to a river on the left side that is named Tas[s]enocogoula [Red River of Louisiana], in which he shows two branches: On the west one are eight villages."[85]

Iberville's Indian-assisted rediscovery permitted the great French cartographers Claude and Guillaume Delisle to publish the first European maps showing the course of the lower Mississippi with reasonable accuracy.[86] See the Introductory Essay and the Map List for discussion of the Delisles and their work. Plates 43 and 47 show Guillaume's 1703 and 1718 maps with their improved depiction of the Mississippi and the Southeast. They should be compared with the earlier Sanson and Du Val maps, Plates 31 and 33.

At the opening of the eighteenth century their colonies in the upper and lower Mississippi Valley and along the Gulf coast saw the French well poised to begin exploration and eventual commercial exploitation of the Southeast in competition with the Spanish and British. The experiences of one French officer—although he did not participate directly in empire-building efforts in the Southeast and was familiar with Indians farther north—are instructive and worthy of mention here. That officer was the Baron de La Hontan, who served in New France in the 1680s and 1690s and published a very popular account of his experiences there titled *New Voyages to North America*, which first appeared in both English and French in 1703.

In a section of his book devoted to "A Short View of the Humors and Customs of the Savages," La Hontan included the following observation:

> They are as ignorant of *Geography* as of other *Sciences*, and yet they draw the most exact Maps imaginable of the Countries they're acquainted with, for there's nothing wanting in them but the Longitude and Latitude of Places: They set down the the True North according to *the pole Star*; The Ports, Harbours, Rivers, Creeks and Coasts of the Lakes; the Roads, Mountains, Woods, Marshes, Meadows, &c. counting the distances by Journeys and Half-journeys of the Warriors, and allowing to every Journey Five Leagues. These *Chorographical Maps* are drawn upon the Rind of your *Birch Tree*; and when the Old Men hold a Council about War or Hunting they're always sure to consult them.[87]

La Hontan's many other illuminating experiences with the Indians cannot be considered here, but this observation does much to confirm that cartographic skills were not restricted to the Indians of any particular region.

THE EIGHTEENTH CENTURY

A young Dutchman who emigrated to Louisiana two decades after its founding by Iberville, encountered and wrote about a truly unique Indian explorer—that is, an Indian who had traveled vast distances to satisfy his curiosity about his own origins. The Dutchman was Antoine Simon Le Page Du Pratz, who spent sixteen years in the French colony along the lower Mississippi River before returning to France in 1734. He is best known as the author of a descriptive "history" of Louisiana that was originally published in Paris in 1758. In 1763 a two-volume edition appeared in English, and in 1774 the one-volume *History of Louisiana or of the Western Parts of Virginia and Carolina* was published in London. Le Page spent about half of his stay in Louisiana living with the Natchez Indians, and what he wrote concerning them is generally acknowledged to be the best and most accurate information available on this interesting group.

Like many Europeans, Le Page was curious about the origins and migration myths of the Indians with whom he dealt. Unsatisfied with what he was being told locally on this score, he "made great inquiries to know if there was any wise old man among the neighbouring nations, who could give me further intelligence about the origin of the natives."[88] Such an expert, named Moncacht-ape, was found living among the Yazoo Indians farther up the river. Le Page described him as "remarkable for his solid understanding and elevation of sentiments." He went on to compare Moncacht-ape "to those first Greeks, who travelled chiefly into the east to examine the manners and customs of different nations, and to communicate to their fellow-citizens, upon their return, the knowledge which they had acquired."[89]

Fortunately for the reader of today, Le Page recorded the account that Moncacht-ape related to him. To our detriment in trying to reach an understanding of the Native Americans, all too few Europeans were as interested in them and as sensitive to their ideas and feelings as was Le Page Du Pratz, the author of the following version of an Indian Argonaut's account of his amazing exploration of the North American continent:

> I had lost my wife, and all the children whom I had by her, when I undertook my journey towards the sunrising. I set out from my village contrary to the inclinations of all my relations, and went first to the Chicasaws, our friends and neighbours. I continued among them several days to inform myself whether they knew whence we all came, or at least whence they themselves came; they, who were our elders; since from them came the language of the country. As they could not inform me, I proceeded on my journey. I reached the country of the Chaouanous, and afterwards went up the Wabash or Ohio, almost to its source, which is in the country of the Iroquois or Five Nations. I left them however towards the north; and during the winter, which in that country is very severe and very long, I lived in a village of the Abenaquis, where I contracted an acquaintance with a man somewhat older than myself, who promised to conduct me the following spring to the Great Water. Accordingly when the snows were melted, and the weather was settled, we proceeded eastward, and, after several days journey, I at length saw the Great Water, which filled me with such joy and admiration that I could not speak. Night drawing on, we took up our lodging on a high bank above the water, which was sorely vexed by the wind, and made so great a noise that I could not sleep. Next day the ebbing and flowing of the water filled me with great apprehension; but my companion quieted my fears, by assuring me that the water observed certain bounds both in advancing and retiring. Having satisfied our curiosity in viewing the Great Water, we returned to the village of the Abenaquis, where I continued the following winter; and after the snows were melted, my companion and I went and

viewed the great fall of the river St. Laurence at Niagara, which was distant from the village several days journey. The view of this great fall at first made my hair stand on end, and my heart almost leap out of its place; but afterwards, before I left it, I had the courage to walk under it. Next day we took the shortest road to the Ohio, and my companion and I cutting down a tree on the banks of the river, we formed it into a pettiaugre, which served to conduct me down the Ohio and the Missisippi, after which, with much difficulty, I went up our small river; and at length arrived safe among my relations, who were rejoiced to see me in good health.

This journey, instead of satisfying, only served to excite my curiosity. Our old men, for several years, had told me that the antient speech informed them that the Red Men of the north came originally much higher and much farther than the source of the river Missouri; and as I had longed to see, with my own eyes, the land from whence our first fathers came, I took my precautions for my journey westwards. Having provided a small quantity of corn, I proceeded up along the eastern bank of the river Missisippi, till I came to the Ohio. I went up along the bank of this last river about the fourth part of a day's journey, that I might be able to cross it without being carried into the Missisippi. There I formed a Cajeux or raft of canes, by the assistance of which I passed over the river; and next day meeting with a herd of buffaloes in the meadows, I killed a fat one, and took from it the fillets, the bunch, and the tongue. Soon after I arrived among the Tamaroas, a village of the nation of the Illinois, where I rested several days, and then proceeded northwards to the mouth of the Missouri, which, after it enters the great river, runs for a considerable time without intermixing its muddy waters with the clear stream of the other. Having crossed the Missisippi, I went up the Missouri along its northern bank, and after several days journey I arrived at the nation of the Missouris, where I staid a long time to learn the language that is spoken beyond them[.] Going along the Missouri I passed through meadows a whole day's journey in length, which were quite covered with buffaloes.

When the cold was past, and the snows were melted, I continued my journey up along the Missouri till I came to the nation of the West, or the Canzas. Afterwards, in consequence of directions from them, I proceeded in the same course near thirty days, and at length I met with some of the nation of the Otters, who were hunting in that neighbourhood, and were surprised to see me alone. I continued with the hunters two or three days, and then accompanied one of them and his wife, who was near her time of lying-in, to their village, which lay far off betwixt the north and west. We continued our journey along the Missouri for nine days, and then we marched directly northwards for five days more, when we came to the Fine River, which runs westwards in a direction contrary to that of the Missouri. We proceeded down this river a whole day, and then arrived at the village of the Otters, who received me with as much kindness as if I had been of their own nation. A few days after I joined a party of the Otters, who were going to carry a calumet of peace to a nation beyond them, and we embarked in a pettiaugre, and went down the river for eighteen days; landing now and then to supply ourselves with provisions. When I arrived at the nation who were at peace with the Otters, I staid with them till the cold was passed, that I might learn their language, which was common to most of the nations that lived beyond them.

The cold was hardly gone, when I again embarked on the Fine River, and in my course I met with several nations, with whom I generally staid but one night, till I arrived at the nation that is but one day's journey from the Great Water on the west. This nation live in the woods about the distance of a league from the river, from their apprehension of bearded men, who come upon their coasts in floating villages, and carry off their children to make slaves of them. These men were described to be white, with long black beards that came down to their breasts; they were thick and short, had large heads, which were covered with cloth; they were always dressed, even in the greatest heats; their cloaths fell down to the middle of their legs, which with their feet were covered with red or yellow stuff. Their arms made a great fire and a great noise; and when they saw themselves outnumbered by Red Men, they retired on board their large pettiaugre, their number sometimes amounting to thirty, but never more.

Those strangers came from the sun-setting, in search of a yellow stinking wood, which dyes a fine yellow colour; but the people of this nation, that they might not be tempted to visit them, had destroyed all those kind of trees. Two other nations in their neighbourhood however, having no other wood, could not destroy the trees, and were still visited by the strangers; and being greatly incommoded by them, had invited their allies to assist them in making an attack upon them the next time they should return. The following summer I accordingly joined in this expedition, and after traveling five long days journey, we came to the place where the bearded men usually landed, where we waited seventeen days for their arrival. The Red Men, by my advice, placed themselves in ambuscade to surprize the strangers, and accordingly when they landed to cut the wood, we were so successful as to kill eleven of them, the rest immediately escaping on board two large pettiaugres, and flying westward upon the Great Water.

Upon examining those whom we had killed, we found them much smaller than ourselves, and very white; they had a large head, and in the middle of the crown the hair was very long; their head was wrapt in a great many folds of stuff, and their cloaths seemed to be made neither of wool nor silk; they were very soft, and of different colours. Two only of the eleven who were slain had fire-arms with powder and ball. I tried their pieces, and found that they were much heavier than yours, and did not kill at so great a distance.

After this expedition I thought of nothing but proceeding on my journey, and with that design I let the Red Men return home, and joined myself to those who inhabited more westward on the coast, with whom I travelled along the shore of the Great Water, which bends directly betwixt the north and the sun-setting. When I arrived at the villages of my fellow travellers, where I found the days very long and the night very short, I was advised by the old men to give over all thoughts of continuing my journey. They told me that the land extended still a long way in a direction between the north and sun-setting, after which it ran directly west, and at length was cut by the Great Water from north to south. One of them added, that when he was young, he knew a very old man who had seen that distant land before it was eat away by the Great Water, and that when the Great Water was low, many rocks still appeared in those parts. Finding it therefore impracticable to proceed much further, on account of the severity of the climate, and the want of game, I returned by the same route by which I had set out; and reducing my whole travels westward to days journeys, I compute that they would have employed me thirty-six moons; but on account of my frequent delays, it was five years before I returned to my relations among the Yazous.[90]

After sharing his remarkable story, Moncacht-ape was rewarded by Le Page with "several wares of no great value, among which was a concave mirror" that "magnified the face to four or five times its natural size." He returned home "highly satisfied" to his own nation. Le Page felt that the Indian's account of the proximity between the western extremity of America and eastern Asia seemed confirmed by a discovery made earlier "in a marsh near the river Ohio." According to Le Page, intact skeletons of two large and two small "elephants" had been found there and implied a link with Asia. He also noted what he termed "a great resemblance" existing between the Indians of America and the "Tartars in the north-east parts of Asia."[91]

Fascinating as the story of Moncacht-ape and his incredible journey may be, it is only fitting to conclude a consideration of the Indian storyteller by acknowledging the possibility that he was, after all, simply a convincing literary invention of Le Page Du Pratz.

In the area of Virginia and the Carolinas the opening decade of the eighteenth century saw several interesting and revealing events that provide insight into the wayfinding and cartographic capabilities of Southeastern Indians. In one of these cases the event was inspired by the hostile actions of one Indian group against another. Unlike most of the "Amerindian Hostilities" referred to in Figure 2, this event had the potential to improve rather than detract from European understanding of the geographic environment of the Southeast. It involved an Indian named Lamhatty, whose map is included here as Plate 43A (see the description for Map 146 in Map List).

Lamhatty was a Towasa Indian whose homeland was on the Florida Gulf coast. As the eighteenth century opened, the Towasa and other Spanish-affiliated Indians in the region were coming under increasing pressure from tribes to the north who were in the economic orbit of the English at Charleston. Particularly insidious in this chapter of Southeastern imperial competition was the vigorous trade in Indian slaves being promoted by the South Carolinians. Creek attacks on the Towasa villages in 1706 and 1707 are seen by Gregory A. Waselkov as "a continuation of the protracted slaving wars instigated by the Carolinians."[92]

Lamhatty entered the annals of history when, during the Christmas season in 1707, he sought refuge at the home of a Virginia frontier settler who took him to Colonel John Walker of King and Queen County. Most of what is known about Lamhatty comes from a manuscript in the collection of the Virginia Historical Society, which takes the form of a map on one side of a sheet of paper with the following text on the reverse:

> Mr. Robert Beverlys Accot of Lamhatty an Indian of Towassa of 26 years of age comeing naked & unarmed into the upper inhabitants on the north side of Mattapany in very bad weather in ye Xt. mass hollidays anno 1707 gives this accot.
>
> The foregoing year ye Tuskaroras made war on ye Towasas & destroyed 3 of theyr nations (the whole consisting of ten) haveing disposed of theyr prisoners they returned again, & in ye Spring of ye year 1707: they swept away 4 nations more, the other 3 fled, not to be heard of. It was at this second comeing that they took Lamhatty & in six weeks time they caryed him to Apeikah from thence in a week more to Jabon, from thence in 5 days to tellapousa (where they use canoes) where they made him worke in ye Ground between 3 & 4 months. then they carryed him by easy Journeys in 6 weeks time to Opponys from thence they were a month crossing ye mountains to Souanouka: where they sold him A party of ye Souanouka's comeing a hunting Northward under the foot of ye mountains took him with them, there were of ye Souanoukas. 6 men 2 women & 3 children, he continewed with them about 6 weeks, & they pitched theyr camp on ye branches of Rapahan: River where they pierce ye mountains, then he ran away from them keeping his course E b S & ESE. Crossing 3 branches of Rapahan:River & thrice crossing Mattapany till he fell in upon Andrew Clarks house which he went up to & Surendered himselfe to ye people they being frightned Seized upon him violently & tyed him tho he made no manner of Resistence but Shed tears & Shewed them how his hands were galled and Swelled by being tyed before; whereupon they used him gentler & tyed ye string onely by one arme till they brought him before Lt. Collo. Walker of King & Queen County where [he] is at Liberty & Stays verry Contentedly but noe body can yet be found that understands his language.
>
> Postscript [torn] after some of his Country folks were found servants [torn] he was sometimes ill used by Walker, became verry melancholly often fasting & crying Several days together Sometimes useing little Conjurations & when Warme weather came he went away & was never more heard of.[93]

And who could blame the deservedly despondent Lamhatty, who had escaped Indian slavery only to gain a cruel white master? As his "Account" is reflected upon, one might wonder about its apparent geographical specificity in light of the fact that no one could be found in the Virginia frontier community who understood his language. Once again the graphic language of cartography was employed to bridge the gulf separating curious whites from a widely traveled Indian.

From a letter written by Lamhatty's captor, Colonel Walker, it is learned that the map on the reverse of Beverley's "Account" was "all his [Lamhatty's] own drawing." Walker explained the Indian's cartographic conventions so that the map could be clearly understood: "ye red line denotes his march, ye black lines, ye Rivers, & ye shaded lines ye mountains, which he describes to be vastly big among some of those Indian Towns."[94] On the face of Lamhatty's map a number of the Indian towns as well as the streams and rivers were named, presumably by Walker attempting to render Lamhatty's identifications into phonetic English. Significantly all of the Indian villages are represented by small circles. As mentioned above, this was a widely practiced convention among Indian cartographers whose work has survived.

The Tuscarora Indians who abducted Lamhatty on the Gulf of Mexico coast were a long way from their homeland on the North Carolina Piedmont. Needless to say they were a major force to be reckoned with as white settlement began to spread into that colony. In 1701 they were visited by a young Englishman who would later become the surveyor general of North Carolina and earn the bitter enmity of the Tuscarora. He was John Lawson, author of *A New Voyage to Carolina . . . and A Journal Of a Thousand Miles, Travel'd thro' several Nations of Indians. Giving particular Account of their Customs, Manners, &c.*, which was first published in London in 1709. Few published works from the eighteenth century approach Lawson's in terms of the quality of his detailed descriptions and appraisals of the Indians of the Carolinas. Regrettably, the maps that appeared with Lawson's book were far less revealing of Carolina's geography, being as they were compilations largely based on earlier published maps. For a discussion of the Lawson maps see the Map List at number 150 ("Lawson 1709") and number 168 ("Lawson 1718").

Lawson first appears in 1700 as a well-educated young Englishman in London and with a yen for foreign travel. Upon meeting "a Gentleman" who had returned from America and assured him of the attractions of Carolina, Lawson abandoned his plans to go to Rome and took ship for Charleston via New York. In December of that year he received a commission to make a reconnaissance survey of the backcountry on behalf of the Lords Proprietors of Carolina.[95] As his journal of that excursion reveals, he was constantly in the company of one or more Indian guides. It is clear also that they were excellent teachers and Lawson their apt and receptive student.

One of those teacher-guides was named Enoe-Will, whom Lawson described as being "of the best and most agreeable Temper that I ever met with in an Indian." Although Lawson could not have realized it at the time, there was a note of foreboding in what he wrote next concerning Enoe-Will, who, "being always ready to serve the English, not out of Gain, but real Affection," was "apprehensive of being poison'd by some wicked Indians." He was therefore, Lawson continued, "very earnest with me, to promise him to revenge his Death, if it should so happen."[96] Obviously the political tensions that would later erupt in the Tuscarora War and Lawson's own death at the hand of those powerful Indians were already present as he began his North Carolina career. Enoe-Will and the neighboring Piedmont Indians threatening him were on the horns of the same dilemma: on the one hand, they had come to require numerous articles of European manufacture, such as firearms, gunpowder, and woolen cloth, which could only be obtained by trade with their white neighbors; but these same neighbors were, on the other hand, their competitors in a growing struggle for control of the land and its resources.

Among his many informative passages Lawson provides an excellent summary of the manner in which the Indians utilized mnemonic aids in their oral record keeping and history. After describing the Indian conception of heaven and hell, he wrote:

> Thus is mark'd out their Heaven and Hell. After all this Harrangue [the Indian conjuror] diverts the People with some of their Traditions, as when there was a violent hot Summer, or a very hard Winter; when any notable Distempers rag'd amongst them; when they were at War with such and such Nations; how victorious they were; and what were the Names of their War-Captains. To prove the times more exactly, he produces the Records of the Country, which are a Parcel of Reeds, of different Lengths, with several distinct Marks, known to none but themselves; by which they seem to guess very exactly, at Accidents that happen'd many Years ago; nay two or three Ages or more. The Reason I have to believe what they tell me on this Account is, because I have been at the Meetings of several Indian Nations; and they agreed, in relating the same Circumstances, as to Time, very exactly; as, for Example, they say, there was so hard a Winter in Carolina, 105 years ago, that the great Sound was frozen over, and the Wild Geese came into the Woods to eat Acorns, and that they were so tame, (I suppose, through Want) that they kill'd abundance in the Woods, by knocking them on the Head with Sticks.[97]

In his account of an elaborate ceremony he witnessed in an Indian town house Lawson provides further insight concerning the ritualization of the Indians' oral record keeping:

> They had Musicians, who were two Old Men, one of whom beat a Drum, while the other rattled with a Gourd, that had Corn in it, to make a Noise withal: To these Instruments, they both sung a mournful Ditty; the Burthen of their Song was, in Remembrance of their former Greatness, and Numbers of their Nation, the famous Exploits of their Renowned Ancestors, and all Actions of Moment that had (in former Days) been perform'd by their Forefathers. At these festivals it is, that they give a Traditional Relation of what hath pass'd amongst them, to the younger Fry. These verbal Deliveries being always published in their most Publick Assemblies, serve instead of our Traditional Notes, by the use of Letters. Some Indians, that I have met withal, have given me a very curious Description of the great Deluge, the Immortality of the Soul, with a pithy Account of the Reward of good and wicked Deeds in the Life to come; having found amongst some of them, great Observers of Moral Rules, and the Law of Nature; indeed, a worthy Foundation to build Christianity upon, were a true Method found out, and practis'd, for the Performance thereof.[98]

In another place Lawson gave an interesting example of the way in which knotted cords, similar to those mentioned above by John Lederer, were employed in the Southeast. An Indian he encountered on his backcountry survey told him that South Carolina's governor "had sent Knots to all the Indians thereabouts, for every Town to send in 10 Skins." Exactly why this levy had been imposed was not explained, but the use of "knots" to emphasize a message appears to have been a fairly common occurence in the region.[99]

But what Lawson observed and recorded concerning the wayfinding skills and cartographic capabilities of the Southeastern Indians he met and traveled with is of more interest here. Under the heading of wayfinding Lawson observed the following:

> They are expert Travellers, and though they have not the Use of our artificial Compass, yet they understand the Northpoint exactly, let them be in ever so great a wilderness. One Guide is a short Moss, that grows upon some trees, exactly on the North-Side thereof.
>
> Besides, they have Names for eight of the thirty two Points, and call the Winds by their several Names, as we do; but indeed more properly, for the North-West Wind is called the cold Wind; the North-East the wet Wind; the South the warm Wind; and so agreeably of the rest. Sometimes it happens, that they have a large River or Lake to pass over, and the Weather is very foggy, as it often happens in the Spring and Fall of the Leaf; so that they cannot see which Course to steer: In such Case, they being on one side of the River, or Lake, they know well enough what Course such a Place (which they intend for) bears from them. Therefore, they get a great many Sticks and Chunks of Wood in their Canoe, and then set off directly for their Port, and now and then throw over a Piece of Wood, which directs them, by seeing how the Stick bears from the Canoes Stern, which they always observe to keep right aft; and this is the Indian Compass by which they will go over a broad Water of ten or twenty Leagues wide.

When traveling overland, Lawson found his Indian guides to be equally clever in reading natural and human clues to find their way.

> They will find the Head of any River, though it is five, six or seven hundred miles off, and they never were there, in their Lives before; as is often prov'd, by their appointing to meet on the Head of such a River, where perhaps, none of them ever was before, but where they shall rendezvous exactly at the prefixt time; and if they meet with any obstruction, they leave certain Marks in the Way, where they that come after will understand how many have pass'd by already, and which way they are gone.[100]

Although there is no evidence of Lawson's having copied any of the symbolic messages he describes in the following extract, a later surveyor-cartographer did publish copies he made in the Southeast. He was Bernard Romans, whose observations are discussed below. Lawson continued:

> Besides in their War expeditions, they have very certain Hieroglyphicks, whereby each Party informs the other of the Success or Losses they have met withal; all of which is so exactly perform'd by their Sylvian Marks and Characters, that they are never at a Loss to understand one another. Yet there was never found any Letters amongst the Savages of Carolina; nor, I believe among any other Natives in America, that were possess'd with any manner of Writing or Learning throughout all the Discoveries of the New-World.[101]

It was in his illuminating discussion of Indian cartographic ability that Lawson struck a second unintended prophetic note in light of his later death at the hands of the Tuscarora Indians. On Indian mapping he observed:

> They will draw Maps, very exactly, of all the Rivers, Towns, Mountains, and Roads, or what you shall enquire of them, which you may draw by their Directions, and come to a small matter of Latitude, reckoning by their Days Journeys. These Maps they will draw in the Ashes of the Fire, and sometimes upon a Mat or Piece of Bark. I have put a Pen and Ink into a Savage's Hand, and he has drawn me the Rivers, Bays, and other Parts of a Country, which afterwards I have found to agree with a great deal of Nicety: But you must be very much in their Favour, otherwise they will never make these Discoveries to you; especially, if it be in their own Quarters.[102]

As mentioned above, John Lawson served as North Carolina's surveyor general and was an active agent in the spread of white settlement onto the lands of the Indians. It was in this role that he fell out of favor with the Tuscarora, whose anti-English faction had grown "weary and tired of the Tyranny and Injustice with which the the whites treated them and resolved to endure the bondage no longer."[103] A distinguished North Carolina historian, Hugh T. Lefler, found it "not surprising that Lawson, who had laid out the town of New Bern and who was the chief surveyor in the colony," was the first victim in the conflict known as the Tuscarora War that threatened the colony's existence in 1711.[104]

Lawson met his execution-style death while in the company of the head of a Swiss land development enterprise, Baron Christoph von Graffenried, for whom he had surveyed and laid out the town of New Bern on the Neuse River in North Carolina. Graffenried survived the ordeal and left a narrative in which he places blame for Lawson's death squarely on the surveyor's own shoulders. The Swiss land developer claimed that Lawson had discouraged his original plan to locate a settlement in Virginia so that he, Lawson, could profit from the North Carolina surveying fees. It seems more likely that Lawson's promoterlike enthusiasm for North Carolina transcended mere personal pecuniary interest as he maneuvered the Swiss toward a settlement venue on the Neuse River. As Graffenried admitted, "It is true that besides the fine speeches of Lawson, it was the first promises of the Lords Proprietors which tempted us to establish ourselves in North Carolina."[105]

Graffenried wrote that the Indians kept the surveyor's execution "very secret," but, a few months later, another informant learned that Lawson had met a horrible death by burning, in the very manner he had previously described as an Indian practice in *A new Voyage to Carolina*. In his book Lawson had written of the Indians' cruelty to their prisoners of war, stressing that their inhumanity in this regard was "a natural failing" in a people he otherwise frequently found admirable. In his summary of their cruelties, Lawson wrote of how pitch-pine splinters were stuck into the bodies of still-living prisoners and then ignited "like so many torches" as the victim was forced to dance around the fire and suffer the blows and buffets of his captors.[106]

The informant in question, Christopher Gale, wrote in a letter dated November 3, 1711, "We are informed that they stuck him full of fine small splinters of torch wood like hog's bristles and so set them gradually afire."[107] Gale did not provide details on the disposition of Lawson's physical remains, but it is probably safe to conclude that Lawson met a fate similar to that he had described in his journal account. There Lawson related how the victimized human torch was derided and punished until he expired, "when every one strives to get a Bone or Relick of this unfortunate Captive."[108]

The colonial history of the Southeast is rich in poignant tales, but the prophetic nature, irony, and sheer horror of John Lawson's life and death cannot help but move one even after the passage of three centuries. No other eyewitness provided such rich insights into the role of the Indians in the exploration and mapping of the region, and no other paid so high a price for what he learned and recorded.

Governor Alexander Spotswood of Virginia was proud of his personal involvement in an exploratory expedition to the Blue Ridge Mountains in 1718. On August 14 of that year he informed his superiors in London of what he termed "a great discovery of the Passage over the Mountains." As will be seen, it was an accomplishment in which he had relied on the services of Indian guides. In his letter, Spotswood related:

> [I] found an easy passage over the great Ridge of Mountains w'ch before were adjudged Unpassable, I also discovered by the relation of Indians who frequent these parts, that from the pass where I was it is but three Day's March to a great Nation of Indians living on a River w'ch discharges itself in the Lake Erie; That from ye Western side of one of the small Mountains, w'ch I saw, that the Lake is very Visible and cannot therefore, be above five days March from the pass aforementioned, and that the way thither is also very practicable, the Mountains to the Westward of the Great Ridge being smaller than those I passed on the Eastern side.[109]

Quite clearly Governor Spotswood placed the most optimistic interpretation on what he was learning from the Indians. It is possible in this instance to attribute this example of "Misunderstanding of Amerindians" to the innate tendency of colonial officials to exaggerate or distort information when it forwarded their political ambitions. Of course it is also possible that he reported naively and honestly what he believed he was being told. In any event, Indian wayfinding skills, including pictographs or blazes on trees, were vital to the success of Spotswood's "great discovery," as is affirmed by one of the governor's associates, John Fontaine, when he writes in his journal: "But we met with such prodigious precipices, that we were obliged to return to the top again. We found some trees which had been formerly marked, I suppose, by the Northern Indians, and following these trees, we found a good, safe descent."[110]

A decade after John Lawson met his grisly death at Indian hands, Bernard Romans was born in the Netherlands. Like Lawson, he left Europe and pursued a surveyor's career in the wilderness of the Southeast. Also like Lawson, he depended on and learned from Indian guides and informants and published an account of his experiences and observations. As Lawson and others before him had, Romans observed the pictographic messages Indians displayed in the country where they traveled, hunted, or fought. Fortunately Romans copied some examples of what he termed "Indian hieroglyphick painting," and these were subsequently printed in his book, *A Concise Natural History of East and West Florida*, published in New York in 1775.[111] A copy of these drawings is included here as Figure 5.

Found by Romans in Choctaw country near the Pasca-

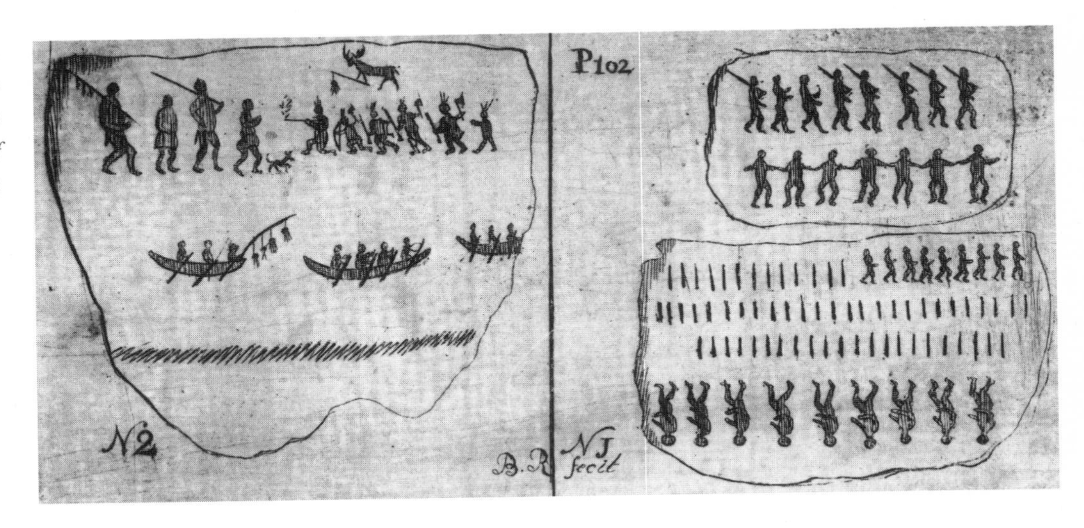

FIGURE 5.
Indian pictographs as shown in Bernard Romans's *Concise Natural History of East and West Florida* (1775)

goula River, the first of these (on the right in Figure 5) was interpreted as follows: "An expedition by seventy men, led by seven principal warriors, and eight of inferior rank, had in an action killed nine of their enemies, of which they brought the scalps, and that the place where it was marked was the first publick place in their territories where they arrived with the scalps."[112] Romans continued, explaining that

> The second [on the left in Figure 5] is a painting in the Creek taste, it means that ten of that nation of the Stag family came in three canoes into their enemies country, that six of the party near this place, which was at Oopah Ullah, a brook so called on the road to the Choctaws, had met two men, and two women with a dog, that they lay in ambush for them, killed them, and that they all went home with the four scalps; the scalp in the stag's foot implies the honour of the action to the whole family.[113]

Romans's statement concerning the Choctaw pictograph having been located at "the first publick place in their territories," agrees with a similar observation made by another eighteenth-century naturalist active in the Southeast. This was Mark Catesby, whose book was published in 1731. Catesby provided the following observation concerning Indian pictographic messages he had seen:

> In their hunting marches, at the entrance of the territories, or hunting grounds of an enemy, the captain, or leader of them chips off the bark from one side of a tree, on which he delineates his own person, with the dreadful hieroglyphick figure . . . which is sometimes a rattle-snake open mouthed, at a corner of his mouth, twisting in spiral meanders round his neck and body, the hero also holding in his hand a bloody Tommahawk. By this menace or challenge is signified that he whose pourtrait is there displayed, hunts in these grounds, where if any of his enemies dare intrude, they shall feel the force of his Tommahawk.[114]

Philip Thicknesse, a somewhat eccentric associate of James Oglethorpe in the early years of Georgia's development, was probably writing about the Creek Indians when he published some reminiscences under the pen name "A. Wanderer" in *The St. James's Chronicle* for April 23, 1786. As his remarks reveal, he saw a positive value in the universality of understanding enjoyed by the Indians' pictographic messages:

> Among the Tribe of Travellers, and Writers relative to the Natives of North America, they uniformly agree that the Indians do all by Memory only; and that our Mode of conveying Information to absent Friends by Writing is to them incomprehensible; but this is true only with Respect to the Natives of Africa. . . . But it is otherwise with the native Indians of America; for they have not only the Art of Writing among themselves, but in some Measure surpass us, for they write to the Understanding of Men of all Nations, while ours is confined to Men of this or that particular country. . . . Before I wandered among the more polished Nations, my first foreign Trip was to America, where walking in the Woods, near the Head of the River Savannah, at Georgia, I found tied to a Tree a Bit of . . . Cedar, on which, in a very uncouth Manner, several Figures were delineated. The first was the Figure of a Canoe, and under it were two Rows (one under the other) of Lines or Notches, thus- [ten pairs of matching lines follow].

The next an Indian Hut, under which were near forty Notches; the third was the Figure of a Man laid down, and under him were eight Notches; fourthly there were two Men one leading the other by a String, and under the captive figure were four Notches; the fourth Mark seemed to represent the Tail of some Animal (a Seale), with two Notches under it; and lastly was the Figure of a Woman and a Child, tied as the Man above; under the the Woman were three Notches, and under the Child thirteen. . . . Ignorant as I then was, and still am, I knew that this Tablet was not placed there without some Meaning, and therefore I brought it away with me, and it fortunately turned out to be my Horn-Book (for I never got much farther in the Business) to learn the Art of Decyphering, as I soon discovered by it that the Upper Creek Indians had been upon a Canoe Expedition against their Enemies;- that they went with ten Canoes, and ten Men in each; that they had surprised an Enemy's Town, with near forty Huts in it; that they had killed eight Men, taken four Prisoners, and brought them away alive; that they had scalped two Men, and brought off three Women and thirteen Children. But after having hastily satisfyed myself that I had read their Letter, I began to suspect I had made too quick a Conclusion; and while I was re-considering the Matter, I happened to turn to the other Side of the Shingle, and there I found one Man down and two Notches under him, which confirmed my first reading; for it showed that they had lost two of their Party.[115]

While in his own writing Bernard Romans was generally critical of the Indians and their lifestyles, he relied on their wayfinding expertise in his surveys. On one occasion he mentioned the help of "three Chactaw savages who extricated us from a great difficulty" and led the party to a river crossing to relieve them of the labor of constructing a raft.[116] Like Captain John Smith long before, Romans relished exciting the Indians' interest and curiosity by demonstrating his navigation instruments. As he wrote in *A Concise Natural History*, for example: "Nothing was more entertaining than the surprize of the savages, at seeing me take observations of the solar altitude, the mercury I used for an artificial horizon, was a matter of great wonder to them, particularly, when i shewed them its divisibility, and the succeeding cohesion of the globules."[117]

On at least one occasion, however, Romans and his fellows encountered a Creek Indian war party, and, as he expressed it, "It was fortunate for me that i had not brought a savage guide with me, which would have exposed us to a volley from those warriors."[118] Not long afterward, he wrote, "Here seems to be the true theatre of the war, for the bark logs are very numerous."[119] When the party camped for the night at a place called "Suktaloosa (i.e.) Black Bluff, from its being a kind of coal," they discovered that they were athwart "a great thoroughfare for warring savages." Romans described the precautions that were called for in such a potentially perilous setting: "We took the usual precaution of large fires, and hanging our hats on stakes." Nor were they ill advised in their efforts, for, as he continued, "in the night we heard the report of small arms."[120]

Clearly the surveying and mapping efforts of Bernard Romans and his associates were influenced to a considerable degree by the "Amerindian Hostilities" swirling around them in the area of present-day Louisiana, Mississippi, and Alabama. This may be the reason why Romans, himself an accomplished cartographer, is silent on the subject of Indian-drawn maps. Recall Lawson's advice that the Indians were often guarded in this respect and would share their maps only with those who were "very much in their Favour . . . especially if it be in their own quarters." It is quite possible that Romans did not enjoy the favor of the Creeks, Choctaws, and Chickasaws and so was never made privy to their maps.[121]

There is no doubt that the mapping efforts of Romans's sometime employer William Gerard De Brahm, surveyor general of the British colonies in the Southeast during the 1760s and early 1770s, were impeded by Indian hostilities. On his large-scale manuscript map titled "East Florida East of the 82nd degree of Longitude from the Meridian of London" De Brahm placed the following caption near the west bank of the Indian River inland from Cape Canaveral:

> Surveyor Gen: Camp March 7, 1765—Thus far the Surveyor General carried on his Survey by land from St. Augustine guided by Sahaykee a Creek Indian, who meeting some Hunting Indians made difficulties and out of fear to disobaye the Nation refused to continue as far as the Haven of Spirito Santo which stop'd the Surveyor General who must have fallen in with many Hunting Ganges of Semiolokee [Seminoles] of which that of the Indian Headman called Cowkeeper was only within one [day's] journey to the Southward of him."[122]

A French military officer named Jean-Bernard Bossu published an account of his travels and experiences with the

Indians of the Southeast during the years from 1751 to 1762.[123] In his book Bossu called attention to the need for aspiring military commanders in America to have "theoretical and practical knowledge of geography and make an effort to know the terrain of the country in which they may be fighting."[124] To drive home his point he related the story of a Spanish expedition against the Missouri Indians. The Spaniards had decided to extirpate the Missouri to make room for a permanent colony that could check France's ambitions in the Mississippi Valley. Unable to field a force large enough to accomplish this goal, the Spanish developed a strategy that hinged on making allies of the Osage Indians, who were described as "neighbors and mortal enemies of the Missouris."[125]

On the long march from Santa Fe the Spanish relied on Indian geographical intelligence, including at least one map. Ignorant as they were of the geography and terrain of the country, the Spanish were easily duped into marching into the hands of the Missouri Indians, whom they had been led to believe were the Osages and their potential allies in the coming campaign. One can only imagine the irony of the scene as the Missouri chief ceremoniously listened to the Spaniards' interpreter explain their desire to form an alliance aimed at the destruction of his own people. As Bossu described, "The great chief of the Missouri Nation, hiding his true feelings, pretended to be very happy and promised to help the Spaniards carry out this pleasant project."[126]

Well and truly duped, the Spanish commander "distributed to the Indians 1,500 rifles and as many pistols, sabers, and hatchets." The Spanish force was destroyed in the massacre that ensued as soon as the now-well-armed Indians were ready. Sometime later the Missouris turned over the unneeded contents of the Spanish baggage train to the nearest French post. Well supplied with mounts taken from the Spanish, the Indians "presented the most beautiful of these" to the French commander. Bossu mentioned also that "they had brought the map which had misdirected the Spaniards and led them to deliver themselves stupidly to their enemy." This must rank as the most dramatic and effective example of an Amerindian map employed to deliberately misinform a hostile enemy.

The need for circumspection during periods of tension with the Southeastern Indians was well understood by John Stuart, George III's superintendent of Indian affairs for the Southeast. In a set of instructions addressed to one of his assistants, David Taitt, traveling to meet with Creek chiefs in 1772, Stuart wrote:

> You are if it is possible to be affected without giving umbrage or raising jealousy in the Indians. Ride thro' all the Indian Villages of the Upper as well of the Lower Creek Nation, and take particular notice of their situations and make such observations as may enable you to draw a plan of the country and of the rivers etc. . . . You will be very particular in your remarks upon the rivers the depths & courses & the distances of the roads & every information which may enable you to lay down the situation of the country, of the Villages & of the roads you shall travel as well from Pensacola to the Nation as from there to Charleston.[127]

In a letter he wrote to Stuart while still in the Creek country, Taitt went into detail concerning the strident complaints of the Indians against abusive white traders and the excessive supplies of rum they were bringing into the nation. More satisfying to Stuart's eyes was his mention of initial success in making the observations required for the preparation of a detailed map of the Creek country. On this score Taitt reported:

> I have made the observations recommended by you in respect to the country from Pensacola to this place [the important Creek town known as Tuckabatchee] and likewise taken the Latitudes of different places: about two miles N by E from crossing Little Scambia I observed in Latitude 31.14 This Town is in Latitude 32.26 and is the farthest any boat or canoe can go up this river on account of falls which begin at the upper end of the Town and at 1½ miles above the Town is one about one hundred feet high. My quadrant is now of no use the sun being got to far to the Northward for observing by land.[128]

Upon completion of his sojourn in the Creek country, Taitt prepared a detailed journal, which Stuart forwarded to London in July, 1772. In that revealing document Taitt spelled out the difficulties he had encountered in making the surveys that are reflected in one of the largest manuscript maps of the colonial Southeast. The map, listed here as Map 440, is known as the Stuart-Gage map of 1773. Titled "A Map of West Florida part of Et: Florida. Georgia part of So: Carolina including . . . Choctaw Chickasaw & Creek Nations," it contains a lengthy synopsis of Taitt's observations on the Creek Indians and their homeland.

At times in fear of his life from bellicose and often drunken Indians, David Taitt lost no opportunity to record detailed descriptions of the terrain as he traveled along the trails and rivers that linked the many Indian towns comprising the Creek nation of that day. His discussions also provide insight into the factional political struggles that were creating such a potentially volatile climate in the Indian country. On March 24, 1772, while at the Creek town known as "Hillabies," Taitt recorded his response to the difficulties posed by an example of what in Figure 2 is termed "Amerindian Hostilities":

> This morning I went to black drink in the square at Socuspoga where I stayed till ten o'clock and then set out for Hillabies. Paya Lucko's son went with me on pretence of being my guide but I suppose this piece of kindness was the effect of jealousy and not intended as any service to me. I made him ride on before and kept my servant between him and me thereby preventing him from seeing me take observations of the course of the path and creeks as we passed.[129]

While he was in the Creek country, Taitt learned that no small part of the Indians' unrest stemmed from reports they had received concerning a Cherokee scheme to satisty their indebtedness to a consortium of traders by signing over title to a large tract of land lying along tributaries of the Savannah River in Georgia. On February 22, 1771, the Cherokee ceded to the traders in return for the voiding of the tribe's debt "a certain tract of land upon Broad River Georgia side, beginning at the mouth of the Kayugas, extending five measures up Savannah River, and Running five measures extending toward the oconies, Viz. five measures long and five measures broad or sixty miles square."[130] As Taitt's superior, Superintendent of Indian Affairs John Stuart, observed, "the Creek Indians and their traders were not inattentive to these transactions: the former claimed part of the ceded land in right of conquest having obliged the Cherokee during the war between them to abandon it."[131]

Long desirous of adding Indian territory to his frontier colony, Georgia's governor, James Wright, intervened and persuaded the tribal leaders and traders that a transfer of Indian land could be legally undertaken only by government. Wright traveled to London and convinced the crown to allow him to treat with the Creeks and Cherokees and effect the transfer of the land to Georgia. The traders would be paid up to the amount of the Indians' indebtedness by the government, which in turn would sell the acreage to the hundreds of frontier families expected to flood in from the north as soon as the Indians surrendered their title.

An Indian congress was convened at Augusta on June 1, 1773, where the Creeks somewhat reluctantly joined the Cherokees in making the desired cession to Georgia. The most colorful account of this important event was penned by the Philadelphia naturalist William Bartram while on a plant collecting tour of the Southeast. Attending as the guest of John Stuart, Bartram noted that the Creeks, "being a powerful and proud spirited people . . . were unwilling to submit to so large a demand, and their conduct . . . betrayed a disposition to dispute the ground by force of arms." He went on to relate that "at length the cool and deliberate counsels of the ancient venerable chiefs, enforced by liberal presents of suitable goods, were too powerful inducements for them any longer to resist."[132]

Eager to observe and collect plant specimens in the Indian country, Bartram joined the party assigned to survey and mark the boundaries of the sprawling cession now referred to as the "new purchase" or "ceded lands." Bartram described the survey group as a "caravan, consisting of surveyors, astronomers, artisans, chain-carriers, markers, guides, and hunters, besides a very respectable number of gentlemen, who joined us, in order to speculate in the lands, together with ten or twelve Indians, altogether to the number of eighty or ninety men, all or most of us well mounted on horseback, besides twenty or thirty pack-horses loaded with provisions, tents and camp equipage."[133]

Bartram and others were most favorably impressed with the promise of the Piedmont soils they found in the broad region stretching from Augusta to modern-day Athens, Georgia. With his characteristic eloquence, Bartram wrote:

> This new ceded country promises plenty & felicity. The lands on the River [Broad River] are generally rich & those of its almost innumerable branches agreeable and healthy situations, especially for small farms, every where little mounts & hills to build on & beneath them rich level land fit for corn & any grain with delightful glittering streams of running water through cain bottoms, proper for meadows, with abundance of waterbrooks for mills. The hills suit extremely well for vineyards & olives as nature points out by the abundant produce of fruitful grape vine, native mulberry trees of an excellent quality for silk. Any of this land would produce indigo & no country is more proper for the culture of almost all kinds of fruits.[134]

Not long after the mounted caravan making up the survey party had entered the "new ceded country" Bartram observed and recorded an event involving, as he termed it, "Indian sagacity" in the survey process. It was an event that, in Bartram's words, "nearly disconcerted all our plans, and put an end to the business." Had the survey been disconcerted and put to an end one of the most remarkable maps included in this study might never have been made. The event Bartram described took place in present-day Oglethorpe County, Georgia, not far from Athens, at a place known to Bartram and his contemporaries as the Buffalo Lick. As Bartram wrote:

> The surveyor having fixed his compass on the staff, and about to ascertain the course from our place of departure, which was to strike Savanna river at the confluence of a . . . certain river, about seventy miles distance from us; just as he had determined upon the point, the Indian Chief came up, and observing the course he had fixed upon, spoke and said it was not right; but that the course to the place was so and so, holding up his hand, and pointing. The surveyor replied, that he himself was certainly right, adding, that that little instrument (pointing to the compass) told him so, which, he said, could not err. The Indian answered, he knew better, and that the little wicked instrument was a liar; and he would not acquiesce in its decisions, since it would wrong the Indians out of their land. This mistake (the surveyor proving to be in the wrong) displeased the Indians; the dispute arose to that height, that the Chief and his party had determined to break up the business, and return the shortest way home, and forbad the surveyors to proceed any farther: however after some delay, the complaisance and prudent conduct of the Colonel made them change their resolution; the Chief became reconciled, upon condition that the compass should be discarded, and rendered incapable of serving on this business; that the Chief himself should lead the survey; and, moreover, receive an order for a very considerable quantity of goods.[135]

It would be easy to interpret Bartram's anecdote as just one more story of how Europeans, trying to cheat Indians out of their lands, were found out through the sagacity of their would-be victims. In this instance such an interpretation would be a mistake. Incredible as it may seem, the "Great Bufloe Lick," where the dramatic exchange between the Indian and surveyor took place, lies immediately adjacent to a prominent magnetic anomaly that is approximately three miles in diameter. This can be seen on the Aeromagnetic Map of Georgia by I. Zietz, F. E. Riggle and F. P. Gilbert, published in 1980 by the U.S. Geological Survey. The size and strength of this anomaly coupled with its proximity to the Buffalo Lick would have influenced the surveyor's compass needle, causing it to point to a direction the Indian knew from experience to be incorrect. We can only conclude that in this instance the surveyor was operating in good faith but had been misled by a faulty reading from his "little wicked instrument."

Contrast this anecdote with the one related by Captain John Smith telling how he saved his life by enthralling the Powhatan Indians with "a round Ivory double compass dyall." Unlike the Powhatans of 150 years before, the Creeks and Cherokees of the late eighteenth century had come to learn something of the compass and how it might be used to their disadvantage in land dealings. Rather than "amazed with admiration" as Smith's Virginia captors had been, Bartram's Indians at the Buffalo Lick displayed a healthy skepticism of the surveyor's compass and what it told in contradiction of their own knowledge.

The map that was prepared from the survey in which Bartram participated is included here as Plate 66C and is described as Map 440B in the Map List. It is titled "A Map of the Lands Ceded to His Majesty by the Creek and Cherokee Indians at a Congress held in Augusta the 1st June By His Excellency Sir James Wright Bart. Captain General Governor and Commander in Cheif [*sic*] of the Province of Georgia The Honorable John Stuart Esqr. Agent and Superintendent of Indian Affairs and the said Indians—containing 1616298 Acres." It was prepared by Philip Yonge from the survey work that he, Edward Barnard, LeRoy Hammond, Joseph Purcell, and William Barnard carried out with the unnamed Indians representing the Creek and Cherokee nations. Because the colonial authorities intended to recoup the expenses incurred in paying off the Indians' indebtedness by selling the land to settlers, more than usual attention was paid to the quality of the soils and other natural features found in the region. T. G. Macfie, a soil scientist, has identified this map by Philip Yonge as being the earliest true soils map prepared in the Southeast.[136]

CONCLUSION

This essay began by stressing the important role played by Native American informants, guides, and mapmakers as the earliest Europeans found their way into the Southeast in the wake of Columbus. To assist in gaining a fuller appreciation of exactly how Native Americans entered into the complex processes that led to the mapping of the region, the theoretical schema presented as Figure 2 was utilized as a sort of intellectual map or set of guideposts pointing the way through a lengthy historical record of selected Indian-white interactions.

At many places in the discussion of that historical record allusions are made to certain diagnostic traits that can serve as clues pointing to Native American authorship or influence in the creation of particular maps. More often than not the maps themselves are attributed to European cartographers or compilers and, lacking overt credits, remain as silent witnesses insofar as their Native American input is concerned.

The systematic interrogation of these silent cartographic witnesses has only begun. It is a research path that may lead to a deeper appreciation of the manifold contributions made by the Southeast's original people to the region's exploration and mapping during the period covered by this study. To assist in that systematic interrogation of the early maps of the Southeast, as well as those of other regions of the continent, I would like to end by presenting a table or guide contrasting the traits of Native American mapping to those of European cartography for the consideration of scholars as well as general readers.

NATIVE AMERICAN MAP DIAGNOSTICS

1. *Distance*
   EUROPEAN: distance between places usually in some unit of constant spatial value (i.e., yards, miles, leagues, etc.)
   NATIVE AMERICAN: separation between places on map often relates to time required to travel from place to place
   Factors to consider:
   A. Season: long vs. short days
   B. Weather: strong headwinds, ice, snow, excessive heat or cold, floods, droughts
   C. Social Attitudes: neighbors friendly or unfriendly—paths between friends well-developed and clear, whereas in enemy territory paths may be avoided or circuitous
   D. Traveler's Age: for children distances may seem greater than for adults
   E. Traveler's Condition: strong or weak, weary or fresh, prisoner or free, fed or hungry
   F. Religion: prayers, chants, charms, or drugs may shrink or stretch perceived distance
2. *Object Size*
   EUROPEAN: objects and areas usually shown consistently (i.e., objects or areas that are large in nature appear on map as larger than similar smaller features
   NATIVE AMERICAN: object size on map may be related to some hierarchy other than size in nature (e.g., power, status, age, veneration)
3. *Bounded Spaces*
   EUROPEAN: great attention paid to boundary details
   NATIVE AMERICAN: areas indicated by simple enclosing shapes (circular or ovoid = "us"; rectangular = "them")
4. *Networks*
   EUROPEAN: asymmetrical or complex networks retain their characteristics to the extent that a particular map allows (e.g., trails distinct from streams)
   NATIVE AMERICAN: asymmetrical or complex networks are frequently represented as approximately symmetrical and uniform
   For example:
   A. Portages or routes between river systems or elements in those systems often not differentiated, especially if well known and frequently used
   B. Rivers depicted as anastomose—a macro scale condition almost never found in nature—resulting from lack of differentiation between waterways and trails in a transport network, etc.
5. *Shapes and Sizes*
   EUROPEAN: map scale allowing, great care exerted to reproduce correct feature shape and relative size as found in nature
   NATIVE AMERICAN: feature caricaturization or shape simplification common, with map size frequently related to position in a hierarchy of value or importance that may be political, spiritual, or secular
6. *Linear Features*
   EUROPEAN: linear features that are irregular in nature are irregular on map and change direction as they do in the landscape
   NATIVE AMERICAN: irregular linear features tend to be regularized as essentially straight or gently curved;

changes in direction can occur at nodes rather than as in nature

7. *Centrality*
   EUROPEAN: centrality dependent on purpose of map
   NATIVE AMERICAN: cartographer's home or present location often enjoys central positioning on map

NOTES

1. J. Mooney provides excellent examples in his highly regarded *Myths of the Cherokee and Sacred Formulas of the Cherokees* [Facsimile reprints from the 19th and 7th Annual Reports of the Bureau of American Ethnology] (Nashville, 1982).

2. L. De Vorsey, "Silent Witnesses: Native American Maps," *The Georgia Review* 46 (Winter, 1992): 709–26.

3. *The Letter of Columbus on the Discovery of America* [A facsimile of the Pictorial Edition, with a New and Literal Translation, and a Complete Reprint of the Oldest Four Editions in Latin] (New York, 1892), 6–7.

4. W. N. Dunwoody, "Guanahani to Saometo to Cuba: Columbus' Track through the Isla de los Lucayos Followed Taino Indian Canoe Routes" (unpublished presentation, annual meeting of the Society for the History of Discoveries, Miami, October 2, 1992).

5. B. de Las Casas, *Historia De Las Indias*, ed. A. M. Carlo and L. Hanke (Mexico-Buenos Aires, 1951), 323–26; S. E. Morison, *Admiral of the Ocean Sea* (New York, 1942), 1:438–42.

6. L. De Vorsey, *Keys to the Encounter: A Library of Congress Resource Guide for the Study of the Age of Discovery* (Washington, 1992), 85.

7. Other Indians were brought to Europe and exploited for their geographical knowledge subsequent to Columbus's first voyage. In 1511, the year that saw the publication of the Martyr map, for example, Juan de Agramonte presented two Indians to King Ferdinand to support his petition for a royal license to "undertake at his own cost a voyage of discovery to a new land." A line in Ferdinand's letter regarding those Indians has led some scholars to believe that they were from the American Southeast rather than the known islands of the Caribbean. The King observed, "Moreover, the Indians Agramonte brought with him were much more sensitive to cold than those of Hispaniola, and we may reasonably trust to God that this will turn out differently from the experience of the Portuguese." In a somewhat strained argument, L. A. Vigneras suggests that Agramonte's Indians were Waccamaws from the coastal "territory that extended for eighty miles between the Cape Fear River and Winyah Bay" in the Carolinas. See Vigneras, "The Projected Voyage of Juan de Agramonte to the Carolinas, 1511," *Terrae Incognitae* 11 (1979): 67–70.

8. L. De Vorsey, "Amerindian Contributions to the Mapping of North America: A Preliminary View," *Imago Mundi* 30 (1978): 74 (fig. 2). A German-language translation of Figure 2 was published by Rainer Vollmar in his *Indianische Karten Nordamerikas* (Berlin, 1981), 13.

9. Anthropologist G. A. Waselkov has brought the best-known surviving examples of Southeastern Indian cartography together in his scholarly essay titled "Indian Maps of the Southeast," in *Powhatan's Mantle: Indians in the Colonial Southeast*, ed. P. H. Wood, G. A. Waselkov, and M. T. Hatley (Lincoln, Nebr., 1989), 292–343.

10. For an informed discussion of the processes involved in map production, see D. Woodward, "The Study of the History of Cartography: A Suggested Framework," *American Cartographer* 1, no. 2 (1974): 101–15.

11. M. R. Hemperley, *Historic Indian Trails of Georgia* (Atlanta, 1989); W. E. Meyer, "Indian Trails of the Southeast," *Forty-second Annual Report of the Bureau of American Ethnology for 1924–1925* (Washington, 1928); and H. H. Tanner, "The Land and Water Communication Systems of the Southeastern Indians," in Wood, Waselkov, and Hatley, *Powhatan's Mantle*, 6–20.

12. For translations of the key documents detailing Ponce de León's activities in the Southeast, see D. B. Quinn, ed., *New American World: A Documentary History of North America to 1612* (New York, 1979), 1:231–47.

13. Ibid., 237.

14. Quinn, *New American World* is also the most accessible source for the documents revealing the Southeastern activities of Lucas Vásquez de Ayllón. See 1:248–73 in this indispensable collection. Emphasis added in quote.

15. Ibid., 248.

16. Ibid., 260.

17. Ibid., 248.

18. Ibid., 265.

19. Ibid., 266.

20. Ibid.

21. Ibid.

22. Ibid.

23. For Martyr's role in the creation of the "Chicora Legend," see P. E. Hoffman, *A New Andalucia and a Way to the Orient: The American Southeast During the Sixteenth Century* (Baton Rouge, 1990), 3–21.

24. Quinn, *New American World* 1:270.

25. Hoffman, *A New Andalucia*, 51.

26. Ibid., 53.

27. Quinn, *New American World* 1:261.

28. Ibid.

29. Ibid.

30. For a concise review of the de Soto entrada route, see C. Hudson, "The Hernando de Soto Expedition, 1539–1543," in *The Forgotten Centuries: Indians and Europeans in the American South, 1521–1704*, ed. C. Hudson and C. C. Tesser (Athens, Ga., 1994), 74–103. The documents bearing on the de Soto expedition can be found in a valuable two-volume set edited by L. A. Clayton, V. J. Knight, and E. C. Moore titled *The De Soto Chronicles: The Expeditions of Hernando De Soto to North America in 1539–1543* (Tuscaloosa, 1991).

31. The translations of the de Soto documents quoted here are from Quinn, *New American World*, vol. 2. This quotation is found

on p. 93. Identical or very similar translations are included in Clayton, Knight, and Moore, *The De Soto Chronicles*.

32. Quinn, *New American World*, 2:102.

33. Ibid., 104.

34. Ibid.

35. Ibid., 110.

36. Ibid.

37. Ibid., 111.

38. Ibid., 113.

39. Ibid.

40. Ibid.

41. Ibid.

42. Hudson, "The Hernando de Soto Expedition," 99.

43. K. O. Kupperman, *Roanoke: The Abandoned Colony* (Totowa, N.J., 1984), 16.

44. This and following quotations from Captain Gabriel Archer's account are taken from his "A Relayton of the Discovery, etc. 21 May–22 June 1607," in *Travels and Works of Captain John Smith . . . 1580–1631*, ed. Edward Arber (Edinburgh, 1910), 1:xl–lv.

45. It appears that Archer is the only authority to employ this Indian toponym. In the judgment of P. L. Barbour, it probably means "Long Hill (or Mountain)." See Barbour, *The Jamestown Voyages Under the First Charter 1606–1609* (Cambridge, 1969), 2:475. In the absence of evidence to the contrary, it appears to have been a good description of the Blue Ridge.

46. The Monacan Indians formed a polity immediately to the west of the Powhatans, who were their traditional enemies. See J. L Hantman, "Powhatan's Relations with the Piedmont Monacans," in *Powhatan Foreign Relations, 1500–1722*, ed. H. C. Rountree (Charlottesville, 1993), 94–111.

47. "Caquassan" or "caquasson" is defined in Barbour, *The Jamestown Voyages*, 2:469, n. 54.

48. For Smith's description of his captivity, see P. L. Barbour, *The Complete Works of Captain John Smith (1580–1631)* (Chapel Hill, 1986), 2:146–52.

49. Ibid., 1:49.

50. Ibid., 55.

51. D. B. Quinn, *The Roanoke Voyages, 1584–1590* (Cambridge, 1955), 1:259.

52. A. Brown, *The Genesis of the United States* (New York, 1890), 396.

53. Ibid., 397.

54. In their "Instructions given by way of Advice," the original Jamestown settlers were directed to spare no effort in finding the shortest way to "the Other Sea." See Barbour, *The Jamestown Voyages*, 1:49–54.

55. Ibid., 2:408–9.

56. Ibid., 413–14. Emphasis added. The scene of Powhatan's dramatic sketching of maps has been identified as Portan Bay in Gloucester County on the north side of the York River in Virginia. See *William and Mary Quarterly* 10 (July 1901): 1–4. In Powhatan's day this was the site of the village known as "Werocomoco" or "Werawocomoco" and was one of his favorite residences.

57. C. Verner, "The First Maps of Virginia, 1590–1673," *The Virginia Magazine of History and Biography* 58 (January, 1950), p.6.

58. Barbour, *Works of Captain John Smith*, 1:150–51.

59. Barbour, *The Jamestown Voyages* 1:52.

60. Ibid., 51.

61. Barbour, *Works of Captain John Smith*, 1:119–90; map, 140–41.

62. For reproductions of these De Bry engravings, see P. Hulton, *America 1585: The Complete Drawings of John White* (Chapel Hill, 1984), 109, 123, 127, and 128.

63. Quoted in Verner, "The First Maps of Virginia, 11.

64. *The True Travels, Adventures and Observations of Captaine John Smith . . .* [From the London Edition of 1629] (Richmond, 1819), 1:183.

65. Quotation from D. Norona, "Maps Drawn by Indians in the Virginias," *The West Virginia Archaeologist* 2 (March 1950): 15.

66. E. W. Andrews and C. M. Andrews, eds. *Jonathan Dickinson's Journal or God's Protecting Providence. Being the Narrative of a Journey from Port Royal in Jamaica to Philadelphia Between August 23, 1696 and April 1, 1697* (New Haven, 1961), 8.

67. The best reprint of Lederer's book and map is the one edited by W. P. Cumming and titled *The Discoveries of John Lederer with Unpublished Letters By and About Lederer to Governor John Winthrop, Jr.* (Charlottesville, 1958). In addition to a facsimile of Lederer's map, it includes a newly drawn map showing his routes and some of the sites he mentioned with their modern names (front endpaper).

68. Cumming, *The Discoveries of John Lederer*, 12.

69. Ibid., 14.

70. Ibid., 20.

71. Ibid., 25.

72. Ibid.

73. Ibid.

74. Ibid., 37.

75. Ibid., 38.

76. Ibid., 37–38.

77. W. P. Cumming et al., *The Exploration of North America, 1630–1776* (New York, 1974), 147.

78. R. G. Thwaites, ed. *The Jesuit Relations and Allied Documents: Travels and Explorations of the Jesuit Missionaries in New France, 1610–1791* (Cleveland, 1896–1901), 59:91.

79. Ibid., 149.

80. M. C. Cross, trans., *Father Louis Hennepin's Description of Louisiana* (Minneapolis, 1938), 80–81.

81. Ibid.

82. J. G. Shea, ed., *Discovery and Exploration of the Mississippi Valley* (Clinton Hall, N.Y., 1852), 204.

83. C. A. Brasseaux, trans. and ed., *A Comparative View of Louisiana, 1699 and 1762* (Lafayette, La., 1981), 60. For a scholarly debate concerning the motivation behind La Salle's reporting the mouth of the Mississippi River to be on the coast of Texas, see L. De Vorsey, "La Salle's Cartography of the Lower Mississippi: Product of Error or Deception?," *Geoscience and Man* 25 (1988): 5–23; and P. H. Wood, "La Salle: Discovery of a Lost Explorer," *The American Historical Review* 89 (April 1984): 294–323.

84. R. G. McWilliams, ed. and trans., *Iberville's Gulf Journals* (Tuscaloosa, 1981), 71.

85. Ibid.

86. L. De Vorsey, "The Impact of the La Salle Expedition of 1682 on European Cartography," in *La Salle and His Legacy: Frenchmen and Indians in the Lower Mississippi Valley*, ed. P. K. Galloway (Jackson, Miss., 1982), 74.

87. J. P. La Hontan, *New Voyages to North America* (London, 1703), 2:13–14.

88. A. S. Le Page Du Pratz, *The History of Louisiana, or of The Western Parts of Virginia and Carolina . . .* [Translated from the French . . . With some Notes and Observations relating to our Colonies] (London, 1744; New Orleans, 1941), 285.

89. Ibid.

90. Ibid., 285–90.

91. Ibid., 290. The fossilized remains of large tusked quadrupeds were mentioned by Jean-Bernard Bossu in his *Travels in the Interior of North America, 1751–1762*, trans. and ed. Seymour Feller (Norman, Okla., 1962), 103–4. Bossu mentioned being given "a molar which weighed about six and one half pounds" by an Indian. Like Le Page, he was led to conclude "that Louisiana is joined to India and that the elephants came here from Asia through the west."

92. The best treatment of the Lamhatty map and account is in Waselkov, "Indian Maps of the Colonial Southeast," 313–20. The quotation is found on p. 317.

93. Ibid., 315–16.

94. Ibid., 314–15.

95. John Lawson, *A New Voyage to Carolina*, ed. H. T. Lefler (Chapel Hill, 1967), xi.

96. Ibid., 62.

97. Ibid., 187.

98. Ibid., 45.

99. Ibid., 48. A Spanish officer sent to spy on the Jamestown colony in 1609 mentioned the use of knotted string as a mnemonic by the Indians of coastal Carolina; see the account of Captain Francisco Fernandez de Ecija in Barbour, *The Jamestown Voyages*, 2:302 and 319.

100. Lawson, *A New Voyage to Carolina*, 213.

101. Ibid., 213–14.

102. Ibid., 214.

103. Ibid., xxxi.

104. Ibid.

105. Ibid., xxxvi.

106. Ibid., 207.

107. Ibid., xxxvi.

108. Ibid., 207.

109. Quoted in C. E. Hatch, *Alexander Spotswood Crosses the Blue Ridge: "A Great Discovery of the Passage Over the Mountains"* (Washington, 1968), 13.

110. Ibid., 47.

111. Romans, *A Concise Natural History of East and West*, is available in a photographic facsimile of the 1775 edition, with an introduction by R. W. Patrick, published in the Floridiana Facsimile and Reprint Series (Gainesville, 1962).

112. Ibid., 102.

113. Ibid.

114. Mark Catesby, *The Natural History of Carolina, Florida, and Bahama Islands* (London, 1731–45), 2:ix.

115. Quoted in R. M. Baine, "Reminiscences of Early Georgia," *Georgia Historical Quarterly* 76 (Winter 1990): 686–87.

116. Romans, *A Concise Natural History*, 309.

117. Ibid., 314.

118. Ibid., 321.

119. Ibid., 324.

120. Ibid., 326.

121. A. J. Pickett, *History of Alabama, and Incidentally of Georgia and Mississippi, From the Earliest Period* (Charleston, 1851), 1:94, states, "Rude paintings were quite common among the Creeks and they often conveyed ideas by drawings. No people could present a more comprehensive view of the topography of a country, with which they were acquainted, than the Creeks could, in a few moments, by drawing upon the ground."

122. For a discussion of De Brahm's Florida mapping activities, see L. De Vorsey, "De Brahm's East Florida on the Eve of Revolution: The Materials for Its Re-creation," in S. Proctor, ed., *Eighteenth-Century Florida and Its Borderlands* (Gainesville, 1975), 78–96.

123. Bossu, *Travels in the Interior of North America.*

124. Ibid., 88.

125. Ibid.

126. Ibid., 90.

127. British Public Record Office, London, Colonial Office manuscripts, C.O. 5-73, folio 56 verso.

128. Ibid., folio 261.

129. Ibid., folios 302–3.

130. British Public Record Office, London, Colonial Office manuscripts, C.O. 5-661, folio 192, quoted by L. De Vorsey, in *The Indian Boundary in the Southern Colonies, 1763–1775* (Chapel Hill, 1966), 162. The area of this proposed cession was delineated on at least two of the maps included in the Map List. The first is Map 323, Georgia Indian Lands ca. 1771 MS, and the second is Map 437, Stuart Georgia 1772 MS.

131. Ibid., 164.

132. William Bartram, *Travels*, ed. Francis Harper (New Haven, 1958), 22.

133. Ibid., 123.

134. William Bartram, "Travels in Georgia and Florida 1773–74: A report to Dr. John Fothergill," annotated by Francis Harper, *Transactions of the American Philosophical Society*, n.s. 33, pt. 2 (November 1943): 144.

135. Bartram, *Travels*, 26.

136. T. G. Macfie, "William Bartram's Observations on Soils, 1773," *Soil Survey Horizons* 33 (1994): 96–102.

List of Maps of the Southeast during the Colonial Period, Including Local Maps and Plans of the Region South of Virginia and North of the Florida Peninsula

REGION COVERED IN THE MAP LIST

This is an annotated map list of regional and local manuscript and printed maps of the Southeast before 1776. It is based on a checklist of the maps found in some of the chief American libraries and of the manuscript map collections in European archives.

General maps which include the entire continent or the British, Spanish, or French possessions in North America are excluded unless they have significance in a study of the historical cartography of the Southeast. Some maps which include areas larger than this study, however, such as Jode 1593, Daniel ca. 1679, Delisle 1703, Delisle 1718, Popple 1733, and Mitchell 1755, are listed because they show characteristic misconceptions of the Southeast, record new discoveries or information for the first time, or influence later regional maps. Local plans and maps limited to areas in the present state of Florida are excluded from this study because their inclusion would considerably enlarge the list and would, except for a relatively few additional maps reported in the last forty years, duplicate the list in P. L. Phillips, ed., *The Lowery Collection: A Descriptive List of Maps of the Spanish Possessions within the Present Limits of the United States, 1502–1820*, by Woodbury Lowery, Washington, 1912. Virginia also, except for regional maps of the Southeast, is omitted; during the colonial period it was frequently included in regional maps of the northern colonies. The problems resulting from inclusion of every local and regional map dealing with Virginia would have increased the problems of the study and the size of this work prohibitively. Limitations of some kind are necessary. The Virginia map lists of Swem, Phillips, and others are available. Mr. Coolie Verner published some excellent bibliographical studies in Virginia cartography.

Those world maps which are fundamental to a study of exploration and discovery along the southeastern coast in the sixteenth century are discussed in the Introductory Essay.

METHOD USED IN THE MAP LIST

In most instances the list is chronological. Maps from the same plate are listed together under the date of first appearance, even though the plate has been changed or the map appears in different books, unless a change of imprint or date on the map makes a different entry advisable. The items below are given as illustrations and do not describe any one map.

Title Tag: Comberford 1657 MS B

Manuscript maps of a given date precede those printed. This title tag indicates a map by or attributed to Comberford, made or dated 1657; it is the second manuscript map listed as by Comberford under this year. The title tag is given for identification; problems of authorship and date are discussed, if necessary, under *Description*.

Title: The south Part of Virginia Now the North Part of Carolina [across face of map]

The location of the title or cartouche on the sheet is not given unless the dedication, engraver, imprint, and map number are in different places on the map.

Size: 18⅝ × 13⅞.

The width and height within neat lines are given in inches unless otherwise noted. If the size is irregular, the bottom and the right-hand neat line measurements are given.

Scale: 1" = ca. 21 miles.

To be read: one inch represents approximately twenty-one miles.

For reduction of German, French, or other measurements to English miles, use is made of the table in Roch-Joseph Julien, *Atlas Topographique et Militaire*, Paris, 1760. Boggs's Natural Scale Indicator is used for maps in which the scale is lacking, varies widely for different parts of the map, or seems faulty. Where degrees are given on such maps, the measurement is made between 33° and 34° North Latitude.

In. Doe, John S. Atlas. London, 1657, No. 55.

Author, title, place of publication, and date of the first edition of the work in which the map is found is given if the map appears in a book. Later editions are indicated by dates under *Copies* in the libraries where they are found; other works in which the same map is found are listed in separate entries under the date of the first appearance of the map.

Inset: The title, size, and scale of insets follow the method used for the main map.

Description. The description usually includes the area shown, features of significance, special bibliographical problems, and historical or biographical items when they are not sufficiently important for the Introductory Essay.

References: Stokes, I. N. Phelps. *The Iconography of Manhattan Island, 1499–1909.* New York, 1915–28, II, 36.

Stokes, *Iconography of Manhattan*, II, 36 (abbreviated form, if work is frequently referred to).

Reproductions: An attempt has been made to list reproductions of maps and to indicate an American library which possesses a photocopy of a manuscript map in a foreign collection.

Copies: Congress 1657,[1] 1662, two copies,[2] 1657–1704,[3] ca. 1684,[4] [1684],[5] separate;[6] University of North Carolina 1662;[7] Yale.[8]

These entries indicate that copies of this map may be found at the Library of Congress, University of North Carolina, and Yale. The Library of Congress has copies of the map in (1) Doe's Atlas, 1657, which is the date of the earliest edition in that library of the atlas which contains the map, (2) date of the next edition having that map in that library (Congress has two copies of this atlas), (3) a copy of the atlas with no imprint date but having maps ranging between the two dates given, (4) a copy of the atlas with no date, probably published about 1684, (5) a copy of the atlas with no date but identifiable as the 1684 edition, (6) a loose copy of the map (either sold separately or detached from the atlas or work in which it is usually found), (7) the University of North Carolina has the 1662 edition of the atlas which contains the map, (8) Yale has no copy of the atlas but has a separately issued or detached copy of the map. The title and date of a book or atlas is usually given as catalogued by the library where it is found, for purposes of identification; significant bibliographical questions are customarily discussed in the *Description.*

LIBRARIES AND COLLECTIONS CONSULTED IN COMPILING THE MAP LIST

Short Citation	Full Name and Location
Congress	Library of Congress, Washington, D.C.
National Archives of Canada	National Archives of Canada, Ottawa, Canada
American Geographical Society	American Geographical Society, University of Wisconsin, Milwaukee, Wisconsin
Boston Public	Boston Public Library, Boston, Massachusetts
John Carter Brown	John Carter Brown Library, Providence, Rhode Island
Charleston Library Society	Charleston Library Society, Charleston, South Carolina
Charleston Museum	Charleston Museum, Charleston, South Carolina
Clements	William L. Clements Library, Ann Arbor, Michigan
Davidson	Cumming Map Collection, Davidson College Library, Davidson, North Carolina
Duke	Duke University Library, Durham, North Carolina
East Carolina University Library	East Carolina University Library, Greenville, North Carolina
Georgia Hargrett Library	Georgia Hargrett Library, University of Georgia Libraries, Athens, Georgia
Harvard	Harvard College Library, Harvard University, Cambridge, Massachusetts
Huntington	Henry E. Huntington Library and Art Gallery, San Marino, California
Kendall	Henry P. Kendall Collection, Kendall Memorial Room, South Caroliniana Library, University of South Carolina, Columbia, South Carolina
Newberry	Newberry Library, Chicago, Illinois
New York Public	New York Public Library, New York, New York
North Carolina State	North Carolina State Library, Raleigh, North Carolina
North Carolina Department of Archives and History	presently Division of Archives and History, Department of Cultural Resources, Raleigh, North Carolina (formerly The North Carolina Historical Commission)
Sondley	Sondley Reference Library, Pack Memorial Library, Asheville, North Carolina
South Carolina Historical Society	South Carolina Historical Society, Charleston, South Carolina
University of North Carolina	Library of the University of North Carolina, Chapel Hill, North Carolina
Yale	Yale University Library, New Haven, Connecticut

For the following map list the author has made an attempt to examine personally and to list all the copies of pertinent maps that are to be found in the foregoing libraries and collections. He has also examined a number of other libraries and collections, such as those of the Massachusetts Historical Society, American Antiquarian Society, University of Virginia Library, The Mariners' Museum, South Carolina Historical Commission, and University of South Carolina Library, for special or unique copies of maps. In addition, he has attempted to list the manuscript maps pertinent to this study which are in the chief European archives. For this purpose map catalogues of European libraries, such works as Hulbert's Crown Collection of maps in the British Library and the Public Record Office, the Karpinski Series of

Reproductions of maps of America in continental archives, and other similar sources have been used. The Geography and Map Division of the Library of Congress has been generous in its cooperation and in obtaining photocopies of manuscript maps in European libraries which it did not previously possess.

Some libraries, such as the John Carter Brown, William L. Clements, University of Virginia, University of North Carolina, Yale University, and the National Archives of Canada, are continuing to add significant items to their cartographical collections of maps of the Southeast. Important items may have been added to such collections between the author's last examination of them and the reader's use of this checklist.

SELECTED BIBLIOGRAPHY

Adams, Randolph G. *British Headquarters Maps and Sketches.* Ann Arbor, Mich., 1928. (British Headquarters maps used during the Revolutionary War period, now in the William L. Clements Library.)

Agee, Rucker. *Maps of Alabama: The Evolution of the State Exhibited in Printed Maps from the Age of Discovery.* Birmingham, Ala., 1955. (A mimeographed list of maps, 1570–1887, in the collection of Rucker Agee, exhibited in the Birmingham Public Library January–February 1955.)

Alden, John R. *John Stuart and the Southern Colonial Frontier . . . 1754–1775.* Ann Arbor, Mich., 1944.

Bagrow, Leo. *Die Geschichte der Kartographie.* Berlin, 1951.

———. *History of Cartography.* Revised and enlarged by R. A. Skelton. Cambridge, Mass., 1964.

Baxter, J. M. "An Annotated Checklist of Florida Maps." Tequesta. Coral Gables, Fla., I, No. 1 (1941), 107–15. (51 maps relating to Florida; 1502–1915.)

The Blathwayt Atlas. Volume 1: The Maps. Edited by Jeannette D. Black. Providence, R.I., 1970. Reproductions of a collection of 48 manuscript and printed maps, gathered for the Lords for Trade and Plantations by William Blathwayt, surveyor and auditor general of the plantations' revenues, about 1683.

The Blathwayt Atlas. Volume 2: Commentary. By Jeannette D. Black. Providence, R.I., 1975.

Calendar of State Papers. Colonial Series. America & West Indies. Various volumes edited by W. Noël Sainsbury, Cecil Headlam, and others. London, 1860–. Vol. I, 1574–1660.

Cartografia de Ultramar. Carpeta II. Estados Unidos y Canada. Toponomia de los Mapas gue la integran relationes de Ultramar. Madrid: Servicio del Ejército, 1953.

Carroll, Bartholomew R. *Historical Collections of South Carolina.* 2 vols. New York, 1836.

Catalogue of the Manuscript Maps, Charts, and Plans, and of the Topographical Drawings in the British Museum. London, 1861. Vol. III, pp. 508–21 (maps of southeastern North America).

Catalogue of the Maps, Plans and Charts in the Library of the Colonial Office. London, 1910.

Catalogue of the Printed Maps, Plans, and Charts of the British Museum. 2 vols. London, 1885.

Catalogue of the Wymberley Jones De Renne Library, Isle of Hope, Near Savannah, Georgia. 3 vols. Wormsloe, Ga., 1931. (L. L. Mackall, librarian 1916–18, supervised card index; Azalia Clizbie compiled the catalogue; the library is now housed in the Hargrett Rare Book and Manuscript Library of the University of Georgia, Athens, Ga.)

Chatelain, Verne E. *The Defenses of Spanish Florida 1565–1763.* Washington, D.C., 1941.

Claussen, M. P., and H. R. Friis. *Descriptive Catalog of Maps Published by Congress, 1817–1843.* Washington, D.C., 1941.

Crane, Verner W. *The Southern Frontier, 1670–1732.* Durham, N.C., 1928.

Cumming, William P. *British Maps of Colonial America.* Chicago, 1974.

———. "Early Maps of the Chesapeake Bay Area: Their Relation to Settlement and Society." In D. B. Quinn, ed., *Early Maryland in a Wider World.* Detroit, 1982.

———. *North Carolina in Maps.* Raleigh, 1966, 1984.

Cumming, William P., R. A. Skelton, and D. B. Quinn. *The Discovery of North America.* London, 1971.

Cumming, William P., S. Hillier, D. B. Quinn, and G. Williams. *The Exploration of North America 1630–1776.* London, 1974.

De Vorsey, Louis. *The Georgia–South Carolina Boundary: A Problem in Historical Geography.* Athens, Ga., 1982.

———. *The Indian Boundary in the Southern Colonies, 1763–1775.* Chapel Hill, N.C., 1966.

———, ed. *The Atlantic Pilot* [a facsimile reproduction of the 1772 edition by William Gerard De Brahm]. Gainesville, Fla., 1974.

———. *De Brahm's Report of the General Survey in the Southern District of North America.* Columbia, S.C., 1971.

Dictionary of American Biography. Edited by Allen Johnson, Dumas Malone, and others. 20 vols. New York, 1928–36.

Dunbar, Gary. *Geographical History of the Carolina Banks*. Coastal Studies Institute, Louisiana State University, Report 8A. Baton Rouge, La., 1956, pp. 245–47, Appendix B: Maps 1529–1807. (18 maps, listed chronologically.)

Fite, Emerson D., and A. Freeman. *The Book of Old Maps*. Cambridge, Mass., 1926.

Florida Library Association. "Maps [of Florida]." *Preliminary Check List of Floridiana 1500–1865 in the Libraries of Florida*. Florida Library Bulletin, II, No. 2 (1930), 15–16. (Short title list of about 100 maps 1520–1865; gives locations.)

Ford, Worthington C. "Early Maps of Carolina." *The Geographical Review*, XIV (April 1926), 264–73.

Freeman, Douglas S., ed. "Confederate War Maps." [Confederate Memorial Literary Society.] Calendar of Confederate Papers. Richmond, Va., 1908, pp. 486–90. (32 maps; 1862–65. "The chief feature of this collection is the series of maps of the late Major-General Jeremy F. Gilmer, Chief of Engineering Bureau, C.S.A.")

Georgia Historical Records Survey. Classified Inventory of Georgia Maps, 1941. East Point, Ga., Georgia State Planning Board, 1941. (Reproduced from typewritten copy; about 2400 titles; 1700–1941; arranged by subject; gives locations, 150 pages.)

Harrisse, Henry. *The Discovery of North America*. New York and London, 1892.

Heaney, H. J., comp. "Check-List of the Elkins Americana, 1493–1869, Now in the Free Library of Philadelphia." In E. Shaeffer, *Portrait of a Philadelphia Collector William McIntire Elkins (1882–1947)*. Philadelphia, 1956.

Hulbert, A. B., ed. *The Crown Collection of Photographs of American Maps*. Series I. 5 vols. Cleveland [1904–8]. Maps in the British Museum. Series II. 5 vols. [Harrow, England, 1909–12]. Maps in the British Museum. Series III [1914–16]. Maps in the Colonial Library, Public Record Office.

John Carter Brown Library. *Annual Report*. Providence, R.I., 1901–66.

John Carter Brown Library. *Annual Reports: Index*. Compiled by Dorothy G. Watts. Providence, 1972.

Journal of the Commissioners for Trade and Plantations. London, 1920–38. From April 1704 to May 1782, in fourteen volumes designated by the periods they cover.

Karpinski, Louis C. *Bibliography of the Printed Maps of Michigan: 1804–1880*. Lansing, Mich., 1931.

———. *Early Maps of Carolina and Adjoining Regions from the Collection of Henry P. Kendall*. Charleston, S.C., 1937.

———. "The Karpinski Series of Reproductions." (Photographic reproductions of manuscript maps illustrating American history found in the chief French, Spanish, and Portuguese libraries and archives, made ca. 1927. The complete series is available in the following libraries: Congress, Clements, Harvard, H. E. Huntington, Newberry (Ayer), New York Public; the French series only, which comprises some 650 photographs, is available in over a dozen other libraries, listed in Karpinski's *Bibliography of the Printed Maps of Michigan*, p. 24.

———. "Manuscript Maps Relating to American History in French, Spanish, and Portuguese Archives." *American Historical Review*, XXXIII (January 1928), 328–30.

Karpinski, Louis C., editor, and Priscilla Smith, compiler. *Early Maps of Carolina and Adjoining Regions Together with Early Prints of Charleston from the Collection of Henry P. Kendall*. Columbia, S.C., 1930.

Laney, F. B., and K. H. Wood. *Bibliography of North Carolina Geology, Mineralogy and Geography with a List of Maps*. North Carolina Geological and Economic Survey. Bulletin 18. Raleigh, 1909. Part II, *List of Maps of North Carolina*, pp. 269–362.

Le Gear, Clara E. *United States Atlases. A List of National, State, County, City, and Regional Atlases in the Library of Congress*. 2 vols. Washington, D.C., 1950–53.

Lewis, C. M., and A. J. Loomie. *The Spanish Jesuit Mission in Virginia 1570–1572*. Chapel Hill, N.C., 1953.

Lowery, Woodbury. *The Spanish Settlements within the Present Limits of the United States: Florida, 1513–1561*. New York, 1901.

———. *The Spanish Settlements within the Present Limits of the United States: Florida, 1562–1574*. New York, 1905.

———. See under P. L. Phillips, ed.

McCormack, Helen G., ed. "A Catalogue of Maps of Charleston Based on the Collection of . . . Alfred O. Halsey." Year Book: City of Charleston, S.C., 1944. Charleston, S.C., 1947 [1948], pp. 178–203. (53 maps; 22 before 1776.)

The Map Collector. Vols. I– . Tring, England, 1971– .

The Map Collectors Circle. Nos. 1–110. London, 1963–74.

Marshall, Douglas, ed. *Catalog of American Maps to Research*

1860 in the William L. Clements Library. 4 vols. Ann Arbor, Mich., 1972.

Meurer, Peter H. *Atlantes Colonienses: Die Kölner Schule der Atlascartographie 1570–1610*. Bad Neustad, 1988.

Nordenskiöld, Adolf E. *Facsimile-Atlas to the Early History of Cartography*. Trans. from the Swedish original by J. A. Ekelof and C. R. Markham. Stockholm, 1899.

———. *Periplus, an Essay on the Early History of Sailing Charts and Sailing-directions*. Trans. by F. A. Bather. Stockholm, 1897.

Owen, T. M. "Maps of Alabama." *Bibliography of Alabama. Annual Report of the American Historical Association*. Washington, D.C., 1897, pp. 1036–43. (75 titles; 1700–1890; alphabetically and chronologically arranged; brief descriptions.)

Paullin, C. O. *Atlas of the Historical Geography of the United States*. Edited by John K. Wright. Washington and New York, 1932.

Penfold, P. A., ed. *Maps and Plans in the Public Record Office. Vol. 2, America and the West Indies*. London, 1974. Maps of the Southeast, pp. 368–506.

Phillips, P. L. *A List of Geographical Atlases in the Library of Congress*. 4 vols. Washington, D.C., 1909–20.

———. *A List of Maps of America in the Library of Congress*. Washington, D.C., 1901.

———. *Virginia Cartography, a Bibliographical Description. Smithsonian Miscellaneous Collections*, Vol. XXXVII. Washington, D.C., 1896.

———, ed. *The Lowery Collection. A Descriptive List of Maps of the Spanish Possessions within the Present Limits of the United States, 1502–1820*, by Woodbury Lowery. Washington, D.C., 1912.

Phillips, Ulrich B. "Maps of Georgia." *Georgia and State Rights. Annual Report of the American Historical Association for the Year 1901*. Washington, D.C., 1902, II, 218–19. (19 maps; 1750–1864; chronologically arranged.)

Powell, William S., ed. *Dictionary of North Carolina Biography*. 4 vols. Chapel Hill, N.C., 1979–91.

Quinn, David B. *The Roanoke Voyages 1584–1590*. The Hakluyt Society, Second Series, CIV–CV (issued for 1952). London, 1955. 2 vols.

———, ed. *New American World: A Documentary History of North America to 1612*. 5 vols. New York, 1979.

Roberts, J. K., and R. O. Bloomer. *Catalogue of Topographic and Geologic Maps of Virginia*. Richmond, Va., 1939. (970 maps, 1782–1939; chronologically arranged; gives locations; extensive notes; indexed.)

Sabin, Joseph, and others. *Bibliotheca Americana. A Dictionary of Books, Relating to America*. 29 vols. New York, 1868–1936.

Sanchez-Aaavedra, E. M. *A Description of the Country: Virginia's Cartographer and Their Maps 1607–1881*. Richmond, 1975. With a portfolio of map facsimiles.

Saunders, William L., ed. *The Colonial Records of North Carolina*. 10 vols. Raleigh, 1886–90.

South Carolina. The Library of the University of. "Maps [Caroliniana]." Author List of Caroliniana in the University of South Carolina Library. Columbia, S.C., 1923, pp. 240–43. (42 maps; 1672–1915; no apparent arrangement. "South Carolina Maps" in the Abney Memorial Library listed on pp. 284–85; 14 maps; 1825–1917; arranged alphabetically by title.)

Stephenson, Richard W. *The Cartography of Northern Virginia: Facsimile Reproductions of Maps Dating from 1608 to 1915*. Fairfax County, History and Archaeology Section, Office of Comprehensive Planning, 1981.

Stevens, Henry, and Roland Tree. "Comparative Cartography Exemplified in an Analytical & Bibliographical Description of Nearly One Hundred Maps and Charts of the American Continent Published in Great Britain during the Years 1600 to 1850." In *Essays Honoring Lawrence C. Wroth*. Portland, Me., 1951, pp. 305–65.

Stokes, I. N. Phelps. *The Iconography of Manhattan Island, 1498–1909*. 6 vols. New York, 1915–28. Vol. II (1916) contains a map list (1500–1700) and reproductions of early maps.

Swanton, John R. *Early History of the Creek Indians and Their Neighbors*. Bureau of American Ethnology, Smithsonian Institution, Bulletin No. 73. Washington, D.C., 1922.

Swem, E. G. *Maps Relating to Virginia in the Virginia State Library and Other Departments of the Commonwealth with the 17th and 18th Century Atlas Maps in the Library of Congress*. Richmond, Va., 1914.

Tennessee War Services Project. Preliminary Inventory of Maps in Tennessee, Published by War Services Project. Sponsored by Tennessee State Planning Commission. Nashville, Tenn., 1942. (106 maps, mostly modern, arranged by counties; descriptive annotations; gives locations; indexed. "Editing of the entries was done by Beatrice O. Cannon.")

Thomassy, M. J. R. Cartographie de la Louisiane (Extrait de la Géologie Pratique de la Louisiane). Nouvelle-Orléans, chez l'Auteur, 1859, pp. 205–26. (42 maps; 1684–1859; gives some locations; bibliographical notes.)

Torres Lanzas, Pedro. *Relación Descriptiva de los Mapas, Planos, &c. de México y Floridas Existentes en el Archivo General de Indias.* 2 vols. Sevilla, 1900. (Chronologically arranged, with an alphabetical index of authors; Volume I, titles 1–319, 1519–1776; Volume II, titles 320–515, 1776–1823.)

Winsor, Justin, ed. *Narrative and Critical History of America.* 8 vols. Boston and New York, 1884–89.

The World Encompassed: An Exhibition of the History of Maps. [Edited by Elizabeth Baer, Lloyd A. Brown, and Dorothy E. Minor.] Baltimore: The Walters Art Gallery, 1952.

Wimberley Jones De Renne Library, Isle of Hope, Near Savannah, Georgia, Catalogue of the. 3 vols. Wormsloe, Ga., 1931. (L. L. Mackall supervised the card index; Azalia Clizbie compiled the catalogue; the library is now in the Hargett Library of the University of Georgia, Athens, Ga.)

Wright, Martin. "A List of Maps and Charts Showing Coastal Louisiana [1500–1954]." *A Geographical and Geological Study of the Louisiana Coast with Emphasis upon Establishment of the Historic Shoreline.* Compiled by the Staff of the Coastal Studies Institute, Louisiana State University, under the supervision of James P. Morgan. Baton Rouge, La., 1955, Appendix A (39 pages). (1500–1954; several hundred maps listed in different categories; older maps, chronologically; topographical survey maps, by serial numbers.)

Map List

01. Waldseemüller 1507

UNIVERSALIS COSMOGRAPHIA SECUNDUM PTHOLOMAEI TRADITIONEM ET AMERICI VESPUCII ALIORUQUE LUSTRATIONES [along bottom of map]

Size: 96 × 54. *Scale:* None.

Description: This large map of the world was printed from twelve wood block engravings each 18 × 24½ inches in size. Only the section devoted to the area of "TERRA ULTERI INCOGNITA," "PARIAS" and the islands lying off the coast is included in Plate 1.

The greatest interest in this map centers on the fact that Waldseemüller and his collaborators used it to forward their argument that the discoveries being made on the western shores of the Atlantic, or as they termed it "Occeanus Occidentalis," should be named in honor of Amerigo Vespucci. In Waldseemüller's book, *Comographiae Introducto*, published with the map, it was stated that: "In the sixth climate toward the antarctic there are situated the farthest part of Africa . . . Zanzibar, the lesser Java and Seula, and the fourth part of the earth, which because Amerigo discovered it, we may call Amerige, the land of Amerigo, so to speak, or America." The southern portion of that "fourth part of the earth" shown on the full Waldseemüller map bears the name "America"—a name that in time came to be applied to the Western Hemisphere's two continents. To further emphasize Vespucci's rank as discoverer, his portrait flanks that of Claudius Ptolemy across a distinctly two-hemisphere, reduced-scale world map in the artistic headpiece to the main map. See L. De Vorsey, "Naming America," in *Keys to the Encounter*, Washington, 1992, pp. 154–59.

After the death of his collaborator, Matthias Ringmann, the Latter, Waldseemüller may have changed his mind regarding Amerigo's claims because his later maps omitted "America" on the South American continent. However, other mapmakers continued to honor Vespucci on their maps to the chagrin of Columbus promoters. It was Mercator, on his 1538 map, who first differentiated between the name South and North America.

Whether or not Vespucci ever sailed along the coast of North America is vigorously disputed; some claim that the Southeast-appearing coastal delineation on the Waldseemüller map reproduced here is imaginary or based upon a misrepresentation of the coast of Cuba or Yucatan. Quite possibly it results from sightings made by an unrecorded voyager to the Florida and Gulf coasts during the period when Spain was actively colonizing Hispaniola. It is also possible that Indians familiar with the region provided some of the geographical intelligence that ultimately went into the map's compilation. Earlier manuscript maps showing coastlines in this region are the "Cantino" planisphere of 1502 and Nicolo Caveri's world chart that probably dates from 1504 or 1505. Photographs of both of these maps as well as the complete Waldseemüller map are included in K. Nebenzahl, *Atlas of Columbus and the Great Discoveries*, Chicago, 1990, plates 11, 13, and 16.

The single surviving copy of Waldseemüller's great wall map was discovered by Joseph Fischer, S.J., in 1901, in the castle of Prince Waldburg-Wolfegg, Württemberg, where it was bound in an old book with a copy of the cartographer's 1516 Carta Marina.

References

Fischer, J., and F. R. von Wieser. *The Oldest with the Name America of the Year 1507 and the "Carta Marina" of the Year 1516, by Martin Waldseemüller (Ilocomilus).* Innsbruck, 1903.

Shirley, R. W. *The Mapping of the World: Early Printed World Maps 1472–1700.* London, 1983, pp. 28–29.

Reproductions

Plate 1 (detail).

Fischer and von Wieser. *The Oldest Map with the Name America of the Year 1507.* Innsbruck, 1903 (facsimile).

Nebenzahl, K. *Atlas of Columbus and the Great Discoveries.* Chicago, 1990, Plate 16, pp. 54–55 (photograph).

Original

Wolfegg Castle, Württemberg, Germany.

---

02. Vespucci 1526 MS

[detail from untitled world map by Juan Vespucci]

Size: 96⅜ × 33⅜. *Scale:* None.

Description: This is a section of a vellum colored manuscript map of the world made in Seville by Juan, the nephew of Amerigo Vespucci. Clearly showing Florida and the adjoining coasts, it appears to record the reports of expeditions led by Juan Ponce de León, Alonzo Alvárez de Pineda, and Pedro de Quexós. Quexós explored the Georgia-Carolina coast for Lucas Vázquez de Ayllón, whose failed attempt to establish a colony there is memorialized by his name on this and a series of maps following Vespucci's. For a listing of early maps showing Ayllón's name associated with the Southeast see L. De Vorsey, "Early Maps and the Land of Ayllón," *Columbus and the Land of Ayllón: The Exploration and Settlement of the Southeast*, ed. Jeannine Cook, Darien, Ga., 1992, Table 2, p. 20.

An old debate concerning the site of Ayllón's failed colony, San Miguel de Gualdape, was revived in 1990 when Paul E. Hoffman published his book *A New Andalucia and A Way to the Orient*, Baton Rouge, 1990. In it he presented an argument for placing the colony somewhere on Sapelo Sound in Georgia as opposed to earlier work which located it on the South Carolina coast. See P. Quattlebaum, *The Land of Chicora: The Carolinas Under Spanish Rule with French Intrusions 1520–1670*, Gainesville, Fla., 1956.

In 1516, ten years before the making of this map, Juan Vespucci had been entrusted with the responsibility of making the padrón real, the official royal chart, for the guidance of Spanish navigators and officials. In 1526 he was a member of the famous Badajoz junta and Pilot of the King. Vespucci had access to the official reports and records of all Spanish explorers; this map is thought to be a copy or preliminary draft of a padrón real being prepared in 1526.

Reproductions

Plate 2 (detail).

Nebenzahl, K. *Atlas of Columbus and the Great Discoveries.* Chicago, 1990, Plate 27, pp. 86–87 (whole map).

Original

Hispanic Society of America. New York.

---

03. da Verrazano 1529 MS

[detail from untitled world map by Gerolamo da Verrazano]

Size: 51 × 102. *Scale:* None.

Description: This map was drawn on vellum by the brother of Giovanni da Verrazano who explored the coast of North America from Georgia to Nova Scotia for France in 1524. When sailing off the hazardous North Carolina Outer Bank barrier islands, Giovanni thought (as he later wrote) that he was observing: "an isthmus a mile wide and about 200 long, in which from the ship, was seen the oriental sea . . . which is the one without doubt, which goes about the extremity of India, China and Cathay." The belief that the continent was pierced by a deep embayment of the Pacific Ocean was further emphasized by the explorer's brother when he drew this map a few years after the voyage. It was an erroneous but persistent feature that was copied for decades on maps of the continent. The "Mare de Verrazano," as the Pacific embayment became known, spawned hopes of an easy route to Asia and played a role in decision making that led to the French attempt to found a colony on the coast of the Southeast in the 1560s, as well as Ralegh's attempts to establish an English colony farther north in the 1580s. See W. P. Cumming, *Mapping the North Carolina Coast: Sixteenth-Century Cartography and the Roanoke Voyages*, Raleigh, 1988, pp. 4–10. A close examination of Plates 8, 11, 15, and 29 will suggest just how persistent the idea of Verrazano's isthmus and arm of the Pacific was in the minds of European geographers and cartographers.

References

Cumming, W. P., R. A. Skelton, and D. B. Quinn. *The Discovery of North America.* New York, 1972, pp. 69–73.

Phillips, P. L., ed. *The Lowery Collection.* Washington, 1912, No. 30, pp. 35–40.

Reproductions

Plate 3 (detail).

Nebenzahl, K. *Atlas of Columbus and the Great Discoveries.* Chicago, 1990, Plate 28, pp. 90–91 (whole map).

Original

Biblioteca Apostolica Vaticana, Vatican City.

---

04. Ribero 1529 MS

[detail from world map prepared in Seville]

Size: 33 × 80. *Scale:* None.

Description: This magnificent map by the Portuguese cartographer, Diego Ribero, was drawn while he was in the service of Charles V of Spain. It is a copy of the official map that was constantly updated by Spain's pilot major, whose duty it was to interrogate all returning overseas navigators. Ribero held that office at the time this map was compiled. Although the land granted to Ayllón in the Southeast is prominently shown, there is no indication of his abortive settlement, San Miguel de Gualdape. The short duration and loss of life in the breakup of the colony probably accounts for this omission. See the discussion of Gualdape's site included in the description of Map 02 by Juan Vespucci, above. A close examination of the coastal waters off the shores of the Southeast reveals a continuous ribbon of fine stippling indicating shoaling or hazardous conditions.

References

Cumming, W. P., R. A. Skelton, and D. B. Quinn. *The Discovery of North America*. New York, 1972, pp. 104–7.

Fite, E. D., and A. Freeman, eds. *A Book of Old Mays Delineating American History*. New York, 1969, pp. 46–49.

Stevenson, E. L. "Early Spanish Cartography of the New World." *Proceedings of the American Antiquarian Society*, XIX (April 1909), pp. 369–419.

Reproductions

Plate 4 (detail).

Cumming, W. P., R. A. Skelton, and D. B. Quinn. *The Discovery of North America*. Plate 115, pp. 106–7.

Nebenzahl, K. *Atlas of Columbus and the Great Discoveries*. Chicago, 1990, Plate 29, pp. 92–95 (whole map).

Original

Biblioteca Apostolica Vaticana, Vatican City.

---

1. De Soto ca. 1544 MS

[endorsed, Mapa del Golfo [*sic*] y costa de la Nueva España desde el Rio de Panuco hasta el cabo de Santa Elena &] | De los papeles que traxeron de Seuilla de Alonso de santta cruz

Size: 23½ × 17⅛. *Scale:* 1" = ca. 120 miles.

Description: This quill-and-ink sketch is an unsigned, undated map of part of southern North America, showing the coast from the vicinity of the Cape Fear River in North Carolina to the Panuco River in Mexico and the interior as far north as the latitude of the Tennessee River. It is also the only extant contemporary map attempting to illustrate the country explored by de Soto (1539–43), a fact which has contributed largely to its fame and frequent reproduction. Many of the 127 names and legends inscribed on it are to be found, though often with different spellings, in the various accounts of the expedition. Most of the coastal names, however, are found on earlier or contemporary Spanish world maps, and locations to the extreme west and in Mexico are the result of information gained from other sources. Indian names on the map such as Ays and Guasco are places in Texas through which Moscoso and the remnants of the expedition passed on their march to the Rio Grande after the death of de Soto on May 21, 1542, and his burial in the Mississippi. On the basis of internal evidence, therefore, the information on the map indicates that it was made after the return of Moscoso to Mexico City in December 1543.

The inscription on the back of the map, which is given above as the title, states that the map was among the papers taken by the royal cartographer, Alonzo de Santa Cruz, from Seville. The title in brackets is that given by Pedro Torres Lanzas, *Relación Descriptiva*, I, 17, No. 1. Santa Cruz, who styled himself archicosmographer to Charles V, aided in editing Spain's master map, the padrón general. A world map by him, dated 1542, is in the Royal Library at Stockholm (reproduced in Erik W. Dahlgren, *Map of the World by the Spanish Cosmographer Alonzo de Santa Cruz, 1542*, Stockholm, 1892); and a sketch of the east coast of North America from a map by Santa Cruz of about 1560 is given in I. N. P. Stokes, *Iconography of Manhattan Island*, II, Plate 18. Dahlgren, op. cit., p. 6, and Franz R. v. Wieser, *Karten von Amerika des Alonso de Santa Cruz*, Innsbruck, 1908, p. viii, both give the date of Santa Cruz's death as 1572, since that is the year in which an inventory of his maps, etc., was made by his successor, Juan Lopez de Velasco. On this basis, Lowery (*Lowery Collection*, p. 78) gives 1572 as the possible date of the de Soto map. B. Boston, "The 'De Soto Map,'" *Mid-America*, XXIII (1941), 242, refers to documents, however, which show that Santa Cruz died November 9, 1567, in Seville. The de Soto map, whether compiled by Santa Cruz himself or not, contains no information concerning the expedition of Tristan de Luna (1559) or the founding of St. Augustine by Menéndez (1565). It is probable that the map was made from information gained from Moscoso or possibly some other member of the de Soto expedition soon after their return; the most commonly given date, ca. 1544, is a reasonable one. It might be called the de Soto-Moscoso-Santa Cruz map.

The fact that there are many more place-names from the latter portions of de Soto's extended march suggests strongly that the map was compiled from memory rather than any sort of day-by-day journal or itinerary. In keeping with this hypothesis, many Indian settlements visited in the first year of the expedition are omitted or incorrectly located.

The spelling and location of Indian settlements differ in many cases from the accounts of the Gentleman of Elvas,

Ranjel, Biedma, and Garcilaso de la Vega, and may be independent of them. B. Boston, op. cit., p. 245, suggests that the author of the map used Biedma and a more complete, non-extant manuscript report by Ranjel as sources. It should be noted, however, that many important places named by Ranjel are missing from the map. There are sufficient similarities between this map and the Ortelius-Chiaves 1584 map (Plate 9), such as in the area shown, in the general topographical outlines and in many of the names, to indicate a definite relationship between the two. A close examination, however, reveals many differences; there are only 56 continental names on the Ortelius-Chiaves map as compared to 127 names and legends on the earlier MS map. The location and spelling of the names often differ, as the following list shows.

East Coast names, reading up from the Martyres

Soto ca. 1544 MS	Ortelius-Chiaves 1584
Los martires	Martyres
florida	La florida
C. de canaveral	C. de Cañarval
R. de corriẽtes	Rio de Coriente
C. delacruz	C. de Cruz
marbaxa	—
R. Seco	Rio Seco
c. d[e] nillo [*or* c ð nillo]	—
C. grueso	C. Grueso
C. de S.ta elena	P. S. Helenae
—	S. Helenae
—	Anames
—	Rio Iordan
—	Rio Canaas

Eastern Interior names, reading east to west
(Names on the de Soto map indicated by number after corresponding names on Ortelius-Chiaves map)

1. Laguna dulce	—
2. chitala	Aymay (4)
—	Catilacheque
3. chalaq	Chalaqua (3)
4. abuymay	Xuala (2)
5. guaqujlla	—
6. cofaq˜ [cotaq˜]	Cafaqui (6)
7. guazullj	Guaxuli (7)
8. canera gae [*or* gai]	Canaraqay (8)
9. capalar	—
10. chiaha	Chiacha (10)

The topographical detail on the MS map is on the whole more accurate than on the printed map. The rivers are more accurately drawn. But the River Seco joins a river flowing into the Gulf of Mexico; on the Ortelius-Chiaves map, another river branches off from the River Seco near its headwaters and empties into the Atlantic north of the mouth of the Seco, an erroneous conception which possibly arose from the heavily broken coastline of Georgia and which is followed by some later sixteenth-century cartographers. It has been argued that the curious way in which some rivers are depicted on these maps is "one of the key diagnostic traits revealing Native American cartographic influence on European maps." Briefly put, this argument maintains that the Europeans misunderstood the manner in which the Indians described or mapped mixed networks such as transportation or travel routes. To the Indian the overall pathway or route system was the important thing and whether a segment was traveled on foot or in a dugout canoe was incidental. Thus rivers on European maps that appear to be forking or joining in ways not found in nature may be the result of some explorer's misunderstanding of cartographic information gained from Indians and his resulting failure to differentiate between river systems and the overland trails which connected them. (See L. De Vorsey, "Silent Witnesses: Native American Maps," *The Georgia Review*, XLVI (Winter 1992), 716–17.) The range of mountains is east-west on the MS map; on Ortelius-Chiaves it runs from southeast to northwest exactly at right angles to the true direction of the Appalachians. Undoubtedly this error of direction, which is found on later maps for over a hundred years, resulted from the report of the de Soto expedition, which entered the mountainous region in western North Carolina and continued in it as they marched westward along the Tennessee River and into northern Alabama.

In the extreme northeastern section of the de Soto map is a large freshwater lagoon or lake, shaped roughly like an "H" and labeled "laguna dulce," and below it is another smaller lake. Harrisse (*Discovery of America*, p. 644) says: "As to the large lagoon on the north-east, it is most likely intended for the Ekanfanoka or Ouaquaphenogaw marsh, between Flint and Oakmulgee rivers in Georgia. The lake south of that lagoon may be the Okeechebe, north of the Everglades in Florida." Paullin (*Atlas of the Historical Geography of the United States*, p. 9) follows Harrisse: "The large lake in the northeast may be the Okefenoke Swamp, in southern Georgia." But there are various indications that these lakes portray a knowledge of the Great Lakes, conveyed to de

Soto's men by the Indians with whom they conversed. They may also be connected with the puzzling appearance of the fabled lake in the region of northern Georgia which is found in many maps of North America until well into the eighteenth century. In the first place, the lakes are not represented as being in the Florida peninsula at all, but to the east of the long row of hills which indicate the Appalachian mountains, at about 35° N.L. In the second place, they are placed east of Chitala on the map. Chitala (Xuala) is in the mountains of western North Carolina. It is possible that the lagoon and smaller lake represent lakes and sounds in eastern North Carolina. The possibility that Chesapeake Bay might be intended also exists. Ortelius-Chiaves (1584) and Cornelius Wytfliet (1597) include cartouches which cover the area where the lakes are shown in the de Soto map. The maps of Florida by Acosta (1698) and Matal (1600) follow Ortelius in having no lake.

Another manuscript map belonging to Santa Cruz and possibly made by him while he was helping to edit the padrón general is found in the Archivo Historico Nacional in Madrid. Without author, title, or date, it is endorsed "Golfo y costa de la nueva esp.[na] De los papeles que trugeron de Seuilla de Alonso de Santta Cruz." It has been reproduced in *Mapas Españoles de América*, Madrid, 1951, Plate VIII, where it is given the erroneous date of 1536. It is also reproduced in Cumming, Skelton, and Quinn, *The Discovery of North America*, p. 172. The map depicts the eastern coast of South and North America from 10° to 42° N.L. It is also reproduced without author or date and with no reference to the endorsement or place of origin, as "Carta de las Antillas, Seno Mejicano y Costas de Tierra Firma y de la America Setentrional," Facsimile R, in *Cartas de Indias*. Publicadas por primera vez el Ministerio de Fomento. Madrid, 1877. Two names on the Florida coast of the map give it an approximate date and also greatly increase its importance. These are "r: de s: agostino" at 29°55' N.L. and "r: de s: matte" at 30°30' N.L. These are within a few minutes of the actual latitude of St. Augustine and San Matheo. The names were given by Pedro Menéndez de Avilés at the time of his attack on the French under Ribaut and are not found in any document or map along this coast before 1565. (See John Gilmary Shea, "Ancient Florida," in Justin Winsor, *Narrative and Critical History of America*, II, 263 and 273.) Since the map was among Santa Cruz's papers at the time of his death on November 9, 1567, the date of this map cannot be earlier than 1565 nor later than 1567. Its significance as an authoritative record of the official Spanish conception of the coast and nomenclature of this period is evident. Starting at the point of Florida and reading northward along the Atlantic coast one finds old names familiar from the time of Ayllón and some new ones: florida, los martiles, tierra de quasto, c: del canaueral, r: de moschites, r: de s: agostino, r: de s: matte, r: de s: joan, r: de s: age, r: de s: andres, r: de guales, r: de tacatecoro, r: de balenas, r: de s: elena, c: de s: elena, r: de medanos, r: dulce, r: de guamas, r: de canoas, r: de nogales, r: de pinos, r: deloro, c: de s: romano, p: del principe, c: de trafalcar, r: del spto: santo, ba: de n[ra] senora [by another hand], r: salado. The names along the Florida coast frequently differ from those given by Chaves in the Quadripartitu[m] en cosmographia pratica [ca. 1539] as listed in Stokes, *Iconography*, II, opp. p. 40; the map also has more and later names. To the interior of the North American continent on the map are a large and a small lake. In the larger lake there is an island with a conventionalized walled city drawn on it; seven rivers flow into or out of the lake. The lake is called "coniuas lacus," and the city "nuevo mesico" (the city of Mexico, also situated in a lake, is shown in its proper place). The latitude is given as "Quere star a 54: grados," though on the latitude scale it would be about 48° N.L. Otherwise the interior of the continent is bare, with only the coastal names.

Possibly the first appearance of the Great Lakes on any printed map is on the great Mercator world map of 1569. On it is a freshwater lake in the middle of the continent, with three smaller connected lakes to the east which flow into the St. Lawrence. On the large lake is the inscription: "Hic mare est dulcium aquarum, cuius terminum ignorari Canadenses ex relatu Saguenaiesum aiunt." This "mare dulce" appears in many subsequent maps, in various forms, for over a half century.

References

Boston, B. "The 'De Soto Map.'" *Mid-America*, XXIII (1941), 236–50.

Harrisse, H. *The Discovery of North America*. New York and London, 1892, pp. 643–44.

Mapas Españoles de América. Madrid, 1951, pp. 53–56.

Phillips, P. L., ed. *Lowery Collection*. Washington, 1912, p. 78, No. 60.

Robertson, J. A., ed. and trans. *True Relation . . . by a Gentleman of Elvas*. Publications of the Florida State Historical Society. No. 11. Deland, Fla., 1933, II, 418–24.

Swanton, J. R., Chairman. *Final Report of the United*

States de Soto Expedition Commission. Washington, 1939, pp. 11, 58–59.

Torres Lanzas, Pedro. *Relación Descriptiva de los Mapas, Planos, &c. de México y Floridas*. Sevilla, 1900, I, No. 1.

Reproductions

Plate 5.

Cumming, Skelton, Quinn. *Discovery of North America*, p. 121.

Harrisse, H. *Discovery of North America*, opp. p. 132.

Mapas Españoles de América, Plate IX.

Paullin, C. O. *Atlas of the Historical Geography of the United States*. Washington, 1932, Plate 12A.

Priestly, H. I., ed. and trans. *The Luna Papers*. Deland, Fla., 1928, I, opp. p. xix.

Robertson, J. A., ed. and trans. *True Relation . . . by a Gentleman of Elvas*. Deland, Fla., 1933, II, opp. p. 418 (with tissue overlay having the names and legends printed).

Congress (Lowery Collection photocopy).

Original

Sevilla. Archivo General de Indias. Indiferente General. Est. 145. Caj. 7. Leg. 8. Ramo 272. (Mapas. Mexico, 1.)

---

2. Gutiérrez 1562

Americae Sive qvartae Orbis Partis Nova Et Exactissima Descriptio avctore Diego Gvtiero Philippi Regis Hisp. Etc. Cosmographo. Hiero Cock Excvda. 1562 [across top of map] | Hieronymus cock excude. cum gratia et priuilegio 1562 [cartouche, bottom left]

Size: 34 × 36¾. *Scale:* 1" = ca. 270 miles.

Description: This is the largest and most detailed map of the New World to be engraved and published up to the time of its appearance. It includes the east coast of North, Central, and South America, and parts of the west coast of Europe and Africa. Diego Gutiérrez (1485–1554), the author, was born in Seville and became a maker of naval instruments and charts. About 1547 or 1549 (cf. Harrisse, *Discovery of North America*, p. 720) he became pilot major, ad interim, by virtue of power of attorney from Sebastian Cabot. In 1550 he made a portolan chart of the Atlantic Ocean, incorporating new information about the east coast of America, and signed it "Cosmographo de Su Majestad." After his death his salary was transferred, October 22, 1554, to his son Diego Gutiérrez, junior.

This map, made by a cosmographer and acting pilot major who had access to the latest Spanish charts and reports of his time, shows clearly the current conceptions and knowledge then available in Spain. It does not, however, show the explorations of de Soto or other recent voyages of discovery by Spanish and French. Although the map was engraved in 1562 by the Flemish Hieronymus Cock, presumably the manuscript original could not have been made later than 1554, when Gutiérrez died. Although Mercator used this map for part of the New World, such as the Gulf coast, in making his great wall map of 1569, he did not use it for his topography of the east coast of North America, which differs from and is better than that of Gutiérrez. The gross misspellings and fanciful coastlines show that the map was engraved primarily for artistic appeal, with little or no knowledgable supervision.

To the interior of southeastern North America are names of Indian tribes reported by Ponce de León, Ayllón, and others, and here accepted as names of regions: Otagil, Calivaz, Apalchen, Avacal, and Mocosa. Along the coast are names given to topographical features during explorations: R. de Iuan ponce and B.ª de Iuã Ponce on the west Florida coast; on the Atlantic coast Rio de corintes, C. de aruz, Marbaxa, R. de terra llana, C. gruedo, C. de S. Elena, R. de S. Elena, Andenes, R. Jordan, C. de S Roman, R. de Canoas, Ananes, Terra llana, R. baxo, P.º del principi, C. de terra falgar, R. de santo spirito, B.ª de s. Maria, R. Salado, R. de la playa, C. de S. Ivan, C. de las aranas. The islands which dot the entrance to Baya de Santa Maria may represent the Outer Banks of North Carolina, although, as C. M. Lewis and A. J. Loomie (*The Spanish Jesuit Mission in Virginia*, pp. 10ff.) show, toward the end of the century the Spaniards were giving this name to Chesapeake Bay. On the other side of the continent, at the southern tip of Lower California, appears "C. Califormia," the first appearance of the name on a map: see Ruth Putnam, *California: The Name*, Berkeley, Calif., 1917.

References

Le Gear, C. E. "Rosenwald Gift of 16th-Century Maps." In Walter W. Ristow, ed., *A la Carte*. Washington, 1972.

The World Encompassed. Baltimore, 1952, No. 229.

Reproductions

Plate 6 (top left section of map, showing North America).

Le Gear, op. cit., opp. p. 18.

Bagrow, Leo. *History of Cartography*. Cambridge, Mass., 1964. Plate 69.

Copies

Congress; British Library.

2A. Parreus French Florida 1562 [1563] MS

Has terras perlustravit, Nicolasis Parreus. Turronensis Caroli Noni Auspicijs Anno D. 1562 [cartouche, upper right] La Terre Francoise Nowellement Decowerte [across face of map, upper center] descripcion de la costa dela florida descubierte por los francese 61 [endorsed]

Size: 15½ × 12⅜ (39.5 × 31.5 cm).

Scale: 1" = ca. 21.1 miles (1:1,400,000).

Description: This map, which extends from Cabo François (Anastasia Island) north to South Edisto River, is a tracing of a French map on thin paper, evidently made by a Spanish spy or scribe. It records the exploration of Jean Ribaut along the coast in May 1562. The names are those given by Ribaut to the islands, bays, and rivers discovered and examined. Finding an excellent harbor at Port Royal Island, he built Charlesfort and garrisoned it with 30 men; he then returned to France for reinforcements. However, he found religious civil war and fled to England. Charlesfort was short-lived; the colonists abandoned it in the summer of 1563, having elected Nicolas Barré, the pilot of Ribaut's expedition, their leader. Becalmed in the small vessel, they began to die and resorted to cannibalism before they were picked up by an English vessel and taken to England. There Barré recovered. In London Barré was interrogated, placed under house arrest, and then allowed to return to France. He probably joined the expedition under Laudonnière or Ribaut that returned to Florida in 1564 and was killed in the massacre of the French by Pedro Menéndez de Avilés in 1565.

There is no Parreus or Parré in contemporary archival records at Tours, though the name Barré, and even Nicolas Barré, is found. Marcel Destombes and Maurice Béguin, conservateur en chef, Archives Departmental d'Indre et Loire, agreed that apparently the Spanish scribe did not see the bottom curve of the B [Barreus] through the tracing paper. A thorough search through Spanish, French, and English archives has failed to reveal how an official map and report is found in the fourteenth volume of notes of an early nineteenth-century Spanish historian.

Le Moyne's map of Florida, published by De Bry in 1591, gave a stylized and confusing representation of the coast explored by the 1562 expedition, which Le Moyne had not accompanied. For a hundred years most European cartographers used French names for the region Ribaut had explored, except for those mapmakers who followed the still earlier Spanish nomenclature. Even after maps with English names for the Carolina coast appeared, the old French names continued to be used on new maps until the middle of the eighteenth century and after. From the beginning, inability to identify Ribaut's nomenclature with the known topography of the region or to find the location of his "Charlesfort" settlement has confused mapmakers and historians. The Parreus map solves most of the problems of identification for nearly fifty names found on early maps and reports along the southeastern coast of North America, such as R. de May (St. Johns River), R. de Seine (St. Marys), and Chenouceau (Battery Creek). The location of Charlesfort was not on Parris Island (the site of the incorrectly located commemorative monument) but on the east bank of Battery Creek near its mouth on Port Royal Island, as it is less exactly shown on Le Moyne's 1591 map.

A description in Spanish of the information on the map written by the scribe or spy who traced the map follows in Navarrete's bound volume, fol. 460; it shows little geographical knowledge or linguistic acumen. It is published in the *Imago Mundi* article listed below, which has an identifying table of this and other maps using the coastal names given by Ribaut.

Reference

Cumming, William P. "The Parreus Map (1562) of French Florida." *Imago Mundi*, XVII (1963), 27–40.

Reproductions

Plate 6A.

Cumming, William P. "The Parreus Map (1562) of French Florida," op. cit., p. 29.

Quinn, D. B., ed. *New American World*. New York, 1979. Vol. II, plate 79.

Original

Madrid. Museo Naval. Navarrete Collection, Vol. XIV, fol. 459, dto. 58.

3. Norment-Bruneau 1565

La Carreline [name in text on page opposite plan]

Size: 3⅝ × 5¾. *Scale:* None.

In: Coppie d'vne Lettre Venant de la Floride enuoyée à Rouen, & depuis au Seigneur d'Eureron: Ensemble le plan & portraict du fort que les Francois y ont faict. A Paris, Pour Vincent Norment & Ieanne Bruneau, en la rue neufue Nostre Dame, à l'Image Sainct Iean l'Euangeliste. 1565, opp. p. [8].

Description: This plan of the fort on "La Riviere de May" (St. Johns River) and the surrounding country is in a letter written by a young Frenchman who accompanied his father on Laudonnière's expedition to Florida. It was sent back to France by ship before Menéndez's attack. The engraving is after a "portraict" drawn by the boy and referred to in the letter. "Le port de entree" of the River May, the adjacent "Bras de Mer," the "sovrse de la Riviere de May," the hills and terrain surrounding the fort, are without topographical resemblance to the actual situation. The plan of the fort is in general similar to that in the ninth and tenth engravings of Le Moyne (1591). "Nous menerẽt," writes the author on the eighth and ninth page of the printed letter, "au mesme ou de present auõs faict nostre fort lequel se nõme le fort de la Carreline, & la on nommé ainsi parceque le Roy a nom Charles, dequel en poueiz veoir le portraict cy apres. Lequel fort est sur ladicte riviere de May, enuiron six lieues dãs la riviere loing de la mer."

The letter but not the plan is reprinted in Henry Ternaux-Compans, ed., *Voyages, Relations, et Mémoires Originaux pour Servir à l'Histoire de la Découverte de l'Amerique*, Paris, 1837–41, pp. 233–45.

Copies

John Carter Brown.

3A. Gerard Mercator 1569

Nova Et Aucta Orbis Terrae Descriptio Ad Usum Navigantium emendate accommodata [along top of map]

Size: 52 × 78. *Scale:* None.

Description: Mercator's engraved world map of 1569 is one of the most important and influential in the history of mapmaking; this importance is based more on the value of its projection than on its geographical content and accuracy. This detail from that map shows the Southeast as it was depicted on the first printed world map drawn on the now famous "Mercator" map projection. Attempting to show the earth's curved surface on flat maps requires ingenious graphic compromise on the part of cartographers, since the three-dimensional globe cannot be accurately depicted on the two-dimensional plane surface of a map. The schemes cartographers employ to produce our flat maps of the curved earth are referred to as "map projections" and involve the sacrifice of certain map qualities to gain others. In simple terms, Mercator mathematically projected the lands and seas he was mapping from a spherical system of latitude and longitude onto the cylinder of paper that forms his map. Along the equator and close to where sphere and cylinder touch the scheme works well, and the resulting map is quite accurate as to direction, distance, shape, and area in the regions shown. Moving from the equator toward the poles, however, results in increasing distortion. This is inevitable because the meridians of longitude that converge at the poles are drawn as straight north-south lines and the poles, which are points, are rendered as straight east-west lines equals to the equator in length. Probably the best known result of this condition is the gross exaggeration found in the area of Greenland, which appears to be larger than South America, although it is only one-ninth the size. Mercator's motivation was to produce a map on which compass directions could be drawn as straight lines and be easily followed by low- and mid-latitude navigators. In this he succeeded. A translation of Mercator's lengthy explanation of the purpose and method of his innovative projection and map is included in E. D. Fite and A. Freeman, *A Book of Old Maps Delineating American History*, pp. 76–80. Edward Wright's explanation of the theoretical principles underlying Mercator's projection appeared in 1599 and led to its increasing use after that date. The entrenched conservatism of navigators, however, slowed its widespread adoption for almost another century.

Mercator depended primarily on Spanish sources for his interpretation of Southeastern geography depicted on the detail included here. The delineation of the coast line, the southeasterly flow of the rivers, and the range of the Appalachian mountains, which are parallel to the shoreline, make this one of most accurate general representations of the region yet to appear.

References

Phillips, P. L., ed. *Lowery Collection*. Washington, 1912, No. 55, pp. 68–74.

Shirley, R. W. *The Mapping of the World: Early Printed World Maps 1472–1700*. London, 1983, pp. 137–42.

Reproductions

Plate 7 (detail).

Nebenzahl, K. *Atlas of Columbus and the Great Discoveries*. Chicago, 1990, Plate 40, pp. 128–29 (whole map).

Original

Bibliothéque Nationale, Paris.

---

4. San Marcos ca. 1578 MS

[Second Spanish Fort at Santa Elena]

Size: 7 × 5 (reproduction). *Scale:* None.

Description: San Marcos, the second Spanish fort at Santa Elena, was built in the fall and winter of 1577–78. It had a central caballero, three cavaliers, and ten pieces of artillery. It was rectangular in structure, unlike the earlier triangular forts at St. Augustine and its predecessors at Santa Elena, the French Charlesfort of 1562, built by Ribaut on Parris Island, and the Spanish San Felipe of 1565–76.

San Felipe was burned by the Indians in 1576. In 1577 Pedro Menéndez Marqués, who was an admiral in the royal navy and had that spring been appointed captain-general and provisional governor of Florida, began the second fort. It was probably finished early in 1578, for Menéndez had expected to be in Havana by Christmas 1577 but was still at that time supervising the building; see J. T. Connor, *Colonial Records of Spanish Florida*, I, 291, 293.

Don Francisco Carreño, governor of Cuba, writing to the king of Spain on February 12, 1578, concerning the importance of a settlement at Santa Elena, stated that more than twenty ships had been wrecked in that same place. Most of the survivors were killed by the Indians; but if peace were made and if those Spanish remaining had not been put to death, they could be ransomed: Connor, op. cit., II, 333. Santa Elena was, next to St. Augustine, the most important settlement along the coast. There were forts at San Pedro, on the island of Tacatacuru or Cumberland Island at the mouth of St. Marys River, and at San Mateo on St. Johns River, but no settlers.

The identification number and the size of the original map in the Archives of the Indies are lacking in the reproduction of the map by Mrs. Connor in the *Colonial Records of Spanish Florida*. It has since been identified. The plan of the fort may have been sent by Balthazar del Castillo y Ahedo, the Visitador (royal inspector) to Florida. In April 1578 Castillo wrote to the king concerning the building of the second fort by Menéndez (Connor, op. cit., II, 43; Archivo General de Indias, est. 54, caj. 1, leg. 34). On October 12–15, 1578, and November 1, 1578, another Visitador, Alvaro Flores, visited San Marcos and made a thorough examination of the fortification, artillery, and the seventy-nine soldiers with their equipment (Connor, op. cit., II, 117; Archivo General de Indias, est. 2, caj. 5, leg. 2/10).

Reproduction

Connor, J. T. *Colonial Records of Spanish Florida, 1570–1580*. Deland, Fla., 1925–30, II, opp. p. 50.

Original

Sevilla. Archivo General de Indias. Mapas. Mexico 4.

4A. Fernandez-Dee 1580 MS

The Cownterfet of Mr Fernando Simon his Sea carte which he lent unto my master at Mortlake. A° 1580. November 20. The same Fernando Simon is a Portugale, and borne in Tercera being one of the Iles called Azores. [upper left at center of the North American continent]

Size: 47¼ × 32. *Scale:* 1" = 218 miles.

Description: This vellum chart of the seacoasts bordering the Atlantic Ocean and of the west coast of the Americas extends from 68° N.L. to 36° S.L., and from California to the Cape of Good Hope. The coastlines with their bays, islands, and shoals are carefully delineated, but no names have been entered on the chart except along the North and Central American coasts. The form and nomenclature show that Fernandez's "Sea carte" derived from a Spanish chart of a type common at the time. The occasional Portuguese spellings (montanhas, tiera lhana) show that the map which "my master's" man was given to copy had been drawn by a Portuguese, possibly not necessarily Fernandez himself.

The map is without identifying title, author, or ornamentation except for the note given above. From it one may infer that the well-known Portuguese master-pilot Simon Fernandez (Simão Fernandes, Ferdinando) took this chart with him on a visit to Dr. John Dee of Mortlake on November 20, 1580, and that Dee borrowed it and had it copied. Dee had an active interest at this time in western voyages of discovery, especially in the Northwest Passage (Map 4B and Plate 8), and Fernandez was probably reporting on his extraordinary three-month voyage of reconnaissance to the coast of North America in the spring of 1580. Fernandez played an important part in the Roanoke voyages; he was master and pilot of Amadas's flagship in 1584 and of Grenville's flagship in 1585. In 1587 Fernandez was master of the flagship "Lion" and was listed as an assistant to the governor of the "City of Raleigh" in Virginia.

Fernandez's map shows graphically the probable reason for the choice of the Carolina Banks area as the location of Ralegh's colony. The only bay shown on the east coast between the Cabo de la Florida and Cabo de Arenas is Bahia de Santa Maria, which on the map is placed at the latitude of Pamlico Sound, to which they directed their course. This is depicted as a large bay with two rows of three islands each across it. It is strategically located for preying on Spanish shipping, and for this Elizabethan sport Fernandez had almost an obsession. With its islands and with the rivers flowing into it the bay appears on the chart to be also the best haven on the coast for concealment from Spanish countermeasures.

What geographic entity the name "Santa Maria" referred to on the Spanish charts is uncertain; a number of modern commentators identify it with Chesapeake Bay. There is little doubt, however, that to Fernandez and the Roanoke colonists it was Pamlico Sound. On the 1585 chart of Virginia, Wococon Inlet is called "the port of saynt maris wher we arivd first" (cf. Plate 10 and Map 6). Governor Lane refers to Ococan Inlet in a letter to Walsingham as "Thys Porte in y[e] Carte by y[e] Spanyardes [Fernandez's chart?] called S[t] Marryes baye" (Quinn, *Roanoke Voyages*, I, 201, n. 10). On Fernandez's chart the entrance of Bahia de Santa Maria is between 35°20' and 35°30'; on White's map (plate 11) the inlet to the south of Wococan Island is at about 35°17' and the inlet to the north is at about 35°43'. The actual latitude of Ocracoke Inlet at present is 35°04'. It was Fernandez who discovered another inlet to the north at 35°50', called after him Port Ferdinando, which was later used by the colonists as the approach to Roanoke Island.

References

Cortesão, A., and A. Teixeira da Mota. *Portugaliae Monumenta Cartographica*. Lisbon, 1960, II, 129–31.

Quinn, D. B. *The Roanoke Voyages: 1584–1590*. London, 1955, I, 98ff.; II, 517ff.

Quinn, D. B. *The Voyages and Colonizing Enterprises of Sir Humphrey Gilbert*. London, 1940, I, 51ff.; II, 239ff.

Taylor, E. G. R. *The Troublesome Voyages of Captain Edward Fenton: 1582–1583*. London, 1959, xxxiii ff.

Reproductions

Cortesão, A., and A. Teixeira da Mota, op. cit., II, Plate 240.

Cumming, Skelton, Quinn. *The Discovery of North America*, p. 174.

Quinn, David B. *New American World*, Vol. III, Plate no. 89.

Original

London. British Library. MS Cott. Rol., XIII, 48.

4B. John Dee ca. 1582 MS

[Northern Hemisphere, from pole to Tropic of Cancer]

Size: 24½ × 19¾. *Scale:* None.

Description: This polar view of the Northern Hemisphere was in the collection of Sir Humphrey Gilbert, an Elizabethan leader in the search for a Northwest Passage. His ownership is testified to by an inscription in his hand that reads, "Humphray Gylbert knight his charte." Below that inscription, in the lower right corner of the map, appears a set of three cabalistic signs identifying the cartographer as John Dee, the philosophical geographer and adviser to colonial expansionists in Elizabeth I's court. The map first came to light in 1928 when it was sold with other materials from the library of one of Sir Walter Ralegh's close associates, Henry Percy, ninth earl of Northumberland. The map's date is in part based upon the paper's watermark, a bunch of grapes with the name Gouault. Similar paper was used in De Jode's atlas of 1578.

Dee's map, like that of Michael Lok which was published in 1582, shows Verrazano's Sea as an arm of the Pacific Ocean almost bifurcating North America. In the area of the Southeast where Lok draws the "R. de May," Dee shows that river linked to a large inland lake south of Verrazano's Sea. A river originating at Mexico City flows into that unnamed inland sea, and another waterway links with the Gulf of Mexico. In Dee's construct the Southeast appears to be divided into two regions or provinces named "Florida" and "Apalchin" separated by "r. meo," presumably Lok's "R. de may," the name later applied to Florida's St. Johns River. With maps like Dee's and Lok's in view it is understandable to read that, a few years later, Governor Ralph Lane of the abortive Roanoke Colony led an expedition up the Roanoke River in North Carolina which, according to Thomas Harriot, "either rises in the Bay of Mexico, or else from very near unto the same, that opens out into the South Sea [the Pacific]."

References

Bishop, R. P. "Lessons of the Gilbert Map." *The Geographical Journal*, LXXII (September 1928), 235–43.

Cumming, W. P. *Mapping the North Carolina Coast.* Raleigh, 1988, pp. 32–42, 127–28.

Heaney, H. J., comp. "Check-List of the Elkins Americana, 1493–1869, Now in the Free Library of Philadelphia," in E. Shaeffer, *Portrait of a Philadelphia Collector William McIntire Elkins (1882–1947)*, Philadelphia, 1956.

Reproduction

Plate 8.

Original

Free Library of Philadelphia (W. L. Elkins Collection, No. 42).

---

5. Ortelius-Chiaves 1584

La Florida. Auctore Hieron. Chiaves. [Cartouche, bottom center] | Cum Priuilegio [Cartouche, top right]

Size: 8⅞ × 6. *Scale:* 1" = ca. 230 miles.

In: Ortelius, Abraham. *Theatrvm Orbis Terrarvm.* Antverpiae, 1584, No. 9.

Description: This map gives the coast southward from Carolina (35° N.L.) along the Atlantic and the Gulf of Mexico to the Mexican coast at the Tropic of Cancer (23½° N.L.); the interior delineation extends somewhat north of the latitude of the Tennessee River (41° N.L.).

The Atlantic coastal names are those commonly found on Spanish maps during the second and third quarters of the sixteenth century but show no knowledge of the French and Spanish settlements and explorations during the third quarter of the century. To the interior are names of Indian settlements first reported in Europe upon the return of members of the de Soto expedition. The area shown, the general configuration, and the names given have similarities to the earlier de Soto map, though the differences in detail are so numerous that any direct dependence of the printed map on the manuscript is problematical. The names of 34 continental rivers, capes, and bays, and of 22 Indian settlements are on the Ortelius-Chiaves, as compared to 52 coastal names and 75 interior names and legends on the de Soto map. The spellings on the printed map show more phonetic similarity to those in the account of the Gentleman of Elvas (and, less frequently, of Biedma) than do those on the de Soto map, which has names not given, or at least not identifiable, elsewhere. Finor in the central Alabama region in the de Soto map is not found in the other accounts; Chitala is spelled Xuala in Chiaves and elsewhere. On the other hand, Ortelius-Chiaves has Catilachegue (Cofitachequi, the important Indian settlement), which is not given on the de Soto map. Since the account of the Gentleman of Elvas was printed by 1557 and the account of Biedma was at least available in the Archivo de Simancas (whence it was later transferred to the Archivo General de Indias in Sevilla), it is at least possible that these original sources were known to the maker of the map. The names (cf. Canaas for Canoas, etc.) are sometimes mutilated; both the draftsman and the engraver were Dutch, copying a map made by a Spaniard (cf. Delanglez, op. cit., pp. 72–73).

Gerónimo de Chaves (other signatures Chiaves, Hieronymus Chauez, H. Chiauez), who designed this map for Ortelius, was born in Seville in 1524. He was the son of Alonso de Chaves, maker of the lost padrón general of 1536 and author of *Opus Quatripartitum en Cosmographia pratica—Espeio de Navegantes* [1539]. Géronimo de Chaves was pilot and successor to Sebastian Cabot in the Chair of Cartography in the Casa de Contratacion in 1552. He graduated as bachelor in mathematics from the University of Seville, and was already noted as a scientist when he published his translation of J. Sacro-Busto's *Sphera del Mondo* in 1545, for which he drew a small chart of America. He was cosmographer to Philip II of Spain and made maps of Andalusia and Sicily for Ortelius, as well as this one of Florida. (See Bagrow, op. cit., p. 56; Harrisse, op. cit., p. 710; Stokes, *Iconography of Manhattan*, II, 40–41.)

Ortelius was, next to Mercator, the greatest geographer of the sixteenth century, and his *Atlas* had a profound influence on early cartography. The map of Florida first appeared in the third supplement or 1584 Additamentum to Ortelius's work, the first edition of which was published in 1570. The map was also included in Ortelius's complete Latin edition of the same year (1584) and was usually included in the Latin editions and in their foreign translations for many years thereafter. An unchanged plate of Florida, having the same accidental markings, was used in all editions.

A complete list of the known editions and translations of the Ortelius Atlas follows: Latin 1570 (four editions), 1571, 1573, 1574, 1575, 1578, 1579, 1584 (two editions), 1589,

1591, 1592, 1595, 1601, 1603, 1607, 1609, 1612; Dutch 1571, 1598; German 1572 (two editions), 1573, 1580, 1602; French 1572, 1574, 1578, 1581, 1587, 1598; Spanish 1588, 1600, 1602, 1612; English 1606; Italian 1608, 1612. (See Bagrow, op. cit., p. 21.)

See de Soto ca. 1544 MS (Map 1 above) and Introductory Essay.

On the same folio are two other maps:

1. Pervviae Avriferae Regionis Typvs. Didaco Mendezio Auctore. *Size:* 8¾ × 13⅛.

2. Gvastecan Reg. *Size:* 8¾ × 6⅞.

References

Bagrow, Leo. "A. Ortelii Catalogus Cartographorum," Dr. A. Petermanns Mitteilungen. XLIII (Ergänzungsheft Nr. 199, 1928), 21, 56.

Delanglez, H. *El Rio Espíritu Santo.* New York, 1945, pp. 71–73.

Harrisse, H. *The Discovery of North America.* New York and London, 1892, p. 710.

Phillips, P. L. *Lowery Collection.* Washington, 1912, p. 85, No. 70.

Reproductions

Plate 9.

Cartografia de Ultramar. Carpeta II, Estadas Unidos y Canada. Madrid, 1953, Plate 47 (with transcribed list of place-names, p. 284).

Cash, W. T. *The Story of Florida.* New York, 1938, I, 6.

Copies

Separates (detached copies from an atlas): John Carter Brown; Clements; Harvard; National Archives of Canada.

Ortelius, Abraham. *Additamentum* III. Antuerpiae, 1584, No. 10.

Latin edition: Congress 1584.

French edition: Anvers, 1585; Congress 1585 (three copies); Harvard 1581 (bound in).

German edition: Congress 1584.

———. *Theatrvm Orbis Terrarvm.* Antverpiae, 1584, No. 9, and subsequent Latin editions; Congress 1579 (two copies, inserted), 1584, 1592, 1595, 1601, 1603, 1609, 1612; American Geographical Society [1584]; John Carter Brown 1591, 1591–92, 1603; Clements 1584, 1592, 1595, 1603, 1612; Harvard 1584, 1592; New York Public 1584, 1585; Yale 1592, 1609; National Archives of Canada.

———. *Théâtre de l'Vnivers.* Anvers, 1587, No. 8, and subsequent editions; Congress 1587, 1588; John Carter Brown 1598; Clements 1587, 1598; Yale 1587.

———. *Theatro de la Tierra Vniversal.* Anveres, 1588, No. 8, Spanish edition; Congress 1588.

———. *Theatro d'el Orbe de la Tierra.* Anveres, 1602, No. 9, Spanish edition; Congress 1602.

———. *Theatrvm Orbis Terrarvm.* Antvverpen, 1598, No. 9, Dutch edition; Congress 1598.

———. *Theatro del Mondo.* Antwerp, 1608, No. 9, and subsequent Italian editions; Congress 1612; John Carter Brown 1608; National Archives of Canada 1612 and a separate.

———. *The Theatre of the Whole World.* London, Printed by Iohn Norton, 1606, No. 9. English edition; American Geographical Society 1606; John Carter Brown 1606; Clements 1606.

---

6. Virginia ca. 1585 MS

A discription of the land of virginia [endorsed]

Size: 15¾ × 12. *Scale:* None.

Description: This anonymous, crudely drawn, but interesting map shows the region surrounding Pamlico and Albemarle Sounds on the Carolina coast. Professor D. B. Quinn (see reference below, II, 846–47) has identified it as "a rough note of the mapping done by White and Hariot in the first phases of discovery. It was most probably sent to England early in September 1585, on the *Roebuck* or *Elizabeth*, possibly with the letter from Lane to Walsingham of 8 September 1585."

The early date of the map and the significance of the legends on it have not been realized until recently because it has been associated with a letter of 1618 from Captain John Smith to Lord Bacon. In this letter Smith said that he was enclosing two maps "to show the difference betwixt Virginia and New England." The New England map is lacking. Apparently during a rearrangement of the Colonial Papers in the Public Record Office some time in the nineteenth century this sketch map of Virginia, also in the Colonial Papers, was arbitrarily placed and numbered with two items from Smith to Bacon. Alexander Brown, in his *The Genesis of the United States*, II, 597, accepted the "Virginia" sketch map as possibly Smith's attempt "to copy from some drawings of our present North Carolina coast." Neither the handwriting on the map nor the endorsement, however, can be identified with that in Smith's letter to Lord Bacon (C.O. 1/1, 42) or his "Description of New England" (C.O. 1/1, 42 [1]), according to a letter of August 4, 1953, from the secretary of the Public Record Office to W. P. Cumming. The information peculiar to the map was not used by Smith in his map of the same area published in 1624; and the legends have no apparent connection with any known expedition from the Jamestown settlement.

Though the technique of the map is not John White's nor the handwriting Thomas Hariot's, the information given, as Professor Quinn (I, 216–17) shows in his careful and extensive footnotes on the legends, can hardly derive from or apply to any other expeditions except those made by Lane's colonists in 1585. The watermark in the paper (entwined columns) is one appropriate to 1585 (Quinn, I, 217, and references there made).

The following transcription of the legends on the map has been checked against those of Brown and Quinn. They follow in roughly a counterclockwise sequence the items on the map, beginning at the upper center.

1. pomaioke: [Lane, Hariot, White, and others reached "The Towne of Pomeioke" on July 12 from Wococon. The lake behind is Lake Mattamuskeet (Paquippe on White's map).]

2. secotan [This village, which Lane and his group reached July 15, is variously located on different early maps; here it is on the north bank of Pamlico River. It is the name given to the whole area back of Pamlico Sound in White–De Bry 1590 and so continued on most maps throughout the seventeenth century.]

3. here is .3. fatham of water: [Soundings were evidently taken as the party went up Pamlico River.]

4. [t]his goithe to a great toune callid nesioke

5. [ ] this to warreã

6. the port of saynt maris wher we arivid first: [This is the inlet of Wococon, the present Ocracoke Inlet. It apparently is the entrance to the Baya de Santa Maria of the early Spanish maps, although by the end of the century the Spaniards applied the name to the Chesapeake Bay.]

7. wococan [Ocracoke Island.]

8. here groith ye roots that diethe read: [Croatoan Island. Quinn, p. 216: probably here Dogwood, *Cornus florida*, which still grows near Cape Hatteras.]

9. the gallis are found here [Quinn, p. 216: probably oak-galls, useful in tanning.]

10. ye king*es* ill: [Roanoke Island; Wingina was the king.]

11. the grase that berither the silke groithe here plentifully:

12. freshe water with great store of fishe:

13. Here were great store of great red grapis veri pleasant:

This manuscript sketch may have some relation to White–De Bry 1590 B, the pictorial engraving which is the second plate in Hariot's *A Briefe and True Report.* This plate of De Bry's, "The Arriual of the Englishemen in Virginia," does not correspond to any of White's British Museum drawings and maps. If, as Quinn suggests, this sketch is an earlier form of a pictorial map, now lost, which White made and which De Bry used for his second plate, the engraving omits the left or southern part and the information given in the legends, and adds the wrecked ships at the inlets and the boatload of Englishmen approaching Roanoke Island.

References

Brown, Alexander. *The Genesis of the United States.* Boston, 1890, II, 596–97.

Quinn, D. B. *The Roanoke Voyages 1584–1590.* The Hakluyt Society, Second Series, CIV. London, 1955, I, 215–17; II, 846–47.

Reproductions

Plate 10.

Brown, A. *The Genesis of the United States*, II, opp. p. 596 (facsimile).

Quinn, D. B. *The Roanoke Voyages*, I, opp. p. 215 (with facsimile and legends on page 215).

Original

Public Record Office. MPG. 584.

---

7. White 1585 MS A

La Virgenia Pars (across face of map)

Size: 14½ × 18½. *Scale:* 1" = ca. 80 miles.

Description: This is a watercolor drawing, with pen outlines, from Chesapeake Bay south to the Cape of Florida. It is ornamented with numerous sailing vessels, dolphins, flying fishes, whales, and other denizens of the sea. The arms of Sir Walter Ralegh are given at the top center.

It is found, together with the next map in this list, in an album of drawings with the title, in the artist's handwriting: "The pictures of sondry things collected and counterfeited according to the truth in the voyage made by Sr Walter Ralegh knight, for the discouery of La Virginea. In the 27th yeare of the most happie reigne of our Soueraigne lady Queene Elizabeth. And in the yeare of o^{r} Lorde God. 1585." This volume was formerly in the library of Lord Charlemont. It was purchased at Sotheby's auction rooms by

Henry Stevens in 1865. Stevens sold it to the British Museum, where it was placed in the Grenville Library in March 1866. It was transferred to the Department of Prints and Drawings in 1905: see L. Binyon, *Catalogue of Drawings in British Museum*, III, 327. It contains, besides the two maps, numerous paintings by the artist; of these, twenty-three are similar to engravings which were cut by De Bry, assisted by his sons and his associate, G. Veen, in De Bry's 1590 edition of Harriot's *A Briefe and True Report*. Since the engravings differ in details from the corresponding drawings and the map in the manuscript, and as more than one of De Bry's engravings has no corresponding painting in this album, it is clear that De Bry had a different set of paintings to work from. A duplicate of the seventh drawing in the album, found in another work in the British Museum, is attributed to Candidus (White); it is reasonable to suppose that he made copies of drawings which would have aroused such lively interest among scientists and friends. D. B. Quinn (*The Roanoke Voyages*, I, 392) has made a tentative but carefully documented table in which he predicates one original set and six other groups of drawings made by White himself.

There are seventy-six watercolor drawings. Henry Stevens, in a letter quoted by P. L. Phillips in *Virginia Cartography*, pp. 15–17, states that after a fire at Sotheby's one volume remained under pressure while saturated with water for three weeks. As a result each drawing became imprinted onto the sheet next to it. The off-sets produced in this way were bound into a second volume. Of the two volumes bound in red morocco which Stevens sold to the British Museum, the first contains these off-prints.

The upper half of this map is very similar to the large-scale map, also in this album, which portrays the coast from Chesapeake Bay to Cape Lookout; but it has fewer names. The inlet to the sound just to the south of Roanoke Island, named Hatrask on the other map and on De Bry's engraving, is here given the name Port Ferdinando. This is, however, not an old Spanish name for the inlet; it was named after Lane's praised and trusted pilot, and back of the dropping of the name lies Governor White's bitter reversal of opinion concerning his character. Ralph Lane, the governor, dates his first letter to Sir Francis Walsingham from "The porte which is called Ferdynando, dyscoverdde by the master and pylotte maggiore of our fleete, your honor's servante, Symon Ferdynando." See R. E. Hale, "Ralph Lane's Letters to Sir Francis Walsingham and Sir Philip Sidney," p. 11. In White's journal of 1587, Ferdinando is constantly charged with treachery to the colony; and to his desertion White ascribes the capital error of staying at Roanoke instead of finding a better place for settlement. Trinety Harbour, to the north of Roanoke on the De Bry engraving, is not shown on either of the manuscript maps.

The coastline in the central part of the map extends from Cape Lookout to Port Royal on almost the same latitude, east-west at around 34°30' N.L. This is correct for Cape Lookout (34°35'); but Port Royal is near 34°15', which places it nearly 150 miles too far north. White compensates for this by raising the River May (St. Johns) two degrees north and elongating the peninsula of Florida, which has the correct latitude at the cape. Mercator, who was followed by De Laët 1625 and others, has this same fault, though he puts the Port Royal–Cape Lookout latitude at 32°30' N.L., and compensates by elongating the distance from Roanoke Island to Chesapeake Bay. A more striking feature of the center of White's map than the east-west direction of the coast is a channel from Port Royal westward to a large body of water, with only its eastern shores visible. Since this is the latitude of Verrazano's Sea, the most probable assumption is that White is suggesting here a passageway to the South Sea. In Le Moyne's map, engraved by De Bry in 1591, appears a sea in the same place, but without the connecting passages. In the British Museum is a world map, Harleian ca. 1544, which has a long, narrow isthmus leading from the Atlantic to a great indentation of the Pacific. This wide bay, strongly reminiscent of Verrazano's Sea, narrows the northern part of the continent to a wide strip on either side of the St. Lawrence River. To the west of the isthmus is "R. de Sª Helene" and to the east is "mer osto." The Harleian map, though based on some form of a Spanish padrón general, is apparently of the French school. Details of its coastal configuration differ from contemporary Spanish maps as well as from White's; but White must have been influenced by a map of this type, as the location of the isthmus to the east of St. Helena and the shape of the bay of the Pacific are similar. The belief, shown in both of these maps, that the Pacific was just over a mountain range near the source of the Atlantic coastal rivers, is found in Ralph Lane's writing and continued for over a hundred years to motivate the expeditions of Governor Berkeley, John Lederer, and other trans-Appalachian explorers of the seventeenth century. (A good reproduction of the Harleian world map is in H. P. Biggar, *The Voyages of Jacques Cartier*, Ottawa, 1924, opp. p. 128; a color reproduction is in Cumming, Skelton, Quinn, *Discovery of North America*, pp. 150–51.)

It is almost certain that for most of the lower, southern

half of his map White was chiefly indebted directly to Le Moyne's original manuscript map. Le Moyne and White were both artists and mapmakers and both were in Ralegh's employ. White could have seen Le Moyne's manuscript map when he drew his own map, for it was still in Le Moyne's possession in 1585. Le Moyne, as late as 1587, according to De Bry's own statement, refused to sell him his drawings and map; it was not until after Le Moyne's death in 1588 that De Bry was able to buy the drawings from Le Moyne's widow. In 1585 White drew his map in which the coastal nomenclature in French was closely similar to that of Le Moyne's map. Sufficient additional detail and differences are on White's map to indicate that he received information possibly from Le Moyne personally.

Where De Bry has translated the nomenclature into Latin for the engraved map of 1591, here is the French original. The interior is sparsely treated except for three or four Indian names, in contrast to the profusion of detail in Le Moyne 1591; but "Montaigne Pallaci" (the Appalachians) is given, with the great lake below it which would haunt the cartography of Europe for generations. The island ("Sarrope" of 1591) in the middle of the lake, is given; the small lake is placed where the Everglades are, and the great lake is in the location of Okefenokee Swamp.

Though the Spanish influence is small on this map, since the far greater and more detailed knowledge of White and Le Moyne have crowded it out, the Spanish names on the west coast of the peninsula of Florida, as well as "C. de S. Helena" and "R. Jordanis," show that the map has used Spanish information where English and French have not supplanted it.

The map is carefully done; it possesses a good deal of detail for its size; it combines whatever knowledge from Spanish and French sources White could add to his own careful surveys. It is an interesting map because it portrays the latest geographical knowledge at that time available to the leader of Ralegh's colonizing ventures.

References

Adams, Randolph G. "An Effort to Identify John White." *American Historical Review* (October 1935).

Binyon, L. *Catalogue of Drawings by British Artists . . . in the British Museum.* London, 1907, IV, 326–37.

———. "The Drawings of John White, Governor of Ralegh's Virginia Colony." *Walpole Society, Annual Volume*, Oxford, 1925, XIII, pp. 19–24 (text), 24–30 (plates).

Brown, A. *Genesis of the United States.* New York, 1890, I.

Bushnell, David I. "John White—The First English Artist to Visit America, 1585." *Virginia Magazine of History and Biography*, XXXV (1927), 419–30.

Cumming, W. P. "The Identity of John White Governor of Roanoke and John White the Artist." *The North Carolina Historical Review*, XV (July 1938), No. 3, pp. 197–203.

Fite, E. O., and A. Freeman. *A Book of Old Maps.* Cambridge, Mass., 1926, pp. 93–95.

Hale, E. E. "Ralph Lane's Letters to Sir Francis Walsingham and Sir Philip Sidney." *Archaeologia Americana*, IV (1860), pp. 3–33.

Lorent, Stefan, ed. *The New World, The First Pictures of America.* New York, 1946, pp. 182ff.

Phillips, P. L. *Virginia Cartography.* Washington, 1896, pp. 3–18.

Quinn, D. B. *The Roanoke Voyages.* The Hakluyt Society, Second Series, CIV. London, 1955, I, 460 n. 2 (excellent discussion of cartographical sources); II, 848.

Reproductions

Plate 11.

Eggleston, E. "The Beginning of a Nation." *Century Magazine*, XXV (November 1882), pp. 66–67.

Fite, E. D., and A. Freeman. *A Book of Old Maps.* Cambridge, Mass., 1926, plate 26.

Hakluyt, R. *The Principal Navigations, Voyages, Traffiques, & Discoveries of the English Nation.* Glasgow, 1904, VIII, p. 400.

Henderson, A. *North Carolina.* Chicago, 1941, I, 30.

Lorent, Stefan, ed. *The New World.* New York, 1946, pp. 186–87.

Quinn, D. B. *The Roanoke Voyages*, I, opp. p. 460.

Taylor, E. G. R., ed. *The Original Writings and Correspondence of the Two Richard Hakluyts.* Hakluyt Society, Second Series, LXXVII, London, 1935, II, opp. p. 414.

Congress (photocopy).

Original

British Museum. Department of Prints & Drawings. 1906-5-9-1(2).

8. White 1585 MS B

Virginea Pars [across face of map]

Size: 18⅞ × 9¼. *Scale:* 1" = ca. 11 miles.

Description: This map shows the coast from Cape Charles to Cape Lookout and west from Cape Hatteras to the mouth of the Roanoke and Neuse Rivers. It embodies the material which the artist, who was sent out by Sir Walter Ralegh in 1585, gained while on his trips of exploration. Many of the towns shown on the map are mentioned in the "Account of the Particulars of the Imployments of the Englishmen left in Virginia," by Master Ralph Lane. (H. S. Burrage, ed., *Early English and French Voyages*, New York, 1906, pp. 243–71. First printed in R. Hakluyt, *Principal Navigations*, London, 1589, III, pp. 254–64.)

This manuscript map and the White Map of Virginia 1590 in De Bry's edition of Harriot's *A Briefe and True Relation* are very similar in area shown, topography, and nomenclature; they stem from a common original. But whereas the MS map has thirty-one names along the coast and rivers, the printed map has thirty-four. A comparison shows that while the MS map has a number of names on the north bank of Albemarle Sound absent in the printed (1590) map, along the Chowan River the reverse is true. The printed map gives more information about the general distribution of the different tribes (Weapemeoc, Chawanook, and Secotan); the MS map indicates only that the four villages to the right of the mouth of the Chowan River are of the Weapemeoc tribe.

The two maps in John White's manuscript volume are of particular interest, since they (and the printed map) portrayed several hundred miles of coast in more detail and with greater accuracy than was done for any other part of the New World for many years to come. When De Bry published Le Moyne's map in 1591, a detailed knowledge of the coast was extended from Port Royal to central Florida, although Le Moyne's work was not as accurate for that area as was White's work in dealing with the Banks area. It was not until Smith's map of Virginia 1612 and Champlain's map of the St. Lawrence River 1612 that there were better surveyed maps of any large section of the North American continent than the combined work of White and Le Moyne. No improvements of any importance were made for the Carolina region in any map until Comberford 1657; and since the Comberford map was not printed or used by other cartographers, the printed maps of the region improved very slowly before the actual settlement of the region almost a hundred years after White.

For the first time on any map appear Chesepiooc Sinus (Chesapeake Bay); Hatrask (Hatteras; not the present Cape, as the map shows, for the inlets and banks of the North Carolina coast have changed greatly since then); Moratuc (a town, but also the early name for the Roanoke River); Roanoac (the island); Croatoan (Croatan, a banks island south of what is now Cape Hatteras); Wokokon (Ocracoke Inlet, though probably not at the same place as the present inlet); and other names. The places visited by the settlers are colored red.

See John White 1585 MS A for a description of the volume in which the map is found, a discussion of the author, and for a list of references. D. B. Quinn (*The Roanoke Voyages*, I, p. 461 n. 1; II, pp. 847–48, 852–72) has made a study of the names and topographical details of this map which is a superb example of modern scholarship.

Reproductions

Plate 12.

Burrage, H. S., ed. *Early English and French Voyages*. New York, 1906, opp. p. 248.

Eggleston, E. "The Beginning of a Nation." *Century Magazine*, XXV (November 1882), 73.

Hakluyt, R. *The Principal Navigations, Voyages, Traffiques & Discoveries of the English Nation*. Glasgow, 1904, VIII, opp. p. 320.

Lorent, Stefan, ed. *The New World, The First Pictures of America*. New York, 1946, p. 186.

Nisser, C., C. Skinner, and W. Wood. *The Pageant of America*. New Haven, 1925, I, p. 160.

Quinn, D. B. *The Roanoke Voyages*, I, opp. p. 461.

Congress (photocopy).

Original

British Museum. Department of Prints and Drawings. 1906-5-9-1 (3).

9. Galle 1587

Novvs Orbis [top left and top right] | Doctiss. et ornatiss. Rich. Hakluyto F.[ilips] G.[alle] S. Cui potius quam tibi Orbem hunc nouum dicassem ? cum tu assiduis eruditisq libris tuis ipsum eundem in dies illustriorem reddas. Eumigitur Vti tua humanitate dignum est accipe, teq[ue] nos Vicissim amabimus. Paris. cal. Maij M.D. LXXXVII. [cartouche, bottom center]

Size: 8 × 6 7/16. *Scale:* 1" = ca. 1,950 miles.
In: Anghiera, Pietro Martire d'. *De Orbe Novo Petri Martyris Anglerii . . . Decades Octo.* Paris, Apvd Gvillelmum Avvray, 1587, opp. p. 1.
Description: This oval map of the Western Hemisphere enclosed by a rectangular ornamental border, often called the Hakluyt–Peter Martyr map, shows a marked advance in geographical knowledge in spite of its small size. The name Virginia appears for the first time on a printed map as "Virginea 1584" to the north of the Ralegh settlement, and "Duhare" and "Chicora," inland from "C. Sta. Helena," are reminiscent of Ayllón's explorations. The general outline of the North American continent follows Mercator's 1569 map, but legends on the continent show English explorations and were furnished by Hakluyt: Cabot's voyage, "Bacallaos ab Anglis 1496"; Frobisher, "Meta incognita ab Anglis Inuenta An. 1576"; and Drake's California, "Nova Albion Inuenta An. 1580 ab Anglis." The map is dedicated to Hakluyt, who edited the work which it accompanies and who dedicated the volume to Sir Walter Ralegh. Formerly F. G. S. was indentified as Francis Gaulte. Quinn and Skelton, in their facsimile edition of Hakluyt's *Principal Navigations*, suggest that the engraver was the Netherlands Filips Galle, an identification now generally accepted.

References

Hakluyt, Richard. *The Principal Navigations; Voiages and Discoveries of the English Nation.* With an introduction by David Beers Quinn and Raleigh Ashlin Skelton. Cambridge, 1965, I, xlviii.
John Carter Brown Library. *Catalogue.* Providence, 1919, I, 311.
Paullin, C. O. *Atlas of the Historical Geography of the United States.* Washington, 1932, p. 9.
Winship, G. P. *Cabot Bibliography.* 1900, p. 162.
Winsor. *America*, III, 42.
The World Encompassed. Baltimore, 1952, No. 203.

Reproductions

Plate 13.
Hakluyt, R. *Principal Navigations.* Glasgow, 1904, opp. p. 272.
Paullin. *Atlas*, Plate 14B (part of Western Hemisphere, enlarged).
Winsor. *America*, III, 42.
The World Encompassed, Plate XLVI.

Copies

John Carter Brown; Boston Athenaeum; Huntington; Johns Hopkins (Garrett); Clements.

---

10. Boazio 1588

The Famouse West Indian voyadge made by the Englishe fleete of *23* shippes and Barkes wherein weare gotten the Townes of S.t Iago: S.t Domingo, Cartagena and S.t Avgvstines the same beinge begon from Plimmouth in the Moneth of September 1585 and ended at Portesmouth in Iulie 1586 the whole course of the saide Viadge beinge plainlie described by the pricked line Newlie come forth by Baptiste B

Size: 21 3/8 × 13 1/8. *Scale:* 1" = ca. 325 miles.
Description: This map, which is based on some padrón general-type chart, shows the coasts bordering on both sides of the Atlantic. Its purpose, as the title indicates, is to show the voyage of Francis Drake; details are sparse, and the only names on the North American continent south of the St. Lawrence River are Norumbega, Virginia, Florida, and S. Augustine. The "pricked line" of Drake's voyage touches the Virginia coast at 39° N.L., above the indented bay which on Spanish maps may represent the Carolina Sounds.

The plate for this map is thought to have been engraved by Boazio in Leyden in 1588 to accompany the 1588 Latin text of Walter Bigges's notable account of the West Indian voyage of Drake, which was published in London in 1589 as "A Summarie and Trve Discourse of Sir Francis Drakes VVest Indian Voyage." The map has usually been described as belonging also to the English text, but recent opinion has been against its being a part of the volume as published. One John Carter Brown Library copy and the Grenville (British Museum) copy of the map have a printed text pasted to the bottom, with the heading: "Sir Francis Drake knight Generall of the whole Fleete of the West Indian voiage in 1585." Several libraries have copies of the map without the appended letterpress explanation, and copies of the explanation without the map are in the New York Public Library and the Society of Antiquaries in London. The date and the provenance of the map are not certain.

Baptista Boazio is thought to have been in the service of Christopher Carleill, vice-admiral of Drake's fleet. He copied several figures of fish and animals from White's drawings; they may both have been on the same ship on the homeward voyage from Virginia. Quinn, *Roanoke Voyages*, I,

34–35; Mary F. Keeler, *Sir Francis Drake's West Indian Voyage 1585–86*, London, 1981, pp. 63, 319 n. 4.

The text of Bigges's "A Summarie and Trve Discourse" is accompanied by four plans of the Spanish towns attacked in America. That of St. Augustine is the earliest printed map of a city within the present limits of the United States.

References

John Carter Brown. *Annual Report*, 1943, pp. 8–18; *Annual Report*, 1948, pp. 17–19. John Carter Brown. Catalogue. Providence, 1919, I. 2. 312. 315.

The World Encompassed. Baltimore, 1952, No. 230.

Reproduction

Cumming, Skelton, Quinn. *The Discovery of North America.* London, 1971, 187.

Copies

John Carter Brown, two copies; Free Library of Philadelphia.

---

11. Hogenberg 1589

Americae et Proximarvm Regionvm Orae Descriptio [cartouche, bottom left] | Per Franc: Hogenberg: A.º 1589. [small cartouche above title cartouche, lower left]

Size: 18⅜ × 13¼. *Scale:* 1" = ca. 740 miles.

In: Bigges, Walter. *Relation Oder Beschreibūg der Rheiss . . . in die . . . Indien gethan Durch . . . Franciscum Drack.* 1589.

Description: This map of the Western Hemisphere has one of the early references to the Ralegh colony at Roanoke; "Virgine" is shown as a city off the coast slightly above 40° N.L. and below "c. de lagus islas." "Montagna S. Johan" is to the interior west of "Virgine." "Augustin" is shown as another city, on the bank of a river that flows just north of the Florida peninsula.

Franz Hogenberg, painter, engraver, and print seller, was born in Malines before 1540, lived briefly in England, and resided after 1570 in Cologne. While living in Antwerp before 1570 he engraved some of the maps for Ortelius's *Theatrum Orbis Terrarum* of 1570; the influence of a map in a later edition of Ortelius, *Theatrum*, the Ortelius-Chiaves Florida of 1584, appears in the southeastern area of Hogenberg's map.

References

John Carter Brown. *Catalogue.* Providence, 1919, I, 2, 315.

LeGear, C. E. "Sixteenth Century Maps." *Library of Congress Quarterly Journal*, VI (May 1949), 21.

Wagner, H. R. *Sir Francis Drake's Voyage Around the World.* San Francisco, 1926, p. 424.

Copies

Congress; John Carter Brown 1589; Clements 1589.

---

12. White–De Bry 1590 A

Americae pars, Nunc Virginia dicta, primum ab Anglis inuenta, sumtibus Dn. Walter Raleigh, Equestris ordinis Viri, Anno Dñi. MDLXXXV regni Vero Sereniss: nostræ Reginæ Elizabethæ XXVII, Hujus vero Historia peculiari Libro discripta est, additis etiam Indigenarum Iconibus [cartouche, top right corner] | Autore Ioanne With Sculptore Theodoro DeBry, Qui et excud. [cartouche, left center]

Size: 16½ × 11⅞. *Scale:* 1" = ca. 25 miles.

In: Harriot, Thomas. *Admiranda Narratio fida tamen, de Commodis et Incolarvm Ritibus Virginiae . . .* , Francofurti ad Moenum, 1590. Plate 1. (Harriot's work is Part 1 of Theodorus De Bry's *Collectiones Peregrinationum in Indiam Occidentalem.* [25 parts] Frankfort, 1590–1634.)

Description: The coast from Chesepiooc Sinus (Chesapeake Bay, at 37° N.L.) to an inlet (New River Inlet?) below Promontorium tremendum (Cape Lookout, at 34°30' N.L.) is given. The inlets along the Banks and the Albemarle and Pamlico Sound area, with the Indian villages in the vicinity, show fairly detailed topographical knowledge; it is clear that the contour of the Banks and the location and number of the inlets have changed pronouncedly since White's day. Although this map lacks the authenticity of White's manuscript map of approximately the same area (Quinn, I, 462), the addition of some names and an improvement in coastal delineation may show increased knowledge gained in 1587–88, after the manuscript map was made.

De Bry's engraving of White's watercolor map is one of the most important type maps in Carolina cartography. Within a period of ten days, De Bry published four editions of Harriot's work, in which White's drawings appear, in Latin, French, English, and German; the engraved map remains essentially unchanged in all editions. Within two years after De Bry's work was published, Jode's map of North America 1593 incorporated information from the map, though not very accurately. Wyfliet 1597, Acosta 1598, Matal 1600, and Quad 1600 followed White's map in their works. Although De Bry's map remained essentially unchanged in all editions, as stated above, a first state contains an erroneous spelling "Ehesepiooc" for the name of the Indian settlement on a river (Lynnhaven) near the mouth of Chesapeake Bay, although the bay itself is correctly spelled "Chesepiooc Sinus." Several early mapmakers, including Wytfliet, Matal, and Norumbega, followed the "Ehesepiooc" spelling. But the engraver, De Bry, soon corrected his plate, incising with his burin a heavy "C" over the still visible "E," as in the copy reproduced herein as plate 14. After 1606, when Hondius included White's information in the Florida-Virginia map he made for Mercator's *Atlas*, most maps of the New World and of this region showed the influence of De Bry's engraving, until the English settlements toward the end of the century gave cartographers a new basis for their work.

For a fuller discussion of this map and for bibliographical references, see the Introductory Essay, White 1585 MS A, White 1585 MS B, and Le Moyne 1591. D. B. Quinn's (I. 462, nn. I and II, 849–50) study of the topography and nomenclature deserve special mention.

Reproductions

Plate 14.

Ashe, S. A. *History of North Carolina.* Greensboro, 1908, I, opp. p. 2.

A Brief Account of Raleigh's Roanoke Colony of 1585. William L. Clements Library, Bulletin XXII, Ann Arbor, Mich., 1935, after p. 18.

Connor, R. D. W. *Beginnings of English America.* Raleigh, N.C., 1907, frontispiece.

Gabriel, R. H., ed. *The Pageant of America.* I, 161.

Hamilton, P. J. *The Colonization of the South.* Philadelphia, 1904, III, opp. p. 49.

Harriot, Thomas. *Narrative of the First English Plantation of Virginia.* London, 1893, following p. 46 (Bernard Quaritch's reprint).

Hawks, F. L. *History of North Carolina.* Fayetteville, N.C., 1859, I, opp. p. 140 (poor facsimile).

Lorant, Stefan, ed. *The New World, The First Pictures of America.* New York, 1946, end papers front and back.

Quinn, D. B. *The Roanoke Voyages*, I, opp. p. 462.

Winsor, J. *Narrative and Critical History of America*, III, opp. p. 124.

Copies

Congress; American Geographical Society (three copies); John Carter Brown (three copies), and 1608 (two copies); Clements (nine copies) and 1608; Duke; Harvard; Mariners' Museum; New York Public (twelve copies); University of North Carolina 1590; Yale 1590.

In: Harriot, T. *A briefe and true report of the new found land of Virginia.* Franckfort, Inprinted by Ihon Wechel, at Theodore de Bry, owne coast and chardges, 1590. Plate I in appendix: "The True Pictures and Fashions of the peple in that Part of America novv called Virginia."

Copies

John Carter Brown; Clements; Folger Shakespeare Library; New York Public.

In: Harriot, T. *Merveilleux et estrange Rapport, toutefois fidele.* Francoforti, 1590. Plate I in appendix: "Les Vrays Pourtraicts, et Facons de vivre du peuple. d'une partie de l'Amerique nouvellement appellee Virginia."

Copies

John Carter Brown; Clements; New York Public (two copies).

In: Harriot, T. *Wunderbarliche doch Warhafftige Erklerung Von der Gelegenheit und Sitten der Wilden in Virginia.* Francoforti, 1590. Plate I in appendix: "Warhafftige Contrafacturen Und Gebreuch der Innwohner der jenigen Landschafft in America welche Virginia ist genennet."

Copies

John Carter Brown; Clements, also 1600 (with small "von" in title); Harvard (two copies, separate atlas); New York Public (four copies) and 1600 (two copies).

In: Harriot, T. *Wunderbarliche doch warhafftige Erklerung Von der Gelegenheit und Sitten der Wilden in Virginia.* Francoforti, 1620. Plate I in appendix: "Warhafftige Contrafacturen Und Gebreuch der Innwohner der jenigen Landschafft in America welche Virginia ist genennet."

Copies

John Carter Brown; Clements (three copies); New York Public (two copies).

In: Bry, Johann Theodor de. *America.* Das ist Erfindung vnd Offenbahrung der newen Welt . . . durch M. Phillipum Ziglerum von Wůrtzburg . . . Franckfurt am Mayn, 1617, after title page.

Copies

John Carter Brown; Clements.

13. White–De Bry 1590 B

Anglorum in Virginiam aduentus [printed above engraving] | T.[heodorus de] B.[ry] 2. [lower right corner]

Size: 8⅞ × 6⅛. *Scale:* 1" = ca. 8 miles.

In: Harriot, Thomas. *Admiranda Narratio fida tamen.* Francoforti, 1590, Plate II in appendix: "Vivae Imagines et Ritvs Incolarvm eivs Provinciae in America, qvae Virginia appellata est."

Description: This semi-pictorial map, of a type very popular at the time, includes the coast from Cape Hatteras to Knott Island in Currituck Sound and shows an Indian village on Roanoke Island with the Englishmen approaching it in a small sailboat from the inlet of "Trinety harbor." Outside the bank islands are two large vessels riding at anchor; at each of the five inlets shown, a sunken ship indicates the dangers of entrance.

Though the printed title above the map differs in the Latin, English, French, and German editions of Harriot's work, the engraving itself remains unchanged. No original corresponding to this is found in White's drawings in the British Museum; but it is very similar to the northern half of the sketch map (Virginia ca. 1585 MS) which is probably related to a White drawing no longer extant.

Reference

Quinn, D. B. *The Roanoke Voyages*, I, 216, 217, 413; II, 849.

Reproductions

Gabriel, R. H., ed. *Pageant of America.* New Haven, 1925, I, 161.

Hamilton, P. J. *The Colonization of the South.* Philadelphia, 1904, III, opp. p. 81.

Hawks, F. L. *History of North Carolina.* Fayetteville, N.C., 1859, opp. p. 88 (poor facsimile).

Henderson, A. *North Carolina.* Chicago, 1941, I, 18.

Lorant, S. *The New World.* New York, 1946, p. 229.

a. [Latin edition]

Copies

Congress, two copies; American Geographical Society, three copies; John Carter Brown, three copies, and 1608; Clements, nine copies, and 1608; Duke; New York Public, twelve copies; Yale.

b. The Arriual of the Englishemen in Virginia | T.B. 2

In: Harriot, T. *A briefe and true report of the new found land of Virginia.* Franckfort, Inprinted by Ihon Wechel, at Theodore de Bry, owne coast and chardges. 1590. Plate II in appendix: "The True Pictures and Fashions of the Peple in that Part of America novv called Virginia."

Copies

John Carter Brown; Clements; Folger Shakespeare Library; Harvard; New York Public.

c. Abord des Anglois en Virginia | T.B. 2

In: Harriot, T. *Merveillevx et estrange Rapport, toutefois fiedle.* Francoforti, 1590. Plate II in appendix: "Les Vrays Povrtraicts, et Facons de vivre dv pevple. d'vne partie de l'Amerique novvellement appellee Virginia."

Copies

John Carter Brown; Clements; New York Public, two copies; University of North Carolina, two copies.

d. Von der ankunfft der Engellender in Virginia. II. | T.B. 2

In: Harriot, T. *Wunderbarliche doch Warhafftige Erklerung Von der Gelegenheit vnd Sitten der Wilden in Virginia.* Francoforti, 1509. Plate II in appendix: "Warhafftige Contrafecturen Vnd Gebreuch der Innwohner der jenigen Landschafft in America Welche Virginia ist genennet."

Copies

John Carter Brown; Clements, and 1600; Harvard; New York Public, four copies, and 1600, two copies; Peabody Museum.

e. Virginia I. *Von der Ankunfft der Engellånder in Virginia* | T.B. 2.

In: Harriot, T. *Wunderbarliche doch warhafftige Erklårung von der Gelegenheit vnd Sitten der Wilden in Virginia.* Franckfurt, 1620. Plate II in appendix: "Warhafftige Contrafecturen Vnd Gebråuch der Innwohner der jenigen Landschafft in America welche Virginia ist genennet."

Copies

John Carter Brown; Clements, three copies; New York Public, two copies.

f. [plate without printed title] | T.B. 2 [bottom right]

In: Bry, Johann Theodor de. *America.* Franckfurt, 1617, p. 228.

Copies

John Carter Brown; Clements.

14. Le Moyne 1591

Floridae Americae Provinciae Recens & exactissima descriptio Auctorè Iacobo le Moÿne cui cognomen de Morgues, Qui Laudonnierum, Altera Gallorum in eam Provinciam Nauigatione Comitat*us* est, Atque adhibitis aliquot militibus. Ob pericula, Regionis illius interiora & Maritima diligentissimè Lustrauit, & Exactissimè dimensus est. Obseruata etiam singulorum Fluminum inter se distantia, ut ipsemet redux Carolo. IX, Galliarum Regi, demonstrauit.

Size: 17¾ × 14⅜. *Scale:* 1" = ca. 82 miles.
In: Le Moyne de Morgues, Jacques. *Brevis Narratio eorvm qvae in Florida Americae Provincia Gallis acciderunt.* Francofvrti, 1591, opp. p. 1.

Description: This important map includes the peninsula of Florida and the surrounding regions from the northern part of Cuba to "Prom Terra falg" or Cape Lookout. It is found in the second volume of De Bry's *Grands Voyages*.

The author, Jacques Le Moyne, was an artist who accompanied Laudonnière on his ill-fated trip to Florida in 1564. He made many graphic drawings of native scenes, a map of the region, and an accompanying narrative. De Bry saw Le Moyne in London in 1587 and attempted to obtain the drawings and papers. But Le Moyne, who was at the time in Ralegh's service, refused to part with them; soon after his death in 1588, however, De Bry purchased them from Le Moyne's widow and published them in 1591. The manuscript map is not extant, but it was in all probability used by John White in making the southern part of his "La Virgenia Pars"; see White 1585 MS A.

The map contains many striking details, frequently erroneous, which were incorporated in other maps for over 150 years. It was Le Moyne's misfortune to have many of his errors incorporated and even exaggerated in Mercator's map of 1606, upon which for half a century much of the subsequent cartography of the region was based.

Le Moyne's coastline is usually correct for latitude, but the shore extends too far east rather than northeast in direction. This caused a striking error in Mercator's map, with a compensating enlargement of the Virginia region; the mistake was corrected somewhat by Jansson 1641 and those who followed him.

Along the top of the map, to the north, extends the shore of a sea, probably Verrazano's Sea. It is unnamed and has no channel connecting it to the Atlantic. A similar body of water is found in Lescarbot 1611 and Seller 1679.

Along the coast are Latin names for rivers and bays, such as Gironda, Garumna, and Charenta, together with a few of the earlier Spanish names. The identification of the rivers has been attempted several times (see the Lowery reference below) but it is doubtful whether Le Moyne had definite knowledge of the number of rivers along the coast himself. The names were given on the first voyage under Ribaut, who in his account makes some reference to their latitude and appearance. Most of the problems of toponymic identification have been solved by a recent analysis of a Spanish spy's tracing on transparent paper of a map made by Captain Nicolas Barré, Ribaut's able pilot on the 1562 voyage. See 2A Parreus French Florida 1562 and Plate 6A. Ribaut's names were eventually superseded by others when the seventeenth-century English settlers arrived; probably the only permanent coastal name first found on Le Moyne's map is "Portus Regalis" or Port Royal.

His placement of "Charlefort" on an island at Port Royal and Carolina (the fort "la Carolina") on the River May are helpful identifications. But the name "Carolina" copied by a later mapmaker and put by Sanson 1656 much farther north, was probably the original source of the later false belief of mapmakers (Delisle 1718, Còvens and Mortier ca. 1730) and even nineteenth-century historians that the whole country was named Carolina by the French.

Le Moyne has several lakes which played a conspicuous part in the later cartography of the Southeast. In the peninsula of Florida is a lake with an island called "Sarrope," which probably represents Lake Okeechobee or one of the lakes in that region. North of Sarrope is a larger lake which may be intended to represent Lake George and which through later mutations of location and size became the great inland lake of the Southeast. Le Moyne locates it slightly southeast of the mouth of "May" (St. Johns River) into which it flows. He calls it "Lacus aquae dulcis" (freshwater lake) and says that it is so large that from one bank it is impossible to see the other side. To the north of the lake, among the "montes Apalatci" (Appalachian mountains) is another large lake, fed by an enormous waterfall. This waterfall may have been inspired by tales of waterfalls in western North Carolina; but it is more likely to depict the legends heard from Indians of the great falls of Niagara. Below this lake is written "In hoc lacu Indigenae argenti grana inveniunt" (In this lake the natives find grains of silver).

Also to be mentioned here are Le Moyne's drawings of scenes and rivers along the coast, which form nine of the forty-two engravings which accompany the work as an appendix. "Arcis Carolinæ descriptio," the tenth picture in this appendix, is of the French fort on St. Johns and was later reproduced in the various editions of the work *America*, issued as "Arx Carolina" by Montanus (1671), Ogilby (1672–75), and Dapper (1673), and in P. vander Aa's *Galerie Agreable du Monde* (1729?).

De Bry made use of part of Le Moyne's map in designing a map to illustrate the voyages of Columbus, which accompanies Benzoni's work in the fourth volume of De Bry's *Grands Voyages*, published in 1594: "Occidentalis Americæ partis, vel, earum quas Christophorus Columbus primu*m* detexit Tabula Chorographica . . ."

A careful study of the Laudonnière expedition and of the various attempted identifications of the rivers on Le Moyne's map is in W. Lowery, *Spanish Settlements within the Present Limits of the United States: Florida, 1562–1674*. See also Phillips, P. L., ed., *Lowery Collection*, pp. 90–92. D. B. Quinn, *The Roanoke Voyages*, I, 460 n. 2(b), lists the names which the original Le Moyne manuscript map, no longer extant, probably furnished White in preparing his map of the southeast Atlantic coast (White 1585 MS A).

An expert and careful examination of the Le Moyne–De Bry 1591 map, Le Moyne's sources personal and derivative, its cartographic qualities, the probable changes made by the engraver, De Bry, and other interesting commentaries is in Dr. R. A. Skelton's essay in Paul Hulton's *The Work of Jacques Le Moyne De Morgues: Hugenot Artist in France, Florida, and England*, London, 1977, Vol. I, Chapter IV, pp. 45–54, to which D. B. Quinn and W. P. Cumming also contributed.

A study of the cartographical influence of the map is found in W. P. Cumming, "Geographical Misconceptions of the Southeast in the Cartography of the Seventeenth and Eighteenth Centuries," *Journal of Southern History*, IV (November 1938), 476–92.

See Introductory Essay.

References

See items referred to above and:

Fite, E., and A. Freeman. *The Book of Old Maps*, pp. 69–70.

Gaffarel, P. L. J. *Histoire de la Floride Françoise*. Paris, 1875, at end of book.

Paullin, C. O. *Atlas of the Historical Geography of the United States*. Washington, 1932, pp. 9–10.

Reproductions

Plate 15.

Fite, E. D., and A. Freeman. *The Book of Old Maps*. Cambridge, 1926, plate 20.

Hakluyt, Richard. *The Principal Navigations, etc.* Hakluyt Society Publications, Extra Series, IX, Glasgow, 1904, opp. p. 112.

La Ronchière. Ch. de. *La Floride Française*. Paris, 1928, I, opp. p. 32. (Le Moyne's map and drawings hand-colored by Daniel Jacomet.)

Lorent, S. *The New World*. New York, 1946, pp. 34–35.

Paullin, C. O. *Atlas of the Historical Geography of the United States*. Washington, 1932, plate 16.

Copies

Congress; American Geographical Society, two copies; John Carter Brown, and [1609]; Clements, four copies; Duke 1591; Harvard (two copies in atlas of De Bry maps); New York Public 1591, four copies, and [1609], four copies; 1634, two copies separate.

In: Le Moyne de Morgues, Jacques. *Der Ander Theyl der Newlich erfundenen Landtschafft Americæ*. Franckfort, 1591, opp. p. 1. (De Bry's German edition of the text of Le Moyne; translation by Oseam Halen.)

Copies

Congress 1591, 1603; John Carter Brown 1591, 1603; Clements; New York Public, three copies, and 1603; Yale; National Archives of Canada.

---

15. Mestas 1593 MS

El fuerte de san ta Elena

Size: 22⅞ × 17. *Scale:* None.

Description: This is a proposed plan in colors, undated and unsigned, for the rebuilding of the fort at Santa Elena. It was brought by Alferez Hernando Mestas to Philip II of Spain, together with three other plans in a letter and deposition dated February 23, 1593, from the governor of Florida, Domingo Martínez de Avendaño. There is no reference to the plan in the letter. The governor recommended the building and manning of the forts for the pacification of the region and defense of the Spanish possessions in Florida. The fort, however, was never built: see V. Chatelain, *The Defenses of Spanish Florida*, Washington, 1941, p. 113, note 12.

The first fortification in this vicinity was the Charlesfort of Ribaut, built on Parris Island in 1562. Le Moyne drew a pictorial sketch of this French fortification which is found in De Bry's edition of Le Moyne, 1591.

References

Connor, J. T. "The Nine Old Wooden Forts of St. Augustine," *Florida Historical Society Quarterly*, IV (April 1926), 171–80.

Phillips, P. L. *Lowery Collection*, p. 92, No. 74.

Torres Lanzas, Pedro. *Relación Descriptiva*, I, 43, No. 46.

Reproduction

Congress (hand-drawn colored reproduction on linen; photocopy).

Original

Sevilla. Archivo General de Indias. Indiferente General. Est. 140. Caj. 7. Leg. 37.

---

16. Jode 1593

Americæ Pars Borealis, Florida, Baccalaos Canada, Corterealis. A Cornelio de Iudæis in lucē edita. [Title in large rectangular cartouche above map] | [Dedication to left of title] Generoso, atq₃ Magnifico Dño, Dño Theodorico Echter, á Mespelbru, Sacr. Caesar. Maiest[ti], et Rever[smo]. Principi, Episcopo Herbipolensi a consilijs primo, &c. Cornelius de Iudæis Antverp. D.D.A.° MDLXXXXIII.

Size: 20 × 14½. *Scale:* 1" = ca. 460 miles.

In: Jode, Gerard de, and Cornelis de Jode, *Specvlvm Orbis Terræ*. Antverpiae, [1593–1613], fol. II.

Inset: View in lower right corner "Incolarū Virginiæ Habitus."

Description: This map, which includes most of North America except Mexico and the extreme western part of the continent, contains many interesting features. It is probably the first map to be influenced by Le Moyne's map for Florida and by White's map for Virginia. It has the French fort "Carolina" situated at the mouth of "R. Mayo" near "S. Helena," and "Charlefort" on the north bank at the mouth of the "R. de Gallo," which joins the "R. Mayo." This combined stream takes its rise near "Xuala" in the mountains of "Apalchē." The "Virginia" names from White's map are erroneously placed above "C. de las Arenas." This puts "Chesepoc Sinus" (Chesapeake Bay) near the actual latitude of Boston; on Jode's map the coastline of "Norombega" (New England) extends eastward from "Chesepoc Sinus," and the St. Lawrence River also runs in an east-west direction. In this Jode merely accentuated the erroneous representation of the New England coast found in the maps of Mercator 1569, Ortelius 1589, and many earlier sixteenth-century cartographers. The mistake of Jode in placing Virginia above C. de las Arenas is followed by Wytfliet 1597, Acosta 1598, Matal 1600, and also by Quad 1600, whose map is a reduced copy of Jode's. Jode (Cornelis de Iudaeis) followed Mercator and Ortelius rather than Le Moyne in situating the headwaters of the great rivers which empty into the Atlantic north of the Florida peninsula to the northwest of their mouths. There are no lakes in the Apalchen mountains, where these rivers take their rise, in Jode's map; but far to the interior, west of the headwaters of the St. Lawrence River, is the "Lago de Conibas," with the island city of "Conibas" situated in its center; a legend states that "hoc mare dulcium aquarum est, cuius terminum ignorari Canadenses aiunt." The freshwater lake with the invisible shore of Mercator-Hondius 1606, though based on Le Moyne's map, may have been influenced by this freshwater lake of the northwest. One of the first appearances of Conibas (which in turn possibly owed its origin to a confused account of one of the Great Lakes related to the Spanish conquerors of New Mexico) is on a manuscript map [ca. 1566] by Santa Cruz: see "Carta de las Antillas, Seno Mejicano y Costas de Tierra-Firme y de la America Setentrional," which is reproduced in Cartas de Indias, Madrid, 1877, plate R (see under de Soto ca. 1544 MS).

In 1578 Gerard de Jode published a work entitled *Specvlvm Orbis Terrarvm*, containing several maps of the New World. Gerard died in 1591, and his son Cornelis completed the enlarged work, *Specvlvm Orbis Terrae*, by the inclusion of several new maps and additional text. The map here described is one of the new maps; it attempts to embody the latest available information. Two of the legends give historical notes on the French and English settlements.

Reproduction

Plate 16 (detail).

Copies

Congress, two copies; John Carter Brown; Clements; Newberry (Ayer); New York Public (Stokes Collection); Yale; National Archives of Canada.

17. Maymi ca. 1595 MS

Planta de la costa de la florida y en que Paraje esta La Guna Maymi y adonde se ha de hacer el fuerte [endorsed on map]

Size: 8¼ × 11^{13}/16. *Scale:* None.

Description: This map, which extends from Cuba, includes the Florida peninsula and the east coast to S elena. The map is anonymous and undated. Pedro Torres Lanzas, *Relación Descriptiva*, Seville, 1900, I, 71, No. 94, gives it the title "Mapa de la Florida y Laguna de Maimi donde se ha hacer un fuerte" and dates it "Siglo 17 (?)."

The Atlantic coastal names, reading down from east to west, with probable identifications, are: S elena [Santa Elena]. ahoya. C. de los baxos [Bahia de los Bajos]. cofonufo. hospogahe [Espogache]. asao. Guadalquini [St. Simon's Sound]. Ballenas. S Pedro [Cumberland Sound]. Sena. S Mateo [St. Johns River]. Sagustin. matancas [Matanzas]. Moysquitos [Mosquito]. cabo de Canaberal [Cape Canaveral]. ays. S iozia. Xega.

This map may have formed part of a letter sent to the home government by Juan de Posada, advising the dismantlement of the forts in Florida, about the year 1595, although the title of the map itself suggests a fort to be built on Lake Miami. See W. Lowery, *The Spanish Settlements within the Present Limits of the United States: Florida, 1562–1574*, New York, 1905, pp. 464–66, and P. L. Phillips, *Lowery Collection*, pp. 100–101, No. 88. For Spanish activities along the coast at this time, see J. G. Johnson, "The Yamassee Revolt of 1597 and the Destruction of the Georgia Missions," *Georgia Historical Society*, VII (March 1923), pp. 44–53.

Reproductions

Hamilton, P. J. *The Colonization of the South*. Philadelphia, 1904, III, opp. p. 32.

Lowery, W. *Spanish Settlements, 1562–1574*, opp. p. 286.

Congress (photocopy).

Original

Sevilla. Archivo General de Indias, Est. 145, Caj. 7, Leg. 7.

18. Wytfliet-Florida 1597

Florida et Apalche [cartouche, upper right] | 16 [upper right corner]

Size: 9 × 11⅜. *Scale:* 1" = ca. 82 miles.

In: Wytfliet, Corneille. *Descriptionis Ptolemaicæ Avgmentvm* . . . , Lovanii, 1597, No. 16.

Description: This map follows Ortelius 1584, although it includes a slightly larger area and has several changes in nomenclature. The country north and east of the great River Seco, which flows in a southeasterly direction from the upper limit of the peninsula of Florida, is called Apalche. The Seco and Sola Rivers join each other to form a great island, as in the Ortelius map. The Florida peninsula itself, which is not given a name, has a rectangular shape, with a bottleneck at the top, unlike the V-shaped form in most of the earlier maps. The work of Wytfliet in which this and the following map are to be found appeared in numerous editions and was highly regarded. In spite of its title, the work owes nothing to Ptolemy in substance or method. The first edition is sometimes called the first atlas of the new world, since it contains no maps except those of the Americas.

See note in the description of Acosta-Norvmbega 1598 (Map 21).

Reference

Phillips, P. L. *Lowery Collection*, pp. 96–97, No. 83.

Reproductions

Plate 17.

Nordenskiöld, A. E. *Facsimile Atlas*. Stockholm, 1889, plate LI g.

Winsor, Justin. *America*, II, 281.

Copies

Found in various editions, 1597–1603: Congress; American Geographical Society; John Carter Brown; Harvard; Kendall; New York Public; National Library of Canada, 1607.

In: Wytfliet, C. *Histoire vniverselle des Indes Orientales et Occidentals*. Douay, 1605, No. 16, between pp. 116–17.

Copies

Found in various editions, 1605–11: Congress; American Geographical Society; John Carter Brown; Charleston Library Society; Duke; Harvard; New York Public; Yale; Clements.

19. Wytfliet-Norvmbega 1597

Norvmbega et Virginia. 1597 [Cartouche, upper left] | 17 [upper right corner]

Size: 11½ × 9⅛. *Scale:* 1" = ca. 80 miles.

In: Wytfliet, C. *Descriptionis Ptolemaicæ Avgmentvm . . . Lovanii 1597*, No. 17.

Description: This map includes the North Carolina coastal region, which is based on the White map 1590. Wytfliet, Matal, and Acosta, in their maps of "Norvmbega," follow the early first state of White–De Bry, since the Indian settlement south of Chesapeake Bay is spelled "Ehesepiooc," an error soon corrected on De Bry's copper plate: see White–De Bry 1590.

Wytfliet's "Florida et Apalche" map is based on the cartography of Ortelius-Chiaves and shows no influence of later discoveries along the coast. It extends to "C. de Arenas" at 38° N.L. Wytfliet's "Norvmbega et Virginia. 1597" begins a little below this point, repeats "C. de las Arenas" at 38°, and then inserts the White topography and nomenclature between 38° and 43° (extremity of the present state of Maine). This is the same error found in the Jode map (1593) and many early general maps. These maps make the coast of Norumbega (New England) extend east-west, latitudinally. This "compensates" for the true northeasterly direction of the coastline to such an extent that on the Wytfliet map Cape Breton Island, which is actually north of Nova Scotia at about 46°, lies on 45° N.L.

Reference

Phillips, P. L. *Virginia Cartography.* Washington, 1896, pp. 18–19.

Reproductions

Stokes, I. N. P. *Iconography of Manhattan Island.* New York, 1915–28, II, Plate 20 (facsimile).

Nordenskiöld, A. E. *Facsimile Atlas.* Stockholm, 1889, Plate LI h.

Copies

Found in various editions, 1597–1603: Congress; American Geographical Society; John Carter Brown; Clements; Harvard; Kendall; New York Public; University of North Carolina.

In: Wytfliet, C. *Historie Vniverselle des Indes Orientales et Occidentales.* Douay, 1605, No. 17, between pp. 120–21.

Copies

Found in various editions, 1605–11: Congress; American Geographical Society; John Carter Brown; Charleston Library Society; Clements; Duke; Harvard; New York Public; Yale.

---

20. Acosta-Florida 1598

Florida et Apalche.

Size: 9½ × 7¼.

Scale: None. No latitude or longitude marks.

In: Acosta, José de. *Geographische vnd Historische Beschreibung der . . . America.* Cölln, 1598, fol. 7.

Description: This is an Ortelius-Chiaves-type map. It follows Wytfliet very closely, though the plate is different and there are a few minor omissions, such as "Tropicvs Cancri" and indications of latitude and longitude within the neat lines of the map, which are found in Wytfliet. In 1588–89 the Spaniard Acosta (ca. 1539–1600) published his treatise *De Natura Novi Orbis*, which may be regarded as a preliminary draft of his celebrated *Historia Natural y Moral de las Indias*, Seville, 1590; this latter work was translated into many European languages. In most editions only a general map of the New World appears; but in the German editions of 1598 and 1605, the map here noted, with twenty-nine others, is also included.

Reference

Meurer, Peter H. *Atlantes Colonienses*, ACO1, p. 50.

Copies

John Carter Brown; New York Public, two copies.

In: Acosta, José de. *America . . . Oder West India.* Vrsel [Wesel], 1605, at end of volume.

Copies

Newberry; New York Public.

---

21. Acosta-Norvmbega 1598

Norvmbega et Virginia

Size: 9⅛ × 7³⁄₁₆.

Scale: None. No latitude or longitude given.

In: Acosta, José de. *Geographische vnd Historische Beschreibung der . . . America.* Cölln, 1598, fol. 8.

Description: This map is like the Wytfliet map 1597.

Joannes Matalius Metellus (1520–97), who had an established reputation in Louvain, at the end of his career in Cologne drew the 20 maps appearing in his and Acosta's volumes noted here and below. Since he died in 1597, it seems possible that he may have drawn his maps prior to or in some kind of collaboration with Wytfliet's. Mr. Richard B. Arkway, a New York map and rare book dealer, advertised a copy of Acosta's rare little volume containing Matal's maps in his Catalogue XXII (1983), No. 1; the title follows: ACOSTA/METELLUS (1598) *Geographische und historische Beschreibung der uberauss grossen landschafft America: Welche auch West-India, und jhrer grosse halben die newe Welt genennet wird. Gar artig vnd nach der kunst in 20 Mappen order Landtafeln verfasset . . .* Köln, Johann Christoffel, 1598. Text is followed by 20 double page maps with text on reverse sides. Part of Arkway's catalogue description follows:

"Although listed under Acosta in Sabin [No. 128], the maps are undoubtedly by Metellus, whose name appears on an edition of 1600 which is composed of the same maps. Metellus' maps are often mistaken as reissues of Wytfliet's. However, significant differences between the two emerge upon examination. Metellus' maps are totally new engravings. Not only are they different in size, design and decorative details, some also employ different delineations with new place-names. Metellus included a significant map of the Pacific and American west coast not found in the Wytfliet. It is the second separate printed map of the Pacific, preceded only by the Ortelius. The world map in the Metellus is completely different than the Wytfliet. The former, based primarily on Mercator, is a much more advanced piece of cartography. Finally, the text in the Metellus is totally different than that in the Wytfliet."

Copies

John Carter Brown; New York Public, two copies.

Reference

Meurer, Peter H. *Atlantes Colonienses*, ACO1, p. 50.

In: Acosta, José de. *America . . . oder West India*. Vrsel, 1605, bound in at end of volume.

Copies

Newberry; New York Public.

21A. Wright 1599

["Thou hast here (gentle reader) a true hydrographical description of so much of the world as has beene hetherto discovered and is comne to our knowledge": portion of text in cartouche in lower right]

Size: 16½ × 25. *Scale:* None.

Description: Known in the literature for many years as the Wright-Molyneux, Hakluyt-Molyneux, or Hakluyt-Wright, this copperplate map of the known world was published in 1599 in Richard Hakluyt's *Principal Navigations* II. As mentioned above (Map 5), it is one of the first maps to be based on Mercator's projection as theoretically described by Edward Wright in his *Certaine Errors in Navigation*, London, 1599. After receiving degrees at Cambridge, Wright, born in Norfolk in about 1558, compiled a distinguished career as a mathematician interested in problems of cosmography and navigation. Expert Helen Wallis noted that of all the maps published by Hakluyt Wright's "was by far the most original . . . and one of the most authoritative of its day." Evidence of its immediate fame is provided by Shakespeare's allusion to it in *Twelfth Night*, in which Maria describes Malvolio as one who "Does smile his face into more lines than is in the new map, with the augmentation of the Indies." Maria is, of course, drawing attention to the many intersecting rhumb lines which crisscross the map.

On the detail of the map showing the Southeast, reproduced here as Plate 18, it can be seen that Wright enlarged Albemarle Sound and Chesapeake Bay unduly and lessened the size of Pamlico Sound. Notable is the prominent placement of "Virginia" across the Southeast while the name "La Florida" appears to designate only the peninsula, hardly a condition that would have met with agreement in Spain at the time.

The early states of this map lack a cartouche with text commenting on Drake's South American discoveries in 1577. The cartouche is located in the Pacific just to the southwest of that continent on the later states.

References

Fite, E. D., and A. Freeman. *A Book of the Old Maps Delineating American History*. New York, 1969, pp. 101–2.

Parsons, E. J. S., and W. F. Morris. "Edward Wright and His Work." *Imago Mundi* III (1938) [1964 reprint], pp. 61–71.

Shirley, R. W. *The Mapping of the World: Early Printed Maps, 1472–1700*. London, 1983, Number 221, pp. 238–39.

Wallis, H. "Edward Wright and the 1599 World Map." In D. B. Quinn, *The Hakluyt Handbook* I, London, 1974, pp. 63–73.

Reproductions

Plate 18 (detail).

Fite, E. D., and A. Freeman. *A Book of Old Maps*, Map 28 (first state), p. 100.

Nebenzahl, K. *Atlas of Columbus and the Great Discoveries*. Chicago, 1990, Plate 50 (second state), pp. 158–59.

22. Matal-Norvmbega 1600

Norvmbega et Virginia.

Size: 9 × 7⅛. *Scale:* None.

In: Matal, Jean. [Joannes Matalius Metellus, Sequanus]. *Speculum Orbis Terrae.* Coloniae Agrippinae, 1600–1602, Part IV: America sive Novvs Orbis. Coloniae Agrippinae, 1600, No. 6.

Description: Although smaller and with differences of design, this is similar to Wytfliet's 1597 map of the same title; see description under Acosta-Norvmbega 1598 (Map 21).

Copies

Harvard; New York Public; National Archives of Canada.

23. Matal-Florida 1600

Florida et Apalche.

Size: 9½ × 7½. *Scale:* None.

In: Matal, Jean. [Joannes Matalius Metellus, Sequanus]. *Speculum Orbis Terrae.* Coloniae Agrippinae, 1600–1602, Part IV: America sive Novvs Orbis. Coloniae Agrippinae, 1600, No. 7.

Description: This is similar to Wytfliet's "Florida et Apalche" 1597, though it differs in several minor ways, such as in the location of Indian settlements.

Reference

Phillips, P. L. *Lowery Collection.* No. 94, p. 105.

Copies

Congress (Lowery Collection); Harvard; New York Public.

24. Quad 1600

Novi Orbis Pars Borealis, America Scilicet, Complectens Floridam, Baccalaon, Canadam, Terram Corterialem, Virginiam, Norombecam, pluresque alias prouincias [seven lines of historical data omitted] Coloniæ laminis Jani buxemechers. [title cartouche across bottom of map]

Size: 11¼ × 9^{5}⁄16. *Scale:* 1" = ca. 710 miles.

In: Quad, Matthias. *Geographisch Handtbuch.* Köln, 1600, Map 78.

Description: This map is based on the Jode map of 1593, although it has numerous minor omissions, changes, and additions. The fort Carolina at "S. Helena" is drawn but the name "Carolina" is not given at the mouth of "R. Mayo," as on the Jode map. Above the Arctic Circle, however, Jode's legend concerning French discoveries, which refers to the building of "Carolina" by Laudonnière in 1565, is retained; and in the last line of the descriptive title the histories published by De Bry are referred to. "Charlesfort" is given on "r. de Gallo." "Florida" is the name given to the region north of the Gulf of Mexico; the area to the northeast has "Apallalei Mō" and "Virginia" is retained. The city of Conibas, situated in the "Lago de Conibas" to the northwest of the St. Lawrence, is given.

Reference

Meurer, Peter H. *Atlantes Colonienses.* QUA6, p. 218.

Copies

Congress; Newberry (Ayer); National Archives of Canada.

In: Quad, M. *Fasciculus Geographicus.* Köln, 1608, No. 82.

Reference

Meurer, Peter H. *Atlantes Colonienses.* QUA7, p. 222.

Copies

Congress; Clements; Newberry (Ayer); New York Public; National Archives of Canada.

25. Tatton 1600

Noua et rece Terraum et regnorum Californiæ, nouæ Hispaniæ, Mexicanæ, et Peruviæ, una cum exacta absolutaq$_3$ orarum Sinus Mexicani, ad Insulam Cubam usq$_3$ Oræq$_3$ maritimæ ad Mare austriacum delineatio, á M. Tattonus celebrem Sydrogeographō edita [cartouche, top center] | Beniamin Wright. Anglus. coelator an.º 1600 [cartouche, top right]

Size: 21¼ × 16¼. *Scale:* 1" = ca. 160 miles.

Description: This map of the southern part of North America and of Central America was engraved by Benjamin Wright, a celebrated English engraver who lived in Holland and Italy and made maps for foreign publishers. Tatton also was an Englishman. Though both designer and engraver were English, that the map was not primarily designed for English users may be supposed by the lack of English legends and omission of Virginia and the Roanoke region from the map. The Florida area on the map follows Ortelius-

Chiaves 1584 closely except for coastal names such as "R. S. Mathio" for the lower reaches of the "R. Seco," which are derived from later Spanish sources. Tatton may have taken the legend placed in the mountain range to the extreme northeast, "Mons Apallaci in quo aurum et argentum est," from Le Moyne 1591. The map is beautifully designed and is a skillful composite of information from various other cartographical documents.

The date 1600 which is on one of the Library of Congress copies of the map has been changed on the other by the erasure of the two zeros and the re-engraving of "16." The 1616 date is also found on the John Carter Brown and Streeter copies. Another map engraved by Benjamin Wright, also dated 1600, similar in technical workmanship and on paper of the same size and crossed-arrow watermarks, is attributed to G[abriel] Tatton as the author: Maris Pacifici, quod uulgo Mar del zur cum regionibus Circumiacentibus, insulisque in eodem passi Sparsis, nouissima descriptio, G. Tattonus Auct. 1600 | Beniamin Wright Anglus coelator. *Size:* 23 × 19. *Scale:* 1" = ca. 425 miles. Although the interest of this map is centered on the Pacific, as the title indicates, the Florida-Carolina coast has considerable detail. Copies of this map are in the Library of Congress, John Carter Brown, and Harvard. Several manuscript maps of America by Gabriel Tatton are extant, including one in the Florence National Library which extends to "Virginia" and "Norumbega" (See *Lowery Collection*, No. 86n.).

References

John Carter Brown. *Annual Report*, 1949, pp. 7–13.
Paullin, C. O. *Atlas*, p. 10.
Phillips, P. L. *Lowery Collection*, Nos. 86, 103.
Wagner, H. R. *The Cartography of the Northwest Coast.* Berkeley, Calif., 1937, II, Nos. 218, 269, 217.
The World Encompassed. Baltimore, 1952, No. 235.

Reproductions

Plate 19 (1616 issue).
Paullin, C. O. *Atlas*, Plate 17E.
The World Encompassed, Plate LI.

Copies

Congress, two copies; John Carter Brown; Harvard.

---

26. Mercator-Hondius 1606

Virginae Item et Floridae Americae Provinciarum nova Descriptio.

Size: 19 × 13½. *Scale:* 1" = ca. 48 miles.
In: Mercator, Gerardi. *Atlas . . . auctus ac illustratus á Iudoco Hondio.* Amsterodami, 1606, No. 143.

Description: Though this map by Hondius is based on two earlier maps, those of White 1590 and Le Moyne 1591, subsequent cartographers usually followed this work rather than the earlier maps. It thus became the most important type map of the region until the Ogilby-Moxon "Discription of Carolina" ca. 1672; and its influence, both direct and indirect, extends into the middle of the eighteenth century.

That part of the territory settled by the French in 1562–65 depends chiefly upon Le Moyne's map. The gold-bearing mountains of Apalatcy, the lake fed by the great waterfall, the lake with the unseen shore, Lake Sarrope, the French names for the rivers, the location of the French fort at Port Royal: these and many other details are based on Le Moyne. However, there are important changes. The great lake with the opposite shore invisible, which in Le Moyne lies in a southwesterly direction from the mouth of the River May, into which it empties, has been moved northward to the foothills of the Apalatcy Montes. This makes the River May flow in a southeasterly direction instead of in the Λ-shaped course given by Le Moyne. Probably in this Hondius was influenced by the other maps of the region—Mercator 1569, Ortelius 1584, Wytfliet 1597—in which the great river Sola flows in a southeasterly direction from the mountains.

However, the River May is Le Moyne's name for St. Johns River in Florida, on which Fort Caroline was built and which flows in a northerly direction from its sources and lakes in Florida until near its mouth, when it turns eastward toward the sea. By putting the great lake and the direction of the flow of the River May to the northwest of its mouth and changing the accompanying topographical features and Indian settlements, Hondius created geographical misconceptions of the region, which lasted for nearly a century and a half. Also he moved the latitude of the mouth of the River May from 29°30' N.L. (St. Johns is 30°25') to 31°20'. This is actually the latitude of the Altamaha in Georgia, the first river above Florida which has its source in the Appalachian range. These factors undoubtedly encouraged the continuance of erroneous beliefs in the minds of subsequent explorers and cartographers. St. Marys River, which lies between St. Johns and the Altamaha, rises in the Okefenokee Swamp. Some writers think that the River May is the St. Marys River and that the great lake is the Okefenokee Swamp: see A. H. Wright, *Our Georgia-Florida Frontier: The Okefinokee Swamp*, Ithaca, 1945, III, 1–16.

Le Moyne was not the only source for information concerning this territory; on the River May is drawn Edelano Island, which is found on many subsequent maps; and nearby is "a lake where gems are found." Ribaut's French names are given for the rivers where Le Moyne uses Latin, and the legend affixed to the map, "Verum nos eam solummodo Floridae partem hic apposuim*us* cujus pleniorem notitiam habemus ex ipso autographo illius qui hanc nomine regis Galliae accuratissime descripsit," may have as a source Laudonnière's *Histoire Notable*, Paris, 1586. The northern part of the map very closely follows the White 1590; however, a few Spanish names appear ("Medano, Hispanis"; "C. S Romano Hispanis").

Cartouches: A large rectangular cartouche containing the title is in the upper left corner; connected to this, on either side, are two ovals purporting to show the differences of buildings and fortifications between settlements in Florida and Virginia: to left, "Civitatum Floridae imitatio"; to right, "Civitatum Virginiae forma." To the lower right is a cartouche which states that the middle meridian of the region is 300° (the degrees of longitude are counted east: Port Royal, which lies near 300° on the map, is 80°40' longitude west from Greenwich), and the latitude extends from 30° to 37° (the actual latitude included in the map is from 29°50' to 39° N).

Gerard Mercator, the great geographer who invented Mercator's projection, died in 1594 after publishing only a few parts of his atlas. His son Rumoldus published a larger collection in 1595 but died in 1600. In 1602 Bernard Brusius published a collection in one volume at Dusseldorf; this was the first Mercator's atlas. But in 1606 Jodocus Hondius brought out an edition of Mercator's atlas, augmented by fifty new maps by himself and others, with the Petrus Montanus text. This edition was the first to contain the map of Virginia and Florida. Many editions in various languages followed; but the map itself remained unaltered, and when it is found as a separate there is no indication of which edition it is from except by a collation of the text on the back with texts in complete atlases. It was also apparently published as a separate with no text on the back.

In the following year, 1607, a small edition of Mercator's atlas was published, with reduced plates. In 1630 a larger reduced edition was published with new and somewhat changed plates. For a list of the different editions of Mercator's atlas, see P. L. Phillips, *A List of Geographical Atlases*, Washington, 1909–20, I, 168–71. For an account of the Mercator-Hondius-Jansson families and a list of maps and atlases published by them, see J. Keuning, "The History of an Atlas. Mercator-Hondius," *Imago Mundi*, IV (1947), 37–62; and J. Keuning, "The Novus Atlas of Johannes Janssonius," *Imago Mundi*, VIII (1951), 71–98.

One of the earliest world maps to show the influence of the Mercator-Hondius is the curious world map of Octavius Pisanus, "Globus terrestris planisphericus" ca. 1610, in which the spectator is supposed to stand on the South Pole, inside the sphere, his head directed toward the center of the earth. See I. N. P. Stokes, *Iconography of Manhattan Island*, New York, 1909–28, II, 134–35 and Plate 21.

The regional maps of Mercator-Hondius 1607, Mercator-Hondius 1630, Laët 1630, Jansson 1642, Sanson 1656, Du Val 1657, Montanus 1671, Aa 1714, Fer 1718, Chatelain 1719, and other maps by these and other cartographers followed the Mercator-Hondius 1606 type; see reference below for full list.

References

Cumming, W. P. "Geographical Misconceptions of the Southeast in the Cartography of the Seventeenth and Eighteenth Centuries." *Journal of Southern History*, IV (November 1938), 476–92; Appendix A (pp. 487–89) gives a list of regional maps influenced by this map.

Keuning, J. "The History of an Atlas, Mercator-Hondius." *Imago Mundi*, IV (1948), 37–62.

Phillips, P. L. *Lowery Collection*. Washington, 1912, p. 115, No. 100.

Reproductions

Plate 20 and Color Plate 2.

Cartografia de Ultramar, II, Plate 38 (with transcription of place-names and legends, pp. 234–35).

Humphreys, Arthur L. *Old Decorative Maps and Charts*. London, 1926, Plate 42.

Smith, Priscilla, compiler. *Early Maps of Carolina . . . from the Collection of Henry P. Kendall*. L. C. Karpinski, editor. 1930, between pp. 16 and 17.

Copies

Separates: Congress, two copies; American Geographical Society; John Carter Brown; Duke; Georgia (De Renne); New York Public, two copies; University of North Carolina; National Archives of Canada.

Latin editions (1606 et seq.): Congress 1607 (No. 143); 1619 (No. 147); 1623, between pp. 367–68; 1630 (No. 162); 1632, opp. p. 725; 1611 (No. 147); John

Carter Brown 1623; Clements 1607, two copies; Harvard 1613, 1638; New York Public 1616, 1623.
French editions (1609 et seq.): Congress 1609, second edition (No. 143); 1619, third edition (No. 152); 1630, opp. p. 663; 1628, opp. p. 698; 1633, Vol. II, p. 698; 1636, opp. p. 663; American Geographical Society 1613, 1633, 1619?; John Carter Brown 1663; Clements 1613, 1619, two copies, 1663; Harvard 1633; New York Public 1628; Yale 1633.
Dutch editions (1634 et seq.): Congress 1634 (No. 174).
English editions (1633 et seq.). *Description:* The title of the first edition is: *Atlas, or A Geographicke description . . . translated by Henry Hexham . . . printed at Amsterdam, by Henry Hondius, And Iohn Iohnson. Anno 1633*, II (1636), between pp. 441–42. The 1636 English edition has the English title and imprint pasted over an earlier title page (a 1633 French Mercator in the Harvard copy), with a second title page having "A Newe Atlas" (as in the Harvard and Huntington copies). The Clements Library copy has a title page for the second volume with "1638 Editio Ultima." The 1641 edition repeats the title of the earlier editions, "*Atlas, or a Geographicke Description.*"
The second volume of the English editions, in which this map appears, is dated 1638 or 1641, except for the University of Virginia copy, with 1636 in both volumes.
Congress; John Carter Brown; Clements (1638); Harvard; Huntington; New York Public; Yale.
In: Mercator, G., Joannes Jansson, H. Hondius. *Atlas Novus.* Amsterdam, 1638, III, Ssss.

Copies

John Carter Brown.
In: Hondius, H. *Nouveau Théâtre du Monde.* Amsterdam, 1639, III, No. 100.

Copies

Congress.
In: Hondius, H., and J. Jansson. *Nouveau Théâtre du Monde.* Amsterdam, 1639–40, III, no. 99 (verso, catchword; ggggg peu).

Copies

Congress.
In: Jansson, Ian. *Le Nouveau Théâtre du Monde ou Novvel Atlas.* Amsterdam, 1639 (verso, catchword: peu).

Copies

Yale.

27. Mercator-Hondius 1607

Virginia et Florida.

Size: 7¼ × 5¾. *Scale:* 1" = ca. 115 miles.
In: Mercator, Gerard. *Atlas Minor . . . à I. Hondio.* Amsterdam, 1607, No. 145.

Description: This map is based on the larger Mercator-Hondius 1606 map. It has omitted a number of names and legends found upon the map of the previous year, the oval cartouches with Indian settlements on either side of the title cartouche, all pictures of animals, sea monsters, and ships, and the cartouches at lower right containing comments on latitude and longitude. The same plate, with an added title printed at the top of the page, was used for the English editions of Mercator; see Mercator-Hondius-Saltonstall 1635.

Although the printed superscriptions differ, the same plate is used in Nos. 33 and 37. This map does not appear in *Atlas Minor* after 1621.

Copies

Latin editions (1607 et seq.). Congress 1607, No. 145; 1609, opp. p. 649; 1610, opp. p. 665 with "Virginia et Florida 665" above the neat line, verso, catchword: sunt; 1621, opp. p. 655; American Geographical Society 1628; Harvard 1613; New York Public 1610; Yale 1610; Clements 1621; National Archives of Canada 1607.
French editions (1608 et seq.), opp. p. 627. Congress 1608, 1613; American Geographical Society 1608?; National Archives of Canada 1608.
German editions (1609 et seq.). Harvard 1609?.

27A. Tindall Virginia 1608 MS

The draught by Robert Tindall of Virginia, anno 1608

Size: 18 × 32. *Scale:* 1" = 2⅘ miles.

Description: This colored chart, the first extant map of Virginia by a Jamestown colonist, shows with unusual degree of accuracy the location of Indian villages on the James and York Rivers observed on the expeditions of Captain Newport in the summer of 1607 on the James and in February 1608 up the York River. Robert Tindall, who was a gunner in the service of Prince Henry, a patron of the Virginia Company, listed fourteen names on the James and five on the York River. Near Cape Comfort at the mouth of "King James his River" is the Indian settlement of "Checkotanke"

(Kecoughtan); farther up the river is Jamestown, with its peninsula location now an island, clearly indicated. Several Indian settlements are marked and named between Chickahominy River and Poetan (Powhatan) below the falls at the present site of Richmond. Below on the chart (and to the north) are "Prince Henry His River" (York) and "Tindall's Point" (now Gloucester Point). When the party reached Powhatan's headquarters on the Pamunkey, the Indian chief attempted to persuade Captain Newport to move his settlement from Jamestown.

For the map's historical importance, see Maurice A. Monk, "The Ethnological Significance of Tindall's Map of Virginia," *William and Mary Quarterly*, XXIII (1941), 371–408.

References

Barbour, P. L. *The Jamestown Voyages under the First Charter 1606–1609.* Cambridge, 1969, I, p. 106.

Brown, Alexander. *The First Republic in America.* New York, 1898 (reprint, 1969), pp. 5–7ff.

Reproductions

Barbour, P. L. *Jamestown Voyages*, I. opp. p. 104 (line drawing).

Cumming, Skelton, Quinn. *The Discovery of North America*, p. 237 (color).

Original

British Library. Cotton MS, Aug. I, ii, 46.

---

28. Zuñiga 1608 MS
[Chart of Virginia]

Size: 28¾ × 22. *Scale:* None.

Description: This is a crude chart of the Chesapeake Bay region, with three rivers to the south of the James River which represent the Neuse, Pamlico, and Roanoke. The chart was sent by Zuñiga, the Spanish ambassador to England, to Philip III with a letter dated September 10, 1608. Alexander Brown (*Genesis of the United States*, I, 184) thinks that it, or the original from which it was copied, was sent to England with Captain Nelson in the summer of 1608 by Captain John Smith, who intended it to illustrate his "A True Relation." If so, it was not used, although details on it are similar to those on Smith's map of Virginia 1612 and both may be derived from a common source. Brown also thinks that Henry Hudson took a copy of this chart with him to Holland in 1608, where he had "letters and charts which one Captain Smith had sent him from Virginia, by which he [Smith] informed him [Hudson] that there was a sea leading into the Western Ocean by the North of the Southern English Colony," about the latitude of forty degrees. Lewis and Loomie (*The Spanish Jesuit Mission in Virginia*, pp. 262–63) suggest Nathaniel Powell, who was one of the cartographers employed by the Virginia Company and who accompanied Smith on many of his excursions, as the author of the chart.

The map has numerous legends. The information given in them, not found on any other map, adds considerably to its interest and value. Those in the North Carolina region show the continued concern over the fate of possible survivors of the Roanoke Colony. The reproduction in this work shows only about a fourth of the map and represents the country south of the James River. The transcription of the names and legends, given below, begins at the south and reads north.

1. pakerakinick

D. I. Bushnell, Jr. ["Virginia from Early Records," *American Anthropologist*, N.S. IX (January 1907), 34] quotes an interesting document, British Museum MS Vol. Z1993, fol. 178ff., which contains instructions to Deputy Governor Thomas Gates: "Here at Peccarecamicke you shall finde four of the English alsoe, lost by Sr Walter Raweley, which escaped from the slaughter of Powhatan of Roanocke upon the first arivall of our Colony and live under the protection of a wiroano called Sepanocan enemy to Powhatan by whose consent you shall never receive them, one of these were worth much laboar and if you find them not, yet search into this countrey it is more probable than towardes the North."

2. Here remay[n] eth 4 men clothed that came from roonock to Ocanahowan

3. rawrotock

4. aumocawnunk

5. morattico

6. machomonchocock

7. Here the King of Paspahegh reported our men to be and wants to go

Paspahegh was the country on the north bank of the James above Jamestown.

8. panawiock

9. roanock

Roanoke Island; the Outer Banks are not given in this rough drawing.

10. chawwone

Chowan River.

11. vttawmussawone

12. ocanahowan

13. nihamaock

An inlet by Knotts Island led to Currituck Sound and up to North Landing River, here called Nihamaock.

14. chissapiack

15. nansamund

16. here Paspahegh and 2 of our men landed to go to Panawiock

Most of these names differ from those on the White and De Bry maps. Pakerakinick was apparently an Indian town on the Neuse, Panawiock south of Albemarle Sound, and Ocanahowan on the Roanoke; on the White–De Bry 1590 map Panauuaioc [?Panawiock] is on the south bank of the Pamlico and Chaunoock is on the Chowan. Several of the names are found in contemporary reports from the Jamestown colony. Captain Smith writes in his *A True Report* (L. G. Tyler, ed., *Narratives of Early Virginia*, New York, 1930, p. 53) that the King of Paspahegh started with two men to search for the Roanoke colonists at "Panawicke beyond Roonok," but "playing the villaine" returned without going further than Werraskoyack. In *The Proceedings of the English Colonies in Virginia* (Tyler, *Narratives of Early Virginia*, pp. 163, 188) two unsuccessful expeditions for the lost colonists are mentioned, one by Michael Sicklemore to Chawonock and another by Nathaniel Powell and Anas Todkill to the Mangoages, the latter probably the same as that mentioned by Smith. William Strachey (*Historie of Travaille into Virginia*, ca. 1616; quoted in Brown, I, 184) refers to a report that the Indians at Peccarecanick and Ochanahoen had two-story stone houses which they were taught to build by the English who escaped the slaughter at Roanoke and that seven of the lost colony, "fower men, two boyes and one yonge mayde," were held at Ritanoe by the weroance Eyanoco.

The best reproduction of the whole map, with useful notes, is in Philip L. Barbour, ed., *The Jamestown Voyages under the First Charter 1606–1609*, Hakluyt Society, second series cxxxvi–vii. Cambridge, 1969, I. 238–40.

References

Barbour, P. L. *Jamestown Voyages*, I, 238–40.

Brown, A. *Genesis of the United States*. Boston, 1890, I, 184–85.

Lewis, C. M., and A. J. Loomie. *The Spanish Jesuit Mission in Virginia, 1570–1573*. Chapel Hill, N.C., 1953, pp. 262–63, 274–75.

Pearce, H. J., Jr. "New Light on the Roanoke Colony." *Journal of Southern History*, IV (May 1938), 148–63.

Weekes, S. B. "The Lost Colony of Roanoke: Its Fate and Survival." *Am. Hist. Asso. Papers*, V (1891), 468.

Reproductions

Plate 21 (southern part).

Ashe, S. A. *History of North Carolina*. Greensboro, N.C., 1908, I, opp. p. 43 (southern part; a tracing).

Barbour, P. L. *Jamestown Voyages*, I, opp. p. 238.

Brown, Alexander. *Genesis of the United States*, I, opp. p. 184.

Bruce, Philip A. *History of Virginia*. American Historical Society, New York, 1924, I, opp. p. 36.

Original

Simancas, Archivo General de Estado. Leg. 2586, folio 148 (M y P IV.66).

---

29. Velasco 1611 MS
[East Coast of North America]

Size: 31 4/8 × 45 5/8. *Scale:* None.

Description: This anonymous English chart of the east coast of North America from Cape Fear (Cape Lookout) to Newfoundland was sent by Don Alonso de Velasco, the Spanish ambassador, to the King of Spain, together with a letter in cipher, dated March 22, 1611. Velasco states in the letter that it is a copy of a map made by a surveyor who was sent by James I to survey the English province in America. The surveyor, according to Velasco, had returned about three months prior to the date of his letter and had incorporated in the map all that he could learn. The date of this copy is therefore properly 1611, though the original survey was made in 1610.

The map is drawn in colors on four sheets pasted together. For the Carolina area many of the names go back to White–De Bry 1590, although several names (Croatamang, Port Fernando, and Mentso) show a knowledge of White's manuscript map. Quinn (see reference below, II, 851–52) thinks that other names may derive from Molyneux, from other unknown maps, and from interested informants such as Hariot. Cape Kenrick, the cape which then existed at what is now Wimble Shoals, is referred to in contemporary narratives but is on no other extant map of the period, though it appears on a number of later seventeenth-century maps; it is probable that this name was furnished by some personal informant. New details in the Roanoke and Chowan Rivers region show an exactness which Captain John Smith gained from the expeditions made south of the James

River. The Chesapeake area has 48 names of Indian villages. Alexander Brown thinks that Smith may have used this map in preparing his 1612 map of Virginia. Stokes (II, 57) thinks that John Daniel may have contributed to its composition. The delineation of the Hudson River is of special interest, since this map is the earliest known document to give the names "Manhata" and "Manhatin."

Coolie Verner ["The First Maps of Virginia, 1590–1673," *The Virginia Magazine of History and Biography*, LVIII (1950), 7] quotes Samuel Adams Drake as contending that the map is too good for the state of geographical knowledge existing in 1610 and concluding that it belongs to a later date.

References

Brown, A. *Genesis of the United States*, I, 455–60.
Fite, C. D., and A. Freeman. *A Book of Old Maps*, pp. 109–11.
Quinn, D. B. *The Roanoke Voyages*, II, 851–52.
Stokes, I. N. P. *Iconography of Manhattan Island*, II, 57, 135–36.

Reproductions

Brown, A. *Genesis of the United States*, I, opp. p. 456.
Cumming, Skelton, Quinn. *Discovery of North America*, pp. 266–67 (color; double page).
Fite, C. D., and A. Freeman. *A Book of Old Maps*, Plate 30.
Quinn, D. B. *The Roanoke Voyages*, II, opp. p. 581 (facsimile of the Outer Banks area).
Stokes, I. N. P. *Iconography of Manhattan Island*, II, C, Plate 22 (frontispiece, photographic reproduction in color).

Original

Simancas, Archivo General de Estado. Leg. 1588, fol. 25.

---

30. Lescarbot 1612

Figure et description de la terre reconue et habitée par les Frªnçois [*sic*] en la Floride et audeça, gisante par les 30, 31, et 32 degrez [Cartouche, bottom right] | De la Main de M. Marc Lescarbot [top center]

Size: 8 1/16 × 6 5/16. *Scale:* 1" = ca. 42 miles.
In: Lescarbot, Marc. *Histoire de la Nouvelle-France*. Paris, 1612, opp. p. 65 (location differs in various copies).

Description: This map shows the coast and the interior from below 30° to 33° N.L. (Cap François to Port Royal) and accompanies an account of the French attempt under Ribaut and Laudonnière to found a Huguenot colony. On it are depicted the great lake of unknown width; the mountain of Palassi, where gold, silver, and copper are found; Indian villages named by their chiefs, although, according to a note on the map, not a thirtieth of the true number is marked; the coastal rivers by their French names; Charlefort at Port Royal; and La Caroline on the River May.

Phillips (*Lowery Collection*, pp. 117–18) quotes a long and detailed note by Kohl on this map. Kohl points out numerous differences from the Le Moyne map 1591, such as the direction of flow and the latitude of the River May (St. Johns), which he states is put half a degree north of Le Moyne's River May and of its true latitude by Lescarbot. "This false picture of Lescarbot was copied by Laët and many other geographers," says Kohl. The accusation of Kohl is unjustified, however, for on the Mercator-Hondius maps, which are clearly the ones actually followed by Laët and others, the mouth of the River May is put at 31°20'; on Lescarbot's at about the true latitude of 30°23'; and on Le Moyne's at 29°50'. In fact, neither in topography nor in nomenclature does Lescarbot show a close dependence upon Le Moyne or Mercator. The maps of Florida by Sanson and Du Val 1659 et ff. and possibly Kocherthal 1709 and Homann 1714 were influenced in their conception of the size, shape, and location of "Apalache" lake by this map.

Lescarbot's work was first published in 1609, with three maps in some copies, but no map of Florida nor reference to it. The 1611 and 1612 printings have this notice, "Av Lectevr": "La figure du Fort de la Floride dit la Caroline, entre la page 66 & la 67." In the 1617 and 1618 printings the notice "Av Lectevr" was changed to "La figure de la Terre de la Floride reconuë & habituée par tes [*sic*] Francois, en la page 65." The change from "Fort" to "Terre" in the notice would indicate that Lescarbot planned at first to draw a design of the fort La Caroline, built by the French on the River May; but this was not done and the map which he drew later was for the whole country. The map is not found in the 1611 edition of *Histoire de la Nouvelle-France*, which was published by Iean Millot of Paris; it is found in copies of the 1612 printing, also published by Millot, and in the printings of 1617 and 1618, published by Adrian Perier of Paris. The date of the map is therefore better given as 1612 than the year 1611, which is the date to which it is assigned by Lowery.

References

Phillips, P. L. *Lowery Collection*, pp. 117–19, No. 102.
Sabin, J. *Bibliotheca Americana*, X (1878), 231, No. 40170.

Reproductions

Plate 22.

Courtenay, W. A. *The Genesis of South Carolina, 1562–1670.* Columbia, S.C., 1907, opp. p. xxxii.

Lescarbot, Marc. *The History of New France.* In Publications of the Champlain Society. Toronto, 1914, XI (Vol. I of Lescarbot's History), end of volume.

Copies

Congress 1618; John Carter Brown 1618; Clements 1618; Harvard 1618; Huntington 1612, 1617, 1618; Newberry 1617, 1618; New York Public 1612, 1618; Yale 1618; National Archives of Canada 1618.

---

31. Bertius 1616

Virginia et Nova Francia [cartouche, lower right] | Descriptio Terræ Novæ. [above neat line of map] | 787 [pagination: top left, outside neat line of map]

Size: 5⅜ × 3¾. *Scale:* 1" = ca. 530 miles.
In: Bertius, Petrus. *Tabvlarvm Geographicarvm Contractvm.* Amsterdam, 1616, p. 787 (opp. p. 779).

Description: This map extends from "Rio May" to Labrador. South of "C de S Roman" the Spanish nomenclature is used: to the north, the White map 1590 nomenclature for "Virginia." Chesapeake Bay is omitted, and "Trinety herbor" (Albemarle Sound) is below "C d las Arenas." Above and to the west of Virginia is the country of Apalchen.

This is a new plate, designed by Hondius; in the 1602, 1606, and 1612 copies of Bertius's work, under different titles, a map "Terra Nova" is used, which shows only Labrador and the land at the mouth of the St. Lawrence River.

Copies

Congress; Clements, two copies.

In: Bertius, Petrus. *La Geographie Racourcie.* Amsterdami, 1618, p. 778.

Copies

Congress; National Archives of Canada.

---

32. Smith 1624

Ould Virginia [across field of map] | A description of part of the adventures of Cap: Smith in Virginia [top right of map] | Graven and extracted out of y^e generall history of Virginia, New England, and Sōmer Iles, by Robert Vaughan. [beneath neat line]

Size: 6⅛ × 5⅛. *Scale:* 1" = ca. 19½ miles.
In: Smith, John. *The Generall Historie of Virginia, New=England, and the Summer Isles . . . London,* 1624, between pp. 20–21 or 40–41.

Description: This map is on a large engraved folding sheet (22" × 10⅝" except in the large paper edition of the *Historie*) which has ten compartments, nine of which are pictures of Indian life and adventures, some of which are influenced by De Bry's engravings of White's drawings, 1590. The tenth compartment, in the lower center of the sheet, contains the Smith map, with "Ould Virginia" written across it. The map shows the coast from "C Henry" to "C feare" (Cape Lookout).

Strangely this map, in spite of the numerous editions and the evident popularity of the work in which it is found, had very little influence on the cartographical development of the region. Although many of the White 1590 names are retained, the author made numerous changes and gave new names to different topographical features, both actual and imagined; acceptance of this map by cartographers would have markedly changed nomenclature along the North Carolina coast. Very few English maps of the Carolina region were made in the next fifty years, however, and the continental mapmakers were not influenced by it. When the flood of English maps of the Carolina region began about 1675, new influences were at work, and Smith's map, which added little to the topographical knowledge of the territory, was not used.

Wilberforce Eames lists eight different issues of Captain John Smith's *A Generall Historie* in Sabin's *A Dictionary of Books Relating to America,* XX (1927), 227; and he gives four different states of the map of "Ould Virginia" as they are tabulated below. Changes of ownership and additional copies acquired by libraries and individuals since Eames's 1927 list are noted below on the basis of replies to enquiries sent out during the summer of 1956. Many copies of *A Generall Historie* lack the maps; only those which have "Ould Virginia" are given here.

A fuller tabulation of copies of *A Generall Historie,* listed under the various dates of publication, may be found in Elizabeth Baer's *Seventeenth Century Maryland,* Baltimore, 1949.

Reproductions

Plate 23.

Cumming, Skelton, Quinn, *Discovery of North America,* p. 281 (with 8 accompanying illustrations).

Smith, John. *The Generall Historie of Virginia. . . .* New York, 1907, opp. p. 208.

Copies

A. First state: with thirty-seven place-names but with no figures of trees. Chapin Library, Williamstown, Mass., 1624; Massachusetts Historical Society 1626; Free Library of Philadelphia 1624 insert; Princeton 1624, 1626; University of Virginia 1624; Walters Art Gallery 1624.

B. Second state: twenty-five new place-names, beginning at the top with "Mountaynes forest," twenty-seven figures of trees added, but before the name of Iames Reeve as printer was inserted.

a. On ordinary paper: Huntington 1624 (large paper ed.); New York Public 1626.

b. On thick paper for use in the large paper issue of 1624: British Museum; Eton College Library, Eton, Bucks.; Huntington, three copies; Lambeth Palace Library, London; New York Public; Pierpont Morgan Library, New York; Princeton, two copies; University Library, Cambridge, England.

C. Third state: the inscription "printed by Iames Reeve" is added to the right of the map in the Pocahontas–John Smith picture; two new place-names are added, "Davers Ile" and "P Barkley"; but "Adams Sound" has not been added above "C Henry." Congress 1626; Columbia University 1626; John Carter Brown 1632; Clements 1624; Princeton 1624; Huntington 1625; Johns Hopkins (Garrett) 1624; Library Company of Philadelphia 1624; Pierpont Morgan Library, New York; Preston Davie, Tuxedo, N.Y., 1626; New York Public 1624, 1632; Henry C. Taylor 1625; University of Virginia 1624; Yale 1626.

D. Fourth state: "Adams Sound" added just above "C Henry." Congress 1624; Boston Public 1627; John Carter Brown 1627, 1631, 1632; Clements 1632; Dartmouth 1624; Duke 1624; Folger Shakespeare Library 1624, large paper edition, 1626, 1631; Princeton 1627, two copies, 1632; Harvard 1632; Huntington 1626, 1627, 1632; Illinois State Historical Society 1624; Lafayette 1624; Library Company of Philadelphia 1627; Newberry 1627, 1632; New York Public 1627, two copies, 1632, three copies; Pierpont Morgan Library 1627, large paper copy; Henry C. Taylor 1632; University of Virginia 1627; William and Mary 1624; Williams (Chapin) 1632.

---

33. Purchas (Hondius) 1625

Virginia et Florida [cartouche, upper center] | Hondivs his Map of Florida [printed above neat line]

Size: 7¼ × 5¾. *Scale:* 1" = ca. 87 miles.

In: Purchas, Samuel. *Purchas his Pilgrimes.* London, 1625, III, p. 869.

Description: This map is a form of the small Mercator "Virginia et Florida" 1607. The map occupies about half the page, the rest of which is part of the printed description of the coast from Florida to Newfoundland. *Purchas his Pilgrimes* was also published uniformly with *Purchas his Pilgrimage*, London, 1626, as in the John Carter Brown copy.

The Widener Collection copy at Harvard has an inscription by Purchas to Sir Robert Heath, his parishioner. This is the only known inscribed copy. A few years later Sir Robert Heath obtained a grant of the land of "Carolana" from Charles I.

This map uses the same plate as No. 27 with different superscription. John Carter Brown 1626 and Newberry 1626 are in Vol. III (1625); only Vol. V is 1626.

Copies

Congress 1625; American Geographical Society 1625; John Carter Brown (with *Purchas his Pilgrimage*, London, 1626); Clements 1625; Harvard 1625, two copies; Newberry 1625, two copies, 1626; Yale 1625, two copies.

---

34. Laët 1630 A

Florida, et Regiones Vicinae

Size: 14 × 11. *Scale:* 1" = 106 miles.

In: Laët, Joannes de. Nieuvve Wereldt. Leyden, 1625, opp. p. 40.

Description: This map is an attempt to put onto one map the information of the Ortelius-Chiaves 1584 map and the Mercator 1606 map. Mercator's names are given for the coast; "Cosa," "Cofachiqui," "Xuala," etc., of Ortelius, which properly are located in or near the Carolinas, have been pushed westward of the Appalachian range and to the eastern branches of the Mississippi River. The Floridian peninsula is called "Tegesta Provinc.," one of the earliest appearances of this name on a map. "Tegesta" for Florida appears on numerous subsequent maps: cf. De Brahm's "The Ancient Tegesta, now Promentory of East Florida," in his *Atlantic Pilot*, London, 1772.

In spite of clear statements of the early explorers that the Mississippi was a single large river, Laët gives a distorted and peculiar conception of three large rivers and three small rivers flowing into a large "Bahia del spiritu Santo." The thesis of J. Delanglez, *El Rio del Espíritu Santo*, New York, 1945, is that this bay and these rivers do not represent the Mississippi River. In the east, surrounded by the "Apalatcy Montes," is a small unnamed lake, to the northeast of the "Lacus Magnus," the great lake of Le Moyne and Mercator.

Kohl and Phillips (see *Lowery Collection*, p. 137) both date this map 1625. It goes more properly under the date 1630 with its companion map, "Nova Anglia, Novvm Belgivm et Virginia," since it is not found in the 1625 edition of Laët's work in the libraries of Congress, Clements, Huntington, New York Public, American Geographical Society, or Harvard. The John Carter Brown Library copy in the 1625 edition of the work is affixed to a stub, though the stub may not be a part of the original binding.

Jansson 1636 and Blaeu 1644 copied this map closely; this and the following map formed the base for Sanson's Canada 1656 and Florida 1656.

Reference

Phillips, P. L. *Lowery Collection*, pp. 136–37, No. 123.

Reproduction

Plate 24.

Copy

John Carter Brown 1625.

In: Laët, J. de. *Beschrijvinghe van West-Indien*. Leyden, 1630, No. 5, opp. p. 137.

Copies

Congress; American Geographical Society; John Carter Brown, two copies; Harvard; New York Public.

In: Laët, J. de. *Novus Orbis*. Lvgd. Batav. 1633, No. 5, opp. p. 95.

Copies

Congress; American Geographical Society; John Carter Brown; Harvard, two copies; Huntington; Kendal; New York Public; Yale.

In: Laët, Jean de. *L'Histoire dv Nouveau Monde*. Leyde, 1640, opp. p. 99.

Copies

Congress; John Carter Brown; Clements; Harvard; Huntington; Yale; National Archives of Canada.

35. Laët 1630 B

Nova Anglia, Novvm Belgivm et Virginia.

Size: 14 × 11. *Scale:* 1" = ca. 110 miles.

In: Laët, Joannes de. *Beschrijvinghe van West-Indien*. Tweede Druck. Leyden, 1630, No. 4, opp. p. 39.

Inset: Bermuda majori mole expressa. *Size:* 3½ × 3¼. *Scale:* 1" = 6¼ millia.

Description: The Carolina coast, which is shown southward to "C. of Feare" (Cape Lookout), is based on White 1590. Dr. L. C. Wroth (John Carter Brown, *Annual Report*, 1943, pp. 23–27) suggests that this map may be the prototype of N. J. Visscher's famous map of the northeastern coast of North America and its hinterland, "Novi Belgii Novae Angliae, nec non Partis Virginiae Tabula . . ."

References

Paullin, C. O. *Atlas*, p. 12.

Stokes, I. N. P. *Iconography of Manhattan*, II, 141, 144–45; VI, 261–62.

Winsor, J. *Narrative and Critical History of America*. III, 381; IV, 417.

Reproductions

Avery, E. M. *A History of the United States*. Cleveland, 1904–7, II, 222.

Paullin, C. O. *Atlas*, Plate 21.

Stokes, I. N. P. *Iconography of Manhattan*. II, C. Pl. 31.

Winsor, J. *Narrative and Critical History of America*. IV, 436; (New England), opp. p. 126.

Copies

Congress; American Geographical Society; John Carter Brown, two copies; Harvard; New York Public.

In: Laët, Joannes de. *Novvs Orbis seu Descriptionis Indiae Occidentalis*. Libri XVIII, Lvgd. Batav., 1633, No. 4, opp. p. 63.

Copies

Congress; American Geographical Society; John Carter Brown; Harvard, two copies; Huntington; Kendall; New York Public; Yale.

In: Laët, Joannes de. *L'Histoire dv Nouveau Monde*. Leyden, 1640, opp. p. 67.

Copies

Congress; John Carter Brown; Harvard; Huntington; Yale; National Archives of Canada.

36. Mercator-Hondius 1630

Virginiae Item Floridae Americae Provinciarum nova Descriptio

Size: $10^{1}/_{8} \times 7^{3}/_{8}$. *Scale:* 1" = ca. 118 miles.

In: Mercator, G. *Atlas*. Amsterdam, 1630, p. 663.

Description: This is a reduced copy of the original Mercator-Hondius "Virginiae Item et Floridae" 1606, with the omission of the two oval cartouches containing pictures of Indian settlements in Florida and Virginia, of figures of boats, monsters, animals, and people, and of several legends. It is, however, fuller than the small Mercator-Hondius 1607, having the title cartouche at upper left, a dugout with three Indians off the River May, and the cartouche with comment on latitude and longitude at bottom center. Of the reduced size of the large Mercator's atlas three editions are known: French editions of 1630 and 1636, and a Latin edition of 1632. The French edition has "De Virginia et la Floride" printed above the neat line of the map; the Latin edition has "Virginia et Florida." The Jansson and Du Sauzet editions have no printed title above the map.

Copies

Congress 1630; John Carter Brown, *United States of America Atlas*, No. 10.

Latin edition: Congress 1632.

In: Jansson (van Waesberge), Joannes. *Atlas sive Cosmographicae Meditationes*. Amsterdam, 1673, No. 174 (verso is blank).

Copies

Yale 1673.

In: Du Sauzet, Henri. *Atlas Portatif*. Amsterdam, 1734, No. 264.

Copies

Congress 1734–[35], 1734–[38]; Clements 1734.

37. Mercator-Hondius-Saltonstall 1635

Virginia et Florida [cartouche, top center] | Virginia 899. [printed, at top of page]

Size: $7^{1}/_{4} \times 5^{3}/_{4}$. *Scale:* 1" = ca. 115 miles.

In: Mercator, G. *Historia Mundi: or Mercator's Atlas . . . Englishᵉd by W[ye] S[altonstall]*. London, 1635, p. 899.

Description: This map has "Virginia [page] 899" printed at the top of the page; otherwise it is similar to, and uses the identical plate of, Mercator's *Atlas Minor*, 1607 (see Map 27, above). Four different issues of this Englished atlas have been examined in this study, with the following differences in imprint:

1. London Printed by T. Cotes, for Michael Sparke 1635 (as printed on the title page of the first Folger Library copy).

2. London Printed by T. Cotes, for Michael Sparke and Samuel Cartwright 1635.

3. London printed for Michaell Sparke, and are to be sould in green Arbowre 1637. Second Edytion. (Ornamental title page; imprint on title page as in 2.)

4. (Title page and ornamental title page as in 3, except for change of date to 1639 on ornamental title page.)

The map of Virginia and Florida is unchanged throughout; the same plate is used as in Nos. 27 and 33.

Copies

Congress 1637, four copies; John Carter Brown 1635, 1637; Clements 1639; Folger 1635 [1], 1635 [2], 1637; Harvard 1635; Huntington 1635, 1637; Mariners' Museum 1637; Newberry 1635, 1637; New York Public 1639; Yale 1635, 1637; National Archives of Canada 1635, 2 copies.

38. Dudley ca. 1636 MS.

[Map of the East Coast]

Size: $19^{1}/_{2} \times 15$. *Scale:* None.

Description: This manuscript map extends from the north part of Florida and the most northern part of the Bahama Islands to Nova Scotia and the mouth of the St. Lawrence. It is No. 43 of the third volume of the Munich MSS of Sir Robert Dudley, styled earl of Warwick and duke of Northumberland. Dudley (1573–1649) lived as an exile in Florence from the early part of the seventeenth century until his death. He collected a large library of works on navigation and many maps which he used in the preparation of his great sea atlas, *Dell Arcano del Mare*, first published in 1646–48, and again in 1661. The large collection of manuscript maps prepared by Dudley, from which A. F. Lucini, his engraver, worked, is now preserved in the Royal Library of Munich. It is contained in three volumes, with 273 numbered leaves, and has the original MSS of the maps of the *Arcano*, as well as unpublished material.

Several maps, on varying scales, which portray the east coast of the North American continent, are contained both in the Munich manuscript and in the *Arcano*.

For a discussion of the date and content of the manuscript and its relation to the published work, see Stokes, *Iconography of Manhattan Island*, II, 145–48. A full description of Dudley's work is found in P. L. Phillips, *List of Geographical Atlases in the Library of Congress*, I, 203–17 (No. 457), and III, 142–46 (No. 3428). A pen-and-ink drawing of one of the Dudley MS maps, "La Florida," is in the Kohl Collection in the Library of Congress. The area of this Kohl sketch extends from Cuba to 31°, to "R. Maio," on the east coast; see P. L. Phillips, *Lowery Collection*, pp. 123–28.

For the two specific maps of the southeast coast in the *Arcano*, see under Dudley 1647; the published form of this map, "Carta seconda Generale del' America" 1646, and the other American maps in Dudley's manuscript atlas and in the *Arcano*, are not listed in this study.

Reproduction

Stokes, I. N. P. *Iconography of Manhattan Island*. New York, 1915–28, II (1916), Plate 36. Plate 37 gives the engraved form of the same map as it appears in Dudley's *Arcano*.

Original

Munich. Kgl. Hofbibliothek.

---

39. Jansson 1636

Nova Anglia Novvm Belgivm et Virginia [heart-shaped cartouche, upper left] | Amstelodami Johannes Janssonius Excudit [cartouche lower right, with scales]

Size: 19⅞ × 15¼. *Scale:* 1" = ca. 110 miles.

In: Jansson, Jan. *Appendix Atlantis oder dass Welt Buchs*. Amsterdam, 1636, fol. Nn (verso, catchword: New).

Description: This is Jansson's reprint of De Laët's map 1630 B on a larger scale and with the omission of the Bermudas inset. The map shows the Atlantic coastline of North America from "Cape Feare" (Cape Lookout) to Nova Scotia; Cape Feare is put at 32° north, which is almost three degrees too far south.

A careful comparison made in Stokes's *Iconography of Manhattan*, II, 144–45, shows that except for a few minor changes of nomenclature and careless mistakes in spelling, Jansson follows De Laët's map 1630 so closely that one may safely conclude that Jansson "copied only from his map, and not from the original from which it was derived. This is confirmed by the text on the back of the map, which is taken nearly verbatim from De Laët."

Jansson probably made this map for his *Appendix Atlantis*, Amsterdam, 16[3]6, and then used it in his English edition of Mercator's *Atlas*, published the same year. The Yale copy of Jansson's *Appendix Atlantis* has the following title: Appendix Atlantis Oder dess Welt Buchs | darinnen viel unterschiedliche | neue | und weitlåuffige Beschreibung | mit deroselben Tabellē | von Teutschland | Franckreich | Niderland | Italien | auch einer und der andern Indien begrieffen: Alles in gute Ordnung gebracht: Anno 1626 [changed with ink to 1636]. This title is pasted over a second title-slip, "Atlas . . . Welt Buchs" [from "Franckreich" on same as above, but no date], which in turn is pasted over a French title, printed on the title page itself, "Appendix Atlantis." The imprint of the Yale copy, "Amsterdam Bey Johan Jansson, Anno 1636," is pasted over the original French imprint.

Three different states of the plate of this map have been noted. In 1647 Jansson changed the title of the map and engraved decorations found on an earlier Blaeu map which has the same title but which does not extend as far south as this map: see Jansson 1647 and Phillips, *Lowery Collection*, pp. 230–31, No. 270.

Reproductions

Karpinski, L. C. *Bibliography of the Printed Maps of Michigan*. Lansing, 1931, Plate II.

Stokes, I. N. P. *Iconography of Manhattan Island*. II, C, Plate 31.

Copy

Yale 1636.

In: Mercator, G., and H. Hondius. *Atlas or a geographicke description of the regions, countries, and kingdomes of the world. . . .* Translated by Henry Hexham. Amsterdam, 1636, II, 44^1–44^2 (second volume has the date 1638 in some copies).

Copies

Congress 1636; Clements 1638; Harvard 1636; Huntington 1636; New York Public (Stokes collection, separate); Yale 1638?; National Archives of Canada 1636.

In: Mercator, G., J. Jansson, and H. Hondius. *Atlas Novus*. Amsterdam, 1638, III, Ssss (verso, Latin text).

Copy

John Carter Brown 1638.

In: Jansson, J., and H. Hondius. *Le Novveau Théâtre du Monde*. Amsterdam, 1639–40, III (1640), No. 96 (verso, catchword: dddd une).

Copies

Congress 1640; National Archives of Canada (separate).
In: Jansson, J. *Nieuwen Atlas.* Amsterdam, 1642–44, II (1642), America, fol. C (verso, catchword: Nievw).

Copy

Congress 1642.

---

40. Vingboons ca. 1639 MS

[Carolina coast]

Size: 28¼ × 19¾. *Scale:* 1" = 22½ miles.

Description: This sheet, which belongs to the Vingboons Atlas in the Library of Congress, extends from "Accowmack" on the eastern shore of Virginia to "Barra de S Matheo" [*sic*] in Florida.

F. C. Wieder, in his *Monumenta Cartographica,* The Hague, 1925, I, 9–10, 89–95, gives a full account of the Vingboons Atlas, which was apparently once in the hands of the great eighteenth-century mapmaking family of van Keulen; it was later broken up and sold to various purchasers. Some of the maps relating to North America came into the possession of Harrisse, who bequeathed them to the Library of Congress. The present map was offered to the Library of Congress by the Argosy Book Shop of New York about 1937, and not until the map had been compared with the maps already in the Library of Congress was it discovered that this was the missing map between the New England–Virginia and the Florida maps in the Library of Congress Vingboons Atlas. Joannes Vingboons's work is carefully and beautifully done, though this map adds nothing to our knowledge of the coast. Vingboons is the draftsman; whether he is also the author is not known.

Wieder, op. cit., IV, p. 118, lists another Vingboons atlas, Vatican Library, No. Reg. Lat. 2105–7, Wereltlycke Verthoningen Der . . . Africa, America, En een Gedeelt Van Europa . . . Door Iohannes Vingboons. Sheet 11 of this atlas has "Powhatan," "Wingandocoa By de Engelsche Genoemt Virginia," with the coastline extending southward to Lat. 31°25' N. There is no title; in the right-hand bottom corner is "I. Vingboons fecit."

Size: 69½ × 48 cm. *Scale:* 30 Duijtsche Mylen = 136 mm.

Wieder lists 242 Vingboons maps; recently four more, which together comprise a chart of the known world, have been found in a private Swedish collection and reproduced in full size, with an introduction: Leo Bagrow, "Vingboons' maps in Sweden," *Anecdota Cartographica,* 2, Stockholm, 1948; reviewed by W. V. Cannenburg, "Four Vingboons Maps Discovered!" *Imago Mundi,* V (1949), 26.

Original

Congress.

---

41. Blaeu 1640

Virginiæ partis australis, et Floridæ partis orientalis, interjacentiumq$_3$ regionum Nova Descriptio.

Size: 19¾ × 15⅛. *Scale:* 1" = ca. 36 miles.
In: Blaeu, W. J., and J. Blaeu. *Le Théâtre dv Monde, ou Novvel Atlas.* Amsterdami, 1638, II (1640), pp. 28–29 (verso, catchword: K ceux; EE).

Description: This map, which covers approximately the same area as the large Mercator-Hondius map of 1606, is based upon that map. There are, however, numerous minor changes. The North Carolina area is correctly reduced in size; the latitude of "C de Trafalgar" or "C oft Feare" is changed to 34° N.L. from 32° in Mercator. This gives a more adequate representation of the South Carolina coast, which was greatly contracted in the Mercator map.

Many other details show that Blaeu attempted to incorporate new geographical knowledge. Although the nomenclature of the Mercator map is closely followed in the southern half of this map, with Le Moyne's names for the coastal rivers, with the numerous Indian names, and with the legends concerning the various lakes, yet for the northern half many new names are added. Chesapeake Bay is changed from a small bay to its proper shape, with "James Town" and many Indian names added to the surrounding country. New names appear along the Carolina coast. P. Grinfild is found in Pamlico Sound; Sandhoeck is at the lower end of two long, curved bays which expand the South Carolina coast. Though these additions to the nomenclature of the coast were not permanent, as the new names given in subsequent English explorations superseded them, nevertheless they show that new information was being gathered. This is the most correct map of this area yet to appear, and it was closely followed by Jansson's map of 1641 and the smaller Montanus 1671; these maps, in turn, influenced other atlases until the end of the century.

Jansson's 1641 map with the same title is so similar to this map by Blaeu that they have commonly been regarded as identical: see P. L. Phillips, *Lowery Collection,* pp. 138–39, and L. C. Karpinski, *Early Maps of Carolina,* p. 24. Numerous small differences, however, show that different plates were used. The parallel lines on the right-hand side of the 1641 map are blacked out opposite the numerals of latitude in the Jansson plate; in the Blaeu plate they are unincised and white. The Jansson plate has an accidental mark below the latitude numeral 34 on the right-hand side; the Blaeu plate has a line radiating from the compass which cuts through the boundary line of the map at the latitude numeral 33 on the right-hand side.

Reproduction
Plate 26.
Copies
Congress 1644; American Geographical Society; Clements 1640, two separates; Georgia (De Renne); Newberry (Ayer); New York Public, four separates; University of North Carolina; Yale 1658?.
In: Blaeu, W. J., and J. Blaeu. *Le Theatre dv Monde.* Amsterdam, 1645, III, Part II, No. 38 (verso blank).
Copy
Congress 1645.
In: Blaeu, J. *Novvs Atlas.* Amsterdami, 1641–42, No. 85.
Copies
Congress 1641–42, No. 85; Harvard 1646–49, II, 29 (verso, catchword: I vnd).
In: Blaeu, J. *Tonneel des Aerdrycx ofte Nieuwe Atlas.* Amsterdami, 1650, II America, No. 9 (verso, catchword: heyt).
Copies
Duke 1650; New York Public 1650; Yale 1650.
In: Blaeu, J. *Tonneel des Aerdriicx.* Amsterdami, 1648–58, II (1658) America, No. 34 (verso, catchword: H sijn).
Copies
Library of Congress 1658; New York Public 1658.
In: Blaev, I. *Atlas Maior.* Amstelaedami, 1662– , XI America, No. 5 (verso, catchword: O monio).
Copies
Congress 1662–65, 1662–72; Harvard 1662.
In: Blaeu, J. *Le Grande Atlas ou La Geographie Blaviane.* Amsterdam, 1663, XII, No. 25 (verso, catchword: K ment).
Copies
Newberry (Ayer) 1663; Yale 1663.
In: Blaeu, J. *Le Grande Atlas ov Cosmographie Blaviane.* Amsterdam, [1667], XII, No. 25 (verso, catchword: K ment).
Copies
Congress 1667; Clements 1667.
In: Blaeu, Joan. *Grooten Atlas.* Amsterdam, 1664– , VII [1665], (verso, catchword: H. sijn).
Copies
Congress 1665; Yale 1665.
In: Wit, F. de. *Atlas Minor.* Amsterdam, 1634–1708, II, No. 73.
Copy
Congress 1634–1708.

---

42. Jansson 1641

Virginiæ partis australis, et Floridæ partis orientalis, interjacentiumq; regionum Nova Descriptio.

Size: 19⅞ × 15¼. *Scale:* 1" = ca. 36 miles.
In: Jansson, Jan. *Nieuwen Atlas.* Derde Deel: Des Nieuwen Atlantis Aenhang. Amstelodami, 1642–44, II (1644), No. 105 (verso, catchword: xxx rivieren).

Description: This map is nearly identical to Blaeu's map 1640 of the same title; see Blaeu's map for identifying differences. J. Keuning, "The Novus Atlas of Johannes Janssonius," *Imago Mundi*, VIII (1951), 75–76, lists this map as No. 623, first appearing in a German appendix or third volume of the *Atlas Novus*, Amstelodami, 1641, a copy of which is in the Stadt- und Hochschul- Bibliothek at Bern, Switzerland.

Jansson's map appeared in many editions of his atlases. The size of the map varies with individual copies from 19¾ × 15 to 19⅞ × 15¼, although apparently the same plate is used. About 1710 Valk and Schenk, using Jansson's plate, added latitude and longitude lines, and affixed their names, probably for use in their *Atlas Contractus*: see Phillips, *Lowery Collection*, pp. 230–31.

Reproduction
Cartografía de Ultramar, II, Plate 40 (with transcription of place-names).
Copies
Congress 1644, two copies; Harvard (two separates); Kendall (two separates); University of North Carolina.
In: Jansson, Joannes. *Novus Atlas, Das ist Welt-Beschreibung.* Amsterdam, 1648–49, III, fol. EEE (catchword, verso: durch).
Copy
Yale 1648–49.

In: Jansson, J. *Nuevo Atlas o Teatro De toto El Mondo.* Amsterdam, 1653, II, No. 42 (verso blank).

Copy

Congress 1653.

In: Jansson, J. *Novus Atlas,* Amstelodami, 1646–49, III, America, fol. Eee (verso, catchword: præse).

Copies

Congress 1649; National Archives of Canada.

In: Jansson, J. *Nieuwen Atlas, ofte Wverelts-Beschrijbenge.* Amstelodami, 1652–53, III, America, fol. Ee (verso, catchword: rivieren).

Copy

Congress 1652–53.

In: Jansson, J. *Novus Atlas.—Atlantis, Pars Quarta.* Amstelodami, 1657, IV, No. 34.

Copy

American Geographical Society 1657.

In: Jansson, J. *Novus Atlas Absolutissimus.* Amstelodami, 1658, VI, fol. EEEee.

Copy

Clements 1658.

In: Jansson, J. *Nieuwen Atlas, ofte Werelts-Beschrijvinge.* Amsterdam, 1657–87, III, America, fol. Ee (verso, catchword: rivieren).

Copy

Congress 1657–62.

In: Jansson, J. *Nouvel Atlas ou Théâtre du Monde.* Amstelodami, 1658, VIII, SSS.

Copy

Clements 1658.

In: Jansson, J. *Atlas Contractus.* Amstelodami, 1666, II, No. 74 (verso, catchword: cipui).

Copy

Congress 1666.

In: Jansson, J. *Atlas Major, sive Cosmographiæ Universalis.* Amstelædami, 1661–75, VIII, No. 421 (verso, catchword: pæse).

Copy

New York Public 1661–75.

In: Seller, John. *Atlas Terrestris.* London, 1676?, verso blank.

Copy

John Carter Brown 1676.

In: Allard, Carel. *Atlas Minor.* Amstelodami, [1696?], No. 141.

Copy

Congress 1696.

A. [ca. 1710: cartouche title unchanged] | Amstelædami, Apud Gerardum Valk. et Petrum Schenk [bottom right]

In: Congress (Lowery Collection), Harvard.

---

43. Jansson 1647

Nova Belgica et Anglia Nova [cartouche, upper left] | Amstelodami Johannes Janssonius Excudit [scale cartouche, bottom right]

Size: 19¾ × 15¼. *Scale:* 1" = ca. 110 miles.

In: Jansson, J. *Novus Atlas: Das Ist: Welt-Beschreibung.* Amstelodami, 1648–49, III (1649), Ccc (verso, catchword: maiz).

Description: This is the second state of Jansson's 1636 map, Novvm Anglia Novvm Belgivm et Virginia. This second state of the map was printed for the first time in the third volume of Jansson's *Atlas* published in 1647 in Amsterdam: see Stokes, *Iconography of Manhattan Island,* II, 144, where a copy of this atlas in the Library of the Royal Dutch Geographical Society is recorded. The heart-shaped cartouche in the upper left of the 1636 map has been erased; in its place a rectangular cartouche with a standing male Indian to the left and a female Indian to the right has been engraved. Below the cartouche two bears have been added, while above is the drawing of a palisaded Indian village, and a few other decorations taken from Blaeu's earlier map of the same title.

The cartouche in the lower right with Jansson's signature and the scales has not been changed, nor have changes in the topography or nomenclature of the map been made. A third state of this map is in the Library of Congress, with changes made by Valk and Schenk about 1710: see Jansson 1636 and Phillips, *Lowery Collection,* pp. 230–31, No. 270.

In the first edition of W. J. Blaeu's *Atlas,* published in 1635, is a map with this same title; but the map represents the country only as far south as the Chesapeake Bay. A reproduction of Blaeu's map is found in Stokes, *Iconography of Manhattan Island,* II, Plate 33.

Reproductions

Plate 25.

Cartografia de Ultramar, II, Plate 22 (with transcription of place-names, p. 146).

Copies

Kendall, separate; Yale 1649.

In: Jansson, J. *Novvs Atlas.* Amstelodami 1646–49, III (1649), America, No. 23 (verso, catchword, Ccc tubuli).

Copies

Congress; Yale; National Archives of Canada.

In: Jansson, J. *Novus Atlas Absolutissimus.* Amstelodami, 1658, VI, CCC.

Copy

Clements 1658.

In: Jansson, J. *Nieuwen Atlas ofte Wverelts-Beschrijbenge.* Amstelodami, 1652–53, III, No. 80 (verso, catchword: E sus).

Copy

Congress 1652–53.

In: Jansson, J. *Nouvel Atlas ou Théâtre du Monde.* Amstelodami, 1658, VIII (verso, catchword: Z tubuli).

Copy

Clements 1658.

In: Jansson, J. *Novus Atlas.* Amstelodami, 1658?, III (verso, catchword: La).

Copy

John Carter Brown 1658?.

In: Jansson, J. *Nieuwen Atlas ofte Werelts-Beschrijvinge.* Amstelodami, 1657–87, III, America, fol. Cc (verso, catchword: Nieuw-).

Copy

Congress 1657–62.

A. [Ca. 1710: same title; in scale cartouche, bottom right, Jansson's signature has been poorly erased, and under Amstelodami is:] Apud G. Valk, et P. Schenk.

Copy

Congress, separate.

---

44. Dudley-Virginia 1647

Carta particolare della Virginia Vecchia è Nuoua. La longitu:ne Comi:ca da l'Isola di Pico di Asores. D'America Carta. III. [cartouche, right center] | A. F. Lucini Fece [bottom right]

Size: 15 × 18⅞. *Scale:* 1" = ca. 15.8 miles.

In: Dudley, Sir Robert. *Dell Arcano del Mare.* Firenze, 1646–48, III (1647), Book 6, No. 101.

Description: This map extends from 35° to 39°15', from "Golfo Pericoloso" and "P:to dell Principe" (Cape Fear) to "C. May" and the New Jersey shore. The Carolina region is based on the White map 1590, through some intermediate source; it has also some of the older Spanish names. For the first time soundings along the banks and in Pamlico Sound are given. In the 1661 edition of the *Arcano*, "L.º 6.º" [Libro Sesto] has been added to the cartouche title under "Carta. III"; the plate is otherwise unchanged.

This map is found in the great work of Dudley, "the first marine atlas in which the maps were drawn on the Mercator projection": see P. L. Phillips, *Lowery Collection*, p. 123. For further notices of this work and the manuscript original, see Dudley ca. 1636.

Reproductions

Plate 27.

Humphreys, A. L. *Old Decorative Maps and Charts.* London, 1926, Plate 56.

Copies

Congress 1647; Clements 1661; Harvard 1647; New York Public 1661; Yale 1661.

---

45. Dudley-Florida 1647

Carta particolare della costa di Florida è di Virginia. La longitu[di][n]e, comi[n]ca da l'Isola di Pico d'Asores. D'America Carta III. [cartouche, upper left] | A. F. Lucini Fece. [bottom right]

Size: 15⅛ × 18¾. *Scale:* 1" = ca. 42 miles.

In: Dudley, Sir Robert. *Dell Arcano del Mare.* Firenze, 1646–48, III (1647), Book 6, No. 102.

Description: The map represents the coastline from 28°30' to 39°15', from "C. Caneuoral" to "C. May" and "C de Pidra Arenas." Thus, since the map extends as far north as the previous map of Virginia, and yet includes the coastline as far south as Cape Canaveral, the scale is much smaller. The northern half is similar to the other Dudley map. From R. Principe (Cape Fear River) southward, the nomenclature is apparently made on the simple plan of drawing enough rivers to list most of the eleven names of the Mercator-Hondius map 1606 and also the Spanish names. Thus, interspersed among the French names first found on the Le Moyne map, from "R. Maio" northward, are Cubacani, Po^to S: Pedro, R. Tacatacoron, R: S: Iago, R. di Galli, R: Paseros, H: Delapas, R. della Cruce, B. dos Bacoas, R. Arboledo, Costa di Canuas, R. di Canuas, Oristan, Oflano, R. Medanos, Caiagua, Baia, C. del Golfo, and R. Grande at Ribaut's Forte Carlo (Charlefort).

L.º 6º [Libro Sesto] has been added to the cartouche of the map in the 1661 edition; the plate is otherwise unchanged. For the manuscript original of this map and other general maps of America in the Dudley MS *Atlas* and in the *Arcano*, consult Dudley ca. 1636 (map 38, above).

Reproduction

Plate 28 (1661 edition).

Copies

Congress 1647; Clements 1661; Harvard 1647; New York Public 1661; Yale 1661.

---

46. Farrer 1650 MS

Ould Virginia 1584, now Carolana 1650

New Virginia 1606 New England 1606 [written across the face of the map]

Size: 10½ × 6½. *Scale:* None.

In: Williams, Edward. *Virgo Triumphans: or, Virginia richly and truly valued; more especially the South part thereof: viz. The fertile Carolana, and no lesse excellent Isle of Roanoke . . . By Edward Williams, Gent. London, Printed by Thomas Harper, for John Stephenson, and are to be sold at his shop on Ludgate-Hill, at the Signe of the Sunne, 1650* (inserted).

Description: This roughly drawn MS map, in pencil, ink, and watercolor, was drawn by John Farrer and is found in his own copy of the first edition of Williams's *Virgo Triumphans*.

Carolana is colored blue, Virginia brown, New England green, and Canada dark green. "The West Sea where Sr Frances Drake was 1577" is over the hills, a few miles beyond the headwaters of the James and Roanoke Rivers; the Hudson and St. Lawrence connect the "West Sea" with the Atlantic. This map is presumably the original drawing for the printed map of Farrer in 1651, though that has many changes and additional legends.

In the address "To the Reader" Williams writes in *Virgo Triumphans* that "The whole substance of it . . . was communicated to me by a Gentleman of merit and quality . . . Mr. John Farrer of Geding in Huntingdonshire." He continues in the first paragraph of the first page of the book:

"The scituation and Climate of Virginia is the Subject of every Map, to which I shall refer the curiosity of those who desire more particular information." By this, in the margin, Farrer has written in ink: "But a map had binn very proper to this Book For all men love to see the country as well as to heare of it and the Eye in this kind is alsoe to be satisfied as well as the Eare. Therefore vnder Correction an Error in not doing it." Farrer's marginalia continue throughout the book, and give many explicit details which show how much more information he had at hand than Williams used.

The chief interest of this map is its evidence of continuance of the belief in the northwest passage and in the closeness of the Pacific Ocean. In the book Williams writes (p. 35) that "twenty-two miles beyond the falls of the James River is a Rock of Chrystall. Three days beyond that is a Hill of Silver Oare. Beyond which, over a ledge of hills, is the Sea of China." A legend on the printed map (see Farrer 1651) states that the Sea of China is ten days' march from the head of the James River.

References

Phillips, *Virginia Cartography*, p. 33.

Quaritch, B. *General Catalogue*, V, 2991 (Catalogue No. 112, Part 2, May 16, 1891, p. 158).

Sabin, *Bibliotheca Americana*, XXVIII, 405 (104193).

Verner, Coolie. "The Several States of the Farrer Map of Virginia." *Studies in Bibliography*. Charlottesville, Va., 1950, III (1950), 281–84.

Winsor, J. *America*, III, 168.

Reproductions

Cumming, Skelton, Quinn. *Discovery of North America*, p. 268 (color).

Waynick, C. *North Carolina Roads and Their Builders*. Raleigh, 1952, p. 8.

Original

New York Public.

---

47. Farrer 1651

A mapp of Virginia discouered to y^e^ Falls, and in it's Latt: From 35. deg: & ½ neer Florida to .41.deg: bounds of new England. [top right] | John Farrer Esq₃ Collegit. Are sold by I. Stephenson at y^e^ Sunn below Ludgate. 1651. [bottom center] | John Goddard sculp: [bottom left]

Size: 13⅞ × 10⅝. *Scale:* 1" = 29½ miles.

Description: This remarkable and fascinating map is apparently the result of Farrer's suggestion of a map to go with Williams's *Virgo Triumphans* and of Farrer's manuscript drawing in his copy of Williams's work: see Farrer 1650. The printed map shows much elaboration over Farrer's rough manuscript draft, though it remains fundamentally the same. The South Sea is ten days' march over the hills and "rich adiacent Valleyes" from the head of James River, and Hudson River is connected by a lake with the "Sea of China"; this is a combination of the error of Verrazano's sea with that of the Northwest Passage. The Muratuk (Roanoke) River flows from the southwest instead of correctly from the northwest, an error which may have misled Lederer and those who were influenced by his map.

The map not only gives shires and county divisions and other details in Virginia and Maryland for the first time on any map but also adds details to the north concerning Dutch, Swedish, and other settlements. Long Island extends north and south instead of east and west. One of the few correct delineations on the map is the unbroken ridge of the Appalachians, extending to the Hudson instead of stopping in the south; but the direction is in error, as it runs directly north and south. Information on the map was apparently derived from various sources available to John Farrer (Ferrar), long an important officer in the Virginia Company; many details show that he drew upon reports and firsthand information not found in the printed maps of the period. Cartographically this map is of interest because it gives numerous place-names for the first time and shows what strange misconceptions of the region a well-informed Englishman of the period could hold; it did not, however, influence later mapmakers. "Carolana," the name given on the map to that part of "Ould Virginia" between the Muratuk (Roanoke) and the Chawanoke (Chowan) Rivers, is the only reference to Heath's grant of 1629 found on an extant contemporary printed map.

The most striking feature of the map, the narrowness of the North American continent, reflects a widely held belief of the time. Farrer had already supported this theory in his *A Perfect Description of Virginia*, London, 1649 (reprinted in Force's *Tracts*, Vol. II): "from the head of James River above the falls . . . will be found like rivers issuing into a south sea or a west sea, on the other side of those hills, as there is on this side, where they run from west down to the east sea after a course of one hundred and fifty miles; but of this certainty Mr. Hen. Briggs, that most judicious and learned mathematician, wrote a small tractate and presented it to the noble Earl of Southampton, then Governor of the Virginia Company in England anno 1623, to which I refer for full information." Winsor (*America*, II, 466, note 4) quotes from Hakluyt's *Purchase his Pilgrimes*, London, 1625, III, 852, where Master Briggs speaks of the South Sea "on the other side of the mountains beyond our falls, which openeth a free and fair passage to China."

John Farrer, who furnished Williams the information about silkworms for his book, was interested in the silk-producing industry of the French and Walloon "Vinerouns" as early as 1623 (*Calendar of State Papers, Col. Series, America and the West Indies*, ed. W. N. Sainsbury, London, 1860, I, 43). His daughter Virginia, who remained a spinster and outlived her father by thirty years, died in 1687. Her interest in the American colonies is shown by the numerous additions to the map after the change of imprint to her name in place of her father's. She seems to have been the inspiration for Samuel Hartlib's book, *The reformed Virginia silkworm, or, a rare and new discovery of a speedy way, and easy means, found out by a young lady in England. She having made full proof thereof in May Anno 1652.* London, 1655.

The map was first engraved for Stephenson's third edition, 1651, of Edward Williams's *Virgo Triumphans: or, Virginia richly and truly valued*, London, 1650. The second edition of *Virgo Triumphans* has a changed title: *Virginia: more especially the South part thereof, richly and truly valued . . . the Second Edition, with the Addition of the Discovery of Silkworms, with their benefit . . . By E. W. gent. London, Printed for T. H. for John Stephenson, at the Sign of the Sun below Ludgate, 1650.* The third edition is known only through the Huth copy of the second edition, which has an inserted page with a different title and an announcement of the map: *Virginia in America, Richly Valued . . . Together with A Compleat Map of the Country from 35. to 41. degrees of Latitude discovered, and the West Sea. London, Printed for John Stephenson . . . 1651.* Stephenson also published in 1651 Edward Bland's *Discovery of Nevv Brittaine*, which described Bland's expedition into "the fertile Carolana," some copies of which have Farrer's map. Two copies of Bland's work, in the British Museum and in Huntington, have the map in an early state without Drake's head, and the British Museum has another Bland with the accompanying map in a later state with Drake's head.

Mr. Coolie Verner (see reference below, p. 282), who has made an excellent study of the different states of the map, assumes from available sources that:

1. Copies of the first and second editions of Williams which contain the map are either late garnered copies or sophisticated ones.

2. Any of the first four states of the map might properly be found with the third edition of Williams, and the third edition would be incomplete without one of them.

3. The copies of the Bland with the Farrer map are accidental or special copies, or sophisticated ones.

4. The plate was obtained by Overton from Stephenson or his successors and was used by him under his own imprint for separately issued (or atlas-bound) copies of the map. The assignment of the year 1667? to the final state of the plate is based on indirect evidence in Plomer, *Dictionary of Booksellers and Printers*, 1641–67, London, 1907, pp. 89–90, 142, 172, and on the fact that after 1668 the Term Catalogues carry notices of Overton's publications with a different imprint address.

P. L. Phillips, in his *Maps of America*, p. 978, mentions a copy of the map having Virginia Farrer's name with the date 1671, an error repeated in Stokes, *Iconography*, II, 154; this is the result of a misreading of 1651 on a copy of the map in the New York Public Library.

Four states of the map, with their chief differences, were noted by Colonel Lawrence Martin, Chief of Division of Maps in the Library of Congress (Sabin, XXVIII, 404, No. 104191). Coolie Verner (see reference below) has identified an earlier state than those listed by Martin in Sabin (cf. Huntington and John Carter Brown copies described below). These five states as listed by Verner, with further identifying differences added, are given below, with the volumes in which they are found.

References

Black, Jeannette D., ed. *The Blathwayt Atlas. Vol. II: Commentary*. Providence, 1975, pp. 141–44.

Harrison, Fairfax. *Landmarks of Old Prince William*. Old Dominion Press, Richmond, Va., 1924, II, 604–5.

Phillips, P. L. *Virginia Cartography*, pp. 30–33.

Sabin, J. *Bibliotheca Americana*, XXVIII, No. 104191.

Stokes, I. N. P. *Iconography of Manhattan*, II, 153–54.

Verner, Coolie. "The Several States of the Farrer Map of Virginia." *Studies in Bibliography*. Papers of the Bibliographical Society of the University of Virginia. Charlottesville, Va., III (1950), 281–84.

Reproductions

Plate 29 (fourth state).

Black. *Blathwayt Atlas. Vol. I: The Maps*, No. 22 (fourth state).

Bland, E. *The Discovery of Nevv Brittaine*. 1651, frontispiece (first state of map from Henry E. Huntington; text from New York Historical Society: Americana series, No. 65; photostat reproductions by Massachusetts Historical Society, Boston, 1923).

Cumming, W. P. *Maps of Colonial America*, p. 2 (first state).

Cumming, Skelton, Quinn. *Discovery of North America*, p. 269 (fourth state).

Fiske, John. *Old Virginia*, II, 12 (fourth state).

Gabriel, R. H., ed. *The Pageant of America*, II, 9 (second state).

Stokes, op. cit., II, plate 47 (third state).

Gabriel, R. H., op. cit., I, 188 (fourth state).

Mathews, E. B. *Maps and Map Makers of Maryland*. Johns Hopkins Press: Geological Survey. Baltimore, Md., 1898, II, 364 (fourth state).

Winsor, J. *America*, III, 465 (fourth state).

Editions:

a. [1651] First State: John Farrer, author; I. Stephenson, publisher; date, 1651, no portrait of Sir Francis Drake at top.

Copies

Huntington; John Carter Brown.

b. [1651?] Second State: "Fort Orang." added, right center, on "Hudsons Riuer."

Copy

New York Public (*Virgo Triumphans*, 1650).

c. [1652?] Third State: "John Farrer Esq$_3$" changed to "Domina Virginia Farrer"; portrait of Sir Francis Drake at top. Added south of the James River, Secotan, Dazamoneak, Waxisquok; Col: Litletons Plan[ta]tion, Majotoks, and Acomac Riv: added on Eastern Shore of Virginia; Cape May correctly substituted for "Cape Iames," which has been engraved at Cape Henlopen, on the south side of "Lord Delaware Bay and Riuer"; Richnek woods, Nanteok, Raritās, Mont Ployden, Eriwoms, and Kildorp added along the Delaware River; Conectacut. Ri:; the passage to the South Sea is still open. Several decorative features have been added: a rampant animal, some trees and mountains.

Copy

New York Public (E. Williams, *Virginia, more especially the South part thereof, richly and truly valued*. London, 1650, frontispiece).

d. [1652?] Fourth State: In the title the word "Falls" changed to "Hills" by making "F" into "H" and putting a dot over the "a"; "Canada flu" separated from the Sea of China by an isthmus which also divides "A Mighty great Lake"; "Rawliana" added as the third name for the Carolina region; "Magna passa" for Albemarle Sound changed to "Rolli passa"; "Anandale C" added in Mary Land; some more decorative mountains engraved.

Copies

New York Public (E. Williams, *Virginia.* London, 1650, frontispiece); Stokes Collection, Map 9; John Carter Brown (*Blathwayt Atlas*, No. 21a); University of Virginia.

e. [1667?] Fifth State: Imprint after "Are sold by" changed to: "Iohn Ouerton without Newgate at the corner of little old Baly"; date erased.

Copies

Congress; Johns Hopkins (Garrett).

---

48. Sanson-Canada 1656

Le Canada, ou Nouvelle France, & c. Ce qui est le plus advance vers le Septentrion est tiré de diverses Relations des Anglois, Danois, &c. Vers le Midy les Costes de Virginie, Nouvlle Suede, Nouveau Pays Bas, et Nouvelle Angleterre Sont tirées de celles des Anglois, Hollandois, &c. La Grande Riviere de Canada ou de St Lawrens, et tous les environs sont suivant les Relations des Francois. Par N. Sanson d'Abbeville Geographe ordinaire du Roy. A Paris. Chez Pierre Mariette Rue S. Iacque a l'Esperance Avecq Privilege du Roy, pour vingt Ans. 1656 [cartouche, bottom right] | I. Somer Sculpsit [bottom right, below cartouche]

Size: 21⅜ × 15¾. *Scale:* 1" = ca. 150 miles.

In: Sanson, N. *Cartes Générales de Tovtes les Parties Dv Monde.* Paris 1658, No. 86; in 1670 edition of same work, II, 31.

Description: This large map, first published in Paris as a loose map, was included in several Sanson atlases and became a type map for the northern region of the United States until the publication of Delisle's map of Canada at the beginning of the eighteenth century. The blank areas to the north and west of the Great Lakes, which are themselves only partially outlined, are indicative of the limits of geographical knowledge at that time. As Dr. J. A. Robinson points out in Paullin's *Atlas of the Historical Geography of the United States*, p. 11, the influence of Champlain on Sanson is evident for the region around the St. Lawrence River.

The boundary lines separating the colonial possessions are of especial interest because of Sanson's official position with the French government. The map extends southward far enough to include most of the South Carolina region, which is called "Floride Françoise." The two lakes of the Mercator map 1606 are present, though they are given no name. "Apalache" is written above the larger lake, and an offshoot of the "Apalatcy Montes" separates "Floride Françoise" from the northern Carolina region, which is made part of "Virginie." The Moscoso–de Soto map nomenclature is still used for the Florida region, though Cosa, Cofachiqui, Xuala, and other Indian settlements near the coast have been placed west of the Appalachian mountain range; the immediate source of this and the next map by Sanson is Laët's 1630 maps.

Reproductions

Plate 30.

Karpinski, L. C. *Maps of Michigan.* Lansing, 1931, Plate III.

Paullin, P. O. *Atlas of the Historical Geography of the United States.* Washington, 1932, Plate 20-B (only lower part reproduced).

Copies

Congress 1658, 1670; Clements 1658, 1664; Yale 1666; National Archives of Canada.

In: Sanson, N. *Cartes Générales de la Geographie.* Paris, 1675, I, No. 23.

Copy

Congress 1675.

In: Sanson, N. [Collection of Maps, Paris, 1640–79, No. 29.]

Copy

Yale 1640–79.

In: Sanson, N. *Tables de la Geographie Ancienne et Nouvelle.* Paris 16[66: numbers erased].

Copy

American Geographical Society 16[66].

In: DuVal, P. *Cartes de Geographie.* Paris, 1676, No. 93.

Copy

Clements 1676.

---

49. Sanson-Floride 1656

Le Nouveau Mexique, et La Floride: Tirée de diverses Cartes, et Relations. Par N. Sanson d'Abbeville Geogr ord[re] du Roy. A Paris. Chez Pierre Mariette, Rue S. Jacque a l'Esperance Avec Privilege du Roy, pour vingt Ans. 1656. [cartouche, top right] | Somer sculp. [bottom right, above neat line]

Size: 21½ × 12¼. *Scale:* 1" = ca. 175 miles.

In: Sanson, N. *Cartes Générales de Tovtes les Parties Dv Monde.* Paris, 1658, No. 87.

Description: This map extends from the latitude of Port Royal on the South Carolina coast westward to California, which is represented as a great island; from northern Mexico the interior of the continent is shown up to Canada and the Great Lakes, of which only Ontario and Erie are given.

This map, together with the accompanying "Le Canade ou Nouvelle France," engraved the same year, exerted a profound influence on the delineation of the North American continent for nearly a century. The division of the continent into the northern and southern halves (excluding Central America for a special map) proved convenient; while the names and topography of Sanson's map were gradually changed with the acqusition of new information, many of Sanson's misconceptions of the region remained in later maps side by side with the inclusion of newer material.

Not all examples of subsequent developments of this type of map are listed in this study, since the Atlantic coastal region is but a small part of the whole map. With the later shift of interest on the part of French cartographers to the Mississippi basin, the eastern region that became Carolina was neglected or relegated to other specific maps.

In the following year Sanson used the Florida section of the map for a special treatment on nearly the same scale, which in turn became a type map for the region; see Sanson's "La Floride" 1657 for a more detailed examination of the sources and characteristics of the southeastern part of this map.

"Floride Françoise" on these two maps is probably found for the first time to designate the Georgia–South Carolina area as part of the French possessions in the New World, though earlier maps, such as Lescarbot's 1611, refer to French settlements there.

Nicolas Sanson d'Abbeville, le père (1600–1667), the designer of this map, published many geographical works and was the founder of a famous family of French cartographers. The first edition of his *Cartes Générales* appeared in 1644. The titles on the same plate were frequently changed, as in the case of the present map (see below).

Reproductions

Plate 31 (right half).

Tooley, R. V. *Maps and Map-Makers.* London, 1952, Plate 77.

a. [Title as given above.]

Copies

Congress 1658; Clements 1658, 1664; Harvard; Yale 1666.

In: Sanson, N. *Tables de la Geographie Ancienne et Nouvelle.* Paris 16[66; numbers erased].

Copy

American Geographical Society 16[66].

b. [Same title to A Paris]: chez l'autheur Re.n avec Privilege du Roy, pour vingt ans. 1656 | Somer Sculp.

Copy

Congress 1670.

c. [Title as above to A Paris]: Chez l'autheu[r] Avec Privilege du Roy, pour vingt Ans. 1656. | Somer Sculp.

In: Sanson, N. *Cartes Générales de la Geographie.* Paris, 1675, I, No. 24.

Copy

Congress 1675.

In: Sanson, N. [A Collection of Maps, Paris, 1640–79], No. 30.

Copy

Yale 1640–79.

---

50. Comberford 1657 MS A

The Sovth Part of Virginia Now the North Part of Carolina [across face of map] | Nicholas Comberford Fecitt Anno 1657 [left center, by scale]

Size: 18⅝ × 13⅞. *Scale:* 1" = ca. 20 miles.

Description: The territory shown on the Comberford map extends from Cape Henry to Cape Fear (the present Cape Lookout, which was often called Cape Fear on the early maps), and the map gives a delineation of the country a short distance up the Chowan, Roanoke, Pamlico, and Neuse Rivers.

The map is on vellum and colored; it is mounted on cedar boards about ⅜" thick. On the front cover is a bookplate of "James Comerford," not having the "b," probably a former owner of the map and descendant of the mapmaker. In the opinion of Mr. Victor H. Paltsits, chief of American History Division and keeper of manuscripts, New York Public Library, the words "Now the North Part of Carolina" were added to the original title in darker ink by a different hand in the seventeenth century. This is confirmed by the duplicate map (see Comberford 1657 MS B), which has only the first line in its title. The Comberford map differs in many ways from the maps which precede it. For the first time many English names which survive in present-day nomenclature are given. The topography is more accurate and detailed than in any previous map of the southeastern coast and indicates that surprisingly careful surveys had been made up the Chowan River branches to Blackwater River at "South Key." This is the present South Quay, Virginia, a few miles below Franklin, Virginia; a legend on the map states that "great sloopes" could come up the river to this point. The survey for the Comberford map was probably made by the 1656 expedition of Dew and Francis or by Roger Green's companions.

Among the names which appear for the first time on the Comberford map are: "Knot Ile" (the present Knott Island, in Currituck Sound); "Machapoungo R" (Pungo River); "Pamxtico River" (Pamlico); "The Neus River" (Neuse). Of special interest is "Lucks Island," for this is apparently the only extant identification of this Banks island which was referred to in the charter of 1663 to the lords proprietors as the northern limit of their grant. "All that territory . . . extending from the north end of the island called Lucke island, which lieth in southern Virginia seas, and within six and thirty degrees of northern latitude" (Colonial Records of North Carolina, I, 21). The inlets which made Lucks Island have since disappeared.

The general excellence of the Comberford map and the number of permanent names which appear on it make it unfortunate that it was not published. Frequently in the Colonial Records one comes across the urgent plea from the lords proprietors for new surveys or new maps; yet it was not until the province became a crown colony that a printed map as good as this was made. Evidently the Comberford map was not known to subsequent mapmakers. The obsolete nomenclature of the White map clung to eighteenth-century maps with a tenacity which would be surprising to one unacquainted with the imitative characteristics of seventeenth- and eighteenth-century cartographers.

Nicholas Comberford (fl. 1646–64) designed numerous charts on vellum, in colors and gold; he lived in London, "neare to the West End of the School House at thee signe of the Platt in Radcliffe," according to one of his portolans now in the Biblioteca Nazionale, Florence, and according to a portolan chart of the Atlantic Ocean, dated 1650, which was advertised for sale in Catalogue 69 of William H. Robinson, Ltd., London. Another Comberford map of the Atlantic coast (British Museum Add. MS 5414, art. 13), dated 1657, does not give any of the new information found in the two maps of the Carolina coast here listed and executed in the same year, nor does a similar map of the Atlantic coast by Comberford, dated 1659, which is in the Houghton Library at Harvard.

Although the specific source of Comberford's 1657 charts of "The South Part of Virginia" remain unknown, a substantial amount of information about Comberford and Batts has been unearthed in the last quarter century, especially during the 1970s. Comberford was an important member of what has come to be called "The Thames School" of chartmakers, described in Tony Campbell, "The Draper's Company and its School of Seventeenth Century Chart-Makers," *My Head is a Map*, ed. by Helen Wallis and Sarah Tyacke, London, 1973, pp. 81–106, and in Thomas R. Smith, "Manuscript and Printed Charts in Seventeenth Century London: The Case of the Thames School," *The Compleat Plattmaker*, ed. by Norman Thrower, Berkeley, Calif., 1978, pp. 45–100. "The Thames School" was active at least through the first two decades of the eighteenth century: see Helen Wallis and William P. Cumming, "Charts by John Friend Preserved at Chatsworth House, Derbyshire, England," *Imago Mundi*, XXV (1971), 81. The charts of the Thames School were on Vellum, usually backed by boards, with distinctive colors and patterns.

Comberford's career is particularly addressed in Thomas R. Smith, "Nicholas Comberford," *Dictionary of North Carolina Biography*, Chapel Hill, 1979, I.119.

A detail on the Comberford map of special interest is "Batts House" near the head of the Albemarle Sound on Fletts Creek (now Salmon Creek), which was built for Batts by Francis Yeardley in 1655 as an Indian trading center. Nathaniel (Nathaniell) Batts is the earliest British landowner and settler south of Virginia for whom there is a recorded deed; he explored the region around the Sounds, and according to George Fox, who visited him, "had been a governor of Roanoke," went by the name of Captain Batts, "and had been a rude, desperate man." Batts's activities were first examined in W. P. Cumming's article on the Comberford map, noted below, and in later accounts summarized in Elizabeth G. McPherson and Herbert R. Paschal, "Nathaniell Batts (ca. 1620–ca. 1679)," *Dictionary of North Carolina Biography*, Chapel Hill, 1979, I.119.

An interesting later version of the Comberford map is in the possession of Paul R. Stoney, a map dealer of Williamsburg, Virginia. It is a somewhat smaller chart, undated, with the same two-line title written across the upper face of the map, but with another title below the neat line, "A Map of the Hunting [? Ground of] y Cherokee." The lettering of the second line written across the upper map face of the manuscript, "Now the North Part of Carolina," differs from the top line, as it does in the New York Public Library copy. There are additional details, however, in a different hand: "Enemy mountains" (top left); "Cherokee Camp" (center right below south key, on a tributary flowing into Chesapeake Bay); "Peter Strong/Homestead" (in blacker ink, near Paquike Lake); a line drawn that corresponds with the boundaries of Wickham Precinct (est. 1705), renamed Hyde County (1712); and two official government seals "GR" with attached blue sheets. The chart's provenance has not been traced, but from the foregoing and other details it appears the map was copied from the New York Public Library rather than the Greenwich original by some draftsman shortly after the Tuscarora War (ca. 1713) and used about 1780 in a loyalist claim for restitution.

See Introductory Essay; see Comberford 1657 MS B.

References

Cumming, W. P. "The Earliest Permanent Settlement in Carolina: Nathaniel Batts and the Comberford Map." *American Historical Review*, XLV (October 1939), 82–89.

Cumming, W. P. *North Carolina in Maps*. Raleigh, 1966, pp. 11–12.

Phillips, P. L. *Virginia Cartography*, pp. 33–34.

Stokes, I. N. P. *Iconography of Manhattan Island*. New York, 1915–28, II, 152.

Reproductions

Plate 32.

Cumming, *North Carolina in Maps*, Plate IV.

Cumming, Hillier, Quinn, Williams, *Exploration of North America*, p. 119 (color).

Henderson, A. *History of North Carolina*. Chicago, 1941, I, 44.

Original

New York Public.

---

51. Comberford 1657 MS B

The Sovth Part of Virginia [across face of map] | Made by Nicholas Comberford Dwelling in Redcliffe Anno 1657 [left center, by scale]

Size: 20 × 15¼ (outside dimensions of border); 18⅓ × 13¹⁄₁₀ (inner dimensions of border).

Scale: 1" = 20 miles.

Description: This map is similar to Comberford 1657 MS A in the area shown, in the topography and legends given, and in many ornamental details. It lacks the second line of the title which was added at a later date to the other map; some of the ornamental details, such as a sea monster and the compass, are located differently, a boat which is being rowed by a man in red and is carrying a passenger has been added in the waters of Pamlico Sound, and a large animal is given to the west of the Dismal Swamp region. This map was offered for sale in Catalogue 69 of William H. Robinson, Ltd., London, S.W. 1, Item 111, and was exhibited by this firm at the New York World's Fair. It has since been purchased by the National Maritime Museum, Greenwich, England.

See Comberford 1657 MS A.

Original

National Maritime Museum, Greenwich, England.

---

52. Sanson 1657 A

Le Canada, ou Nouvelle France, &c. Tirée de diverses Relations des Francois, Anglois, Hollandois, &c. Par N. Sanson d'Abbeville Geographe ord[re] du Roy. Avecq Privilege pour Vingt Ans. A Paris Chez l'Auteur. 1657. [cartouche, bottom right]

Size: 12⅛ × 8⅜. *Scale:* 1" = ca. 210 miles.

In: Sanson, N. *L'Amerique en Plvsievrs Cartes*. Paris, 1657, No. 2.

Description: This is a small edition of the large Sanson map of 1656.

a. [same title to A Paris]: Chez P. Mariette, rue S. Iacques a l'Esperance avec Privil. du Roy pour vingt ans. [in cartouche]

In: Sanson, N. *L'Amerique en Plvsievrs Cartes Novvelles*. Paris, 1662, No. 2.

Copies

Congress 1662, 1667?; John Carter Brown 1662; Clements 1667?; Harvard 1667?; New York Public 1662; National Archives of Canada 1657, 1667.

53. Sanson 1657 B

La Floride, Par N. Sanson d'Abbeville Geogr ord.re du Roy A Paris Chez l'Auteur Avecq Privil pour 20 Ans. 1657. [Cartouche, bottom center]

Size: 10 × 7. *Scale:* 1" = ca. 185 miles.
In: Sanson, N. *L'Amerique en Plvsievrs Cartes.* Paris, 1657, No. 4.

Description: The coastline of this map extends from "Secotan" in "Virginia" to "Panuco" in Mexico; the interior is shown to 40° N.L. In many respects this map gives a new and influential interpretation of the region, though in general outline it follows the de Laët map for its names, topographical interpretation, etc.; it has the Chaves nomenclature west of the "Apalachi Montes" and the Le Moyne material, including the two lakes and French names for coastal rivers. However, it has added a few contemporary influences; the region north of "R. Iordan" is called "Virginie"; "Floride Françoise" occupies roughly the present area of Georgia and South Carolina. At a fork of the river entering at "Port Royal" is "Caroline," a confusion of the Port Royal "Charlefort" with the French fort "Caroline" at St. Johns River, which probably had something to do with the idea later in the century and in the eighteenth century that the whole Carolina region was called "Caroline" by the French before the English occupation. The map is a good example of the method used by both French and English official geographers (cf. Delisle 1718 and Moll 1715) to increase the claim of their countries to territory in the New World by enlarging its boundaries on their maps. "L. Erie" is in the upper right-hand corner five or six degrees lower than it should be, with several lakes below it, where the "laguna dulce" and the smaller lake were on the de Soto ca. 1544; this tends to substantiate the theory that the lakes in the earlier map refer to Indian rumors of the Great Lakes.

This map was included in a number of later atlases and had a strong influence on the subsequent cartography of the region.

See Sanson 1683.

a. [Title as given above.]

Copies

Congress 1657, separate; John Carter Brown 1657; Clements 1667?; Harvard 1657; New York Public 1657; Yale 1657.

b. [Title unchanged to A Paris]: Chez P. Mariette, rue S.t Iacques a l'Esperance Avec Pri. pour 20 ans.
In: Sanson, N. *L'Amerique en Plvsierrs Cartes Novvelles.* Paris, 1662, No. 3.

Copies

Congress 1662, 1667?; John Carter Brown 1662; Harvard 1667?; New York Public 1662; National Archives of Canada.

54. Du Val 1659

La Virginie Par P. Duval Geogra. du Roy. 1659 [cartouche, right center] | La Virginie et les Isles Bermvdes [across top of map]

Size: 4⅞ × 3⅞. *Scale:* 1" = ca. 210 miles.
In: Du Val, P. *Le Monde Terrestre.* Paris, [1661?], No. 7.
Inset: Isles Bermudes. *Size:* ½" × ½".

Description: This map is similar to the 1660 map by Du Val, with a slightly different title; it may be from the same plate. It extends from the mouth of "May R" to "Nle Svede," "N. Holande," and the south shore of the New England coast. The two Mercator 1606 lakes are given. The territorial lines bounding the English possessions include roughly the present area of North Carolina, Virginia, and Maryland; the name given to this region is "Apalchen."

The same plate, with many changes, is used in No. 56; Dr. Coolie Verner has identified four states of this plate.

Copy

New York Public 1661?.

55. Du Val ca. 1659

Floride.

Size: 4⅝ × 3¾. *Scale:* 1" = ca. 350 miles.
In: Du Val, P. *Le Monde Terrestre.* Paris, [1661?], No. 8.

Description: This map is very much like Sanson's "La Floride 1657," though smaller; Du Val evidently drew upon his father-in-law's map as a model. It is in the same volume with Du Val's "La Virginie 1659" and is of approximately the same date; this map may be from the same plate as Du Val's 1660 map with a slightly different title.

It has the two lakes; the French fort "Caroline ou Charlesford" at Port Royal (33° N); "Tegesta presqu' Isle" for the Floridian peninsula; and "Lac Erie" at about 40° N.L. In 1669? Du Val published a map based on Lescarbot, but in this map he followed the new information embodied in Sanson's map.

Du Val published another map from a different plate in 1679 which is very similar to this one. The 1679 map has a different title and minor changes, such as "Noueau Mexique" to the extreme left of the map instead of "Noua Mexico," which is on this map.

Copy

New York Public 1616?.

---

56. Du Val–Virginie 1663

Carte de La Virginie Par P. Duval Geogra. du Roy. A Paris [cartouche, right center] | [at top, with longitude line] La Virginie et les Isles Bermudes

Size: 4⅞ × 3⅞. *Scale:* 1" = ca. 210 miles.

In: Du Val, P. *Le Monde ov La Géographie Vniverselle.* Paris, 1660, I, No. 7.

Inset: Isles Bermudes. [top right] *Size:* ½" × ½".

Description: This map extends from the mouth of "May R" to "N. Holande" (including Long Island). The two Mercator lakes are given, and "Apalchen" is given in the region now occupied by North Carolina. Slight changes have been made on the plate in the 1682 state; "N. Holande" and "al Manhate" have been added to "N. Bas Pais" and "Nouuel Amsterdam" of the 1670 plate.

Copies

Congress 1670, 1682, two copies; American Geographical Society [1660 engraved but 1663 printed]; John Carter Brown 1676; Clements 1682; Harvard 1682, two copies; Kendall 1682; Yale 1682.

---

57. Du Val–Floride 1663

La Floride Par P Du Val G.O.D.R.

Size: 4⅝ × 3¼. *Scale:* 1" = ca. 350 miles.

In: Du Val, P. *Le Monde ov La Geographie Vniverselle.* Paris, 1660, I, No. 8; between pp. 54–55; 1682, opp. p. 40.

Description: See Du Val's "Floride" ca. 1659. The legend at Port Royal, "Caroline ou Charlesford," which is on the plate in the 1660 edition, was erased in 1682 and "Charlesfort" engraved over the partially obliterated marks.

Reproduction

Plate 33.

Copies

Congress 1670, 1682, two copies; American Geographical Society [1660 engraved but 1663 printed]; John Carter Brown 1676; Clements 1682; Harvard 1682, two copies; Kendall 1682; Yale 1682.

---

58. Shapley 1662 MS

Discouery made by William Hilton of Charles towne In New England marriner from Cape Hatterash Lat: 35:30" to y^e^ west of Cape Roman in Lat: 32:30" In y^e^ yeare 1662 and layd Down in the forme as you see by Nicholas Shapley of the towne aforesaid Nouember:1662.

Size: 11 × 12. *Scale:* 1" = ca. 20 miles.

Description: This is a rude outline sketch on paper of the exploration of the Cape Fear River by William Hilton. It is the first map to show any detail of the Cape Fear region, giving the soundings along the coast from "C Romana" to "C Hatterash," with detailed drawing and nomenclature of the Cape Fear River to a distance about seventy-five or eighty miles from the mouth. It has been identified by W. C. Ford as being copied entirely in John Locke's hand from an original by Shapley now lost. Shapley is probably the one by that name who was clerk of the writs in Charlestown, Mass., in 1662 and who died there in May 1663 (Winsor, *America*, V, 337). The William Hilton referred to in the map came to Plymouth, Mass., as a child with his father in 1623; he died in Charlestown, Mass., in 1675. The map is the only record of the voyage made by Hilton in 1662 in the interest of some intending settlers from the Barbadoes and New England. In 1663 a William Hilton, presumably the same person referred to by Shapley in this map, made another voyage which was described in his "A Relation of a Discovery lately made on the Coast of Florida . . . London, 1664," in turn the apparent basis for the Horne map of 1666. There are differences in nomenclature between the Shapley map and the Horne map which have been analyzed in parallel columns, together with references to the Blathwayt maps (see Lancaster 1678 MS A) in Ford's "Early Maps of Carolina" pp. 267, 269.

References

Ford, Worthington C. "Early Maps of Carolina." *Geographical Review*, XVI (1926), 264–73.

Proceedings. Mass. Hist. Soc., Series I, XX (December 1883), 402 (minutes; reported by Mr. Hallam, the secretary).

Salley, A. S., ed. *Narratives of Early Carolina*. New York, 1911, pp. 31–61 (Hilton's Relation).

Winsor, J. *Narrative and Critical History of America*, V, 337.

Reproductions

Courtenay, W. A. *The Genesis of South Carolina*. 1907, opp. p. 4.

Hulbert, A. B. *Crown Collection*, Series I, Vol. V, Map 30.

Ford, W. C., op. cit., p. 265.

Winsor, J., op. cit., p. 337.

Original

British Library. Add. MS. 5415. g. 4.

---

59. Moxon 1664

Americae Septentrionalis Pars [in upper field of map] | London Sold by Joseph Moxon 1664 [in cartouche in upper left corner]

Size: 12 15/16 × 14 3/4. *Scale:* 1" = ca. 200 miles.

Description: This map shows the east coast of North America from Yucatan and the West Indies up to Labrador. It is probably the first engraved map to give the name "Carolina" to indicate the province of the lords proprietors. The first charter was granted to the lords proprietors in the previous year; "Carolina" is written across the interior, south of "Roanoke R" and north of "C of Faire."

"Probably it is the first map showing Manhattan Island published after the British occupation, and the first on which the name New York appears. A pencil note on the back of this chart suggests that it may be 'Part of a Large Map of the World, 10 ft × 5 newly corrected price 50s. June 1670 . . .'" See I. N. P. Stokes, *Iconography of Manhattan Island*, New York, 1909–28, II, 156. Only coastal names are given, except for some of the Great Lakes, as it is a hydrographic map. The names for the Georgia–South Carolina coastal region are based on Le Moyne 1591, and those for North Carolina on the White map 1590.

The only known copy, in the Stokes Collection, is on permanent exhibit in the New York Public Library.

Reproduction

Stokes, I. N. P. *Iconography of Manhattan Island*. New York, 1909–28, II, Plate 50.

Copy

New York Public Library.

---

60. Horne 1666

Carolina Described.

Size: 8 3/4 × 5 7/8. *Scale:* 1" = 47 miles.

In: A Brief Description of the Province of Carolina on the Coasts of Floreda . . . Together with a most accurate Map of the whole Province. London, Printed for Robert Horne, 1666, frontispiece.

Description: This map, which shows the Carolina coast (31° to 37°), is probably the first printed map of Carolina so entitled, though Moxon's map 1664 already had the name Carolina on it. The designer of this map may have had Shapley's map 1662, or a copy of it, as the nomenclature along "Charles River" (Cape Fear River) found on the Shapley map is in general followed. However, much new information, not on the Shapley map, is apparently taken from the records of a voyage made by William Hilton in 1663 which were published the following year (see Introductory Essay).

"Charles Town," the short-lived settlement on the west bank of the Cape Fear River established in May 1664 by the Barbadian planters, is given; the similarity in name to the later settlement on the west bank of the Ashley River caused some confusion to later cartographers. "Cape Fear" is for the first time given to the cape which still bears that name, while "C Hope," after Shapley's "Hope," is given to Cape Lookout, the Cape Fear or Faire of the earlier maps. "Colleton Ile" is at the mouth of "Albemarle Riv."; these names show the influence of the Proprietorship and appear for the first time on any printed map. At Port Royal is "Charles Fort," the French settlement of one hundred years before; "S. Hellens" Island, "R. Jordan," and "R Grandy" are given.

The soundings along the coast are not dependent on Shapley's map and are not found in Hilton's printed *Relation*. Decorative detail, including ships and various animals indigenous to the country, is delineated.

The authorship of *A Brief Description of Carolina* is unknown, though it is sometimes attributed to Robert Horne, since the title states that the pamphlet was printed for him. Horne published manuals and books on trade, and sold products of the patent-medicine variety. The pamphlet was undoubtedly published with a view to obtaining new settlers for the lords proprietors' land. It is now very rare and the map itself still rarer, as it is usually not found in the extant copies. The map was closely followed by Blome (1672), and the area, which shows the coastline but very little of the interior, is the same as that in numerous maps during the rest of the century (cf. Morden, Thornton, Sanson, and their imitators).

See Introductory Essay. See Shapley 1662 MS. See Lancaster 1679? MS A.

References

A Brief Description. Ed. John T. Lanning. Charlottesville, Va., 1944, pp. 1–12.

Ford, W. C. "Early Maps of Carolina." *Geographical Review*, XVI (1926), 264–73.

Reproductions

A Brief Description. Ed. John T. Lanning. Charlottesville, Va., 1944, opp. p. 13.

Bryant, W. C., and S. W. Gay. *A Popular History of the United States*. New York, 1878, II, opp. p. 285 (poor and inaccurate facsimile).

Craven, W. F. *The Southern Colonies in the Seventeenth Century, 1607–1689*. [Baton Rouge], 1949, opp. p. 324.

Ford, W. C. "Early Maps of Carolina." *Geographical Review*, XVI (1926), 268.

Gabriel, R. H., ed. *The Pageant of America*. I, 266.

Hawks, F. L. *History of North Carolina*. Fayetteville, N.C., 1859, II, opp. p. 42.

Henderson, A. *North Carolina*. Chicago, 1941, I, 46.

Copies

John Carter Brown 1666; Huntington 1666; University of Virginia 1666; Clements 1666.

---

61. Du Val 1665

La Floride Francoise Dressee sur La Relation des Voiages que Ribaut, Laudonier, et Gourques y ont faits en 1562. 1564. et 1567. Par P. Du Val, Geographe du Roy.

Size: 9 × 6¼. *Scale:* 1" = ca. 66 miles.

In: Du Val, P. *Diverses Cartes et Tables*. Paris, [1665], Pt. 2, No. 5.

Description: This map is based on, and is very similar to, Lescarbot's map 1611 of the same region. It is found in the section of Diverses Cartes entitled "Cartes Pour les Itineraires et Voyages Modernes"; the later geographical knowledge shown in Du Val's earlier map of this region, which was taken from Sanson's maps, is not here included.

Reproduction

Courtenay, William A., ed. *The Genesis of South Carolina, 1562–1670*. Columbia S.C., 1907, opp. p. xxxii.

Copy

Congress [1669?], separate.

---

62. Anon. ca. 1670 MS

[Albemarle and Pamlico Sounds]

Size: 15 × 12. *Scale:* None.

Description: This is a rude outline chart of the North Carolina coast bordering Pamlico and Albemarle Sounds. To the north of the mouth of the Roanoke River is Governors Plantation; to the west of Mackay Creek on the south shore of Albemarle Sound is Governors Plantation. Since "Albemarle River" is given on the map and that name was not given to the sound until about three years after the First Charter in honor of one of the proprietors, the statement in the British Museum catalogue that it was "drawn about 1660" is erroneous. It is a very early map, probably made soon after the first governor's arrival, and gives valuable indications of the location of several early settlements.

Reference

Cumming, W. P. "The Earliest Permanent Settlement in North Carolina." *Am. Hist. Rev.*, XLV (1939), 85–86.

Reproduction

Hulbert, A. B. *Crown Collection*. Cleveland, 1908, Series I, Vol. V, Map 29.

Original

British Library. Add. MS. 5027. a. 59.

---

63. Culpeper 1671 MS A

Culpepers Draught of Ashley Copia vera [on back of map]

Size: 23½ × 18. *Scale:* 1" = 1 mile.

Description: This is a chart of the Ashley and Cooper Rivers, showing the location of settlers at Charles Town and the land owned at that time. Charles Town was then situated on the west bank of the Ashley River; it was not until 1680 that the site was moved to Oyster Point. Though the map gives much detail and numerous legends, it is crudely drawn and the topography is frequently incorrect. A note states that Culpeper understands the lords proprietors already have received information concerning the soundings of the rivers.

Culpeper, Surveyor General of the colony, had agreed with the Council at Charles Town on August 25, 1671, to lay out the land in lots. See Collections of the South Carolina Historical Society, Charleston, 1897, V, 332, note 1. This draft was apparently sent by the *Blessing*, which sailed September 1, 1671; Culpeper sent a draft by that boat and promised a "perfecter" one later: CSCHS, V, 355.

Although Ogilby did not use the map as the basis for his Ashley and Cooper Rivers inset, which was copied by many subsequent cartographers, apparently he took several names from it for the inset.

A long table, listed A to Z, is given at the bottom left of the map, which is printed in full in CSCHS, V, 339–40. See the note under the Ashley-Cooper map 1671. See Charles Town ca. 1725 MS.

References

McCormack, Helen G. "A Catalogue of Maps of Charleston." *Year Book 1944: Charleston, S.C.* Charleston, 1947 [1948], p. 180.

Penfold, P. A., ed. *Maps and Plans in the Public Record Office.* London, 1974, p. 494.

Reproductions

Collections of the South Carolina Historical Society. Charleston, 1897, V, front. Cf. p. 339.

Courtenay, W. A., ed. *The Genesis of South Carolina.* Columbia, 1907, opp. p. 125.

Original

London. Public Record Office, MPI 13.

---

64. Culpeper 1671 MS B

Culpepers Draught of the L[ds] P[rs] Plantacon, Carolina 1671.

Size: 30 × 18. *Scale:* 1" = 5 chains.

Description: This is a chart of two plots of land, one of 340 acres laid out according to warrant by Governor West for Lord Ashley, Sir George Carteret, and Sir Peter Colleton, and the other of 160 acres for Sir Peter Colleton.

The chart shows the star palisade and is bounded on the north by "the Landing place" (on Old Town Creek), Mr. Samuel West, and Mr. Jno. Mavericke; on the west by "land not taken up"; on the south by Mr. Jos Dowden and Jas. Jones; and on the east by "Marshes of Ashley river."

References

Calendar of State Papers. Colonial Series. America and West Indies, 1669–1671. London, 1889, p. 282.

Collections of the South Carolina Historical Society. Charleston, S.C., 1897, V, 371, Section 668.

Penfold, P. A., ed. *Maps and Plans in the Public Record Office.* London, 1974, pp. 491–92.

Original

London. Public Record Office, MPI 406.

---

65. Locke 1671 MS

Map of Carolina [16]71 [endorsed by John Locke]

Size: 21½ × 17¾. *Scale:* 1" = ca. 80 miles.

Description: This sketch, drawn in pencil with many additions in ink, includes the southern and eastern part of the North American continent from Yucatan in Mexico (17° N.L.) to the head of Chesapeake Bay (39° N.L.), and the islands off the coast from Jamaica to the Bahamas. A dotted line across the map shows the southern limits of the Carolina grant, and "Carolina" is written across the continent from Texas eastward. "The Rickohockans" (Appalachian mountains) extend in a great U from Virginia down to the headwaters of "R. May" at "Ushery Lake," swinging northwest beyond "The great lake" (Lake Erie). The general topography derives from a Spanish or Dieppese-type chart, on which English nomenclature and geographical information have been superimposed.

The chief interest of this map lies in the heavily penned names added in the eastern region, most of which are derived from John Lederer's account of his several expeditions. It may have been used by Locke and the lords proprietors to keep up with current knowledge about the proprietary. A comparison with the Lederer map of 1672 shows that, though the two maps have many similarities, the engraver of the printed map probably did not use this manuscript map as a basis for the printed map; the topographical details differ and more names are given on the Lederer map for the area which it shows.

See Introductory Essay.

References

Cumming, W. P. "Naming Carolina." *North Carolina Historical Review*, XXII (January 1945), 41.

Penfold, P. A., ed. *Maps and Plans in the Public Record Office.* London, 1974, p. 339.

Reproductions

Plate 35 (right half).

Congress (photocopy).

Original

London. Public Record Office. MPI 11.

---

66. Ashley-Cooper 1671 MS

[Ashley and Cooper Rivers]

Size: 26⅜ × 21½. *Scale:* 1" = 1⅓ miles.

Description: This map in colors of the Ashley and Cooper Rivers, with the adjacent territory, is the first careful survey of the vicinity of Charles Town. It is beautifully drawn and may have been the "perfecter" draft promised the lords proprietors by Culpeper September 1, 1671. See *Collections of the South Carolina Historical Society*, V, 355. This survey evidently is the basis for the inset of the Ashley and Cooper Rivers found in the Ogilby-Moxon map ca. 1672 and from that map copied by other cartographers for many years. The similarity of design, soundings, and nomenclature shows that it is the source for the Ogilby-Moxon map, though the addition by Moxon of three names, "Store Cr," "Ittivan R," and "Ittivan C," shows the use of Culpeper's rough but more detailed draft of the same year. W. Noel Sainsbury assigns this map to the year 1671; see *Calendar of State Papers, America and West Indies, 1669–1774*, London, 1850– , VII (1889), 282.

See Culpeper 1671 MS and Charles Town ca. 1725 MS.

Reference

Penfold, P. A., ed. *Maps and Plans in the Public Record Office*. London, 1974, p. 492.

Original

London. Public Record Office. MPI 14.

---

67. Montanus 1671

Virginae partis australis, et Floridae partis orientalis interjacentiumq₃ regionum Nova Descriptio.

Size: 14 × 13⅜. *Scale:* 1" = ca. 51 miles.

In: Montanus, Arnoldus. De Nieuwe en Onbekende Weereld: of Beschrvving van America en't Zuid-land. Amsterdam. By Jacob Meurs, Boekverkooper en Plaetsnyder, 1671, between pp. 142 and 143.

Description: This map, which includes the area from northern Florida to the York River in Virginia, is a copy of Blaeu's 1640 or Jansson's 1641 map. Different only are the cartouche to the bottom right consisting of a bison's skin held up by two Indian youths and the scale at the top left, above a waterfall and a pool in which several Indians are bathing.

The work by Montanus in which this map appears is one of a series under the general editorship of a Dr. Olfert Dapper, printed by Jacob Meurs of Amsterdam. The copyright for this volume was obtained by Meurs on July 28, 1670. Meurs evidently had an agreement with John Ogilby permitting publication of an edition in English of the work. Ogilby makes no reference to Montanus or to Meurs in the English edition, though much of the work is a direct translation; however, he has significant additions and enlargements, particularly concerning the British possessions and colonies. In general, Ogilby used the plates and maps made for the Dutch edition by Meurs, although he also added a few new maps. In the John Carter Brown Library, New York Public Library, and Harvard Library are editions of Ogilby with the date 1670 on the title page. This apparent priority of date has caused confusion concerning authorship and relative sequence of publication. Actually, Ogilby, like Meurs, had intended to publish the work in 1670; he had difficulty in getting information about the British colonies, as we know from a letter to John Locke from Peter Colleton. Its publication was announced in June 1670 for the following January; in November 1670 it was announced as in "a good forwardness"; and it was finally published November 3, 1671 (see Arber, *Term Catalogues*, I. 45, 50, 63, 94; III. 227; Stokes *Iconography of Manhattan*, II, 262). In 1673 the High German edition, the authorship of which is attributed to Olfert Dapper, appeared, although the text is a translation of Montanus and the engravings are from the plates, by then somewhat worn, of Meurs. The German work was privileged in 1670 but was delayed for the Dutch edition and then for the Ogilby edition.

For the complicated question of the date of the first lords proprietors map, which was engraved by Moxon and used in later issues of Ogilby's *America*, see Ogilby ca. 1672.

This map is in three states. Aa (Map 205, below) made extensive changes, adding details from Lederer and changing the cartouche. Covens-Mortier (Map 195, below) further changed the cartouche.

References

Cumming, W. P. "Geographical Misconceptions of the Southeast in the Cartography of the Seventeenth and

Eighteenth Centuries." *Journal of Southern History*, IV (November 1938), 477–92.
Stokes, I. N. P. *Iconography of Manhattan*, II, 262.

Reproduction

Tooley, Ronald V. *Maps and Map-Makers*. London, 1949, opp. p. 116.

Copies

Congress, separate; American Geographical Society; John Carter Brown; Clements, separate; Harvard, two copies; Kendall; Mariners' Museum; New York Public Library, two copies; University of North Carolina; Yale.

In: Ogilby, John. *America: being an accurate description of the New World*. London, 1671, between pp. 112–13.

Copies

John Carter Brown; Harvard; Kendall; New York Public, two copies; National Archives of Canada.

In: Dapper, Olfert. *Die Unbekante Neue Welt, oder Beschreibung des Welteils Amerika*. Amsterdam, Bey Jacob von Meurs, 1673, between pp. 164–65.

Copies

Congress; American Geographical Society; John Carter Brown; New York Public, two copies; Yale.

---

68. Lederer 1672

A Map of the Whole Territory Traversed by Iohn Lederer in His Three Marches. [across top of map] | Cross Sculpsit [bottom left, below scale]

Size: 8 × 6⅝. *Scale:* 1" = ca. 40 miles.

In: Lederer, John. *The Discoveries of John Lederer, In three several Marches from Virginia, To the West of Carolina, And other parts of the Continent: Begun in March 1669, and ended in September 1670. Together with a General Map of the whole Territory which he traversed. Collected and Translated out of Latine from his Discourse and Writings, By Sir William Talbot Baronet*... London, Printed by J. C. for Samuel Heyrick, at Grays-Inne-gate in Holborn. 1672, opp. p. 1.

Description: This map, which covers the area now included in Virginia and North Carolina, purports to show the topography of the regions covered by a German explorer in the journeys which he made from the falls of the James River westward to the Virginia Blue Ridge and southwest in the Carolina Piedmont. It is an important and influential map, for it provided new data suitable for the incorporation in maps for a region which had been cartographically barren and which was receiving special attention in promotional literature during the decade following the map's appearance. Its information, which contained misleading details, was quickly adopted by Ogilby and by other English and continental mapmakers. Some of the errors continued to be retained in new maps until the middle of the eighteenth century. See Introductory Essay.

References

Alvord, C. W., and L. Bidgood. *The First Explorations of the Trans-Allegheny Region*. Cleveland, 1912, pp. 131–71.
Ashe, S. A. "Was Lederer in Bertie County?" *North Carolina Booklet*, Raleigh, N.C., XV, No. 1 (July 1915), 33–38.
Carrier, Lyman. "The Veracity of John Lederer." *William and Mary Quarterly*, 2nd Series, IX (1939), 435–45.
Cumming, W. P. "Geographical Misconceptions of the Southeast in the Cartography of the Seventeenth and Eighteenth Centuries." *Journal of Southern History*, IV (November 1938), 476–92 (Appendix B gives a list of maps influenced by Lederer).
Cumming, W. P., ed. *The Discoveries of John Lederer*. Charlottesville, Va., 1958.
Cunz, Dieter. "John Lederer, Significance and Evaluation." *William and Mary Historical Magazine*, XXII (April 1942), 175–85.
Cunz, Dieter. *The Maryland Germans*. Princeton, N.J., 1948, pp. 30–39.
Harrison, Fairfax. *Landmarks of Old Prince William*. Richmond, Va., 1924, II, 606–7.
Lederer, John. *The Discoveries . . . with an explanatory introduction by H. A. Rattermann*. Antiquarian Reprints, Cincinnati, 1879.
Mooney, James. *Siouan Tribes of the East*. Washington, 1894, pp. 34–70 passim.
Ratterman, H. A. "Der Erste Erforscher des Allegheny Gebirges—Johannes Lederer." *Der Deutsche Pioneer*, VIII (January 1877), 399–407.
Rights, D. L. *The American Indian in North Carolina*. Durham, N.C., 1942, pp. 62–66.
Rights, D. L. "The Trading Path to the Indians." *North Carolina Historical Review*, VIII (1931), 403–26.
Shipman, F. W. "John Lederer." *Dictionary of American Biography*, XI, 91.
Thomas, Cyrus. "Was John Lederer in Either of the

Carolinas?" *American Anthropologist.* N.S., V (1903), 724–27.

Thomas, Cyrus, and J. N. B. Hewitt. "Xuala and Guaxale." *Science.* N.S., XXI, No. 544 (June 2, 1905), 863–67.

Winsor, J. *Narrative and Critical History of America,* V, 338–40.

Reproductions

Plate 36.

Alvord, C. W., and L. Bidgood. op. cit., opp. p. 139.

Bartholemew, J. G. *A Literary & Historical Atlas of America.* New York: Everyman's Library, p. 125.

Courtenay, W. A. *The Genesis of South Carolina.* Charleston, S.C., 1907, opp. p. 134.

Cumming, W. P., ed. *The Discoveries of John Lederer.* Charlottesville, Va., 1958, frontispiece.

Fiske, J. *Old Virginia and Her Neighbors.* Boston, 1900, II, 291.

Gabriel, R. H., ed. *Pageant of America,* II, 8.

Harrison, Fairfax. op. cit., I, opp. p. 157.

Lederer, John. *The Discoveries.* [Reprint.] Introduction by H. A. Rattermann. Published by Oscar H. Harpel, Cincinnati, 1879, frontispiece.

———. [Reprint.] [Walker, Evans, and Cogswell Co., Charlestown, S.C., 1891.] Frontispiece.

———. [Reprint.] G. P. Humphrey, Rochester, N.Y., 1902, frontispiece.

Winsor, J. op. cit., V, 339.

Copies

Congress; Apprentice's Free Library, Philadelphia; Boston Public; Chapin (Williams College); Archibald Craige, Winston-Salem, N.C.; Free Library of Philadelphia; Harvard; Haverford College; Huntington; John Carter Brown; Johns Hopkins (Garrett); Newberry (Ayer); New York Historical; New York Public; Princeton; University of Minnesota; University of Pennsylvania; University of Pittsburgh; University of Virginia; William L. Clements; Oxford (Bodleian); British Library, four copies (and two without map); Edinburgh University; Royal College of Physicians, London.

---

69. Blome 1672

A Generall Mapp of Carolina. Describeing its Sea Coast and Rivers. London. Printed for Ric. Blome.

Size: 9¾ × 7⅛. *Scale:* 1" = ca. 47 miles.

In: Blome, Richard. *A Description of the Island of Jamaica; With the other Isles and Territories in America, to which the English are Related.* London, 1672, opp. p. 126 (opp. p. 56 in the 1678 edition).

Description: This map is copied from the Horne 1666 map, though it is not as well engraved. The drawings of indigenous plants and animals found on the Horne map have been omitted; [Cape] "Carteret" and "Ashly Riv" have been added; and a cartouche with the eight coats of arms of the lords proprietors, added to the lower left corner of the map, has the following legend: "To ye R^{t} Hon^{rbl} ye Ld Proprietors of ye Country of Carolina, this Mapp & Treatise where in their Hon^{es} are concerned, is Humbly dedicated by Ric: Blome."

"The map as a whole reflects the imperfect knowledge available in London at the time when it was made. It will be seen that the compiler has fallen into marked confusion of mind respecting the position of Charles Town. The first settlement of that name had been located some twenty or thirty miles up the Cape Fear River. The Charles Town founded in 1670 was placed on the west side of Ashley River and before long transferred to the present position between the Ashley and Cooper Rivers. Blome's map indicates 'Ashly Riv.' near his Charles Town but gives the latter a position near Cape Fear and not far from the old site on the Cape Fear River. The proper position of the name Ashley would be against the river lying between 'C. Romano' and 'R. Grandy' (the North Edisto). Another point deserving attention is that Charles Fort, the short-lived Huguenot establishment, is set on a large island east of 'S. Hellen's,' the two islands being of about the same size; whereas, . . . St. Helena is a large island east of Broad River, while Charles Fort was on a small island formed by Broad River, Port Royal River, and Pilot Creek, lying southwesterly from St. Helena." (Note by J. Franklin Jameson, in A. S. Salley, ed., *Narratives of Early Carolina, 1650–1708,* New York, 1911, v–vi.) The position of Charles Town was the result of confusion on the part of the compiler of this map as is stated in the note; however, its location is copied unchanged from the earlier Horne map 1666, which antedates the Charles Town of 1670, and "Ashly Riv." was added to the coast near it. Blome published another edition of this work in 1678 on large paper; the same plate, however, was used for the map.

The John Carter Brown Library has a copy of *A True Description of Carolina,* London, 1682, which has Blome's 1672 map bound with it; but there is no connection between the broadside and the map. According to a MS note in the book, they were probably bound together in the modern binding, together with a copy of Ogilby-Moxon ca. 1672 and a copy of Gascoyne 1682. The two latter maps have since been removed from the book and are kept as separates.

Although Blome does not use Lederer's map, the text of the work includes a summary of the explorations; this probably is the first use made of Lederer's *The Discoveries* in English promotion literature.

Reproductions

Plate 34.

Salley, A. S., ed. *Narratives of Early Carolina.* New York, 1911, frontispiece.

Copies

Congress 1672, 1678; John Carter Brown 1672, 1678; Clements 1678; Harvard 1672; Huntington 1678; Kendall 1672; New York Public 1672, 1678; Yale 1672, 1678.

---

70. Ogilby-Moxon ca. 1672

A New Discription of Carolina by Order of the Lords Proprietors [cartouche, top right] | James Moxon Scul [bottom right]

Size: 21½ × 16½. *Scale:* 1" = 26 miles.

In: Ogilby, John. *America.* London, 1671, between pp. 212–13.

Insets: [Ashley River and Cooper River] *Size:* 7 × 5. *Scale:* 1" = 3 miles.

Description: This map of Carolina from the coast to the Appalachians has the latest information which Ogilby was able to obtain from the lords proprietors, and is often called The First Lords Proprietors' Map. It was substituted by Ogilby for the Blaeu 1640-type map by Montanus which is found in some of the earlier copies of Ogilby's volume and in the continental editions of the work. John Locke apparently furnished the promotional account of Carolina which appears on p. 212 of the volume and also contributed some of the new nomenclature, which is heavily interspersed with names of the lords proprietors.

Ogilby relied heavily on Lederer 1672 for information concerning the interior, and it was chiefly through this popular map in Ogilby's work that Lederer's misconceptions became so quickly disseminated and so widely copied. Hilton's and Sandford's reports of the coast are also used. The inset is based on Ashley-Cooper 1671 MS, with some names taken from Culpeper 1671 MS.

Although in 1682 appeared the much-improved Gascoyne map (the second lords proprietors' map), in the same year a Carolina promotional tract by the proprietors' secretary, Samuel Wilson, was published with the Ogilby Map.

A wretched misrepresentation of this map, based upon a copy in the New York Public Library with the erroneous date 1671 pencilled in, was first given in *The Charleston Year Book*, 1880, p. 240, and has since been reproduced. See P. J. Hamilton, *Colonization of the South*, opp. p. 236.

See Introductory Essay.

References

Cumming, W. P. "Geographical Misconceptions of the Southeast." *Journal of Southern History*, IV (1938), 484–91.

———. "Naming Carolina." *North Carolina Historical Review*, XXII (1945), 34–42.

Ford, W. C. "Early Maps of Carolina." *Geographical Review*, XVI (1926), 264–73.

Karpinski, L. C. *Early Maps of Carolina.* Camden, S.C., 1937, pp. 26–28.

Stokes, I. N. P. *Iconography of Manhattan.* New York, 1915–28, II, 262.

Reproductions

Plate 37.

Avery, E. M. *A History of the United States.* Cleveland, 1904–7, IV, 9.

Courtenay, W. A. *The Genesis of South Carolina.* Columbia, S.C., 1907, opp. p. 128.

Hawks, F. L. *History of North Carolina.* Fayetteville, N.C., 1859, II, opp. p. 53.

Henderson, A. *North Carolina.* Chicago, 1941, I, 49.

Copies

John Carter Brown, separate; Kendall 1671; New York Public 1671, two copies, separate; Sondley 1671; Yale 1671, two copies.

In: Wilson, Samuel. *An Account of the Province of Carolina.* . . . London, 1682, front.

Copies

Congress 1682; John Carter Brown 1682; Clements ca. 1672.

71. Culpeper 1672/3 MS

Plott of y^e Lords Prop[rietors'] plon[tation] 44½ Acres lond [endorsed by John Locke]

Size: 18¾ × 14½. *Scale:* 1" = 132 feet.

Description: The following note is given in the lower left-hand corner of the map: "Carolina This plott Represents the shape and forme, of the Cleare plantable Land belonging to the Lords proprietors of this province, whereon Colo Joseph West now liveth, which at Request of the said Colo Joseph West I have measured and surveyed and find it to containe fforty foure acres and one halfe of Land or neare thereabouts, scituate and being neere Charles Towne In the abouesaid province, butting and bounding as by the plott appeares, performed March 7th 1672 and Certified

John Culpeper Survey[or] gen^le"

This is a survey of the plantation of Governor Joseph West, situated on the right bank of the Ashley River, about two miles north of its junction with the Cooper River. West was appointed governor of the colony upon the death of Sayle, September 1670; he was removed by the lords proprietors April 19, 1672, shortly after this survey. West was again appointed governor from 1674 to 1682, and during 1684–85. In 1674 he was appointed one of the Landgraves of the province. See E. McCrady, *The History of South Carolina under the Proprietary Government 1670–1719*, New York, 1897, pp. 115, 138, 717, 719, and the authorities there cited. The map shows the buildings, the design of the formal garden, and the boundaries of the plantation. On the back of the map, in addition to the endorsement, is given the date: 7 March 1672/3.

References

Collections of the South Carolina Historical Society. V, 421.

Penfold, P. A., ed. *Maps and Plans in the Public Record Office*. London, 1974, p. 494.

Sainsbury, W. Noel, ed. *Calendar of State Papers, America and the West Indies 1669–1674*. London, 1860, VII (1889), 472 (& 1045).

Original

London. Public Record Office. MPI 12.

72. Herrman 1673

Virginia and Maryland As it is Planted and Inhabited this present Year 1670 Surveyed and Exactly Drawne by the Only Labour & Endeavour of Augustin Herrman Bohemiensis [cartouche, bottom left center] | Published by the Authority of his Ma:^ties Royall Licence and particuler Priviledge to Aug. Herrman and Thomas Withinbrook his Assignee for fourteen yeares from the year of our lord 1673 [bottom, to left of cartouche] | Sold by John Seller. Hydrographer to the King at his Shop in Exchang^e ally in Cornhill. London [under cartouche, with Herrman's portrait, bottom right center] | W: Faithorne Sculp:^t [bottom, to right of Herrman cartouche]

Size: 32 × 37 (printed on four sheets).

Scale: 1" = 7½ miles.

Description: This notable map of Virginia and Maryland extends only a short distance into Carolina. It embodies the results of what is probably the best surveying in the colonies during the seventeenth century, although for scope and detail it is not superior to the unpublished MS map which the South Carolina Surveyor General Mathews completed a few years later. Much of the information on Mathews's map was used by Crisp in his map of 1711. These two maps—Herrman's and Crisp's—complement each other and give, as Mr. Lawrence C. Wroth has pointed out, a cartographical representation from Maryland to Florida unrivalled at that time by the other colonies.

Detail in the Carolina region of the map is sparse; Herrman has "Part of Roanock River by others relations," gives "Durands" (George Durant) between "Chowann River" and "Pashpetank," and "C. Stephen" and "The Falls" of the "Moratuck R," a tributary of the Chowan River.

The British Library copy lacks Seller's imprint, which is pasted on the John Carter Brown Library copy under Herrman's portrait. John Carter Brown's *Blathwayt Atlas* has a colored vellum MS of the Herrman map; the printed map is apparently a later version, as it has additional legends, with some correction in detail. According to Mlle. Foncin, conservateur en chef du Départment des Cartes et Plans in the Bibliothèque Nationale, two copies of the Herrman map are in that library, one of them in the Collection d'Anville and the other in the collection of the Service Hydrographique de la Marine.

References

Baer, Elizabeth. *Seventeenth Century Maryland: A Bibliography*. Baltimore, 1949, pp. 75–76.

Capek, Thomas. *Augustine Herrman of Bohemia Manor*. Prague, 1930.

John Carter Brown. *Annual Report*, 1929–30, pp. 10–15.
Keuthe, J. L. "A Gazetteer of Maryland, A.D. 1673." *Maryland Historical Magazine*, XXX (1935), 310–25.
Phillips, P. L. *The Rare Map of Virginia and Maryland by Augustine Herrman*. Washington, 1911.
Scisco, L. D. "Notes on Augustine Herrman's Map." *Maryland Historical Magazine*, XXXII (1938), 343–51.

Reproductions

John Carter Brown Library, [J.C.B. copy]. Providence, 1941.
Phillips, P. L. op. cit. [Facs. of B.M. copy].
Stevens, B. F. [Facsimile of B.M. copy]. London, ca. 1896.

Copies

John Carter Brown; Congress; Pepys Collection, Magdalene College, Cambridge; Bibliothèque Nationale, Paris, two copies.

---

73. Morden-Berry 1673–77?

A New Map of the English Plantations in America. both Continent and Ilands, shewing their true Situation and distance, from England or one with another, By Robert Morden, at the Atlas. in Cornhill nere the Royal Exchange, and William Berry at the Globe. between York House and the New Exchange in the Strand, London.

Size: 21 × 17⅛. *Scale:* 1" = 140 English miles.
Inset: A half circle, with diameter of 6¼", in upper right corner, showing the North Pole and the northern regions of Europe and North America.
Description: The map extends from the equator to 44° N.L., including Maine; though it covers a large area, its detail for the English colonies is unusually good for a map of this period.

The active mapmaking partnership of Morden and Berry lasted, according to notices of publication in the Term Catalogues (I, 151ff.), between 1673 and 1677. However, the notice on the map at Panama that "In December 1670 Panama was taken by the English and Kept 28 Dayes" probably indicates a date not long after 1673. This is probably one of the earliest maps to use the material given in Lederer's map; it agrees with Ogilby ca. 1672 as against Speed 1676 in having "Eruco R" as a tributary to Roanoke River.

Copies

Congress; American Geographical Society (Collection of Maps, 1649–1700, No. 43); John Carter Brown, a separate, *Blathwayt Atlas*; Kendall; New York Public.

---

74. Roggeveen ca. 1675

Caerte vande Cust van Florida tot de Verginis Streckende van Cabo de Canaveral tot Baya de la Madalena [cartouche, top left] | 25 [to right of scale cartouche, lower right]

Size: 21 × 16⅜. *Scale:* 1" = 37 miles.
In: Roggeveen, Arent. Het Eerste Deel Van het Brandende Veen . . . West-Indien. Amsteldam, Uytgegeben door Pieter Goos, [1675], No. 25.
Description: This hydrographic map shows very little influence of French and English explorations; the coastal names are Spanish. Neither the Le Moyne nor the White nomenclature is used, although other Dutch cartographers, such as Blaeu and Jansson, had introduced the later names some decades before.

Another map in the Roggeveen Atlases, entitled "Pascaerte vande Virginies Van Baÿe de la Madelena tot de Zuÿdt Revier | 26," shows the coast from Pamlico Sound, which is much enlarged, to a place on the coast north of the entrance of the Delaware River.

Reference

Phillips, P. L. *Lowery Collection*, pp. 152–53.

Copies

Congress 1675; New York Public 1675, two copies.
In: Roggeveen, Arent. *The First Part of the Burning Fen.* Amsterdam, 1675, No. 28.

Copy

Harvard 1675.
In: Roggeveen, Arent. *Tourbe Ardante Illuminant toute la Region des Indes Occidentales.* Amsterdam, 1676?, No. 25.

Copy

New York Public 1676?
In: Roggeveen, Arnoldo. *La Primera Parte . . . India-Occidental.* Amsterdam, 1680, No. 25.

Copies

Congress; New York Public.

75. Seller ca. 1675

A Chart of the West Indies from Cape Cod to the River Oronoque, By John Seller. Hydrographer to the King. at the Hermitage Stairs in Wapping. London.

Size: 20⅜ × 16¾. *Scale:* 1" = 147 miles.

In: Seller, John. *Atlas Maritimus, or the Sea-Atlas*. London, 1675, No. 40.

Description: This map is one of the earliest to show the influence of the Lederer lake, desert, and nomenclature in the Carolina region. Seller continually improved his atlases, making new maps which included the latest geographical information available. In the preface of this 1675 edition of his *Sea-Atlas* Seller calls it "the first essay of this nature that hath been compleated in England."

Seller's earlier atlases in the 1670s show a steadily increasing knowledge of explorations and settlements on the Carolina coast in the West Indies charts. Therefore, although in spite of the 1675 imprint this atlas has charts dated 1676 and 1677, the date of the map is probably 1675 or soon after. The National Archives of Canada has dated their copy of the atlas [1677–79]. The William L. Clements Library has a map with all the recent coastal and Lederer information but with a different title: "A General Chart of the West India's from Cape Cod to The River Oronque. By John Seller, John Colson, William Fisher, James Atkinson, and John Thornton." This is in a copy of John Seller's *Atlas Maritimus*; or *Sea Atlas*, London, 1675, No. 39.

Seller's *Atlas Terrestris*, noted below, is a collection of plates from Jansson, F. de Wit, Robert Morden, and others, together with Seller's own plates. The shop addresses of Seller given on the plates, compared with those recorded in the Term Catalogues, make probable a date of 1676 or 1677 for this atlas.

Reproduction

Plate 38.

Copies

Congress 1675; Harvard 1675.

In: Seller, John. *Atlas Terrestris*. London, [1676?], No. 60.

Copy

John Carter Brown 1676?

76. Speed 1675

Carolina

Size: 3⅜ × 4⅞. *Scale:* 1" = 140 miles.

In: Speed, John. *An Epitome of Mr. John Speed's Theatre of the Empire of Great Britain and of His Prospect of the Most Famous Parts of the World*. London, 1675, opp. p. 250.

Description: This small map follows Ogilby's map ca. 1672 and is the first imitation of it which can be definitely dated. On account of its small size, many details are omitted, such as the names of Indian villages in the south, the marking of Lederer's route, and the inset of Ashley and Cooper Rivers. That it is a copy of Ogilby and not of the large Speed (1676) is shown by agreement with the former in some details, such as the name of Eruco River, which are omitted in Speed.

The map is found in *A Prospect of the Most Famous Parts of the World*, London, 1675, an addendum to the *Epitome*. The accompanying text is the same as that of the large *Prospect* published in 1676.

In the Term Catalogues for February 1675 (I, 202, 229) the small volume in which this map appears is advertised together with Speed's *Theater* (1676), and was evidently prepared along with it. But apparently the small volume, some copies of which have a 1675 imprint, was actually published earlier than the larger Speed *Atlas*.

Copies

Congress 1675; Clements 1676; Kendall 1675; University of North Carolina 1676; Yale 1676.

77. Speed 1676

A New Description of Carolina. Sold by Tho: Bassett in Fleetstreet. and Ric: Chiswell in S.t Pauls Churchyard. [cartouche, bottom left] | Francis Lamb sculp. [bottom right]

Size: 20⅛ × 14¾. *Scale:* 1" = 26 miles.

In: Speed, John. *The Theatre of the Empire of Great Britain*. London, 1676, No. 25, between pp. 49–50 of *A Prospect of the most Famous Parts of the World*, London, 1676.

Description: This map is based upon Ogilby's ca. 1672 map and follows it closely. Some of the names, such as Eruco (River), which are given on Ogilby and even on the small Speed 1675, are omitted; the cartouche is found in the place of the inset of the Ashley and Cooper Rivers of the Ogilby map.

On the back of the map is "The Description of Florida" (p. 49) and "The Description of Carolina" (p. 50). The Carolina page gives a full synopsis of Lederer's expedition into Carolina; this widely circulated map must have done much to spread the knowledge of Lederer's explorations further than his own pamphlet would have done. It also serves to explain the paths of Lederer's journey; this clarification is lacking in Ogilby, who delineates Lederer's route but refers to him neither on the map nor in the text. The Speed map was also sold as a separate, as in one of the John Carter Brown copies, which has no writing on the back of the sheet.

The map is found, not in Speed's *Theatre*, which contains only British maps, but in an addendum entitled *A Prospect of the Most Famous Parts of the World*, in which a number of new maps were added.

John Speed (1552?–1629) was one of the foremost cartographers of seventeenth-century Britain and is notable for his English county maps and the beauty of hand-coloring given to them. His atlases have now become quite rare, owing to the custom of breaking them up to sell individual decorative maps. The writer has seen a number of beautifully colored copies of Speed's "Carolina" privately owned, and has a copy in his own collection.

Speed had twelve sons, but they apparently did not continue his business long after his death; the 1676 atlas was edited and augmented by E. Phillips. The Speed *Atlas* had been out of print since the Great Fire; a reprint was announced in February 1675 (Edward Arber, *The Term Catalogues, 1668–1709 A.D.*, I, 202, 229) and was published with added maps in the following year. For a detailed description of this edition of Speed's *Atlas*, see Thomas Chubb, *Printed Maps in The Atlases of Great Britain and Ireland*, London, 1927, pp. 37–40.

References

Cumming, W. P. "Geographical Misconceptions of the Southeast in the Cartography of the Seventeenth and Eighteenth Centuries." *Journal of Southern History*, IV (November 1938), 489.

Ford, W. C. "Early Maps of Carolina." *Geographical Review*, XVI (April 1926), 271–72.

Reproductions

Color Plate 4.

Cartografia de Ultramar, II, Plate 42 (with transcription of place-names, pp. 248–49).

Copies

Congress 1676, two separates; John Carter Brown, two separates; Charleston Library Society; Charleston Museum; Clements; Duke; Harvard 1676; Huntington 1676; New York Public 1676, three separates; North Carolina Department of Archives and History, two separates; Yale 1676, two copies; University of North Carolina (Southern Collection), separate; National Archives of Canada, three separates; University of South Carolina; Davidson.

---

78. Blathwayt ca. 1679 (1664–83) MS A
The Mapp of Carolina [endorsed]

Size: 17¾ × 29⅜. *Scale:* 1" = 12½ miles.
In: Blathwayt Atlas, No. 18.

Description: The title tag of this unsigned, undated vellum manuscript chart of the Carolina coast has been changed from "Lancaster 1679 c. MS A" in earlier editions of this work to "Blathwayt ca. 1679 (1664–83) MS A." This and the following three charts of the Carolina coastal area, all in the *Blathwayt Atlas*, are so similar in decorative coloring, technique, and other features that they have in the past been ascribed to James Lancaster, who signed his name and date (1679) to one of the four charts (MS D). With the identification by modern scholars of a "Thames School" of chart makers, members or apprentices of the Drapers' Company with distinctive patterns of execution, Jeannette D. Black, later curator of maps in the John Carter Brown Library, concluded that no date or authorship of any of the ten manuscript charts of the "Thames School" can be ascribed to any except the one signed by James Lancaster (Jeannette D. Black, *The Blathwayt Atlas, Vol. II: Commentary*, pp. 15–22).

This map extends from below Hilton Head at Port Royal Sound to Roanoke River (Albemarle Sound) and shows the discoveries made by William Hilton in 1662–64 during his three voyages along the Carolina coast. The explorations which Hilton made on his third voyage at Port Royal and St. Helena Sound and up the Cape Fear River for about fifty miles are given in detail, as well as soundings along the entire coast. The rest of the map gives only general information. The similarities to the Horne (1666) map are numerous and significant; but enough small differences in nomenclature and contour occur to posit a common original, probably a sketch of the voyage which was made by Hilton or one of his companions and which came into the possession of the lords proprietors or was available to them.

The Blathwayt Atlas in the John Carter Brown Library, in

which this and the three following manuscript maps are found, as well as other printed maps which are listed elsewhere in this work, is a collection of maps of unusual cartographical importance concerning the period of discovery and settlement of the English in North America. The latest dated map in the atlas, John Thornton's "New Mapp of the World," is printed and dated 1683. Although some of the maps were made earlier than 1675 (cf. Farrer's "Virginia" 1651), it probably contains no map made later than 1683. When Blathwayt retired from the Board of Trade in 1710, it is possible that he took the atlas with him to his home, Dyrham Park, believing it contained only out-of-date maps. We know that the Board continued to acquire maps for its reference library. William Blathwayt, the original owner, was clerk and then surveyor and auditor general of plantation revenues; these maps were collected in an attempt to gather the latest information about the colonies. Blathwayt may have ordered some of the manuscript maps, such as these of the Carolina coast, especially made for him from original information at his disposal. The atlas is a large folio volume in contemporary binding of rough calf, containing thirteen manuscript and thirty-five printed maps of the period 1670–83. Most of the maps relate to North America, although some are of the West Indies, South America, Africa, and India.

Ten of the thirteen manuscript maps in the atlas are attributed to members of the Thames School, as they are all on vellum, use the same four colors and gold with yellow borders, and have the same type of craftsmanship and style (cf. Black, *The Blathwayt Atlas: Commentary*, p. 15). The execution of the four vellum maps showing parts of the Carolina coast was probably ordered by the same person at the same time. In this connection a letter to Blathwayt from Sir Peter Colleton, one of the lords proprietors, may be pertinent, although his "passages" refer to Culpeper's Rebellion, not geography. See Black, *Blathwayt Atlas: Commentary*, p. 137 fn. 2. It shows that Colleton was supplying information about Albemarle.

> Sr,
>
> The bearer hereof will give you a Narrative of the passages of Albemarle as they have apeared to the Proprietors by letters & informations of p*er*sons come from thence with which I should have waited on you myself but that I am so extreamely ill of the gout that I am not able to stand wch I hope will excuse
>
> Yo^r humble servant
> P Colleton
>
> this 9th of February 1679[80]
> [*Colonial Records of North Carolina*, I, 286]

Map No. 20 has detailed drawings of three entrances to Albemarle Sound as well as readings of the soundings in the three channels; no other seventeenth-century printed or manuscript map has this particular feature drawn with such care. Map No. 21 has drawings of houses, showing the location of settlers along Albemarle Sound.

William Blathwayt (1649?–1717) was appointed one of the secretaries of Sir William Temple at the Hague in 1668 and thereafter was sent to Rome, Stockholm, and Copenhagen. In 1676 Blathwayt was assistant to the clerks of the Privy Council committees of Trade and Foreign Plantation (*Calendar of State Papers for 1675–1676*, London, 1893, p. 380 (sect. 889). On May 18, 1680, the Privy Seal issued a patent "for erecting and establishing an office of general inspection, examination, and audit of all and singular accounts of all moneys arising or accruing or which shall arise or accrue to the King from any of his foreign dominions, Colonies, and Plantations in America. . . . The chief officer of said office is to be hereby styled the surveyor and auditor general of all his Majesty's revenues arising in America, and William Blathwayt is hereby appointed thereto with full power . . . to determine all accounts of all such rents, revenues, prizes, fines, escheats, forfeitures, duties, and profits whatsoever" (quoted from G. A. Jacobsen, *William Blathwayt*, New Haven, Conn., 1932, p. 158, note 16). In August 1683 Blathwayt purchased from Matthew Locke the post of Secretary of War (at that time not much more than a clerkship), an office which he held until 1704 and which grew considerably in importance during his tenure. In 1689 he became clerk of the Privy Council and later acted as Secretary of State under William III during the campaign in Flanders. He was a commissioner of Trade and Plantations from 1696 to 1706. Blathwayt retired from public service in 1710 and died at Dyrham Park, Gloucestershire, in the great house which he had completed on his wife's estate. At Dyrham Park many of his letters and documents are preserved; among them was the atlas, which was in the possession of a descendant, the late Robert Wynter Blathwayt (1850–1936), until shortly before it was sold by Sotheby, Wilkinson, and Hodges after advertisement in their catalogue of April 15, 1910. It was purchased by the John Carter Brown Library in 1912 and is one of the great cartographical treasures not only in that institution but also in this country. See the article on William Blathwayt in the *Dictionary of National Biography*, V,

206; see also Stephen S. Webb, "William Blathwayt; Imperial Fixer: Mudelling Through to Empire, 1689–1717," *William and Mary Quarterly*, 3rd ser., XXVI (1969), 373–415; and Jeannette D. Black, *Blathwayt Atlas, Vol. II: Commentary*, Providence, R.I., 1975, pp. 3–32.

References

Black, *Blathwayt Atlas: Commentary*, pp. 118–24.

Ford, W. C. "Early Maps of Carolina." *Geographical Review*, XIV (1926), 268.

John Carter Brown. *Annual Report*, 1911–12, p. 12.

Reproductions

Jeannette D. Black, ed. *Blathwayt Atlas: The Maps*. Providence, R.I., 1970, Map No. 20.

Davidson (photocopy).

Congress (photocopy).

Black, *Blathwayt Atlas: The Maps*. Providence, R.I., 1970, Map No. 18.

Original

John Carter Brown.

---

79. Blathwayt ca. 1679 (1662–83) MS B

[Carolina]

Size: 20⅞ × 24⅛. *Scale:* 1" = ca. 11 miles.

In: Blathwayt Atlas, No. 19.

Description: This map, showing the coast from "C. Hatteras" in the north to "C. Romana" in the south, is meagre of detail except for the Cape Fear region. It is based upon the explorations of William Hilton up the Cape Fear River in 1662 and is apparently a copy of the Shapley 1662 map, which it follows to the smallest detail; the outline of the coast and the soundings along it, the branches and names along the Charles River (Cape Fear River), and details such as anchor marks at the mouth of the river are identical. The names on this map are given in W. C. Ford, "Maps of Early Carolina," *Geographical Review*, XV (1926), 267. A report by Hilton and his fellow voyagers is given in Black, *Blathwayt Atlas: Commentary*, pp. 127–31.

See Shapley 1662 MS and Blathwayt ca. 1679 (1664–83) MS A.

Reference

Black, *Blathwayt Atlas: Commentary*, pp. 125–33.

Reproductions

Congress (photocopy).

Davidson (photocopy).

Black, ed. *Blathwayt Atlas: The Maps*, Providence, R.I., 1970, Map No. 19.

Original

John Carter Brown.

---

80. Blathwayt ca. 1679? (1677) MS C

Carolina [Across face of map; also so endorsed]

Size: 25¼ × 22¼. *Scale:* 1" = 22 miles.

In: Blathwayt Atlas: The Maps, No. 20.

Description: This map extends from St. Augustine to Cape Henry. It is particularly detailed for the entrances to Albemarle Sound and in the vicinity of the Ashley and Cooper Rivers, as well as for soundings along the whole Carolina coast.

The information for the map was gathered after 1674, for "Lady Yeamans house" is drawn to the north of Charleston Harbor, opposite "Silezuant I" (Sullivans Island). Governor Yeamans died in 1674. Information was gathered before 1680, for Charles Town is still on the left bank of the Ashley River. Near it is "Governors house" on Governor West's land (see Culpeper 1673).

This map is the most detailed general map extant up to that time and evidently incorporates the latest information obtainable. The detail is fuller than in the Gascoyne (1682) Second Lords Proprietors' Map for the Albemarle region but has far less complete detail for the Ashley-Cooper River area, for which Gascoyne had new information sent by Mathews. Apparently Gascoyne did not use the data on this map, or used it sparingly, as the soundings along the coast frequently differ and the contours of the coastline are not alike. However, Thornton-Morden-Lea ca. 1685, Thornton-Fisher 1698, and the Chart of Carolina ca. 1700 follow the soundings along the coast, particularly north of Cape Fear River, which are on this map; Thornton's information came from a common original, if it was not taken directly from this map.

Jeannette D. Black (*Blathwayt Atlas: Commentary*, p. 134) considers that the handwriting of the anonymous craftsman who made this chart differs from that of James Lancaster, to whom it was formerly attributed. She believes it is similar to the lettering of two charts of Virginia/Maryland in the *Blathwayt Atlas* (Nos. 16 and 17) and also that of British Library, Add. MS 13970b, a map of Carolina almost identical to the Blathwayt chart, listed hereinafter as No. 99. Several copies of this map may have been ordered to be made at about the same time.

Reference
Jeannette D. Black. *Blathwayt Atlas: Commentary.* Providence, R.I., 1975, pp. 134–36.
Reproductions
Congress (photocopy).
Davidson (photocopy).
Black, *Blathwayt Atlas: The Maps*, No. 20.
Original
John Carter Brown.

---

81. Lancaster 1679 MS D

Carolina [across face of map] | Made by James Lancaster: Anno Domi: 1679 [bottom center, below scale] | North Part of Carolina [endorsed]

Size: 25¼ × 19. *Scale:* 1" = 5 miles.
In: Blathwayt Atlas: The Maps, No. 21.

Description: This map is of Albemarle Sound and its immediate vicinity. No soundings are given; but the names of rivers flowing into the Sound are recorded, and twenty-three small houses are drawn along the Sound, near some of the tributaries, and on Colleton Island, to show the location of settlers or settlements. The irregular placement of the houses indicates that they were drawn as the result of specific information and not for decorative purposes; and contemporary records confirm this impression by giving evidence that certain of the settlers were located, about fifteen years after the granting of the Carolina charter, in several places where the houses appear on this map. Unfortunately few names of settlers are listed with the houses; the detailed information about the location of individual landowners sent from Charles Town concerning the Ashley-Cooper region and recorded with increasing minuteness in contemporary maps has no parallel for Albemarle Sound and its vicinity.

Pertinent comments on this chart by Jeannette D. Black (*Blathwayt Atlas: Commentary*, pp. 137–40) include comments on the copyist James Lancaster of the Thames School of chart makers, and on Captain John Whitty, an important early trader and settler whose house and name are at the mouth of the Pasquotank River. Ms. Black (p. 140) suggests that the original sketch for this chart may have been made by Captain Whitty between 1663 and his death by or before 1667; Whitty was in active communication with Sir Peter Colleton, one of the proprietors and owner of Colleton [Collington] Island.

Reproductions
Congress (photocopy).
Davidson (photocopy).
Jeannette D. Black. *Blathwayt Atlas: The Maps.* Providence, R.I., 1970, No. 21.
Original
John Carter Brown.

---

82. Daniel ca. 1679

A Map of ye English Empire in ye Continent of America: Viz Virginia Mary Land Carolina New York New Iarsey New England &ct. by R: Daniel Esqr [cartouche, bottom center] | sold by R. Morden at y^e^ Atlas in Cornhill neer y^e^ Royal Exchang & by W. Berry at y^e^ Globe neer Charing Cross London. [cartouche, bottom left center] | W. Binneman sculpsit [to right of Morden-Berry cartouche] | Licensed by R. l'Estrange Esqr [to left of Berry-Morden cartouche]

Size: 23½ × 19⅝. *Scale:* 1" = ca. 42 miles.
Inset: Caroliniae [*sic*] Pars. *Size:* 5⅞ × 7⅞.
Scale: 1" = 85 miles.

Description: The large map extends from the Chesapeake to the St. Lawrence River. The inset extends from "S. Augustine" to the Chesapeake Bay.

The Carolina inset has full Lederer (1672) material, with the omission of the actual line of Lederer's march. The Indian names from Le Moyne are omitted in the southern part. By the Lederer "Sauana" is the following legend: "This larg Sauana Lies as Nilus land from May to Sep^t^ vnder Water & from Sep^t^ to May perpetually green stock'd with dears Variety of beasts wild Turkies & Other fowles Innumerable."

The date given for this map is usually 1679, for a map with approximately the same title is listed under that year in E. Arber, *The Term Catalogues*, London, 1903, I, 372.

About 1685 a similar map, with Daniel's and Berry's names and the "Licensed by R. L'Estrange, Esqr" deleted, was published by Morden, with "Pennsilvania" added in the place of "by R: Daniel, Esqr" See under Morden ca. 1685. It is slightly different in size, probably because of the use of a different paper; there are differences in the title, and numerous additions of detail, such as "Lake Hurons"; "Lake Erius or Felis," ("Lake Ontarius" has been changed to "Lake Hurons" and two other lakes inserted), "Pennsilvania," "Philadelphia," "Manhatans I.," and others. The strangest aberration in this later Morden issue is that the Potomac River flows from Lake Erius southward, while from the same lake a river flows northward through Lake Ontarious to the St. Lawrence. In spite of these differences, however, Morden's map is apparently made from the same plate as the Daniel map, for there are numerous identical "accidental" scratchings on both. The Carolina insert is unchanged. A comparison of the two maps with the copy of the Herrman map of Virginia 1673 indicates that the Daniel map utilized the Herrman map, as shown by the names "Red lyon," etc., the eastward bend at the head of Chesapeake Bay, and the crook in the Potomac River; the Morden issue of the map with an extension of the head of Chesapeake Bay from 39°45' to 41°25', reflects the fight William Penn began, even before he came to America, to obtain a port on the Chesapeake. See Morden ca. 1685.

References

Baer, E. *Seventeenth Century Maryland.* Baltimore, 1949, pp. 116–17.

Stevens, H., and R. Tree. "Comparative Cartography . . . One Hundred Maps." In *Essays Honoring Lawrence C. Wroth.* Portland, Me., 1951, p. 319.

Reproduction

Stokes, I. N. P. *Iconography of Manhattan Island.* New York, 1915–28, II (1916), Plate 51.

Copies

John Carter Brown; New York Public (Stokes Collection); British Library.

---

83. Sanson 1679 A

Canada, sive Nova Francia &c. Per N. Sanson

Size: 12 × 8¼. *Scale:* 1" = ca. 210 miles.

In: Sanson, N. *Geographia Exactissima.* Frankfort, 1679, America, p. 17.

Description: This map is similar to Sanson 1657 A but has Latinized the names.

Copies

Yale; National Archives of Canada.

---

84. Sanson 1679 B

Florida. Per N. Sanson.

Size: 10 × 7. [10⅛ × 7⅛] *Scale:* 1" = ca. 185 miles.

In: Sanson, N. *Geographia Exactissima. Die Ganze Erd-Kugel.* Frankfort, 1679, America, p. 23.

Description: This map is similar to Sanson 1657 B but has Latinized nomenclature.

Copy

Yale.

---

85. Du Val 1679 A

Virginia [in cartouche—right center] | Virginia et Insulæ Bermudes. [across top of map]

Size: 4⅞ × 3⅞. *Scale:* 1" = ca. 210 miles.

In: Du Val, P. *Geographiæ Universalis Pars Prior.* Nurnberg, 1679, between pp. 60–61.

Inset: Isles Bermude[s]. ["S" lost in folds of engraved "cloth."] *Size:* ½" × ½".

Description: See Du Val's La Virginie 1660 (Map 56, above), upon which this map is based, though there are slight changes on the map and the title is Latinized. Dr. Coolie Verner has identified two states of the plate in this German edition.

Copies

Congress 1685?; John Carter Brown 1681, 1690; Yale 1679, 1681, 1694.

---

86. Du Val 1679 B

Florida

Size: 4⅝ × 3¼. *Scale:* 1" = ca. 350 miles.

In: Du Val, P. *Geographiæ Universalis Pars Prior. Das ist: Der allgemein Erd-Beschreibung.* Nurnburg, 1679, No. 9, between pp. 84–85.

Description: This map is like Du Val's Floride ca. 1559, except for the Latinization of names.

Copies

Congress 1685?; John Carter Brown 1681, 1690; Yale 1679, 1681, 1694.

87. Seller 1679

Florida

Size: 3¾ × 2⅛. *Scale:* None.

In: Seller, J. *Atlas Minimus*. London, 1679, No. 90.

Description: This is a very small copy of the popular Sanson's Florida 1657 type map, extending from "R. de May" to Mexico. The lakes are given and there is an interesting late example of Verrazano's sea, which is given at the top of the map, but which is not shown in the general map of "The English Empire in America" in the same volume.

The page opposite the map gives this comment: "Florida is Divided into these V Parts: The Peninsula (Cabo de Florida); The Spanish Florida (St. Matheo, St. Augustino, Quilata); The French Florida (Charles Fort, Carolina); The Kingdom of Apalachites (Melilot); the Isles of Bermudes lying opposite to the Eastern Shoar."

Copies

Congress 1679; Yale 1679, two copies of map in same volume.

88. Morden 1680

Carolina Virginia Mary Land & New Iarsey by Robt. Morden.

Size: 4⅛ × 4¾. *Scale:* 1" = ca. 158 miles.

In: Morden, Robert. *Geography Rectified: or, A Description of the World . . . Illustrated with above Sixty New Maps*. London, Printed for Robert Morden and Thomas Cockeril. At the Atlas in Cornhill, and at the three Legs in the Poultrey over against the Stocks-market. 1680, p. 379.

Description: This map shows the east coast from 30° to 40° N.L. (Bay St. Matheo to the northern boundary of Maryland). It has all of the Lederer material. At the bottom of the page "A Description of Carolina" is followed by seven lines of the printed descriptive text.

Copies

Congress 1680; John Carter Brown 1680; Sondley 1680; Yale 1680; Clements 1680.

89. Thornton [1673]

A New Mapp of the north part of America from Hudson Straights commanly call'd the Nor west Passage Including Newfoundland New Scotland New England Virginia Mayland & Carolena Made and sold by John Thornton at the signe of England Scotland & Ireland in the Minories.

Size: 21⅛ × 16¼. *Scale:* 1" = ca. 98 miles.

Inset: [Atlantic coast from 28°20' to 41° N.L.: "B. d Masquetoe" to "Staten I."]

Description: The large map extends from "Staten I" to 64°30' N.L. The inset shows the Carolina coast in some detail; it has "Ashley's Lake," "Savana," "Deserta Arenosa." "Charlestowne" is written below Ashley River. This map, formerly dated tentatively "ca. 1680," was advertised for sale in the *London Gazette* for 20–24 February 1673. Jeannette D. Black (*Blathwayt Atlas: Commentary*, pp. 48–53) says that it is therefore the earliest map known to have been published by John Thornton, the first printed map to show the results of activity by the Hudson Bay Company, and the second printed map to contain (in its inset) the features of Lederer's discoveries published in 1672.

Reproduction

Jeannette D. Black, *Blathwayt Atlas: Commentary*, Vol. I, No. 5.

Copies

John Carter Brown; Clements; Royal Geographical Society; National Archives of Canada.

90. Moore 1681

Florida

Size: 4½ × 3¾. *Scale:* 1" = ca. 340 miles.

In: Moore, Sir Jonas. *A New Geography with Maps to each Country, Third Book*. London, 1681, No. 50 (opp. p. 25).

Description: This map is like Sanson's 1657 map, except that some phrases are anglicized: "unknown lands," "golfe of Mexico or of New Spain." A map of "New Mexico" is on the same sheet. This work in the Library of Congress copy is bound in with Sir Jonas Moore, *A New Systeme of the Mathematicks*, London, 1681.

Copies

Congress; New York Public.

91. Keulen 1682?

Pas Kaart Van de Kust van Carolina Tusschen C de Canaveral en C Henry Door C. I. Vooght Geometra T Amsterdam By Iohannis van Keulen Boek en Zee Kaart verkoper aande Niewen brugh Inde Gekroonde Lootsman Met Privilegie voor 15 Iaren [cartouche, right center] | 18 [lower left corner] | 18 [lower right corner; added after 1690]

Size: 23⅛ × 20⅜. *Scale:* 1" = ca. 28 miles.
In: Keulen, J. van. *The Great and Newly Enlarged Sea Atlas or Waterworld.* Amsterdam, 1682–86, III, No. 11.
Inset: De Kust van Carolina Tusschen Riviers Mondt en C de S. Romano in't groodt. *Size:* 8⅜ × 5¼. *Scale:* 1" = ca. 6 miles.

Description: This is a hydrographic chart with nothing except coastal names given; it extends from 28°10' (below "C. Canaveral") to 37°40' ("de Bay van Cheseapeke off Bahia de Madre de Dios"). The Le Moyne names for the rivers south of Cape Fear and sparse use of the White map 1590 nomenclature for the northern half of the map show that little use of recent information was made; Keulen gives the old Spanish names for several of the Capes. "Charles Towne" is on the west bank of the "Rio Grande" (Ashley River) in both the large map and in the inset. Numerous soundings off the coast are given from Cape Hatteras northward and in Chesapeake Bay. This map, unchanged, appears in Keulen atlases for nearly fifty years. The date of the map is uncertain; but since the English edition of J. van Keulen's atlas, which is evidently based on an earlier Dutch atlas, is 1682–86, it is probable that this map appeared by 1682. The Loots map 1706 is based on this map. The van Keulen family was one of the chief publishers of nautical works in the latter half of the seventeenth century. *The English Great and Little Sea Torches*, as well as the French *Flambeaux de La Mer*, were copied from van Keulen's great work. See P. L. Phillips, *A List of Geographical Atlases*, III, 177.

Reproductions
Plate 40 and Color Plate 5.
Copies
Congress 1682–86; Kendall; Davidson.
In: Keulen, J. van. [De Lichtende Zeefakkel. Amsterdam, 1681–96,] II, No. 36.
Copy
Congress 1681–96.
In: Keulen, J. van. *De Groote Nieuwe Vermeerderde Zee-Atlas ofte Waterwerelt.* Amsterdam, 1695, No. 109.
Copy
Congress 1695.
In: Vooght, Claes Jansz. *La Nueva, y Grande Relumbrante Antorcha de la Mar.* Amsterdam, J. van Keulen, 170 [*sic*; 1700?], No. 88.
Copy
Congress 1700.
In: Loon, Jan van, and Claes Jansz Vooght. *De Nieuwe Groote Lichtende Zee-Fackel.* Amsterdam, J. van Keulen, 1699–1702, IV (1701), No. 18, between pp. 40–41.
Copies
Congress 1701; Yale 1687.
In: Voogt, C. J., and Gerard van Keulen. *Le Nouveau et Grand Illuminant Flambeau de la Mer.* . . . Amsterdam, 1713, IV, No. 18.
Copies
Harvard 1728; Mariners' Museum 1713.

91A. Mathews 1682 MS

Carolinass
This Platt Represents the shape and forme of five Thousand Eight Hundred Acres of Land, which According to a warr[r]. to me Directed Under the hands of the Hon[ble] Joseph Morton Esq[z] and the rest of the Deputyes of the Right Honourable y[e] Lords Proprietors beareing date the thirteenth of November 1682. I haue admeasured and Laid out vnto M[r]. John Smith and Ann his wife scituate lying and beeing in a collony of Colleton County not yet Distinguesht by any Name or Number and Butting and bouding Easterly vpon the ffresh River of Edistoh, and Westerly Vpon Lands not yet run out, and Northerly and Southerly Vpon Lands not yet run out, and Marked with such Trees and other Markes as doth Appeare in this Platt. Certifyed this Twenty seaventh day of January anno 1682 Mau: Mathews Surv[r]. Genll— [cartouche, right center]

Size: 18⅛ × 20½ (irregular).
Scale: 1" = 20 chains (1,320').

Description: This vellum manuscript plat of a tract of land on Edisto River is attached to a warrant for the survey. The warrant is registered and attested by Edward Mayo on June 21, 1683. This grant presumably refers to the land belonging to "Mr. Smith" just above New London on the Edisto River, shown on Mathews 1685 MS and Crisp 1711 (see Plate 44). "Here is a revealing record of the activities of Maurice Mathews as a surveyor in Carolina. The nature of the document suggests that the competence of Mathews in the larger survey of 1685 was based upon intimate first-hand knowledge of the small components which made up the whole": John Carter Brown. *Annual Report*, 1945–47, p. 37.

Reference

John Carter Brown. *Annual Report*, 1945–47. Providence, 1958, 37–38.

Original

John Carter Brown.

---

92. Gascoyne 1682

To The Right Honorable Will. Earle of Craven, Pallatine and the rest of ye true and absolute Lords and Proprietors, of the Province of Carolina, This Map is humbly Dedicated by Ioel Gascoyne.

A New Map of the Country of Carolina. With it's Rivers, Harbors, Plantations, and other accomodations. don from the latest Surveighs and best Informations, by order of the Lords Proprietors.

Sold by Ioel Gascoyne at the Signe of the Plat nere Wapping old Stayres. And by Robert Greene at the Rose and Crowne in y^e^ middle of Budge Row. [Two lines crudely erased from plate, probably referring to another bookseller] London.

Size: 22⅝ × 19¼. *Scale:* 1" = ca. 20 miles.
Insets: 1. A Perticular Map for the going into Ashly and Cooper River. *Size:* 8¾ × 4¾. *Scale:* 1" = ca. 2½ miles.

2. A Table of the names of such Settlements as are upon Ashly, and Cooper Rivers. & other adjacent places.
In: A True Description of Carolina. London: Printed for Joel Gascoin and Robert Greene . . . [1682]. (4 pp.)

Description: This map, called the Second Lords Proprietors map, extends from "Mosquetos" below St. Augustine to "C. Henry." It is the most accurate representation of the Carolina region yet to appear: the Le Moyne–Mercator conception of the Southeast is gone; the coastal detail from Cape Henry south to Port Royal and the soundings off the coast show greatly increased knowledge over the Ogilby map; the Lederer influence is entirely absent and in its place is an innocuous and rather non-committal row of parallel rivers running southeastward to the sea; and the Charles Town area is heavily dotted with names and plantations, which is evidence of excellent work done by the surveyor general of the southern settlement, Maurice Mathews. North of the Cape Fear River, the map was formerly thought to be based on information gathered by Governor Henry Wilkinson, who was instructed by the lords proprietors in 1681 to "be sure as soon as you can to send home the mapp of the County mended by your owne or frds: experience": *Colonial Records of North Carolina*, I, 338. Apparently, however, Wilkinson did not send information for the Gascoyne map back to England, for he never reached Carolina. He was arrested by the English government a few days before leaving for his post in North Carolina; see C. M. Andrews, "Captain Henry Wilkinson," *South Atlantic Quarterly*, XV (July 1916), 216–22.

No more careful or accurate printed map of the province of Carolina as a whole was to appear until well into the eighteenth century than the Gascoyne map and its imitators. Perhaps its rather unimaginative accuracy militated against it; not only continental geographers but even the mapmakers preferred the varied detail of Lederer's great lake and desert to the comparative barrenness of Gascoyne. The map in Ogilby's *America* was much more widely distributed and available than Gascoyne's map, which was part of the promotion effort of the proprietors in 1682. It was sold as a separate and also apparently accompanied a little four-page pamphlet, *A True Description of Carolina*, which in turn was an abridgement of R. F.'s *The Present State of Carolina with Advice to Setlers*, London, 1682. *A True Description*, therefore, could have been published either in 1682 or thereafter. From the tone of its text, however, it could not have been published long thereafter.

The date of the map is 1682, since "A New Map of Carolina . . . Sold by R. Green and by J. Gascoigne" was published in Michaelmas Term (November) 1682 (Arbor, *Term Catalogues*, I, 513). This year is given as its date also at the conclusion of "To the Reader" in T. Ashe's *Carolina*, London, 1682.

The following pertinent note is furnished by Mr. Lawrence C. Wroth:

"*A True Description* comprises four leaves, printed on one side only, with title at the top of leaf one. Laid side by side the breadth would be approximately the same as that of the Gascoyne map, i.e., 23½ inches from neat line to neat line. Our Blathwayt copy with its margins measures 25½ inches wide. There is a physical evidence along the top edge of our copy of *A True Description* that it once was pasted down. If it had been issued as a broadside its title would in ordinary usage have been centered at the top and not set in the upper left-hand corner as it would stand if these four leaves were put together as a full sheet broadside rather than as an oblong piece in four equal columns. These circumstances—blank versos of each leaf, evidence of pasting down, and identity of breadth—together with the nature of the contents lead me to make the suggestion that the map was published in some instances with *A True Description* in four columns pasted along its bottom edge. This of course would be in accord with common practice. I think the map in such cases could usually be purchased either with or without the printed description."

Samuel Wilson, secretary of the lords proprietors, in a bill made out to them on May 10, 1683, itemizes 2/3/—"For a Plate of y[e] map of Carolina & printing 2000," and 11/—"to p[d] Mr. Gascoyne for the map of Carolina." See *Colonial Records of North Carolina*, I, 344.

Samuel Wilson also published a promotion pamphlet, *An Account of the Province of Carolina*, London, 1682, for which he used Ogilby's map ca. 1672. Apparently the new Gascoyne map was not yet ready when he wrote and published his pamphlet, although information in his text, such as the changed location of Charles Town, clearly post-dated the information on the Ogilby map, by that time nearly a decade old. Among the maps which followed the Gascoyne map are Thornton-Morden-Lea ca. 1685; Sanson ca. 1696; Thornton-Fisher 1698; Carolina ca. 1700; Moll-Oldmixon 1708; Verelst 1739 MS.

Three states of the Gascoyne map have been identified by Jeannette D. Black (*Blathwayt Atlas: Commentary*, pp. 145–47).

1. Two lines in the lower part of the cartouche beginning with "and by" evidently gave the name and address of a third bookseller. These lines have been erased in all recorded copies; no example of this state is known.

2. A second state (with two erased lines) has a long narrow lake near the Westo Indian settlement, and at the head of the Ashley River is "Earle Shaftesbury Signory Call'd Wadwad Malow." Only one copy of this state is known, British Library, Add. MS 71903 (3).

3. All other known copies have the two lines erased, relocated and broadened the contour of the lake, and erased the two words "Wadwad Malow" after "Earle Shaftesbury Signory Call'd." There are many other small changes, chiefly in the relocation of headwaters of rivers and of the "Apalation Mountaines."

The suggestions of Jeannette D. Black (*Blathwayt Atlas: Commentary*, pp. 147–48) that these geographical changes or improvements were made from information provided orally by Dr. Henry Woodward, are reasonable and probable. Dr. Woodward had returned to London for the first time since 1666, when he had been left by Sandford to learn the language and customs and make friends with the Indians. Most of the revisions are in areas which Woodward had explored, the country of the Westo Indians and the "Indian Trading Path" to Virginia.

See Introductory Essay.

References

Black, *Blathwayt Atlas: Commentary*, pp. 144–48.

Ford, W. C. "Early Maps of Carolina." *Geographical Review*, XVI (1926), 264–73.

Karpinsky, L. C. *Early Maps of Carolina*. Camden, S.C., 1937, pp. 26–31.

Reproductions

Plate 39.

Black, ed., *Blathwayt Atlas: The Maps*, Map 23.

Smith, P. *Early Maps of Carolina . . . from the Collection of Henry P. Kendall*. 1930, Plate 10, opp. p. 16.

Copies

British Library (3), second state; Congress; John Carter Brown 1682, *Blathwayt Atlas* Map 23; University of South Carolina; Kendall; Yale; Clements 1682.

---

93. Seller 1682

Carolina Newly Discribed By Iohn Seller

Size: 5 1/2 × 4 1/4. *Scale:* 1" = ca. 91 miles.

In: Seller, John. *Atlas Maritimus: or a Sea Atlas*. London, 1682, No. 21.

Description: This map of the Ogilby ca. 1672 type is apparently copied from Speed's small map of Carolina 1675. It gives the Lederer nomenclature but omits the actual lines indicating his march, as does the small Speed map; unlike the Speed, it is elaborately ornamented with trees.

This map is found in many of Seller's smaller Atlases; it may also have appeared as one of the maps to accompany the text of *A New System of Geography* which was announced in the *Term Catalogues* for February 1685.

Reproduction

Craven, W. F. *The Southern Colonies in the Seventeenth*

Century, 1607–1689. [Baton Rouge], Louisiana State University Press, 1949, opp. p. 330.

Copies

Harvard; Johns Hopkins (Garrett); Massachusetts Historical Society; Clements.

In: [Wilson, S.] *An Account of the Province of Carolina.* London, 1682, frontispiece (this is an insert; properly a copy of the Ogilby-Moxon map should accompany this book).

Copy

New York Public 1682.

In: Seller, John. *A New System of Geography.* London, [1685?], No. 39.

Copies

Congress 1685?; Clements 1682.

In: Seller, John. *A New System of Geography*, London, 1690, No. 43.

Copies

Congress ca. 1690; Yale 1694.

In: Seller, John. *Hydrographia Universalis*, [London, 1690?], No. 24.

Copy

Congress [1690?].

In: [Seller, John]. *Atlas Terrestris.* London, [1706?], II, No. 42.

Copy

Congress [1700?].

---

94. Spanish 1683 MS

Mapa De la Ysla de la Florida [top right] | Mapa de la Florida remitido por el Marq.[es] Governador Juan Marquez Cabrera con carta de 28[de] Junio de 1683 [endorsed]

Size: 36 3/16 × 24 3/4. *Scale:* 1" = ca. 42 miles.

Description: This map includes the entire peninsula of Florida and the east coast up to "Rio do Nogales," on which is situated "Puerto y Poblacion de S. Iorge de la Nacion Inglesa." This refers to Charles Town; the Ashley is the St. George's Bay of the Spaniards. Because of the English-Spanish treaty of 1670, the Spaniards did not consider the southward movement of the English along the coast justifiable. "The first immigrants had not yet settled on the Ashley when the Spaniards appeared, giving them notice that they must fight for its existence. In 1686 they destroyed the settlement under Lord Cardross [on Port Royal Island] and ravaged the country nearly to the fortifications of [Charles Town]." E. McCrady, *The History of South Carolina under the Proprietory Government, 1670–1719*, New York, 1897, p. 684.

This map has a great deal of detailed information concerning the location of Spanish settlements immediately north of the Florida peninsula, on both east and west coasts. In 1683 the pirate Agramont destroyed the remaining Spanish missions north of St. Marys River on the east coast. This map shows six missions north of St. Johns, including three on Guadalquini and one as far north as Santa Elena. The Scots settlement near Port Royal is not shown, nor the English settlements which continued to expand southward from Charles Town soon after the Scots settlement was destroyed; this map illustrates conditions and gives Spanish names to the coastal region before 1683. It was made in 1683, as a result of the revived strategic importance of the region to the Spaniards. Chatelain (see reference below) dates it 1680–1700.

Such details as the location of the Spanish missions and the Spanish coastal names make this the most informative and generally useful map of Spanish Florida since the latter part of the sixteenth century.

Reproductions

Plate 40A.

Cartografia de Ultramar, II, Plate 48 (with transcriptions of place-names, p. 286).

Chatelain, Verne B. *The Defenses of Spanish Florida, 1565–1763.* Washington, 1941, Map 7.

Karpinski Series of Photographs, No. 163.

Congress (photocopy).

Original

Madrid. Minist. de la Guerra 9a.2[a]-a.10.

---

95. Mallet 1683

Floride [cartouche, top center] | De L'Amerique. Figure CXXVIII [at top of page] | 295 [top right, within border] | T iiij [below neat line to right]

Size: 4 1/2 × 6. *Scale:* 1" = ca. 280 miles.

In: Mallet, A. M. *Description de L'Univers.* Paris, 1683, V, 295.

Description: This map has very little detail; "Virginie" is above the country of "Apalche ou Apalache," which is above the two Mercator-type lakes. "Tegesta" is the name given to the Florida peninsula. The map is apparently based on one of Sanson's. The same map, from a different plate, is in the Frankfort 1686 edition; see Mallet 1686 and Mallet 1719.

On page 297 of the same volume is the engraving of the harbor, town, and triangular fort of "St. August. de Floride."

Copies

Congress; John Carter Brown; Clements; New York Public; Yale; National Archives of Canada.

---

96. Sanson-Canada 1683

Le Canada, ou Nouvelle France, &c. Tirée de diverses Relations des. Francois, Anglois, Hollandois, &c. Par N. Sanson de Abb. Geogr. ord.re du Roi. | [right, below neat line] A.d. Winter Sculp:

Size: 11⅞ × 8³⁄₁₆. *Scale:* 1" = ca. 210 miles.

In: Sanson, N., and Sanson, N., fils. *L'Europe, [l'Asie, l'Afrique, l'Amerique] en Plusieurs Cartes.* Paris, 1683. L'Amerique, No. 2.

Description: This is an exact copy of Sanson-Canada 1679, from a slightly smaller plate. In 1700 additions of latitude and longitude lines, as well as ornamental trees, were made on the plate of this map. See also Sanson's "Le Canade" 1657.

Copies

Congress; Clements.

In: Sanson, N. *Geographische en Historische Beschryvingh.* Utrecht, 1683, opp. p. 558.

Copies

Congress; Yale.

In: Luyts, J. *Introductio ad Geographiam. Trajecti ad Rhenum, 1682*, between pp. 694–95.

Copy

Congress 1692.

In: Sanson, N., et Fils. *Description de Tout L'Univers.* Amsterdam, 1700, Amerique, between pp. 10–11.

Copies

Congress 1700; Harvard 1700; Newberry 1700; Yale 1700.

In: Du Sauzet, Henri. *Atlas Portatif.* Amsterdam, 1734, No. 260.

Copies

Congress, two copies; Clements.

---

97. Sanson-Florida 1683

La Floride Par N. Sanson d. 'Abbeville Geogr. Ord.re du Roÿ. [cartouche, bottom center]

Size: 9⅞ × 6¹⁵⁄₁₆. *Scale:* 1" = ca. 205 miles.

In: Sanson, N., and Sanson, N., fils. *L'Europe [l'Asie, l'Africae, l'Amerique,] en Plusieurs Cartes.* Paris, 1683, L'Amerique, No. 3.

Description: This map is a close copy of Sanson's "La Floride" 1657, although it is slightly smaller and has a single neat line (the 1657 map has double marginal lines) which becomes double by an engraver's error at the bottom of the map at longitude 280°.

In 1700 and in subsequent use of the engraved plate, latitude and longitude lines have been added and trees have been engraved for decorative purposes. This same plate was used, with a Dutch title engraved instead of the French, in Phérotée de La Croix's *Algemeene Weereld-Beschryving*; see under Sanson 1705.

Copies

Congress 1683; Clements 1683.

In: Sanson, N. *Geographische en Historische Berschryvingh der Vier bekende Werelds-Deelen.* Utrecht, by Johannes Ribbius, 1683, No. 3, opp. p. 572.

Copies

Congress 1683; Yale 1683.

In: Luyts, J. *Introductio ad Geographiam.* Trajecti ad Rhenum, 1692, between pp. 702–3.

Copy

Congress 1692.

In: Sanson, N., et Fils. *Description de Tout L'Univers.* Amsterdam, 1700. Amerique, opp. p. 14.

Copies

Congress; Harvard; Newberry; Yale; Clements.

In: Du Sauzet, Henri. *Atlas Portatif.* Amsterdam, 1734, No. 261.

Copies

Congress 1734, two copies.

---

98. Thames School ca. 1684 MS A

Carolina [across face of map]

Size: 22 × 16. *Scale:* 1" = ca. 23 miles.

Description: "A colored map, on vellum, of the coast of the Carolinas and Georgia, from Cape Henry southward to St.

Augustin; drawn about 1680 (by William Hack), without a scale, but about 3 inches to a degree." British Museum *Catalogue of Maps and Drawings*, London, 1861, III, 513.

This chart, formerly listed under the title tab "No. 98 Hack ca.1684 MS A," is a product of the Thames School of chart makers; the older ascription to Hack was made before the recent identification of a school of mapmakers, a subdivision of the Drapers' Guild, with shops near the Thames River in or below London. This school or sub-guild, with its system of masters and apprentices, produced charts and maps from about 1590 to 1720, using similar techniques of craftsmanship, coloring, etc. See references in Jeannette D. Black, *Blathwayt Atlas: Commentary*, pp. 15–24, and authorities there cited.

This chart is closely similar to No. 80 (Map 80 in the *Blathwayt Atlas*); "the hand is the same, the coloring almost identical; the outlines, the soundings, the latitudes, and the longitudes are practically identical" (*Blathwayt Atlas: Commentary*, p. 134). Ms. Black continues with a detailed list of small differences, including the size of the vellum parchments, and concludes that the two may have been ordered for different lords proprietors at the same time.

Reproductions

Hulbert, A. B. *Crown Collection*. Cleveland, 1904–8, Series I, Vol. V (1908), map 34.

Congress (colored photocopy).

Original

British Library. Add. MS.13970.b.

---

99. Thames School ca. 1684 MS B

North Carolina [across face of map]

Size: 24 × 18. *Scale:* 1" = 4½ miles.

Description: This chart shows the coastal and sound area from Cape Henry to "Ococock Inlet." It gives the depth of the water in Pamlico Sound from Hatteras Inlet to the southern end of Roanoke Island, the soundings on both sides of the island and from the inlet which is opposite Roanoke up to Point Durant in Albemarle Sound. Though Wimble 1738 and Mouzon 1775 give channel depths from Ocracoke to Neuse River which are not parallelled in the surveys recorded on the Hack map, this seventeenth-century map gives more soundings from Hatteras Inlet north than have been noted in any other map before the nineteenth century. The location of the inlets—two opposite Roanoke Island, one to the north of Colleton Island, and Hatteras and Ocracoke entering Pamlico—differs from that shown in Gascoyne 1682. The inlets shown are similar to those in No. 80 Blathwayt ca. 1679 MS C, which calls Hatteras Inlet Roanoke Inlet and calls one of the inlets opposite Roanoke Island the "Old Inlet"; No. 80 also shows at "Conetto Inlet" a channel (absent in No. 99) which leads to the Albemarle region, and has many more soundings outside the Banks.

It has one of the earliest uses of the name "North Carolina" to denote that part of the province to the north of the Cape Fear River.

Ms. Black (*Blathwayt Atlas: Commentary*, p. 136) comments that this chart has much of the navigational information, such as soundings of the inlets and in the sounds, that the proprietors had asked for in 1676. She suggests that the answers shown thereon were so discouraging that they may have contributed to the proprietors' loss of interest in promoting transoceanic trade to northern settlements and turned their attention to the better harboring at Charlestown.

Reproductions

Plate 41.

Hulbert, A. B. *Crown Collection*. Series I, Vol. I, No. 29.

Original

British Library. Add. MS 5415.g. 6.

---

100. Hack 1684 MS

Carolina [across face of map] | Made by William Hack, at the Signe of Great Britaine & Ireland, neare New Staires, in Wapping, Anno Domini, 1684 [cartouche, bottom right]

Size: 21¼ × 25¼. *Scale:* 1" = ca. 26 miles.

Inset: [The Ashley and Cooper Rivers]. *Size:* 7 × 5. *Scale:* 1" = 3 miles.

Description: This manuscript map, heavily colored on vellum, is drawn from the Ogilby First Lords Proprietors Map ca. 1672; the Lederer details are given to the interior, and the inset is like that on Ogilby's map, though somewhat simplified.

R. A. Skelton (*Geog. Journ.* CXXV (1959), 263) notes another Hack map, also a copy of Ogilby ca. 1672, in the Hack atlas in King George III's Maritime Collection in the British Library.

William Hack was a clever map and chart maker "who might have been forgotten except that he travelled in fast company"; he sailed with the notorious buccaneer Bartholomew Sharpe on a marauding expedition along the coast of South America in 1679–82. Landing on the east coast at Porto Bello, which they sacked, the members of the expedition marched across to the Pacific, spreading devastation and terror as they went. On the west coast they captured the Spanish treasure ship "Rosario," which had on it a Spanish pilot's atlas of the west coast. For the English this volume was a possession of great value, as the Spanish zealously guarded their cartographical information. Hack made a copy of the volume: see *Description of a Mappemonde by Juan Vespucci and of a Buccaneer's Atlas by William Hack*, London, 1914, published by Bernard Quaritch. For a more extended account of the cartographical and piratical activities of William Hack, see "William Hack and the South Sea Buccaneers," Chapter V in Edward Lynam's *The Mapmaker's Art*, London, 1953, pp. 101–16.

Reproductions

Hulbert, A. B. *Crown Collection*. Series II, Vol. III, No. 36.
Congress (colored photocopy).

Original

British Library. Add. MS. 5415. g. 5.

---

101. Mathews ca. 1685 MS

A Plat of the Province of Carolina in North America. The South part Actually Surveyed by Mr Maurice Mathews. Ioel Gascoyne fecit.

Size: 62 × 50. *Scale:* 1" = 6 miles.

Insets: 1. A Particular Draught for ye going in to Ashly and Cooper Rivers.

Size: 17½ × 13. *Scale:* 1" = ca. 1¼ miles.

2. A Table of Names of ye severall settlements upon [?Colleton], Stone, Ashly, & Cooper Rivers. [Two hundred sixty-one names given: some are illegible.]

Description: This large colored map extends from Cape Henry in the north down the coast to the mouth of the Westo (Savannah) River. The coastline above Charles Town has not been changed from the Gascoyne map of 1682. However, from Charles Town down the coast to Port Royall, the map shows evidence of later surveys and gives much fuller information concerning the settlements. From the thirty-three listed in Gascoyne's 1682 map, the Table of Names has increased to over two hundred and fifty, with the location of every one on the map indicated. No other comparable area in the Carolinas was as carefully drawn before the middle of the following century. This is an important type map, as it is the basis for Thornton-Morden 1695, Sanson 1696 B, and Crisp 1711.

The interior of the country, as in Gascoyne's 1682 map, is devoid of significant information; ornamental hills and trees abound and the rivers stop long before they reach the "Apalatian Mountaines." The rivers, however, correctly run down to the southeast, and the Appalachian range extends at right angles to the rivers toward the northeast. The road from Charles Town to "Ye old fort, Savana Towne and fort" on the Savannah River, and on up the river to the Indian country, is shown.

That the date of the map cannot be earlier than 1684 is shown by "Scotts Settelment" on Port Royal Island. Lord Cardross settled it in 1684; the Spaniards attacked and destroyed the colonists there in 1686. See V. Crane, *Southern Frontier*, pp. 26–31, and original documents there cited. The *ad quem* date of the map is 1695, for by 1694 Mathews was dead and in the following year the Thornton-Morden map, based on the Mathews MS map, was published. See Introductory Essay.

Reference

Karpinski, L. C. *Early Maps of Carolina*. Camden, S.C., 1937, p. 31.

Reproduction

Congress (colored photocopy).

Original

British Library. Add. MS. 5415.24.

---

102. Spanish ca. 1685 MS

[Florida and West Indies]

Size: 30½ × 18¼. *Scale:* 1" = 125 miles.

Description: This manuscript map shows many Spanish names along the coast, which extends northward to "C d Trafalgar" (Cape Fear). It has "Florida" as a river or bay in the vicinity of St. Johns River and "London" at "Sta elena." This latter is New London, later Willtown, on the South Edisto River in Colleton Precinct. The map is dated 16— by the Biblioteca Nacional. It is certainly not earlier than 1685, when the Spaniards were considering active measures against the English settlements; the settlement of New London on the Edisto had been established only a few years before. See E. McGrady, *The History of South Carolina under the Proprietary Government 1670–1719*, New York, 1897, p. 198.

Reproduction
Congress (photocopy).
Original
Madrid. Biblioteca Nacional. Min. For. Af.

---

103. Morden ca. 1685

A Map of ye English Empire in ye Continent of America Viz Virginia Maryland Carolina New York New Iarsey New England Pennsilvania [cartouche, bottom center] | Sold by R: Morden at y^e^ Atlas in Cornhill neer y^e^ Royal Exchange London. [cartouche, bottom left center] | W: Binneman sculpsit [to right of Morden cartouche]

Size: 22½ × 19⅝. *Scale:* 1" = ca. 42 miles.
Inset: Caroliniae [*sic*] Pars. *Size:* 5⅞ × 7⅞.
Scale: 1" = 85 miles.
Description: See Daniel 1679.

The Henry E. Huntington Library copy of the map has printed texts pasted on either side and below the map, making a sheet 32¼" × 26¼" in size. These marginal additions give a running account of the English Empire, including a specific account of the different provinces. Carolina, which is given a special heading, has a promotional blurb encouraging immigration, derived in part from Ogilby's *America.* In an italicized Preface to the marginalia is an attack on the Dutch and the French, who in their maps and geographies claim English possessions under the titles of New Netherlands and Nova Francia. Blaeu and Sanson are especially condemned. In several libraries, including the Huntington Library, one finds a facsimile made by Henry Stevens of the Daniel 1679 map with the printed marginalia. Concerning this facsimile and concerning the date of the Morden ca. 1685 map, Mr. Lyle H. Wright, head of the reference department of the Huntington Library, has written me the following note: "I surmise that Mr. Stevens laid the British Museum 1679 map over our 1685, and the text which is pasted to our map was transposed to the 1679. The date we use is based on Stevens's knowledge of Daniel's maps—he sold the British Museum their copy of 1679 in 1897. He says when he offered the map to us in 1924, 'I assign the date of about 1685 to this state, from the inclusion of Pennsylvania and by reason of sundry other cartographic details.' Phillips attributes the date 1690? to the map but without the information that Stevens had; therefore I am inclined to accept the latter's date."

Reproductions
Adams, J. T., ed. *Album of American History.* New York, 1944, I, 252.
Karpinski, L. C. *Bibliography of the Printed Maps of Michigan.* Lansing, Mich., 1931, opp. p. 100 (only the northern part is reproduced, from the copy in the Michigan Historical Commission archives).
Copies
John Carter Brown; Huntington (with accompanying text); Johns Hopkins (Garrett); Michigan Historical Commission.

---

104. Thornton-Morden-Lea ca. 1685

A New Map of Carolina. By Iohn Thornton at the Platt in the Minories, Robert Morden at y^e^ Atlas in Corn-hill, And by Philip Lea at the Atlas & Herculus in the Poultry. London.

Size: 17⅞ × 21¾. *Scale:* 1" = 21 miles.
Insets: 1. A Perticular Map for the going into Ashley and Cooper Rivers. *Size:* 5¼ × 4. *Scale:* 1" = 3⅕ miles.
2. A Table of the names of such Settlements as are upon Ashley and Cooper Rivers & other adjacent Places. [Thirty-three names]
Description: This map of Carolina extends westward to "The Apalatian Mountains" and is based chiefly on Gascoyne's map 1682 or the original from which that was made. The details of the rivers, the list of thirty-three settlements and their corresponding numbered location on the map, and other information concerning the interior are alike. A few of the names of scattered settlements near the Ashley and Cooper Rivers have been omitted and the inset of that region is less detailed. The coastline and the soundings, however, differ from the Gascoyne map and agree minutely with the greater detail and accuracy of the Lancaster 1679 MS C; that manuscript map or the original on which it was based was also used.

The date 1685? is given as about the time when the combined names of John Thornton, Robert Morden, and Philip Lea are found in the various publications listed in the *Term Catalogues* for 1683–85.

This map is found in three different states. About 1690

Lea bought the map, erased Thornton's and Morden's names from the cartouche, and added his own address. Above the Charleston Harbor inset he added "Ashley & Cooper River" as a title. About 1700 he used this inset as a small separate map for his *Hydrographia Universalis*. In 1705 Lea died, but his widow carried on his business. George Willdey (or Willday), who bought up old plates about 1725 to 1735, purchased this one, revised the imprint, and published the map otherwise unchanged.

Reproductions

Color Plate 6.

Ashe, S. A. *History of North Carolina*. Greensboro, N.C., 1908, I, opp. p. 145 (upper part of map only).

Charleston, S.C., City of. *Year Book*. Charleston, 1886, p. 280 (Lea edition).

Gabriel, R. H., ed. *The Pageant of America*. I, 268 (inset only, from New York Public Library copy).

Copies

Congress; John Carter Brown; Huntington, two copies; Kendall; South Carolina Historical Society.

A. A New Map of Carolina. By Philip Lea at the Atlas and Hercules in Cheapside London ["Ashley & Cooper River" added above Charleston Harbor inset]

Copies

John Carter Brown, Binder's Atlas, No. 104, separate; Charleston Museum; Duke; Kendall; New York Public; University of North Carolina; Yale; Clements.

B. A New Map of Carolina Sold by Geo: Willdey at the Great Toy, Spectacle, Chinaware and Print Shop, ye Corner of Ludgate Street near St Pauls London [with added title of Charleston Harbor, as in the Lea state]

Copies

Congress; John Carter Brown.

---

105. Crouch 1685

The English Empire in America By R. B. [center, extreme right] | London. Printed for Nath. Crouch [below neat line]

Size: 2¾ × 4⅞. *Scale:* None.

In: B[urton], R[obert], [pseudonym for Nathaniel Crouch]. *The English Empire in America*. London, printed for Nath. Crouch, 1685, front.

Inset: New Found Land. *Size:* 1" × 1".

Description: This map extends from northern Florida to Maine. In the Carolina region "L. Ashley," and "Sauana," and the desert of Lederer's map appear. It has many ornamental designs for so small a map, such as hunting and warring Indians, animals, and mythological beasts.

The work in which this map is found is attributed to Nathaniel Crouch, who used R[obert] B[urton] as his pseudonym. See J. Sabin, *A Dictionary of Books Relating to America*, New York, 1868–1936, III (1980), 162, who lists seven editions of the work.

Passages in the book are also found in Richard Blome's *Description of the Island of Jamaica*, London, 1672. The same dedication "To the Reader," an untranslated English poem, the initial chapters on the discovery of America given in this book by Crouch, and in fact practically the entire text in translation are also found in *Englisches America*, Leipzig, 1697, attributed to Richard Blome on the title page. Parts of the chapters on the different colonies in *The English Empire in America* are found in Blome's *The Present State of His Majesties Isles and Territories in America*, London, 1687, though the latter has added several passages, interpolated from Blome's *Jamaica*, concerning the islands, and has also added material, such as Charles I's grant to the lords proprietors of Carolina, which are not in either of the earlier works. Blome dedicates *The Present State of His Majesties Isles and Territories in America* to James II expressly as his own work. Thus it is clear that there was a good deal of cross-borrowing without acknowledgment, as was common at the time, in these books. But in the 1728 edition of *The English Empire in America* the title page attributes the work to Richard Burton, not R. B.; the evidence, however, is in favor of Nathaniel Crouch as the author. The ascription of *Englisches America* to Richard Blome is probably due to a confusion of R. B. with Blome's initials by an unauthorized translator or a piratical publisher. In 1688 Richard Blome's *L'Amerique Angloise*, a translation of his *Present State*, appeared in Amsterdam in a French translation. Blome died in 1705; Nathaniel Crouch died about 1725.

Three different plates are used for this map, with different titles and other small changes.

1. In the 1685, 1692, 1698, and 1711 editions of *The English Empire in America*, "London. Printed for Nath. Crouch" is at the bottom, while right above this line an Indian is resting, a stick upon his knee.

2. In Blome's *Englisches America* (1697) the title is "Das Englische America," the line at the bottom is omitted, and the Indian is drawn without a stick. See under Crouch 1697.

3. In the 1728 edition, the title in the cartouche is the

same as in the 1685 edition; the Crouch line is omitted, and the figure is holding a stick up in the air. The engraving for this map is poorly done. See under Crouch 1728.

Copies

Congress 1692; John Carter Brown 1685, two copies, 1702; Boston Public 1698; Harvard 1685, 1711; Huntington 1685, 1692; Newberry 1685; New York Public 1685, 1698; Yale 1685, 1692; National Archives of Canada 1728; Clements 1739.

---

106. Mallet 1686

Floride [cartouche, top center] | das Landt Florida Fig: 18 [above neat line at top of page]

Size: 4½ × 5⅞. *Scale:* 1" = ca. 276 miles.

In: Mallet, Allain M. *Description de L'Univers.* Francfourt, chez Jean David Zunner, 1686, V, between pp. 174–75.

Description: This is similar to Mallet 1683, though it is slightly smaller and from a different plate. A similar map, possibly from the same plate, is given under Mallet 1719.

Copy

John Carter Brown 1686.

---

107. Morden 1687

A New Map of Carolina By Robt. Morden

Size: 4⅛ × 4¼. *Scale:* 1" = ca. 158 miles.

In: [Blome, Richard,] *The Present State Of His Majesties Isles and Territories in America* . . . , London, 1687, opp. p. 150, or p. 182.

Description: This small map has much of the Lederer material as given by Ogilby ca. 1672. It extends from "Bay st. Matheo" to latitude 40½°, including Virginia and Maryland. The map accompanies "A Description of Carolina," pp. 150–63, and a printing of the Second Charter (1665) to the lords proprietors, pp. 163–82.

Reproductions

Avery, E. M. *A History of the United States.* Cleveland, 1904–7, III, 24.

Hamilton, P. J. *The Colonization of the South.* Philadelphia, 1904, III, opp. p. 336.

Winsor, J. *A Narrative and Critical History of America.* Boston, 1886, V, 341.

Copies

Congress; American Geographical Society; John Carter Brown; Charleston Library Society; Huntington; Kendall; New York Public; North Carolina State Library; Yale; Clements.

---

108. Blome 1688

Nouvelle Carte de la Caroline par R. Morden. [cartouche, bottom right] | P. 195 [outside neat line, upper right]

Size: 4⅛ × 4¾. *Scale:* 1" = 158 miles.

In: Blome, R. *L'Amerique angloise, ou Description des Isles et terres du Roi d'Angleterre.* Amsterdam, 1688, p. 195.

Description: This map is based upon Morden's A New Map of Carolina 1687 but has differences of title, size, and nomenclature. It has "Hilfon Head" instead of the "Hilfton Head" of Morden.

The book is a translation of Blome's *The Present State of His Majesties Isles and Territories in America*, London, 1687.

Though the title has been changed, the contents of Blome's *Description* (see below) is the same as that of Blome's *L'Amerique Angloise.*

Copies

Congress, separate; John Carter Brown 1688, two copies; Huntington 1688; New York Public 1688; National Archives of Canada 1688.

In: Blome, R. *Description des Isles et Terres Que l'Angleterre possede en Amerique.* Amsterdam, 1715, opp. p. 195.

Copies

John Carter Brown 1715; Huntington 1715.

---

109. Morden 1688

A Map of Florida and ye Great Lakes of Canada By Robt. Morden [cartouche, bottom right] | Page. 73 [within upper right border]

Size: 5 × 5⅛. *Scale:* 1" = ca. 360 miles.

In: Morden, R. *Geography Rectified.* London, 1688, p. 557.

Description: This map has only the large lake emptying into the R. May remaining of the Lederer influence; it includes all of Florida, the five Great Lakes, and the Mississippi. On the same page begins the printed account "Of Florida." Before 1693 "Of Florida" and "557" are printed above the plate; in 1693 and thereafter the page number is changed to "587."

Reproduction

Karpinski, L. C. *Maps of Michigan*. Lansing, Mich., 1931, opp. p. 110.

Copies

Congress 1688, 1693, 1700; New York Public 1693; Yale 1700.

In: Morden, R. [*Atlas Terrestris*. London, 1688?,] No. 73.

Copies

Congress 1700; New York Public 1688?; Mariners' Museum 1702?

110. Morden 1688

A New Map of Carolina By Robert Morden [cartouche, bottom right] | A Description of Carolina 559 [printed at top of page; pagination changed to 589 in 1693 and 1700; changed to "Page 74," top left, within neat line, in *Atlas Terrestris*]

Size: 5 × 4⅞. *Scale:* 1" = ca. 75 miles.

In: Morden, Robert. *Geography Rectified*. London, 1688, p. 559; in later editions, 589.

Description: This is a new map with no Lederer material, though the title is similar to Morden's earlier map 1687. The area shown extends from 31° ("R. May") to 36°10' (northern Carolina boundary). It is a simplified form of Thornton-Morden-Lea ca. 1685. A printed "Description of Carolina" begins on the same page. In the 1680 edition of *Geography Rectified* is a map on a smaller scale, entitled "Carolina Virginia Mary Land & New Iarsey by Robt. Morden," which extends from 30° to 40° N.L.

Copies

Congress 1688, 1693, 1700; John Carter Brown 1688, 1693, 1700; Harvard 1688; Kendall; New York Public 1693; Yale 1700; National Archives of Canada 1688.

In: Morden, R. *Atlas Terrestris*. London [1688?], No. 74.

Copies

Congress 1700; New York Public 1688?; Mariners' Museum 1702?

110A. Thornton–Fisher 1689

A New Mapp of Carolina by John Thornton at ye Platt in ye Minories and by Will: Fisher at ye Postorn Gate on Tower hill London

Size: 20½ × 16½. *Scale:* 1" = 18 miles.

In: *The English Pilot*. London, 1689, Fourth Book, No. 17.

Inset: A Large Draught of Ashly and Coopers River.

Size: 7⅞ × 5½. *Scale:* 1" = 3 miles.

Description: This hydrographic map, which gives only coastal names and soundings, extends from "Westo Inlett" in South Carolina to "C. Charles" in Virginia. In topography and in soundings along the coast it follows closely the Thornton-Morden-Lea map ca. 1685. However, in the Ashley and Cooper Rivers inset and south of Charles Town, it shows evidence of more careful soundings which indicate that the makers of this map used the original Maurice Mathews MS ca. 1685, as the soundings follow the Mathews readings closely. This same map, apparently using the same plates, is found in the fourth book of the *English Pilot*, which contains American maps, for 60 years, including the 1749 edition. About 1729 everything in the cartouche except "A New Mapp of Carolina" was erased; the erasure from the metal plate was done rather carelessly, however, and markings of the original title are still visible. The inset remains unchanged. In 1767 a copy of this map was made for the Dublin issue of the *English Pilot* from a slightly smaller plate.

The dominance of the Dutch sea atlases, even for charts of the English coasts, had been unquestioned since the appearance of Lucas Jansz Waghenaer's *Mariner's Mirrour* in the 1580s and of its numerous successors. In the seventeenth century "wagoners" became an anglicized generic name for sea atlases. But the training of chart makers under the apprentice system of the "Thames School" developed a growing skill and body of knowledge that could be used for the needs of a spreading colonial system.

In 1669 John Seller proposed "a sea Waggoner for the whole World" which eventually resulted in the production of *The English Pilot* in five books. Seller found the enterprise more than he could accomplish alone and enlisted the support of others. Eventually the publication was taken over by a partnership between William Fisher, a successful London bookseller and printer, and John Thornton, a cartographer who had been trained in the Thames School, as the chart makers of the Draper's Guild are now called. In 1689 appeared *The English Pilot: Book IV*, with charts of America.

Some of the charts were old Dutch plates that had been bought by Seller and others; Thornton erased the Dutch cartouche and inscribed an English title with the imprint of Fisher and Thornton. In spite of the obsolescence of information on many of the charts, *The English Pilot: Book IV* was quickly accepted in the English colonies and was regarded as the bible of the American sailor. New charts were added or substituted for old ones; eighteen charts were in the first (1689) edition but a total of sixty-four charts are found in the dozens of editions published between 1689 and 1794. The information on some of the charts covering the same area and included in decades of successive editions is quite contradictory.

The work remained a low-budget operation; the "New Mapp of Carolina" continued with no changes, except for a cartouche imprint to show new owners of the publication, until it was finally dropped after 1749. Richard Mount, partner of the firm after the death of his father-in-law, William Fisher, published James Wimble's map of the North Carolina coast in 1738 separately but made no changes in the *English Pilot.* Charles Town and the South Carolina coast, however, which drew shipping and commerce, were sufficiently important to result in the addition of two maps in the 1706 and subsequent editions of the *Fourth Book*: "A Large Draught of Port Royal" and "A Large Draught of South Carolina from Cape Roman . . ."

For further discussion of the complicated history of *The English Pilot*, see the works listed below under *References.*

References

Cumming, William P. *British Maps of Colonial America.* Chicago, 1974, pp. 39–45, 94–95.

Verner, Coolie. *A Cartographical Study of the English Pilot: The Fourth Book.* Charlottesville, 1960.

Verner, Coolie. *The English Pilot: The Fourth Book, 1689.* Amsterdam, 1967.

Reproductions

Color Plate 7.

Verner, Coolie. *The English Pilot: The Fourth Book, 1689.* Amsterdam, 1967. (A facsimile edition with a helpful introduction and a list of charts in 37 editions.)

Copies

British Library 1689; Harvard 1689, 1698, 1706; Congress 1706, 1721, American Maps, No. 39, separate; John Carter Brown 1706, separate; Charleston Museum; North Carolina Department of Archives and History.

In: Keulen, Johannes van. *The Great and Newly Enlarged Sea Atlas or Waterworld.*

Copy

Congress 1682?

A. [1729] A New Mapp of Carolina

In: The English Pilot. London, 1729, Fourth Book, No. 17.

Copies

Congress 1737, 1745, 1749; American Geographical Society 1742; John Carter Brown 1745, 1749; Clements 1751; Harvard 1729, 1742; Kendall; New York Public 1742; University of Virginia 1737, two copies.

110B. de Leon 1690 MS

Viage que el año 1690 hizo el Gouernador Alonso de Leon desde Cuahuila hasta la Carolina Prouincia habilitada de Texas y otras naciones al Nordeste de la Nueua España. [cartouche, top left center]

Size: 23 × 23¼. *Scale:* 1" = 7 Spanish leagues.
Inset: Descripcion exacta del Lago de S. Bernado y del Todos los Santos que neuuamente se hallo este año de 1690. *Size:* 8¼ × 8¼. *Scale:* 1" = 3½ Castilian leagues.

Description: At the left side of the sheet is a long legend corresponding to the lettering on the map and giving the distances and directions of the journey. "Carolina" is at the top right of the map; the journey stops there at the "Poblacion de los Texas" on the "Rio d S Miguel."

Reproduction

Gil Munilla, Roberto. "Politica Española en el Golfo Mexicano," *Anuario de Estudios Americanos [Sevilla]*, XII (1955), opp. p. 524 (ref.: p. 565).

Original

Sevilla. Archivo General de Indias. Mexico 88 [est. 61, caj. 6, leg. 21(3)].

---

111. Müller-Virginia 1692

Virginia

Size: 3 × 2½. *Scale:* 1" = ca. 300 miles.
In: Müller, J. U. *Kurtz-bündige Abbild- und Vorstellung Der Gantzen Welt.* Ulm, 1692, No. 90.
Inset: I. Bermudes. *Size:* ¾" × ⅜". [Lower right]
Description: This map is very much like Du Val's Carte de

La Virginie 1682, although it differs in several details. It is of the Mercator-Hondius type, with the two lakes and "Apalachen" across present North Carolina. The title is printed below the neat line of the map.

Copies

Congress; Clements.

---

112. Müller-Florida 1692

Florida

Size: 3 × 2½. *Scale:* 1" = ca. 550 miles.

In: Müller, J. U. *Kurz-bündige Abbild- und Vorstellung Der Gantzen Welt.* Ulm, 1692, No. 91.

Description: This map is a miniature copy of Sanson's Florida 1657, with the same area, though with some difference in detail. "Charlesfort" instead of Sanson's "Carolina" is given at Port Royal, and only one lake appears. The title is below the neat line of the map; the lower part of the page gives a written description of Florida and Virginia.

Copies

Congress; Clements.

---

113. Archdale ca. 1695 MS A

A platt of ye Towne [endorsed] | A: Shewes the fort on Ashley and Cooper Rivers B: Shewes the fort on Captaine Daniells Cricke The pricked Lines shewes the Streets and Lotts [legends written in lower right center]

Size: 15 × 19¼. *Scale:* None.

In: Papers relating to ye Province of Carolina, principally whilst John Archdale Esq: was Governour & Commander in chief of ye Province. anno. 1694. 1695 &c. with a Draught of ye Town, Mapps of ye Forts, Rivers, Coasts &c. Fol. 48.

Description: This is a rough pen-and-ink drawing of Charles Town, indicating the squares of the town which had been laid at Oyster Point, and the two forts which had been built. The titles to this and the following maps are in Governor Archdale's own hand; it is probable that at least some of the maps themselves are drawn by him.

The manuscript volume which contains this and the following four maps has, on the first page, below the title, and in the same hand, the information that it was bought at an auction of the "Manuscripts of ye Late Mr Granger" January 25, 1732–33. Below, in a different hand, is written: "De la Bibliotheque de la Chevaliere D'eon." While most of the papers and letters in this volume are of the years 1694 and 1695, some of the items are dated later, extending into the first decade of the eighteenth century. Governor Archdale died in 1717.

Reproductions

Duke, and University of North Carolina (photocopies); South Carolina Archives (microfilm).

Original

Congress Manuscript Division.

---

114. Archdale ca. 1695 MS B

The Draughts of the Forts [endorsed]

Size: 14⅞ × 12. *Scale:* 1" = 39 feet.

In: Archdale, J. Papers relating to ye Province of Carolina. Fol. 48.

Description: This is a pen-and-ink plan of the two forts at Charles Town referred to in the legend on the MS map of Charles Towne in the Archdale papers (fol. 48): see Archdale ca. 1695 MS A.

Reproductions

Duke, and University of North Carolina (photocopies); South Carolina Archives (microfilm).

Original

Congress Manuscript Division.

---

115. Archdale ca. 1695 MS C

[Charles Town harbor and neighboring coastline]

Size: 8¼ × 13. *Scale:* None.

In: Archdale, J. Papers relating to ye Province of Carolina. Fol. 50.

Description: This is a rough draft of the Harbor of Charles Town and the coastal region from "Stoenoe River" to "Cap Roman": see Archdale ca. 1695 MS A.

Reproductions

Duke, and University of North Carolina (photocopies); South Carolina Archives (microfilm).

Original

Congress Manuscript Division.

116. Archdale ca. 1695 MS D

The Mapp of Ashley & Cooper rivers: &: [endorsed]

Size: 16½ × 13. *Scale:* 1" = 4 miles.

In: Archdale, J. Papers relating to yͤ Province of Carolina. Fol. 51.

Description: This pen-and-ink map gives the location of the plantations of twenty-one settlers. It is not nearly as detailed as the Thornton-Morden ca. 1695. The map shows Charleston and the reaches of the Ashley and Cooper Rivers for a distance of about twenty miles above the town. See Archdale ca. 1695 MS A.

Reproductions

Duke, and University of North Carolina (photocopies); South Carolina Archives (microfilm).

Original

Congress Manuscript Division.

117. Archdale ca. 1695 MS E

A rough draught of North Carolina coast to Santee John [repeated five times] Archdale [writing partly obliterated]

Size: 14⅝ × 18½. *Scale:* None.

In: Archdale, J. Papers relating to yͤ Province of Carolina. Fol. 52.

Description: This is a crude pen-and-ink sketch of the coastline and main rivers from Albemarle Sound and its tributaries down the coast to the Santee River in South Carolina. Only the mouths of the rivers are sketched in, and the coastline and Banks islands are drawn by one line with considerable inaccuracy. See Archdale ca. 1695 MS A.

Reproductions

Duke, and University of North Carolina (photocopies); South Carolina Archives (microfilm).

Original

Congress Manuscript Division.

118. Thornton-Morden ca. 1695

To the Right Honorable William Earl of Craven; Palatine, Iohn Earl of Bath. George Lord Carteret. Anthony Lord Ashley. S.ʳ Iohn Colleton Barr.ᵗ Thomas Archdale, Thomas Amy. and the Hieres of Seth Sothell, Esq.ʳˢ This New Map of the Cheif Rivers, Bayes, Creeks, Harbours, and Settlements, in South Carolina Actually Surveyed is humbly Dedicated by John Thornton. & Rob.ᵗ Morden. [cartouche, upper left] | Sold by Iohn Thornton in the Minories and Robert Morden in Cornhill: London [lower right, above scale]

Size: 22¼ × 18⅞. *Scale:* 1" = 2⅛ miles.

Description: This important map is largely based on the South Carolina area in the large manuscript map of Mathews ca. 1685. It extends from South Edisto to Sewee Harbor.

It agrees with the Mathews manuscript in the location of over 250 plantations or settlers, as well as in the topography and soundings, of that section which it covers; thus it is a rich and authentic source of information for the condition of South Carolina settlements at this time. There are a few additions not found on the Mathews map, such as "The French Settlemt" on the Santee, "Indian Settlements" on Kayawah Island, and "Edestow Settˡemᵗˢ" on the Edisto, which show later or at least independent information.

It was copied by other mapmakers. Sanson made a similar map with almost no changes except that some of the legends are in French: see Sanson ca. 1696. Crisp's map of 1711 used this map or the original Mathews manuscript, although Crisp added numerous later settlements and enlarged the area shown, evidencing independent surveys. Herman Moll's Dominions map of 1715 includes most of the map for one of its insets, using the area covered by this map but including items from Crisp not found on it. Moll's later Map of the Province of Carolina 1730 still covers approximately the same area, though many of the names are omitted.

This map was presumably published in 1694 or 1695, as may be determined by the names of the lords proprietors listed in the title dedication. It is not earlier than 1694, since the heirs of Seth Sothel (who died in North Carolina in 1694 intestate and without heirs) are mentioned. John, earl of Bath, became a proprietor April 24, 1694, and John Colleton succeeded his father Peter, who died in April 1694. It was not published later than 1695, since George Lord Carteret, listed in the title, died in that year and was succeeded by his son John.

Reproduction

Plate 42.

Copies

Congress; John Carter Brown, Binder's atlas, No. 105, separate; Charleston Museum; New York Public;

University of South Carolina; Clements; British Public Record Office.

---

119. Morden-Brown ca. 1695

A New Map of the English Empire in America viz Virginia Mary Land Carolina Pennsylvania New York New Jersey New England Newfoundland New France &c by Robert Morden [cartouche, right center] | Sold by Robt. Morden at the Atlas in Cornhill, and by Christopher Brown at ye Globe near the West end of St. Pauls Church: London [below cartouche] | I. Harris sculp: [bottom right]

Size: 23½ × 19⅞. *Scale:* 1" = ca. 92 miles.
In: Wit, F. de. *Atlas Maior.* Amsterdam, [1706?], II, No. 72.
Insets: 1. The Harbour of Boston or Mattathusetts [*sic*] Bay [right center, to left of Cartouche] *Size:* 3 × 3⅛. *Scale:* 1" = 9 miles.
2. A Generall Map of the Coasts & Isles of Europe, Africa and America [lower right corner] *Size:* 12 × 8¾. *Scale:* 1" = ca. 550 miles.

Description: This map extends from the Cape of Florida to Hudson Bay and from Newfoundland to the Mississippi. In spite of its scope and size it has a good deal of information for the Carolina coastal region.

Its special interest lies in its continued use of the Lederer lake, savanna, and desert, and in its striking delineation of a trident-shaped formation for the Appalachian mountain range, with the handle extending deep into Florida, and the three prongs, separating in western North Carolina, stretching west to the Mississippi, north through the present state of Michigan, and northeast into Pennsylvania. In the Michigan peninsula the range has a broad plateau running throughout its length, with the following legend: "On the top of these mountains is a Plaine like a Terras Walk aboue 200 miles in length."

This map is similar in many ways to the map first published by Daniel about 1679 and later published with Morden's imprint about 1685. See Daniel ca. 1679.

The plate for Morden-Brown's map was used, with a changed title and imprint, by Senex with the date 1719 and was included in his *A New General Atlas*, London, 1721. See Senex 1719.

Robert Morden was in business from 1671 to 1702, and Christopher Browne from 1691 to 1707, according to entries in the *Term Catalogues*. The date of this map is probably near the turn of the century. See H. R. Plomer, *A Dictionary of Printers and Booksellers*, London, 1922, pp. 54, 210.

Sanson copied this map closely in his "Carte Nouvelle d L'Amerique"; see Sanson ca. 1700.

Delisle did not use the Morden-Berry map for his "Carte du Mexique et de la Floride" 1703, but drew items from it or from Sanson's version for his famous "Carte de la Louisiane" 1718.

Reference

H. Stevens and Tree. "Comparative Cartography . . . One Hundred Maps," p. 319.

Copies

Congress 1706?; John Carter Brown, Binder's atlas, No. 110; National Archives of Canada; Clements 1695?; British Public Record Office.

---

120. Sanson ca. 1696 A

Carte Generale de la Caroline Dresse sur les Memoires le Plus Nouveaux par le Sieur S[anson] A Amsterdam. Chez Pierre Mortier, Libraire, Avec Privilege de Nos Seigneurs les Etats

Size: 22½ × 18½. *Scale:* 1" = ca. 20 miles.
In: Sanson, N. *Atlas Nouveav Contenant Toutes Les Parties du Monde.* Paris, 1696, No. 22.
Inset: [Ashley and Cooper Rivers.]
Size: 5⅞ × 4⅜. *Scale:* 1" = 3⅓ miles.

Description: This is based on the Thornton-Morden-Lea map ca. 1685, even to the smallest topographical detail, except that the name and legends are sometimes Gallicized: "Lac," "Compagne Agreable" for "Pleasant Valley and Lake." The table of settlers has been omitted. "Charle Ville ou Charles Town" is on the Cape Fear River (see Blome 1672), as well as Charles Town on the Ashley and Cooper Rivers. This map and the following one by Sanson are found in many editions of the French *Neptune*, made for the King of Portugal, the King of England, the King of France, etc.

Reproductions

Color Plate 9.
Cartografia de Ultramar, II, Plate 44 (with transcription of place-names, pp. 254–55).

Copies

Congress, several separates; John Carter Brown 1696, 1700, separates; Charleston Museum, several separates; Clements, two separates; Harvard, two separates; Huntington, separate; Kendall, separate; New York Public (General Atlas); North Carolina Department of Archives and History, separate; University of North Carolina, separate; University of South Carolina, separate; Yale, several separates.

In: Ablancourt, N. P. d'. *Le Neptune Francois.* Amsterdam, Pierre Mortier, 1700.

Copy

Harvard 1700.

In: De Fransche Neptunus. Amsterdam, 1693–1700, II, 29.

Copy

Congress 1693–1700.

In: Suite du Neptune Francois. Amsterdam, Pierre Mortier, 1700.

Copy

Huntington 1700.

In: Sanson, N. *Nieuwe Atlas.* Amsterdam, Pieter Mortier, ca. 1700, No. 85.

Copy

American Geographical Society ca. 1700.

In: Iaillot, H. *Atlas Nouveav.* London, Sold by David Mortier, Bookseller in the Strand near ye Fountain tavern att the Seing of Erasmus S'Head, ca. 1700, I, 22.

Copy

Harvard 1700.

In: Wit, F. de. *Atlas Minor.* Amsterdam, [1634–1708], II, No. 75.

Copy

Congress 1634–1708.

In: Atlas [1722?; no title page, maps by Jaillot, de Wit, Mortier, etc.]

Copy

Yale 1722?.

In: Leth, Hendrick de. *Atlas.* Amsterdam, [1735].

Copy

Yale 1735.

In: Cóvens, J., and C. Mortier. *Atlas Nouveau.* Amsterdam, [1683–1761], IX, 57.

Copy

Congress 1683–1761.

In: Iaillot, Hubert. *Atlas Nouveav.* Amsterdam, ca. 1757, III, 96.

Copy

Clements ca. 1757.

---

121. Sanson ca. 1696 B

Carte Particuliere de la Caroline. Dresse sur les Memoires les plus Nouveaux par le Sieur S[anson] A Amsterdam. Chez Pierre Mortier, Libraire, Avec privilege de Nos Seigneurs les Etats.

Size: 23 5/8 × 19. *Scale:* 1" = ca. 2 1/2 miles.

In: Sanson, N. *Atlas Nouveav Contenant Toutes les Parties du Monde.* Paris, 1696, No. 23.

Description: This map shows the coast from the South Edisto River to the Santee River and gives the names and locations of the plantations around Charles Town. It is a copy of the Thornton-Morden ca. 1695 map, though the cartouche is different and the legends are in French. It is frequently found in atlases with Sanson's "Carte Generale de la Caroline," and was probably published along with it.

See Sanson ca. 1696 A and Thornton-Morden ca. 1685.

Reproductions

Color Plate 10.

Cartografia de Ultramar, II, Plate 43 (with transcription of place-names, pp. 251–52).

Copies

John Carter Brown, 1696, 1700; Charleston Museum, separate; Clements, separate; Harvard, two separates; Kendall, separate; New York Public, separate, General Atlas, 1700–1707; University of North Carolina, separate; University of South Carolina, two separates; Yale, separate.

In: Ablancourt, N. P. d'. *Le Neptune Francois.* Amsterdam, Pierre Mortier, 1700.

Copy

Harvard 1700.

In: De Fransche Neptunus. Amsterdam, 1693–1700, II, No. 30.

Copy

Congress 1693–1700.

In: Suite du Neptune Francois. Amsterdam, Pierre Mortier, 1700.

Copy

Huntington 1700.

In: Iaillot, H. *Atlas Nouveav.* London, Sold by David Mortier, Bookseller in the Strand near ye Fountain Tavern att the Seing of Erasmus S'Head. Ca. 1700, I, 23.

Copy

Harvard ca. 1700.

In: Wit, F. de. *Atlas Minor.* Amsterdam, [1634–1708], II, No. 74.

Copy

Congress 1634–1708.

In: Atlas [1722; no title page; maps by Jaillot, de Wit, Mortier, etc.]

Copy

Yale 1722?

In: Leth, Hendrick de. *Atlas.* Amsterdam, [1735].

Copy

Yale 1735.

In: Cóvens, J., and C. Mortier. *Atlas Nouveau.* Amsterdam, [1683–1761], IX, 58.

Copy

Congress 1683–1761.

In: Iaillot, H. *Atlas Nouveau.* Amsterdam, ca. 1757, III, 97.

Copy

Clements ca. 1757.

In: Sanson, N. *Nieuwe Atlas.* Amsterdam, Pieter Mortier, ca. 1700, No. 86.

Copy

American Geographical Society ca. 1700.

---

122. Crouch 1697

Das Englische America

Size: 2¾ × 4⅞. *Scale:* None.

In: [Crouch, N.?] *Richardi Blome Englisches America.* Leipzig, 1697, opp. p. 1.

Inset: New Found Land. *Size:* 1 × 1.

Description: This map extends from northern Florida to Maine. The Carolina region has the Lederer lake, savanna, and desert. For its relationship to the map entitled "The English Empire in America," of which it is a copy, see Crouch 1685.

Although Richard Bome's name is on the title page as the author and though the Library of Congress index card states that the text of the book is largely a translation of Blome's *The Present State of His Majesties Isles and Territories in America,* London, 1687, actually the work is a translation of R. B.'s *The English Empire in America,* with a few scattered additional sentences. R[obert] B[urton] was the pseudonym of an energetic London writer and publisher, Nathaniel Crouch (1632?–1675?). Apparently the German translator or publisher pirated Crouch's work and was misled by the initials to attribute it to Blome, who had written several works on the English empire in America.

Copies

Congress 1697; John Carter Brown 1697.

---

123. Thornton-Fisher 1689

A New Mapp of Carolina By John Thornton at y^e^ Platt in y^e^ Minor^ies^ and by Will: Fisher at y^e^ Postorn Gate on Tower hill London.

Size: 20½ × 16½. *Scale:* 1" = 18 miles.

In: The English Pilot. London, 1689, Fourth Book, No. 17.

Inset: A Large Draught of Ashly and Coopers River.

Size: 7⅞ × 5½. *Scale:* 1" = 3 miles.

Description: See no. 110A, Thornton-Fisher 1689. The Carolina chart is not found in some copies of the first edition [1689] of *The English Pilot: Book IV.*

Reproduction

Color Plate 7.

Reference

Verner, Coolie. *A Carto-Bibliographical Study of the English Pilot, Fourth Book. . . .* Charlottesville, Va., 1960.

Copies

Congress 1737, 1745, 1749; American Geographical Society 1742; John Carter Brown, 1745, 1749; Clements 1751; Harvard 1689, 1729, 1742; Kendall; New York Public 1742; University of Virginia 1737, two copies; British Library, 1689; Oxford, All Souls 1689; Massachusetts Historical Society 1689; Pennsylvania Historical Society 1689.

---

124. English Possessions 1699 MS

A Map of the English Possessions in North America and New Foundland as it was presented and Dedicated to his most Sacred Maj:^sty^ King William 1699.

Size: 15½ × 12. *Scale:* None.

Description: This map represents the Atlantic coast from Newfoundland to North Carolina just north of the Cape Fear River. The Carolina coast and interior are almost void of names; but along "Albemarle R" are fourteen red dots indicating settlements, with two or three dots for settlements on "Noratoke R" [Roanoke River] and Chowan River. The interior is decorated with mountains, and with trees colored in green.

Original

Congress.

---

125. S.C. Coast ca. 1700 MS

[Coast from Port Royall to Charles Town]

Size: 15¾ × 8⅞. *Scale:* 1" = 4 miles.

Description: This anonymous colored vellum manuscript map of the South Carolina coast between Charles Town and Port Royal is apparently the result of new soundings taken along the coast. Although neither the soundings nor the settlements shown are as full as on the Mathews ca. 1685 or Thornton-Morden-Lea ca. 1695, it was probably made a good deal later. Houses are drawn at places not shown as inhabited in 1685. Charles Town extends up the Cooper River a mile from the Point, although Johnson's Fort is not given. The map shows a fort and some buildings on Port Royal Island.

Reproductions

Hulbert, A. B. *Crown Collection.* Series I, Vol. V, No. 31.

Congress (colored photocopy).

Original

British Library. Add. MS. 5414. 16.

---

126. French ca. 1700 MS

[Carolina]

Size: 24⅜ × 20½. *Scale:* 1" = ca. 38 miles ("A Scale of Fifty English Leagues").

Description: The coastline shown extends from Cape Canaveral to Cape Henry. The map has much of the material found on the Ogilby map, particularly near the sea coast. It is a heavily colored hydrographic map. The English names (cf. scale) indicate that it was made in England or copied from an English map. It has "Ashley Lake" but none of the other Lederer details; very little is given except what is on the sea coast.

Reproduction

Karpinski series (photocopy).

Original

Paris. Service Hydrographique de la Marine. Archives. 137.1.1.

---

127. Carolina ca. 1700

A Chart of Carolina

Size: 17⅛ × 21. *Scale:* 1" = ca. 35 miles.

Inset: Ashley & Cooper River. *Size:* 7⅛ × 5⅜.

Scale: 1" = ca. 2⅓ miles.

Description: This hydrographic chart shows the coast and sounds along the coast from Cape Canaveral at 28°15' to Cape Henry at 37°. The soundings along the coast follow closely those of the Thornton-Morden-Lea map ca. 1685 as far south as Port Royal, the southern limit of the soundings on the 1685 map; from there on the markings are taken from some other source. The few details given for the interior, such as "Pleasant Valley and Lake" with a drawing of the lake on a branch of the River May, also appear to follow the 1685 map. The inset, which is unusually detailed for both soundings of Charles Town Harbour and the names of the settlers in the vicinity, is based on that part of the Thornton-Morden South Carolina ca. 1695 map which it delineates.

Reproduction

Congress (photocopy).

Copy

Kendall.

---

128. Lea 1700?

Ashley & Cooper River [at top of map] | A Perticular Map for the going into Ashley and Cooper Rivers [upper right]

Size: 5⅛ × 5⅛. *Scale:* 1" = 3⅕ miles.

In: Lea, Philip. *Hydrographia Universalis.* London, [1700?], No. 115.

Description: Lea apparently made this map by using the inset from the large "A New Map of Carolina" ca. 1685; the accidental markings on the plate are the same. Apparently Lea inked only the corner inset of the large plate, running off the copies on small sheets of paper; the left-hand part of this copy is not clearly impressed for lack of ink.

When Lea bought the map plate from Thornton and Morden (see Thornton-Morden ca. 1685), he changed the imprint and added "Ashley & Cooper River" above the inset.

Copy

Congress 1700.

129. Sanson 1700

Carte Nouvelle de L'Amerique Angloise contenant La Virginie, Mary-Land, Caroline, Pensylvania, Nouvelle Iorck, N: Iarsey N: France, et Les Terres Nouvellement Decouertes Dresse sur les Relations les Plus Nouvelles Par le Sieur S[anson] a Amsterdam Chez Pierre Mortier Libraire Avec Privilege de nos Seigneurs les Etats

Size: 35⅜ × 23½. *Scale:* 1" = ca. 95 miles.
In: Le Neptune Francois. Amsterdam, Pierre Mortier, 1700.
Description: See Morden-Brown ca. 1695 and Senex 1719. That Sanson copied the Morden-Brown map is shown by various untranslated phrases from the English map, such as "Mines of Iron" on the Ohio River and "Copper Mine" on Lake Illinois near the present site of Chicago. It is not usually found in the Sanson or Jaillot atlases and probably sold as a separate.

Reproduction
Color Plate 8.
Copies
Harvard, separate; Clements, separate.
In: Jaillot, H. *Atlas Nouveav.* London, Sold by David Mortier Bookseller in the Strand near y[e] Fountain Tavern att the Seing of Erasmus S' Head, ca. 1700, I, No. 25.
Copies
Congress; Harvard.

130. Wells 1700

A New Map of the most Considerable Plantations of the English In America Dedicated to His Highness William Duke of Gloucester [cartouche, top left] | Sutton Nichols sculp [bottom right, within latitude line]

Size: 18⅝ × 14. *Scale:* 1" = ca. 52 miles.
In: Wells, Edward. *A New Sett of Maps.* Oxford, 1700, No. 42.
Insets: 1. Carolina. *Size:* 5¼ × 5⅛.
2. New Scotland. *Size:* 4¾ × 6⅝.
3. I. of Jamaica. *Size:* 7 × 3.
4. Bermudaz. *Size:* 3½ × 3¼.
5. I. of Barbados. *Size:* 3⅛ × 3⅛.
Description: The large map extends from "Part of Carolina" (35°) to northern Maine (45°). Very little information south of New York is given except along the sea coast. The inset for Carolina has the usual river and cape names; it has "Charles T." and the "Ashley Lake" flowing into "May R," the latter a survival of the Lederer material. This atlas continued to include the map in numerous editions until as late as 1738.

Copies
Congress 1700, two copies, 1701?, 1718, [1719?], 1722, 1738, [1738?], separate; American Geographical Society 1714?, 1720?, two copies; John Carter Brown 1722?, 1738?; Clements 1700; Harvard 1738?; Kendall; New York Public 1726?, 1719?, 1738?, separate; Sondley 1700?; University of North Carolina; Yale 1700, 1722.

131. Delisle ca. 1701 MS

Carte des environs du Missisipi. Par G. de l'Isle Geogr. Donné par M[r] d'Iberville en 1701.

Size: 40 × 28¾. *Scale:* 1" = ca. 70 miles.
Description: This map represents most of the area now in the United States south of Lake Erie and east of New Mexico. Across the map from South Carolina to the Mississippi is drawn a path, with "Chemin que tiennent les Anglois de la Caroline pour venir aux Chicachas." Thus the map is notable for its early representation of the Carolina Trading path from Charlestown to "la Mobile R." (upper branches of the Tombigbee in northern Mississippi) and west to the intersection of the Arkansas River with the Mississippi River. This French map shows this trading path ten years before it is shown on the earliest English printed map (Crisp 1711); it was two decades before the Barnwell map (1721) showed Capt. Welch's trader's route to the Mississippi established in 1698.

On the "R. d'Oubache ou Akansea Sipi" (here the Tennessee River) is the legend "Route que les Francois tiennent pour se rendre a la Carolinne." The Indian settlements are given on the "Riviere des Caskinampo" (Little Tennessee River). The conception of the Alabama rivers and Appalachian range is sketchy; coastal Carolina is based on Sanson's Carte Générale de la Caroline ca. 1696. This map contains much detail also for the Indians along the Red River and upper Rio Grande.

Delisle used this map, with many changes, for his printed map: see Delisle 1703. For a discussion of the sources used in the Mississippi River area in this map and for the part of Claude Delisle, the father of Guillaume, in its preparation, see the references to Delanglez below.

This map is described in *Catalogue Général des Manuscrits des Bibliothèques Publiques de France.* Bibliothèques de la Marine Par Cu. de La Roncière, Paris, 1907, p. 229.

References

Crane, V. W. *The Southern Frontier*, p. 349.

Delanglez, J. *El Rio del Espíritu Santo.* New York, 1945, pp. 9, 134–44.

Delanglez, J. "The Sources of the Delisle Map of America, 1703." *Mid-America*, XXV (1943), 275–98.

Phillips, P. L. *Lowery Collection*, pp. 219–22.

Reproductions

Congress (photocopy).

Karpinski series (photocopy).

Original

Paris. Bibliothèque du Dépot des Cartes et Plans de la Marine. 4040. C. 4.

---

132. Moll 1701 A

The English Empire in America, Newfoundland, Canada, Hudsons Bay, &c. in plano. Herman Moll Fecit.

Size: 7 × 8½. *Scale:* 1" = ca. 420 miles.

In: Moll, H. *A System of Geography.* London, 1701, II, 161.

Description: Carolina is compressed; very little but coastal names are given for the Carolina region. The map extends all the way from Cuba to Greenland. The first two maps attributed to Moll, entitled "America" and "Europe" respectively, appeared in Sir Jonas Moore's *A New Systeme of Mathematicks Containing . . . A New Geography . . .*, in 1661. See J. N. L. Baker, "The Earliest Maps of H. Moll," *Imago Mundi*, II (1937), 16.

Copies

Congress; John Carter Brown; Harvard; Yale, two copies; Clements.

In: Moll, H. *Atlas Manuale.* London, 1709, No. 37.

Copies

Congress 1709, 1723; John Carter Brown 1723; New York Public 1709; Yale 1709.

In: Moll, H. *The Compleat Geographer.* London, 1723, p. 195; in some copies found in Part II, entitled *Thesaurus Geographicus*, is the date 1722.

Copies

John Carter Brown; New York Public; University of North Carolina; Yale.

---

133. Moll 1701 B

Mexico, or New Spain. Divided into the Audiance of Guadalayara, Mexico, and Guatemala. Florida.

Size: 7¼ × 6½. *Scale:* 1" = ca. 460 miles.

In: Moll, H. *A System of Geography.* London, 1701, II, 178.

Description: In this map, "Palasi Lake," the smaller northern lake of the Mercator type maps, empties into the Ohio River, though it is near to the headwaters of the Santee; "May Lake" empties into "R. May." Comparatively little detail is given for the Carolina region in this map, which includes Central America and the northern shore of South America.

Copies

Congress; John Carter Brown; Clements; Harvard; Yale, two copies.

In: Moll, H. *Atlas Manuale.* London, 1709, No. 38.

Copies

Congress 1709, 1723; John Carter Brown 1723; Clements 1723; New York Public 1709; Yale 1709.

In: Moll, H. *The Compleat Geographer.* London, 1723, p. 214.

Copies

John Carter Brown 1723; New York Public 1723; Yale 1723.

---

134. Delisle 1702 MS

Carte du Canada et du Mississipi Par Guillaüme Del'Isle de l'Academie Royale des Sciences 1702

Size: 31 × 21. *Scale:* 1" = 110 miles.

Description: This map extends from Cuba to Hudson Bay, including part of Mexico. Though it is smaller in size and represents two or three times as much area as does Delisle's ca. 1701 MS map, with consequent loss of detail, it is not merely a simplification. It gives an improved conception of the lower Mississippi River as the result of new information brought back by d'Iberville's expedition (1698–1700). Though it lacks the interesting legends concerning the French and English traders' routes which give such historical interest to the 1701 map, it shows the English trading route from Charles Town to the upper Cherokees, and it adds the names and locations of Indian tribes which were used in Delisle's printed map of 1703. Not all of Delisle's additions in this map are improvements: "Lac d'Ashley" flows into the Ashley River and "Lac de Theomi," here the lower lake of the Mercator-Hondius 1606 type, flows into the River May. He includes some Lederer (1672) names, though not as many as in the 1703 printed map.

Nicolas de Fer, in his *L'Atlas Curieux*, Paris, 1700–1704, has a map of the Mississippi (1701) and of North America (1702) in which he makes use of d'Iberville's reports. A copy of this atlas is in the Library of Congress. For a discussion of earlier manuscript maps of the Mississippi River environs by Nicolas de Fer and by Claude and Guillaume Delisle, see J. Delanglez, *El Rio del Espíritu Santo*, New York, 1945, pp. 9, 134–44.

Reproductions

Congress (photocopy); Karpinski series (photocopy).

Original

Paris, Ministère des Affaires Érangères. Service Géographique.

---

135. Müller 1702 A

Virginia [across face of map] | V. c[ärtlein; fifth card of America; at top of page, above map, which occupies half the page]

Size: 3 × 2½. *Scale:* 1" = ca. 350 miles.

In: Müller, J. U. *Neu-aussgefertigter Kleiner Atlas*. Ulm, 1702, America, p. 168, map 5.

Inset: I. Bermudes (lower right corner). *Size:* ¾ × ⅜.

Description: This very small map is somewhat similar to Du Val's "La Virginie" 1659. It gives the name "Apalchen" to the North Carolina region and "Florida" to the South Carolina region. According to the written description below the map on the same page, "Virginia" comprehends all the territory from Florida to "Canada," which is written in the Pennsylvania region. "L. May" is about the size of "Lac Erie," also shown on the map.

Copy

Yale 1702.

---

136. Müller 1702 B

[Florida] [across face of map] | VI. c[ärtlein; above map]

Size: 3 × 2½. *Scale:* 1" = 600 miles.

In: Müller, J. U. *Neu-aussgefertigter Kleiner Atlas*. Ulm, 1702, America, p. 170, map 6.

Description: This map is of the Sanson "La Floride" 1657 type, with "Florida Hispanica" and "Florida Francica." The large lake and "Charles fort" are given in the Carolina area.

Copy

Yale 1702.

---

137. Delisle 1703

Carte du Mexique et de la Floride des Terres Angloises et des Isles Antilles du Cours et des Environs de la Riviere de Mississipi. Dressée sur un grand nombre de memoires principalem^{t} sur ceux de M^{rs} d'Iberville et le Sueur. Par Guillaume Del'Isle Geographe de l'Academie Royale des Sciēces. A Paris Chez l'Auteur Rue des Canettes pres de St Sulpice avec Privilege du Roy po^{r} 20. ans. 1703 [cartouche, bottom left] | C Simonneau fecit [below cartouche]

Size: 25½ × 18¾. *Scale:* 1" = ca. 150 miles.

In: Delisle, G. *Atlas*, [1700–1704], No. 26.

Description: This map extends from the northern coast of South America to the Great Lakes and from New England to, and including, New Mexico. In spite of the large area shown, the map has much detail, and it influenced subsequent continental mapmakers profoundly in their delineation of the Mississippi Valley and, to a lesser extent, of the southeastern region.

It is striking to find that maps of North America did not show the interior Mississippi River drainage system in any realistic form until in the seventeenth century. Sixteenth- and early-seventeenth-century cartographers such as Diego Gutiérrez in 1562, Gerard Mercator in 1569, Corneille Wytfliet in 1597, Gabriel Tatton in 1616, Nicolas Sanson in 1656, and Pierre Du Val in 1660 (see plates in this volume), all perpetuated the myth of an east-west-trending mountain system to the north of the Gulf of Mexico. Had such a barrier existed it would have made a river with the Mississippi's size and flow an impossibility.

Conjecture, misunderstood Indian reports, and incomplete exploratory forays as a basis for portraying the Mississippi's general location and course should have ended in 1682 with La Salle's successful descent from the Illinois River to the Gulf of Mexico. Surprisingly, however, the maps based on La Salle's reports were even more distorted and inaccurate than many that antedated his exploration. What the La Salle–based maps showed was a Mississippi displaced far to the west to join the Gulf of Mexico on the Texas coast south of present-day Galveston. It was this 1703 map produced by Delisle two full decades after La Salle's discovery that first correctly located the Mississippi within North America. During the 1980s, historian Peter Wood and geographer Louis De Vorsey published conflicting arguments explaining the background and motivations which led to such an incredible time lag between La Salle's momentous geographical discovery and its appearance on maps.

"Caroline" is divided into several subdivisions, such as "C. d'Albermarle," "Vieille Virginie," "Comte de Clarendon" for northern Carolina, and "Chaouenons," "Colleton," and "Craven C" for southern Carolina. The Lederer (1672) names of Indian villages, the desert, and the savanna are apparently taken from Daniel's 1679 map, together with the inscription on the savanna, which is translated "Plaine couverte d'eau." This was misread by the engraver of the Cóvens-Mortier 1722 plate of this map as "Flame Couverte d'eau." The great lake of the southeast is put west of the Appalachians but flows through a pass into the "R. de Wallea." At "Baye St. Mateo" is the town "S. Mathieu aux Anglois."

On this map the upper lake of the Mercator 1606 type is transferred to the west of the Appalachians, is given a peculiar crescent shape, and flows through the Apalachicola River into the Gulf. The names found along the Apalachicola River derive from the discredited tales of the seventeenth-century Englishman Brigstock: see Schröter 1753.

Along the Tennessee River, together with the names of some of the Indian tribes, is a legend indicating that it was a route for French traders: "Route que les Francois tiennent pour se rendre à la Caroline." This legend, with many other details, is found on Delisle's 1701 MS map; but the path of the English Indian traders found on the same MS map is not given on the printed map. Delisle also uses some details from his preliminary study of the previous year (Delisle 1702 MS) which are not given on the 1701 MS map. He adds still other items not found on either MS map.

Besides the different states of this map listed below and the new plate issued by Cóvens and Mortier in 1722, several other mapmakers followed this map closely. Of these the most frequently found in eighteenth-century atlases is Homann's "Regni Mexicani seu Novae Hispaniæ Floridæ, Novæ Angliæ, Carolinæ, Virginiæ et Pennsylvaniæ" (*Size:* 22 1/8 × 18 5/8). Seutter's "Mappa Geographica Regionem Mexicanam et Floridam," in M. Seutter's *Atlas Novus*, Augsburg, 1745, II, 39 (Congress copy), follows Homann's map but adds four insets of cities and harbors. Tobias Conrad Lotter made a map, copies from Homann or Seutter but larger: see P. L. Phillips, *Lowery Collection*, No. 328, *Wymberley Jones De Renne Georgia Library Catalogue*, III, 1201, and Kendall Collection XLIV.c. "Carte . . . du Mexique et la Floride" in Chatelain's *Atlas Historique*, Amsterdam, 1719, is based upon Delisle's map: See Chatelain 1719.

References

Delanglez, J. *El Rio del Espíritu Santo*. New York, 1945, pp. 9, 134–44.

Delanglez, J. "The Sources of the Delisle Map of America, 1703." *Mid-America*, XXV (1943), 175–98.

De Vorsey, L. "The Impact of the La Salle Expedition of 1682 on European Cartography." *La Salle and His Legacy*, ed. Patricia K. Galloway (Jackson: University Press of Mississippi, 1982), pp. 60–77; and "La Salle's Cartography of the Lower Mississippi: Product of Error or Deception?" *Geoscience & Man*, XXV (1988), 5–23.

Wood, P. "La Salle: Discovery of a Lost Explorer," *The American History Review*, LXXXIX (1984), 294–324.

Reproduction

Plate 43.

Copies

Kendall; Yale 1700–1704.

In: Julien, R. J. *Le Théatre du Monde*. Paris, 1768, I, 58.

Copy

Congress 1768.

———A. Carte du Mexique et de la Floride des Terres Angloises et des Isles Antilles du Cours et des Environs de la Riviere de Mississipi. Dressée Sur un grand nombre de memoires principalem.t sur ceux de M.rs d'Iberville et le Sueur. Par Guillaume De l'Isle Geographe de L'Académie Royale des Sciēces. A Paris Chez l'Auteur sur le Quai de l Horlóge Privilège du Roy po.r 20 ans. 1703. [cartouche, bottom left] | C Simonneau [below cartouche]

In: Delisle, C. *Atlas de Géographie*. [Paris, 1700–1712], No. 46.

Copies

Congress 1700–1712, separate; Georgia (De Renne); Harvard.

In: Jaillot, C. H. A. *Atlas François*. Paris, 1695[–1720], No. 15.

Copies

John Carter Brown 1695–1720.

In: [A Collection of Maps, Paris, ca. 1710], No. 107.

Copy

Clements 1710.

———B. [Title same as in A; in 1745 Buache imprint added:] Ph. Buache P. G d. R. d. l'A. R. d. S. Gendre de l'Auteur. Avec Privilege de 30 an. 1745 [below neat line, right]

In: Delisle, G., and P. Buache. [*Atlas Géographique & Universel.* Paris 1700–1762], No. 76.

Copies

Congress 1700–1762, 1700–1763, two copies; Clements 1700–1763.

In: [Bellin, N.] *Cartes et Plans de l'Amerique*. 1745, No. 26 [atlas with title and date in ink].

Copy

Congress 1745.

In: Homann Heirs. *Atlas Compendiarivs*. Norimbergae, 1752[–55], No. 60.

Copies

Congress 1752–55.

---

138. Crisp 1704

A Plan of Charles Town from a survey of Edwd Crisp in 1704 [top left] | Engraved by James Akin [below title]

Size: 11⅜ × 9 (of reproduction). *Scale:* 1" = 660 feet.

Description: The title, size, and scale are taken from the engraving by James Akin which is found in Ramsey's *History of South Carolina*. The original of this map was undoubtedly similar to Edward Crisp's later 1711 inset of Charles Town in his large map, though the 1711 plan includes a larger area (Iohnsons Fort), has more details, and gives slight differences in spelling. This later 1711 map was widely copied in later plans of Charles Town, though the 1704 plan seems to be the basis for other maps, such as the inset in Moll's "Dominions" 1715. The 1704 plan, as engraved by Akin, gives the fortified area, the location of different landowners' houses, and a table of references.

The following note on this plan by J. Franklin Jameson is taken from A. S. Salley, ed., *Narratives of Early Carolina 1650–1708*, New York, 1911, p. vi: "The original cannot now be found. It is perhaps identical with a map which Dr. Ramsay describes in his *History* (II, 262) as having been preserved among the papers of the distinguished family of Prioleau. Some doubt surrounds the origin of the map. Mr. Salley finds a record in South Carolina, of date 1716, reciting a grant that had previously been made to Edward Crisp of London, but finds nothing further to identify him with South Carolina [see Introduction and Crisp 1711]. He signalizes two errors of fact in the 'References' which are placed beneath the map. N is marked as Keating L. Smith's Bridge (wharf). There was no Keating L. Smith at that time; the owner was Keating Lewis. W [behind Governor Landgrave Smith's house] is indicated as the scene of the first rice patch in Carolina; but Mr. Salley considers this to have no historical foundation. In general, however, the plan is correct."

James Akin has been identified as a Philadelphia artist whose earliest work was for Drayton's *View of South Carolina*, 1802: see Helen G. McCormack, "A Catalogue of Maps of Charlestown," *Year Book of Charleston, S.C.*, 1944, Charleston, 1947 [1948], p. 189, note 35.

Reproductions

Gabriel, R. H., ed. *The Pageant of America*. New Haven, 1926, III. 55.

Ramsey, David. *History of North Carolina*. Charleston, 1809, II, opp. p. 3.

Reps, John W. *Town Planning in Frontier America*. Princeton, 1969, p. 227.

Salley, A. S., ed. *Narratives of Early Carolina 1650–1708*. New York, 1911, opp. p. 364.

---

139. Sanson 1705

Florida zoo als het van de Spaanschen en Franschen wordt bezeten, door N. Sanson, geog^r Ord^re du Roÿ.

Size: 9⅞ × 7. *Scale:* 1" = ca. 185 miles.

In: Phérotée de La Croix, A. *Algemeene Weereld-Beschryving*. Amsterdam, 1705, III, between pp. 340–41.

Description: For this map the plate of the Sanson 1683 map of Florida has been used, though all except the last line of the title has been changed. The map has longitude and latitude marks and additional trees which were added to the plate in 1700.

See Sanson 1683.

See P. L. Phillips, *Lowery Collection*, p. 224; De Renne Wymberley Jones Georgia Library Catalogue, III, 1195 (erroneously states this to be an exact copy of Sanson's 1657 map).

Copies

Congress 1705; Georgia (De Renne).

---

140. Loots 1706

Caerte vande Cust Carolina Tusschen B. de S. Matheo en C. Henry op Niew Verbeetert door Ioannes Loots mel [*sic*; for *met*] Voor 15 Iaaren A° 1706

Size: 20¾ × 16. *Scale:* 1" = ca. 28 miles.

In: Jacobsz, Theunis, and John Loots. *The Fifth Part of the new great Sea-Mirrour*. . . . Translated . . . by Ericus Walton. Amsterdam, 1717, between pp. 46–47.

Inset: Le Partie du Carolina Grand point. *Size:* 6½ × 7½. *Scale:* 1" = ca. 3 miles.

Description: "A° 1706" may have been a later addition; it puts the title line out of symmetry and the lines are heavier. This hydrographic map is based on van Keulen 1682, though it does not cover as large an area. It extends from "B. St Paule" (30°20') to "C Henry" (36°50'). The inset, which is below the cartouche on the left of the map, extends from "Charles Towne" to "C. de S. Romano."

Copy

Congress 1717.

---

141. Aa 1706 A

De Vaste Kust Van Chicora Tussen Florida on Virginie Door Lucas Vasquez d'Ayllon En Andere, van Hispaniola Besterend. [cartouche] | Uytgeroerd to Leyden door Pieter vander Aa med Privilegie [below neat line, left] | Alonzo d'Ojeda [below neat line, right] | Pag: 89 [within neat line, upper right]

Size: 8⅞ × 6 (with ornamental border: 13¼ × 8½). *Scale:* 1" = ca. 210 miles.

In: Aa, P. vander. *Naaukerige Versameling der Gedenk-Waardigste Reysen na Oost en West-Indiën*. Leyden, 1707, IX (1706), between pp. 88–89.

Description: This map, which includes the islands of the Greater Antilles, Florida, and the east coast to Cape Hatteras (35° N.L.), illustrates the territory covered in the expeditions of Ayllón.

The country of Chicora covers roughly the area of South Carolina and Georgia; "Chicora," an Indian town, is placed on the "R. Jordan"; Duare is a smaller country to the south of Virginia. Apalache is a land to the north of "Lac Grande," which is the great lake of the Southeast found on an unnamed river which flows down to "C. de S. Helena."

Another map which is usually found in the same volumes which contain this map and which illustrate the voyages of Franciscus de Garay is "Scheeps togt Van Jamaica"; "Chicora" is given on this map also in the South Carolina region.

Reproductions

Florida Historical Pageant. Official Program. Jacksonville, 1922, p. 29.

Mann, F. A., ed. *Sunny Lands*. Saint Augustine, Florida, I (January 1900), No. 1, p. 17.

Copies

John Carter Brown 1706; Harvard 1706; New York Public 1706.

In: Gottfried, Johann L. *De Aanmerkenswaardigste en Alomberoemde Zeeen Landreizen . . . van de Cost-en Westindiën*. Leyden, 1706–27, IV (location in volume varies).

Copies

John Carter Brown 1706; Clements 1706; New York Public 1706; Yale 1706.

In: Aa, Pieter vander. *Cartes Des Itineraires & Voïages Modern*. Leiden, [1707], No. 94.

Copy

Clements ca. 1710.

———A. [Same title in cartouche; number in upper right within neat line erased; below map, within border:] Terre-Ferme de Chicora, entre la Floride et la Virginie, décrite par Lucas Vasquez d'Ayllon sur les Voyages de Don Alonso d'Ojeda, et d'autres qui y ont navigé de l'Ile Hispaniola présentement mise en lumiere par Pierre vander Aa, à Leide. Avec Privilege.

In: Aa, Pieter vander. *Atlas Nouveau et Curieux Des plus Célèbres Itinéraires*. Leide, [1714], II, No. 33.

Copies

Congress 1714; John Carter Brown 1714.

142. Aa 1706 B

'T Amerikaans Gewest van Florida Door Ferdinand de Soto Nader Ontdekt en Groot deels Bemagtigd. | Uytgeroerd te Leyden door Pieter vander Aa met Privilegie [below neat line] | Ferdinand de Soto. | [above neat line, upper right] Pag: 1

Size: 9 × 6. *Scale:* 1" = ca. 210 miles.

In: Aa, Pieter vander, *Naaukeurige Versameling der Gedenk-Waardigste Zee en Land-Reysen na Oost en West-Indiën.* Leyden, 1707, XIV (1706), opp. p. 1.

Description: This is a map of the Sanson (1657) type, extending from "Lac Erie" to "Tampico" in Mexico. The whole country is called Florida; the eastern coast has "Virginia," "Apalache," and "Tegesta Prov." (for the Florida peninsula); "Carolina" and "Chicora" are smaller regions in the country of "Apalache." The City of "Carolina" is shown at Port Royal, and the "L. Grande" is given. Though the volume is dated 1707, the separate journeys, each with its own title page, are dated 1706.

An earlier edition of P. Richelet's translation of Vega's *Histoire* was published in Paris by G. Nyon in 1709 (the title page of the second volume has 1707) without the map. A copy of this edition is in the Harvard College Library.

Copies

John Carter Brown 1706; Harvard 1706; New York Public 1706.

In: Gottfried, Johann L. *De Aanmerkenswaardigste en Alomberoemde Zeeen Landreizen . . . van de Oost-en Westindiën.* Leyden, 1706–27. VII, opp. p. 4 (in the de Soto section, which is dated 1706).

Copies

John Carter Brown 1706; Clements 1706; Kendall 1706; New York Public 1706; Yale 1706.

———A. [Same title in cartouche; no number above neat line]

In: Aa, Pieter vander. *Cartes Des Itineraires & Voïages Modernes Leide*, [1707], No. 63.

Copies

Clements ca. 1707; Harvard ca. 1710.

———A. [Same title in cartouche; below map, within border:] La Floride, Grand Païs de l'Amerique Septentrionale, plus avant decouverte et presque toute conquise par Ferdinand de Soto en 1534. tirée de ses Mémoires, et de tous ceux qui ont paru jusqu'à présent, nouvellement rendue publique par Pierre vander Aa, à Leide, Avec Privilege.]

In: Aa, P. vander. *Atlas Nouveau.* Leide, [1714], II, No. 32. (The ornamental border around the map measures 11¼ × 8½.)

Reproduction

Cash, W. T. *The Story of Florida.* New York, 1938, I, 3 and 12.

Copies

Congress 1714, two copies; John Carter Brown 1714.

———B. [Title only; no imprint, page number, or border]

In: Vega, Garcilasso de la. *Histoire de la Conquête de la Floride.* Leide, Pierre vander Aa, 1731, II, front.

Copies

Congress; American Geographical Society; John Carter Brown; Clements; Harvard; New York Public; Sondley Reference.

143. Aa 1706 C

Zee en Land Togten der Franszen Gedaan na, en in't Americaans Gewest van Florida, allereerst door Ioh, Pontius Ontdekt. [cartouche, bottom right] | Uytgeroerd te Leyden door Pieter vander Aa met Privilegie [under neat line] | Pontius | Pag: 21 [above neat line, upper right]

Size: 9⅛ × 6¹⁄₁₆. *Scale:* 1" = ca. 65 miles.

In: Aa, Pieter vander. *Naaukeurige Versameling der Gedenk-Waardigste Zee en Land-Reysen na Oost en West-Indiën.* Leyden, 1707, XVI (1706), opp. p. 21.

Description: This map is of the Mercator-Hondius type, with a few later influences. There is no waterfall into the smaller, upper lake. "Chicola" is a town on the "R Iordaan." The two French forts built by Ribaut in 1562 and Laudonnière in 1564 are both indicated by the title "Karel Slot."

The map is found in volume 16 of Aa's *Naaukerige Versameling*, in a section having a title page "Versheyde scheepstogten na Florida, door Pontius, gedaan in het jaar 1562, en vervolgens. Te Leyden, P. vander Aa, 1706." The general title page for the volume has the date 1707.

Pontius in the map title refers to Ponce de León.

The 1714 state of this map has a large ornamental border, measuring 11⅛ × 8½.

Copies

John Carter Brown 1706; Georgia (De Renne) 1706; Harvard 1706; New York Public 1706.

In: Gottfried, Johann L. *De Aanmerkenswaardigste en Alomberoemde Zeeen Landreizen . . . van de Oost-en Westindiën.* Leyden, 1706–27, VII (location in volume varies).

Copies

John Carter Brown 1706; Clements 1706; New York Public 1706; Yale 1706.

———A. [Same cartouche title, but with no numbering above neat line.]

In: Aa, Pieter vander. *Cartes Des Itineraires & Voïages Modernes.* Leide, [1707], No. 66.

Copies

Clements ca. 1707; Harvard ca. 1710.

———B. [Same title in cartouche; no numbering above neat line; below map within border:] Voyages par Mer et par Terre des François dans la Floride, prémierement découverte par Jean Pontius, dressez sur ses Mémoires, et perfectionez par un grand nombre d'autres plus recens, de nouveau mis au jour par Pierre vander Aa, à Leide, Avec Privilege.

In: Aa, P. vander, *Atlas Nouveau.* Leide, [1714], II, No. 31.

Copies

Congress 1714, two copies; John Carter Brown 1714.

---

144. Thornton 1706

A Large Draft of South Carolina from Cape Roman to Port Royall. By John Thornton Hydrographer at the Signe of England Scotland and Ireland in the Minories London

Size: 21½ × 17⅜. *Scale:* 1" = 5 miles.

In: The English Pilot. London, 1706, Fourth Book, No. 16, between pp. 24–25.

Description: This hydrographic map gives the soundings and indicates some of the settlements near the coast. It apparently uses some information taken verbatim from the Mathews MS ca. 1685 and not used before, such as "These Creeks are Passable for Canoes." Many details of the map, however, are evidently drawn from new information and new soundings along the coast.

The same plate, with the same accidental markings, was used for the maps by Samuel Thornton and by W. and I. Mount and T. Page, listed below. For further description of the *English Pilot, Book IV* see under No. 110A, Thornton-Fisher 1689 and references there given.

Copy

Congress 1706.

———A. A Large Draft of South Carolina from Cape Roman to Port Royall by Sam[l] Thornton Hydrographer at the Signe of England Scotland and Ireland in the Minories London.

In: The English Pilot. London, 1706, Fourth Book, between pp. 24 and 25.

Copy

John Carter Brown 1706.

———B. A Large Draft of South Carolina from Cape Roman to Port Royall Sold by W. & I. Mount & T. Page, on Tower Hill London.

In: The English Pilot. London, [1755?], No. 14.

Copies

Congress 1755?, 1758, 1759, 1760, 1767, 1773, 1775; John Carter Brown 1758; Charleston Museum; Harvard 1765, 1770, 1773, 1778, separate; New York Public 1764; University of Virginia 1760; Yale 1760; Clements, 1758.

---

145. Thornton–Port Royall 1706

A Large Draught of Port Royall Harbour in Carolina. By Iohn Thornton at the Platt in the Minories London.

Size: 9⅝ × 16⅜. *Scale:* 1" = ca. 3⅕ miles.

In: The English Pilot. London, 1706, Fourth Book, No. 15.

Description: A comparison of the topography and soundings on this map with the *English Pilot* "New Mapp of Carolina" 1689 shows that this map is based on entirely new coastal information. The map listed below is from the same plate as this, but Iohn Thornton has been changed in the title to Sam[l] Thornton. The Port Royall Chart appears only in the 1706 *English Pilot.* See Coolie Verner, *The English Pilot, 1689*, Amsterdam, 1967, p. xix.

Copy

Congress 1706.

———A. A Large Draught of Port Royall Harbour in Carolina By Sam[l] Thornton at the Platt in the Minories London

In: The English Pilot. London, 1706, between pp. 23 and 24.

Copy

John Carter Brown 1706.

---

146. Lamhatty 1708 MS

With "Mr. Robert Beverley's Acco.[t] of Lamhatty" [on reverse of sheet].

Size: 12 × 10½. *Scale:* None.

Description: This map appears to be a contemporary copy of an original drawn by Lamhatty, a Towasa Indian, for Colonel John Walker, a resident of King and Queen County, Virginia, in 1708.

According to a letter written by Walker, an unarmed Lamhatty had sought refuge at a frontier settler's cabin on January 3, 1708. After being brought to Walker's on the following day and showing no hostility, he was befriended by the Virginian, who encouraged him to recount the story of his capture nine months earlier on the Florida Gulf coast by a Tuscarora war party. To make clear the route of his long odyssey Lamhatty prepared a map that Walker described as "all of his own drawing." He went on to explain the Indian's cartographic conventions by pointing out that "ye red line denotes his march, ye black lines, ye Rivers, & ye shaded lines ye mountains."

Virginia historian and frontier land speculator Robert Beverly probably prepared this copy of Lamhatty's map, and his account of the Indian's captivity added details absent from Walker's letter. Work by twentieth-century anthropologists has revealed Lamhatty's map to be a rich source of rare ethnohistorical information on the Indian societies of the Southeast in the opening years of the eighteenth century. David L. Bushnell, who first mentioned the map in print, and Gregory A. Waselkov attempted to identify the many Indian villages, rivers, and other features shown on Lamhatty's map.

References

Bushnell, D. I., Jr. "The Account of Lamhatty." *American Anthropologist*, N.S. (October–December 1908), X, 568–74.

Stout, W. W. "Lamhatty's Road Map." *The Southern Quarterly*, II (April 1964), pp. 247–54.

Swanton, J. R. *Early History of the Creek Indians.* Washington, 1922, pp. 13, 130, 138.

Waskekov, Gregory A. "Indian Maps of the Southeast." *Powhatatan's Mantle*, eds. Peter H. Wood, Gregory A. Waselkov, and M. Thomas Hatley, Lincoln, 1989, pp. 313–20.

Reproductions

Plate 43A.

Bushnell, D. I., op. cit., between pp. 570–71 (Plate XXXV).

Volmar, R. *Indianische Karten Nordamerikas.* Berlin, 1981, p. 48.

Waselkov, G. A., op. cit., Figure 2, pp. 296, 314.

Original

Virginia Historical Society, Richmond, Va. (Lee Family Papers, MS 1L51f677.)

---

147. Low 1708 MS

[Plan of the town of Low Wickham, Precinct of Pasquotank]

Size: 23⅝ × 20⅞. *Scale:* None.

Description: At top center: "Land in the Possession of Emanuel Low Esq[r]"

Below, in cartouche: "North Carolina Surveyed at the Instance and Request of Emanuel Low Esq[r] One of the True and Absolute Lords Proprietors of Carolina a parcell of Land containing 350 [?] Acres designed for a Township by the Name of Low Wickham lying on the S[o] W[t] side of the Mouth of New Begun Creek in the Precinct of Pasquotank Beginning at a Red Oak by the creek side running thence N[o] 35° E[t] [120] Pole to a White Gum, then S[o] 59 F[t] 30 Pole to a White Oak, then N[o] 50 E[t] 60 Pole to a Red Oak by the side of the Pasquotank River Thence down the River to the Creeks mouth and up the [creek] to the First Station Certified this 20th of [Sep]tember Anno D[ni] 1708 By [illegible] Gener[ll]"

At the top right of the map, in ink: "Presented to the State Library Col. R B Creecy of Elizabeth City, N.C. May 7th 1888."

On August 3, 1716, Emanuel Low petitioned that his legal title to a tract of land of 1,006 acres at the mouth of New Begun Creek be granted him. According to his petition the original patent for the land was taken out in his name, before he had become an inhabitant of North Carolina, by his father-in-law, John Archdale, then governor of Carolina: *Colonial Records of North Carolina*, II, 242. In 1711 and 1712 Low was engaged in Cory's Rebellion and was at first not allowed bond when others were released, because he had seized and hidden the records of the Land Office. *CRNC*, II, 864, 873, 792.

Original

Unfortunately this vellum map disintegrated during an unsuccessful attempt to preserve it by lamination.

148. Moll-Oldmixon 1708

Carolina By Herman Moll Geographer Note that y[e] plantations are marked thus ΙΙ [cartouche, bottom right] | Vol. I. pag 325 [upper left, above neat line]

Size: 6 1/8 × 6 7/8. *Scale:* 1" = ca. 70 miles.

In: Oldmixon, J. *British Empire in America.* London, 1708, I, opp. p. 325; opp. p. 456 in 1741 edition.

Description: No new information is on this map, which follows in general the Gascoyne 1682 map. On the same folding sheet is another map: "A Map of y[e] Island of Bermudos, Divided into its tribes, wth the Castles, Forts, &c. By H. M." In 1741 and thereafter the volume and page reference are deleted. In Bowles's *Atlas Minor*, London, ca. 1781, "53" was engraved in the top right corner of the adjacent Bermudos map. The same map is found, with a different title, and from a different plate, in the Dutch translation, *Het Britannische Ryk in Amerikâ*, Amsterdam, 1721, I, opp. p. 248; see Moll-Kyser 1721. There are no maps in *Brittanisches Amerika*, 1710, a German translation of Oldmixon; two general maps, but no Carolina maps, are in the German edition of 1756.

Copies

Congress 1708, separate; American Geographical Society 1708, 1741; John Carter Brown 1708, 1741; Clements 1708; Harvard 1708; New York Public 1708, only Bermuda part in second copy; North Carolina State 1708; Sondley 1741; University of North Carolina 1741; Yale 1708, 1741.

In: Moll, H. *Forty-two New Maps of Asia, Africa, and America.* London, 1716, No. 42.

Copy

Congress 1716.

In: Atlas Geographicus: or, A Compleat System of Geography. London, 1711–17, VI (1717), opp. p. 688, 692, or 694.

Copies

Congress 1717; John Carter Brown 1717; Clements 1717; New York Public 1717; Sondley 1717; Yale 1717.

In: Moll, H. *Bowles Atlas Minor.* London, for C. Bowles, [1718?], No. 53.

Copies

Congress 1781?, 1790?; Clements 1781?; New York Public 1781?.

149. Kocherthal 1709

Virginia Nord Carolina Sud Carolina [across face of map]

Size: 7 3/16 × 6 1/16. *Scale:* 1" = ca. 70 miles.

In: Kocherthal, Josue von. *Aussführlich und umständlicher Bericht Von der berühmten Landschafft Carolina In dem Engelländischen America . . . Vierter Druck*, Franckfurt, 1709, opp. p. 6.

Description: This map extends from Maryland to the mouth of "May Fl." It has "Nord Carolina" and "Sud Carolina," with a very large "Apalache Lac" to the west of "Sud Carolina." "Ahssli Fl" flows down past "Carls Thon" to "C. Feare"; this is a confusion of the Charles Town in South Carolina with the short-lived settlement of the same name on the Cape Fear, abandoned in 1667. The same error was first made on the 1672 map of Carolina by Blome.

The origin of the peculiar geography of Kocherthal's map is not clear, but it may be derived from an earlier map by Visscher. Nikolas Visscher (1649–1709) has a large map of eastern North America with the title "Nova Tabula Geographica complectens Borealiorem Americæ partem; in qua exacte delineatæ sunt Canada sive Nova Francia, Nova Scotia, Nova Anglia, Novum Belgium, Pensylvania, Virginia, Carolina, et Terra Nova, cum Omnibus Littorum. Pulvinorumque Profunditatibus. Amstlodami [*sic*] â Nicolao Visscher. Cum Privilegio Ordinum Generalium Fœderati Belgii. [across top of left-hand sheet; same title in French on right-hand sheet]. | Luggardus van Anse Schulp. [bottom center]. *Size:* Two sheets, each 17 5/8 × 23 1/4. *Scale:* 1" = ca. 60 miles. This map, which is found both loose and in several different atlases, has the large "Apalache Lac," the town of "Carolina" at Cape Feare, and the much enlarged North Carolina coastline which is found in Homann's "Virginia, Marylandia, et Carolina," 1714 and (except for the town of "Carolina") in Kocherthal's map. Visscher, the heir of a line of mapmakers founded in 1616, died in 1709. His wife carried on the business until 1717, when the stock passed into the hands of P. Schenk: see Feil, J., *Über das Leben und Wirken des Geographen Georg Matthaeus Visscher*, Wien, A. Pichler's witwe und sohn, 1857. Thus, while Visscher atlases were sold after that date, 1709 is the latest Visscher himself could have made this map. Kocherthal's map shows no great skill or originality; Homann's has added a great deal of later information about the German migrations in Pennsylvania and down the Valley of Virginia. Therefore, it is probable that Kocherthal and Homann borrowed the lower section of this rather common map by Visscher (which he perhaps engraved in the first years of the eighteenth century) and changed details according to their own new information. If this is ture, Visscher apparently obtained his information from some earlier map, following the old fallacy of the great lake flowing into the "R. May," and adding a mélange of misapplied information from later sources: "Clarinon" and "Iourdain" are the upper branches of the Cape Feare; the "Waitree or Winae Riv." and the "Seyne River" join at the town of "Sante" north of "C Romanie" and west of "Craven County." Visscher apparently used some unidentified English map for his authority; such phrases as "The Sea of Virginny" and "The Part of Hudsons Bay" indicate at least an ultimate English source.

Kocherthal's book is partly based on "Erster anhang aus Richard Blome, Englischen America," pp. 40–65, for its description of Carolina. *Englisches America*, Leipzig, 1697, may have been written by Nathaniel Crouch, alias Robert Burton: see under Crouch 1685. Kocherthal's work was an optimistic account of Carolina and of the fortunes of the German emigrants there. Two years later an attack on this book by Anton Wilhelm Boehme, *Das verlangte nicht erlangte Canaan*, published at Frankfort and Leipzig, 1711, contained contrary accounts of the life of the Germans in Carolina and Pennsylvania and bitingly questioned the truth of Kocherthal's statements. Boehme was the pastor of the German Court Chapel of St. James: see *Pennsylvania German Society Proceedings and Addresses*, VII (1896), 47ff.

Kocherthal issued his volume first in 1706, though he did not go to New York before 1708; the work proved so popular that by 1709 it had reached its fourth edition. Not only does Kocherthal's book appear to have been one of the most powerful inducements in the great German immigration toward Pennsylvania, starting with 50 under Kocherthal in 1708 and increasing to 10,000 without a leader by 1709; it is the only book mentioned by the Swiss immigrants to New Bern under von Graffenried. See Todd, V. H., *Christoph von Graffenried's Account of the Founding of New Bern.* Publications of the North Carolina Historical Commission, Raleigh, 1920, 13–14.

Copies

Congress 1709; John Carter Brown 1709, two copies; Duke 1709; Historical Society of Pennsylvania 1709.

---

150. Lawson 1709

To His Excellency William Lord Palatine; The most Noble Henry Duke of Beaufort; The Right Hono^ble^. Iohn, Lord Carteret; The Hono^ble^ Maurice Ashley, Esq., Sr John Colleton Baronet; Iohn Danson, Esq; And the rest of the True and Absolute Lords Proprietors of Carolina in America This Map is Humbly Dedicated by Io^n^ Lawson Surveyor General of North Carolina. 1709 [cartouche, bottom right] | Iohn Senex sculpsit [below cartouche]

Size: 12 × 15. *Scale:* 1" = ca. 33 miles.
In: Lawson, J. *A New Voyage to Carolina.* London, 1709, opp. p. 60. [From: Stevens, Capt. John]. *A New Collection of Voyages and Travels.* London, 1708–9.

Description: The great seal of Carolina appears above the title cartouche in the lower right corner of the map. Though the coastline extends from below "St. Augustin" at 29° to "Chisapeak Bay" at 37°, very little of the South Carolina region except the seacoast is given. "R. May" is shown with "A Pleasant Valley with a Lake" at the latitude of Cape Fear; but neither its headwaters nor those of "S. Matheo R." are shown. Thus the question of the existence of the great lake of Apalache or Ushery is avoided. To the interior of the North Carolina region unnamed hills dot the western portion, with vague notes, such as "Hilly Land," "Rich Land," "Marble Rocks & Free stone"; but in this part of the map no new features of importance are given, in spite of Lawson's title of "Surveyor General of North Carolina." Numerous topographical similarities show that Lawson relied chiefly on the Thornton-Morden-Lea map (ca. 1685). For the South Carolina part Lawson omitted many details of settlements found in the Thornton map; his contribution to the cartographical knowledge of the region is found in detailed names of creeks, rivers, and settlements in the vicinity of Pamlico Sound. "Bath" is named for the first time on a printed map, and the shape, soundings, and sand bars of Pamlico Sound are given. It cannot be said that Lawson's map added much information of value, though it retained several old inaccuracies and included some new ones.

His book, first issued under the title *A New Voyage to Carolina*, in John Stevens's *A New Collection of Voyages and Travels*,

London, 1708–10, and dated May 1709, was reprinted as a separate volume in the same year, 1709, and again in 1711. Though the 1711 edition refers to the plate and map in the title, the copies examined in the Yale Library, the New York Public Library, and the Kendall Collection have no map. A collected edition of Stevens's *A New Collection of Voyages and Travels* was published in 1711, as in the Newberry Library copy; the date on the title page of Lawson's *A New Voyage to Carolina* in this edition, however, remains 1709. Mr. Sabin said that he had seen two copies of a large paper edition of Lawson's *A New Voyage*: see Thomas L. Bradford, *The Bibliographer's Manual of American History*, edited and revised by S. V. Henkels, Philadelphia, 1907, II, 311. The John Carter Brown Library has the Edmund Paley dedication copy of Stevens's *Collection*, 1711, in a large paper edition; Wake Forest College, North Carolina, also has a large paper edition. The map itself is unchanged. The work was published under a new title, *The History of Carolina*, in 1714 and 1718. A new map was made for the 1718 edition, printed after Lawson's death; it is a strange retrogression to the old errors of Lederer. German editions of Lawson's *History of North Carolina* appeared in 1712 and 1722. The German edition of 1712 has a map similar to that of the 1709 edition, but from a different plate.

Lawson reached Charles Town in August 1700, and soon thereafter began an expedition to the interior of the colony. After he had followed the Indian Trading Path to the site of Hillsborough, N.C., he turned eastward to the white settlements of the colony. In 1705 he was one of the persons who secured the incorporation of the town of Bath. He was made surveyor general of North Carolina in 1708; in 1709, while in London, he met the Swiss colonizer, Christopher de Graffenried. Lawson aided in the migration of 600 Palatines to New Bern, N.C. In September 1711, while on an exploring expedition, he was seized by the Tuscarora Indians, tortured, and killed, though his companion de Graffenried escaped.

For a bibliographical collection of various editions of Lawson's work, see James C. Pilling, *Bibliography of Iroquoian Languages*, Bur. of Amer. Ethn., Washington 1886, No. 6, pp. 107–9.

See also under Lawson 1718 and Brickell 1737.

Reproductions

Avery, E. M. *A History of the United States*. Cleveland, 1904–7, III, between pp. 224 and 225.

Hawks, F. L. *History of North Carolina*. Fayetteville, N.C., 1859, II, opp. p. 104. (This facsimile of only the North Carolina coastal section of the map is also found in S. H. Ashe, *History of North Carolina*, Greensboro, N.C., 1908, opp. p. 169.)

Indians of North Carolina. 63rd Congress, Senate Document No. 677. Washington, 1915, opp. p. 100.

Lawson, John. *History of North Carolina*. Edited by F. L. Harriss. Richmond, Va., 1937, frontispiece.

Lawson, John. *A New Voyage to Carolina*. Edited by H. T. Lefler. Chapel Hill, N.C., 1967, p. xxxviii.

Copies

John Carter Brown 1709, 1711 large paper edition; Newberry 1711; New York Public 1709; University of North Carolina; Wake Forest College 1711 large paper edition.

In: Lawson, John. *A New Voyage to Carolina*. London, 1709, opp. p. 60.

Copies

Congress 1709; John Carter Brown 1709; Archibald Craige, Winston-Salem, N.C., 1709; Clements 1709; Duke 1709; North Carolina State 1709; Yale 1709.

In: Lawson, J. *The History of Carolina*. London, Printed for W. Taylor . . . and J. Baker, 1714, opp. p. 60.

Copies

Congress 1714; Archibald Craige, Winston-Salem, N.C., 1714; Newberry 1714.

In: Lawson, J. *The History of Carolina*. London, Printed for T. Warner, 1718, front.

Copies

Congress 1718 (insert); University of North Carolina 1718 (?insert).

---

151. Crisp [1711]

A Compleat Description of the Province of Carolina in 3 Parts. 1st The Improved part from the Surveys of Maurice Mathews & Mr John Love. 2ly the West part by Capt Tho. Nairn 3ly A Chart of the Coast from Virginia to Cape Florida. Published by Edw. Crisp. [Title in two lines above neat line of map.] | To His Excellency William Lord Craven Palatine The Most Noble Henry Duke of Beaufort The Rt Honble Lord Carteret. The Honble Maurice Ashley Esq Sr Iohn Colleton Baronet Iohn Danson Esqr And the rest of the True and Absolute Lds & Proprietors of the Province of Carolina This Mapp is Humbly Dedicated By Edw. Crisp [in cartouche, top center] | This Flourishing Province Produces Wine, Silk, Cotton, Indigo, Rice, Pitch, Tar, Drugs, & other Valuable Comodityes. Sold at the Carolina Coffee House inn Birchen Lane London. [in cartouche, below scale] | Engraven by Iohn Harris in Bull-head Court Newgate Street London. [below cartouche] [This is the main map, running diagonally from lower left to upper right.]

Size: 38¾ × 30. *Scale:* 1" = 1½ miles.
Insets: 1. A Map of South Carolina Shewing the Settlements of the English, French, & Indian Nations from Charles Town to the River Missisipi by Capt. Tho. Nairn. [upper left corner] *Size:* 16½ × 8¾. *Scale:* 1" = 120 miles.

2. The Town and Harbour of St Augustin [inset upper left] *Size:* 7½ × 5¾. *Scale:* 1" = 1¾ miles.

3. A Plan of the Town & Harbour of Charles=Town [inset right center] *Size:* 10½ × 11 (irregular shape, with lettered and numbered references a–w and 1–15). *Scale:* 1" = 650 feet.

4. A New Chart of the Coast of Carolina and Florida from Cape Henry to the Havana in the Island of Cuba Described by Cap Tho. Nairn and others [inset bottom right] *Size:* 22 × 10½. *Scale:* 1" = 42 miles.

Description: This large map, with its detailed central part giving the names and location of nearly 300 owners of land in the inhabited region of South Carolina, with its insets of Charles-Town and St. Augustin, with its chart of the seacoast from Virginia to Cuba, and with its delineation of the Southeast by Captain Nairne, is one of the most important maps in the cartography of the region.

From the central map of South Carolina came Moll 1715 (Carolina inset), Moll 1730, Homann 1730; from Crisp's Charles-Town inset, or its original, the no longer extant Crisp 1704, were derived most of the maps of Charles-Town for the next forty years; Nairne's inset of the Southeast was the type map for Moll 1715 (southeastern inset) and Georgia 1732, used in the promotion literature of the Georgia Colony. The conception of southern Florida as an archipelago, found in the two insets by Nairne, continued to be found occasionally in maps for the next half century, as in the Scacciati-Pazzi 1763 map. The general value of the Crisp map lies in its representation of the extent of settlements and the expansion of knowledge of the back country. The map is printed on five sheets of varying size, including the long title strip at the top, which are pasted together.

It is undated. P. L. Phillips, *Lowery Collection*, p. 236, assigns the date 1711 to the map for the following reasons:

"Firstly, because William, Lord Craven to whom it is dedicated was chosen Lord Palatine on jan. 16, 1707–8 and died oct. 9, 1711.

Secondly, because in the original m.s. minute book of the Proprietors of North Carolina, vol. 7 N.C., preserved in the Public record office, London, appears the following entry: 'June 1711 ordered that mr E. Crisp be paid 10 guineas for his map of Carolina and the draught of Port Royal. Order'd that six hundred acres of land be directed to be admeasured and set out for the said Crisp in South Carolina, signed by the Lord Carteret for himself and the Palatine.'"

See Introductory Essay.

References

John Carter Brown. *Annual Report*, 1951, pp. 10–17.
Karpinski, L. C. *Early Maps of Carolina*. Charleston, 1937, pp. 32–33, No. 25.
McCormack, Helen G. "A Catalogue of the Maps of Charleston," Charleston, S.C., *Year Book 1944*. Charleston, 1947 [1948], pp. 185–86, No. 11.
Moore, Alexander, ed. *Nairne's Muskhogean Journals: The 1708 Expedition to the Mississippi River*. Jackson, 1988.
Phillips, P. L. *Lowery Collection*, Washington, 1912, pp. 235–36.

Reproductions

Plate 44 (Charles Town inset); Plate 45 (southeastern inset).
Congress (photocopy).

Copies

Congress; John Carter Brown; Kendall; Huntington; British Public Record Office.

---

152. Ochs 1711

Die Provinz Nord und Sud Carolina. [cartouche, bottom right] | Ioh. Hen. sculp. Tiguri [below neat line, right]

Size: 8½ × 10⅜. *Scale:* 1" = 50 miles.
In: Ochs, Johann Rudolf. *Americanischer Wegweiser*. Bern, 1711.

Description: This map, which appears in a guidebook for prospective Swiss settlers coming to America and especially to Carolina, is based on Lawson's 1709 map. The legends are in German and are found chiefly at the foot of the mountains, as they are in the original map: "kein Schleckt Land hier," "Eysen Minen," "Marmor Steinfels Vnd Ledig Stein."

Copies

John Carter Brown; Duke, New York Public; Historical Society of Pennsylvania, Philadelphia; University of North Carolina.

---

153. Lawson-Visher 1712

Die vornehmste Êigenthums Herren und Besitzer von Carolina. Lord Craven. Herzog von Beavfort. Lord Carteret. Ashley, Esq₃. Colleton, Baronet. Danson, Esq₃. etc. etc.

Size: 15 × 12. *Scale:* 1" = ca. 33 miles.

In: [Lawson, John] *Allerneuste Beschreibung der Provintz Carolina . . . aus dem Englischen übersetzet durch M. Visher.* Hamburg, 1712, opp. p. 1.

Description: This is a copy of the Lawson 1709 map.

Copies

Congress; John Carter Brown; Duke; Harvard; New York Public; University of North Carolina.

---

154. Fort Nohucke ca. 1713 MS

[Colonel Moore's attack on Fort Nohucke]

Size: 25¼ × 16½. *Scale:* None.

Description: An unsigned, undated map. Colonel James Moore's successful attack on the hostile Tuscaroras under Chief Hancock at Fort Nohucke (Nocherooka in the plan) began on March 20, 1713. The friendly Tuscaroras under Tom Blount aided in the siege, which was completely successful. Eight hundred of Hancock's followers were captured and many were slain.

The map gives a plan of the Indian fort and vicinity, together with legends and a full account of the fight. The original treaty of September 1712 between Thomas Pollock, President of the Council in North Carolina, and Tom Blount, a chief of the friendly Tuscarora element among the Indians, is now in the Hall of History, Raleigh, N.C.

Reproduction

Gabriel, R. H., ed. *The Pageant of America*, VI, 52.

Original

South Carolina Historical Society.

---

155. Aa 1713

La Floride, Suivant les Nouvelles Observations de Messrs de l'Academie Royale des Sciences, etc. Augmentées de Nouveau. À Leide, Chez Pierre vander Aa. Avec Privilege.

Size: 11½ × 8¾; with border in Gueudeville, 15⅛ × 10¼; in Aa's *La Galerie Agréable du Monde*, 16 × 13⅛.

Scale: 1" = ca. 147 miles.

In: Gueudeville, N. *Le Nouveau Théâtre du Monde.* Leiden, 1713, No. 49.

Description: This map is probably based upon Delisle's "Carte du Mexique de la Floride . . . 1703," using only the part of Delisle's map which was usually included in the Florida maps. The large lake first found in Le Moyne's 1591 map is called "Lac Grande"; but it has no outlet, for "R de Wallea," which is the name given to "R. May" in some of the larger general North American maps of the period, does not touch it. A semicircular lake is at the head of Apalachicola River, in western Georgia or Alabama.

The cartouche is in the lower center; to the left is a smiling lion, to the right, two Indians. This same map, apparently from the same plate but with the title "La Louisiane," is found in Cóvens and Mortier's *Nouvel Atlas*, Amsterdam, [1735?], No. 96. See Cóvens-Mortier ca. 1735.

Reference

Phillips, P. L., ed. *The Lowery Collection*, p. 258.

Reproduction

Color Plate 11.

Copies

Congress; Duke; University of North Carolina; Yale; Clements, No. 43.

In: Aa, P. vander. *Nouvel Atlas*. Leiden, [1714], No. 96.

Copy

Congress 1714.

In: Aa, Pierre vander. *La Galerie Agréable du Monde.* Leiden, [1729], No. 20.

Copies

Congress 1729; Harvard 1729; New York Public 1729; Yale 1729.

---

156. Homann 1714

Virginia Marylandia et Carolina in America Septentrionali Britannorum industria excultæ repraesentatæ a Ioh. Bapt. Homann S.C.M. Geog. Norimbergæ

Size: 23 × 19¼. *Scale:* 1" = 36 miles.
In: Homann, J. B., *Atlas Novus Norimbergæ*, 1714, No. 27.

Description: This map, which extends from South Carolina to Connecticut, is very similar to Visscher's "Nova Tabula . . . Borealiorem Americae" (see Kocherthal 1709), though it has more recent and fuller information concerning settlements of German emigrants. It has the very large "Apalache Lacvs" in western South Carolina. On the west bank of the Cape Fear River, where Kocherthal has "Carls Thon," Homann has a settlement named "Carolina." For much of the North Carolina region he has made use of Gascoyne's 1682 or Thornton's ca. 1685 map, giving many Indian names and marine soundings which are not in Visscher or Kocherthal. Homann evidently made a serious attempt in this map to aid prospective emigrants and indicate German settlements: "Germantown Teutsche Statt" is at the headwaters of the "Rappahanock," and the detail in Pennsylvania, New York, and Connecticut is unusually full. As Homann's map extends from "C Romanie" to southern Connecticut, it includes more of the seaboard than does Kocherthal's map.

This map first appears in Homann's *Atlas Novus*, 1714, where it is listed in the register in the Yale copy. The popularity of the map is evidenced by the number of atlases it appears in throughout the century and by the number of separate copies which are extant. Below "Norimbergæ" in the cartouche, about 1730 and thereafter, is found the line "Cum Privilegio Sac. Cæs. Majest."

Reproduction
Plate 46.

Copies
Congress; American Geographical Society 1714, two separates; John Carter Brown (Binder's Atlas), separate; Duke; Harvard 1714; Kendall; New York Public, two separates; University of North Carolina, two separates; Yale 1714, three separates; Clements, separate.

In: Ottens, R. *Atlas Maior.* Amstelædami, [1641–1729], VII, 103.

Copies
Congress 1641–1729; New York Public 1729?

In: Homann, J. B., and Delisle, G. [*Atlas*, 1699–1739.]

Copies
Yale 1699–1739.

In: Doppelmaier, J. G. *Atlas Novus Coelestis*, Norimbergæ, 1742, No. 158.

Copy
American Geographical Society 1742.

In: [Bellin, N.] *Cartes et Plans de l'Amerique*, 1745, No. 19 [in an atlas with title and date in ink and with many MS maps by Bellin dated 1739].

Copy
Congress 1745.

In: Homann heirs. *Atlas Geographicvs Maior.* Norimbergae, 1759– , No. 143.

Copies
Congress 1759–81, 1759–84, two copies; Newberry (Ayer), 1759[–81].

In: Homann heirs, *Atlas Homannianus.* [Amsterdam, 1731–95], IV, 384.

Copies
American Geographical Society 1731–96.

---

157. Indian Villages ca. 1715 MS

[Map of North and South Carolina and Florida]

Size: 22 × 15. *Scale:* 1" = 30 miles.
Inset: A Map of New France Containing Canada, Louisiana &c. in N^th^ America. According to the Patent granted by the King of France to Monsieur Crozat, dated the 14th of Sep. 1712 N.S. and registered in the Parliament of Paris the 24th of the same Month. By H. Moll Geographer. [This printed map is affixed to the lower left corner of the MS; a separate copy of the same printed map is conjecturally dated 1717?]

Description: This is a somewhat damaged map, showing the distribution of the Indian tribes in the Southeast about the year 1715. Under the location of each of the tribes is written, in a cursive hand, the number of men the tribe possessed. Under "A French Fort" (Fort Toulouse) at the fork of the Coosa and Tallapoosa Rivers is written in a later hand "Since ye warr," which refers probably to the Yamassee War of 1715; on the Barnwell 1721 map is a statement that for twenty-eight years before 1715 this was the uninterrupted location of an English factory. Also at Muscle Shoals is written "A French Fort," and, in a later hand, "Since ye Warr." One of the earliest delineations on a map of the English trading path to the Mississippi is given, with the legend "Formerly ye Common road of ye English Indian Traders." The titles "North Carolina" and "South Carolina" extend across this map from the Mississippi to the Atlantic. The trading paths or roads in South Carolina westward are shown. Crane thinks that this map may have been the engraver's copy for Moll's 1720 map of North America, though it is more detailed. See V. Crane, *Southern Frontier*, p. 350.

In southern Florida is recorded, with explanatory comments, the route of an Indian slave-hunting and exterminating expedition made by Captain Thomas Nairne of Charles Town with his company of Carolina Indians. In 1705 Nairne wrote, "we have these two . . . past years been entirely kniving all the Indian Towns in Florida which were subject to the Spaniards." (Verner Crane, *The Southern Frontier*, 1929, p. 81.) This same expedition was mentioned by Herman Moll on his 1720 map of North America. It is suggested that Moll utilized this manuscript map or one like it along with other sources in compiling that map. Moll's primary purpose was to counter the French territorial claims embodied in Delisle's 1718 "Carte de la Louisiane et du Cours du Missisipi" (see Map No. 170, below). Moll titled his cartographic response "A New Map of the North Parts of America claimed by France" and included it in his *The World Described*. The portion of Moll's map covering the Southeast is reproduced in Cumming, Hillier, Quinn, and Williams, *The Exploration of North America*, p. 94, which also reproduced in full the legend describing Nairne's expedition.

Swanton (in a note on his facsimile of the map) states that "the disposition of the Indian tribes indicates that it was prepared shortly before the outbreak of the Yamassee War, in 1715."

Reference

Penfold, P. A., ed. *Maps and Plans in the Public Record Office. 2. America and West Indies*. London, 1974, 340.

Reproductions

Plate 46A.

Hulbert, A. B. *Crown Collection*, Series III, Plates 13–16.

Swanton, J. R. *Early History of the Creek Indians and their Neighbors*. Washington, 1922, Plate 3 at end of the book (inaccurate copy).

Original

London. Public Record Office. C.O. 700, Carolina no. 3.

158. Moll 1715

A New and Exact Map of the Dominions of the King of Great Britain on ye Continent of North America. Containing Newfoundland, New Scotland, New England, New York, New Jersey, Pensilvania, Maryland, Virginia and Carolina. According to the Newest and most Exact Observations by Herman Moll Geographer [top center] | To the Honourable Walter Dowglass Esqr Constituted Captain General and Chief Governor of all y^e^ Leeward Islands in America by her late Majesty Queen Anne in y^e^ Year 1711. This Map is most Humbly Dedicated by your most Humble Servant Herman Moll Geog^r^ 1715 [cartouche, lower center] | Sold by H. Moll over against Deverux Court in the Strand. [lower left]

Size: 23⅞ × 40. *Scale:* 1" = 50 miles.
In: Moll, H. *The World Described*. London, 1709–20, No. 8.
Insets: 1. [Southeastern North America.] "The Design of this Map is to shew the South Part of Carolina, and the East Part of Florida, possess'd since September 1712 by the French and called Louisiana; together with some of the principal Indian Settlements and the Number of the Fighting Men According to the account of Capt T. Nearn and others." *Size:* 6 × 7. *Scale:* 1" = 155 miles. This inset is a modification of Thomas Nairne's inset in the Crisp 1711 map of South Carolina and settlements of the English, French, and Indian nations; additional information is added in this inset in the 1730 edition and in the "Georgia" edition.

2. A Map of the Improved Part of Carolina With the Settlements &c. By Her. Moll Geographer. [With a list of Planters on the right side.] *Size:* 12⅛ × 10½. *Scale:* 1" = 5½ miles. Though this inset is very similar to the upper right-hand section of the Mathews-Love section of the Crisp map, the area included differs slightly; various small differences in nomenclature and wording show that this is a new map by Moll, based not only on Crisp but also upon Mathews's original map and other more recent information. Like the Crisp map, it has much detailed information concerning settlements in South Carolina between Ashpo Island and the Santee River. A reproduction of this inset and of Charles-Town (No. 4 below) is found in E. M. Avery, *A History of the United States*, Cleveland, 1904–7, III, between pp. 226 and 227, and in W. A. Courtenay,

Genesis of South Carolina, Charleston, 1907, opp. p. 178.

3. A Map of the Principal Part of North America. *Size:* 4⅝ × 4½. *Scale:* 1" = ca. 1,135 miles. A map on a very small scale, extending from the equator to Hudson Bay and westward from the Atlantic to "California I."

4. A Draught of y^e^ Town and Harbour of Charles-Town. *Size:* 4¾ × 5¾. Scale 1" = 75 yards ("Paces.") With a table of explanatory numbered identifications. This is similar to the Charles Town inset in Edward Crisp's map (1711) and is probably based upon it. A reproduction of this draft is given in R. H. Gabriel, ed., *The Pageant of America*, I, 267.

5. The Cataract of Niagara. *Size:* 9⅛ × 8⅛. *Scale:* None. This striking picture of Niagara is also called "The Beaver Inset" because of the number of beavers in the foreground and the lengthy explanation of their activities. For this picture Moll followed very closely an inset in Nicolas de Fer's "Carte de la mer du Sud & de la mer du Nord où se trouvent, les costes d'Amérique, d'Asie, d'Europe & d'Afrique, située sur ces mers . . . Par N. de Fer . . . 1713." *Size:* 75 × 47. One of the most notable features of this great map, of which there is a copy in the Library of Congress, is the "Saut de Niagara." *Size:* 17 × 8. Moll, in copying this inset, reversed the features so that what is to the left on Fer's inset is to the right on Moll's.

Description: This beautifully designed map, with its insets, gives a great deal of information about the Carolina region. The main map gives the Atlantic coastline from "St. Augustin in Louisiana" to "The Great Bay of the Esquimaux" in Labrador. Though several of the insets were apparently suggested by Nairne's insets in Crisp's 1711 map, Moll did not depend entirely upon them for his information. For later changes in the Carolina inset Barnwell's 1721 map was probably used. Four different issues of this map were known to H. N. Stevens (see P. L. Phillips, *A List of Geographical Atlases*, Washington, 1909–20, III, 256), corresponding to a, d, e, and g below. Henry Stevens and Roland Tree, "Comparative Cartography," *Essays Honoring Lawrence C. Wroth*, Portland, Me., 1951, pp. 343–44, note five issues, corresponding to b, c, e, f, and g below. The issues published after 1732 have "Georgia" engraved on the plate in Inset I [Southeastern North America], although "1715" is not erased from the map imprint. The map usually appears in Moll's *The World Described*, in which the earliest of the series of large two-sheet maps is dated 1709, although in some editions of the atlas maps are dated as late as 1736.

While no definitive study of Moll or of his cartographical publications has been made, R. V. Tooley, *Maps and Map-Makers*, London, 1952, p. 55, lists a number of Moll's atlases, including *A New and Complete Atlas*, 1719 (26 maps), which is the subtitle given to *The World Described* on the sheet with the table of contents which accompanies that atlas. The British Library lists an undated Moll atlas with the title "Athlas Royal." About 1735 George Grierson, whose place of business was "at the King's Arms, and Two Bibles in Essex-Street," published an edition of *The World Described* from new plates, with twenty-eight maps. Mr. Cecil Byrd, associated director of the university libraries, Indiana University, who has examined a large number of Moll atlases, has been generous in sharing bibliographical information which he has collected.

More recently Dennis Reinhartz included a "Selected Cartobibliography of the Works of Herman Moll Depicting the American Southwest" as an appendix to his essay "Herman Moll, Geographer: An Early Eighteenth-Century European View of the American Southwest." Reinhartz's essay is included in the volume he and Charles C. Colley edited, *The Mapping of the American Southwest*, College Station, 1987, pp. 79–83.

Here may be mentioned another map in Moll's *The World Described*, No. 17: "To the Right Honourable John Lord Sommers . . . This Map of North America According to y^e^ Newest and most Exact Observations . . . B. Lens delin. G. Vertue Sculp." *Size:* 38 × 23. *Scale:* 1" = ca. 250 miles. This undated map (ca. 1715) has ten insets of North American harbors, among them one of the Ashley and Cooper Rivers. On the Tennessee River is the note: "The Road usually Taken from Caroline to Canada is by this River," one of several details which show the influence of Delisle's "Carte du Mexique et de la Floride" 1703.

See Introductory Essay.

Reproductions

Color Plate 12.

Adams, J. T., ed. *Album of American History*. New York, 1944, I, 198 (South Carolina and Charles Town insets).

Campbell, T. *Early Maps*. New York, 1981, 36.

Cartografia de Ultramar, II, Plate 19 (two plates; issue D; transcription of place-names, pp. 115–23).

Copies

Listed under the different editions of the map.

A. 1715. "Sold by H. Moll over against Deverux Court

in the Strand." [On the Southeastern North America inset: "Cherecies 3000 men."]

Copies

Congress 1709–20; Indiana University, three separates; Clements.

B. 1715. "Sold by H. Moll over against Devereux Court without Temple Bar." See H. Stevens and Roland Tree, "Comparative Cartography," p. 34. No example noted in this study.

C. 1715[1726?]. "Sold by H. Moll and by I. King at ye Globe in ye Poultrey near Stocks Market." [No reference to the Indian King who visited England in 1730; "Georgia" not yet inserted instead of "350 Men" under "Yamesee" in Inset 1.]

Copies

John Carter Brown (Binder's atlas, No. 24); Kendall; Mariners' Museum (in H. Iaillot, *Atlas Nouveav*); Indiana University, three separates.

D. 1715[1732?]. Moll's name and address deleted and the name of I. King added. Inset 1 has "Cherecies 300 men, one of ye Kings of this Nation was in England in 1730." No copy of this has been noted in the collections examined. H. N. Stevens suggests this state appeared soon after Moll's death in 1732 (see above, under inset I).

E. 1715[post-1732]. "Printed and Sold by Tho: Bowles next ye Chapter House in St Pauls Church-yard, John Bowles at the Black Horse in Cornhill, and by I. King at ye Globe in ye Poultrey near Stocks Market." With the enlarged Cherokee note and "Georgia" in Inset 1.

Copies

Congress (American Maps I. 12), separate; John Carter Brown; Clements 1709–20, separate; Duke; Indiana University, three separates; New York Public; Yale, two separates; Clements.

F. 1715[1735?]. Another issue from an entirely new plate. Dedicated to Luke Gardiner; imprint "Sold by Geo. Grierson. . . ." See Stevens and Tree, "Comparative Cartography," p. 344.

Copies

Indiana University, three separates.

G. 1715[post-1735]. "Printed and Sold by Tho: Bowles next ye Chapter House in St Pauls Church-yard; John Bowles & Son at the Black Horse in Cornhill, and by I. King at ye Poultrey near Stocks Market." With the enlarged Cherokee note and Georgia in Inset 1, and "& Son" after John Bowles; except for imprint, the plate is the same as (E).

Copies

Congress 1709–20; John Carter Brown; American Geographical Society; Harvard; Indiana University, three separates.

---

159. Indian Wars ca. 1716 MS

[Carolina]

Size: 29 × 20. *Scale:* 1" = ca. 16 miles.

Description: This sketch map shows in detail the route of the forces sent in the years 1711, 1712, and 1713 from South Carolina to the relief of North Carolina, and of the forces sent from North Carolina in 1715 to the assistance of South Carolina. It also shows the controverted bounds between Virginia and Carolina and extends on the coast from the Savannah River to Cape Charles, Virginia. In the lower right-hand corner is a Table of Explanations of the different routes.

Reproductions

Hulbert, A. B. *Crown Collection.* Series III, Plates 17, 18.

Winsor, J. *Narrative and Critical History of America*, V, 346.

Congress (photocopy).

Original

London. Public Record Office. C.O. 700, Carolina No. 4.

---

160. Graffenried 1716 MS A

Nord Carolina [across face of map at top] | ad pag 6 [top, left corner]

Size: [reproduction] 5 × 9. *Scale:* None.

Description: This map and the two following maps were published, together with a French version of Graffenried's account of the Swiss settlement at New Bern, by the German-American Historical Society at Philadelphia; Professor Albert B. Faust of Cornell University edited the manuscript. It was written by Graffenried himself in 1716, and is a careful revision of the earlier sketches, contained in the so-called A and B manuscripts. MS C is the only one containing maps.

The text of MS C is given, with notes, in Professor Faust's edition, noted below.

For this first map, which extends from Cape Henry to Cape Fear, Graffenried probably used a Gascoyne 1682 or Thornton-Morden-Lea 1685? map as a guide. But there are many new names for the islands and inlets along the coast which appear here for the first time: Golfe de Sampee, I. de Boque, I. de Topsail, Drum Passage, etc. They do not appear on a printed map until Moseley 1733.

A letter to W. F. Mülinen, Bern, Switzerland, asking the size and scale of the original maps, was returned unopened. A later attempt by the present writer to find Mr. Mülinen and examine the MSS during a visit to Bern in the winter of 1945 was unsuccessful.

Since that time the North Carolina Department of Cultural Resources, Division of Archives and History, has obtained a microfilm of many of Graffenried's papers including his 1716 manuscript maps described here as "MS A, B, and C": see Cain, B. T., ed. *Guide to Private Manuscript Collections in the North Carolina State Archives*, 3rd rev. edition, Raleigh, 1981, pp. 450–51.

Reproduction

Faust, Albert B. "The Graffenried Manuscript C." *German American Annals*, N.S. XII (March–October 1914), opp. p. 70.

Original

Private library of Wolfgang Friedrich Mülinen, Bern, Switzerland.

---

161. Graffenried 1716 MS B

Baronie de Bernberi [top of map] | Plan de la Colonie Suisse & Palatine on Nord Caroline etablie en Octobre 1710 par le B: le Graffenried [bottom of map]

Size: 5 × 7 [reproduction]. *Scale:* None.

Description: A plan of the Neuse and Trent Rivers, from the town of New Bern at their confluence up toward their sources, with the location of houses on the banks.

Reproduction

Faust, Albert B. "The Graffenried Manuscript C." *German American Annals*, N.S. XII (March–October 1914), 68.

Original

Private library of W. F. Mülinen, Bern, Switzerland.

---

162. Graffenried 1716 MS C

Plan der Schwijtzereschen Coloney In Carolina angefangen im October 1710 dürch Christophe von Graffenried und Frants Lüdwig Mishul

Size: 6¾ × 8½. *Scale:* None.

Description: This is a more detailed map than the preceding one, with many explanatory legends.

A reproduction of this plan, with the printed title below of "Anlage der Stadt Neu-Bern 1710 Nach einem Plane der Bibliothek von Mülinen" (*Size:* 13½ × 17 inches) is in *Neujahrsblatt herausgegeben vom Historischen Verein des Kantons Bern für 1897*, Bern, 1896. A copy of this Swiss publication, with the map, is in the North Carolina Room of the University of North Carolina Library.

Reproduction

Faust, Albert B. "The Graffenried Manuscript C," *German American Annals*, N.S. XII (March–October 1914), 104.

Original

Private library of W. F. Mülinen, Bern, Switzerland.

---

163. Le Maire 1716 MS A

Carte Nouvelle de la Louisiane et païs circonvoisins dressée sur les lieux pour être presentee a S M.te T.C. par Le Maire pretre parisien et mission.e apostolique MDCCXVI

Size: 32½ × 16⅞. *Scale:* 1" = ca. 110 miles.

Description: This map covers the southern part of North America and the regions bordering on the Gulf of Mexico, though the coastal borders only of Mexico and South America are shown.

This map, or some copy of it, was used by Delisle in preparing his famous 1718 map, in the title of which Delisle records his indebtedness to Le Maire's Memoires. Though Delisle draws on Le Maire for many items, such as "Route de M.r Saint Denis en 1714" in Texas, he omits a number of interesting and significant legends about the trading paths of the English. From "Yasous" on the Mississippi across Alabama is "Chemin des Anglois," Captain Welch's 1698 route; from "Appalache ou R Chaoünons" is "Chemin ordinaire des Anglais"; from Charleston another path goes to "Habitation Angloise" on the "Jourdain ou Santi" and thence the "Chemin des Tcheraqui" leads to the Over-the-Hills settlements of the Cherokees.

The measurements are those of the Clements photocopy.

See Le Maire 1716 MS B which, in spite of slight differences in the title, is very similar to this map. The Library of Congress has a photocopy of another MS map by Le Maire of this date which is in Paris (Service Hydrographique. 138 bis - 1–6). "Delineabat F. le Maire P.P. missionaire Apostolic anno 1716." *Size:* 44½ × 12⅛. It has names and soundings along the Gulf Coast from Mexico to the Florida Peninsula.

See Delisle 1718.

Reproduction

Clements (photocopy).

Original

Paris. Dépot des Cartes et Journ. de la Marine. 4044.C.46.

---

164. Le Maire 1716 MS B

Carte nouvelle de la Louisiane et Päys Circonvoisons Dressée sur les lieux pour estre presentée a sa Majesté tres Chretienne Par F. le Maire Prestre parisien et missionnaire Apostolique. 1716

Size: 23¼ × 18⅝. *Scale:* 1" = ca. 150 miles.

Description: See Le Maire 1716 MS A.

Reproduction

Karpinski series of reproductions.

Original

Paris. Bibl. Nat. Ge. D. 7883.

---

165. Vermale 1717 MS

Carte Generale de la Louisiane ou du Micicipi, dressée sur les plusieurs mémoires et dessinée par le Sr Vermale, cy deuant cornette de dragons. 1717 [title in blank space to left of map]

Size: 36 × 23. *Scale:* 1" = ca. 79 miles.

Description: This manuscript map has many details found on Delisle's 1718 "Carte de la Louisiane," such as the route across Texas taken by Monsieur S. Denis in 1716, the fort built by Bienville on the Red River, and the names of Indian settlements in the Mississippi-Alabama region. There are numerous differences between the two maps; but the many similarities indicate a common source or the borrowing of one from the other. For the Carolina coastal region Vermale's map shows no similarity to any Delisle map; the information is antiquated and the names of the rivers, St. Jean, St. Dominique, St. Philipe, are strange. The "1717" which is under the title of the map is written in a different hand and with different ink.

Reproductions

Marcel, G. *Reproduction de cartes et de globes* . . . Paris, 1893, Plate 25.

Congress (photocopy).

Original

Paris. Bibl. du Ministère de la Marine. 4044.C.11.

---

166. Montgomery 1717

A Plan representing the Form of Setling the Districts, or County Divisions in the Margravate of Azilia.

Size: 12⅜ × 12. *Scale:* 1" = 2 miles ("The 116 Squares . . . are, Every one a mile on Each Side," p. 11).

In: Montgomery, Sir Robert. *A Discourse Concerning the design'd Establishment of a New Colony to the South of Carolina, In the Most delightful Country of the Universe.* London, 1717, opp. p. 10.

Description: This is a map of the abortive "Margravate of Azilia," which shows Montgomery's visionary plan for a walled, militarily secure settlement pattern to be established in the region south of the Savannah River. Drawing on the recent Yamasee War debacle and near collapse of South Carolina, Montgomery correctly focused on the fatal vulnerability of the Carolina colonists resulting from what he termed, "a want of due Precaution in their Forms of settling, or rather, to their settling without any form at all." Projected for a hotly contested coastal region formerly occupied by Spain, Azilia was not to be defended by "building here and there a Fort, the fatal Practice of America." To the contrary, Montgomery's plan was to make the "whole Plantation one continued Fortress" by enclosing habitations and farmlands by "Military Lines, impregnable against the Savages." The would-be Margrave did not specify where within the debatable land lying between the Savannah and Altamaha Rivers would be found the required "level, dry, and fruitful Tract of Land" needed for his "just square of twenty Miles Each Way." Nor, unfortunately, did Montgomery publish any general map of the area he proposed to settle.

On June 19, 1717, Sir Robert Montgomery, baronet of Skilmorlie, received from the lords proprietors a grant of land between the rivers Altamaha and Savannah. In 1718 he stated before the council that he had raised £30,000 among his friends, for his project. The assumption of the government of the province of South Carolina by the crown threw obstacles in the path of colonization in 1720, and Sir Robert died in 1731 without executing his plans. An abstract of the grant to Sir Robert Montgomery for the Margravate of Azilia is found in *A Description of the Golden Islands*, London, 1720, a pamphlet of forty-five pages, in which the land between the Savanna, and Altamaha is indicated as the limits of the grant; but the pamphlet deals specifically with the four "Golden Islands" of "St. Symon, Sapella, Santa Catarina, and Ogeche." This pamphlet contains a letter from Colonel John Barnwell, of Carolina, who is "generally once a Day at the Carolina Coffee-house in Birchinlane." A copy of this pamphlet is in the John Carter Brown Library.

Though Azilia never materialized, the name is found on numerous later maps of the Carolinas: Moll's A New Map of the North Parts of America Claimed by France . . . 1720, Barnwell's MS map ca. 1721, Moll's Carolina 1729, Homann's Dominia 1737, Herbert-Hunter 1744 MS.

References

Andrews, C. M. *The Colonial Period of America History*, II, 226, and references there listed.

Crane, V. "Projects for Colonization in the South, 1684–1732." *Mississippi Valley Historical Review*, XII (1925), 23–35.

Crane, V. *Southern Frontier*, pp. 210–14.

De Vorsey, L. "Oglethorpe and the Earliest Maps of Georgia." *Oglethorpe in Perspective: Georgia's Founder After Two Hundred Years*, eds. Phinizy Spalding and Harvey H. Jackson. Tuscaloosa, 1989, pp. 25–29.

Reese, T. R. *The Most Delightful Country of the Universe. . . .* Savannah, 1972, p. 12.

Reproductions

De Vorsey, L. "Oglethorpe and the Earliest Maps of Georgia," 26.

Force, Peter. *Tracts*. Washington, 1836–46, I, opp. p. 8.

Gabriel, R. H., ed. *The Pageant of America*. New Haven, 1925, I, 271.

Montgomery, Sir Robert. *A Discourse* . . . [reprint]. *Magazine of History*, IX, Extra number, 33 (1914), frontispiece.

Montgomery, Sir Robert. *A Discourse concerning the design'd Establishment of a New Colony to the South of Carolina*. Rochester, G. P. Humphrey, 1897, opp. p. 10.

Reps, John. *Town Planning in Frontier America*, 236.

Copies

Congress; John Carter Brown, three copies; Clements; Georgia Hargrett Library; Harvard; Newberry.

---

167. Maule-Roanoke 1718 MS

This Plane as here delineated and Layed Out represents the Island of Roan-Oak in North Carolina Containing Twelve Thousand Acres of Land and Marsh. As Surveyed Anno 1718 By Wm Maule, Sury^{r} Gen^{ll}

Size: 12¼ × 29½. *Scale:* 1" = 2,640 feet.

Description: This is a crude pen-and-paper drawing which gives numerous legends concerning houses, landings, trees, and topographical features. It shows also the nearby mainland and the Outer Banks.

The last numeral in the date may be a six, making the date of this map 1716; but the unusual formation of the number, making two completed adjoining circles, and the apparent reference to this map in 1729 in Moseley's copy (q.v.), giving 1718 as the date of the survey, makes 1718 the probable date of the map.

Original

North Carolina Department of Archives and History.

---

168. Lawson 1718

A Map of the English Plantations in America

Size: 7¼ × 7. *Scale:* 1" = ca. 320 miles.

In: Lawson, J. *The History of Carolina*. London, T. Warner, 1718, opp. p. 1.

Description: This is entirely different from the map in other editions of Lawson's *History*; it has Lederer's Lake, Savanna, and Desert, though they are not named. It is a retrogressive map.

The Library of Congress and University of North Carolina copies of this volume have the Lawson 1709 map, probably as inserts.

Copies

John Carter Brown 1718; North Carolina State 1718; Yale 1718.

169. de Fer 1718

Partie Méridionale de la Rivière de Missisipi, et ses environs, dans l'Amérique Septentrionale. Mis au jour par N. de Fer. Géographe de sa Majesté Catolique 1718 [across top, below neat line] | À Paris dans l'isle du Palais sur le Quay de l'Orloge à la Sphère Royale avec Priv.e du Roy 1718. [across bottom of map, above neat line]

Size: 24¾ × 18. *Scale:* 1" = ca. 80 miles.
In: Aa, P. vander. *La Galerie Agreable du Monde.* Leiden, 1729–50, I, No. 72.

Description: This map represents nearly the same area as Delisle's 1718 map, from 22° to 41½° north, and the coast from Charleston, in Carolina, to Rio de Garde in Mexico. But while Delisle's map makes a striking advance in its portrayal of the North American continent, de Fer is retrogressive.

The Newberry Library holds a rare first state of this map which is dated 1715. While the geographical details appear to be identical, none of the decorative features, such as the ships in the Gulf of Mexico and Indian villages, is present. The western part has most of the information given in Delisle's 1703 map; in the southeast de Fer goes back to Sanson's general map of Carolina, ca. 1696, giving the lake with "Vallée Agreable." West of the Appalachians are the two Mercator lakes, as in Delisle 1703. Lake Erie is the only Great Lake shown.

This map uses the bottom left plate only of a large map of North America by de Fer made for the Compagnie d'Occident in 1718, but with the title "Partie Meridionale . . ." of the bottom section concealed by the overpasted top sheet, or added to the plate. The imprint remains the same. The title of the large map, a copy of which is in the Library of Congress, is: Le Cours du Missisipi ou de S.t Louis . . . Dressée sur les Relations et Memoires du Pere Hennepin et de M.rs de la Salle, Tonti, Laontan, Ioustel des Hayes, Joliet, et le Maire & par N de Fer Geographe de sa Majesté Catolique Tous ces Memoires Relations et decouvertes se sont faites depuis 1681 jusques en 1717, qui est lannée de l'Establissement de la Compagnie doccident et pour laquelle cette Carte a été Dressée A Paris Chez l'Auteur Isle du Palais à la Sphère Royale 1718. [cartouche center left] | La France Occidentale dans L'Amerique . . . 1718 [cartouche lower right]. *Size:* 41¾ × 38. *Scale:* 1" = ca. 80 miles. There is an inset at the top left of Louisiana and the mouth of the Mississippi.

A [1715]

Copies

Newberry.

B [1718]

Copies

Harvard 1729–50.

C [1718]. [same title: imprint at bottom changed to:] A Paris chez J. F. Benard sur le Quay de l'Orloge à la Sphère Royale avec Priv.e du Roy 1718

Copies

Congress; Clements; Kendall.

170. Delisle 1718

Carte de la Louisiane et du Cours du Mississipi Dressée sur un grand nombre de Memoires entrautres sur ceux de M.r le Maire Par Guill.aume Del'isle de l'Academie R.le des Scien.ces [across top of map below neat line] | A Paris Chez l'Auteur le S.r Delisle sur le Quay de l'Horloge avec Privilege du Roy Juin 1718 [bottom center, above scale]

Size: 25⅝ × 19⅛. *Scale:* 1" = ca. 95 miles.
In: Delisle, G. *Atlas.* Paris, 1700–1762.
Inset: Carte Particuliere des Embouchures de la Rivie. S Louis et de la Mobile. *Size:* 6 × 5. *Scale:* 1" = ca. 40 miles.

Description: This notable map, extending from Long Island to the Rio Grande and southward to the lower half of the Florida peninsula, was first made as a separate by Delisle for the Compagnie d'Occident.

Its importance for the Carolina area lies in its political purpose; it circumscribes and lessens the western boundaries of the English colonies and states that "Caroline" was first discovered, named, settled, and possessed by the French and that Charles Town is the French "Charles fort." For the angry English reaction to these claims, see the Introductory Essay.

For the Carolina region, extending into present Tennessee and Kentucky, Delisle's map is sparse of detail and inaccurate. The erroneous conception of the Appalachians, with one range extending into the Michigan peninsula, he apparently borrowed from the map of Morden-Brown ca. 1695 or Sanson ca. 1700. It has, however, the first attempt on a modern map to trace the route of Hernando de Soto; the chief locations on the course which Delisle gives are surprisingly close to the identifications made by recent investigators: see John R. Swanton, ed., *Final Report of the United States De Soto Expedition Commission*, Washington, 1939. It is interesting to note that although Swanton located the important Indian town of Cofitachequi on the Savannah River, Delisle's map places it ("Cutifaciqui") to the east of "R. Sante ou Jourdain," in what is now east central South Carolina. Possibly basing his choice for this site on information gained from an informant or unpublished source, Delisle is in essential agreement with this portion of the de Soto route reconstruction argued by Charles Hudson and his associates: see Charles Hudson, M. T. Smith, and C. B. DePratter, "The Hernando De Soto Expedition: From Apalachee to Chiaha," *Southeastern Archaeology*, 3 (1984), pp. 65–77. In other segments, however, he departs widely from Hudson's reconstructed route. The routes of Alonso de Leon (1689), St. Denis (1713–16), and other western explorers are also traced. F. Le Maire, to whom Delisle expresses his especial indebtedness for information used in the map, was a French missionary who in 1717 published his *Mémoire* and in the previous year made two manuscript maps of the Mississippi valley.

It is for the Mississippi valley, particularly the Gulf area, that the cartography of this map is notable for employment of new information, wealth of detail, and relative accuracy. Kohl says that "this map is the mother and main source of all the later maps" of the Mississippi (P. L. Phillips, ed., *Lowery Collection*, p. 230); and more recent scholars substantiate its historical importance (C. O. Paullin, *Atlas of the Historical Geography of the United States*, p. 12). Almost twenty years earlier Claude Delisle, the father of Guillaume, had begun his study of the Mississippi with special instructions for information which he gave to Iberville's expedition. See "Lettre de Mr. Delisle à Mr. Cassini, sur l'embouchure de la riviere de Mississipi," in [Jean I. Bernard's] *Recueil de voiages au Nord*, Amsterdam, 1715–38, III (1715), pp. 257–67. This letter was first published in *Journal des Sçavans*, May 10, 1700, pp. 201–6, and an earlier manuscript draft is in the Service Hydrographique, Paris, 115–20; No. 17 B. The best study of the historical cartography of the Mississippi River, including the contributions of the Delisles, is that of Father Delanglez: see under the reference list below.

The importance of the map is shown not only by its influence on subsequent regional delineation but by the number of impressions of the map itself and by its direct imitators. See Senex 1721, Cóvens-Mortier ca. 1730, and Bernard 1734.

Homann, the most indefatigable mapmaker of his time in Germany, made a plate based on Delisle's map but covering a slightly larger area which is found in many Homann atlases and in at least three different issues: "Amplissimæ Regionis Mississipi Seu Provinciæ Ludovicianæ a R. P. Ludovico Hennepin Francise Miss. in America Septentrionali, anno 1687 detectæ, nunc gallorum Coloniis et Actionum Negotiis toto Orbe Celeberrimæ. Nova Tabula-Edita, A Io Bapt. Homanno, S.C.M. Geographo, Norimbergæ." Susan M. Reed (see reference below, p. 214) considers Homann's map "A mere perversion" of Delisle, though actually it follows him closely, translating the legends and giving de Soto's route.

A beautifully executed small copy (16½ × 12½) of Delisle, without the inset, is "Novissima Tabula Regionis Lvdovicianæ Gallice dictæ La Louisiane," found in D. Kohler's *Bequemer Schul und Reisen-Atlas*, Nürnberg, C. Weigeln, ca. 1734, a good reproduction of which is in Fite and Freeman's *Book of Old Maps*, Map 46, p. 176. A good historical note accompanies the reproduction, though the editors are confused in their bibliography of the map and make no reference to the original of Delisle. In 1744 Bellin published a map which omitted de Soto's route, eliminated several outdated references, and introduced new information: "Carte de la Louisiane cours du Mississipi et Pais Voisins Dediée à M le Comte de Maurepas, Ministre et Secretaire d'Etat Commandeur des Ordres du Roy. Par N. Bellin Ingenieur de la Marine, 1744," in N. Bellin, *Cartes et Plans de L'Amerique*, [Paris], 1745, No. 21. In 1750 Bellin made a similar map, with revisions and corrections, dedicated to Rouillé, secretary of state. A German edition of Bellin's map is entitled "Karte von Louisiana, dem Laufe des Mississipi, und den Benachbarten Laendern Durch N. Bellin Ingenieur de la Marine. 1744. | No. 11," in *Allgemeine Historie der Reisen zu Wasser und zu Land*, Leipsig, bey Arkstee und Merkus, Vol. XIV (1756), opp. p. 308.

Among manuscript copies of the Delisle map, the most interesting, in view of Governor Burnet's attack on the map in his letter to the lords of trade and plantation (see Introductory Essay), is in the British Public Record Office: Colonial Office Library, North America, No. 1 (*Size:* 23 × 18; *Scale:* 1" = ca. 75 miles), with the legends translated into English but without title or date. It is dated [ca. 1720] by the Colonial Office. A reproduction is in A. B. Hulbert, *Crown Collection*, Series III, Plates 1–4.

Another manuscript copy is reproduced by W. H. Milburn, in *The Lance, Cross, and Sword*, New York, N. B. Thompson Publishing Company, 1892, pp. 74–75, with an interesting note. Edmund J. Forstall, in French's *Historical Collections of Louisiana*, Part II, refers to this same map, attributing to it erroneously the date 1700. The map is ornamented with animals, ships, and with one phrase in the lower right-hand corner, inexplicable because it alone is in English: "De Soto landed 31 May 1539."

See Introductory Essay.

References

Delanglez, J. *El Rio del Espíritu Santo*. New York, 1945, pp. 3, 9, 136–37, 142–44.

Delanglez, J. "The Sources of the Delisle Map of America," *Mid-America*, XXV (1943), 275–98.

Fite, E. D., and A. Freeman. *A Book of Old Maps*. Cambridge, 1926, pp. 177–79.

Karpinski, L. C. *Bibliography of the Printed Maps of Michigan*. Lansing, 1931, pp. 133–34.

Phillips, P. L., ed. *Lowery Collection*. Washington, 1912, pp. 229–30, No. 269; p. 239, No. 288.

Reed, Susan M. "British Cartography of the Mississippi Valley in the Eighteenth Century." *Mississippi Valley Historical Review*, II (1915), 213–24.

Reproductions

Plate 47.

Cartografia de Ultramar, II, Plate 99 (of first issue; transcription of place-names, pp. 382–86).

Karpinski, L. C. *Bibliography of the Printed Maps of Michigan*. Lansing, 1731, Plate XI.

Paullin, C. O. *Atlas of the Historical Geography of the United States*. Washington, 1932, Plate 24.

———A [1718].

Copies

Congress; Clements, two separates; Harvard; Kendall; New York Public 1741?; Yale 1700–1762.

In: Jaillot, C. H. A. *Atlas Françoise*. Paris 1695–[1720], No. 16.

Copies

John Carter Brown 1699–1720.

———B [1745]. [Title and plate unchanged] | Ph. Buache P.G. d R. d. l'A. R. d. S. Gendre de l Auteur. Avec Privilege du 30 an. 1745 [added in bottom right corner]

In: Delisle, G., and P. Buache. [*Atlas Geographique et Universelle*— Paris, 1700–1763], No. 76.

Copies

Congress 1700–1763, three copies; Clements 1700–1763.

———C [1782]. [Buache imprint unchanged:] Carte de la Louisiane et du Cours du Mississipi Avec Les Colonies Anglaises. Revue, Corigee et considerablem^{t} augmentee en 1782. Par guill^{aume} Delisle de l Academie R^{le} de Scien^{ces} [across top] | A Paris chez Dezauche successeur des Sieurs De l'Isle et Buache, Rue des Noyers prés la Rue des Anglis. [bottom center, above scale]

In: Delisle, G., and P. Buache. *Atlas Géographique et Universel*. Paris, Chez Dezauche, 1781–[84], II, 131.

Copies

Congress 1781–84; John Carter Brown (Binder's Atlas, No. 39); Clements; Newberry (Ayer).

---

171. Mallet 1719

Floride [cartouche, top center] | das Landt Florida Fig: 18 [above neat line]

Size: 4½ × 5⅞. *Scale:* 1" = ca. 276 miles.

In: Mallet, A. M. *Beschreibung des gantzen Welt-Kreises*. Franckfurt am Mayn, 1719, J. A. Jung, VIII, opp. p. 346.

Description: The maps in this atlas are similar to those in Mallet's French Atlas of 1683. The title and names are in French, and the maps may be from the 1686 plates. German titles are given outside the border. See Mallet 1683 and Mallet 1686.

Copies

Congress; Yale.

---

172. Senex 1719

Most humbly Inscrib'd to Hewer Edgly Hewer of Clapham Esqr &c [top of map] | A New Map of the English Empire in America Viz., Virginia, Mary Land, Carolina, Pennsylvania, New York, New Iarsey, New England, Newfoundland, New France, &c. Revis'd by Io^{n} Senex. 1719. [cartouche, right center] | I. Harris sculp. [bottom right]

Size: 23⅜ × 19¾. *Scale:* 1" = ca. 92 miles.

In: Senex, John. *A New General Atlas*. London, 1721, opp. p. 237.

Insets: 1. the Harbour of Boston or Mattachusetts [*sic*] Bay. *Size:* 3 × 3⅛. *Scale:* 1" = 9 miles.

2. A Generall Map of the Coasts & Isles of Europe, Africa and America. *Size:* 12 × 8¾. *Scale:* 1" = ca. 550 miles.

Description: See Morden-Brown ca. 1695 and Sanson ca. 1700. For this map Senex used the plate of the Morden-Brown map, as the accidentals and erasures show; he "Revis'd" it very slightly.

Reproduction

Color Plate 13.

Copies

Congress 1721, separate; American Geographical Society 1721, separate; John Carter Brown 1721; Clements 1721; Harvard 1721; Kendall 1721; New York Public 1721; University of North Carolina; Yale 1721, separate.

---

173. Chatelain 1719

Nouvelle Carte de la Caroline.

Size: 3 × 2¼. *Scale:* None.

In: Chatelain, Henri A. *Atlas Historicus.* Amsterdam, 1705–20, VI, 1719, No. 26, p. 100.

Description: This small map, on a large sheet with seven maps and descriptive writing about parts of North America, is apparently drawn from a Blome-Morden map with the same title. Although small, it has in full detail all the Lederer 1672 material. In 1732–39 a second edition of the same work was published. Also contained in the sixth volume of both editions are two large maps, covering a much wider region but giving detailed information for the Southeast:

1. "Carte de la Nouvelle France . . . Tom: VI N° 23 Pag: 91." *Size:* 17⅜ × 16⅝. This map extends from Cuba to Hudson Bay. The Le Moyne lake still appears, though unconnected to any river, and lying beyond the Appalachian range.

2. "Carte contenant le Royaume du Mexique et la Floride Dressez sur les observations & sur les Memoires les plus Nouveaux. Tom: VI, No. 27. Pag: 101." *Size:* 20½ × 16. This map, extending from the equator to 45° N.L. and based largely on Delisle's 1703 map, has the lake, the desert, the savannah, and the towns of the Lederer 1672 map.

Reproduction

Color Plate 14.

Copies

Congress 1719, 1732; American Geographical Society 1732; John Carter Brown 1732; Charleston Library Society 1719; Clements; Harvard 1719; New York Public 1732; Yale 1719.

---

174. Hughes ca. 1720 MS

A Map of the Country adjacent to the River Misisipi—Copy'd (by A[lexander] S[potswood]) from the Original Draught of Mr Hughes

Size: 18 × 14½. *Scale:* 1" = 70 miles.

Description: This map has a strong resemblance to the Ortelius-Chiaves 1584 map in its general outlines, which are rather crude. It represents the lower half of North America from Charles Town, S.C., to the Rockies. It is properly of the date 1713, since Hughes drew the original in that year, but this copy was made by Lieutenant Governor Spotswood of Virginia about 1720. See A. Spotswood, *Official Letters*, ed. R. A. Brock, Virginia Historical Society Collections, new series, Richmond, 1882–85, II, 331.

The interest of this map is twofold. It shows the route of exploratory missions of Price Hughes, the ardent colonizer, toward the Mississippi, and is significant in view of its date and its authorship. See V. Crane, *The Southern Frontier*, pp. 99–107. In the second place, it shows more correctly than any earlier map the northeast-southwest direction of the valleys as well as of the mountain ranges of the lower Appalachian system: "the East Ridge" (Blue Ridge), "The Highest Ridge" (Smoky Mountains, Clinch Mountain ranges), and the intermediate mountain plateaus.

Reference

Crane, V. *The Southern Frontier*, p. 349.

Original

London. Public Record Office. C.O. 700, Virginia no. 2.

---

175. Delisle ca. 1720

Carte de la Louisiane et du Cours du Mississipi Dressée sur un grand nombre de memoires entrau tres sur ceux de Mr. le Maire Par Guill.aume De l'Isle de l'Academie R.le des Sciences

Size: 16⅛ × 14. *Scale:* 1" = ca. 100 miles.

Description: This map is very similar to Delisle's 1718, on which it is based. The area shown is somewhat more circumscribed, and the inset showing the mouth of the Mississippi is omitted.

Copy

Clements.

176. Delisle-Louisiana 1720

A New Map of Louisiana and the River Mississipi. [cartouche, bottom center]

Size: 8¾ × 9¹³⁄₁₆. *Scale:* 1" = ca. 150 miles.
In: [Smith, Dr. James?] Some Considerations on the Consequences of the French Settling Colonies on the Mississipi. London, 1720, after title page.

Description: This is an English edition of that part of Delisle's 1703 map which shows Louisiana, Florida, and Carolina. It has the Lederer (1672) Indian towns, the "Plains coverd w.th Water," and "The Road the French take to go to Carolina."

The pamphlet in which the map is found was at first attributed to Richard Beresford, the surveyor general of Carolina, by Justin Winsor (*Narrative and Critical History of America*, V, 76) but later to Dr. James Smith, judge advocate of Massachusetts and Carolina (Winsor, *Mississippi Basin*, p. 141). The author recounts the troubles of an advocate-general in New England and shows no firsthand acquaintance with the southern frontier such as Beresford had.

Reproductions

Some Considerations on the Consequences of the French Settling Colonies on the Mississippi. Reprint by the Historical and Philosophical Society of Ohio, with a Preface by Beverly W. Bond, Jr. [Chicago, 1928].

Winsor, J. *The Mississippi Basin.* Boston, 1895, pp. 142–43.

Copies

Boston Public 1720; John Carter Brown 1720; Clements 1720, 1726.

177. Altamaha River ca. 1721 MS

A Map or Plan of the Mouth of Alatamahaw River with the adjacent Lands.

Size: 18¼ × 16⅛. *Scale:* 1" = 2 miles.

Description: A careful survey, without soundings, of the coast for about thirty miles, and of the course of the river inland for another thirty, from the mouth of the Altamaha in Georgia. Fort King George is given, with roads from it northeast to "Palachicolas" and northwest "to the Okanees." The handwriting of the legends is very like Barnwell's. At the continued insistence of Colonel John Barnwell, the English started the construction of Fort King George on the Altamaha in 1721. See John T. Lanning, "Don Miguel Wall and the Spanish Attempt against the Existence of Carolina and Georgia," *No. Car. Hist. Rev.*, X (July 1933), 186. V. Crane (*Southern Frontier*, p. 350) attributes this, the following two maps, and King George's Fort 1722 MS to Col. John Barnwell. Jeannine Cook details the authentic modern reconstruction of Barnwell's blockhouse in her well-illustrated *Fort King George: Step One to Statehood*, Valona, Ga., 1990.

Reference

Maps and Plans in the Public Record Office Vol. 2: America and West Indies. London, 1974, p. 409. (This book lists the maps of Georgia in the Colonial Office on pp. 405–12.)

Reproductions

Congress (photocopy); Georgia Archives (photostat).

Original

London. Public Record Office. C.O. 700, Georgia no. 1.

178. Barnwell 1721 MS

The Northern B[ra]nch of Alatama River which joyns y.e main River 3 miles higher up [. . .]

Size: 22¾ × 18. *Scale:* 1" = 110 feet.

Description: This map is dated August 29, 1721, and signed "For His Excellency General Nicholson By his Ser.t J. Barnwell."

In the upper right corner is a drawing or inset of a fort or block house, "the front of Alatama plank'd house," which is located on the main map at the juncture of the branch river with a creek. There are many explanatory legends, all in Barnwell's hand, as "Ten acre field, sandy, good for potatoes, left by the Huspaw People (belonging to the Yamasees) when they heard in 1715 that the English were coming to attack them."

Reference

Crane, V. *Southern Frontier*, pp. 233, 237, 350.

Reproductions

Congress (photocopy); Georgia Archives (photostat).

Original

London. Public Record Office. C.O. 700, Georgia no. 2.

179. St. Simon's 1721 MS

A Chart of S! Simon's Harbour. September 2d 1721

Size: 22¾ × 18. *Scale:* 1" = 1¼ miles.

Description: This shows the harbor at the mouth of the Altamaha River; the writing is in Barnwell's hand.

See Barnwell 1721 MS.

Reproductions

Congress (photocopy); Georgia Archives (photostat).

Original

London. Public Record Office. C.O. 700, Georgia no. 3.

180. Beaufort 1721 MS

Mapp of Beaufort in South Carolina 1721.

Size: 29 × 21. *Scale:* 1" = 5 chains.

Description: This is a survey of Beaufort, and of land belonging to Capt. Woodward, to Col. Barnwell, and to Mr. Hazzard, possibly drawn by Col. Barnwell.

Reproductions

Hulbert, A. B. *Crown Collection*, Series III, Nos. 19–20.

Original

London. Public Record Office. C.O. 700, Carolina no. 5.

181. Herbert 1721 MS

South Carolina

The Ichnography or Plann of the Fortifications of Charlestown and the Streets, with the names of the Bastions quantity of acres of Land, number of Gunns and weight of their shott, By his Excellency^s Faithfull & Obedient Servt John Herbert Oct^br 27 1721 [top left]

Size: 23 × 18. *Scale:* 1" = 132 feet.

Reproduction

Hulbert, A. B. *Crown Collection of Photographs of American Maps*. London [1914–16], Series III, No. 21.

Original

London. Public Record Office. C.O. 700, Carolina no. 6.

182. Senex 1721

A Map of Louisiana and of the River Mississipi by Iohn Senex [across top of map] | This Map of the Mississipi Is Most humbly Inscribed to William Law of Lawreston Esq: [cartouche, bottom right]

Size: 22½ × 19¼. *Scale:* 1" = ca. 137 miles.

Reproduction: Georgia Archives (photostat).

In: Senex, John. *A New General Atlas*. London, 1721, opp. p. 248.

Description: This is "a most impudent plagiarism" from Delisle, for it is closely based on his 1718 map, showing de Soto's route, the route of Mr. S. Denis in 1713–16 across Texas, etc. The Carolina part is rather vacant and does not have Delisle's reference to the French priority of settlement. The map does not represent as large an area as Delisle's, for it extends eastward only to the eastern shore of Virginia and does not include New York.

Copies

Congress; American Geographical Society; John Carter Brown, Binder's atlas, No. 12; Clements; Harvard; Kendall; New York Public; Yale.

183. Moll-Kyser 1721

Carolina door Herman Mol. de Plantasien zyn aldus gemerkt [**II**]. | J. Kyser F[ecit] | I. Deel. Blatz 249. [at top right, outside neat line]

Size: 5⅞ × 6⅝. *Scale:* 1" = ca. 70 miles.

In: [Oldmixon, John.] *Het Britannische Ryk in Amerika*. Amsterdam, 1721, I, opp. p. 249.

Description: This map is a copy of the Moll-Oldmixon "Carolina" 1708 from a different plate, with Dutch legends; it is found in the Dutch translation of Oldmixon's work by R. G. Wetstein.

The "Island of Bermudos" map is on the same folded sheet.

Copies

Congress 1721; John Carter Brown 1721; Clements 1721; Kendall 1721; Yale 1721. Also in the 1727 edition. Copy: British Library, 1727.

184. Barnwell ca. 1721 MS
[Southeastern North America]

Size: 52¼ × 31⅛. *Scale:* 1" = ca. 18 miles.

Description: The area shown extends from Cape Charles in Virginia southward to Cape Canaveral in Florida, and westward to the Mississippi. This large unsigned, undated manuscript gives the location of French, Spanish, Indian, and English settlements; the chief trading paths to the Indians, including the route of Captain Welch to the Mississippi River in 1698 and of Hughes in 1715; the routes taken and the location of battles in the various military expeditions of the South Carolinians during the first decades of the eighteenth century; and numerous explanatory legends about the quality of the land, the location of trading posts, the size of Indian tribes, and details of marauding forays.

The map was used in making Popple 1733, Catesby 1731, Bull 1738 MS, Bull-Verelst 1739 MS, Barnevelt ca. 1744 MS, and Mitchell 1755. Through the Popple 1733 and Mitchell 1755 maps, which were quickly and frequently copied by other mapmakers in England and on the continent, the information on this map was widely if somewhat tardily disseminated. It is a mother map of the first rank. Verner Crane, who makes frequent use of it in his work, calls it "A notable map, based on reports of Indian agents, Indian 'censuses,' etc.; the first detailed English map of the southern frontier extant." Cf. *The Southern Frontier*, p. 350.

For its ascription to Colonel John Barnwell and for the date, see the Introductory Essay.

Reproductions

In earlier editions of this work, Plate 48.

South Carolina Archives (photostats).

Reference

P. A. Penfold, ed. *Maps and Plans in the Public Record Office.* London, 1974, No. 1998, p. 341.

Original

London. Public Record Office. C.O. 700, North American Colonies General no. 7.

184A. Barnwell-Hammerton 1721 MS
[Southeastern North America]

Size: 54 × 31⅛. *Scale:* 1" = ca. 18 miles.

Description: This is another copy of the Barnwell map described above. There are a number of differences including an elaborate decorative but uncompleted cartouche not found on that Public Record Office copy. Although incomplete, the cartouche contains the name "Hammerton," presumably William Hammerton who died in Charleston in 1732. See *Charleston Wills* III, p. 38. Hammerton was a mariner in that city who was re-commissioned as chief Naval Officer by South Carolina's governor, Sir Francis Nicholson, in 1721. The cartouche also contains the date "1721." This version of the map was first identified and examined by W. P. Cumming in 1969 while it was in the library at Chatsworth House, the estate of the duke of Devonshire in Derbyshire, England. See W. P. Cumming, *British Maps of Colonial America*, p. 12.

In 1978 the map was sold at auction in New York and acquired by Yale University's Center for British Art. Before the map left Chatsworth House it was carefully examined by G. M. Lewis, director of the Amerindian and Inuit Maps and Mapping Programme of the University of Sheffield. His unpublished notes of that examination have been invaluable in preparing the present description. The map was drawn on one sheet made up of seven panels of paper. The panel including portions of coastal Virginia and North Carolina appears to have been inserted as a correction by whoever prepared the original map. Black, red, yellow, and blue-gray inks were used in the drawing, and the paper bears a watermark. Lewis found the watermark (a strasbourg bend and lily) to be very similar to No. 437 in W. A. Churchill, *Watermarks on Paper*, Amsterdam, 1935. Lewis differed from Cumming in his opinion as to whether Hammerton drew the map. To Lewis's eye the map appeared to have been drawn by "a trained cartographer/draftsman/copyist." Among the differences between this map and the version in the Public Record Office, Lewis called attention to an inscription located north of the Roanoake River on both. On the Public Record Office map the inscription reads, "Remnants of Tuskaroroes Settled here," whereas on the copy now at Yale it reads, "Remnants of Tuskaroroes Settled here 1720."

The Hammerton version may be the earliest extant form of John Barnwell's map. It has "Carolina" displayed prominently across present Alabama and Mississippi, whereas the Public Record Office copy carries the name "Carolana." In 1722 the son of Dr. Daniel Coxe published a map and pamphlet supporting the Coxe claim to the 1629 grant of "Carolana" to Sir Robert Heath. The "Carolana" spelling shown on the Public Record Office version of the Barnwell map may have been inspired by Coxe's map and spelling. See Map 190 below for a description. In the area of present Georgia, the Barnwell-Hammerton map lacks the "Margravate of Azilia" and "Fort King George" designations found on the Public Record Office version. On the Barnwell-Hammerton the fort is indicated by an unnamed symbol, and in his correspondence Barnwell does not appear to have referred to the fortification on the Altamaha River as "Fort King George" until 1722. Generally speaking the Public Record Office map seems to embody many more details which would weigh in favor of expansionist policies being forwarded by the South Carolinians, especially as against the French in the Southeast. For information on Azilia and Fort King George see Maps 166, 177, and 178.

There is another, and presumably later, copy of Barnwell's map in the De Renne Collection in the University of Georgia Hargrett Library. It is described below as Map 255 Barnevelt ca. 1744 MS.

Reproductions

Plates 48, 48A–D.

Hatley, Tom. *The Dividing Paths.* New York, 1993, Figure 1, facing p. 146 (South Carolina area).

South Carolina Archives (photographs).

Reference

W. P. Cumming. *British Maps of Colonial America.* Chicago, 1974, p. 12.

Original

Yale Center for British Art.

---

185. King George's Fort 1722 MS

A Plan of King George's Fort at Allatamaha, South Carolina Latitude 31°12. North.

Size: 11 × 8. *Scale:* 1" = 300 feet.

Description: Attached to the plan is a copy of the colors of the Swiss Company, endorsed: "Carolina. Copy of the colours belonging to the Swiss Company, given them by The Missisipi Company in France | Reced from Mr Young Agent for South Carol[ina] | Reced Decr 4th. 1722."

Reproductions

Congress (photocopy).

Georgia Archives (photostat).

Hulbert, A. B. *Crown Collection.* Series III, No. 132.

Original

London. Public Record Office. C.O. 700, Georgia no. 4.

---

186. Stollard 1722 MS

[Plan of Fort King George and part of the Altamaha River]

Size: 17 × 22. *Scale:* 1" = 23 feet [for fort scale]; 1" = 1 mile [for the river].

Inset: An Abstract of the Journall of the Voyage from Fort King George in South Carrolina to St Simons Island & Barr in the Elizabeth Sloop Capt Stollard Commander. [The Abstract is dated August 1722.]

Size: 8 × 13½.

Description: Fort King George was started in 1721 for protection against possible Spanish reprisals or French encroachments. See Altamaha River ca. 1721 MS. In addition to the large-scale plan of the fort, there is a sketch map of the river for several miles below its site. Bearings and distances for the river passage are listed in eighteen consecutive "reaches" or segments in the Abstract.

Reproductions

Georgia Archives (photostat).

Hulbert, A. B. *Crown Collection.* Series III, Nos. 133, 134.

Congress (photocopy).

Original

London. Public Record Office. C.O. 700, Georgia no. 5.

---

187. Fort King George–St. Simon's Island ca. 1722 MS

[A. Fort King George. B. Part of St. Simon's Island, Georgia.]

Size: 16½ × 20. *Scale:* 1" = 24 ft. [for the fort]; 1" = 1 mile [for an outline drawing of St. Simon's Island].

Reproductions

Congress (photocopy).

Georgia Archives (photostat).

Original

London. Public Record Office. C.O. 700, Georgia no. 6.

---

188. Fort King George ca. 1722 MS A

The Ishnography or Plan of Fort King George

Size: 18 × 12. *Scale:* 1" = 20 feet.

Description: The fort is shown to occupy a point of land

where a marsh-bordered stream captioned "The Back River" joins the "Alatamahaw River." The landward side of the point is defended by a blockhouse and ditch-rampart and a "landing" is indicated on the Altamah River side. Within the defended area three rectangles are identified as "Barracks 80 feet long by 14 wide," "Barracks 60 feet long by 14 wide," and "Hospital 30 foot long." Outside the defended area the land is described as "An Old Indian field formerly a Town," and, closer to the fort, "The Turnip Patch."

Inset: The Profil of the ffort King George by a Scale of ten feet in an inch.

Reproductions

Georgia Archives (photostat).
Hulbert, A. B. *Crown Collection.* Series III, Nos. 135, 136.
Congress (photocopy).

Original

London. Public Record Office. C.O. 700, Georgia no. 7.

---

189. Fort King George ca. 1722 MS B

[A plan of Fort King George]

Size: 22 × 18. *Scale:* 1" = 32 feet.

Description: At bottom center is the "Explaination," its items labelled A to K. Items A to F describe the fort and its immediate surroundings. Item G, however, indicates "a river that discharges itself into the maine River that that [*sic*] coms within Land from St. Augusteen beyond St. Simons inlett." Additionally item K informs that, "Indian fields . . . run all along the side of ye River G."

Reproductions

Georgia Archives (photostat).
Hulbert, A. B. *Crown Collection.* Series III, No. 137.
Congress (photocopy).

Original

London. Public Record Office. C.O. 700, Georgia no. 8.

189A. Beaufort ca. 1722

Plan of Beaufort Town

Size: 15 × 9½. *Scale:* None.

Description: This plan of Beaufort, North Carolina, is laid off into 106 lots, with streets named. "James Green" appears on the right side. The plat is in the Secretary of State's Papers, Land Patent Book No. 7 (1707–40).

Original

North Carolina State Department of Archives and History.

---

190. Coxe 1722

Map of Carolana and of the River Meschacebe.

Size: 21½ × 16¼. *Scale:* 1" = ca. 118 miles.

In: Coxe, Daniel. *A Description of the English Province of Carolana.* London, Printed for B. Cowse, at the Rose and Crown in St. Paul's Church Yard, 1722, opp. p. 1.

Inset: A Map of the Mouth of the River Meschacebe. [bottom right corner] *Size:* 6¾ × 7⅜. *Scale:* None.

Description: This map shows the lands claimed by Dr. Daniel Coxe (1640–1730), who had procured the assignment of Sir Robert Heath's patent to Carolana (1629) by 1692. Dr. Coxe, who was physician to Charles II and Queen Anne, played an important part in early colonial history. He was nominally governor of West Jersey (1687–92) and was assisted in his various claims by his son Daniel (1673–1739), who came to America in 1702. In 1722 the younger Coxe published *A Description of the English Province of Carolana*, which contained memoirs of traders and explorers collected by his father, and set forth what is believed to be the first printed plan for a political confederation of the North American colonies. See *Dictionary of American Biography*, New York, 1928–30, IV, 485.

Apparently to avoid conflict with the established settlements of the lords proprietors only the country west of the settled portion of Carolina was claimed by Coxe. "The King and Council [had] declared Heath's charter void; but as it had not been legally so adjudged, Coxe's descendants obtained a recognition of their rights from the Board of Trade, and received from the Crown in 1768, in lieu of their claim to Carolina, 100,000 acres of land in the interior of New York": E. McCrady, *The History of South Carolina under the Proprietary Government, 1670–1719*, New York, 1897, p. 56, note 1. See also C. W. Alvord and L. Bidgood, *The First Explorations of the Trans-Allegheny Region*, Cleveland, 1912, pp. 229–49; F. C. Melvin, "Dr. Daniel Coxe and Carolana," *Mississippi Valley Historical Review*, I (1914), 257–62; G. D. Scull, "Biographical Notice of Doctor Daniel Coxe, of London," *Pennsylvania Magazine of History*, VII (September 1883), 317–38.

This is one of the very few maps that have "Carolana" on them; see also Farrer 1650 MS, Farrer 1651, and Barnwell ca. 1721 MS.

Reproduction

Winsor, J. *A Narrative and Critical History of America*, V, 70 (facsimile of part of the map).

Copies

American Geographical Society; John Carter Brown; National Museum of the American Indian, New York; New York Public.

In: [The same.] London, Printed for A. Bettesworth, at the Red Lyon in Pater-Noster-Row. 1726, opp. p. 1.

Copies

Clements 1726; Duke 1726.

In: [The same.] London, Printed for Edward Symon, against the Royal Exchange in Cornhill, 1727, opp. p. 1 (no reference to map in title page of volume).

Copy

John Carter Brown 1727.

In: [The same.] To which is added, A large and accurate Map of Carolana, and of the River Meschacebe. Printed for and sold by Olive Payne, at Horace's Head in Pope's Alley, Cornhill, opposite the Royal Exchange, 1741, front.

Copies

Congress; John Carter Brown; Clements; Charleston Library; Harvard; Kendall; New York Public; North Carolina State; Yale.

---

191. Delisle 1722

Tabula Geographica Mexicæ et Floridæ &c [across top of map, above the neat line] | Carte du Mexique et de la Floride des Terres Angloises et des Isles Antilles du Cours et des Environs de la Riviere de Mississipi. Dressée Sur un grand nombre de Memoires principalement sur ceux de M.rs d'Iberville et le Sueur Par Guillaume De l'Isle Geographe de l'Academie Royale des Sciences. A Amsterdam chez Jean Covens & Corneille Mortier Avec Privilege. 1722 [in cartouche, lower left] | I. Stemmers Senior Sculp. [below cartouche]

Size: 23½ × 18½. *Scale:* 1" = 150 miles.

In: Delisle, G. *Atlas Nouveau*. Amsterdam, 1730, No. 48.

Description: This is the same as Delisle's 1703 map; but it has a slightly different title, is smaller, and is from a different plate.

Reproduction

Color Plate 15.

Copies

Congress 1730, 1733, three copies; 1741; Clements 1733, three copies; Harvard 1733, 1741, ca. 1745.

In: Delisle, G. [*A Collection of Maps, 1722–1774*], No. 37.

Copies

Congress 1722–74.

---

192. Southeastern Indians ca. 1724 MS A

A Map Describing the Situation of the several Nations of Indians between South Carolina and the Massissipi [*sic*] River, was Copyed from a Draught Drawn & Painted upon a Deer Skin by an Indian Cacique: and Presented to Francis Nicholson Esq.r Governour of Carolina.

Size: ca. 57 × 45 (deerskin shaped). *Scale:* None.

Description: This large map states that it has an origin similar to the two others that follow, but it differs so markedly in detail and method of drawing that there is little apparent connection. It differs in the placement of the tribes and in the tribes mentioned; it extends its information beyond the Mississippi River while the other two restrict theirs to the Carolina area.

All three maps are valuable for the information they give concerning the location and names of the Indian tribes at this early period. The other two maps give the name and location of the Indian tribes near the coast; this map gives names and tribes from the Cherokees to the Mississippi and slightly beyond, including the Alabama and Mississippi state region.

The date of these maps has usually been given as about 1730. But Sir Francis Nicholson died in 1729. He was provisional governor of South Carolina from 1721 to 1729; however, he returned to England in 1724 and Arthur Middleton administered the government in South Carolina from 1724 until Robert Johnson's appointment in 1729. The original deerskins referred to in the titles of the maps would probably, therefore, have been made and presented by 1724, and these copies were probably made soon after Governor Nicholson's return to England in April 1724. In his innovative analysis of this map Gregory Waselkov chose to depart from the Public Record Office date and placed it at "circa 1723." He also omitted the phrase "& Painted" from the title information found on the map. See his "Indian Maps of the Southeast," *Powhatan's Mantle*, p. 324.

References

Penfold, P. A., ed. *Maps and Plans in the Public Record Office*. London, 1974, No. 2002, p. 342.

Wood, P. H., G. A. Waselkov, and M. T. Hatley, eds. *Powhatan's Mantle: Indians in the Colonial Southeast*, Lincoln, 1989, pp. 320–29.

Reproductions

Congress (photocopy).

Wood, P. H., G. A. Waselkov, and M. T. Hatley, eds. *Powhatan's Mantle: Indians in the Colonial Southeast*. Lincoln, 1989, Figure 4, p. 297.

Original

London. Public Record Office. C.O. 700, North American Colonies General no. 6(2).

---

193. Southeastern Indians ca. 1724 MS B

A Map Describing the Situation of the several Nations of Indians between South Carolina and the Massissipi [*sic*] River, was Copyed from a Draught Drawn & Painted upon a Deer Skin by an Indian Cacique: and Presented to Francis Nicholson Esq.r Governor of Carolina.

Size: ca. 44 × 32 (deerskin shaped). *Scale:* None.

Description: This map is numbered with the preceding map and has a similar title; but it is like the British Library map which follows, showing the location of the chief Indian tribes in North Carolina, South Carolina, and in the southern Appalachian mountain range. This and other maps of the Southeast, drawn by Indians or copied from their drafts, are reproduced and analyzed by Gregory A. Waselkov in his essay, "Indian Maps of the Colonial Southeast," appearing in Peter H. Wood, Gregory A. Waselkov, and M. Thomas Hatley, eds., *Powhatan's Mantle: Indians in the Colonial Southeast*, Lincoln, 1989, pp. 292–343.

Reference

P. A. Penfold, ed. *Maps and Plans in the Public Record Office*. London, 1974, No. 2001, p. 342.

Reproductions

Plate 48E.

Gabriel, R. H., ed. *The Pageant of America*. New Haven, 1929, II, p. 22.

Congress (photocopy).

Wood, P. H., W. A. Waselkov, and M. T. Hatley, eds. *Powhatan's Mantle*, Figure 3, p. 297.

Original

London. Public Record Office. C.O. 700, North American Colonies General no. 6(1).

---

194. Southeastern Indians ca. 1724 MS C

This map describing the scituation of the Several Nations of Indians to the N W. of South Carolina, was coppyed from a Draught drawn & painted on Deerskin, by an Indian Cacique and presented to Francis Nicholson, Esq.r, Governour of South Carolina by whom it is most humbly Dedicated to His Royal Highness George Prince of Wales.

Size: 44 × 32 (deerskin shaped). *Scale:* None.

Description: Similar in detail to the preceding map.

Reference

Wood, P. H., G. A. Waselkov, and M. T. Hatley, eds. *Powhatan's Mantle*, pp. 320–24.

Reproductions

Hulbert, A. B. *Crown Collection*. Series II, Vol. III, No. 37.

Winsor, J. *Narrative and Critical History of America*, V, 349.

Original

British Library. Additional MS 4723 (formerly Sloane MS 4723).

---

195. Charles Town ca. 1725 MS

A Platt of Charles Town

Size: 18 × 18 (irregular). *Scale:* 1" = 338 6/13 feet.

Description: This ground-plan of the land around Oyster Point divides the land between the Ashley and Cooper Rivers into about 300 numbered lots. It is thought to be a copy of the lost "Grand Model" referred to in grants and conveyances. The Grand Council of Carolina, meeting in the second year of the colony at Charles Town on the west bank of the Ashley, on April 22, 1672, ordered John Culpeper, the surveyor general, to lay out 12,000 acres of land between the Ashley and Cooper Rivers. On July 27, 1672, the council ordered Culpeper to lay out a town on Oyster Point. Culpeper fled the settlement at Charles Town the following year. Though the original of this plan has been ascribed to Culpeper, no records exist to show that Culpeper made the survey he was ordered to make, nor is the plat identifiable as a copy of any "Model" by Culpeper. The earliest recorded grant to a lot specified on the plan is February 3, 1678, and the settlement did not move from the west bank of the Ashley to Oyster Point until 1680. See Helen G. McCormack, "A Catalogue of Maps of Charleston," *Charleston Year Book*, 1944, p. 184.

This plat was removed by General Wilmot G. DeSaussure from some papers marked for destruction in the Charleston City Hall and is preserved in the South Carolina Historical Society archives. It is accompanied by a larger sheet, 34 × 25, listing the first grantees of town lots. The list, written in the same hand on the same kind of paper as the plat, bears the date 1725, which is therefore presumably the date of the plat.

The copy of another early (lost) plan of Charleston may be mentioned here: "A Plat of Charlestown containing — acres and sixteen perches with lots Market Place Burying Place Whart & Streets pr Stephen Bull a true copy Daniel H. Tillinghast, Sur. Gen." *Size:* 15¾ × 9½. After Maurice Mathews's removal as surveyor general in 1684, Stephen Bull's name appears as surveyor general to 1688. Daniel H. Tillinghast's term of office was in 1809, which is therefore the date of this copy, which is in the John McCrady Collection of Plats, Book I, p. 8, Office of the Registrar of Mesne Conveyance, Charleston, S.C. Bull's plat corresponds very closely to the 1725 plat listed above and is probably, according to Helen G. McCormack, op. cit., p. 185, from another copy of the lost original "Model."

References

McCormack, Helen G. "A Catalogue of Maps of Charleston." *Year Book* 1944, City of Charleston, S.C. Charleston 1947[1948], pp. 184–85.

Smith, Henry A. M. "Charleston—The Original Plan and the Earliest Settlers." *South Carolina Historical and Genealogical Magazine*, IX (1908), 11–27 (also gives references to other earlier surveys referred to in the records).

Reproductions

McCormack, H. G. Op. cit., opp. p. 186.

Smith, H. A. M. Op. cit., opp. p. 12.

Original

South Carolina Historical Society.

---

196. Fort King George 1726 MS

A Plan of Fort King George as it's now Fortifyed, 1726.

Size: 23½ × 18. *Scale:* None.

Description: There is an accompanying list of references from A to Z.

Reproduction

Georgia Archives (photostat).

Original

London. Public Record Office. M.P.G. 13.

196A. Popple 1727 MS

A Map of English and French Possessions on
the Continent of North America. 1727. H. Popple
[in cartouche at right center]

Size: 26¼ × 20⅜ (from photocopy).
Scale: 1" = ca. 116 miles.

Description: This colored manuscript may be Henry Popple's original compilation from which his famous engraved map was later developed (see map 216). Popple had no particular reputation as a cartographer prior to the appearance of his engraved map in 1733. He was, however, a member of a family of bureaucrats with close associations with the Board of Trade. On April 18, 1727, the same year in which this map was completed, Henry was named clerk there to fill the "vacancy occasioned by Mr. Hoskins dismissed."

Reproduction

Congress (photocopy).

Original

British Library. Add. MS. 23, 615, fol. 72.

---

197. Virginia–North Carolina Line 1728 MS A

We the underwritten Commissioners appointed
for laying out and settling the Boundary betwixt the
governments of Virginia & North Carolina. Pursuant
to His Majesty's Direction and Assent of the Lords
Proprietors of Carolina having in the months of March
and April last past caused a due West Line to be run
from the North shore of Currituck Inlet to Black Water
River, thence down the same according to it's [*sic*]
Meanders to the mouth of Notaway River altering the
Latitude Southward forty four chains, thence a due

West Line again crossing Meherrin River three times. And having now caused the said due West Line to be continued from thence to a Chestnut Oak on the Southside of the Southern Branch of Roanoke River after crossing the same four times according to the above Plan. We do hereby mutually agree that the same shall be and remain so far the Dividing Line betwixt the two Colonys. Done at the Camp on the South Branch of Roanoke River this seventh day of October Anno. 1728 [Carolina:] C. Gale T. Lovick E. Moseley W. Little [Virginia:] W Byrd R. Fitzwilliams W. Danbridge [Boundary Commissioners]

Size: 29 × 20½. *Scale:* 1" = 5 miles.

Description: A map showing the boundary between Virginia and North Carolina and signed by the commissioners of the two states at their camp on the south branch of Roanoke River on October 7, 1728. It is divided into two parts with one above the other. The map was received in London "with Majr. Gooch Lt. Govr. of Virginia's letter dated 26th March 1729" and "Read 3rd Jun 1729" where it was noted, "The Virginia Commissrs (being deserted by the Carolina Commissrs.) proceeded much higher up into the Country and sent another Plan of the whole Survey."

For the account of the colorful survey party that created this map and the two that follow there is no better source than William Byrd's *Histories of the Dividing Line Betwixt Virginia and North Carolina*, New York: Dover Publications, 1967. This is an unabridged republication of the work first published in 1929 by the North Carolina Historical Commission and has a new introduction by Percy G. Adams.

Reproductions

North Carolina Department of Archives and History (photocopy).

Congress (photocopy).

Original

London. Public Record Office. C.O. 700, Virginia no. 3.

---

198. Virginia-Carolina Line 1728 MS B

A Map of the Boundary between Virginia and Carolina laid out by his Majestys Order and Assent of the Lords Proprietors of Carolina in the Year 1728. Delineated by Wm Mayo.

Size: 55½ × 12. *Scale:* 1" = 5 miles.

Reproductions

Boyd, W. K., ed. *William Byrd's Histories of the Dividing Line.* Raleigh, 1929, opp. p. 322 (MS. C. 1).

Congress (photocopy).

Original

London. Public Record Office. C.O. 700, Virginia no. 5.

---

199. Virginia-Carolina Line 1728 MS C

[1. Copy of 1728 MS A above. 2. Copy of 1728 MS B above.]

Size: 40 × 7. *Scale:* 1" = 5 miles.

Description: The signatures on this copy are not original, and the two parts of the map have been joined to make the line of Lat. 36°31' N continuous. The dedication: "To His Excellency John Lord Carteret Palatin and the rest of the true and absolute Lords Proprietors of Carolina This Delineation of the Boundary betwixt the Province of Carolina & the Colony of Virginia so far as it is run & agreed unto by the Commrs. of both Governments is humbly Presented By Your Lords^s^ Most Obedient and Most Humble Servant E. Moseley Surveyor Genl of North Carolina." The further endorsement states: "Reced & read July ye 8th 1729."

Reproductions

Boyd, W. K., and P. G. Adams, eds. *William Byrd's Histories of the Dividing Line Betwixt Virginia and North Carolina.* New York, 1967, end.

Congress (photocopy).

Original

London. Public Record Office. C.O. 700, Virginia no. 5(1).

---

200. Moll 1728 A

A Plan of Port Royal = Harbour in Carolina with the Proposed Forts, Depth of Water &c. Latitude 32^D^—6 North. According to an Origenal Draught by H. Moll. [top left] | 1728 [upper left, under legend below title]

Size: 10⅝ × 7⅞. *Scale:* 1" = 4½ miles.

In: Moll, H. *Atlas Minor.* London, 1729, No. 52.

Description: This was one of the new maps made for Moll's *Atlas Minor,* which superseded an earlier atlas with thirty-two maps published about 1727. See Moll's introduction, quoted by P. L. Phillips, *A List of Atlases in the Library of*

Congress, Washington, 1909–20, I, 322–23. The *Atlas Minor* was issued with T. Templeton's *A New Survey of the Globe* in the early editions.

A caption just below the title box provides the following favorable endorsement for the area now occupied by a major U.S. Marine Corps training facility: "Port Royal River lies 20 Leagues from Ashley River S.W. it has a bold Entrance 19 or 20 Foot Low-water, the Harbour is large, safe, commodious and runs into ye best Country in Carolina. Here ye Air is always cleer and agreeable to Europian Constitutions. It Flows 9 Foot."

In the 1729 edition of Moll's atlas, under a legend describing the harbor, is the date 1728, which is deleted in the 1732 and subsequent issues. The number of the map, 52, is added to the top right corner in the issue of 1732 and thereafter.

Reproduction

Color Plate 16.

Copies

Congress 1729, 1732?, 1736?, 1763?, two copies; American Geographical Society 1736?; John Carter Brown 1732?; Duke 1736?, three copies; Harvard 1732?; Huntington, two separates with the date 1728 unerased; New York Public 1745, two separates; South Carolina Archives and History, separate; Clements 1736.

---

201. Moll 1728 B

Florida Calle'd by y^e French Louisiana &c. By H. Moll Geographer 1728

Size: 10½ × 7¾. *Scale:* 1" = ca. 165 miles.

In: Moll, H. *Atlas Minor*. London, [1728?], No. 54.

Description: This is one of the early printed maps giving the roads and trading paths from Carolina westward. It follows in general conception and in many details Moll's "A New Map of the North Parts of America claimed by the French . . . 1720" for that area which the later map shows.

Beginning with the 1732 edition and thereafter, "1728" and the erroneous apostrophe in "Calle'd" are deleted, "Georgia" and "54" (the number of the map, placed above the neat line to the right) and "Cherekeys 30 Vill." are added.

Copies

Congress 1729, 1732?, 1736?, 1763?, two copies; American Geographical Society 1736?; John Carter Brown 1732?; Duke 1736?, three copies; Harvard 1732?; New York Public 1728?, 1745?, two separates.

---

202. Crouch 1728

The English Empire in America By R. B.

Size: 2⅞ × 5⅞. *Scale:* None.

In: Burton, Robert. *The English Empire in America*. Sixth Edition. London, Printed for A. Bettesworth and J. Batley, 1728, front.

Inset: New Found Land. *Size:* 1 × 1.

Description: Though this is a map from a new plate with a different title, it is a copy of a map which appeared in several earlier editions of this work. See Crouch 1685. The Harvard 1728 and Brown 1729 copies lack the map.

Copies

John Carter Brown 1728; Clements 1739; Harvard 1739.

---

203. Maule-Moseley 1729 MS

This Plan represents Roanoke Island containing Twelve thousand acres of Land & Marsh, as it was surveyed. Anno 1718 By W. Maule Surveyr. Genl. A true Coppy By E Moseley Survr. Genl. 1729

Size: 17¾ × 9 (photocopy). *Scale:* None.

Description: This is a simplified copy of the Maule 1718 MS map, with fewer legends and only slight topographical information.

Reproduction

North Carolina Department of Archives and History (photocopy).

Original

Owned by John Wood, Edenton, North Carolina.

---

204. Gascoigne-Swaine ca. 1729 MS

A Plan of Port-Royal, South Carolina. Copy'd from a plan of Capt. John Gascoigne by Francis Swaine [Title in top left corner: "Remarks" and "Explanations" below]

Size: 25 × 32. *Scale:* 1" = 1 mile.

Description: Under "Remarks" is written "Variation of the Compass 1729 - 1°: - 24' E." This chart is so dilapidated and fragmentary as to be almost illegible. Although the *Catalogue of Maps, Plans, and Charts in the Library of the Colonial Office*, London, 1910, North American Colonies, p. 9, gives the date 1729 to this manuscript, the style of drawing, with its great care and detail, suggests that it may have been prepared in the 1770s as an engraver's copy of the original 1729 plan by Gascoigne.

If this is an engraver's copy, it may have been prepared for Jefferys and Faden's "A Plan of Port Royal in South Carolina. Survey'd by Capn. John Gascoigne . . . Engraved by Jefferys and Faden. Geographers to the King [across top, below neat line, followed by explanations]. *Size:* 22¾ × 27¹⁵⁄₁₆. *Scale:* 1" = 1 mile.

Jefferys and Faden's chart was apparently published in 1776, not ca. 1773, the date usually assigned, as can be seen by the reference in Le Rouge's edition mentioned below. Jefferys and Faden follow the MS with great exactness, though Swaine's manuscript copy represents a slightly larger area, including the whole of Port Royal Island.

In Sayer and Bennett's *North American Pilot*, London, 1777, Pt. II, No. 9, is a later edition of the same chart with corrected soundings and additional information, such as "Twenty-five Families on this Island" on Trench's Island and "Littleton Fort or New Fort" on Port Royal Island or Beaufort. The Jefferys-Faden imprint is absent, and near the top right, under title and notes, is: "Printed for R. Sayer and J. Bennett map & chart-sellers' No. 53 Fleet Street, as the Act directs 15 May 1776." A later issue is unchanged except for the imprint: "Published 12th May 1794 by Laurie & Whittle. 53 Fleet Street, London," in Laurie and Whittle's *North American Pilot*, London, 1800, Pt. II, No. 16. An Admiralty chart, apparently using the Gascoigne-Swaine MS, but giving new soundings, was published in 1777 with the following title across the top of the chart: "Port Royal in South Carolina taken from Surveys deposited at the Plantation office | Publish'd as the Act directs by J. F. W. Des Barres Esq. Augst. the 10th 1777 [bottom right]. *Size:* 22⅛ × 28. *Scale:* 1" = 1 mile. This is found in copies of the third volume of the *Atlantic Neptune*.

A French edition of the Jefferys and Faden chart, including "D'awfoskee," was engraved by Petit and published by order of M. de Sartine in 1778: "Plan de Port Royal et de la riviere et d'etroit D'awfoskee Levé par le Capt. John Gascoigne . . . 1778." *Size:* 23 × 34¼. *Scale:* 1" = 1½ miles.

A somewhat smaller edition is found in G. L. Le Rouge, *Pilote American Septentrionale*, Paris, 1778, No. 23: "Port Royal Dans la Caroline Meridionale Levé par le Capt. Gascoigne Publié a Londres en 1776 Traduit A Paris Chez Le Rouge Rue des Augustins 1778." *Size:* 22⅝ × 24½. *Scale:* 1" = 1 mile. The absence of "Fort Littleton or New Fort" (which was built in 1757) shows that Le Rouge based his map on Jefferys-Faden, not on the revised Sayer-Bennett. Copies of all these variations of the Gascoigne map are to be found in the Library of Congress. Henry Stevens and Roland Tree, "Comparative Cartography," *Essays Honoring Lawrence C. Wroth*, Portland, Me., 1951, p. 352, note four issues of the Port Royal map: (a) Jefferys and Faden [1773], (b) Sayer and Bennett 1776, (c) Sayer 1791, and (d) Laurie & Whittle 1794.

About 1776 Jefferys and Faden published a companion piece to the Port Royal chart, also based on Gascoigne's 1729 surveys, with the title "A Plan Of the River, and Sound of D'Awfoskee, in South Carolina, Survey'd by Captain Gascoigne." [Top left of chart, above "Remarks"]. *Size:* 17¾ × 25½. *Scale:* 1" = 1 mile. Copies are found in the Harvard College Library and Charleston Museum. Sayer and Bennett (1776), Laurie and Whittle (1794), De Sartine (1777), and Le Rouge (1778) also published editions of the D'Awfoskee chart. See Introductory Essay.

Reproduction

Congress (photocopy).

Original

London. Public Record Office. C.O. 700, Carolina no. 7.

---

205. Aa 1729

Partie Meridionale de la Virginie, et la Partie Orientale de la Floride, dans L'Amerique Septentrionale suivant les Mémoires les plus exacts de ceux qui les ont decouvertes. de nouveau mises en lumiere par Pierre vander Aa, Marchand Libraire À Leide a Amsterdam chez J. Cóvens et C. Mortier.

Size: 13⅞ × 11¾. *Scale:* 1" = ca. 39 miles.

In: Aa, P. vander. *La Galerie Agreable du Monde*. Leiden [1729], LXIII, No. 17 (21 in "Table des Figures," Yale copy).

Description: This peculiar map shows the coast from Baye S. Matheo to the eastern shore of Virginia (30° to 38°) and gives detail for the interior garnered indiscriminately from various sources.

Apparently Aa based his work on a Mercator-Hondius type map, probably a Jansson, Blaeu, or Montanus, and added to it as much detail as he could utilize from an Ogilby-Moxon type map. The Lederer material, consequently, is abbreviated and distorted. It has "Plaine decouverte d'Eau" for the savanna region, just above the smaller lake with the great waterfall. The great lake is given no name. The names in the extreme western part of the map are based on Delisle and are derived eventually from the strange and discredited tales of the seventeenth-century Englishman Brigstock: see Delisle 1703 and Schröter 1753. "Chez J. Cóvens et C. Mortier" is not on the Yale copy of the map. The map has been removed from the Library of Congress set of vander Aa's work.

The work by Aa in which this map occurs is one of the rarest and most magnificent of its kind that have been produced, in sixty-six volumes with over 3,000 maps and views, some by Aa and others from plates of many of the great European mapmakers of the preceding half century. Aa stated that only 100 copies of the great work were made: See P. L. Phillips, *A List of Atlases*, III, 288.

Reproductions

Plate 49.

Cumming, W. P. "Geographical Misconceptions of the Southeast in the Cartography of the Seventeenth and Eighteenth Centuries." *Journal of Southern History*, IV (1938), opp. p. 492.

Copies

Congress, separate; Harvard 1729–50; Newberry (Ayer) 1729; New York Public 1729; Yale 1729.

In: Cóvens, J., and C. Mortier. *Atlas Nouveau.* Amsterdam, [1683–1761], IX, No. 56.

Copies

Congress 1683–1761.

---

206. Moll 1729

Carolina. By H. Moll Geographer 1729

Size: 10¾ × 7⅞. *Scale:* 1" = 75 miles.

In: Moll, H. *Atlas Minor.* London, 1729, No. 51.

Description: This map, which extends from "The South Bounds of Carolina" to "C. Charles" in Virginia, shows the location of Indian tribes in the Carolinas. It gives the chief roads or trading routes westward from Charleston, and many islands along the coast are identified by name for the first time on a printed map. The destruction of "St. Maria de Palaxy" on the Gulf is noted, as are Col. Barnwell's defeat of the Indians in North Carolina in 1712 and Col. Craven's victory in 1716. The earlier states of the map have "Azilia" in South Carolina for the much-advertised but abortive Margravate of Sir Robert Montgomery (see under 1717). The 1736 edition of the map was copied by Homann in 1737 as one of the four divisions of his frequently published "Dominia" map. In later impressions of Moll's map "Azilia" was changed to "Georgia."

A great deal of new information concerning names is found in this work of Moll, though it shows little topographical improvement over his previous maps. Five different impressions of this map are found in Moll's *Atlas Minor* and in Salmon's *Modern History*, London, 1739.

Reproduction

Plate 50.

A. Title as given above, with "1729," and with "Azilia" on the map in South Carolina.

Copies

Congress 1729; Duke.

B. The title as given above, with erasure of date "1729" and with "51" at top right.

Copies

Congress 1732; John Carter Brown 1732?; Duke 1732?; Harvard 1732?; Sondley.

C. Title as in (B) and with "according to the last charter" added to original "The South Bounds of Carolina."

Copies

Congress 1736, 1736?, two separates; American Geographical Society 1736?; Duke 1736?, three copies; New York Public 1745?, two separates; Sondley; University of South Carolina 1745?; Clements 1736.

D. Title as above, with "51" and "according to the last charter," but with "Vol. 3 p. 589" at top left, within neat line.

In: Salmon, Thomas. *Modern History.* London, 1739, III, opp. p. 589.

Copies

Clements 1739; Yale 1739.

E. Title as in (C), with "Vol. 3 p. 562" at top left, within neat line. This was taken from Samuel Simpson, *The Agreeable Historian*, London, 1746, vol. III, p. 562.

Copies

Congress; British Library.

F. Title changed: North and South Carolina, Georgia & part of E. Florida By H. Moll Geographer. ("Azilia" erased, "Georgia" substituted; volume and page reference erased.)

Copies

Congress 1763?, two copies (of Moll's *Atlas Minor*); Yale.

G. Title as in (B) lacking "according to the last charter," "Georgia," "Mantances I.," "Mantances Inlet.," and "Sault R." Added just outside the lower line indicating longitude are the capital letters "A" (center), and "H" (three at right-hand corner). Probably engraved for the pirated edition of Moll's *Atlas Minor*, published in Dublin by G. Grierson in 1740.

Copy

University of North Carolina.

---

207. Hunter 1730 MS

[Cherokee Nation and the Traders' Path from Charles Town via Congaree]

Size: 25⅞ × 16½.

Scale: Varies greatly for different parts of the map.

Description: This pencil-and-ink drawing shows the Indian trading path crossing the branches flowing into the Congaree (Santee) from the south, crossing many branches of the Savannah River near its head, and finally reaching the Tellico Plains region of the Overhill Cherokee in Eastern Tennessee. The Indian villages along the route are indicated and named.

The map has no title, but it has the following certification: "This represents the Charecke Nation by Col[l] Herberts Map & my own Observations with the path to Charles Town, its Course & (distance measured by my watch) the Names of y[e] Branches, Rivers & Creeks, as given them by y[e] Traders along that Nation May 21, 1730, certified by me George Hunter."

The map, with its numerous notes and comments, is a rich mine of information for contemporary trading conditions. This Congaree route became important after the Yamassee and Creek attacks made the southern trail more exposed; the building of Fort Congaree in 1718 gave added protection to it. See V. Crane's *The Southern Frontier*, pp. 128–30, 350, and A. S. Salley, ed., "George Hunter's Map of the Cherokee Country and the Paths thereto in 1730," *Bulletin of the Historical Commission of South Carolina*, No. 4, Columbia, S.C., 1917. See Herbert-Hunter 1744 MS.

Reproduction

Salley, A. S. Op. cit., frontispiece.

Williams, S. C. *Early Travels in the Tennessee Country, 1540–1800*. Johnson City, Tenn.: Watauga Press, 1928, opp. p. 114 (upper part).

Original

Congress (Faden Collection, No. 6).

---

208. Cóvens-Mortier-Delisle ca. 1730

Carte de la Louisiane et du Cours du Mississipi Dressée sur un grand nombre de Memoires entr' autres sur ceux de M[r] le Maire Par Guill[me] de l'Isle de l'Academie R[le] des Sciences [above neat line] | a Amsterdam Chez Jean Cóvens et Corneille Mortier Geographes. [bottom center, above scale]

Size: 23⅜ × 17⅛. *Scale:* 1" = ca. 85 miles.

In: Delisle, G. *Atlas Nouveau*. Amsterdam, 1730, No. 47.

Inset: Carte Particuliere de Embouchures de la Riviere S. Louis et de la Mobile [bottom right]. *Size:* 5⅞ × 4¾. *Scale:* 1" = ca. 42 miles.

Description: See Delisle 1718; this is a small edition of the same map.

References

Phillips, P. L. *Lowery Collection*, No. 269.

Wymberley Jones De Renne Georgia Library Catalogue, III, 1195–96 (dated erroneously [1718]).

Copies

Congress 1730, 1733, three copies, 1741; Clements 1733, three copies; Georgia (De Renne); Harvard 1733, 1741, ca. 1745; Kendall; Yale 1730, 1742, 1757.

In: Sanson, N., and H. Iaillot. *Atlas Nouveav*. Amsterdam, ca. 1760, III, 98.

Copies

Clements 1760.

---

209. Moll 1730

A Map of the Province of Carolina, Divided into its Parishes, &c. according to the latest accounts. 1730. By H. Moll Geographer.

Size: 15¼ × 14. *Scale:* 1" = 31 miles.

In: Humphreys, David. *An Historical Account of the*

Incorporated Society for the Propagation of the Gospel in Foreign Parts. London, 1730, p. 81.

Inset: A Map of y[e] most Improved Part of Carolina.

Size: 7 3/8 × 5 1/8. *Scale:* 1" = 8 miles.

Description: Though this map does not extend as far to the interior, the information on it is very similar to that in Moll's Carolina map of the preceding year. The inset is similar to the "Improved Part of Carolina" inset in Moll's Dominions map 1715, though it does not have the names of settlers.

Reproductions

Winsor, J. *Narrative and Critical History of America*, V, 351 (inset only). Comment, Ibid., V, 348.

Copies

Congress; John Carter Brown, two copies; Charleston Library Society; Clements; Duke; Harvard; New York Public; Sondley; University of North Carolina; Yale.

---

210. Catesby [1734]

A Map of Carolina, Florida and the Bahama Islands with the Adjacent Parts.

Size: 23 3/4 × 17 1/4. *Scale:* 1" = ca. 85 miles.

In: Catesby, Mark. *The Natural History of Carolina, Florida, and the Bahama Islands.* 2 vols., London, 1731–32 (variously placed).

Description: This map is not an original work by the otherwise versatile Catesby. It is based largely on Henry Popple's cartographic opus, "A Map of the British Empire in America with the French and Spanish Settlements adjacent thereto" (Plate 55). For readers of his *Natural History* it provided, in the words of Catesby scholar G. F. Frick, "a good representation of the better English ideas about the geography of North America in the 1730's and early 1740's." See G. F. Frick and R. P. Stearns, *Mark Catesby: The Colonial Audubon*, Urbana, 1961, p. 71.

The apparent contradiction of Catesby's use of a map published in 1733, or two years after the first volume of his *Natural History* appeared, has been a source of puzzlement, if not confusion. Nor has that confusion been lessened by the sometime placement of the map in volume II rather than volume I. The clue to the explanation of the placement and dating of Catesby's map is provided in the changed wording found on the title pages in later editions of *Natural History*. The title page in the 1731 edition includes the statement, "To the *Whole* is Prefixed a New and Correct Map of the Countries Treated of" (emphasis added). In later editions the statement has been changed to "To Which is Prefixed, a New and Correct Map of the Countries; with Observations on Their Natural State, Inhabitants, and Productions." What appears to have occurred is that original subscribers to Catesby's *Natural History* were assured of a map with the *whole*, or completed, work and received their maps bound in volume 2 when it finally appeared in 1743. Later purchasers were in turn provided with "a New and Correct Map" properly "Prefixed" at the beginning of their first volume of Catesby's *Natural History*. In some cases where the volumes have been rebound the map may, of course, be found at the beginning of the first volume of the early edition.

While never published as originally intended, a version of Catesby's map was destined to become famous 250 years after he helped prepare it. Showing the area from Cuba to the North Carolina Outer Banks and west to Mobile Bay, this version forms an inset continuation of the eastern seaboard of the United States on a Rawlinson copperplate discovered in Oxford's Bodleian Library in 1986. See Pearce S. Grove, "The Discovery at Oxford," *The Map Collector* 56 (Autumn 1991), p. 12. M. B. Pritchard and V. L. Sites have mounted a convincing argument supporting the conclusion that Catesby was directly involved with the etching of the inset map for his friend the well-known Virginian. See their *William Byrd II and His Lost History: Engravings of the Americas* (1993). Although Catesby's contributions to the project took place in 1737, Byrd's book was never published nor has his manuscript been found. In 1988, however, the National Geographic Society used an image of the Rawlinson copperplate with Catesby's map to grace the cover of their centennial publication, *Historical Atlas of the United States.* Regrettably, the details of the plate's provenance were at that time uncertain, so Catesby failed to receive the credit he was due.

Copies

Congress 1731, 1754, 1771, two separates; American Geographical Society 1731; John Carter Brown 1731, 1771; Charleston Library Society 1771; Duke 1754; Georgia, Hargrett Library 1771; Harvard 1731, 1754, 1771; Kendall 1754, separate; New York Public 1754, two copies, 1771; University of North Carolina 1754; Yale 1743; Clements 1731.

211. Georgia 1732

[Southeastern North America]

Size: 7³⁄₈ × 5¹⁄₄. *Scale:* 1" = ca. 190 miles.

In: Some Account of the Designs of the Trustees For Establishing the Colony of Georgia in America. London, [1732], sheet preceding page 1.

Description: This map, extending from "Carituck R" to "Cape Florida" and west to the "Messesipi River," is usually considered the first map with the name "Georgia" upon it and is a close copy of the Nairne inset in the Crisp 1711 map. The four-page pamphlet in which it first appears is in two editions, the second edition having a different set of engravings. In the "A" edition the lower half of page four has the map in the De Renne copy; the lower half of the same page is blank in the John Carter Brown copy, the "B" edition, but has the map on the sheet preceding the first page of the text. The map itself is known in at least three states. In the De Renne copy is an inscription concerning the Carolina Indians, which is similar to a legend in the Nairne inset and which is written across the peninsula of Florida. In the John Carter Brown copy this inscription has been partially obliterated from the plate. In the state of the map found in Smith's *Sermon* and Martyn's *Reasons*, both published the following year (1733), additional erasures have been made on the plate. See the note by L. C. Wroth in Verner W. Crane's "The Promotion Literature of Georgia," *Bibliographical Essays: A Tribute to Wilberforce Eames*, Cambridge, Mass., 1924, p. 287, note 18; L. L. Mackall, "The Wymberley Jones De Renne Georgia Library," *Georgia Historical Quarterly*, II (June 1918), 74–76; *Catalogue of the Wymberley Jones De Renne Georgia Library at Wormsloe*, Isle of Hope near Savannah, Georgia, Wormsloe, 1931, I, 18.

In his essay "Oglethorpe and the Earliest Maps of Georgia," Louis De Vorsey explains the role of these erasures to the printer's plate as a part of Oglethorpe's clever use of maps to help insure the success of the Georgia colonial venture. See P. Spalding and H. H. Jackson, eds., *Oglethorpe in Perspective: Georgia's Founder After Two Hundred Years*, Tuscaloosa, 1989, pp. 30–34.

Reproductions

De Renne Georgia Library Catalogue, I, 19 (first state).

Ibid., p. 49 (from Smith's *Sermon*).

De Vorsey, L. "Oglethorpe and the Earliest Maps of Georgia," in Spalding and Jackson, eds. *Oglethorpe in Perspective*. Tuscaloosa, 1989, pp. 23–43 and p. 31, Figure 2.4.

Gabriel, R. H., ed. *The Pageant of America*. I, 272 (Martyn's *Reasons*).

Winsor, J. *Narrative and Critical History of America*. V, 365 (Martyn's *Reasons*).

Copies

John Carter Brown 1732; Georgia (De Renne) 1732.

In: Smith, Samuel. *A Sermon Preach'd before the Trustees for Establishing the Colony of Georgia in America . . . In the Parish Church of St. Augustin, On Tuesday February 23, 1730/31.* London, 1733, opp. p. 34.

Copies

Congress 1733; Duke 1733; Georgia (De Renne) 1733.

In: Martyn, Benjamin. *Reasons for Establishing the Colony of Georgia.* London, 1733, p. 1.

Copies

Congress 1733; John Carter Brown 1733, three copies; Duke 1733; Georgia (De Renne) 1733, three copies; Huntington 1733, two copies; Kendall 1733; Massachusetts Historical Society 1733; Newberry 1733; New York Public 1733, two copies.

212. Carolina-Georgia post-1732 MS

Carte de la Caroline du nord et du sud

Size: 13³⁄₈ × 12³⁄₄. *Scale:* 1" = ca. 68 miles.

Description: This map was apparently drawn from an English original after 1732 and before the French and Indian Wars. It is based on some Lederer-type map (1672) of the late seventeenth century, as it has "Lac d'Ushley," "Savane de 80 lieues de long," and other Lederer details. Above "R. May" is written "Nouvelle Georgia."

Reproductions

Karpinski series of reproductions.

Original

Paris. Service Hydrographique de France. C. 4044 Etats Unis. Cartes Générales. No. 40.

213. Beaufort ca. 1733 MS

Beaufort S° Carolina

Size: 22¹⁄₂ × 18. *Scale:* 1" = 200 feet.

Description: This plan has no nomenclature except for the

identifying title written in the upper right corner and a half-obliterated legend at the bottom right: "Beaufort, 200 feet in an Inch." Only the plan of the streets is given. This is probably the plan referred to in a letter from Governor Johnson to the duke of Newcastle, dated March 30, 1733: "Mr. Yonge will lay before your Grace a plan to the Town of Beaufort on Port Royal River, and has directions to assist Mr. Fury in solliciting H. M. that that port and harbour may be fortified and secured" (from Cecil Headlam and A. P. Newton, eds., *Calendar of State Papers, Colonial Series, America and West Indies*, 1733, London, 1939, p. 68, No. 86).

In 1721 an act providing an assize court at Beaufort was posted; in 1722 a free school was established there (Edward McCrady, *The History of South Carolina under the Royal Government*, pp. 44, 46).

Reproduction

Congress (photocopy).

Original

London. Public Record Office. C.O. 700, Carolina no. 9.

---

214. Byrd 1733 MS

My Plat of 20,000 acres in N° Carolina. Survey'd in September 1733, by Mr. Mayo, being 15 miles long, 3 Broad at the W. End, & one at the Est.

Size: 14½ × 9¾. *Scale:* None.

In: Byrd, William. *A Journey to the Land of Eden.* [1733].

Description: This map shows the tract of land, located along and south of the Virginia–North Carolina line near the junction of the Dan and Irvine Rivers, which Byrd bought from the North Carolina commissioners in 1728. They, in turn, had received it as a recompense for their services as surveyors. In September and October of 1733, Byrd laid out his purchase and wrote his *Journey to the Land of Eden*. He called the tract Eden because of its supposed fertility; no doubt the pun on the name of Governor Eden also pleased him. This possession of William Byrd is not to be confused with the "new gefundenes Eden," which he sold to the Helvetian Society in 1736. See Eden 1737.

The manuscript volume in which the map occurs is described in T. H. Wynne's edition of Byrd's works, listed below, which also has a reproduction of the map. *A Journey to the Land of Eden* has been published also in *The Westover Manuscripts*, ed. E. Ruffin, Petersburg, Va., 1841; in *The Writings of Colonel William Byrd of Westover in Virginia Esqr.*, ed. J. S. Bassett, New York, 1901; and in Mark van Doren's edition, New York, 1928.

In 1743, Byrd took out an additional patent from the North Carolina government for 6,000 acres to include the rich site of the abandoned Indian Saura Town. The map is found in the Westover Manuscripts owned by George Harrison of Brandon and now in the custody of a New York Trust Company (see Boyd's edition of the *Histories*, p. xvi). The MS volume in which it is found has been fully described in Wynne's edition, p. xvii.

Reproductions

History of the Dividing Line and other Notes, from the Papers of William Byrd. Ed. Thomas H. Wynne, Richmond, 1866, II, p. 37.

William Byrd's *Histories of the Dividing Line.* Ed. W. K. Boyd. Raleigh, 1929, opp. p. 168.

William Byrd's Histories of the Dividing Line Betwixt Virginia and North Carolina. Eds. W. K. Boyd and P. G. Adams.

Original

Westover Manuscripts (see above).

---

215. Crenay 1733 MS

Carte De partie de la Louisianne qui comprend le Cours du Missisipy depuis son embouchure jusques aux Arcansas celuy des rivieres de la Mobille depuis la Baye jusqu'au Fort de Toulouse: des Pascagoula de la riviere aux Perles, le tout relevé par estime. Fait a la Mobille en Mars 1733 par les soins et recherches de Monsieur le Baron de Crenay Lieutenant pour le Roy et commandant a la Mobille. [cartouche and scale, top left]

Size: 44½ × 30⅜. *Scale:* 1" = ca. 10 miles.

Description: This large map of the territory between the Chattahoochee and the Mississippi Rivers was compiled by Baron de Crenay, commandant of the post at Mobile. It is a beautifully made map and has much detailed information concerning the location of Indian tribes and trading paths, information important in the intense rivalry between the French and English.

Reproductions

Hamiton, Peter J. *Colonial Mobile.* Boston, 1910, p. 190.

Karpinski series of reproductions.

Swanton, J. R. *Early History of the Creek Indians and Their Neighbors.* Washington, 1922, Plate 5.

Original

Paris. Ministère de Colonies. No. 1, Louisiane.

215A. Wimble 1733 MS

A Large and aisect Drafe [exact Draft] of the Sea Cost of N^{o}. Carolina with the true Latt^{d}. of ye Harbours from Sholote inlet to Coretuck with a true Draft of Cape faire Rever the depth of Water and the Land Marks of Sailing into Cape faire Rever the Govennors' pint on ye w^{t}. [west side] of the Rever Gest apesid [just opposite] with the ball heed Roils and And Leads vou over the Barr in 3 fath^{m}. water as you may See by ye line Dran [Drawn] A : B [Top center; no cartouche]. Drafted and Layed out in AP^{l}. 16:1733 By James Wimble wo as yausd [who has used] the Coast and trad [traded] this 12 yeare past Desiering the same mav Be in Prent for the safe Gard of Shipeing and trad j Wimble [bottom left; no cartouche]

Size: 31¾ × 12¾. *Scale:* 1" = 3 nautical miles.

Description: This manuscript chart by a practical navigator named James Wimble, although crude in its execution, contains more hydrographic detail than any chart of the area covered available until that time. It shows the entire complex coastline of North Carolina, with depth soundings in the sounds and offshore as well as the names of islands and ship passages through the sounds. While not primarily concerned with interior settlements, Wimble shows "New Carthage town" with "Wimbl Castelell located across the lower Cape Fear River from "Browmswick [Brunswick] town." In a letter with the chart, Wimble discussed tidal currents as well as pilotage landmarks and important anchorages, with the request that his work be delivered to "M^{r}. Mount On Tower Hill London." Mount was a leading map and chart publisher of the day who had one of Wimble's later maps engraved for sale in 1738. See Map 241, "Wimble 1738," below. Significantly, "Wilmington" is shown on the printed chart where Wimble had originally indicated the site of "New Carthage" and "Wimbl Castele" on his manuscript.

Wimble was born and spent his youth on the Sussex coast. His baptism is recorded on January 31, 1696 [1697], in the parish register of the Hollington Church-in-the-Wood, where births and deaths of Wimbles continued to be recorded into the nineteenth century. To judge by his spelling, his formal schooling was rudimentary at best; but he did receive practical training as a seaman and sailed his own vessel to the West Indies in 1718. Sometime before 1722 Wimble had acquired land on the island of New Providence in the Bahamas. Having lost his own ship he was actively engaged as a sailing master on the sloops of others transporting cargoes between Jamaica and the Carolinas. In 1723 Wimble received a grant of 640 acres at the head of "Scuppernong River," probably located near present-day Columbia, North Carolina. Marrying and starting a family in the locality of today's Tyrell County, North Carolina, Wimble set up a distillery. In 1726 he appears to have sold his North Carolina distillery and removed to Boston, where he continued his maritime activities, frequently sailing to North Carolina and British West Indies ports. A small marginal box on the 1738 printed chart contains the note: "Sold by W. Mount & T. Page on Tower Hill and J. Hawkins in Fenchurch Street London and the Author at Boston in New England." Wimble did not abandon his interests in North Carolina but continued to acquire land there, and in 1733, he was actively engaged in the promotion of New Carthage, the settlement that ultimately became known as Wilmington. Wimble's house and garden site, then occupied by his son, was shown on the map of Wilmington completed by Claude Joseph Sauthier in December 1769. See Map 378 "Sauthier-Wilmington 1769 MS," below. In the period marked by conflict with Spain known as the War of Jenkins Ear, Wimble turned to privateering to recoupe earlier losses. In a heated battle with a large Spanish ship "mounting thirty guns and 250 men and passengers," he suffered the loss of his left arm "about 5 inches from my body." In spite of "loosing a great deal of blood," Wimble survived to assure his monarch, "If His Majesty would trust me with a 20 gun ship the Spaniards should well pay for it." Not one to be beached by the loss of an arm, Wimble continued his privateering assaults on Spanish shipping until, according to a report, he lost his vessel "on Cuba." Whether he fell victim to Spanish guns, unmarked reefs, or stormy weather is not recorded; but James Wimble, the mariner-cum-distiller–land speculator, and author of North Carolina's 1733 coastal chart was never heard from again. For citations to the sources of Wimble's biographical details, see the two articles by William P. Cumming entitled "The Turbulent Life of Captain James Wimble" and "Wimble's Maps and the Colonial Cartography of the North Carolina Coast," appearing in *The North Carolina Historical Review* 46, nos. 1 and 2, pp. 1–18, 157–70, respectively.

Reproduction

Cumming, W. P. "Wimble's Maps and the Colonial Cartography of the North Carolina Coast," *The North*

Carolina Historical Review 46, no. 2, between pp. 164 and 165 (photograph printed with Map 241 on reverse).

Original

Oxford. Bodelian, Rawlinson MS. D. 708, f. 207r.

---

216. Popple 1733 A

A Map of the British Empire in America with the French and Spanish Settlements adjacent thereto. by Henry Popple. C. Lempriere inv. & del. B. Baron sculp. [cartouche, bottom left, sheet 17] | To the Queen's Most Excellent Majesty this Map is most humbly Inscribed by Your Majesty's most dutiful, most Obedient, and most Humble Servant Henry Popple [lower right, sheet 15 | Sold by S. Harding on the Pavement in St. Martins Lane, and by W. H. Toms Engraver in Union Court near Hatton Garden Holborn

	£	s.	d.
Price in Sheets	1	11	6
Bound	1	16	6
On Rollers & Coloar'd	2	12	6

[below neat line, left, sheet 17] | London Engrav'd by Willm. Henry Toms 1733. [below neat line, right, sheet 20]

Size: 102 × 104 (in twenty sheets varying in size; the Carolina sheet, No. 10, is 26¾ × 19⅜).
Scale: 1" = 40 miles.
Insets: Twenty-two insets, including: The Town and Harbour of Charles Town in South Carolina [sheet 12] *Size:* 6 × 4¾. *Scale:* 1" = 1½ furlongs (ca. ⅕ mile).

Description: Extending from the Yucatan peninsula and Jamaica northward to Hudson's Bay and from the Bay of Fundy westward slightly beyond the Mississippi, with twenty-two detailed insets of harbors, towns, and views, Popple's great map is impressive in conception and elaborate in detail, if at times faulty in execution. For the Southeast he relied chiefly on Barnwell's map ca. 1721. By 1727 he had already made a full-size draft of the area with the title "A Map of the English and French Possessions on the Continent of North America. 1727. H. Popple" (see Map 195A). This large draft uses Barnwell's MS map sparingly. The small printed key map which accompanies the large printed map of 1733 includes much more Barnwell detail. The large printed map itself draws heavily on Barnwell for Indian settlements to the interior and for coastal names of the Carolina region, which are given more fully than in any previously printed map. On the other hand, the wealth of explanatory notes and legends on the Barnwell map is not used; and the direction of the rivers, flowing almost straight east to the Atlantic, and the Appalachian Mountains, which are the poorest features of the Barnwell map, are unfortunately followed closely by Popple. Professor Verner W. Crane (*Southern Frontier*, p. 351) writes that Popple used Barnwell's map for the South, but "not very intelligently." The North Carolina region, largely dependent on Barnwell, is rather bare of detail; Moseley's 1733 map evidently appeared too late for use.

The area west of the Allegheny Mountains is largely based on Guillaume Delisle's "Carte de la Louisiane et du Cours du Mississipi" (Map 170; Plate 47), published in Paris in 1718 and itself an important contribution to the cartography of the Mississippi valley. Professor L. C. Karpinski (*Maps of Michigan*, p. 137) states that the Great Lakes are better delineated in Popple than on the Mitchell 1755 map (Plate 59).

In New York and south of the Great Lakes the "carrying places" for strategic and important portages between the river systems show the use of Cadwallader Colden's "A Map of the Countrey of the Five Nations" [1724]. A few names appear on Popple's map which do not appear on earlier maps, such as "Uncas" in southeastern Connecticut denoting the disputed territory claimed by Uncas, chief of the Mohegan Indians.

Popple has a lengthy legend on sheet 20 of his map, quoting a commendatory opinion of Professor Edmund Halley, the Oxford professor of astronomy, and stating that the map was undertaken with the approbation of the lords commissioners of trade and plantations. In 1755 the English commissaries who had been negotiating boundaries with France pointed out that, although the lords commissioners had approved Popple's undertaking the map, they had never approved its execution, and that the map had no authority behind it. The adverse criticism of the English in 1755 was apparently motivated by features on Popple's map which told against their claim. See *The Memorials of the English and French Commissaries concerning the Limits of Nova Scotia or Acadia*, London, 1755, pp. 277–78.

For professional skill in the compilation, engraving, and publication of his map, Popple relied on Clement Lempriere, described in the cartouche as the designer and craftsman ("C. Lempriere inv. & del."). According to his obituary dated 9 July 1746 in *The Gentleman's Magazine*, XVI, Lempriere was "an ingenious gentleman, draughtsman to the office of Ordnance, and Capt. of a marching Reg. of foot" (p. 383). A number of manuscript maps still preserved in the Public Record Office and British Library testify to Lempriere's qualifications as a surveyor and cartographer of considerable ability. A map of Bermuda, first published in 1738, with longitudes based on observations of the moon in 1722 and 1726, was jointly dedicated to the governor, Alured Popple, by C. Lempriere and W. H. Toms, one of the engravers of the Popple map. Other leading names in the London map publishing world appear on the map as well.

The map was issued with a number of different imprints. The title and imprint given above is the fullest though probably not the earliest. Another impression omits everything below the neat line at right on sheet 17 and has below the neat line on sheet 20: "London Engrav'd by Will.m Henry Toms 1733." A copy of this impression is in the Library of Congress. Another impression is similar to this but has "London Engrav'd by Will.m Henry Toms & R. W. Seale, 1733." Two copies are in the Library of Congress. This form does not have the maps numbered 1 to 20, although the plates are the same. Another impression differs in having below the neat line of sheet 17: "Sold at Stephen Austen's Book Seller in Newgate Street & By Tho.s Willdey at the great Toy Shop in S.t Pauls Church Yard, London." The John Carter Brown Library has this impression, and an imperfect copy of it is in the Library of Congress. The old John Carter Brown Library Catalog lists the Popple map under both 1732 and 1733; the 1732 date is followed in Sabin XV, 75250, and Winsor V, 81. But there seems to be no evidence that the large Popple map was published before 1733, and the present John Carter Brown card catalog lists it under 1733.

There is no doubt that the plate for sheet 10, the Carolina-Georgia area on the large Popple map, was completed and sheets pulled before 1734, when James Edward Oglethorpe, following his first year in Georgia, arrived in London. The coastal configurations on those original sheets were highly generalized and inaccurate and, of course, lacked any evidence of Britain's newest colony, Georgia. Exactly how and when Oglethorpe passed his geographic knowledge and surveys to Popple and his team of engravers is not certain, but the evidence of the transfer can be observed on several extant large Popple maps. One thing is certain, however, Oglethorpe lost no opportunity to employ cartography in publicizing and promoting Georgia. For details and illustrations of three evolutionary stages through which Popple's Carolina-Georgia map passed see L. De Vorsey, "Maps in Colonial Promotion: James Edward Oglethorpe's Use of Maps in 'Selling' the Georgia Scheme," *Imago Mundi* 38 (1986), pp. 35–45.

The map was used widely by later cartographers; within ten years there were Dutch (Cóvens-Mortier ca. 1737) and French (Le Rouge ca. 1742) versions of the map, and Bowen (1747) and other English mapmakers relied on it for regional maps of the English possessions in America. Popple lacked the ability and professional skill of Delisle, Mitchell, and Evans, but on the whole animadversions on his map have probably been too severe.

Little is known of Henry Popple. Twenty years before the publication of his map, he had been cashier to Queen Anne; thereafter he served as agent to several English regiments. He died in Bordeaux on September 27, 1743.

On November 7, 1721, when Mr. John Spencer was appointed clerk of the Board for Trade and Plantations, Mr. Henry Popple was promised the next vacancy. On April 18, 1727, Henry Popple was appointed clerk, but he resigned the post on April 19, 1727, for another "inployment." On the same day that he resigned, Mr. Alured Popple was confirmed as secretary of the Board. He had succeeded his father, William Popple, Esq., deceased, a previous secretary, after April 19, 1722. See *Journal of the Commissioners for Trade and Plantations, 1718–1722*, London, 1925, pp. 327, 350, and Ibid., 1722–28, London, 1928, pp. 328, 346, 359.

References

Paullin, C. O. *Atlas of the Historical Geography of the United States*, p. 13.

Phillips, P. L. *Lowery Collection*, pp. 265–67.

Winsor, J. *Narrative and Critical History of America*, V, 81.

Reproductions

Cartografia de Ultramar, II, Plate 1 (in six plates; transcription of place-names, pp. 12–21).

Popple, H. *A Map of the British Empire in America with the French and Spanish Settlements Ajacent Thereto.* [A full scale facsimile with introductory notes by William P. Cumming and Helen Wallis] Harry Margary, Lympne Castle, Kent, 1972.

Copies

Congress, five copies; American Geographical Society; John Carter Brown; Clements; Georgia (De Renne), Carolina sheet; Harvard; Huntington, two copies; Kendall, Carolina sheet; New York Public; Yale; Public Record Office. C.O. 700, America North and South no. 11.

217. Popple 1733 B

America Septentrionalis [above neat line] | A Map of the British Empire in America with the French and Spanish Settlements adjacent thereto by Hen. Popple [cartouche, bottom left] | To the Queens most Excellent Majesty This Map is most humbly Inscribed by Your Majesty's most Dutiful, most Obedient and most Humble Servant Henry Popple [cartouche, lower center right] | W. H. Toms Sculp. [below neat line, right]

Size: 19 × 19½. *Scale:* 1" = 175 miles.

Insets: Twenty-two insets and view of various harbors and sights, including: Charles Town. *Size:* 1⅜ × 1. *Scale:* 1" = 1 mile.

Description: This is the one-sheet key map for the great Popple map, crossed and numbered to indicate the location of the sheets on the large map. The key map is like the large Popple, except for numerous omissions and abbreviations of names in the interior and in titles, as in the Charles Town inset.

At least four states of this key map are known, all with the title as given above; one state has nothing below the neat line at the bottom left, one has "Sold by the Proprietors S: Harding on the Pavement in St. Martins Lane, and W. H. Toms Engraver in Union Court near Hatton Garden Holborn. Price 2 shills.," another state has "Sold at Stephen Austen's Book seller in Newgate Street & by Thos. Willdey at the great Toy Shop in St. Pauls Church Yard London. Price 2 shillings," and a fourth has "Printed and Sold by D. Steel Bookseller lower end Minories Little Tower Hill 1775."

There is an exact duplicate of this map, in the state and with the title given at the beginning of this item, in British Library Add. MS. 35907. A comparison of a photostat of this map in the Library of Congress with a printed copy shows accidental markings alike on both: since the British Library copy also has "W. H. Toms sculp." and the lines for the twenty large sheets, it is apparently the engraver's original copy.

For the relation of this map to the large Popple and to other maps, see Popple 1733 A.

Reproductions

Plate 55 (detail).

Avery, E. M. *A History of the United States.* Cleveland, 1904–7, III, between pp. 4 and 5.

Paullin, C. O. *Atlas of the Historical Geography of the United States*, Plate 27.

Popple, H. *A Map of the British Empire in America with the French and Spanish Settlements Ajacent Thereto.* [A full scale facsimile with introductory notes by William P. Cumming and Helen Wallis] Harry Margary, Lympne Castle, Kent, 1972.

Copies

Congress, five copies; American Geographical Society; John Carter Brown, two copies; Clements; Huntington, two copies; Yale, two copies; Public Record Office, MPI 303, and F.O. 925/1562 (with part of sheet ii missing).

218. Moseley 1733

To his Excely. Gabriel Johnston, Esq., Captain General & Governour in Chief in and over His Majesty's Province of North Carolina in America. This map of the said Province is most humbly Dedicated and Presented by your Excellency's most obedient Humble servant Edward Moseley [lower left corner] | A New and Correct Map of the Province of North Carolina. By Edward Moseley, late Surveyor General of the said Province. 1733 [lower right corner] | J. Cowley, Sculp. Sold at the Three Crowns on Fan-Church Street, over against Mincing Lane London [below title, lower left-hand corner]

Size: 57⅛ × 45¼. *Scale:* 1" = 5 miles.

Insets: 1. Port Brunswick or Cape Fear Harbour. *Size:* 6⅜ × 6⅝. *Scale:* 1" = 2½ miles.

2. Port Beaufort or Topsail Inlet. *Size:* 6¼ × 6⅝. *Scale:* 1" = 2 miles.

3. Ocacock Inlet. *Size:* 6 × 6⅝. *Scale:* 1" = 4¼ miles.

4. Explanation.

5. Directions for Ocacock Inlet.

Description: This very rare map extends from "Cape Carteret or Cape Roman" to "Part of Virginia" (33°12' to 36°37'). It gives a great number of names of settlers and plantations on Albemarle Sound and the rivers flowing into it, on Pamlico River, on Neuse River and its tributaries, on Cape Fear River and its tributaries, and in the vicinity of Waccamaw Lake. The delineation of the coast and nomenclature along it provide the first detailed and accurate cartographical survey of the North Carolina coastal area.

Moseley's map is one of the most important type maps in the history of North Carolina cartography, directly influencing Wimble 1738, Collet 1770, Mouzon 1775, and other less important maps, for the area Moseley covers. It is the first approach in the history of early North Carolina cartography to the detailed survey for the region around Charles-town made by Mathews nearly fifty years before in his general map of Carolina ca. 1685. Unfortunately, while the Thornton-Morden ca. 1685, Sanson ca. 1696, Crisp 1711, and Moll maps 1715, 1729, 1730, etc., multiplied in convenient form the full details of Mathews's survey. Moseley's map never was, so far as is known, published in convenient or inexpensive form; and no later map attempted to give in full the detailed information concerning settlers found on Moseley's map. Wimble's map, which appeared five years later (1738), shows fuller knowledge of soundings, inlets, and locations of settlements along the coast; but Moseley's map gives much more information, the result of his many years as surveyor general, and extends that information farther away from the coast. Wimble's map represents only the region bordering on the ocean and the sounds.

Although Moseley attempts to portray the interior of the country, which Wimble does not, the paucity of information still evident in Moseley's map in this respect is striking. Up the Cape Fear River about fifty miles in a direct line from its mouth are two "Welch Settlements," and about eighty-five miles from its mouth is a "Palatine settlement." In view of the sparseness of knowledge about North Carolina settlements on the Cape Fear River at this period, an anomymous pamphlet, *A New Voyage to Georgia*, published in London in 1735, is of special interest because it adds to the information given by Moseley. Its author describes in some detail a trip to the Cape Fear River country made by him in 1734; he visited a Nathaniel More, who lived "sixty miles at least from the bar"; and he states that there are persons at least forty miles farther up. See *Collections of the Georgia Historical Society*, Savannah, 1840, II (1852), 56–57.

Except for the items mentioned, between the coastal region and the "Charokee Mountains" in the northwest corner of the map Moseley gives only rivers, their branches, the location of Indian villages, and a few scattered legends. The "Indian Trading Road from the Cataubos and Charokee Indians to Virginia" is shown from its crossing of the "Sapona or Yatkin River" (near the present town of Salisbury, North Carolina) to its crossing of the Roanoke near the Virginia line. A legend between Deep River (incorrectly made a branch of the Pee Dee instead of the Cape Fear River, on Moseley's map) and Haw River begins "This Country abounds with Elk & Buffaloes at the distance of about 150 miles from the sea" and describes the abundance of game and the fertility of the soil. This legend may have had an appreciable influence in promoting the influx of settlers into the Piedmont section of North Carolina during the next twenty or thirty years.

As early as 1709 there are references in the Colonial Records to Moseley's participation in surveying the North Carolina–Virginia line; and this map of 1733 was probably used in the dispute over the North Carolina–South Carolina boundary line, a controversy then at its height. The fullness of information concerning settlements around Lake Waccamaw (spelled "Waggomau" on map) and on the banks of the Cape Fear River lends support to this view, as that was the focal point of the dispute between the two provinces.

A map made before 1730 by Edward Moseley was used at the meetings of the commissioners for trade and plantations at Whitehall, in London, during the conferences of the commissioners with Governor Burrington of North Carolina and Governor Johnson of South Carolina, which began January 8, 1730. In the dispute which arose over the exact lines of the boundary, Governor Johnson published a rejoinder in *The South Carolina Gazette* of Saturday, November 4, 1732, to an earlier communication from Governor Burrington, in which he refers to Moseley's map:

"Governor Burrington, and Myself were summonded to attend the Board of Trade (ca. January 8, 1730), in order to settle the Boundaries of the two provinces, Governor Burrington laid before their Lordships Col. *Moseley's* Map, describing the Rivers of Cape Fear and Wackamaw, and insisted upon Wackamaw River being the Boundary, from the Mouth to the Head thereof, &c." (Quoted from A. S. Salley, *The Boundary Line Between North Carolina and South Carolina, Bulletin of the Historical Commission of South Carolina*, No. 10, Columbia, S.C., 1929, p. 5.)

In May 1731, Moseley wrote to the duke of Newcastle that "I am preparing a large Map of this Province for his Majesty's view, drawn from several Observations I collected when I was surveyor general of this Province and many helps I have received from several Gentlemen of this and the neighboring Governments" (*Colonial Records of North Carolina*, III, p. 137).

Deficient as it was except for the coastal region and in spite of the rapid settlement westward, little more adequate surveying was done for many years after Moseley's map appeared, as a letter from Governor Johnston of North Carolina to the secretary of the Board of Trade in February 1751 shows. Johnston writes: "I am sorry I cannot transmitt any other map of this Province than that of the late Colonel Moseley's of which there is one in your office. It is very deficient, expecially in the back settlements, many thousand persons having sat down there since that map was published" (*Colonial Records of North Carolina*, IV, pp. 1073–74).

Edward Moseley, who died in 1749, was surveyor general as early as 1710, succeeding John Lawson (*Colonial Records of North Carolina*, I, p. 747). Other surveyors general followed him; but in 1723 he was again appointed and remained in the office until about 1730, when he was succeeded by Joseph Jenoure (*Colonial Records of North Carolina*, II, p. 503; III, pp. 204, 209, 217–18). Moseley was one of the notable characters in the early history of Carolina. By 1705 he was a member of the Council and landowner; in 1714 he was licensed to practice law and became the foremost lawyer in the Province; he was commissioner for the North Carolina–Virginia, North Carolina–South Carolina, and Granville–North Carolina boundary lines. He was Public Treasurer, chief Baron of the Exchequer, and Associate Justice of North Carolina, maintaining a reputation for impartial justice and of hatred for oppressive government (*Colonial Records of North Carolina*, IV, pp. xi–xiii). See also J. T. Shinn, "Edward Moseley: A North Carolina Colonial Patriot and Statesman," *Southern Historical Association Publication*, III (1889), No. 1.

A careful description of Moseley's map appeared in the *North Carolina University Magazine*, Raleigh, second series, II (1853), 1–4, in which the writer of the article states that the copy he has examined belongs to Hugh Williamson Collins, Esq., of Edenton, and that he has "good reason to suppose that this is the only copy in North Carolina, and probably in the Union" (p. 2).

In 1822 Major T. Roberdeau of the topographical office of the Department of Engineers made, from an original copy of the map, a tracing which is now preserved in the National Archives at Washington (H 47 Roll). A photostat of this tracing is in the Library of Congress. Major T. Roberdeau's tracing omits some of the details of the original map, especially along the coast.

A copy of Moseley's map, possibly the one referred to by Governor Johnston in 1751 as being in the office of the Board of Trade (see above), is now in the British Public Record Office. C.O. 700, Carolina no. 11. It is in damaged condition.

There is a reference to another map by Moseley under the Helvetische Societät card in the Library of Congress: New und exacte carte von Soud & North Carolina, Virginia, Mary Land und Pensilphania, 1736. E. Moseley (*Size:* 16 × 12½). Phillips suggests that this map was probably taken from "Neu-Gefundenes Eden. Oder ausführlichen Bericht von Süd und Nord Carolina, Pensilphania, Mary Land & Virginia. In Truck verfertiget durch Befel der Helvetischen Societät, 1737." But this is not the title of the map in the John Carter Brown Library copy of Neu-Gefundenes Eden. The map has been missing for some years from the Library of Congress.

A photocopy of a survey of 640 acres granted to Richard Pase on the "Moratoke R^r" next to land belonging to James Binford, which is "Certified & returned this 22nd Day of October Anno Dom^o 1706 pr Edwd Moseley Sur^r Gen." is in the North Carolina State Department of Archives and History.

Of the three known extant copies of Moseley's very rare 1733 map, the best preserved is in the East Carolina University Library, Greenville, North Carolina. The two copies in England's Public Record Office and Eton College are in poor condition. The copy once owned by the late Henry P. Kendall of Camden, South Carolina, and Boston appears to have been lost or accidentally destroyed after his death.

Reproductions

Plates 50A, 51–54.

Cumming, W. P. *North Carolina in Maps*. Raleigh, 1966, Plate VI.

Waynick, C. *North Carolina Roads and Their Builders.* Raleigh, 1952, opp. p. 74 (large folded reproduction of Roberdeau's tracing).

Congress (photocopy).

Copies

East Carolina University; Eton College (Mann Atlas, vol. III); Public Record Office. C.O. 700, Carolina no. 11.

---

219. Oglethorpe-Jones-Gordon 1734

A View of Savanah [*sic*] as it stood the 29th of March, 1734 [above engraving] | To the Honble the Trustees for establishing the Colony of Georgia in America This view of the town of Savanah is humbly dedicated by their Honours Obliged and Most Obedient Servant, Peter Gordon. Vue de Savanah dans la Georgia [below engraving; French title on same line and to left of "Peter Gordon"] | P. Gordon Inv. [below engraving, to left] | P. Fourdrinier Sculp. [below engraving, to right]

Size: 21⅞ × 15⅞. *Scale:* None.

Description: This is an arresting high oblique perspective view of Savannah as it was originally laid out, in 1733, by James Edward Oglethorpe, the founder of Georgia. Set in a clearing in a seemingly endless forest that stretches to the horizon, the townsite is on a high bluff overlooking the Savannah River, shown busy with seagoing ships and small craft. It is the view an observer might have had by looking due south from a vantage point above Hutchinson Island, which is identified and named in a numbered legend at the foot of the engraving. The town is laid out according to a perfectly rectangular plan of straight streets intersecting at right angles and uniform building lots grouped in modules around four large open squares. Many of the already completed buildings and structures are numbered and identified in the legends. These include, "The stairs going up," and "The Crane . . ." for allowing people and cargo to gain access to the town from the river bank below. Also shown are "Mr. Oglethorpe's Tent," "The publick Mill, The publick Oven, The Parsonage House," and "The Fort," to name but a few of the facilities shown as completed along with a large number of small houses set in fenced yards.

The plan bears a curious resemblance to the square divisions of the Margravate of Azilia in the map which illustrates Sir Robert Montgomery's *A Discourse Concerning a New Colony to the South of Carolina*, London, 1717 (Map 166).

A few privately printed lithographed copies of this view were made on heavy paper from the original engraving in the British Library for George Wymberley Jones De Renne about 1876.

Thanks to the prominence of his name in the dedication of this engraving to the Georgia Trustees, Peter Gordon has long been given undeserved credit for its authorship. Research into the historical accuracy and provenance of "A View Of Savanah" has shown that Gordon, an upholsterer and Savannah's first bailiff, merely served to deliver to London the town plan from which the perspective view was prepared when he was forced to return home for medical treatment. The author of Savannah's original town plan or plat was Noble Jones, and the artist who converted it to a geometrically correct perspective view was his relative George Jones.

In the production of this famous engraving there were actually five principal actors with various roles: Oglethorpe, Gordon, Noble Jones, George Jones, and Fourdrinier. Oglethorpe, whom the Trustees had placed in charge of all propaganda for the colony, surely created the idea, secured the plat and details from Savannah's surveyor Noble Jones, and deputized Gordon to carry these to George Jones, thence to the Trustees, and finally to the engraver Fourdrinier, all to await his final approval before printing. See Rodney M. Baine and Louis De Vorsey, "The Provenance and Historical Accuracy of A View of Savannah as it Stood the 29th of March, 1734," *Georgia Historical Quarterly* 72 (1989), pp. 784–813.

References

Coulter, E. M., ed. *The Journal of Peter Gordon, 1732–1735*. Athens, 1963.

Winsor, J. *A Narrative and Critical History of America*. V, 369, note 1.

Reproductions

Adams, J. T., ed. *Album of American History*. New York, 1944, I, 220.

Gabriel, R. H. *The Pageant of America*. New Haven, 1925, I, 272.

Howell, Clark. *History of Georgia*. Chicago, 1926, I, opp. p. 72.

"A View of Savannah as it Stood the 29th of March, 1734." [Full-scale facsimile], Historic Urban Plans, Ithaca, New York.

Copies

Congress; Boston Public; Duke; Georgia Hargrett Library; Historical Society of Pennsylvania, Philadelphia; Clements.

219A. George Jones 1734 MS

His Majestys Colony of Georgia in America [pasted to map top center] By George Jones [at bottom centered between explanatory legends]

Size: 26⅛ × 15¼ (top neat line missing). *Scale:* None.

Description: This sketch is the original drawing on which was based the world-famous engraving "A View of Savanah as it stood the 26th of March, 1734." Until 1989 it had been identified incorrectly as a copy of Fourdrinier's engraving rather than its model. See Leonard L. Mackall, ed., *Catalogue of the Wymberly Jones De Renne Georgia Library 3*, Wormsloe, Georgia, 1931, p. 1279. In terms of its content it is almost identical to the engraving.

George Jones appears to have been a cousin of Georgia's first land surveyor, Noble Jones, who arrived with Oglethorpe at the colony's founding. There is no evidence that George ever visited Georgia. His sketch was doubtlessly based on a plat and drawing, augmented by information supplied by Gordon and Oglethorpe, sent to him in London by his cousin Noble in the care of Peter Gordon. See Rodney M. Baine and Louis De Vorsey, "The Provenance and Historical Accuracy of A View of Savanah as it Stood the 29th of March, 1734," *Georgia Historical Quarterly* 72 (1989), pp. 784–813.

Reproduction
Plate 55A.
Original
Georgia Hargrett Library.

---

220. Bernard-Delisle 1734

Carte de la Louisiane et du Cours du Mississipi Dressée sur un grand nombre de Mémoires entrau tres sur ceux de M.r le Maire Par Guill.aume De l'Isle de l'Academie R.le des Sciences. [above map, below neat line]

Size: 16 × 14. *Scale:* 1" = ca. 98 miles.
In: Bernard, Jean F. (publisher). *Recueil de Voyages au Nord*. Amsterdam, 1725–38, V (1734), opp. p. 1.
Description: This map extends from 28° to 46° N.L. (Cape Canaveral to the southern shores of Lake Superior) and from the Atlantic coast near Charleston, S.C., to the upper reaches of the Rio Grande. As the title indicates, this is a copy of Delisle's map of similar title (1718). See Delisle 1718 and Bernard-Delisle 1737.

Copies
Clements; Yale.

---

221. Charlottenbourg ca. 1735 MS

A Plan of the Township of Charlottenbourg on the South side of the River Alatamaha in Georgia

Size: 17¾ × 12½. *Scale:* 1" = 130 [perches?].
Description: There are three similar versions of this colored drawing of a town plan. The best known includes straight streets lined with houses, a church, an obelisk, the governor's house, etc., laid out beside the Altamaha River. There is no evidence that such a town was ever surveyed or settled. The fact that it was described as "on the South side of the River Alatamaha" is curious because at that time Georgia's southern boundary was marked by the Altamaha. In a technical sense "the South side" of the river could have been construed as being within South Carolina until 1763, when Georgia's southern boundary became the St. Marys River. It is for this reason that Janice G. Blake suggested that Charlottenbourg might have been a settlement scheme projected in South Carolina. See J. G. Blake (comp.), *Pre-Nineteenth Century Maps in the Collection of the Georgia Surveyor General Department*, Atlanta, 1975, p. 37. Tracings of all three plans were prepared for Wymberley Jones De Renne, the Savannah collector. They are now found in the University of Georgia Hargrett Library in Athens, Georgia.

References
Catalogue of the Wymberley Jones De Renne Georgia Library. Wormsloe, Georgia, 1931, III, 1200.
Reproductions
Hulbert, A. B. *Crown Collection*. Series III, No. 138.
Georgia State Archives (Photostat). Congress (photocopy).
Original
London. Public Record Office. C.O. 700, Georgia no. 9(1)&(2) and 10.

---

222. Albemarle ca. 1735 MS

[Tract of Land in Albemarle County]

Size: 17¼ × 13¼. *Scale:* "100 Pole to an Inch."
Description: This is a plat of several five-hundred-acre tracts, laid out in rectangles, with the drawings of houses, the names of their owners, and the names of persons to whom land was sold by Cullen Pollock: David Stell, M. Shellet, Jos. Sharp, John McCracken, Wm. Creige, Hans Wingle, Hollowell [Ro?]sbury, Ja.s Kirkpatrick, Mulhollan. Cullen Pollock was the oldest son of Thomas Pollock, president of the Council and acting governor in 1713–14. Thomas came to Carolina about 1682 and died August 20, 1722 (North Carolina Colonial Records, II, pp. 279, 460). In 1720 Cullen Pollock petitioned for a confirmation of his right to a tract of land north of Moratock River. After his father's death he was a large landowner, paying quit rents in 1729–32 for over 1,200 acres of land in Bertie, Edgecombe, and Tyrrell Precincts and also possessing land on the Neuse and Trent Rivers. He died about 1752. See North Carolina Colonial Records, II, p. 389; IV, pp. 53, 71; XI, p. 116; XXII, pp. 243, 245, 246.

Original
Loan to Hall of History, North Carolina Department of Archives and History.

223. Gray 1735 MS

A Map of the Division Line between the Province of North & South Carolina containing thirty miles along the Sea Coast from the Mouth of Cape fear River, and from thence continued on a NW Course for Sixty-two miles and one Quarter. Surveyd in the presence Robert [Halton: name torn off] Mathew Rowan & Edward Moseley Esq.[rs] Comissioners [ca. six letters torn off]ted on the part of North Carolina by W. Gray Surveyor

Size: 95 × 30 [irregular]. *Scale:* 1" = 1 mile.

Description: This survey shows the coast from the mouth of the Cape Fear River to Little River, where the survey began; thence the line goes across Waccamaw River, Little Pedee River, etc., until it reaches Beaverdam and Crow Creeks.

The governors of North and South Carolina appointed three commissioners each to run out the line, which had been agreed upon after much dispute, in April 1735. In May 1735 the commissioners met and began their survey. They found that the line should begin at Little River, which was thirty miles from the mouth of the Cape Fear River. They then separated, agreeing to meet on September 18, 1735. On that date the North Carolina commissioners arrived and began their survey. In October 1735 the South Carolina commissioner arrived, checked the work which had been done, and agreed to it. When they had surveyed the northwest course for sixty-two and a quarter miles, they stopped because they had received no pay. See *Colonial Records of North Carolina*, III, pp. 84, 115; V, pp. 381–82, and M. L. Skaggs, *North Carolina Boundary Disputes Involving Her Southern Line*, Chapel Hill, 1941, pp. 40–44.

In the South Carolina Historical Commission Archives is a map with dates 1735 and 1737 giving the North Carolina–South Carolina line in a N.W. direction for eighty-six-plus miles to the supposed intersection with 35° N.L.: Chart No. 1, A.

Original

London. Public Record Office. C.O. 700, Carolina no. 8.

224. Cóvens-Mortier ca. 1735

La Louisiane, Suivant les Nouvelles Observations de Messrs de l'Academie Royale des Sciences, etc Augmentees de Nouveau. A Amsterdam chez Cóvens et Mortier. | 96 [below neat line, to right]

Size: 11⅝ × 8¾. *Scale:* 1" = ca. 147 miles.

In: Cóvens, J., and C. Mortier. *Nouvel Atlas*. Amsterdam, [1735?], No. 96.

Description: This is the same map as Aa's "La Floride" 1713, having a different title but possibly using the same plate. See Aa 1713.

Copy

Congress 1735?

225. McCulloh 1736 MS

North Carolina. Mr. McCulloh's Draught of Land that he proposes to settle in North Carolina. Rec[d] Read Febry 17: 1735/6 [Endorsed]

Size: 18 × 17½. *Scale:* 1" = ca. 40 miles.

Description: This colored manuscript shows the North Carolina coastline; to the interior a rectangle is drawn around the high bluffs of the northeast branch of the Cape Fear River. Farther to the interior, another rectangle is drawn between the Neuse and Cape Fear Rivers, east of the juncture of the Haw River and Deep River. This second proposed place of settlement east of "The Haw Fields" is, therefore, probably in or near lower Wake County, south of Raleigh.

The protracted, complicated, and bloody disputes over the land claimed by McCulloh and his settlers lasted until after the Revolutionary War: see *Colonial Records of North Carolina*, V, pp. xxxii–xxxv.

Reproduction

North Carolina Department of Archives and History (photocopy).

Original

London. Public Record Office. C.O., M.P.G. 268.

226. Jenner 1736 MS

North Carolina

Draught of the Tract of Land desired by Mr. Jenner for the Switzers in North Carolina. Recd Read Febry 17. 1735/6 [Endorsed]

Size: ca. 19½ × 13½. *Scale:* 1" = 10 9/10 miles.

Description: This map extends northward from 35° to 36°45' and westward from slightly east of the juncture of the "Indian Roade to Virginy" and the Roanoke River at the Virginia–North Carolina state line to the Blue Ridge Mountains.

"The Line of the Land Petitioned for" begins at the juncture of the Roanoke River and the Old Indian Trading Path and the North Carolina–Virginia state line; it extends in a diagonal southwesterly line to 35°55'; then it runs westward to the mountains. Roughly, therefore, it includes that part of Carolina north of Durham, High Point, and Lenoir.

In 1736 Samuel Jenner purchased a large tract of land from William Byrd in Virginia. But colonists of the Helvetian Society in Switzerland, for which Jenner was agent, went to Carolina instead. See Eden 1737. See also *Colonial Records of North Carolina*, IV, pp. 166–68.

Reproduction

North Carolina Department of Archives and History (photocopy).

Original

London. Public Record Office, M.P.G. 169.

---

227. Frederika 1736 MS

A Plan of Frederika a Town in the Plantation of Georgia in the province of Carolina as layd out by Mr Oglethorpe 1736

Size: 11½ × 14⅝. *Scale:* None.

Description: The plan of the town, laid out in 84 small rectangles, occupies the top of the sheet. It accompanies a plan of the fort of Frederica. A legend to the right of the title gives the width of the streets, which are labled A through D on the plan. In 1736 James Edward Oglethorpe, Georgia's founder, personally oversaw the choice of Frederica's site and early construction to guard the colony's southern frontier against a perceived threat from the Spanish in Florida. When the tide of imperial competition ended and Frederica lost its military significance, the town went into a gradual decline and remains today as a National Historic Monument.

Reference

Colonial Records of the State of Georgia, 21, pp. 75–76.

Reproductions

Georgia State Archives (photostat).

Original

John Carter Brown.

---

228. Frederica Fort 1736 MS

The Fort at Frederica in Georgia as layd down by a Swiss Engineer facing [the] principal Street of that Town.

Size: 11½ × 14 5/16. *Scale:* 1" = 40 [feet?].

Description: At the top right of the sheet is this note: "1736 brought over by Mr Saml Hodgkinson of Spalding who went thence over thither at the Instance of & with his Excellency General Oglethorpe the Governour to assist in cultivating a manufacture of Hemp & Flax there wherein he had here been a considerable dealer, [one or two words illegible] brought a Specimen of the growth of Georgia."

The "Swiss Engineer" mentioned in the title may be Samuel Augspourguer of Canton Berne, Switzerland, who "aged 42 and upwards" came to Purrysburg, S.C., in 1734. Two years later he moved to Georgia.

In 1736 Auspourger was appointed "Ingenier of the Fort and Land Surveyor" at Frederica with a salary of 3 shillings a day.

Reference

Colonial Records of the State of Georgia, 5, p. 70.

Reproduction

Georgia State Archives (photostat).

Original

John Carter Brown.

---

229. Arredondo 1737 MS A

Plano de la entrada de Gvaliqvini Rio de S.n Simon situado à 31 Grados, y 17 min.s de latitude Septentrional [upper right] | Le dió à la S.ria de Indias d.n Manual Joseph de Ayala [at bottom]

Size: 18 × 15¾; with title and explanation, 24 × 18.
Scale: 1" = 700 feet.

Description: This plan was sent to the governor of Cuba, Ayala, by Arredondo, the royal army engineer who the governor had sent to Florida to make a report on fortifications and policy. The plan was probably included with the report to the governor dated January 22, 1737, extracts in translation of which are given in P. L. Phillips, *Lowery Collection*, pp. 270–74. It shows "Fuerte Frederico," "Isla de Operavanas ó Ballenas," "Rio que corre a la Barra de Ballenas," etc.

To the left of the plan is an "Explicacion," lettered A to P. See Arredondo 1742 MS.

References

Phillips, P. L. *Lowery Collection*, p. 270, No. 343.

Reproductions

Congress (drawing). Karpinski series of reproductions.

Original

Madrid. Dept. de la Guerra. L.M. 8a. 1a.-a. No. 43.

---

230. Arredondo 1737 MS B

Plano de la Entrada de Gvalqvini Rio de San Simon, situado en 31 grados de altura del Polo Septentrional. Explicacion [Items A to P; cartouche with title and explanation in top right corner] | Havana 15 de Mayo de 1737. Dn Antonio de Arredondo [top left, below scale]

Size: 28¾ × 17¼ (with explanation).

Scale: 1" = ca. 700 feet.

Description: This plan of the entrance to the Altamaha River and the vicinity was transmitted by Arredondo with a letter of the same date. Antonio de Arredondo, a royal engineer in the Spanish army, was sent to St. Augustine by the Governor of Cuba to inspect the fortifications and to recommend measures for the best possible defense.

A tracing of this plan is in the Georgia Hargrett Library De Renne Collection (see *Catalogue*, III, 1200), and a reproduction of the tracing is given in the Georgia Historical Society *Collections*, VII, Part III (1913), 71.

References

Phillips, P. L. *Lowery Collection*, p. 275, No. 349.

Torres Lanzas, Pedro. *Relación Descriptiva*, I, No. 131.

Reproductions

Bolton, H. E. *Arredondo's Historical Proof.* Berkeley, Calif., 1925, opp. p. 192.

Georgia State Archives (photostat).

Original

Sevilla. Arch. Gen. de Indias. Est. 87. Caj. 1. Leg. 2 (1).

---

231. Cóvens-Mortier-Popple ca. 1737

Carte Particulière de L'Amerique Septentrionale [above neat line of map] | A Map of the British Empire in America with the French, Spanish and Hollandish Settlements adjacent thereto by Henry Popple - . - at Amsterdam Printed for I. Cóvens and C. Mortier [cartouche bottom left] | T. Condet s. [below neat line, right]

Size: 19 × 19⅜. *Scale:* 1" = ca. 175 miles.

In: Delisle, G. *Atlas Nouveau.* Amsterdam, [1741?], II, No. 27.

Inset: Twenty-two views and plans, including Charles Town. *Size:* 1½ × 1. *Scale:* 1" = 1⅛ miles.

Description: This is exactly like the Popple 1733 key map except for the title. Cóvens and Mortier also published a Popple of this size without the insets; a copy is in the Library of Congress. See Popple 1733 B.

Copies

Congress 1741?, separate; Harvard 1745.

In: Cóvens, J., and C. Mortier, *Atlas Nouveau.* Amsterdam [1686–1761], IX, No. 40.

Copies

Congress 1686–1761.

---

232. Eden 1737

Eden in Virginia Von der Helvetischen Societet Er Kaufte 33400 Jucharten Land A° 1736

Size: 13⁷⁄₁₀ × 15⁸⁄₁₀. *Scale:* 1" = ¾ mile.

In: [Helvetischen Societät]. Neu-gefundenes Eden. Oder: aussfuehrlicher Bericht Von Sud- und Nord-Carolina, Pensilphania, Mary Land | & Virginia . . . Helvetischen Societät, 1737, between pp. 96–97.

Inset: In the upper left-hand corner is a plan of a city, with streets, parks, etc. *Size:* 3¹⁵⁄₁₆ × 4. *Scale:* None.

Description: The area of "Eden," the land purchased, lies wholly in what is now Halifax County, Virginia, although the map itself includes some of the adjacent land in Virginia and North Carolina. The map accompanies a tract (pp. 96–202) in *Neu-gefundenes Eden* entitled "eine kurtze Beschreibung von Virginia," which is a translation by Samuel Jenner of Bern, agent in America of the Helvetian Society, of a description of Virginia. Jenner states that the description was made for him by Wilhelm Vogel, who is William Byrd II, author of the dividing line histories.

According to the deed of purchase as published by Jenner, the section of Byrd's land ceded is twenty-six miles in length and from three to seven miles in width. It is situated near the island of Acconeechy, also part of Byrd's extensive holdings in this region (he died owning 179,440 acres). Starting at Hycolotte Creek (the present Hyco River) at its juncture with the Dan ("Roanock River"), the boundaries of the tract run along the Dan to Deer Creek (Lawson's Creek), up Deer Creek, and south from it to Buffalo Creek (Powell's Creek), which flows eastward to its juncture with the Hyco-otte at the Dividing Line, and thence back to the Dan.

In spite of Jenner's payment of £3,000 toward the purchase of the land, the Swiss never arrived; "somebody deluded them to Carolina," declared Byrd, "where many of them, I understand, are dead. They would probably have had better luck at the place I intended for them." See Jenner 1736. Professor Beatty, in his introduction (see below), gives a very interesting account of Byrd's various attempts to persuade settlers to come to his Eden. Another boat-load of Switzers perished in shipwreck; Byrd then turned to the Scotch-Irish from Pennsylvania, though he likened them to Goths and Vandals. Then he attempted to colonize directly from Scotland; the group with which he was negotiating, however, also went to North Carolina. Finally, he turned his hopes to German emigrants, but he died with his colonization schemes unsuccessful.

The lack of a map in the Library of Congress copy of the volume has resulted in a curious error in the introduction by Professor Richard C. Beatty to his re-translated edition of the Jenner "translation": see William Byrd's *Natural History of Virginia or the Newly Discovered Eden.* Edited by R. C. Beatty and W. J. Mulloy, Richmond, Va., 1940. Professor Beatty confuses this Eden of 33,400 acres in Virginia with the 20,000 acres in North Carolina purchased by Byrd in 1728 from the North Carolina Commissioners (see Byrd 1733). Jenner's Eden is not the tract of land which lies "at the junction of the Dan and Irvine Rivers"; it lies farther east, between the present location of South Boston, Virginia, and the Dividing Line of North Carolina and Virginia. Professor Beatty must have seen a copy of one of the reproductions of Byrd's 1733 map; the fact that there are two maps of land called Eden, both areas owned by Byrd, and both in accompanying tracts attributed to him, undoubtedly is confusing, especially if only one of the maps has been seen.

The map of "Eden in Virginia" has an additional interest in that it contains a plat of a proposed township section very much like that found in Robert Montgomery's *Discourse concerning the design'd establishment of a new colony to the south of Carolina.* . . . London, 1717, and of the same sort approached in the building of Savannah and afterwards adopted in the development of the northwest.

Although the Library of Congress does not have this map in its copy of Jenner's *Neu-gefundenes Eden*, on the Library of Congress card which lists the volume there is a reference to a loose map in the Library of Congress Geography and Map Division with the title: "New und exacte carte von Soud & Nord Carolina, Virginia Mary Land und Pensilphania 1736; E. Moseley" (*Size:* 16⅛ × 12⅜). The Library of Congress card suggests that the map may have come from a copy of *Neu-gefundenes Eden.* Diligent search has not found this map in the Library; no other copy of such a map by Moseley is known. It is doubtful whether the map belongs to the book, though the spellings "Soud" and "Pensilphania," found in the title of *Neu-gefundenes Eden*, may indicate that it was published by Jenner or the Helvetian Society. The British Library copy of *Neu-gefundenes Eden* has this map as a frontispiece.

References

Besides the works mentioned in the description, and references listed therein, see J. Sabin, *Dictionary*, XIII, No. 52362; Clayton-Torrence, *Trial Bibliography of Colonial Virginia*, Richmond, 1908, No. 143; L. C. Wroth, "Notes for Bibliophiles," Herald Tribune Books, November 16, 1941.

Reproductions

Avery, E. M. *A History of the United States.* Cleveland, 1904–7, III, opp. p. 238.

Copies

John Carter Brown 1737; Clements.

---

233. Homann Heirs 1737

Dominia Anglorum in America Septentrionali Specialibus Mappis Londini primum a Mollio edita nunc recusa ab Homannianis Hered. Juncta est Mappulæ D facies ejus regionis quam Coloni Salisburg incolunt [on left page, top] | Die Gross-Britannische Colonie-Lænder in Nord-America, in accuraten Special-Mappen nach den Londone. Originalien getreulich mitgetheilt und herauss gegeben von Homæñischen Erben. [on right page, at top]

Size: 21¾ × 19½.
Scale: 1" = (differs for different maps).
In: Homann, J. B. *Grosser Atlas.* Nürnberg, 1737, No. 148.
Insets: This sheet consists of four maps in separate compartments, with the general title as given above; titles to the separate maps are as follows:

A. New Fovndland, od. Terra Nova, S. Lavrentii Bay, die Fisch-Bank, Arcadia, nebst einem Theil New Schotland [upper left of sheet] *Size:* 10⅜ × 7¾. *Scale:* 1" = ca. 92 miles.

B. New Engelland, New York, New Yersey, und Pensilvainia [upper right] *Size:* 10⅜ × 7¾. *Scale:* 1" = ca. 70 miles.

C. Virginia und Maryland [lower left] *Size:* 10⅜ × 7¾. *Scale:* 1" = ca. 22 miles.

D. Carolina nebst einem Theil von Florida [lower right] *Size:* 10¾ × 7¾. *Scale:* 1" = ca. 93 miles.

Description: As the Latin title states, these four maps are taken from Moll's maps. The original of each of the four compartments is found in his *Atlas Minor*, London, 1729. Carolina is based on the second state of the map, with changes made about 1736; it has "Azilia" for that part of South Carolina between the Altamaha and the Savannah Rivers and "The South Bounds of Carolina according to the last Charter." "According to the last Charter" is not found in Moll's 1732 impression but is found in the 1736 impression. Various legends on the Moll map have been transcribed, with translations at times in German, elsewhere in Latin.

The chief addition to the original map has been made in the vicinity of the Golden Islands: "Ebenezer of the Salzb." has been put on "Savana R." and "Fort Col Angile" [Argyle] on the "Howgeche R." The Urlsperger Tracts, describing this region, had already roused interest in Germany.

Reproduction

Color Plate 17.

Copies

Congress 1737; American Geographical Society; John Carter Brown, two copies; Georgia Hargrett Library; Harvard; Kendall (Carolina map); Yale 1737, separate; Clements, 2 separates.

In: Homann, J. B. *Atlas Novus*. Norimbergae, [1755?], No. 31.

Copy

Harvard 1755?

In: [Bellin, N.] *Cartes et Plans de L'Amerique*. 1745, No. 17 [in an atlas with title and date in ink and with many MS maps by Bellin dated 1739].

Copy

Congress 1745.

In: Homannischer Atlas. Nuernberg, 1747–[57], No. 94.

Copies

Congress 1747–[57].

In: Homann heirs. *Atlas Geographicvs Maior*. Norimbergae, 1759–[81], I, No. 141.

Copies

Congress 1759–[81], 1759–84, two copies; American Geographical Society, 1759– ; Sondley Reference 1759– .

In: Homann heirs. *Atlas Mapparum geographicorum*. Norimbergae, 1728–[93], No. 95.

Copies

Congress 1728–93.

---

234. Bernard 1737

New map of Georgia Amsterdam for I. F. Bernard 1737 [lower right-hand corner]

Size: 10⅛ × 7⅜. *Scale:* 1" = ca. 137 miles.

In: [Bernard, Jean Frederic, ed.] *Recueil de Voyages au Nord*. Amsterdam, 1725–38, X (1738), opp. p. 470.

Description: This map is apparently an enlarged reproduction of the map published in Martyn's *Reasons for Establishing The Colony of Georgia in 1733*. The engraver has made numerous omissions and has made errors because of his lack of knowledge of English: "no Inhabitants from tience [hence] to the Point of Florida."

The Georgia Hargrett Library copy is found at the end of the ninth volume of the set, which was published in 1737.

Copies

John Carter Brown 1738; Georgia Hargrett Library 1737; Yale 1738.

---

235. Delisle-Bernard 1737

Carte de La Louisiane et du Cours du Mississipi Dressée sur un grand nombre de Memoires entrau tres sur ceux de Mr. le Maire Par Guill.aume De l'Isle de l'Academie R.le des Sciences.

Size: 16¼ × 14. *Scale:* 1" = ca. 98 miles.

In: Garcilasso de la Vega, called the Inca. *Histoire des Incas*. Amsterdam, chez Jean Frederic Bernard, 1733–37, II, opp. p. 1.

Description: This is a copy of Delisle 1718, reduced both in size and area represented. It illustrates Jean Baudoin's translation of Garcilasso's *L'Histoire de la Conquête de la Floride*, which occupies the second volume of *Histoire des Incas*.

Reproduction

Winsor, J. *Narrative and Critical History of America*, II, 294–95 (erroneously dated 1707 by Winsor).

Copies

Congress, separate; John Carter Brown; Harvard; New York Public; Clements 1720 and 1737.

---

236. Brickell 1737

A Map of North Carolina

Size: 8¾ × 7¼. *Scale:* 1" = ca. 40 miles.

In: Brickell, John. *The Natural History of North-Carolina.* Dublin, 1737, opp. p. 1.

Description: The map has little new material; it is much like the Lawson 1709 map, although it is a poorer and more inaccurate piece of work. The coastline extends from Cape Carteret to the latitude of the Dismal Swamp; the interior is almost void of information except for the "Charokee Mountains."

The book itself has been attacked as an almost verbal transcript of Lawson's *New Voyage to Carolina* (1709); see *North American Review*, XXIII, 288–89. However, only half of the book depends upon Lawson's narrative; Brickell both added much from his own observation, according to Stephen B. Weeks, and systematized Lawson's own account; see Weeks's introduction to the reprint noted below.

Reproductions

Avery, E. M. *A History of the United States.* Cleveland, 1904–7, III, 348.

Brickell, John. *The Natural History of North Carolina.* Reprinted by Authority of the Trustees of the Public Libraries, Raleigh, 1911, front.

Copies

Congress; American Geographical Society; John Carter Brown; Archibald Craige Collection, Winston-Salem, N.C.; Clements; Duke; Harvard; Kendall; New York Public, two copies; Sondley Reference.

236A. Gascoigne [1737] MS

A True Copy of a Draught of the Harbour . . . with its approaches, Soundings, anchorages, shoals, place names, proposed forts

Size: 11⅞ × 17⅛. *Scale:* 1" = 3⅓ miles.

Description: This manuscript chart was sent to the Admiralty by Captain James Gascoigne with "some remarks on the Very Dangerous Errors with which the Public Sea Draughts do generally abound." On May 1, 1738, the Georgia Trustees wrote to Sir Robert Walpole "setting forth the good services of Capt. James Gascoigne, royal navy, to Georgia." In their letter the Trustees mentioned that Gascoigne in the sloop *Hawk* had been "indefatigable in viewing all the Southern Inlets . . . and ha[d] continually Cruized upon the Coast, or staid with, assisted, and Protected the Southern Settlements. . . ." For his invaluable services the Trustees urged that Gascoigne be advanced to the command "of one of His Majesty's Ships of War." See *The Colonial Records of the State of Georgia*, 29, p. 259.

Reference

Maps and Plans in the Public Record Office America and West Indies. London, 1974, No. 2863, p. 496.

Original

London, Public Record Office. MPI 274.

236B. Ouma–De Batz 1737 MS

Nations Amies Et Enemies Des Tchikachas

Size: 7¼ × 12. *Scale:* None.

Description: This map was copied from an Indian original by Alexandre De Batz, the French engineer-draftsman, in 1737. It is not a map in the usual sense of the word but rather an Indian view of geopolitical and military conditions persisting in the Southeast at that time. In this sense it is similar to the maps collected by the governor of South Carolina in the early 1720s (see Maps 192, 193, and 195). The Choctaw Indians were allied with the French in a bitter but unsuccessful campaign against the Chickasaw in the Spring of 1737. In an effort to gain intelligence, the French sent an Alabama Indian, named Captain of Pakana, into the Chickasaw country. While there, Pakana interviewed Mingo Ouma, a leader in the Chickasaw peace faction, who gave him this map painted on a skin. P. K. Galloway raises the possibility that Captain of Pakana redrew Mingo Ouma's original map, basing her suggestion on a comment in a cover letter written on October 24, 1737, by a French officer: "the explanations that the Captain of Pakana has made himself and from which he has drawn the two maps enclosed herewith which I take liberty of sending" (see note 27, p. 154,

Vol. IV, *Mississippi Provincial Archives, French Dominion, 1729–1748*). As G. A Waselkov points out, however, both the explicit statement in the map legend and the map's overall structure argue for Mingo Ouma's authorship. See his "Indian Maps of the Colonial Southeast," *Powhatan's Mantle*, p. 332. This map places the Chickasaw Nation ("Toute la Nation Tchikachas") prominently at the center of its field of coverage, exactly what would be expected in a map prepared by the Chickasaw Mingo Ouma. For a full description of the Indian groups shown and a translation of De Batz's lengthy French explanatory text on this map, consult G. A. Waselkov's essay mentioned above.

Reproductions

Galloway, P. K., ed. *Mississippi Provincial Archives French Dominion, 1729–1748*, facing p. 142.

De Villiers, M. "Note Sur Deux Dessinees Par Les Chikachas En 1737," *Journal de la Societe Des Americanistes De Paris*, n.s. 13 (1921), Planche II.

Waselkov, G. A. "Indian Maps of the Colonial Southeast." In *Powhatan's Mantle*, edited by P. H. Wood, W. A. Waselkov, and M. Thomas Hatley, Figure 5, p. 298.

Original

Paris, Archives Nationales, AC, C13A, 22, fol. 67.

236C. Pakana–De Batz 1737 MS

Plan Et Scituation Des Villages Tchikachas. Mil Sept Cent Trente Sept.

Size: 7¼ × 12. *Scale:* 1" = 1 league.

Description: Like the map described above (Map 236B) this map was copied from an Indian original by the French engineer-draftsman Alexandre De Batz in September of 1737. The author of the original was an Alabama Indian known as Captain of Pakana, who was acting as a French intelligence agent-cum-emissary in their war against the Chickasaw Indians in that year. At first glance this map by Pakana and the one by Mingo Ouma, described above, share many similarities in that both are dominated by constellations of linked circles. In Mingo Ouma's map the circles are intended to represent geopolitically significant groups spread far and wide across the Southeast, from the English in Carolina to the French in Mobile and the Arkansas or Quapaws west of the Mississippi, to the Huron and Iroquois in the far north. On the other hand, this map by Pakana is drawn at a much larger scale and shows only the eleven villages in the Chickasaw country that he visited in his spying mission for the French. Like the ethnic groups shown in Ouma's map, the Chickasaw villages appear as a set of linked circles connected by paths. In addition to showing their fortified villages Pakana also showed the pattern of Chickasaw cultivated fields, or as De Batz identified them in his legend, "Deserts." A village occupied by Natchez Indians who had taken refuge with the Chickasaws following a French campaign aimed at their extinction is also shown. For a discussion of this unique example of American Indian cartography consult G. A. Waselkov, "Indian Maps of the Colonial Southeast," in *Powhatan's Mantle*, edited by P. H., W. A. Waselkov, and M. Thomas Hatley, pp. 332–34.

Reproductions

Galloway, P. K., ed. *Mississippi Provincial Archives French Dominion, 1729–1748*, facing p. 154.

De Villiers, M. "Note Sur Deux Dessinees Par Les Chikachas En 1737," *Journal de la Societe Des Americanistes De Paris*, n.s. 13 (1921), Planche I.

Wood, P. H., G. A. Waselkov, and M. Thomas Hatley, eds. *Powhantan's Mantle: Indians in the Southeast*, Figure 6, p. 299.

Original

Paris, Archives Nationales, AC, C13A, 22, fol. 68.

---

237. Bull 1738 MS

This Chart was transmitted by Col.o Bull (President and Comander in Chief of South Carolina) with his Representation to the Board of Trade, dated the 25.th of May 1738. Rec.d July ye 27th 1738 [bottom left corner]

Size: 37 × 29¼. *Scale:* 1" = ca. 30 miles.

Description: This map, covering the whole Southeast, is a copy of Barnwell ca. 1721, though numerous names and legends are omitted. The southern half of Florida, which is not included in Barnwell, is given not as a peninsula but as an agglomeration of islands of various sizes, evidently a following of the Nairne inset on the Crisp map 1711. The substitution of "The Colony of Georgia" for Barnwell's "Margravate of Azilia" is the only significant change on the map; specific recent developments, such as the establishment of Savannah, the location of which Bull in person had helped Oglethorpe to choose, and Fort Frederika, already built by May 1738, are not given.

The name of the actual copyist of the map is not known. Obviously it was not drawn by Bull, for he would have made changes and improvements; probably no contemporary Carolinian was more eminently qualified to add recent information to the map. William Bull was the son of Stephen Bull, who came with the first colonists under the lords proprietors. William Bull became a commissioner of Indian trade and later, as president of the council, he became lieutenant governor upon the death of Broughton in November 1737. In the absence of a royal governor in residence, he administered the government of the province of South Carolina ably, from time to time, until 1743. He remained lieutenant governor until his death in 1755. See E. McCrady, *History of South Carolina under the Royal Government*, pp. 38, 176–77, 299.

Reproductions

Chatelain, V. *The Defenses of Spanish Florida*. Washington, 1941, Map 8.

Original

London. Public Record Office. C.O. 700, Florida no. 2.

---

238. Ruiz 1738 MS

Plano y Perfil del nuevo Fortin de S.n Francisco de Pupo, Situado en la orilla del Norte del Rio S.n Juan siete leguas del presidio de S.n Agustin de la Florida [top right] | S.n Agustin de la Florida y octubre 25 de 1738 Dn Pedro Ruiz de Olàno. [bottom left]

Size: 16¼ × 13. *Scale:* 1" = 3 tuesas.

Description: This military engineer's plan shows a small tract of land on the bank of the St. Johns River in Florida at the crossing of the main trail between St Augustine and Fort St. Marks. The crossing was defended on the west by Fort St. Francisco de Pupo, shown in plan and cross-sectional views. This small defense work was opposite better known Fort Picolata, the site of which is near the present-day community of Picolata in St. John's County, Florida. For a discussion of the important Indian congresses held at Fort Picolata during the British Period of Florida's colonial history see, L. De Vorsey, *The Indian Boundary in the Southern Colonies, 1763–1775*, pp. 186–98.

Reproductions

Cartografia de Ultramar. Carpeta II. Madrid, 1953, Plate 64 (with transcription of place-names, p. 326).

Original

Madrid. Servicio Geografico del Ejército. LM.-S.a - 1.a - a. - Núm.26.

---

239. Virginia-Carolina Line ca. 1738

Virginiae Pars Carolinae Pars [across the face of the map]

Size: 11⅛ × 7⅞ (reproduction). *Scale:* 1" = ca. 7½ miles.

Description: The lower part of this engraving of the boundary line between Carolina and Virginia run in 1728 has figures of an Indian, birds, fish, snake, and opossum, with the location where they were observed during the survey marked.

In July 1737 William Byrd wrote to Peter Collinson, then in London, that he was expecting to finish his *History of the Dividing Line* during the following winter. As Byrd expected to describe some of the wild animals of the frontier, he asked Collinson to make arrangements for some plates. For Byrd's correspondence with Collinson, see J. S. Bassett, ed., *Writings of Colonel William Byrd*, New York, 1901, p. lxxix; also *Histories of the Dividing Line*, ed. William K. Boyd, Raleigh, 1929, xv–xvi. The Rawlinson copperplate was made for the History which Byrd intended to publish.

Collinson showed Byrd's work to the naturalist Mark Catesby, who had been Byrd's guest while in Virginia in 1712. On June 27, 1737, Byrd wrote to thank Catesby for his compliments on the manuscript. See Marion Tinling, *The Correspondence of William Byrd*, Charlottesville, Va., 1977, Vol. II, p. 519. In her contribution to the article titled "Discovery of the Rawlinson copperplate maps of the Americas and their related prints," Helen Wallis suggests that etched figures of the snake with the squirrel (no. 2) and the oppossum (no. 3) decorating the boundary line map were executed on the plate by Catesby. See *The Map Collector* 56 (Autumn 1991), p. 16 (with a photograph of the map print found in Holkham Hall in Norfolk, England).

The North Carolina State Department of Archives and History has a photocopy of a pull which was made from the copperplate.

Original

Oxford. Bodleian Library. Rawlinson, Copperplate No. 29.

240. Keith 1738

A New Map of Virginia humbly Dedicated to y^e^ Right Honble Thomas Lord Fairfax. 1738.

Size: 8¼ × 13. *Scale:* 1" = ca. 38 miles.

In: Keith, Sir William. *The History of the British Plantations in America . . . Part I, Containing the history of Virginia.* London, 1738, p. 38.

Description: About a third of the map, which extends from the mouth of Cape Fear River to Philadelphia, is devoted to "North Carolina," for which section Popple's 1733 map is apparently the source.

Of Sir William Keith's projected history, only this first volume, which relates almost wholly to Virginia, was published.

Copies

Congress; American Geographic Society; John Carter Brown; Harvard; New York Public; Yale.

241. Wimble 1738

To His Grace Thomas Hollis Pelham Duke of Newcastle Principal Secretary of State and one of His Majesties most Honourable Privy Council, &c. This Chart of his Majesties Province of North Carolina With a full & exact description of the Sea-coast, Latitudes, Capes, remarkable Inlets, Bars, Channels, Rivers, Creeks, Shoals, depths of Water, Ebbing & Flowing of the Tides, the generally [*sic*] Winds Setting of the Currents, Counties, Precincts, Towns, Plantations, and leading Marks, with directions for all the navigable Inlets; are Carefully laid down and humbly dedicated, by Your Grace's most humble, most dutiful, & most Obedient Servant, James Wimble. Engrav'd and Publish'd according to Act of Parliament Anno 1738. [cartouche, top left] | Sold by W. Mount & T. Page on Tower Hill and J. Hawkins in Fenchurch street London, and the Author at Boston in New England [bottom right] | I. Mynde [lower right-hand corner]

Size: 37¾ × 22¼. *Scale:* 1" = 7½ miles.

Inset: "Directions to Sail into all the Navigable Inlets in this Chart." [lower right corner]. *Size:* 8 × 9⅜. *Scale:* None.

Description: This elegant and refined chart appears to owe but little to James Wimble's 1733 manuscript chart discussed above (Map 215A). Even a cursory comparison quickly reveals that the chart's London publishers, William Mount and Thomas Page, compiled this dedicated chart from sources in addition to Wimble's original manuscript. It has been speculated that the appearance of Moseley's map in 1733 may have discouraged the immediate publication of Wimble's manuscript chart of that year. Wimble had continued gathering information for an improved chart which he probably carried to London in 1737, when he presented a plea for compensation for the loss of his ship before the lords commissioners for trade and plantations. He had written to the duke of Newcastle the year before to request an appointment and that he be allowed to dedicate his map to the duke. In that letter, dated July 10, 1736, he mentioned that it would cover "ye ole Coste of the ole Colleny," and "be newly taken in a plaine draft and woold becompleated in a month times to ye advantage of navigation." Clearly Wimble continued his surveys in the period following 1733, but regrettably, they do not appear to have survived. The printed chart includes the name and location of numerous settlers, though there are not as many itemized settlements shown as on the Moseley 1733 map (Map 218). Wimble's is a navigation chart rather than a map of the colony and thus shows only those colonists settled close to the rivers and inlets with no great attention being paid to the interior parts. Wilmington is indicated for the first time on any large map; Moseley has "Watson," the original landowner, on his map where by 1733 Newton and where by 1735 Wilmington was planned. The chart's information around the Cape Fear River region is especially detailed: Wimble plantations on the Black River (a branch of the Cape Fear now known as South River) and the New River are shown, as well as those of Governor Burrington and Moseley on the "N. W. River" (Cape Fear River), and of many others. Thus while many of the names of settlers on the Moseley are not repeated, Wimble adds much new information concerning settlement along the coast. On the New River eighteen settlements, with names of seven of the more important landowners, are noted. Careful as this chart is and expensive as its preparation must have been, there is no colonial record of official approval or aid having been extended to Wimble's worthy efforts. A letter written to the Bishop of London from Richard Marsden on March 13, 1737, has the postscript: "Capt. Wimble intreats your Lordship to accept this map of North Carolina"; see *Colonial Records of North Carolina*, IV, p. 245. Marsden was a minister of the Society for the Propagation of the Gospel serving the Cape Fear River parish while it was part of the diocese of the Bishop of London. Wimble's chart did much to clarify and make less hazardous the routes of entry for North Carolina coastal navigation and thus encouraged the commerce and trade between the province and Old and New England as well as the West Indies. Together with Moseley's map it must have played a significant part in the rapid increase in the trade in naval stores and general development of North Carolina that took place in the late 1730s and 1740s.

References

Cumming, W. P. "The Turbulent Life of Captain James Wimble" and "Wimble's Maps and the Colonial Cartography of the North Carolina Coast," *The North Carolina Historical Review* 46 (1969–70), pp. 1–18, 157–70.

Reproductions

Cumming, W. P. "Wimble's Maps and the Colonial Cartography of the North Carolina Coast," *The North Carolina Historical Review* 46 (1969–70), between pp. 164 and 165 (photograph printed with Map 215A on reverse).

Copies

Congress; John Carter Brown; Kendall; North Carolina State Department of Archives and History, three copies; University of North Carolina.

---

242. Verelst 1739 MS

New Map of the Country of Carolina, With its Rivers, Harbours, Plantations, and other accommodations. Done from the latest Surveys and best Informations. By order of the Lords Proprietors. [cartouche, bottom left]

Size: 23⅜ × 19¼. *Scale:* 1" = 21 miles.

In: [Verelst, Harman.] *Some Observations on the Right of the Crown of Great Britain to the North West Continent of America.* [1739], opp. p. 50.

Description: This is a pen-and-ink drawing based on the Gascoyne 1682 printed map, with the insets; it contains no new information. It is found in the same manuscript volume with Bull-Verelst 1739 MS.

It is placed in the volume opposite copies of Charles II's grants to the lords proprietors in the historical account of the territory given in the defense of Great Britain's legal right to the country.

See Bull-Verelst 1739 MS.

References

John Carter Brown. *Annual Report,* 1945–46. Providence, 1946, p. 21.

Reproductions

Francis Edwards, Ltd. Catalogue 594. London, 1936, opp. p. 10.

Original

John Carter Brown 1739.

---

243. Bull-Verelst 1739 MS

This Chart was transmitted by Col.º Bull (President & Comander in chief of South Carolina) with his Representation to the Board of Trade, dated the 25.th of May 1738. Rec.d July ye 27.th 1738 [lower left]

Size: 36⅜ × 28¾. *Scale:* 1" = ca. 32 miles.

In: [Verelst, Harman.] *Some Observations on the Right of the Crown of Great Britain to the North West Continent of America.* [1739], opp. p. 68.

Description: This pen-and-ink map is a duplicate of the Bull 1738 map in the Public Record Office and was copied in the same or the following year. It is found in a folio volume of over 105 pages in a manuscript containing Verelst's *Some Observations.* It is the complement to Arredondo's *Historical Proof* of 1742. An account of the Verelst volume and its historical background is given by Mr. Lawrence C. Wroth in John Carter Brown Library, *Annual Report,* 1945–46, Providence, 1946, pp. 14–23.

Harman Verelst was accountant to the Trustees for Establishing the Colony of Georgia, and this volume was "Humbly Presented to His Grace the Duke of Newcastle His Majesty's Principal Secretary of State by His Grace's Most Obedient humble servant Harman Verelst 16. April 1739" [p. 17]. The manuscript volume contains, besides a historical examination of the legal claims of Britain to the Carolina-Georgia area, copies of numerous contemporary depositions. Among these is a long Representation or report from William Bull to the commissioners for trade of 25 May 1738, which includes the map.

See Verelst 1739 MS.

References

Francis Edwards, Ltd. Catalogue 594. London, 1936, p. 10.

John Carter Brown. *Annual Report,* 1945–46, pp. 20–21.

Original

John Carter Brown.

244. Roberts-Toms 1739

The Ichnography of Charles-Town at High Water. [top center] | G: H: Delin: W: H: Toms sculp^{t}. [bottom right] | Published According to Act of Parliament June 9. 1739. by B: Roberts and W: H: Toms. [below alphabetized list at bottom of sheet]

Size: 21½ × 14⅜. *Scale:* 1" = 300 feet.
Inset: A Sketch of the Harbour. *Size:* 5½ × 8.
Scale: 1" = 2 miles.

Description: This detailed plat of Charles Town gives the location of each house, street, and wharf. The main houses are identified in an alphabetized list below the neat line, measuring 21" × 2¾". The title is given on a scroll at the top center of the plan; at the top right is a dedicatory cartouche, "Honorabili Carolo Pinckney Armigero . . . D. D. Servus Devotissimus G. H."

Reproductions

Avery, E. M. *A History of the United States.* Cleveland, 1904–7, III, between pp. 334–35.
Charleston Year Book. Charleston, S.C., 1884, frontispiece.

Copies

John Carter Brown; Kendall.

245. Desbruslins 1740

Plan de la Ville et du Port de Charles Town dans la Caroline Méridionale Aux Anglois

Size: 6 × 4. *Scale:* 1" =¼ mile.
In: Aa, P. vander. *La Galerie Agreable du Monde.* Leide, [1729–50], II, No. 10.

Description: This chart of Charles Town is found on a large sheet with five other plans taken from a French edition of Popple's 1733 map. At the bottom right of the large sheet is "Desbruslins Sculp." and at the bottom center, the date, 1740. The map was probably published by Buache, for whom Desbruslins engraved.

This is not similar to the Cóvens-Mortier map of Charles Town in their reduction of Popple's map.

Copies

Harvard 1729–50.

246. Georgia [1735]

A Map of the County of Savannah. [cartouche, top center] TFL fec [bottom right]

Size: 14¼ × 15½. *Scale:* 1" = 3 miles.
In: Urlsperger, Samuel. *Der Ausführliche Nachrichten . . . Halle, 1735.* Viertes Stück, opp. p. 2073.

Description: This is a map of Georgia after its first year of settlement. It was published in Halle, Germany, in the 1730s and 1740s by Samuel Urlsperger with the monumental collection commonly known as the "Salzburger Tracts." It was during Georgia's first year that groups of German protestants, who became known collectively as Salzburgers, began to arrive in the colony. The inscription "TFL fec" on the map has led some experts to believe that it was engraved by the well-known German mapmaker Tobias Conrad Lotter. This in spite of the middle initial on the plate being F rather than C. While not conclusive, there is evidence suggesting that the plate was in fact engraved in England and thus may have no connection with the Lotter firm. James Edward Oglethorpe, Georgia's cartographically astute founder, is known to have been forwarding maps to Germany in an effort to encourage displaced Germans to make the colony their new home during this period. Wherever its compilation and engraving took place, "A Map of the County of Savannah" is truly remarkable. It clearly shows the geography of the farms, settlements, villages, roads, forts, and outposts that surrounded Savannah. Shown adjacent to Savannah is Georgia's Utopian cadastre, made up of a regularly sized rectangular town, garden, and outlying farm lots. This map is found variously located in the Urlsperger Tracts, to accompany Tract 4, "Kurze Nachricht von Georgien." It is also bound in at the end of a re-issue of the *Journal of Samuel Urlsperger* first printed in 1735, but with a title page "*Ausfuührliche Nachricht Von den Salzburgischen Emigranten . . .*," Halle, 1744.

In the 1980s a heretofore unknown version of this map, with the title "Nova Georgia," was discovered in Oxford's Bodleian Library among a collection of copperplates forming part of the famous Rawlinson Collection. It is marked III and follows the "Map of Roanoke, Blackwater, and other rivers in Virginia and Carolina," which is attributed to William Byrd II (see Map 197 above). Pritchard and Sites persuasively argue that the "Nova Georgia" and a number of the other Rawlinson plates were originally prepared sometime just after 1737 to illustrate a book on Virginia's history and natural conditions, which Byrd was planning to publish. See Pritchard, M. P., and V. L. Sites, *William Byrd and His Lost History: Engravings of the Americas*, Williamsburg, Virginia, 1993. No manuscript of Byrd's projected work has come to light, and it was believed that, except for a limited number produced by the Bodleian in 1988, prints had never been pulled from the related plates. In 1991, however, Dr. Helen Wallis discovered a number of original prints, including "Nova Georgia," in the library of one of England's finest stately homes, Holkham Hall, seat of the earl of Leicester. See Wallis, H., "Discovery at Holkham Hall," *Colonial Williamsburg* 15 (1993), pp. 49–55. The "Nova Georgia" map shares many similarities with "A Map of the County of Savannah," and they appear to be based on a common original draft or survey. "Nova Georgia" is the smaller of the two, measuring only 8" × 8⅞"; and it contains fewer details as well as features and place-names. The only name exclusive to "Nova Georgia" is applied to a stream to the northwest of Savannah called "Westbrook." Immediately notable differences include Nova Georgia's sparser and more regular pattern of tree symbols and more generalized cadastral pattern south of Savannah.

Reference

De Vorsey, L. "Oglethorpe and the Earliest Maps of Georgia," in P. Spalding and H. H. Jackson, eds., *Oglethorpe in Perspective: Georgia's Founder after Two Hundred Years*. Tuscaloosa, Alabama, 1989, pp. 40–43.

Reproductions

Plate 55B.

Adams, J. T., ed. *Album of American History*. New York, 1944, I, 220.

Georgia Historical Society. *Collections*, VIII (1913), opp. p. 96.

Historic Urban Plans. Ithaca, New York (full scale facsimile).

Jones, C. C. *The History of Georgia*. Boston, 1883, I, opp. p. 148.

Winsor, J. *Narrative and Critical History of America*, V, 373 (right half only).

Copies

Congress 1740, 1744; Georgia Hargrett Library 1735; Harvard 1740, 1744; New York Public Library 1740; Clements, 1740.

246A. Thomas 1740 MS

A MAP of the ISLANDS of St. SIMON and JEKYLL with the PLANS and PROFILS of their FORTIFICATIONS as Proposed by the late John Thomas Engineer and design'd to be Executed under his Directions for the Deffence and Security of the said Islands and Town of FREDERICA: most Humbly Dedicated To HIS GRACE The DUKE OF ARGYLL & GREENWICH by HIS GRACE's most Humble and most Dutifull Servant. John Thomas. 1740. [cartouche, top center]

Size: 42½ × 29½. *Scale:* 1" = 1 mile.

Description: This detailed colored manuscript map, compiled by the military engineer Captain John Thomas and his son, was purchased in London in 1892 by the Chicago collector Edward E. Ayer. Ayer presented it as a gift to the Jekyll Island Club, where it hung until the island was purchased by the State of Georgia, at which time the Jekyll Island Club presented it to the Fort Frederica Association. The map now hangs in the museum at the National Park Service Fort Frederica National Monument. Thomas was a retired veteran when ordered to Georgia to plan fortifications there. The map of the islands is accompanied by four inset plans showing aspects of Thomas's elaborate fortification scheme. There are also lengthy blocks of explanatory text. En route to England with his son, a sub-engineer, the elder Thomas died in Charleston in 1739. Before he could render his father's surveys and plans in a finished form, young Thomas was appointed as an engineer to the military force being sent to the West Indies. It was not until 1743 that he was able to present the Georgia Trustees with "a neat plan of the fortifications there" and receive the twelve guineas they awarded him for its preparation. See *The Colonial Records of the State of Georgia*, V, p. 706. There are several of Thomas's fortification sketches in the British Library's King George III's Topographical Collection. Heretofore incorrectly dated, they are listed below. See the entry for Map 318A; it is particularly interesting because it mentions evidence of Thomas's possible plan to engrave this map.

Reference

Blake, J. G. *Pre-Nineteenth Century Maps in the Collection of the Georgia Surveyor General Department*. Atlanta, Georgia, 1975, pp. 44–48.

Reproduction

Georgia State Archives (reduced photostat).

Original

Fort Frederica National Monument, Georgia.

---

247. Jekyll Sound 1771

Plan du Port de Gouadaquini now called Jekil Sound in the Province of Georgia in North America Latitude 31°: 13' North.

Size: 14⅝ × 10⅝. *Scale:* 1" = ca. ⅙ mile.
Description: This map is identical to 402 below.

References

Museum Book Store, Catalog 93, Item 393.

Reproductions

Georgia Historical Society. *Collections*, VII, Part III (1913), opp. p. 66.

Copies

Georgia Hargrett Library; Huntington.

---

248. Seale 1741

Georgia [printed vertically across face of map from Ogechee R. to beyond Florida boundary] | R. W. Seale sculp. [bottom right, below neat line]

Size: 11⅜ × 14¾. *Scale:* 1" = ca. 17 miles.
In: [Martyn, Benjamin]. *An Account Showing the Progress of the Colony Georgia in America from its First Establishment.* London, 1741, frontispiece.
Inset: Great S^t. Simons Isle [written across inset]
Size: 3½ × 6. *Scale:* 1" = 3 miles.

Description: This map primarily shows developments in Georgia but extends to the north as far as "Charles T." and "Dorchester." It is interesting in that it shows roads connecting South Carolina with several Georgia settlements including "Fort Augusta," "Ebenezer," "Savannah," and "Darien," where the former South Carolina outpost Fort King George is indicated as "IF. K. George." The map is based on older surveys because it shows the Salzburger settlement Ebenezer in its original inland site. In 1736, Ebenezer's German-speaking settlers were allowed to move to a better situation on the Savannah River, where they founded New Ebenezer. A single line of dots following the Carolina bank of the Savannah River traces the boundary of that colony with Georgia; the river is described in Georgia's royal charter as "the most Northern Stream of a River there commonly called the Savannah." See L. De Vorsey, *The Georgia–South Carolina Boundary: A Problem in Historical Geography*, Athens, Georgia, 1982, p. 23. Georgia's southern boundary was originally at the Altamaha River (Alatamaha on the map), which is shown here flowing into the sea adjacent to an unnamed St. Johns River, far south of its true mouth. The row of dots indicating the Georgia-Florida boundary commences at the mouth of this incorrectly located Altamaha River. The rectangular land division plan associated with Savannah is shown here in a more simplified form than found on "A Map of the County of Savannah" (Map 246, above). R. W. Seale, the map's engraver, had a good opportunity to become acquainted with the spatial aspects of the Georgia project when he assisted W. H. Toms in engraving the great Popple maps (Maps 216 and 217, above). One of the curious features included on this Georgia map is the isolated portion of a stream shown to the west, "Couanouchie R." It is captioned, "R. undiscover'd" and terminates at the letter O in Georgia.

Most copies of *An Account* (1741) do not have the map and it may be an insert; the copy in the Hargrett Library at the University of Georgia appears to have been folded in half but never bound. It contains the handwritten inscription, "N.B. a fathom is 6 feet," following the printed note in the bottom left corner. John Carter Brown Library has two copies of the 1741 pamphlet, but only one contains the map. The New York Public Library has no map in this nor in the 1742 reprint of *An Account* sold by Jonas Green at his printing office in Annapolis. See Sabin 45000 and *Catalogue of the Wymberley Jones De Renne Georgia Library*, I, 90; III, 1202. The University of Georgia's version of the map was reprinted in volume I of *Detailed Reports on the Salzburger Emigrants Who Settled in America . . .* , edited by George Fenwick Jones, Athens, 1968, p. viii. According to the editor's caption printed under the map, "it has been reproduced in several publications; including Urlsperger's *Ausführlichten, 13te Continuation, Erster Theil* (Halle and Augsburg, 1747)."

Under the year 1741 may be mentioned "A View of the the Orphan House . . ." at Bethesda, with an accompanying plan, which is found in George Whitefield's *An Account of Money Received and Disbursed for the Orphan-House in Georgia*, London, 1741: reproduced in J. T. Adams, ed., *Album of American History*, New York, 1944, I, 227. See also Whitefield, 1768. It is interesting to note further that the authorship of "A View Of the Orphan-House" is given in the lower right corner of the engraving as "N. Jones del." This was doubtlessly Noble Jones, Georgia's first official surveyor. The orphan-house view is drawn as an oblique perspective in the manner of the Jones-Gordon 1734, "A View of Savanah as it stood the 29th of March, 1734" (Map 219, above).

Reproductions

Georgia Historical Society. *Collections*, VII, Part III (1913), opp. p. 7.

Howell, Clark. *History of Georgia.* Atlanta, 1926, I, opp. p. 88.

Jones, George Fenwick, ed. *Detailed Reports on the Salzburger Emigrants Who Settled in America. . . .* Athens, Georgia, 1968, I, p. viii.

Copies

Congress; John Carter Brown; Georgia Hargrett Library; British Public Record Office, MPG 360.

---

249. Popple-Cóvens-Mortier ca. 1741

The Town and Harbour of Charles Town in South Carolina.

Size: 5¾ × 4½. *Scale:* 1" = 1,056 feet.
In: Delisle, G. *Atlas Nouveau.* Amsterdam, ca. 1741, II, No. 28e.

Description: This is one of fifteen insets on a large folio page which gives the plans of nineteen harbors and forts in America. It is sheet five in the six-sheet form of the Cóvens-Mortier-Popple map. The signature "I. K. & s." is at the bottom right of this fifth sheet. Across the top of the sheet is written "Les Principales Forteresses Ports &c. de L'Amerique Septentrionale."

Copies

Congress ca. 1741; John Carter Brown ca. 1741, ca. 1745; Harvard ca. 1741; New York Public ca. 1741.
In: Cóvens, J., and C. Mortier. *Atlas Nouveau.* Amsterdam, 1684–1761, IX, No. 45.

Copies

Congress 1684–1761.

---

250. Savannah 1741

A View of the Town of Savanah [*sic*] in the Colony of Georgia, in South Carolina. Humbly inscribed to his Excellency Gener! Oglethorpe. Published Oct! yᵉ 1st [*sic*] 1741. Reference [A to P]

Size: 22¼ × 15½. *Scale:* None.

Description: At first glance this appears to be a fairly close copy of Jones-Gordon 1734 "A View of Savanah as it stood the 29th of March, 1734" (Map 219, above). Like that earlier engraving it is an arresting high oblique perspective view purporting to show Georgia's first settlement and major town. It is, however, significantly different than the 1734 view and omits one of the most distinctive elements in Savannah's original plan. Rather than showing correctly the uniquely designed squares and Trust blocks around which Savannah's residential wards and tythings were grouped, this unsigned engraving depicts a simple grid of intersecting straight streets and lanes. Some of the houses shown on the Jones-Gordon view are missing here also. In the list of explanatory references printed at the foot of this 1741 version there also are found interesting differences. Reference letter P, for example draws attention to "The Wood Covering the Back & Sides of the Town with several Vistas cut into it." Some of these vistas can be discerned in the forest of high trees shown surrounding the town on three sides and stretching to the horizon. Vistas such as these allowed visual signaling between Georgia's dispersed settlements in time of emergency. By 1741 Savannah boasted a total of six squares with surrounding wards, two more than shown in this out-of-date view. See Rodney M. Baine and Louis De Vorsey, "The Provenance and Historical Accuracy of a View of Savanah as it Stood the 29th of March, 1734," *The Georgia Historical Quarterly* 72 (1989), pp. 784–813.

Reproductions

Cassell's *History of the United States.* Edmund Ollier, ed. New York, 1875–77, I, 487.

Gay, S. H., and W. C. Bryant. *Popular History of the United States.* New York, 1876–98, III, 140.

Jones, C. C. *History of Georgia*, I, 121.

Winsor, J. *Narrative and Critical History of America*, V, 368.

Reference

Catalogue of the Wimberly Jones De Renne Georgia Library, III, 1279.

Copies

Georgia Hargrett Library.

---

251. Charles Town 1742 MS

Ciudad de Carolina [Charles Town] por la latitude de 32 grades 35 minutos

Size: 14⅛ × 14⅛. *Scale:* 1" = ca. 1⅛ miles.

Description: This is a careful survey of the harbor of Charleston, possibly taken from some English source, with soundings and anchorages; it was evidently made for use in a possible attack on the city by the Spaniards. The title, a table of references A to K, and scale are given in an enclosure to the left of the map. The date is given by Torres Lanzas, *Relación Descriptiva*, I, 104.

References

Torres Lanzas, Pedro. *Relación Descriptiva*, I, 104, No. 139.

Reproductions

Congress (tracing).

Original

Sevilla. Archivo General de Indias. Est. 87. Caj. 1. Leg. 24.

---

252. Ayala ca. 1742 MS

Mapa de la Costa de la Florida desde el Cabo Cañaberal salida de la Canal de Bahama hasta la Carolina con sus Bassas y fondos de ellas y el terreno usurpado por los Yngleses

Size: 24½ × 15¾. *Scale:* 1" = 14½ miles.

Description: An anonymous undated manuscript of the Florida-Carolina coast, in colors, with the title above the map and a detailed legend below, giving the names of seventy rivers, entrances, etc., which are indicated by numbers on the map itself. The coastline is shown in detail from Charles Town to Cape Canaveral. On the upper right of the map is written in a different hand: "Le dió a la S^{ria} de Yndias d^{n} Manuel Joseph de Ayala Arch.^{ro} de ella."

Reproductions

Cartografia de Ultramar, II, Plate 50 (with a transcription of the legend and a description of the map on pp. 289–91).

Original

Madrid. Servicio Geografico del Ejército. 9a. 2a. a. Núm. 8.

252A. Zathalin 1742

Plano y Descripcion Del Puerto, De Gval Qvini Citvado, 25. Legvas Alnor^{E} De La Florida Enaltvra De 31 Gra^{s} 13 M^{s} Con Las Brasas de Brasas de Fond. Cascajo En La Canal, Y Lama Svelta En El Pverto [cartouche lower left] Delineado Y Lebrantado, Por Gabriel Mvñoz, Se Gvndo Piloto D La R^{l}; Arm^{a}; D S:M:A: de 1742 D D Cadoad | Alexano Zathalin [cartouche lower center]

Size: 20½ × 28¼. *Scale:* 1" = 8,650 feet.

Description: This is a carefully prepared pen-and-ink manuscript showing the Spanish fleet entering St. Simon's Sound on the coast of Georgia. English fortifications and establishments on St. Simon's Island, on Jekyll Island, and at Frederica are identified by numbers which are explained in a key block on the map's left-hand edge. Shown as "camino de federico" is the road along which a Spanish force was advancing when engaged by General James Edward Oglethorpe's men at the battle of Bloody Marsh—regarded by many as a turning point in the conflict with Spain.

Original

Clements.

---

253. Arredondo 1742 MS

Descriptio Geographica, de la part que los Españoles poseen Actualmente en el Continente de la Florida, del Del [*sic*] Dominio en que estan los Yngleses con legitimo Titulo solo en Virtud del Tratado de pases del año 1670 y de la Jurisdicion que indevidamente an Ocupado despues de dicho Tratado, en que se Manifestan las Tierras que Vrsurpan; y se definen los limites que deven prescrivirse para Vna y Otra Nacion en Conformidad del derecho de la Corona De España [title to right of map, above adjacent table of explanations]

Size: 14⅞ × 16½; 27⅜ × 16½, with appended explanatory sheet. *Scale:* 1" = ca. 65 miles.

Description: The map represents an area extending from northern Cuba to Maryland and westward to the Mississippi River. It portrays the Spanish claims in the "debatable land" between Spanish Florida and the English settlements. It is found with and forms part of the extensive historical document, *Demonstración Historiographica*, completed in 1742 by the Spanish royal army engineer Antonio de Arredondo, which Professor H. E. Bolton has edited and published as *Arredondo's Historical Proof of Spain's Title to Georgia*, Berkeley, Calif., 1925.

The map is on the left half of a large sheet; on the right

half is the map title and under it an itemized list, A to Z, or historical and geographical notes. For the Spanish nomenclature of the Florida peninsula and the Atlantic coast north to Cape Fear, this map has as much detail as any known extant drawing of the region. It shows the provinces of Guale, Orista, and Chicora, the coastal lands with their Spanish settlements and missions, the limits of the Spanish and English possessions, and the route of de Soto (1539–43). Peter J. Hamilton, in *Colonial Mobile*, Boston, 1910, pp. 553–56, prints the whole text of the right-hand sheet with the historical notes, which he took from a copy of the map made by de la Puente in 1765.

Arredondo was a diplomat and soldier as well as an engineer; he played an important part in the Spanish-English borderland disputes of the period. He was sent to Georgia from Havana in 1736 to protest to Oglethorpe against the Georgia colony. His mission was ineffective, and for the coming struggle with England he spent the next several years preparing fortifications, writing reports, and drawing maps. In 1742 he was chief-of-staff in the campaign against Georgia by the Spaniards. In the spring of that year, before the campaign began, he wrote a reasoned defense of the Spanish rights and position at the request of Güemes y Harcasitas, captain-general of Havana, who directed the attack against the English. This report is the *Demonstración Historiographica*, referred to above; the text has been edited, with a translation and introduction by Professor H. E. Bolton (see below). The introduction and translation, without the text, has been published as *The Debatable Land*, Berkeley, Calif., 1925.

See Arredondo 1737 MS and P. L. Phillips, *Lowery Collection*, pp. 270–75, for earlier plans and maps drawn by Arredondo. For duplicates of this 1742 map and their reproductions, see Arredondo-Martinez 1765 MS and Arredondo–de la Puente 1765 MS. Several maps of this period in Servicio Geográfico del Ejército, with transcriptions of place-names, are reproduced in *Cartografía de Ultramar*, II, Plate 50, the coast from Cape Canaveral to 33° N.L.; Plate 51, from Cayo Biscaino to St. Augustine; Plate 62, Gualiquini; Plate 63, the attack of the Spanish fleet on the English at Gualiquini (Fort Frederica); Plate 52, the Florida Keys (Los Martyres).

References

Phillips, P. L. *Lowery Collection*, pp. 342–43, No. 497.

Torres Lanzas, Pedro. *Relación Descriptiva de los Mapas, Planos, &c. de México y Floridas Existentes en el Archivo General de Indias*. Sevilla, 1900, I, 104–5.

Reproductions

Plate 56 (from a certified copy in the Library of Congress, made in 1914 by Jose Luis Gomes).

Bolton, Herbert E. *Arredondo's Historical Proof of Spain's Title to Georgia*. Berkeley, Calif., 1925, opp. p. 112.

Congress (photocopy).

Original

Archivo General de Indias. Sevilla. Est. 86. Caj. 5. Leg. 24.

---

254. Le Rouge–Popple 1742

Amerique Septentrionale Suivant la Carte de Pople faite à Londres en 20 feuilles [across top, below neat line] | A Paris Par et Chez le S.r le Rouge Ingenieur Géographe du Roÿ rue des grands Augustins vis-a-vis le panier fleurÿ avec Privilege du Roy 1742 [lower left corner, in cartouche]

Size: 18⅞ × 19⅞. *Scale:* 1" = ca. 190 miles.

In: Le Rouge, G. L. *Atlas*. Paris, 1740–47, No. 6.

Inset: Charles Town and other insets. *Size:* 1½ × 1. *Scale:* 1" = ca. 3½ miles.

Description: This is one of the reduced "key" type copies of Popple's map, made in France by Le Rouge, with eighteen other insets of towns and harbors in North America. In 1755 Le Rouge made another map, probably for aid in discussion of boundaries in the peace negotiations in progress between France and England: "Canada et Louisiane Par le S.r le Rouge . . . 1755." *Size:* 24¼ × 19⅝. *Scale:* 1" = ca. 71 miles. This map gives more detail than the Popple map for the Carolina region.

Le Rouge's 1756 French edition of Mitchell's "Map of the British and French Dominions in North America . . . 1755" is found in his *Atlas Général.*

Copies

Congress 1740–47, 1741–48; Newberry, separate.

In: Le Rouge, G. L. *Atlas Général.* Paris, 1741–62, No. 140.

Copies

Congress 1741–62.

254A. Eyre 1743 MS

A MAP OF the Colony of Georgia in America. Presented to Robert Eyre Esq.r By His Most Obedient humble Serv.t Thomas Eyre. 1743 [lower right corner, in cartouche]

Size: 28½ × 20¼. *Scale:* 1" = ca. 25 miles.

Description: This is a finely drawn colored manuscript map showing the area between the Atlantic coast and the Mississippi River from just south of St. Augustine to Bull's Island northeast of Charleston. In addition to showing many of the roads and settlements established in Georgia during its first decade under the administration of the Trustees, the map contains information on the Southeastern Indians. Indian groups shown include the "Cherrokee Indians," on the "Toogolo," "Keewee," "Chatooga," "Tanassee," "Watogo," and "Quanassee" Rivers. The "Ogeasees or Lower Creek Indians" are placed on the "Chattahouchee R.," while the "Talapoosies or Upper Creek Indians" are located on the "Ockfuskee R." Farther west the "Coosas" are on the river of the same name with "French Ft. Albamas" located where the Coosa and Ockfuskee Rivers join. The westernmost Indians shown are the Chickasaw and Choctaw, both on rivers with the same names. The Indian and European settlements are linked by trails indicated by fine dotted lines. It is interesting to note that Eyre drew the trails connecting with the French fort with much more closely spaced dots than those leading to English outposts and settlements. This may have been his way of indicating that the French-oriented trails were more hazardous. In Spanish Florida Eyre showed "Ft. Diego," "Ft. Moosa," "Ft. Picolata," and "Ft. de Poopa" as well as the fort at St. Augustine.

Thomas Eyre served in Georgia as "A Cadet in General Oglethorpe's Regiment" and was sent on missions to the Cherokee Nation, including one to prevent the distribution of rum there. In 1741, the Georgia Trustees requested that Thomas Eyre be employed to make maps of the two divisions of Georgia, "the County of Savannah" and "the County of Frederica." Robert Eyre, to whom this map is dedicated, was elected one of the Georgia Trustees on March 21, 1734. The Eyres were of New House in Wiltshire, England, and the dedicatee appears to have been the son of Sir Robert Eyre (1666–1735), who served as the Lord Chief Justice. Now found in the Fellowes Collection in the Norfolk Record Office, Norwich, England, this map is accompanied with listings of the inhabitants of Savannah and Frederica as well as other information about the colony supplied to Robert Eyre in 1743 by one Thomas Jones.

References

Colonial Records of the State of Georgia, XXX, pp. 7, 138, 182, 196, 297.

Original

Fellowes Collection, Norfolk Record Office, Norwich, England.

255. Barnevelt ca. 1744 MS

The Original of this Map was drawn by Col. Barnevelt who Commanded several Expeditions against the Indians in the Time of the Indian War as also Served under Col: Moore in all his Expeditions in the said War. It is highly Approved of by Leiutenant [*sic*] Governour Bull who is allowed to be the best Iudge [*sic*] of Carolina and the Indian Countrys round it, of any Person, now in the Province. [In cartouche upper left] Mississippi and Carolina Barnvelt. [Manuscript label posted on reverse]

Size: 53¼ × 31. *Scale:* 1" = ca. 19½ miles.

Description: This map is a copy of Barnwell ca. 1721, which it follows in size, in the area represented, and in innumerable details and legends. The copyist lacked knowledge of South Carolina history and terrain, as he shows by such errors as Barnevelt and Barnvelt for Barnwell, and Combetree R. for the Combahee River. He also omits a few notations and legends found on Barnwell's very detailed map. Along the Atlantic coast the boundaries and names of South Carolina counties are lacking, and below the Savannah River, where Barnwell put "Margravate of Azilia," the copyist omitted that title and some other names, without adding "Georgia."

The map is more than a bare reduplication of Barnwell's work, however. Several letters are placed from the Altamaha River north to the Pedee and show the location of new settlements and townships along the coast. In the bottom right corner of the map the "Explanation" tabulates the list of developments as follows:

A The Township of Purrysburg
B The Town of Savanna in Georgia
C Fort Augusta in Georgia
D The Swiss Township call'd Savanna or New Windsor
E Orangeburg Township on Ponpon R.
F Amelia Township on Santee

G Saxe Gotha Township on Santee
H Frederick or the Waterbee Township on the N.E. of the River
I Lands Reserved by the Gov.r and Council on the N.E. Side of Santee and Wateree Rivers for the North Britons
K Williamsburg the Irish Township
M New Darien 15 Miles above Frederica
N 200,000 Acres laid out to Col. Selwyn
O Welsh Tract on Pedee
P Kingston or Wacomaw Township
Q Queensbury on Pedee

References

Catalogue of the Wymberley Jones De Renne Georgia Library, III, 1198.

Reproductions

Georgia Historical Society. *Collections*, VII, Part 1 (1909), opp. p. 38 (part only).

Congress (photocopy).

Original

Georgia Hargrett Library.

---

256. Herbert-Hunter 1744 MS

A New Mapp of His Maiestys Flourishing Province of South Carolina Showing ye Settlements of y' English French and Indian Nation. In Herbert. 1725. [cartouche, lower right-hand corner] | True Copy from an Original done by Colonel John Herbert deceased late Commissioner of the Indian Trade who often was in these Nations which Original in his own hand Writing He has shewn me as a just & correct Mapp thereof. Certified this 11th Day of June 1744. By George Hunter, Surveyor General. [upper left-hand corner]

Size: 32¼ × 27¾. *Scale:* 1" = ca. 23½ miles.

Description: The area represented extends northward from Cape Carnaveral [*sic*] to the Cape Fear River and its upper branches and westward from the Atlantic coast to the "Point of Mobile," or from 28°15' to 38°45' N.L. and approximately from 77° to 89° West Long. The map shows the location of the Indian tribes in North and South Carolina, Alabama, and eastern Tennessee. As the title states, this map is a copy of a map by John Herbert drawn in 1725, which is apparently no longer extant. This copy and therefore Herbert's original derive from Barnwell ca. 1721. Most of the legends and much other detail found on Barnwell's map are lacking; the names of tribes and English settlements and the details of rivers and their branches, so far as they are given, are similar to Barnwell. The population of the Indian tribes is unchanged and "Margrvate [*sic*] of Azilia" is written between the Altamaha and Savannah rivers. There is a lack of additions showing new settlements or changes in the twenty years preceding 1744, for King George is located far up the "Altamaha" River where the Oconee and Ocmulgee Rivers meet to form that stream.

P. L. Phillips stated that in his opinion Herbert's 1725 original of this map was the source of George Hunter's map of the Cherokee country, 1730. See A. S. Salley, "George Hunter's Map of the Cherokee Country," *Bulletin of the Historical Commission of South Carolina*, No. 4, Columbia, 1917. Hunter's 1730 map, however, represents a smaller area and gives more detail for that area; in contrast to the general representation of the Southeast in the Herbert map, Hunter's 1730 pencil and pen-and-ink sketch shows little but the branches of the Santee and upper Savannah Rivers and the Indian settlements of eastern Tennessee. More recently Professor Marvin T. Smith has argued that the map in the University of Georgia Hargrett Library is the original Herbert map of 1725 with additions made by Hunter in ca. 1744. Smith bases his assertion on the wording found in the cartouche and the fact that there are two distinctly different handwritings and two different legends on the map. He also states that the Indian town names of interest to him were all in the hand that matched the writing in the 1725 title cartouche. See his *Historic Period Indian Archaeology of Northern Georgia*, University of Georgia Laboratory of Archaeology Series Report No. 30, April 1992, p. 42.

References

Catalogue of the Wymberley Jones De Renne Georgia Library, III, 1198.

Reproductions

Congress (photocopy).

Original

Georgia Hargrett Library.

---

257. Carteret-Toms 1744 MS

Map of Lord Carterets ⅛ part of Carolina Engraved by W H Toms in Union Court near Hatton Garden Holbourn, Aug. 1st 1744

Size: 11¼ × 10½. *Scale:* 1" = 9½ miles.

Description: The dividing line for the Granville District began at a cedar post six miles south of chickinacomack Inlet, sixty miles south of the Virginia line at 35°34', and was carried west to Bonner's Field, near Pamptico River, northwest of Bath Town, in 1743. In 1746 the line was extended to the Salisbury Road beyond Coldwater Creek (near present-day Kannapolis on U.S. Route 29. The line can still be traced by the northern or southern boundary of several counties. In 1774 the line was extended to the Blue Ridge Mountains. See *Colonial Records of North Carolina*, V, lv–lxi, and E. Merton Coulter, "The Granville District," *James Sprunt Historical Collections*, Chapel Hill, N.C., 1913, XII, No. 2, pp. 35–36.

See Carteret-Toms 1744 and Carteret 1746 MS.

Reproductions

North Carolina Department of Archives and History (tracing on linen).

Original

London. Public Record Office. Colonial Office Library. 5. Vol. 324, p. 37.

---

258. Carteret-Toms 1744

[Map of the eighth part of Carolina granted to John Lord Carteret] Engraved by W. H. Toms in Union Court near Hatton Garden Holbourn, Aug^st 1744

Size: 11½ × 11. *Scale:* 1" = 9½ miles.

In: Grant and Release of One Eighth Part Carolina, from His Majesty to Lord Carteret. London, 1744, opp. p. 18.

Description: On the map is written above the indicated lines: "Dividing Line between Virginia and Carolina made in the Year 1728," and "Dividing Line for the ⅛ Part of Carolina set out and marked 1743 | Lat. 35°34'." All the other heirs of the lords proprietors sold their claims and charter to the province in 1729, except Lord Carteret, later Earl Granville, who held one-eighth part bordering on Virginia.

Copies

Congress; John Carter Brown.

---

259. Bellin 1744

Carte Des Costes De La Floride Françoise Suivant les premieres découvertes. Dressée par N. Bellin Ing^r. de la Marine. [above map, within outer bordering lines] | 39 [outside neat line, to left]

Size: 5¾ × 8⅛. *Scale:* 1" = ca. 65 miles.

In: Charlevoix, P. F. X. de. *Histoire et Description Generale de la Nouvelle France.* Paris, Chez Nyons Fils, 1744, I, opp. p. 24.

Description: This map, which shows the coast from St. Augustine (30° N.L.) to Cape Fear (34½° N.L.), attempts to locate the Spanish, French, and English settlements along the coast. It gives Ribaut's names for the rivers as well as the contemporary ones. A curious error is found at "Charles Town," correctly located at the mouth of the Cooper River; the Ashley River is called "Pouhetan or R. James," and "James Town" is on its north bank. There are other errors of location.

This same map, from a different plate and with German translations of the title and legends, is found under Bellin 1756. The *Lowery Collection*, pp. 289–90, lists several Bellin maps of the Gulf coast in this work of Charlevoix.

References

Phillips, P. L. *Lowery Collection*, No. 377, p. 289.

Reproductions

Charlevoix, P. F. X. de. *History and General Description of New France.* Trans. by John Gilmary Shea, New York, 1900 (first translation, 1866), I, opp. p. 134.

Coulter, E. M. *A Short History of Georgia.* Chapel Hill, 1933, opp. p. 5.

Lowery, W. *The Spanish Settlements within the Present Limits of the United States: Florida, 1513–1561.* New York, 1901.

Copies

Congress, separate; American Geographical Society; Harvard; New York Public; Yale.

In: The same title but a different imprint. Paris, Chez Didot, Quai des Augustins, à la Bible d'or, I, opp. p. 36.

Copies

Clements.

In: The same. Paris, Chez la Veuve Ganeau, I, opp. p. 36.

Copies

Congress; Clements; Harvard; New York Public; Yale, two copies.

In: The same. Paris, Chez Pierre-François Gifford, I, opp. p. 36.

Copies

Harvard; New York Public.

In: The same. Paris, Chez Rollin Fils, I, opp. p. 36.

Copies

John Carter Brown.

In: [Bellin, N. *Cartes et Plans De L'Amerique 1745.*] No. 22 ["39" erased; in an atlas with title and date in ink and many MS maps by Bellin dated 1739].

Copies

Congress 1745.

---

260. Moll 1744

Carolina Von Herman Moll Geographe NB.
Die Pflantzung sind also gezeichnet İI | Page 656
[at top left, outside neat line]

Size: 6⅛ × 7. *Scale:* 1" = 70 miles.

In: Oldmixon, J. *Das Britische Reich in America . . . uebersetzet von Theodor Arnold.* Lemgo, 1744, opp. p. 574.

Description: This is a copy of the Moll-Oldmixon map of 1708, with German legends on the map and on a different plate. The "Bermudos" map is on the same folded sheet. The book is a translation by Theodor Arnold of the 1741 edition of Oldmixon's work.

Copies

John Carter Brown 1744; Clements 1744; Sondley 1744.

---

261. Carteret 1746 MS

This Plan sheweth part of the Southern Boundary of the Lands Granted the 17th day of September 1744 by His Majesty King George the Second To the Rt. Honble John Lord Carteret now Earl Granville, as the same was laid out and marked in the Months of March and April 1746 by Eleazer Allen, Matthew Rowan, William Forbes, and George Gould Esqrs Commissioners on the part of His Majesty, and Edward Moseley and Roger Moore Esqrs commissioners on the part of Earl Granville.

Size: 28 × 8 and (second sheet) 20 × 10.

Scale: 1" = 5 miles.

Description: This was certified as a true copy on January 10, 1746, by the commissioners with J. Hawks as witness. The following statement is also found on the map: "This is to certify that the above Plan has been compared and Examined with the original lodged in the Secretarys office of this Province. No. Carolina Sec. office July 5th 1765 Benn Heron Sec."

The survey extends 240 miles from the coast to the Rocky River and the Catawba Indian Trading Path.

See Toms 1744.

Reproductions

Hulbert, A. B. *Crown Collection.* [London, 1914–16], Series III, Plates 22–24.

Congress (photocopy).

Original

London. Public Record Office. C.O. 700, Carolina 15(1) and (2).

---

262. Bowen ca. 1747

The Town and Harbour of Charles Town in South Carolina.

Size: 5¾ × 4⅝. *Scale:* 1" = ca. 1½ furlongs.

In: Bowen, E. *Atlas of the World.* London, 1744–47, No. 73.

Description: This is a copy of the Popple-type plat of Charles Town, with the large fort and small Johnson's Fort to the south of the Town on the west bank of the Ashley. It is one of ten "Draughts and Plans of Principal Towns and Harbours in America and the West Indies, Collected from the best Authorities. By Emmanuel Bowen" on the same sheet (*Size:* 17¼ × 14). The plans derive eventually from the insets on the great Popple map of 1733, but are probably influenced more directly from the Cóvens-Mortier-Popple sheet of fortresses, ca. 1740.

The Bowen sheet of "Draughts and Plans" is found in some copies of the third volume of Bowen's *A Complete System of Geography*, London, 1747, with "No. 105" below the neat line to the left.

See the next map for copies of Bowen's *Geography*.

Copies

Congress (Charles Town inset only); American Geographical Society 1747; Charleston Library Society (Charles Town inset only).

263. Bowen 1747

A New & Accurate Map of the Province of North & South Carolina, Georgia, &c. Drawn from late Surveys and regulated by Astron[l]. Observat[ns] By Eman. Bowen. [cartouche, top left] | N[o]. 76. [below neat line, at right]

Size: 17 × 13½. *Scale:* 1" = 40 miles.

In: Bowen, E. *A Complete System of Geography*. London, 1747, II, between pp. 642–43.

Description: This is a copy of the Carolina sheet of Popple's great twenty-sheet map of North America (1733), with a separate title, reduced in size.

Some copies of the third volume of this atlas have a sheet with drafts and plans of Towns and Harbours in America and the West Indies, by Emanuel Bowen, including: "The Town and Harbour of Charles Town in South Carolina."

Copies

Congress; John Carter Brown; Charleston Museum; Clements; Duke, two separates; Georgia Hargrett Library; Harvard; Kendall; New York Public; Sondley Reference; University of North Carolina, three separates.

In: Bowen, E. [*Atlas of the World*. London, 1744–47], No. 64.

Copies

American Geographical Society 1747.

In: Bowen, E. *A Complete Atlas*. London, 1752, No. 58.

Copies

Congress; American Geographical Society; John Carter Brown; Yale.

264. Lotter 1747

Georgia [vertically across face of map] | T. C. Lotter sculps Aug[ustae] Vind[el] [below neat line, to left]

Size: 8¾ × 19⅝. *Scale:* 1" = ca. 18 miles.

In: Urlsperger, Samuel. *Ausführliche Nachricht von den Saltzburgischen Emigranten*. Halle und Augsburg, 1735–52, III (1747), opp. p. 72.

Insets: 1. Great St. Simon's Isle. *Size:* 3½ × 6. *Scale:* 1" = 3 miles.

2. [Below the map but on the same sheet is a picture of the method of operating water wheels for irrigations, grain-mills, etc.] *Size:* 8¾ × 4⅜.

Description: This is a map of the coast from "St. Augustin" to "Charles T." Under the English names are written German equivalents. It is a copy, with few changes, of the Seale map (1741) (see Map 248, above).

This map is usually found in the first part of the thirteenth continuation of *Ausführliche Nachrichten*. It is sometimes found on a large, heavy single sheet of paper with the plan of Ebenezer: see Urlsperger-Seutter 1747.

Reproductions

Corry, J. P. *Indian Affairs in Georgia, 1732–1756*. Philadelphia, 1936, p. 82.

Gabriel, R. H., ed. *The Pageant of America*. New Haven, 1925, I, 273 (St. Simon's Isle inset).

Hamilton, P. J. *The Colonization of the South*. Philadelphia, 1904, III, opp. p. 316.

Winsor, J. *Narrative and Critical History of America*, V, 379.

Copies

Congress, separate; John Carter Brown; Georgia Hargrett Library; Huntington; Kendall.

265. Urlsperger-Seutter 1747

Plan Von Neu Ebenezer verlegt von Matth. Seutter, Kayser[l]. Geogr. in Augspurg

Size: 19½ × 11¾. *Scale:* 1" = 225 feet.

In: Urlsperger, Samuel. *Aussführliche Nachricht von den Saltzburgischen Emigranten*. XIII Dreyzehenten Continvation . . . Erste Theil. Halle und Augsburg, 1735–52, III (1747), opp. p. 72.

Description: This plan of Ebenezer, the German settlement on the Savannah River, is laid out in ordered lots, forming a large square on the bank of the Savannah River. The town itself is divided into four smaller squares, each with its own marketplace in the center, surrounded by lots and public buildings. That this ordered plan was partially executed is shown by the actual survey of Ebenezer taken about 1770 by De Brahm (Map 412, below).

This plan appeared first in the 1747 tract; the Urlsperger tracts continued to be published at intervals from 1735 to 1752.

The Library of Congress, Kendall Collection, H. E. Huntington, University of Virginia, and Georgia Hargrett Library have an edition of this map on a single large heavy sheet of paper (24" × 21"), which also has the map of Georgia (1747) engraved by T. C. Lotter and the diagram of a water mill. Apparently Seutter sold copies of this large sheet separately. See *Catalogue of the Wymberley Jones De Renne Georgia Library*, Wormsloe, Georgia, 1931, II, 1203.

See references to the Urlsperger tracts and to this map in J. Winsor, *Narrative and Critical History of America*, V, 396.

Reproductions

Avery, E. M. *A History of the United States*. Cleveland, 1904–7, III, between pp. 334 and 335 (colored plate).

Gabriel, R. H., ed. *The Pageant of America*. New Haven, 1925, I, 274.

Hamilton, P. J. *The Colonization of the South*. Philadelphia, 1904, III, opp. p. 304.

Copies

Congress, separate; John Carter Brown; Georgia Hargrett Library; Huntington (two copies); Kendall; U. of Virginia.

---

266. Turner 1747

Map Nº I. Note that What are called by the following Names in this Map ** Hudson's River Delaware River New York Albany New Castle ** Were in the Dutch time Called ** Noordt Rivier Zuydt Rivier Nieuw Amsterdam Fort Orangie Fort Casimir. ** Note also that These Names in this Map ** Bridlington Burnigat Little Egg ** Are commonly Called ** Burlington Barnagat Little Egg Harbour ** Colours . . . [title and legends in cartouche, right center] | Engraved & Printed by James Turner near the Town House Boston [below cartouche]

Size: 12 × 15¼. *Scale:* 1" = ca. 38 miles.

In: A Bill in the Chancery of New Jersey. New York, Printed by James Parker, 1747, opp. p. 124.

Description: This map shows the Atlantic coast from Cape Hatteras to Boston Harbour. Nothing new is given in the Carolina area; only "Carolina," "Roanoke Inlet," and "Cape Hatteras" are found below Virginia.

This map is of little interest for the region south of Virginia. It is, however, a work of bibliographical and human interest because of the persons involved in producing it. Benjamin Franklin suggested (incorrectly, as it turned out) that copper plates for the three maps in the pamphlet would cost less than forty hand-drawn colored copies. There is good evidence, though not absolute certainty, that Lewis Evans designed this map for James Turner, thus bringing together these two for their first joint production. See George J. Miller, *The Printing of the Elizabethtown Bill in Chancery. In:* Board of General Proprietors of the Eastern Division of New Jersey, Pamphlet Series No. 1. Perth Amboy, N.J., 1942, pp. 12–13, 17.

Copies

Congress; John Carter Brown; Clements; Harvard.

---

267. Bowen 1748

A New Map of Georgia, with Part of Carolina, Florida and Louisiana. Drawn from Original Draughts, assisted by the most approved Maps and Charts. Collected by Eman: Bowen, Geographer to His Majesty. [across top of map, below neat line] | Vol. II, p. 323. [top right corner]

Size: 18⅞ × 14¼. *Scale:* 1" = 38½ miles.

In: Harris, John. *Navigantium atque Itinerantium Bibliotheca. Or, A Complete Collection of Voyages and Travels*. London, 1744–48, II (1748), opp. p. 323.

Description: This map extends from about 28½° to 35° N.L.; the coast is shown from "Charles Town" to the "Missisipi," excluding southern Florida. It is somewhat similar to other maps of this period by Bowen but is on a larger scale; the coastal settlements in South Carolina, the Indian tribes friendly or hostile to the English, and the chief trading paths are shown. Indian territories are shown by fine dotted lines. The Georgia roads and settlements are particularly full.

Harris first published his collection of voyages in 1705. For the 1744–48 and the 1764 editions a new chapter was added, giving a history of Georgia. It was for this added chapter that the map was made.

Reproductions

Color Plate 18.

Bolton, H. E. *Arredondo's Historical Proof of Spain's Title to Georgia*. Berkeley, Calif., 1925, opp. p. 208.

Historic Urban Plans. Ithaca, New York (a colored facsimile of the 1748 map).

Hodler, T. W., and H. A. Schretter. *The Atlas of Georgia*. Athens, Georgia, 1986, endpapers.

Copies

Congress 1764; John Carter Brown 1764; Clements 1748; Duke 1748; Georgia Hargrett Library 1748 (and two separates), 1764 (and one separate);

Harvard 1748, 1764; Kendall Collection; New York Public 1748, two copies; Yale 1764.

267A. Hamar 1748 MS

Map of the entrance to Port Royal Harbour.

Size: 15¼ × 21¾. *Scale:* 1" = 1 mile.
Inset: Plan of H.M.S. Adventure's careening creek.
Size: 9½ × 11½. *Scale:* 1" = 100 fathoms (600 ft.).
Description: This colored manuscript map was enclosed in a letter sent to the British Admiralty by Captain Joseph Hamar of the *Adventure*. The letter was dated May 18, 1748. Soundings, bottom characteristics, and currents, as well as anchorages, are indicated. On land, trees and buildings are shown in relief. Tables and notes on tides are included.

Reference

Maps and Plans in the Public Record Office, 2. America and West Indies. London, 1974, No. 2864, p. 496.

Original

British Public Record Office. MPI 430.

---

268. Virginia–North Carolina Boundary 1749 MS

A Plan of the Line between Virginia and North Carolina, from Peters Creek to Steep Rock Creek, ran in the Year of our Lord 1749. by Joshua Fry. Peter Jefferson Will.m Churton. Dan.l Welden done by a Scale of five Miles to an Inch.

Size: 16¾ × 13. *Scale:* 1" = 5 miles.
Description: Also on the map, below the title in the upper right corner, is written the following note: "Rec.d June ye 24.th 1751. With Gov.r Johnston's Letter dated the 15th of Feb.ry 1750."

This survey of ninety miles was used in Fry & Jefferson's map of Virginia (1751). Of this survey W. L. Saunders writes in the Prefatory Notes to the *Colonial Records of North Carolina* (1886, IV. xiii): "Governor Johnston says 'they crossed a large branch of the Mississippi [New River] which runs between the ledges of the mountains, and nobody dreamed of before.' It so happens, however, that no record of this survey has been preserved, and we are today without evidence, save from tradition, to ascertain the location of our boundary for ninety miles."

Steep Rock is about 25 miles southeast of Abingdon, Virginia. According to William M. Darlington, in the preface to his edition of *Christopher Gist's Journals*, Pittsburgh, 1893, p. 22, Steep Rock Creek is "the White Top or Laurel Fork, of Holston River, in the present county of Grayson, and in 1779 Dr. Thomas Walker and Daniel Smith continued the survey of the line from Steep Rock Creek to the Kentucky River."

See Fry-Jefferson ca. 1749 MS.

Reproduction

North Carolina Department of Archives and History (photocopy).

Original

London. Public Record Office, MPG 361.

---

269. Fry-Jefferson ca. 1749 MS

This is a Plan of the Line between Virginia and North Carolina which was run in the Year 1728 in the Spring and Fall from the Sea to Peter's Creek by the Honorable William Byrd, William Dandridge and Richard Fitzwilliams Esquires Commissioners and M.r Alexander Irvine and M.r William Mayo Surveyors; and from Peter's Creek to Steep Rock Creek was continued in the Fall of the year 1749 by Joshua Fry and Peter Jefferson [cartouche, upper right corner]

Size: 74½ × 10⅞. *Scale:* 1" = 5 miles.
Description: This is a pen-and-ink drawing on cloth, in six parts pasted together, in the handwriting of Peter Jefferson. It is not the original (see Virginia–North Carolina Boundary 1749 MS), but was probably made soon after the survey. As the title indicates, the map includes the whole boundary line, and not only the part surveyed by Fry and Jefferson.

Reference

Thurlow, C. E., and F. L. Berkeley. *The Jefferson Papers*. Charlottesville, Va., 1950, pp. 1–2, No. 6.

Original

University of Virginia.

---

270. Hyrne 1749 MS

A New and Exact Plan of Cape Fear River from the Bar to Brunswick. By Edward Hyrne 1749.

Size: 12¼ × 15⅛. *Scale:* 1" = 1⅔ miles.
Description: This carefully surveyed map of the channel of

the Cape Fear River from its mouth to the town of Brunswick was twice printed; see Hyrne-Jefferys 1753 and Hyrne-Sayer 1768.

Reproductions

Congress, North Carolina Department of Archives and History (photocopies).

Original

British Library. Add. MS. 31981. B.

---

271. Robert de Vaugondy 1749

La Floride divisee en Floride et Caroline Par le S.r Robert de Vaugondy Fils de M.r Robert Geog. ordin. du Roi. Avec Privilege 1749 [cartouche, bottom left] | 189 [top right, above neat line]

Size: 6⅞ × 6½. *Scale:* 1" = ca. 125 miles.

In: Robert de Vaugondy, Gilles. *Atlas Portatif, Universel et Militaire*. Paris, 1748–[49], No. 189.

Description: This map gives a great deal of detail for its size, particularly concerning the location of Indian tribes north of the Gulf and west of the Appalachians. Florida has the narrow, almost V-shaped quality peculiar to some of the French and Italian maps of this period, though in this map only the lower half is made into a series of islands. Central and upper Florida has a plateau or range of mountains which extend from the Appalachians. The political bias of the map is shown by the area included in "Florida," which extends along the Altamaha River south and west of "Georgia" to a latitude above 35° N.

The American Geographical Society, the New York Public Library, and the Georgia Hargrett Library have undated copies of the *Atlas Portatif* with this map, on which the date is erased and the number outside the neat line, upper right, is 93. The Library of Congress has a separate copy of the map with the date and with the number "61" at top right.

This map was designed by Didier Robert (1726–86), son of Gilles Robert de Vaugondy (1688–1766), of the famous mapmaking family of Sanson; Gilles was a nephew of Pierre Moulart Sanson, who in turn was a nephew of one of the sons of Nicolas Sanson. Gilles succeeded to the collection of maps and to the business in 1730 and was geographer to the king until his death in 1766. Didier Robert was made Geographer Royal in 1760. The atlases of Gilles and Didier Robert were among the most highly esteemed and widely used in Europe in the latter half of the eighteenth century. Their maps were influential; but in making them they were apparently often indebted to the work of Bellin and Anville.

Copies

Congress 1748–49, separate; American Geographical Society ca. 1774; Georgia Hargrett Library; Huntington; New York Public 1748–49, ca. 1774; University of North Carolina.

---

272. Bowen 1749

A Map of the British American Plantations, extending from Boston in New England to Georgia; including all the back settlements in the respective provinces as far as the Mississipi. By Eman: Bowen Geog.r to His Majesty. | Tho.s Bowen Sc:

Size: 11 × 8¾. *Scale:* 1" = 100 miles.

In: The London Magazine, XVIII (July 1749), opp. p. 308.

Description: The territory covered by the map is sufficiently indicated in the title; the location of the Indians, especially west of the Blue Ridge range, is one of the fullest in a printed map up to this time.

Copies

Congress; Clements.

---

273. Carolina Rivers ca. 1750 MS

[Sketch Map of the Rivers Santee, Wateree, Saludee, &c., with the road to the Cuttauboes.]

Size: 23 × 11. *Scale:* 1" = ca. 7 miles.

Description: A roughly drawn draft of the trading path from the conjunction of the Wateree and Congaree Rivers, which forms the Santee River, northward along the Catawba to "Cattauboes or Nassau Towne" near "the fishing creek," just below the present boundary line between North and South Carolina.

"Cattauboes Towne" is shown as a large palisaded fortified settlement about ten miles above the smaller "Wateree Towne" on Fishing Creek (the present Sugar Creek). On the north side of the fortified "Cattauboes Towne" is "The Gate to Virginia road." To the north of the settlement is "Sugar Town," and to the south along the Catawba River are "Waterree Chickens," "Sugar Ditts," "Waxau Town," and "Waterree Old town."

At upper left, "36" in large numbers is twice written

across the face of the map. The chief value of the map lies in its location of Indian settlements on the Catawba and Congaree Rivers about the middle of the eighteenth century.

Reproductions

Hulbert, A. B. *Crown Collection.* [London, 1914–16], Series III, Plates 25–26.

Congress (photocopy).

Original

London. Public Record Office. C.O. 700, Carolina no. 16.

---

274. Juan de la Cruz ca. 1750

Plano de Charles Town Capital de la Carolina merid^{l}. con vn Mapa de la Costa que media entre esta Ciudad y la de S. Agustin de la Florida en America. | D. Juan de la Cruz fecit. [Lower left, without neat line]

Size: $7\frac{7}{8} \times 5\frac{1}{2}$. *Scale:* None.

Insets: There are two parts to this sheet:

1. The plan of Charles Town. *Size:* $5\frac{1}{2} \times 5\frac{1}{2}$. *Scale:* 1" = $\frac{4}{5}$ mile.

Description: This is a later development of the Crisp 1711 or Popple 1733 map, with "F^{te} Johnsons" on the south side of the Ashley River; this particular map has French names and is possibly based on the French map of Desbruslins 1740.

2. The Carolina coast from "Charles Town" to "S. Augustin." *Size:* $2\frac{1}{4} \times 5\frac{1}{2}$. *Scale:* 1" = ca. 45 miles.

Description: This map has Georgia, Savanah Town, Josephs Town, and Frederika. It is very similar to, and copied from, the Seale 1741 map or the Urlsperger Tracts 1747 map (Map 264, above).

Copies

Kendall.

---

275. Savannah ca. 1779 MSS

[Town of Savannah]

Size: Each of three: $14\frac{1}{2} \times 18$. *Scale:* 1" = 400 feet.

Description: These three pen-and-ink drawings are listed in J. R. Sellers and P. M. Van Ee, *Maps and Charts of North America, 1750–1789*, as maps number 1582, 1583, and 1584. Numbers 1582 and 1583 appear to be preliminary sketches with 1584 being the final watercolored plan of Savannah's fortifications. They were incorrectly dated in earlier editions of this work. All three show the distinctive plan of Savannah's streets and open squares enclosed within a defense work. The final plan, number 184, includes the roads leading into the town, the wharves on the riverside, as well as outlying buildings and strongpoints. Sellers and Va Ee identify the fortification at the southeast corner of Savannah as "Fort Prevost." The only clue to who prepared these plans is found on the reverse of number 1584 where, in an apparently eighteenth-century hand, is written "Mr. M Blimes."

Sellers and Van Ee describe several other Savannah fortification plans that appear to be related to these. They may record stages in the improvement of that city's Revolutionary War defense works. See pp. 343–45 in their guide cited above.

Originals

Congress.

---

276. Condor ca. 1750

Plan of the Harbour of Charles Town, South Carolina [across face] | T. Condor [below neat line] | Various plans and Draughts of Cities Towns [below neat line; remainder of phrase cut off]

Size: $3\frac{7}{8} \times 5$. *Scale:* 1" = $2\frac{2}{3}$ miles.

Description: This plan was cut out from a larger sheet of plans, probably deriving from the insets of the Popple 1733 map.

Copies

Charleston Library Society.

---

277. Delisle 1750

Carta Geographica Della Florida Nell'America Settentrionale

Size: $17\frac{1}{8} \times 12\frac{15}{16}$. *Scale:* 1" = ca. 97 miles.

In: De L'Isle, Guglielmo. *Atlante Novissimo.* Venice, 1740–50, II (1750), No. 41.

Description: This map extends from "Panuco" in Mexico to "Filadelfia" at 40°. This is the same territory shown in the Ortelius (1584) and numerous other early maps; but most of the sixteenth-century names have disappeared, and a large number of later names of rivers, towns, Indian villages, and political divisions are given. It is one of the latest maps to give Lederer's nomenclature for the piedmont area east of the Appalachian Mountains; it indicates the town of Lederer's 1672 map and gives "Deserto," "Pianura coperta d'acqua," and a "Lago Grande."

"Apalache" is the name given to the region west of Georgia and east of Appalachicola River. The English colonies are restricted to the territory east of the Appalachian Mountain range, which is pushed close to the Atlantic coast. The semi-circular lake at the head of the Apalachicola River is an indication of the use of early eighteenth-century French maps. "Virginia Vecchia" is found in the "Carolina," south of "Contea Albemarle." This map is retrogressive in many details, for it reincorporates numerous geographical misconceptions which Delisle discarded in his other excellent maps of the Southeast published at this time and many years before, as in his "Carte de la Louisiane" of 1718. "Georgia," given as part of "Carolina," indicates contemporary political knowledge which is in marked contrast to the generally anachronistic features of the map.

Copies

Congress; American Geographical Society; Kendall; Yale.

---

278. Haig 1751 MS

[Indian Country, Western Carolina]

Size: 29 × 21. *Scale:* 1" = 10 miles.

Description: This map begins at Saxagotha (about 123 miles W.N.W. from Charleston, S.C., according to a note by George Hunter) and gives the rivers, trading paths, and Indian villages of up-country South Carolina and western North Carolina. It shows such locations as "Toxawa" and "Conastee," the Little Tennessee, the French Broad, and in general the territory west of Asheville.

On the upper right of the map is the following legend: "S° Carolina 1751. This Map is exactly Copied from one done by Cap[t] George Haig, for his Excellency Governour Glen, and now in his possession. Mr. Haig was carried off and Murdered by French and Northern Indians several years ago, but I am well acquainted with his hand Writing and manner of drawing, he having been for several years one of my Deputy Surveyors of this Province. George Hunter Sur. Gen[l]."

In the lower left is the explanatory note "N.B. Mr. Haig should have said 93 miles from Amelia Township, Saxagotha lies above 30 miles further up, and by my observations when I was Twice at Amelia and Saxagotha, the course was WNW instead of NW as Mr. Haig has laid it down. George Hunter Sur: Gen[l]." "May 27th 1752 Received from James Crokatt Esq. Agent for ye Province of South Carolina Without any letter S.G."

On the left side of the map a number of Cherokee towns are identified with the names of their "Headman" and number of "Gun Men" in each.

References

Crane, Verner W. *Southern Frontier*, p. 351.

Reproductions

Hulbert, A. B. *Crown Collection*. Series III, Plates 27–30.

Original

London. Public Record Office. C.O. 700, Carolina no. 17.

---

279. Tybee Island ca. 1751 MS

[Tybee Island and Cockspur Island]

Size: 21½ × 16¾. *Scale:* 1" =½ mile.

Description: This untitled manuscript is unsigned and undated. It shows Tybee Island with its lighthouse and "mark tree" used by navigators making an entrance into the Savannah River. Buildings shown nearby may indicate the pilot's residence. A small rectangle with an X inside may indicate Fort George, a small defense work located on the southeast tip of Cockspur Island. It was built by De Brahm in 1761 (see Maps 329A and 416, below). This map has been attributed to Henry Yonge but it does not resemble any of his known work.

Reference

Sellers, J. R., and P. M. Van Ee, comps. *Maps and Charts of North America in the Library of Congress*, No. 1591, p. 345.

Original

Congress.

---

280. Yonge 1751 MS

A Plan of the Inlets, & Rivers of Savannah & Warsaw in the Province of Georgia. Performed by order of the President & Assistants this 1st. June Anno Dom$_3$. 1751 p[er] Henry Yonge Surv[r] [cartouche]

Size: 35 × 38¾. *Scale:* 1" = ½ mile.

Description: Henry Yonge, although much less well known than his contemporary William Gerard De Brahm, was a gifted surveyor-cartographer. Yonge prepared this large and detailed chart of the lower Savannah River and Sound at the request of Georgia's first General Assembly. Acting more as a fact-finding rather than a true law-making body, they were concerned about the deteriorating condition of the navigation channel giving Savannah access to the sea. Not only did Yonge complete an amazingly detailed hydrographic survey of the river's large and dynamic tidal estuary; he went further and included the first large-scale river improvement scheme for the lower Savannah. This can be seen by Yonge's placement of a number of small arrow-shaped symbols at strategic points in the river's course to the sea. In his "Explanation" legend it is pointed out that where these symbols appear "are shoals and narrow Places in the River Savannah, which if stoped up would probably open the main Chanel." A slightly different symbol indicates "Proper Place[s] for a Buoy in Savannah River." The decorative qualities of Yonge's Savannah River map are on a par with its impressive engineering and scientific content. Both his title and legend are framed in artistically penned cartouches, and the offshore waters, marked "Ocean," are occupied by two sailing ships, one in the act of firing a broadside. The large compass rose and scale block are similarly artistic in their rendering.

Reference

Hawes, Lilla. "Proceedings of the President and Assistants in Council of Georgia, 1749–1751." *Georgia Historical Quarterly* 35 (1951), pp. 323–50.

Reproductions

Plate 58A.

De Vorsey, L. *The Georgia–South Carolina Boundary*, Figure 11, pp. 71–72 (black and white reduced photograph).

Original

Congress.

---

281. Fry-Jefferson 1751 [1753]

A Map of the Inhabited part of Virginia containing the whole Province of Maryland with Part of Pensilvania, New Jersey and North Carolina Drawn by Joshua Fry & Peter Jefferson in 1751 [cartouche in lower right corner] | To the Right Honourable, George Dunk Earl of Halifax First Lord Commissioner; and to the Rest of the Right Honourable and Honourable Commissioners, for Trade and Plantations. This Map is most humbly Inscribed to their Lordship's, By their Lordship's Most Obedient & most devoted humble Servt. Thos. Jefferys. [bottom right, below picture in cartouche] | Engrav'd and Publish'd according to Act of Parliament by Thos. Jefferys Geographer to His Royal Highness the Prince of Wales at the Corner of St. Martins Lane, Charing Cross London [bottom right, above neat line]

Size: 48½ × 30½. *Scale:* 1" = 10⅓ miles.

Description: This map includes, in addition to other provinces indicated by the title, North Carolina as far south as 35°45', approximately to the latitude of the present cities of Raleigh and Statesville. It extends west to the present northwest corner of North Carolina, showing the Iron Mountain range beyond New River. In its western reaches this map gives more information for North Carolina than does any other map until Collet, nearly twenty years later. Its influence was felt immediately in subsequent maps; even continental mapmakers soon showed the west branches of the Yadkin and Catawba Rivers and the Brushy Mountains of the Blue Ridge. This is the first printed map that portrays adequately the parallel direction of the Appalachian mountain ranges, with the valleys between the ridges running northeast-southwest, though unfortunately this information does not extend far into North Carolina.

Much of the content of this map was based on the surveys and experiences of its Virginia-based compilers Joshua Fry and Peter Jefferson. Both were land surveyors with an abundance of firsthand experience throughout the frontier and settled regions of their home colony. In the words of another Virginia cartographer, better known as the third president of the United States, Thomas Jefferson, his father, Peter, and Joshua Fry "possessed excellent materials for so much of the country as is below the blue ridge; little being then known beyond the ridge." Joshua Fry had emigrated from England many years before and flourished in his adopted home. In 1738 he had joined Robert Brooke and John Mayo in forwarding to the House of Burgesses a proposal "to make an exact survey of the colony, and publish a map there of, in which shall be laid down the bays, navigable rivers, with the soundings, counties, parishes, towns, and gentlemen's seats." Regrettably the Burgesses rejected this ambitious undertaking. It was not until 1750 in response to an urgent request from the Board of Trade that Virginia's colonial government moved to commission the preparation of a general map. In his letter transmitting the completed map to his superiors in London Governor Burwell described Joshua Fry as "a Gentlemen very eminent for his skill in mathematicks, who was formerly a Professor of it in our College of William and Mary, and since his quitting that Place has retired to the back Settlements in Order to raise a Fortune for his family." Peter Jefferson had become closely connected with the former mathematics professor in 1745 when he was appointed as deputy county surveyor of Albemarle under Fry. From that time onward their careers were inextricably linked.

In 1755 or soon after, a new edition of the map was published, unchanged in title but with numerous improvements of the western half of the map, especially to the west of the Blue Ridge. The mountain ranges are added to, the Ohio River and its tributaries are improved, and the main roads throughout Virginia have been added. Lake Erie is erased from the upper left section in the 1755 edition. The Yougyougain (Youghiogheny) River flows northwest and not south into the New River; this revised state shows an increased knowledge of the Ohio Valley region. In North Carolina the improvements are many, especially for roads, ferry crossings, numerous settlements, and names of additional settlers. This second edition delineates the upper "Catawba River" and its tributaries, probably for the first time on any map, and shows the "Head of the Yadkin," with that river breaking through several parallel ranges to emerge from the Blue Ridge above the "Brushy Mountains." One of the most interesting roads drawn is that from Pennsylvania through the Valley of Virginia to the Moravian "Unitas" on "Gargals Cr."; called "The Great Road from the Yadkin River," this is the Carolina Road over which tens of thousands of settlers drove their wagons south during the third quarter of the century. This 1755 edition is usually called the Dalrymple edition, for in the upper left sheet of the map a table of distances between different towns in the state and in the colonies is given by John Dalrymple: "These Distances with the Course of the Roads on the Map I carefully collected on the Spot and entered them in my Journal from whence they are now inserted. J. Dalrymple, London Jan.y y.e 1.st 1755." According to an advertisement of the map in *The Gentleman's Magazine*, XXV (January 1755), p. 47, Dalrymple was captain of the Virginian regiment. This review also gives the information, not on the map, that the cartouche (of an eighteenth-century tobacco warehouse and wharf) was designed by Hayman and that the map was engraved by Grignion.

Delf Norona (vide infra, pp. 27–30) has shown that the map could not have been published before 1752, although as the title states, it was "Drawn by Joshua Fry & Peter Jefferson in 1751." Presumably it was submitted to Colonel Burwell with Fry's report in May 1751; but owing to Burwell's delay it was not forwarded until August 21, 1751, and was not received in London until November 30 or December 9. It was presented to the commissioners of the Board of Trade on March 10 and 11, 1752. These dates are fixed by endorsements on Burwell's letter of transmittal and Fry's report, preserved in the Public Record Office. Delf Norona suggests plausibly that Lord Halifax, pleased with the wealth of new material on the map, ordered it to be engraved and published by Thomas Jefferys. Norona further shows that the original MS extended farther west than the engraved map, and Jefferys used information from it for the map in his reprint of George Washington's *Journal*, London, 1754.

Coolie Verner, in his checklist appended to Dumas Malone's *The Fry & Jefferson Map of Virginia and Maryland*, Princeton, N.J., 1950, pp. 13–19, dates the printing of the first state as before January 1755 and probably in 1754. He bases this on a letter of August 9, 1755, from Rev. James Maury to Moses Fontaine, which discusses changes in the plates of the map and which is printed in Jacques Fontaine's *Memoirs of a Huguenot Family . . . from the Original Autobiography of The Rev. James Fontaine, and Other Family Manuscripts . . . By Ann Maury*, New York, 1853, and on the advertisement of the 1755 state in *The Gentleman's Magazine*, XV (January 1755), 47.

Mr. John Cook Wyllie of the Alderman Library, University of Virginia, and Miss L. Mulkearn of the Library of the University of Pittsburgh have independently called the attention of the author to the following notice by M. Maty, *Journal Britannique*, XII (Septembre–Decembre 1753), 427–28:

"On vient de publier une nouvelle Carte très détaillée & gravée avec beaucoup d'élégance, de la Virginie. Elle a été contruite sur les desseins de Mrs. Fry & Jefferson, qui ont arpenté le pais en 1751. Outre la Province principale on y voit encore celle de Maryland, & une partie de la Pensylvanie, du nouveau Jersey, & de la Caroline Septentrionale. Cette Carte qui consiste en quatre feuilles réunies, & qui se vend une demi guinée, a été gravée & publiée par Mr. Jefferys Geographe de S. A. R. Monseigneur le Prince de Galles."

Since this notice of the Fry-Jefferson map, written by M. Maty during the fall of 1753 or not long before, refers to it as "une Nouvelle Carte," the actual publication of the map probably occurred some time during the year 1753.

The 1755 "Dalrymple edition" has extensive changes on

all four plates of the map, with longitudinal degree marks inside the upper and lower borders from 65° to 72°19' W. of London. A correction of this error of ten degrees, to 75°19'–82°19', was made, probably later in 1755, and makes the third state of the map which has been noted. Mr. Verner suggests 1761? as the date for a fourth state of the map, since Christoph Daniel Ebeling, in "Neue Karten von Amerika," *Amerikanische Bibliothek, Erstes Stuück*, Leipzig, 1777, No. 16, p. 137, notes that he received a copy of this map from Jefferys with the date "1751" changed by pen to "1761." The reference to Jefferys as "Geographer to his Royal Highness the Prince of Wales" is stricken from the title in this state; the Prince of Wales became George III in 1760.

After the 1755 additions to the plates there are no changes on the face of the map, though the imprint was changed about 1761, and a further title change was made in 1775. The New York Public Library Bancroft copy of the rare first edition has "Most" inserted in ink before and above "Inhabited" in the title.

H. Stevens and R. Tree, "Comparative Cartography," pp. 360–61, in their list of different states of the map, note two states ("b" and "c" in their list) which are not recorded here. Their "a" is the first edition. The "b" has an unchanged title but adds incorrect degree marks (65°19' to 72°19' West of London). The third ("c") is another pre-Dalrymple state with the degree marks corrected to "75°19' to 82°19'." However, since the copies of the first Dalrymple 1755 state ("B" in the list below) which have been examined in this study have uncorrected degree marks, the records of Stevens and Tree are almost certainly confused. It is extremely unlikely that the degree marks would have been re-engraved incorrectly if they had been corrected for a previous issue.

The French editions of the Fry and Jefferson map by Le S[r] Robert de Vaugondy 1755 (with five states noted) and by Le Rouge 1777 are not listed here because they omit the Carolina area given on the English plates of the map.

See Fry-Jefferson 1775 for the 1775 and 1794 states of this map.

See Introductory Essay.

References

De Vorsey, L. "Introductory Notes," *North America at the Time of the Revolution* [A Collection of Eighteenth Century Maps], Part II (1974).

Fite, E. D., and A. Freeman. *A Book of Old Maps*, pp. 243–45.

Garrison, H. S. "Cartography of Pennsylvania Before 1800," *Pennsylvania Magazine of History and Biography*, LIX (1935), pp. 274–75.

Harrison, Fairfax. *Landmarks of Old Prince William*. Richmond, Va., 1924, II, pp. 629–34.

Malone, Dumas. *The Fry & Jefferson Map of Virginia and Maryland*. Princeton, N.J., 1950 (with Mr. Coolie Verner's "checklist"), pp. 13–19.

Mathews, E. B. *Maps and Mapmakers of Maryland*. Baltimore, Maryland Geological Survey, 1898, II, 391–94.

Norona, Delf, ed. "Joshua Fry's Report on the Back Settlements of Virginia (May 8, 1751)," *The Virginia Magazine of History and Biography*, LVI (January 1948), 22–41.

Phillips, P. L. *Virginia Cartography*, p. 48.

Sanchez-Saavedra, E. M. *A Description of the Country: Virginia's Cartographers and Their Maps 1670–1881*, pp. 26–34.

Stevens, H., and R. Tree. "Comparative Cartography." *Essays Honoring Lawrence C. Wroth*. Portland, Me., 1951, pp. 301–2.

Swem, E. G. *Maps Relating to Virginia*. Richmond, 1914, pp. 61, 65, 70; Nos. 175, 189, 234.

Reproductions

Plate 57 (1751 [1753]); Plate 58 (1755): bottom left part of both editions.

Malone, Dumas. *The Fry & Jefferson Map*. Princeton, N.J., 1950 (full-scale reproduction of first state).

Congress (photocopies of New York Public and University of Virginia holdings of the first state).

A. [1751 (1753): first edition, with title and imprint as given above.]

Copies

New York Public; University of Virginia.

B. [1755 revision; same title and imprint except for change of plate to include "most" before "Inhabited" in second line of title; Dalrymple 1755 note and many topographical changes and additions on all four plates; the addition of (incorrect) degree notations within neat lines: 65°19' to 72°19' West of London.]

Copies

Congress; John Carter Brown; Yale.

C. [1755? state; no changes except for correction of degree marks inside upper and lower borders of ten degrees, 65°19'–72°19' W. of London to 75°19'–82°19'.

Copies

Congress; Clements; Harvard.

D. [1761? state; title as in B and C; only imprint changed, bottom right, above neat line, to:] | Printed for Robt Sayer at No 53 in Fleet Street, & Thos Jefferys at the corner of St Martins Lane, Charing Cross, London.

In: Jefferys, T. *A General Topography of North America and the West Indies.* London, 1768, Nos. 54–57.

Copies

Congress 1768, separate; Boston Public 1768.

---

282. Fontaine 1752 MS

Map of the Virginia and North Carolina dividing line

Size: 4 × 1½. *Scale:* None.

Description: This rough draft of the survey is found at the head of a letter of Peter Fontaine, Jr., dated Lunenburg, Virginia, 9th July, 1752. In the 1749 survey of the Virginia–North Carolina line, Rev. Peter Fontaine accompanied the Virginia commissioners as chaplain and surveyor; the map is found in a letter from him to a relative.

Reproductions

[Fontaine, James]. *Memoirs of a Huguenot Family.* Trans. and ed. by Ann Maury. New York, 1853, p. 356.

———. Reprinted from the original edition of 1853. New York, [1907], p. 356.

Hawks, F. L. *History of North Carolina.* Fayetteville, 1858, II, opp. p. 102.

Original

University of Virginia.

282A. De Brahm 1752 MS

A Map of Savannah River beginning at Stone-Bluff, or Nexttobethell, which continueth to the Sea; also, the Four Sounds Savanāh, Hossabaw, and St: Katharines with their Islands likewise Newport, or Serpent River, from its mouth to Benjehova bluff. Surveyed by William Noble of Brahm late Captain Ingenier unter his Imperial Majesty Charles the VII. [across upper part of map]

Size: 53½ × 25½. *Scale:* 1" = 80 chains.

Description: This colored manuscript map was presented to the Georgia Trustees within the first year of De Brahm's arrival in the colony. In his covering letter, dated March 24, 1752, De Brahm described it as a "map of that part of Georgia, which I have had an opportunity of Surveying since my arrival here; which I flatter myself will speak in my behalf, and be more satisfactory and agreeable than anything I could say in a long and tedious letter." See *The Colonial Records of the State of Georgia,* 26, pp. 374–48. De Brahm had served in the Imperial Army as a captain engineer before renouncing Roman Catholicism and joining the ranks of other German-speaking Protestant refugees seeking better fortunes in Georgia. His map is carefully drawn and colored according to an "Explanation of the Colours" legend [lower left corner]:

Deep Red Signifies Height [high] Land.
Pale Red ..... Sandbanks, unter [under] water.

Verdy grease or Saxon Green—Water.
Ditto mixed with Sea Green—Pond.
Ditto mixed with brown—River Swamp.
Grass Green—Oak Land.
Yellow—Pine Barren.
Brown—Swamp.
Sea Green—Marsh.
Black—Bluffs and Mountains

In 1754 De Brahm and Henry Yonge were named as Georgia's two surveyors general of land in the colony's royal administration. When South Carolina's surveyor general died in 1755, De Brahm received an interim appointment to the office, which he held for almost a year until a permanent appointee assumed the position. In 1764 he was appointed surveyor general of the new British colony, East Florida, as well as the Southern District of North America, posts he held until the final phases of the Revolutionary War. The map covers the Savannah River Valley to a point some miles below the site of Augusta and the coastal islands and inlets from the river's sound south to St. Catherines Island. Many settlements and plantations are identified.

References

De Vorsey, L., ed. *De Brahm's Report of the General Survey in the Southern District of North America.* Columbia, S.C., 1971.

Sellers, J. R., and P. M. Van Ee. *Maps and Charts of North America and the West Indies, 1750–1789.* Washington, D.C., 1981, pp. 339–40.

Reproduction

Georgia Archives (photosat).

Original

Congress.

---

283. Warner 17[5]3 MS

[Edisto River] A Plan of Twenty different Tracts of Land, containing in the whole 20,000 acres. Situate on Hollows of McTiers Creek,—on Boggy Gully Creek,—on Gitty's Swamp Creek,—& Deans Swamp, called Holmer's Camp Branch & c[a]. all Waters of Edisto River, in the Forks of said River, & on the Main Ridge Road to Charleston. Survey'd in the Name of James Coalter (except 1000 Acres in the Name of W[m] Morris) the whole granted to James Coalter. P[r] Tho[s] C. Warner Dep[y] Surv[r] In the month of Nov[r] 17 [hole in map] 3.

Size: 35½ × 26¼. *Scale:* 1" = ca. 27 chains.

Description: This is a large colored manuscript survey, with much topographical detail. The date is that given by the Division of Maps, Library of Congress; part of what looks like a "5" is still visible in the date.

Original

Congress.

---

284. Hyrne-Jefferys 1753

A New and Exact Plan of Cape Fear River from the Bar to Brunswick, by Edward Hyrne, 1749. [in cartouche, upper left] | Published according to Act of Parliament March 20th 1753 by T. Jefferys Geographer to His Royal Highness the Prince of Wales at the Corner of Martin's Lane Charing Cross [across bottom of map, within neat line]

Size: 12½ × 15¼. *Scale:* 1" = 1⅔ miles.

Description: This same map, with a different imprint, is found in T. Jefferys, *A General Topography of North America and the West Indies*, London, 1768, No. 63; see under the date 1768.

Copies

Congress; Duke; Harvard; Kendall; Public Record Office, C.O. 700, Carolina no. 18.

---

285. Schröter 1753

Gegend der Provinz Bemarin im Königsreich Apalacha A. Haupt-Stadt Melilot B. die grosse Kirche C. Pallast des Parakusse D. Berg Olani E. Sonnen Tempel F. Gestalt d. empfindz: Pflanze u. ihrer Blume [across top, above the picture] | Am. Hist: Th. II, pag. 592 [top right, above line]

Size: 11⅝ × 7⅞. *Scale:* None.

In: Schröter, Johann Friedrich. *Algemeine Geschichte der Länder und Völker von America . . . Nebst einem vorrede Siegmund Jacob Baumgartens.* Halle, 1753, II, opp. p. 592.

Description: This is not so much a map as an idealized landscape, with alpine mountains in the background, the walled group of buildings called Melilot in the title, a river springing out of a cave in a hillside, at the top of which is the sun-temple, which looks like a medieval Rhine castle, and a number of very formally and heavily attired individuals, apparently nobles, strolling over the landscape.

The "map" is based on the extraordinary supposed adventures of a mysterious Englishman, Bristock (Bristok, Bristol, Brigstock), who visited Apalacha and the "Apalachites" in 1653. Bristock, who had spent many years in the West Indies and knew several Indian languages, apparently published several ephemeral tracts, which were first reported by Abbé Rochefort in his *Histoire Naturelle et Morale des Illes Antilles de l'Amerique*, Rotterdam, 1658, II, 331–53. Schröter's "Nachricht von den Apalachiten," in his *Geschichte*, pp. 562–98, is a translation of Rochefort. Rochefort's work was translated in Davies's *History of the Caribbee Islands*, London, 1666, pp. 228–49, which contains new independent information not in Rochefort; other accounts of Bristock are in Ogilby's *America*, London, 1671, Oldmixon's *The British Empire in America*, London, 1708, and elsewhere.

According to "Bristock," the Apalachites inhabited the country of Apalacha (33°25' N.L. to 37° N.L.) and originated from Mexico. Their bitter enemies were the Cofachites, with whom they fought for centuries; finally, by a religious stratagem, the Apalachites drove the Cofachites to Florida, whence they emigrated to the West Indies. According to Davies's account (op. cit., p. 245) several English families, escaping from the Indian massacres in Virginia in 1621, were shipwrecked on the coast of Florida and, after passing through the provinces of Matica, Amana, and Bemarin, at last settled among the Apalachites. There, according to Bristock, in 1653 they had laid the foundations of a small colony—"they have built Colledges in all those places where there are Churches"—to instruct the children of the Apalachites in true piety.

Although parts of the Bristock story received credence from as noted a student as D. G. Brinton (*Notes on the Floridian Peninsula*, Philadelphia, 1859, pp. 94–96), John R. Swanton (*Early History of the Creek Indians*, Washington, 1922, p. 118) dismisses the Indian ethnology as essentially a fabrication.

Cartographically, the Bristock account had some influence. As early as 1679 John Seller makes the Kingdom of the Apalachites, with its capital Melilot, one of the five provinces of Florida in his *Atlas Minimus*. In 1703 the famous French geographer Delisle, in his "Carte du Mexique et de la Floride," places the "Pays des Cofachi" to the west of the Appalachian mountain range in upper Alabama, with Amana and Bemarin to the south. The numerous maps of other European cartographers who followed Delisle's 1703 map during the next quarter of a century have these same names, together with the lakes and other topographical peculiarities that accompany them.

Copies

Congress 1753.

---

286. Dobbs–Fort Johnston 1754 MS

A Plan of Fort Johnston with some Alterations. No. 2.

Size: 9¼ × 13 (reproduction). *Scale:* 1" = 30 feet.

Description: A plan of Fort Johnston on the Cape Fear River accompanying a letter from Governor Arthur Dobbs of North Carolina to Lord Halifax, dated November 20, 1754, "relative to the present State of that Colony, with an acc[t] of the State of the Fort built at Cape Fear." The letter is printed in the *Colonial Records of North Carolina*, V, pp. 157–60.

Reproduction

North Carolina State Department of Archives and History (photocopy).

Original

London. Public Record Office. C.O. 5/297 c. 63.

286A. Yonge 1754 MS

An Exact=Plan of GEORGE=TOWN. so Named by Patrick Graham Esq[r] President of the Province of Georgia in Hon[r] to his Royal Highness George Prince of Wales, etc [a] Surveyed & Delineated by order of the President & Council of Georgia, this 9th Day of May 1754. By Henry Yonge Surv[r] [cartouche center right]

Size: 20¾ × 29. *Scale:* 1" = 400 feet.

Description: This pen-and-ink and watercolor town plan shows a pattern of straight streets and open squares carefully laid out on a neck of land formed by a large oxbow meander of the "Great Ogeche [Ogeechee] River" south of Savannah. On February 4, 1755, Governor Reynolds indicated his satisfaction that the name of the "town lately laid out, at a Place commonly called the Elbow on great Ogechee River" would be called "Hardwicke." See *The Colonial Records of the State of Georgia*, VII, p. 101. The new name honored his relative, Philip Yorke (1690–1764), earl of Hardwicke, and Lord High Chancellor of England. There were a number of unsuccessful attempts to make Hardwicke the capital of Georgia in the next several years. An attempt was made to re-establish the town through incorporation in the post–Civil War period, but today the name Hardwick designates only a residential community outside of a sprawling modern Savannah.

References

Sellers, J. R., and P. M. Van Ee, *Maps and Charts of North America and the West Indies, 1750–1789*. Washington, D.C., 1981, no. 1572, p. 341.

Original

Congress.

---

287. Wilmington ca. 1755 MS

Plan of Wilmington Scituate on the East side of the North East Branch of Cape Fear River. Agreeable to the Original Survey.

Size: 27½ × 19½. *Scale:* 1" = 17 poles.

Description: This mutilated and faded sheepskin plan, preserved in the North Carolina State Department of Archives and History, has the following legend written above the town: "This Plan represents the Plan of Wilmington as laid out by the Order of the Proprietors of the [illegible] in the year 1755 [1735?] At which time it was agreed by the said proprietors to begin the survey of their [part ?: illegible] whereon lyes the Threshold of the said door of the said House near the River now possessed by Mr. Hugh Blaning." The legend continues, giving the breadth, direction, and names of several streets.

In the year 1731 John Maultsby and John Watson took out warrants for land on the Cape Fear River about fifteen miles above Brunswick, at that time the chief port in that part of the province. Persons settling on Maultsby's land in 1732 called the settlement New Liverpool; those settling on Watson's land in 1733 named it Newton. In 1734 Governor Johnston arrived from England; he espoused the cause of Newton as against Brunswick, bought land near it, and held the Court of Exchequer and meetings of the Council there in 1735. He wanted to have the town named Wilmington, after his patron, the earl of Wilmington. But it was not until 1739, by a typically high-handed method of overriding the members of the Council, that he succeeded in achieving his purpose with the aid of William Smith, chairman of the Council. With the flourishing of Wilmington came the decay of Brunswick, which dwindled to such an extent that by the Revolutionary War it was abandoned. See J. Sprunt, *Chronicles of the Cape Fear River 1660–1916*, Raleigh, 1916, pp. 45–49; and *Colonial Records of North Carolina*, IV, v, xix, pp. 235, 448, et seq.

Hugh Blaning, referred to on the map, is possibly a son of the Justice of the Peace with the same name who lived in Bladen County and died in 1751; see J. Bryan Grimes, *Abstract of North Carolina Wills*, Raleigh, 1910, p. 33. The date of the map itself, however, is uncertain.

Original

North Carolina Department of Archives and History.

---

288. De Brahm 1755 MS A

The Profile of the whole Citadelle of Frederica . . . [top right of sheet, followed by explanatory text] | Profile to the Projects for Savannah & Hardwick [center, right] | Profile to the Redoubt & Ye Blockhouse [center left]

Size: 18½ × 11½. *Scale:* 1" = 300 feet.

Description: This sheet gives a series of four colored profiles and three ground plans for fortifications at Frederica, Hardwick, and Savannah. Endorsed in pencil: "Georgia | Plans and Elevations necessary in Georgia | Rec.d with Govr. Reynolds letter of 5 January 1756." A full bibliographical description of a copy of this map in the Hargrett Library is found in *Catalogue of the Wymberley Jones De Renne Georgia Library*, III, 1205–6.

At the bottom of the sheet is written: "N.B. all thes ground Plans are laid down by a Scale of 300 Feet in an Inch by William D^e Brahm."

Reference

De Vorsey, L. *De Brahm's Report of the General Survey.* Columbia, S.C., 1971, pp. 14–15.

Reproductions

Adams, J. T., ed. *Album of American History.* New York, 1944, I, p. 224.

Hulbert, A. B. *Crown Collection of Photographs of American Maps.* Cleveland, [1914–16], Series III. Nos. 139–40.

Gabriel, R. H., ed. *The Pageant of America*, VI, p. 57. Savannah plans from Georgia Hargrett Library copy.

Georgia Historical Society. *Collections*, VII, Part III (1913), opp. p. 11.

Congress (photocopy).

Original

London. Public Record Office. C.O. 700, Georgia no. 11.

---

289. De Brahm 1755 MS B

A Map of the inhabited Part of Georgia laid down to Shew the Latitudes & Longitudes, of y^e Places that are proposed to be Fortified, in order to Iudge of there communications by William d^e Brahm [above neat line of map] | N.º 15 [upper left corner of sheet]

Size: 3½ × 5¾. *Scale:* 1" = 28 miles.

Description: This small colored map was compiled by De Brahm at the request of Georgia's governor, John Reynolds, who was anxious to prepare the colony for defense against a feared French-inspired Indian attack. It was sent to London on January 5, 1756, along with the set of fortification plans described above (Map 288) to illustrate De Brahm's unsigned "Representation of the forts and Garrisons Necessary for the Defense of Georgia. . . ." The map features Coxspur Island, Savannah, Augusta, and Fort Argyle, as well as other settlements. Dashed red lines indicate the trails connecting the strategic points to be fortified. De Brahm's scheme was to provide the Georgians with maximum flexibility in warding off the feared attacks. Fortunately they never occurred.

An early but undated tracing of this map was preserved in the Wymberley Jones De Renne Georgia Library and is now in the collection of the Hargrett Library at the University of Georgia (see *Catalogue of the Wymberley Jones De Renne Georgia Library*, III, 1206).

Reference

De Vorsey, L. *De Brahm's Report of the General Survey.* Columbia, S.C., 1971, pp. 14–16.

Reproduction

Congress (photocopy).

Original

London. Public Record Office. C.O. 700, Georgia no. 12.

---

290. De Brahm 1755 MS C

Plan of a Project to Fortifie Charlestown Done by Desire of his Excellency the Governour in Council by William de Brahm Captain & Ingeneer in the Service of his Late IMP MAJ Charles VII.

Size: 35 × 40. *Scale:* 1" = 400 feet.

Description: This large colored plan of Charleston, with the principal streets, buildings, and landmarks, has fortifications extending across the neck of land upon which the town is built at a distance of four miles from Oyster Point. Several profiles and plans of fortifications are on the sheet. This is one of the largest and fullest maps of Charleston to this time.

Grandiose would be an understatement in describing the line of fortifications De Brahm proposed for the neck of land occupied by Charleston. His drawings are in the style of Vauban, Europe's most esteemed military engineer and designer of impregnable fortifications. After studying his scheme, South Carolina's governor allowed that "the List of utensils given in by Mr. De Brahme was so large that it would take the greatest part of £50,000 to Purchase them." Needless to say, a far more modest plan was agreed upon, and De Brahm proceeded to direct construction of "Ramparts, forming regular Bastions, detach'd or joined with Curtains." See L. De Vorsey, *De Brahm's Report*, pp. 17–18. An item of potential archaeological interest is a small diamond-shaped revetment captioned "Indian Fort" near a marshy creek emptying into the Cooper River.

See Introductory Essay regarding the South Carolina Historical Society volume containing references to De Brahm in the *Journal of the Charles Town Commissioners of Fortifications.*

See De Brahm 1757 MS and De Brahm [1773] MS A.2, 3, and 4.

Reproduction

Congress (photocopy).

Original

London. Public Record Office. C.O. 700, Carolina no. 14.

---

291. Anon. ca. 1797

Carte de la Caroline Méridionale et Septentionale et de la Virginie

Size: 16⅞ × 12⅝. *Scale:* 1" = 46 miles.

Description: This map is properly dated ca. 1797 and is identical to Map 447, below, but with "Tardieu-Valet" trimmed off. Its coverage extends from Port Royal, South Carolina, to Philadelphia, Pennsylvania. It is found in Edme Mentelle and P. G. Chandlaire, *Atlas Universel*, Paris, 1797–1801, no. 135.

Copies

Kendall; Congress 1797–1801, 1797–1807.

---

292. Catesby-Seligmann 1755

Carolinae Floridae nec non Insularum Bahamensivm cum partibus adjacendibus delineatio ad Exemplar Londinense in lucem edita a Ioh. Michael Seligmann. Norimbergae. A° 1755.

Size: 23 × 16⅞. *Scale:* 1" = ca. 85 miles.

In: Catesby, M. *Die Beschreibung von Carolina . . . Uebersetzt von d. Georg Leonhard Huth.* Nuernberg, [1755], map at end of book.

Description: See Catesby 1731.

Copies

Congress 1755; John Carter Brown 1755; Clements 1755; Harvard 1755; Kendall; University of North Carolina 1755; Georgia Hargrett Library.

In: Catesby, M. *Histoire Natvrelle de la Caroline.* Nvremberg, chez les Heritiers de Seligmann, 1770, end of book.

Copies

Congress 1770.

In: Lotter, T. C. *Atlas Géographique*. Nuremberg, 1778, No. 98.

Copies

Congress 1778.

---

293. Mitchell 1755

A Map of the British and French Dominions in North America with the Roads, Distances, Limits, and Extent of the Settlements, Humbly Inscribed to the Right Honorable the Earl of Halifax, and the Other Right Honorable The Lords Commissioners for Trade & Plantations, By their Lordships Most Obliged, and very humble servant Jn.º Mitchell [cartouche, bottom right] | Tho: Kitchin Sculp. Clarkenwell Green [bottom right, below cartouche] | Publish'd by the Author Feb.ry 13.th 1755 according to the Act of Parliament, and Sold by And: Miller opposite Katherine Street in the Strand. [bottom right center, below neat line]

Size: 76¼ × 53. *Scale:* 1" = ca. 32½ miles.

Inset: A New Map of Hudson's Bay and Labrador from the late Surveys of those Coasts [inset at top left of map]. *Size:* 11½ × 10⅝. *Scale:* 1" = 160 miles.

Description: Most experts would agree with the conclusion of the John Mitchell entry in *The Dictionary of American Biography* where it is stated: "Without serious doubt Mitchell's is the most important map in American history." This is because it was the official map employed by British and American diplomats in framing the definitive Treaty of Peace of September 3, 1783, the document which legitimized the full and independent membership of the United States in the world family of nation states. In the words of John Adams, a signatory for the Americans, "We had before us, through the whole negotiations, a variety of maps, but it was Mitchell's map upon which was marked the whole of the boundary lines of the United States." Known as the "King George" or "red line" map, the British negotiator's annotated copy is similarly esteemed by the New York Historical Society. More recently the Mitchell map's historical significance was emphasized by the innovative analytical treatment it received in Lester Cappon's *Atlas of Early American History*, Princeton, New Jersey, 1976, p. 58.

Mitchell's map covers the seaboard of North America from southern Labrador and Newfoundland, to Florida and Texas, and extends inland to what is now Oklahoma, Kansas, Nebraska, and South Dakota. Though rich in topographic, hydrographic, and settlement detail, it was to have its greatest impact as a political map illustrating the many boundaries dividing the land that would soon be in heated contest, first by Europe's great imperial powers and their Red allies, and then by Britain's disaffected and rebellious colonists. For the southern region, Mitchell used the 1755 revised edition of Fry-Jefferson's map north of the Granville line. He used Barnwell ca. 1721 MS for the area to the south, though he had access to other maps and reports of the Board of Trade and of the British Admiralty, as he states in his second edition. Mitchell's full use of names and legends on Barnwell ca. 1721 MS is especially helpful, since in several places the Barnwell MS is now almost illegible.

Mitchell's map met with a very favorable reception in England as well as on the continent of Europe and in the American colonies. A surprisingly large number of editions and plagarisms of it were issued over a period of three or more decades. The considerable cartobibliographic problems involved with these many editions and printings are major and lie beyond the scope of this study since no appreciable changes are found on the plate for the area south of Virginia. Col. Lawrence Martin, longtime Chief of the Library of Congress Geography and Map Division, identified seven English impressions; two Dutch editions with English titles; ten French impressions, several with titles and notes in German as well as French; and two Italian piracies published in Venice. Martin's interest in the Mitchell map was the subject of an informative essay by Walter W. Ristow that was augmented with a "Table for Identifying Variant Editions and Impressions of John Mitchell's Map of the British and French Dominions in North America," compiled by Richard W. Stephenson. See W. W. Ristow (comp.), *A la Carte: Selected Papers on Maps and Atlases*, Washington, D.C., 1972, pp. 102–13. The cartobibliographers Henry Stevens and Roland Tree included a section on the Mitchell map in their *Comparative Cartography*, which was republished in R. V. Tooley, *The Mapping of America*, London, 1985, pp. 86–87. The Library of Congress has copies or photocopies of most of the different editions of the map. Only copies of the first impressions of the first edition which have been noted in this present study are listed below. The second issue of the first edition has "Millar" and "Katharine" in the imprint changed from the incorrect spelling "Miller" and "Katherine Street." The reproduction in this work is from the copy of the second issue in the John Carter Brown Library.

See Introductory Essay.

References

Berkeley, E. B., and D. S. Berkeley. *Dr. John Mitchell, The Man Who Made The Map of North America.* Chapel Hill, North Carolina, 1974.

De Vorsey, L. *North America at the Time of the Revolution: A Collection of Eighteenth Century Maps*, II, Lympne Castle, England, 1974.

Fite, E. D., and A. Freeman. *A Book of Old Maps.* Cambridge, Mass., 1926, pp. 181–84, 292–93.

Jones, G. W. "The Library of Dr. John Mitchell of Urbanna," *The Virginia Magazine of History and Biography*, 76 (1968), pp. 441–43.

Miller, Hunter, ed. *Treaties and Other International Acts of the United States of America.* Washington, 1933, III, pp. 328–56 (p. 328, n. 1, has a list of Colonel Martin's publications concerning the map).

Winsor, J. *Narrative and Critical History of America*, V, p. 1983.

Reproductions

Plate 59 (southeastern section).

De Vorsey, L. *North America at the Time of the Revolution* (full scale, 1775 Jefferys and Faden edition).

Fite, E. D., and A. Freeman. *A Book of Old Maps*, Plate 47 (British Museum "George III" copy; fourth English edition).

Miller, H. *Treaties and Other International Acts*, III, folder at end of volume (Steuben-Webster copy; fourth English edition).

Senate Document No. 431, 25th Congress, 2nd session, serial 318, Map A (Dashiell's 1830 map, based on Map A of the treaty of 1827 between Great Britain and the United States).

Copies

(of first impression, first edition) Clements; Yale.

---

294. Baldwin 1755

A Map of Virginia, North and South Carolina, Georgia, Maryland with a part of New Jersey, &c. [above neat line] | Printed for R. Baldwin in pater Noster Row 1755 [below neat line]

Size: 10½ × 8⅜. *Scale:* 1" = 105 miles.

In: The London Magazine, XXIV (1755), p. 312.

Description: The map extends from 28° to 40°20' N.L. (from Cape Canaveral to Philadelphia) and westward to the Mississippi. It has much information concerning towns and fortifications and is an attempt to give detail helpful in following the war against the French.

A more general map by I. Hinton, in which the use of Mitchell's map is discernible, appeared in the *Universal Magazine*, XVII (1755), opp. p. 145: A Map of the British and French Settlements in North America. Univ. Mag. I. Hinton Newgate Street [cartouche, bottom right], *Size:* 14⅞ × 10⅞. *Scale:* 1" = ca. 110 miles. *Inset:* Fort Frederick at Crown Point built by the French in 1731 [bottom center]. This map covers the same area as the Mitchell map; it shows more detail than many of the small generalized maps of the British colonies in America printed about this time. "Unitas [Fratrum]," for the Moravian settlement, is shown north of Lord Granville's line; the boundary lines of all the southern provinces extend straight west without consideration of French claims.

Copies

Congress, three separates; Clements; Georgia Hargrett Library; Harvard; New York Public, two separates; University of North Carolina.

---

295. Robert de Vaugondy 1755

Partie de l'Amerique Septentrionale, qui comprend le Cours de l'Ohio, la N^lle^ Angleterre, la N^lle^ York, le New Jersey, la Pensylvanie, le Maryland la Virginie, la Caroline. Par le Sr Robert de Vaugondy Géographe ordinaire du Roi. Avec Privilège 1755.

Size: 24½ × 18¾. *Scale:* 1" = 45 miles.

In: Robert de Vaugondy, G., and son. *Atlas Universel.* Paris, 1757–[58], No. 104.

Inset: Supplément pour la Caroline. *Size:* 8½ × 7.

Scale: 1" = 45 miles.

Description: The inset, at the top left of the sheet, gives a detailed map of the Georgia and South Carolina coast from 31° to 34° N.L. The large map, which is on the same scale, begins at 34°, just below Cape Fear, and extends to northern New England. The information in the map and in the inset is apparently based upon Mitchell's map of the same year. In later editions of Robert's atlas, usually called "Le Grand Vaugondy," the final line of the title, "Avec Privilège 1755," has been erased.

Under this date may be mentioned "Carte des Prétensions des Anglois dans l'Amérique Septentrionale Suivant leurs Chartres Tant sur les Possessions de la France que sur celles de l'Espagne," in Gt. Brit. Commissioners for Adjusting the Boundaries for the British and French Possessions in America. Mémoires des Commissaires du Roi. Paris, 1755, IV, iii. On this map lines running across North America illustrate the various British claims to the continent; Yale has issued photocopies from the copy in its library.

Reproduction

Color Plate 19.

Copies

Congress 1757, 1757–86, 1783–99, 1792?, 1793?; American Geographical Society 1757; John Carter Brown 1757, "United States of America Atlas"; Clements 1757; New York Public 1757; Harvard 1757; University of North Carolina; Yale 1757.

---

296. Anville 1755

Canada Louisiane et Terres Angloises Par le S.r d'Anville de l'Académie R.le des Inscriptions et Belles-Lettres, et de celle des Sciences de Petersbourg, Sécretaire d S. A. S. M.gr Le Duc d'Orléans Novembre MDCCLV Sous le Privilège de l'Académie. Chez l'Auteur, aux Galeries du Louvre. [cartouche, top left] | G. De-la-Haye [bottom center]

Size: in four sheets: 22 × 18⅞; 22¾ × 18⅞; 25⅜ × 15½; 22⅛ × 18⅛. *Scale:* 1" = 15 miles.
In: Anville, Jean Baptitse B.d.' [*Atlas General.* Paris, 1727–80], Nos. 32–34.
Inset: Le Fleuve Saint-Laurent Représenté plus en détail que dans l'étendue de la Carte. *Size:* 22⅛ × 18½ (irregular shape). *Scale:* 1" = ca. 10 miles.
Description: This map is based on that of Mitchell, which appeared earlier in the same year. Omission of legends, of references to English factories in the disputed trans-Allegheny region, and inclusion of the St. Lawrence River inset and some increased detail from French sources of information, show differences from Mitchell 1755.

References

Phillips, P. L. *Lowery Collection*, pp. 305–7, No. 408.

Copies

Congress 1727–80, 1727–86, 1743–80; Clements 1755, separate.

---

297. Le Rouge 1755

Charles-Town, Capitale de la Caroline.

Size: 4½ × 6⅞. *Scale:* 1" = 200 yards.
In: Le Rouge, G. L. *Recueil des Plans de l'Amérique Septentrionale.* Paris, 1755, No. 8.
Description: A map of the two forts at Charles Town.

Copies

Congress 1755, two copies; John Carter Brown 1755; Kendall; University of South Carolina; Clements.

---

298. Upper Creek Indians ca. 1756 MS

[Indian Tribes of the Tombigbee, Alabama, region]

Size: 18½ × 19⅝. *Scale:* 1" = ca. 25 miles.
Description: This is a French map of the Upper Creek region. It distinguishes the towns of the allies of the English and French but is sparse of detail except for the rivers, the Indian villages and their names, and the number of warriors each possesses. The descriptive title of this map given by the Bibliothèque Service Hydrographique is: "Carte de la rivière des Charaquis ou Grande Rivière, avec les tribes Chouenons, Alibamons, Talapouches, Abekas, Caonitas, Charakis, qui tiennent soit pour les Anglais, soit les Français."

The date is that given by Crane (see below), who, however, incorrectly places this map in the Bibliothèque Nationale.

Reference

Crane, V. *Southern Frontier*, p. 351.

Reproduction

Congress (photocopy).

Original

Paris. Bibl. Serv. Hydrographique. C.4044–55.

---

299. Georgia–Florida Coast ca. 1777 MS

Sketch of the Country between the River Alatamaha and Musqueta Inlet, containing part of Georgia and East Florida; laid down from Information on a scale of eight Miles to an Inch. N:B: From St. Mary's to Smirna is taken from a Plan of the deceased Mr. Hurd's.

Size: Two pasted sheets. Size of upper sheet: 12½ × 17; lower: 16⅞ × 12½. *Scale:* 1" = 8 miles.
Description: Because of an inscription on its reverse, this

map is incorrectly dated 1756 in the *Catalogue of the Wymberley Jones De Renne Georgia Library*, III, 1207. The date given here is based on evidence internal to the map. "Fort McKintosh," shown on the "Great Satilla River," was destroyed in 1777, and "Fort Howe" on the "Alatamah River" was known as Fort Barrington prior to the winter of 1776. This map may have served as military intelligence in the Revolutionary War activities taking place in the Georgia–East Florida theater. An inset in the bottom left contains a "Plan of the Town and Harbour of Augustine" with depth sounding in the approach channel and harbor.

References

Blake, J. G. *Pre-Nineteenth-Century Maps in the Collection of the Georgia Surveyor General Department*. Atlanta, Georgia, 1975, no. 215, p. 105.

Hawes, L. M. "Letter Book of Lachlan McIntosh, 1776–1777," *Georgia Historical Quarterly* 38 (1954), p. 253.

Original

Georgia Hargrett Library.

---

300. Claud Thomson [1786?] MS

An Actual Survey of the Lands of Mess.[rs] Elliots, Squires. Situated upon Great Ogechee River, a little below the Ferry. Consisting in the whole of 3405.4.[r].4.[r]. Acress of Land. Laid down by a Scale of ten Chains to an Inch.

Size: 35¾ × 23½. *Scale:* 1" = 660 feet.

Description: This survey plat was incorrectly dated [1756?] in earlier editions of this work. Little is known about its author, Claud Thomson, and the third figure of the date is illegible on the original. Thomson entered the historical record in 1786 when he began soliciting surveying work in Savannah's newspaper the *Gazette of the State of Georgia*. In a notice that appeared on February 2, 1786, he announced that he had "been regularly bred to that business." Thomson soon became Chatham County surveyor and in 1787 he announced his plan to publish a map of backcountry Georgia. Some months later Thomson published his "warmest thanks" to subscribers but was forced to regretfully announce the abandonment of the project due to their small number.

Reference

Cadle, Farris W. *Georgia Land Surveying History and Law*, pp. 121–22; 149; 153.

Original

National Archives, Washington, D.C.

---

301. Linares 1756 MS

PLANO I. Descripcion de la Costa, desde el Cavo de Cañaveral, hasta cerca de la boca de la Vir[ginia], contando, Costa de Florida, Georgia, y Carolinas del S, y N, con todos sus Puertos, Este-n [about five letters obliterated] letas, Baxos, Islas, y Rios; segun las vlti- [about seven letters obliterated] icias, hata oy Octubre de 1657. COPIADO Por Juan Linares, Pilotin de Numero de la Real Armade: baxo la Correccion D[n] Joseph Francisco Badaraco Maestro Delineador por S. M. en dicha Real Escuela | Esc La Real Escvela De Navegcion En el Departamen[to] de Cadiz 1756. [in cartouche]

Size: 20 × 29. *Scale:* 1" = ca. 15 miles.

Description: This map, colored in green, extends from Cape Canaveral to a place along the coast between Cape Fear and Cape Lookout. There are detailed soundings between Cape Canaveral and "S.[n] Agustin."

Reference

Seller, J. R., and P. M. Van Ee. *Maps and Charts of North America and the West Indies, 1750–1789*, No. 1382, p. 297.

Original

Congress.

---

302. Seale–North Carolina ca. 1756

Virginia, North Carolina, and Parts adjacent; from the latest Discoveries [cartouche, right center] | R. W. Seale sculp [bottom right, below neat line]

Size: 9⁷⁄₁₆ × 7⅜. *Scale:* 1" = ca. 85 miles.

In: A New and Complete History of the British Empire in America. [London?, ca. 1756], III, opp. p. 163.

Description: According to a manuscript note in the first of the three volumes in the Harvard College Library, this work was published in numbers of twenty-four pages each, without title page. The work was left in an uncompleted state, which accounts for its being so little known. It was commenced about 1756. The pamphlets on Carolina may not have been published before 1758. The title is taken from the first page of the first volume.

Both this and Map 303 are based on Mitchell 1755, though with the omission of numerous legends and topographical details.

References

Winsor, J. *Narrative and Critical History of America*, V, 350.

Copies

Harvard ca. 1756; Clements, 1756.

---

303. Seale–South Carolina ca. 1756

South Carolina, Georgia, and Parts adjacent; from the latest Improvements. [cartouche, bottom center] | R. W. Seale sculp [bottom right, below neat line]

Size: 9 7/16 × 7 3/8. *Scale:* 1" = ca. 85 miles.

In: A New and Complete History of the British Empire in America. [London?, ca. 1756], III, opp. p. 265.

Description: This map includes northern Florida and the Indian country westward to the Mississippi, as well as South Carolina and Georgia proper, and is a simplification of the Mitchell map 1755 for the region which it covers. For the volume in which it is found, see the preceding map.

Copies

Harvard ca. 1756; Clements, 1756.

---

304. Reynolds 1756 MS

A Map of Georgia laid down to shew the Latitude and Longitude of the Places that are proposed to be Fortified in order to Judge of their Communication.

Size: 7 3/4 × 12 [size of sheet]. *Scale:* 1" = ca. 28 miles.

In: Reynolds, [John]. *A Representation & Estimate of what is necessary to be done for the Security & Defence of His Majesty's Province of Georgia.* 1756.

Description: This manuscript map is very similar to Map 289, above. It is found in a five-page manuscript (apparently in De Brahm's handwriting) signed but not written by Governor Reynolds and enclosed in a letter from him dated 23 July 1756, and found among the Loudoun Papers. Early in 1756 John Campbell, fourth earl of Loudoun, was appointed commander-in-chief of the British forces in America. The Loudoun documents contain correspondence from all the colonial governors.

References

Huntington Library Bulletin, I (1931), 70–72.

Original

Huntington.

---

305. Bellin 1756

Karte von den Küsten des Französischen Florida Nach den ersten Entdeckungen entworffen von N. Bellin Ingr. de la Marine. [above map, within outer border lines] | No. 3 [above neat line, right]

Size: 8 × 5 3/4. *Scale:* 1" = ca. 65 miles.

In: Allgemeine Historie der Reisen zu Wasser und Land. . . . Leipzig, Arkstee & Merkus, 1756, XIV, opp. p. 16, No. 3.

Description: This is a copy of the map by Bellin in Charlevoix's *Histoire et Description de la Nouvelle France*, Paris, 1744, I, opp. p. 24: See Bellin 1744.

References

Phillips, P. L. *Lowery Collection*, p. 289, No. 375.

Copies

Congress 1756, separate.

---

306. Bonar 1757 MS A

A Draught of the Creek Nation. [top center] | Taken in the Nation May 1757 by William Bonar. [lower center]

Size: 19 × 14 3/4. *Scale:* None.

Description: Professor Crane describes this as an "important map of towns on Chattahoochee and Alabama and branches, with trading paths." On the border are vignettes of "A Hot House," "A Publick Square," "A Junker Yard," Indian weapons, a fort, a warrior, and a squaw. The map shows the Creek Indian heartland, along the Coosa, Talapoosa, and Chattahoochee Rivers, in what is today the state of Alabama. Although distances are badly distorted, the map gives the names and locations of many Creek towns as well as the two chief routes into the area from South Carolina and Georgia. Fort Toulouse indicated as "French Fort" was built in 1716. A plan of it is included as one of the border vignettes. Bonar had gained entry to the fort by posing as a packhorseman, but he was found out and arrested. He was rescued and set free by a party of pro-English Creek Indians.

See Bonar 1757 MS B.

References

Crane, Verner W. *The Southern Frontier*, p. 351.

De Vorsey, L. "Early Maps as a Source in the Reconstruction of Southern Indian Landscapes," in C. M. Hudson, ed. *Red, White and Black: Symposium on Indians in the Old South*, Athens, 1971, pp. 19–21.

Reproductions

Plate 59E.

Clements (photocopy).

Cumming, W. P. *British Maps of Colonial America.* Chicago, Illinois, p. 58.

Original

London. Public Record Office. C.O. 700, Carolina no. 21.

---

307. Bonar 1757 MS B

A Draught of the Upper Creek Nation, taken in May 1757. 1280 Gun Men, 27 Towns. [across top of map, within neat line]

Size: 13¼ × 12¼. *Scale:* None.

Description: This pen-and-ink sketch is like Bonar 1757 MS A, though without the elaborate side pictures that adorn that map.

Original

Clements (Gage).

---

308. De Brahm ca. 1757 MS

The Plan and Profile of ffort Loudoun Lattitude 36°. 7.m Projected by William De Brahm his Majestys Surveyor Genl. in Georgia.

Size: 15¾ × 11½. *Scale:* 1" = 100 feet.

Description: This plat is a well-executed pen-and-ink drawing with the bastions, palisade retrenchments, and other features common to the star-shaped forts of the period and found in slightly varying form in many of the subsequent plans for fortifications listed later as by De Brahm.

Fort Loudoun was part of Britain's policy of cementing the Cherokee Indians to their cause in the controversy known as the French and Indian War (1755–63). In the terms of a treaty signed in May of 1755, South Carolina agreed to build a fortification designed to protect the Overhill Cherokee towns from a surprise attack by the French and their Indian allies. Governor James Glen took pains to explain to the Indians that the fort would be built along the Little Tennessee River "near the place where enemies are most frequently discovered," and that it would serve as a refuge "to receive your women and children" in the event of such attack. William Gerard De Brahm was charged with designing and supervising Fort Loudoun's construction in that remote river valley in the mountains of what is now eastern Tennessee. Governor Glen had specified that it be built close to the "towns it is to protect," and, for the benefit of the garrison who would man it, "in a good air and near good water not very near any Eminence and not far from good corn land or from a convenient range for their cattle." In his "Report on the General Survey" that he presented to King George III, De Brahm included an account of his activities at Fort Loudoun as well as a plan and map of its setting (Maps 409 and 410, below). See Louis De Vorsey, *De Brahm's Report of the General Survey*, Columbia, S.C., 1971, pp. 18–24, 100–105.

Under this date may be entered a plan of Sunbury, Parish of St. John, Georgia, a town which was founded on the river Medway by a grant from George II in 1757: "Plan of the Town of Sunbury—containing 3430 feet in Length from North to South & 2230 feet in Breadth on the North." Across the top are numbered plots, at bottom the specifications. This drawing is reproduced in C. C. Jones, *The History of Georgia*, Boston, 1883, opp. p. 449.

References

Alden, John. *John Stuart and the Southern Colonel Frontier.* Ann Arbor, 1944, pp. 58, 60.

Original

Huntington.

---

309. De Brahm–Charles Town 1757 MS

Copy of Mr. De Brahms Plan for fortifying Charles Town, South Carolina, as now doing, with additions and Improvements July 1757.

Size: 16½ × 21. *Scale:* 1" = 400 feet.

Description: Unlike his 1755 "Plan of a Project to Fortifie Charleston . . ." (Map 290, above), this manuscript plan is restricted to the city's defensive walls and bastions. Detailed profiles of the chief installations are shown below them and are partially colored as is the main plan. The inset measures 15¼ × 5½; scale: 1" = 80 feet. Students of Charleston's early development will find the attention given to the townsite's topography and drainage patterns particularly valuable.

References

De Vorsey, L. *De Brahm's Report of the General Survey.* Athens, Georgia, 1971, pp. 17–18, 99–100.

McCormack, H. G. "A Catalogue of Maps of

Charleston," in *Year Book of Charleston, S.C., 1944,* Charleston, S.C., 1947 [1948], pp. 190–91, No. 21.
See Introductory Essay.

Reproduction

Congress (photocopy).

Original

London. Public Record Office. C.O. 700, Carolina no. 20.

---

310. De Brahm 1757

A Map of South Carolina and a Part of Georgia. containing the Whole Sea-Coast; all the Islands, Inlets, Rivers, Creeks, Parishes, Townships, Boroughs, Roads, and Bridges: As Also, Several Plantations, with their Proper Boundary-Lines, their Names and the Names of their Proprietors. Composed From Surveys taken by The Hon. William Bull, Esq. Lieutenant Governor, Captain Gascoign, Hugh Bryan, Esq; And the Author William De Brahm, Surveyor General to the Province of South Carolina, one of the Surveyors of Georgia, and late Captain Engineer under his Imperial Majesty Charles VII. Engrav'd, by Thomas Jefferys, Geographer to his Royal Highness the Prince of Wales [cartouche, bottom right] | To the Right Honourable George Dunk, Earl of Halifax First Lord Commissioner; and to the rest of the Right Honourable the Lords Commissioners, of Trade & Plantations. This Map is most humbly Inscribed to their Lordships, By their Lordships most Obedient & most devoted Humble Servt. William de Brahm [bottom, below Explanation, and to left of cartouche] | London. Published According to Act of Parliament by T. Jefferys. Octr. 20 1757. [below neat line, left center]

Size: Four sheets, each 24 × 26½. *Scale:* 1" = ca. 5 miles.
In: Jefferys, T. *A General Topography of North America.* London, 1768, Nos. 59–62.

Description: This map shows the coast from the North Carolina boundary line southward to St. Marys River in Georgia and extends westward to the Indian country. For the first time, for any large area in the southern colonies, a map possesses topographical accuracy based on scientific surveys. For the coastal region and up the larger rivers as far as the settlements extend, great care and detail in surveying is evident. Farther to the interior and away from the rivers information is meager, small streams and branches are not given, and the course of the rivers in the Piedmont region is inaccurate. The actual amount of topographical information given in the low country, however, is impressive; swamps, marshes, and the winding of tidal channels and rivers are delineated. The shape of the coastal islands is reasonably correct and identifiable; no longer, as in the earlier maps, are they symbolic blobs to indicate that islands are there. Along the North Carolina–South Carolina boundary line "the Nature of the Land in this Course" is shown by different types of shading, offering the nearest approach to a soil map for many years to come. For some reason only the first edition of De Brahm's map has this feature.

An explanatory legend to the left of the dedication text aids the user of this map by explaining symbols and shadings used to indicate "Towns," "Forts," "Churches & Chapels," "Houses," "Roads," "Swamp Lands," "Marsh Lands," "Oak Lands," and "Pine Lands."

De Brahm first revealed his intention to publish a map of South Carolina and Georgia by way of a public notice printed in the *South Carolina Gazette* on October 1752, just over a year after his arrival in America. In the notice he solicited subscriptions to the project which was to appear at the large scale of three miles to one inch. In the next few years he would travel and survey extensively in the settled regions as well as the frontiers of both colonies.

For many years De Brahm's South Carolina 1757 continued to be the basis of later maps, such as Cook 1773, Mouzon 1775, and Stuart's 1780 edition of De Brahm's map. Professor L. C. Karpinski, *Early Maps of Carolina and Adjoining Regions from the Collection of Henry P. Kendall, 1937,* p. 37, writes: "De Brahm is the first strictly scientific cartographical expert to practice his art in the Carolinas. For his activity in America after 1751 and involved work in Georgia and East Florida, the title of Surveyor General of the Southern District of North America was well deserved."

Of the previous surveys by Gascoigne, Bryan, and Bull that De Brahm refers to in his title, unfortunately but few records remain (see Swaine-Gascoigne 1729 MS). Lieutenant Governor Bull wrote to the Board of Trade from Charleston a number of years later, on December 17, 1765, of "a new map of this Province which I made last year & had the honor of presenting to your Lordships Predecessors with my other Dispatches sent in the Hillsborough Packet last winter": A. S. Salley, *The Boundary Line between North Carolina and South Carolina, Bulletins of the Historical Commission of South Carolina,* No. 10, Columbia, S.C., 1929, p. 22. This map has not been found or identified.

A French edition of De Brahm's map was published in G. L. Le Rouge, *Atlas Amériquain Septentrional*, Paris, 1778, Nos. 19–20: Caroline Meridionale et Partie de las Georgie Par Chev.r Bull Gouverneur Lieutenant, le Captaine Gascoign, Chev.r Bryan et de Brahm, Arpenteur Général de la Caroline Merid.le et un des Arpenteurs de la Georgie, en 4 Feuilles, a Paris Chez le Rouge, Ingénieur Geographe du Roi, rue des grands Augustins avec Privilege du Roi. 1777. *Size:* 51 × 40½. *Scale:* 1" = 1½ marine leagues. At the bottom right of this map are insets giving the course of the Hudson River by C. J. Sauthier. Copies of this map are found at Congress, Brown, Harvard, South Carolina Historical Society, and elsewhere.

The 1780 edition of the map, listed below, has many changes. Along the coast are soundings, absent in the 1757 edition, new plantations and roads, and names of islands or streams added or changed. To the interior the additions are much more marked; scores of names along the rivers and roads have been added, together with the names of new precincts and parishes. Most noticeable is the greatly increased knowledge of the minor creeks and streams, which form a heavy dendritic pattern over the whole map, where in the 1757 edition the spaces remained bare except along the coast and up the main rivers. Most, though not all, of these additions and changes can readily be traced back to Cook's 1773 map of South Carolina.

References

De Vorsey, L., ed. *De Brahm's Report of the General Survey.* Columbia, S.C., 1971, pp. 15–16, 31–33.

De Vorsey, L. "William Gerard De Brahm, 1718–1799," *Geographers Biobibliographical Studies*, 10 (1986), pp. 41–47.

Sellers, J. R., and P. M. Van Ee. *Maps and Charts of North America and the West Indies, 1750–1789*. Washington, D.C., 1981, no. 517, pp. 326–27.

Reproductions

Plates 59A–D.

Bloom, Sol, Director General, United States Constitution Sesquicentennial Commission, Washington, D.C., 1938 (separate sheet, reproducing Faden's 1780 edition of map).

Cartografia de Ultramar, II, Plate 45 (in two sheets; 1777 French edition, with Sauthier's Chart of the Hudson River; with transcription of place-names, pp. 257–62.

A. 1757 [Title as given above].

Copies

Congress 1768, two copies, two separates; Boston Public 1768; John Carter Brown; Charleston Museum, two separates; Clements; Duke, two separates; Georgia Hargrett Library; Huntington; New York Public; University of South Carolina; Public Record Office.

In: Faden, W. *The North American Atlas*. London, 1777, Nos. 32–33.

Copies

Congress 1777.

B. 1780 [Stuart-Faden edition: title unchanged up to "William De Brahm, Esq.r"]: Surveyor General of the South.n District of North America, Republished with considerable Additions, from Surveys made & collected by John Stuart Esq.r His Majesty's Superintendant of Indian Affairs. By William Faden, Successor to the late T. Jefferys, Geographer to the King. Charing Cross, 1780. [cartouche, bottom right] | To the Right Honourable Lord George Germaine, First Lord Commissioner, and to the rest of the Right Honourable the Lords Commissioners of Trade & Plantations. This Map is most humbly Inscribed to their Lordships, By their Lordships most Obedient & most devoted Humble Serv.t William Faden. [bottom center, to left of cartouche] | London. Published as the Act directs. by W.m Faden, Charing Cross, June 1.st 1780 [below neat line, left center]

In: Jefferys, T. *The American Atlas*. London, 1776, No. 24a (inserted).

Copies

Congress 1776; John Carter Brown; Clements; Georgia Hargrett Library; Kendall; University of South Carolina; Public Record Office.

---

311. Bellin 1757

Carte de la Caroline et Georgie. Pour servir à l'Histoire Générale des Voyages. Tirée des Auteurs Anglois par M. B. Ing de la Marine 1757. [cartouche, bottom right] | Tom XIV. in 4.o N.o 11 [below neat line, left] | Tome 14 in 8.o Page 148 [below neat line, right]

Size: 11¼ × 7⅜. *Scale:* 1" = ca. 61 miles.
In: *Histoire Générale Des Voiages*. Paris, 1746–61, XIV (1757), opp. p. 564.
Description: This map probably used some of the contem-

porary maps by E. Bowen. It attempts to show the course of the Tennessee River in western North Carolina as well as the coastal rivers east of the Blue Ridge, and has more information for the territory west of the Blue Ridge, extending into Tennessee and Kentucky, than is usual in maps of this period. In some copies of this map, as in those at Harvard and Clements, the writing below the neat line has only "Tom XIV" to the left, "No. 11" to the right.

Professor Karpinski attributes this map to M. Bonne; but in the "Advertissement" of Volume I, p. xvii, the translator refers to "l'Auteur de ces belles Cartes . . . M. Bellin, Ingenieur de la Marine, Garde du Dépôt Royal des Plans & des Cartes." This is Jacques Nicolas Bellin (1703–72), the author of numerous works on geography and hydrography.

This same map, with the 1757 date present, is found in La Harpe, J. F. de, *Abrégé de l'Histoire Générale des Voyages*, Paris, 1780 (and in the later edition, Paris, 1820), No. 47.

A later Italian map based on and similar to the 1757 map is found at the Charleston Library Society, John Carter Brown Library, New York Public Library, and University of North Carolina: "Carta della Carolina e Giorgia. Tratta dalle Carte Inglesi da Mr. Bellin. Ing. della Marina," in J. N. Bellin, *Teatro della guerra marittima*, Venezia, 1781, No. 31, opp. p. 11.

This same volume of *Histoire Générale* (1757) has another map, "Carte de la Floride, de la Louisiane, et Pays Voisins . . . 1757," opposite page 570, which includes South Carolina and Georgia on a smaller scale.

Copies

Congress; John Carter Brown 1757; Charleston Museum; Clements; Harvard 1757; Kendall Collection; University of North Carolina; Yale 1757.

---

312. Hess ca. 1758 MS

Fort Littleton at Port Royal [written within the plan of the fort] | Em: Hess Lieut. in the Royl. Amn. Regt [bottom left]

Size: 16⅞ × 11⅛. *Scale:* 1" = 30 ft.

Description: Fort Lyttelton, about three miles from the present site of Beaufort, South Carolina, was erected in 1758 against Indian and Spanish invasions, and was named for Governor William Henry Lyttelton, who was active during the Indian incursions. This is also the site of Stuart Town, where some Scot Covenanters, with Lord Cardross's aid, attempted to settle but were massacred by the Indians at Spanish instigation in 1686.

Emanuel Hess has not been identified; it is possible that this map was drawn during the Revolution.

Original

Huntington.

---

313. Bellin 1758

Karte von Carolina und Georgien, zur all gemeinen Geschichte der Reisen. Aus den englaendischen Nachrichten von M. B. Ing. de la Marine 1757 | No. 19 [upper right, outside neat line]

Size: 10 × 7¼. *Scale:* 1" = ca. 65 miles.

In: Allgemeine Historie der Reisen zu Wasser und zu Land. Leipzig, 1758, XVI, No. 19.

Description: This map (31° to 37°) is a copy of Bellin's 1757 map with legends and names in German. "Land der Apalachen" is south of "Fl. Alatahama" and "Georgien."

Copies

Congress 1758; University of North Carolina.

---

314. Gibson 1758

Carolina and Georgia [cartouche, top left] | 44 [lower right corner]

Size: 3¾ × 2½. *Scale:* 1" = 230 miles.

In: Gibson, J. *Atlas Minimus.* London, 1758, No. 44.

Description: This map extends west to "River Mississipi" and "Pt. of Louisiana." It is on too small a scale to give much detail. No. 43 of the *Atlas Minimus* is "The English and French Settlements in Nth. America." The 1774 edition of the Atlas was "Revised, corrected, and improved, By E. Bowen, Geographer to His Majesty"; the map is unchanged.

Copies

Congress 1758; John Carter Brown 1774; Clements 1758.

---

315. Lopez 1758

La Florida

Size: 3³⁄₁₆ × 4⁹⁄₁₆. *Scale:* 1" = ca. 175 miles.

In: Lopez de Vargas Machuca, Thomas. *Atlas Geographico*

de la America Septentrional y Meridional. Madrid, 1758, opp. p. 31.

Description: This map gives detail only for the peninsula and the rivers and Indian settlements in the western Georgia and Alabama region.

Copies

Congress; John Carter Brown.

---

316. Sepp 1758

Carte de la Louisiane, Mary-land, Virginie, Caroline, Georgia, avec une Partie de la Floride A Amsterdam chez Covens & Mortier, 1758 [above neat line of map] | C. Sepp, sculpsit. [bottom left corner]

Size: 23⅛ × 15⅜. *Scale:* 1" = ca. 45¾ miles.

In: Cóvens, J., and C. Mortier. *Atlas Nouveau.* Amsterdam, [1683–1761], IX, No. 51.

Description: This map, which extends from northern Florida to New Jersey and westward from the New Jersey coast to the Mississippi, is apparently based even to the smallest detail upon the bottom sheet of the three parts of Anville's "Canada Louisiane et Terres Angloises . . . MDCCLV."

It shows the Indian settlements between the Appalachians and the Mississippi. It gives the Granville Line in North Carolina but not the division between North and South Carolina.

Copies

Congress 1683–1761; Harvard.

In: Delisle, G. [*Collection of Maps*, Amsterdam, 1722–74], No. 36.

Copies

Congress 1722–74.

---

317. Auspouger ca. 1739 MS

[A Map of St. Simon's and Jekyll Islands on the coast of Georgia.]

Size: 36 × 19. *Scale:* 1" =⅝ mile.

Description: The islands have been very carefully surveyed; neither this nor the following MS (Map 318) has a title. The British Library credits Samuel Auspouger as the compiler of this unfinished map. It was incorrectly dated in earlier editions of this work. One of the map's most distinctive features deals with the vegetation of these barrier islands. Auspouger distinguishes very carefully between the marshy areas subject to frequent inundation and the forested and presumably higher areas.

Reproduction

Congress (photocopy).

Original

British Library. King George III's Topographical Collection. cxxii. 70.

---

318. Thomas ca. 1738 MS

[A plan of St. Simon's and Jekyll Islands, showing the position of the forts, redoubts, etc.]

Size: 36 × 19. *Scale:* 1" = ⅝ mile.

Description: Incorrectly dated in earlier editions of this work, the artistically drawn but unfinished cartouche and scale blocks suggest that it is a fair copy of a compilation being prepared for engraving and publication. Surveys by Samuel Auspouger, Captain John Thomas, and others could have been used as sources. The British Library currently ascribes authorship of this potentially valuable map to Captain John Thomas. Important features include detailed cadastral plans for the area centered on Frederica as well as the southern tip of St. Simon's Island and a plantation on Jekyll. See Map 246A.

Reproduction

Congress (photocopy).

Original

British Library. King George III's Topographical Collection. cxxii. 71.a.

318A. Thomas ca. 1738 MS

[Plans and profiles of the forts and redoubts in St. Simon's and Jekyll Islands, on a large scale; unfinished: 4' × 1'8", with rough details]

Size: 47½ × 19; 13⅛ × 11¼; 9 × 6. *Scale:* None.

Description: Three drawings that appear to be associated with fortifications designed by Captain John Thomas for the defense of the Georgia coast. Thomas was a military engineer who worked there in 1738–39. The item filed in the British Library as King George III's Topographic Drawing. cxxii. 7/d is notable. It is inscribed "The Plan of the Plan" and appears to be a preliminary sketch for a large map that would have included the manuscript survey by Captain Thomas that is described above as Map 246A.

Reproduction
Congress (photocopy).
Original
British Library. King George III's Topographical Collection. cxxii. 71b, c, and d.

---

319. Thomas ca. 1738 MS

2e Plan dune Batterie & d'une Redoute avec vn Corps de Garde construit en forme de Redoute a Machecoulis pour y servir de Retranchement a vne Garde de 20 hommes ou plus

Size: 12 × 18. *Scale:* None.

Description: Another plan for St. Simon's Island that was incorrectly dated in earlier editions of this work.

Reproduction
Congress (photocopy).
Original
British Library. King George III's Topographical Collection. cxxii. 73.

---

320. St. Simon's ca. 1760 MS

Plan Dun Fort convenable en quelque endroit que ce soit dune Cote maritime.

Size: 24 × 11. *Scale:* 1" = 32 toises (205 feet).

Description: A proposed fort for St. Simon's Island, off the Georgia coast. The date given for this map may be incorrect.

Reproduction
Congress (photocopy).
Original
British Library. King George III's Topographical Collection. cxxii. 74.

---

321. St. Simon's ca. 1760 MS

[Fort and redoubt for St. Simon's Island]

Size: 22 × 10½. *Scale:* 1" = 32 toises (140 feet).

Description: This plan is very much like King George III's Topographical Collection MS. cxxii. 74, though on a smaller scale; it has a long technical "Explanation." The date given for this map may be incorrect.

Reproduction
Congress (photocopy).
Original
British Library. King George III's Topographical Collection. cxxii. 75.

---

322. Thomas ca. 1738 MS

Plan dun petit Fort pour L'Isle de St Andre capable de contenir outre les Magazins, des Barraques pour 200 Hommes de Garnison, & environ 20 Chambres de Surplus pour quelques autres Habitants [across top of plan, followed by lettered references] | St Andrews at Camberland [*sic*] Island [penciled at bottom, in later hand]

Size: 15 × 12. *Scale:* 1" = 33 toises (210 feet).

Description: A Plan for a fort to be built on Cumberland Island, Georgia, prepared by Captain John Thomas. See Map 246A. The date given here may be incorrect.

Reproduction
Congress (photocopy).
Original
British Library. King George III's Topographical Collection. cxxii. 76.

---

323. Georgia Indian Lands ca. 1771 MS

Unpurchased Land the Whole Extraordinary good for, Hemp Tobacco, Indigo, Wheat &.&- [top right] | Rich Wheat and Hemp Land Unpurchased Land [top center] | The Indian Hunting Lands [center left]

Size: 20 × 29. *Scale:* 1" = 10 miles.

Description: This colored manuscript sketch was incorrectly dated ca. 1760 and erroneously attributed to William Gerard De Brahm in earlier editions of this work. It is a very generalized depiction rather than a map based on or compiled from surveys. There are a number of captions describing the fertility of Georgia's Piedmont soils as well as the most important Indian trading paths. Important upcountry settlements such as Wrightsborough, Queensborough, and Augusta are shown, as well as Mount Pleasant, Ebenezer, Fort Barrington, and Savannah nearer the coast. There are copies found in both the British Library and British Public Record Office. The P.R.O. copy was removed from a collection of memorials and letters supporting the plan of Governor James Wright to purchase a large tract of Indian land. For details on Georgia's enlargement through the purchase of a large tract of interior land from the Creek and Cherokee Indians, see L. De Vorsey, *The Indian Boundary in the Southern Colonies, 1763–1775*, Chapel Hill, North Carolina, 1966, pp. 161–72. The British Library version of this map is included in Additional MSS. 14,036, where it is folio f. The Library of Congress Geography and Map Division holds a photocopy.

Reproduction

Congress (photocopy).

Original

London. Public Record Office. MPG 20.

---

324. Wachau ca. 1760 MS

Tractus I der Wachau in Nord Carolina unter 36 Gr. noerdlicher Breite [top left] | Explicatio . . . [top right]

Size: 13 × 18¼. *Scale:* 1" = ½ mile.

Description: This unsigned map, identified by Miss Adelaide Fries as the careful work of the Moravian surveyor Christian J. Reuter, shows the land in the vicinity of "Die Grosse Iohanna" (the present Old Town Creek, flowing by Bethabara). The manuscript is one section of a larger chart, as the lengthy "Explicatio" refers to sections not shown on this sheet. This map may have been drawn later than 1760; see Reuter 1766 (Map 353), with biographical references and a list of other maps by Reuter there given.

Original

University of North Carolina.

---

325. Yonge–De Brahm 1760 MS

A Plan of the Islands of Sappola, containing 9520 acres. That is to Say, the main Island 770 acres. The Islands called black Beard ..... 1600 [acres]. six small Islands of little Sappola ..... 220 [acres]. Acres .... 9520 [upper right] Georgia. Pursuant to a warrant from his Excellency Henry Ellis Esqr. Captain General & Governor in Chief in & over his Majesty's said Province to us directed bearing date the 2d day of September 1760. We have caused to be admeasured & laid out unto Grey Elliot Esqr. the Islands of Sappola . . . Certified the 30th day of September in the year of our Lord seventeen hundred & Sixty—By Henry Yonge Will De Brahm Suvrs. Gen. [lower left]

Size: 29 × 23½. *Scale:* 1" = 1,980 feet.

Description: In James Edward Oglethorpe's early dealings with the Creek Indians he agreed that they should retain a number of the barrier islands along the coast. As the Indians declined in numbers their use of the islands also fell off and title to them passed to an unscrupulous Englishman who had married Oglethorpe's Creek Indian interpreter, known variously as Mary Musgrove, Mary Matthews, or Mary Bosomworth. After long and contentious wrangling with the Bosomworths the colony agreed to pay Mary 2,100 pounds sterling "provided so much should be produced from the Sale of Ossabaw Sapala." On July 24, 1759, the Georgia Council ordered the surveyors general, Henry Yonge and William Gerard De Brahm, to "make an actual Survey of the Islands Ossabaw and Sapala." In the same session they provided for the sale of the islands to pay off the Bosomworth's claim. See *The Colonial Records of the State of Georgia*, VIII, pp. 85–88, 307–8. On their plat Yonge and De Brahm noted the location of a number of buildings as well as a "Spanish Fort," and "Indian Town." Sand dunes marshes, uplands, and "large Savanna of fresh Water," as well as a large number of witness and corner trees are identified. Another copy of this survey is in the collection of the Surveyor General Department in the Georgia State Archives in Atlanta. See J. G. Blake, comp., *Pre-Nineteenth-Century Maps in the Collection of the Georgia Surveyor General Department*, No. 148, p. 69.

References

Catalogue of the Wymberley Jones De Renne Georgia Library, III, 1208–9.

Reproduction

In the 1970s this plat was redrawn and printed for distribution by the Office of the Vice President for Research at the University of Georgia.

Original

Georgia Hargrett Library.

---

326. Bonne ca. 1760

Carte de la Louisiane et de la Floride. Par M. Bonne, Ingenieur - Hydrogphe de la Marine. [lower center] | Liv. XVI. No. 46. [above neat line of map]

Size: 8¼ × 12⅝. *Scale:* 1" = 100 miles.

In: Bonne, R. [A collection of Amsterdam maps, ca. 1760].

Inset: Supplement [showing part of the Missouri River].

Size: 2⅛ × 1¾.

Description: This map, which shows Florida, Georgia, and South Carolina westward to the "Mississipi," also shows the mountainous western parts of North Carolina and Virginia in detail. It was probably made in the 1760s, judging from its detailed knowledge of Indian settlements.

Copies of this map are to be found in the Library of Congress, Yale, and elsewhere in Rigobert Bonne's *Atlas de toutes les parties connues du Globe Terrestre*, Genève, J. L. Pellet, 1780. See P. L. Phillips, *Lowery Collection*, p. 397.

Reproduction

Color Plate 20.

Copies

Congress; Yale ca. 1760, separate.

---

327. Kitchin 1760

A New Map of the Cherokee Nation with the Names of the Towns & Rivers. They are Situated on N° Lat. from 34 to 36 [cartouche, lower left] | Engrav'd from an Indian Draught by T. Kitchin. [below neat line] | For the London Mag: [above neat line]

Size: 8¾ × 6⅝. *Scale:* 1" = 16 miles.

In: The London Magazine, XXIX (1760), opp. p. 96.

Description: This map shows the Indian tribes at the headwaters of the French Broad, Little Tennessee, and Savannah Rivers. It is probably partly based on Mitchell's 1755 map, though it has some additional information. It is one of the earliest printed maps of Western North Carolina. On it is written the following legend: "NB. Col. Pawley wrote in 1746, That there was a Fall ¼ Mile long 12 miles below Uforsee to which the French Boats Might come & from thence transport what they please to any Town, over the Hills. Mr. Kelly a Trader Said the French Boats came up formerly to great Uforsee." The account accompanying the map is headed "Successful Expedition of Governor Lyttelton against the Cherokee Indians, with an accurate Map of their Country." In addition to reviewing the cause and results of Lyttleton's punitive incursion, the article stresses the military strength of the Cherokee and the importance securing them to the British cause in the backcountry competition with France.

Reproductions

Drake, S. G. *Early History of Georgia, embracing the Embassy of Sir Alexander Cumming to the Country of the Cherokees.* Boston, 1982, frontispiece (reprinted from *The New-England Historical and Genealogical Register*, July 1872).

Winsor, J. *The Mississippi Basin.* Boston, 1895, pp. 272–73.

Copies

Congress, separate; Clements; Duke, separate; Georgia Hargrett Library; Harvard; New York Public, two separates; Yale.

---

328. Stuart ca. 1761 MS

A Map of the Cherokee Country. [cartouche, lower left] | John Stuart fecit. [below cartouche]

Size: 25 × 19. *Scale:* 1" = 16 miles.

Description: This interesting colored manuscript map of what is now western North Carolina and eastern Tennessee shows the upper reaches of the Savannah and Tennessee ("Tannasse") Rivers and gives a rather detailed delineation of the Appalachian mountain range between southern Virginia and northern Alabama. "Ft. Loudoun" is at the lower center part of the map, with the note "in 35 d. 15° N° Latitude 450 computed miles from Charlestown." There is a table of the different tribes of the Cherokee nation, with the number capable of bearing arms in each. Stuart had been a member of the Fort Loudoun garrison that was forced to surrender to the Cherokees in 1760. He was fortunate in escaping the massacre that followed the English surrender. One of the trails shown on the map is captioned "The Road by which Capt. Stuart escaped to Virginia."

Professor John Alden, in his *John Stuart*, Ann Arbor, Michigan, 1944, p. 365, dates the map ca. 1761 but writes that he considers it may have been drawn by Stuart any time between 1761 and 1764. He was not appointed Superintendent of Indian Affairs in the Southern District until 1762.

References

De Vorsey, L. "The Colonial Southeast on 'An Accurate General Map,'" *The Southeastern Geographer* 6 (1966), pp. 20–32.

Reproductions

Gabriel, R. H., ed. *The Pageant of America.* New Haven, VI (1927), 109 (simplified facsimile); II (1929), 23 (detailed facsimile of the center, including Fort Loudoun region).

Hulbert, A. B. *Crown Collection.* Harrow, England, 1909–12, Series II, Vol. III, Nos. 34–35.
Congress (colored photocopy).

Original

British Library. Add. MS. 14036. fol. e.

328A. Cherokee Country 1761 MS

Sketch of the Cherokee Country and march of the Troops under Command of Lieut.n Col.o Grant to the Middle and Back Settlem.ts-1761 [left side above scale bar]

Size: 41½ × 13¾. *Scale:* 1" = 3 miles.

Description: This carefully drawn military sketch map covers the mountainous area stretching from the Cherokee town of Keowee and its adjacent Fort Prince George to the Tennessee River to below where it is joined by the Hiwassee River [Heywassee on map]. Although there is no indication of the overall topography of the area, mountain symbols do line certain segments of the trail Grant followed in his punitive raid against the Cherokee Indians. However, major rivers, streams, and trails are clearly shown with the locations of Grant's encampments as well as the many Cherokee towns. A caption along the trail from Fort Loudoun to Great Tellico states: "Here the Garrison of Fort Loudon was attack'd."

This map was part of the archive of Jeffery, first baron of Amherst, commander-in-chief of British forces in America. It was bequeathed to the Royal United Services Institution by his heirs and was purchased with other maps by the British Museum in 1968.

Original

British Library. Add. MS. 57714.16.

328B. Bullitt [1762?] MS

The Number of Farm's Improv'd by British Planters the Colony Virginia, on these Branches of the Mississippi at the Comencment of Hostilities Between the French their Indians and British Colonies. the Number of souls subsisted by sd.d Farms at a Low Estimation 1500- Jos: Bullitt Capt. of ye Provincial Regt. of Virga - [under bar scale lower right] | Not having an opportunity to Take the Latitude of places Represented in this Sketch. Refer to Evan's Map as the Most Correct for those Places he Touches on. [under compass rose lower left]

Size: 29¼ × 62½. *Scale:* 1" = 5 miles.

Description: This large ink and colored wash manuscript map is on paper and, like Map 328A, was formerly among the Amherst maps in the Royal United Services Institution in London. It presents a historically valuable view of the Cherokee Indian territory west of the Blue Ridge, stretching from the area near present-day Roanoke, Virginia, southwest to the Cherokee Overhill towns on the Little Tennessee River, south of present-day Knoxville, Tennessee. This map would be particularly useful in any attempt to analyze the activities of the Virginia forces in the region during the period of warfare that saw the surrender of the garrison at Fort Loudoun in August 1760. See Maps 328, 349, 401A, and 409. Among the places shown along the trail to the "Bigg Island" of the Holston River are "Post Amherst," "Post Fauquire," "Post Chiswell," "Stephens Spring," "Post Atticully Formerly Stalnakers," "Walkers Survery of Ten Thousd. acres of Land in the year 1746," and "Post Robinson." A caption on the map, located downstream from the "Bigg Island," states: "The courses and Distances from this Down from Lt. Timberlakes observations Both of the River & Road." Lt. Henry Timberlake's experiences while delivering the articles of peace agreed to by Indian and white negotiators at the Great Island of the Holston in November 1761 are described in his book, *The Memoirs of Lieut. Henry Timberlake*, London, 1765. In a footnote on page 67 of his reprint of Timberlake's *Memoirs*, S. C. Williams provides evidence that Sir Jeffrey Amherst was to receive "a fair copy" of Timberlake's "draught of the courses and bearings" of the river he traversed.

Reference

Williams, S. C., ed. *Lieut. Henry Timberlake's Memoirs 1756–1765*, Marietta, Georgia, 1948.

Original

British Library. Add. MS. 5771.6.

---

329. Gibson 1762

A New & correct map of the Provinces of North & South Carolina, Georgia, & Florida [cartouche, bottom right] | J. Gibson sculp.t [bottom right, within neat line]

Size: 11 × 14. *Scale:* 1" = ca. 75 miles.
In: The American Gazetteer. London, 1762, I, under "Carolina."
Description: This map extends from 26° to 36½° N.L., in-

cluding the region south of the Virginia line and west to the Mississippi. Numerous legends give information concerning English settlements and treaties; roads and the location of Indian tribes are very fully portrayed. It derives largely from Mitchell 1755.

Copies

Congress; American Geographical Society; John Carter Brown; Clements; Harvard; Kendall; New York Public; Sondley Reference; Yale.

329A. De Brahm 1762 MS

Above is the Map of Coxspur Island in the Enbuschure [*sic*] of Savannah River . . . [caption below] | Above is Ichnography [*sic*] & Orthography of the redoubted Caponiere . . . [caption below] | projected & Directed by W. G De Brahm Late Cap:tn Ingeneer [both on single sheet].

Size: 6¾ × 4½. *Scale:* 1" = 40 chains (2,640 feet).

Size: 9 × 7½. *Scale:* 1" = 20 feet.

Description: This small map and fortification plan was forwarded to the Board of Trade by Georgia's governor, James Wright, with his letter of June 10, 1762. It shows "Part of Tybe Island" with its lighthouse overlooking the "Sound of Savannah River," and "Coxspur Island" with its blockhouse fortification, Fort George (see Maps 339, 416, and 430). In his letter Wright stated that the fort was "now erecting at Cockspur Island." Sandbanks and flats are shown constricting navigation to south and north channels on either side of Cockspur Island. The longitude of Tybee Light is given as 79° and 37' west from London.

References

Blake, J. G. *Pre-Nineteenth-Century Maps in the Collection of the Georgia Surveyor General Department*, No. 153, p. 72.

De Vorsey, L. *De Brahm's Report of the General Survey in North America*, pp. 158–60.

Maps and Plans in the Public Record Office, 2, No. 2385, p. 409.

Reproduction

Georgia Archives (photostat).

Original

London. Public Record Office. C.O. 5/648 f. 144.

329B. Powder Horn 1762 MS

A New Map of Carolina and Likewise a plan of Ye Cherokee Nation Conqurd by the arme Commandt by Lieutt Col Jas Grant 1762 [engraved in decorative cartouche near center of horn]

Size: Single engraved cow horn. *Scale:* None.

Description: This unique scrimshaw-style powder horn map shows South Carolina from the Charleston waterfront to the Cherokee Country in the interior. It was probably prepared by or for a British officer attached to James Grant's punitive expedition which destroyed fifteen Cherokee towns during the early summer of 1761. The powder horn was an heirloom in the family of Thomas and Vanessa Doyle of Dun Laoghaire in the Republic of Ireland before its presentation to the Museum of the Cherokee Indian in Cherokee, North Carolina, in 1976. On the large end of the horn, about 10 percent of the map is devoted to what appears to be a historically accurate depiction of Charleston's waterfront skyline. Beyond Charleston the trail on the map leads to "Monks Corner," where Grant's army spent three weeks getting into fighting trim before proceeding to link up with the Carolina Provincials and a battalion of Royal Scots at prominently marked "Congarees." From the Congarees the trail leads to "Fort Ninety-six," where the final invading force was assembled. From the frontier fort, named from the fact that it was ninety-six miles from Charleston, the two-mile-long column began its march toward the Cherokee settlements. Shown and named on the powder horn map are "Fort Prince George," "Estatoe Old Field," "Field of Battle," "Etchoee," "Tassie," "Nockase," "Watoga," "Ayoree," "Cowee," "Usinah," "Cowichhe," "Cowhee Gap," "Stickoee," "Kittoa," "Tuckorithe," and "Tassatee." The name of "Alejoy" town is given, but its site is not, probably because Grant's force never reached it. Dawena Walkingstick prepared a reconstructed map of the invasion route which is a valuable aid in reading the powder horn map. Her map appeared in D. H. King, "A Powder Horn Commemorating the Grant Expedition Against the Cherokees," *Journal of Cherokee Studies* I (Summer 1976), p. 35.

References

Powder Horns In the Southern Tradition, Exhibition Catalogue by the Museum of Florida History, 1985.

Reproductions

King, D. H. "A Powder Horn Commemorating the

Grant Expedition Against the Cherokees." *Journal of Cherokee Studies* I (Summer 1976), 14 photographs.
Evans, E. R., and D. H. King. "Historic Documentation of the Grant Expedition Against the Cherokees, 1761." *Journal of Cherokee Studies* II (Summer 1977), p. 274, Figure 2.

Original

Museum of the Cherokee Indian.

---

330. Anon. ca. 1763 MS

[Carolina and Granville, Spanish, and French claims]

Size: 53½ × 36¼. *Scale:* 1" = 16 miles.

Description: This large sketch map has little detail; it outlines heavily in color the boundaries of "Louisiana, French," west of the Appalachians, "Spanish, Florida," including east and west Florida and a neck running up back of the English colonies to a point near North Carolina's southwestern boundary, "Georgia," and "South Carolina," which joins North Carolina at a line running west of "Cataboe" land on the 35° N.L. Across 35°30' N.L. runs the Granville line, with the following notation: "By the above Line Earl Granville claims near a third of what was formerly deemed the whole of N^o & S^o Carolina & near a sixth of such part of said provinces as are now in dispute with France and Spain." The map indicates clearly the various territorial claims being made at that time.

Reproduction

Congress (photocopy).

Original

London. Public Record Office. C.O. 700, Carolina no. 23.

---

331. Yonge–De Brahm's Georgia 1763 MS A

A Map of the Sea Coast of Georgia and the inland parts thereof extending to the Westward of that part of Savannah called broad River Including the several Inlets, Sounds, Creeks, Rivulets, Towns, Roads Forts & most remarkable places therein; performed at the request of His Excellency James Wright Esqr Capt General & Governor in Chief of the said Province, the 20th day of August 1763. And in the third year of His Majestys Reign. By Henry Yonge Wm D^e Brahm Surv^r Genl

Size: 23½ × 15¼. *Scale:* 1" = ca. 11 miles.

Description: This map extends from 30°30' N.L. to 34°30' N.L. and from 79°30' W.L. to 82° W.L.

It and the following map were compiled for Georgia's royal governor, James Wright, as he prepared to meet with other governors and the leaders of the southern Indian nations at the momentous conference held in Augusta in October 1763. See *Journal of the Congress of the Four Southern Governors, and the Superintendent of That District, with the Five Nations of Indians, At Augusta, 1763*, Charles-Town, South Carolina, 1764. Even a brief comparison reveals that Henry Yonge was the cartographer who compiled this map showing Georgia's historic boundaries with the Indians. On the original a yellow line is used to show that part of the colony "ceded by the Indians to His Excellency General Oglethorpe in the year 1739," while a red line indicates the large interior tract Wright was planning to treat for with the Indians.

The next map, now in the Clinton Collection, is almost identical but was drawn by Yonge's co–surveyor general, William Gerard De Brahm.

Reference

De Vorsey, L. *De Brahm's Report*, pp. 30–31, 267.

Reproduction

Congress (photocopy).

Original

British Library. Add. MS. 14036. g.

---

332. Yonge–De Brahm's Georgia 1763 MS B

A Map of the Sea Coast of Georgia & the inland parts thereof extending to the Westward of that part of Savannah called broad River including the Several Inlets, Rivers, Islands, Sounds, Creeks, Rivulets, Town, Roads, Forts & most remarcable places therein, performed at the request of his Exc^{lly} James Wright, Esqr. Captn Genl & Governor in Chief of the Said Province the 20th day of August 1763 & in the third year of the Reign of His Majesty King George the III By Henry Yonge J W G De Brahm Sur^y Genl

Size: 22⅞ × 12½. *Scale:* 1" = 11½ miles.

Description: See Map 331.

References

Brun, C., comp. *Guide to the Manuscript Maps in the William L. Clements Library.* Ann Arbor, 1959, No. 635, p. 152.

Reproduction

Congress (photocopy).

Original

Clements (Clinton Collection 329).

---

333. Wright 1763 MS

A Map of Georgia and Florida Taken from the latest and most accurate Surveys. Delineated & drawn by a Scale of 69 English miles to a Degree of Latitude by Thomas Wright, 1763.

Size: 37 × 16. *Scale:* 1" = ca. 19 miles.

Description: Practically nothing is shown on this map to the north of the Savannah River; between the Savannah and Altamaha Rivers the parishes, settlements, trading paths, and small creeks are shown in considerable detail. The Florida peninsula is very narrow and is broken into islands at the south.

"The great Swamp Owquaphenogaw" (Okefenokee) correctly empties into St. Marys River, which flows into the Prince William Sound north of St. Johns River. On the map this swamp has a curious similarity in position to the great inland lake of the seventeenth-century cartographers; it is possible that Okefenokee Swamp contributed to the error and helped perpetuate it.

Thomas Wright (ca. 1740–1812) was educated at the Royal Mathematical School of Christ's Hospital, London, during the period 1755–58. Just how and when he came to Georgia is not known, but he may have arrived just after leaving the Royal Mathematical School. In 1764 he was appointed principal assistant to Samuel Holland, the surveyor general of the Northern District of North America. Wright's later career was spent as surveyor general of Prince Edward Island, 1773–1812, and in several other government offices in the colony. Wright's papers are located in the National Archives of Canada.

Wright's may be the first map to show the boundaries of Georgia's parishes as they were given in the 1758 act that created them. See *The Colonial Records of the State of Georgia*, XXVIII, pp. 258–72. The provisions of that act resulted in the unilateral extension of Georgia's territory toward the interior since there is no indication that the Indians were consulted beforehand. For a discussion of Georgia's colonial boundaries with the Indians see L. De Vorsey, "Indian Boundaries in Colonial Georgia," *The Georgia Historical Quarterly*, LIV (Spring 1970), pp. 63–78.

A pen-and-ink map, supposedly copied from this original, is in the De Renne Collection, Hargrett Library, Athens, Georgia; the upper part has been reproduced in C. Howell's *History of Georgia*, Atlanta, 1926, I, opp. p. 168.

Reproductions

Plate 60 (northern portion).

Hulbert, A. B. *Crown Collection of Photographs of American Maps*. [Cleveland, 1914–16], Series III, Nos. 141–43.

Original

London. Public Record Office. C.O. 700, Georgia no. 13.

---

334. Jefferys 1763

Florida from the Latest Authorities. By T. Jefferys, Geographer to His Majesty. [bottom left]

Size: 14 × 15⅛. *Scale:* 1" = ca. 68 miles.

In: Roberts, William. *An Account of the First Discovery, and Natural History of Florida*. London, 1763, frontispiece.

Description: Interior settlements and rivers are given in some detail northward to the Santee River. The map includes "North Carolina," but there is no information for that region except the names of the three Capes: Fear, Lookout, Hatteras. The area shown extends west to Mississippi. Most of Florida is crisscrossed with rivers, lakes, waterways, and bays.

Jefferys, besides his other labors, edited the notes of Roberts, adding to them this general map, which he states he collected and digested with great care from a number of French and Spanish charts taken from prize ships. Among the several plans in this book is one of St. Augustine and its harbor, showing the depth of water, the site of the sea wall, etc.

Reference

Brinton, D. G. *Notes on the Floridian Peninsula*, p. 88.

Copies

Congress; Boston Athenaeum; Duke; Georgia Hargrett Library; Harvard, two copies; New York Historical; New York Public; Yale; Clements.

In: Jefferys, T. *A General Topography of North America and the West Indies*. London, 1768, No. 65. Same as Map 366.

Copies

Congress 1768; Boston Public 1768.

335. Scacciati-Pazzi 1763

Carte Rappresentante la Penisola della Florida [cartouche, lower center] | Andrea Scacciati sc [below neat line, left] | Giu. Pazzi scrise. [below neat line, right]

Size: 8¼ × 8⅞. *Scale:* 1" = ca. 92 miles.

In: Il Gazzattiere Americano. Livorno, 1763, I, opp. p. 191, No. 30.

Description: There is very little material for North Carolina, but the Florida, Georgia, South Carolina, and Alabama regions have contemporary information concerning the location of Indian tribes. The title page states that the information for the gazetteer was taken from English sources. The southern part of the Florida peninsula is broken into an archipelago.

Copies

Congress; John Carter Brown; Yale.

336. Gibson 1763

A Map of the New Governments, of East & West Florida [cartouche, top right] | Gent. Mag [upper right, outside neat line] | J. Gibson Sculp. [lower right, within neat line]

Size: 9⅞ × 7½. *Scale:* 1" = ca. 95 miles.

In: Gentleman's Magazine, XXXIII (1763), opp. p. 552.

Inset: Plan of the Harbour and Settlement of Pensacola. *Size:* 4½ × 3¼. *Scale:* 1" = 6½ miles.

Description: This map extends northward from Cuba (23°) to "Savanah Sound" (32°) and westward from "Savanah Sound" to the "Mississipi." Florida south of the St. Johns River is delineated as an archipelago, with scores of intersecting channels. Gibson was the map engraver for *The Gentleman's Magazine*, as Kitchin was for *The London Magazine.*

Copies

American Geographical Society, separate; Clements; Duke; Harvard; New York Public, separate; Yale.

336A. Proclamation Boundaries 1763

The British Governments in N.th America laid down agreeable to the Proclamation of Oct.r 7. 1763 [cartouche upper right] | J. Gibson Sculp: [lower right]

Size: 7¼ × 9¼. *Scale:* 1" = 240 miles.

In: The Gentleman's Magazine, XXXIII, London (1763), p. 612.

Inset: Bermuda or Summer Islands. *Size:* 3¾ × 2¾. *Scale:* 1" = 3⅓ miles.

Description: The Royal Proclamation of October 7, 1763, was the most significant and far-reaching policy statement for the administration the American colonies ever issued. To provide governments for the territories won from France and Spain and awarded to Britain by the Peace of Paris, it brought into being the new British colonies of Quebec, East Florida, West Florida, and Grenada. In addition to the creation of new colonies, far-reaching changes in the boundaries articulating the map and landscape of the Southeast were spelled out. The debated territory between the Altamaha and St. Marys River was made part of Georgia to clear up any lingering confusion there. Even more important, however, is the heavy line shown on this map as forming the boundary between West and East Florida and extending up the Flint River to follow the watershed of the eastern Appalachian and "Allegany" Mountains to bound on the west Georgia, the Carolinas, Virginia, Pennsylvania, New York, and New England (Plate 60A). This was what became known as the Proclamation Line. It was the main purpose of the omnibus Royal Proclamation of October 7, 1763, and was designed to forestall future Indian wars like Pontiac's "Conspiracy" that had raged earlier in the year. The eastern watershed of North America, in the language of the proclamation, would form "a western boundary . . . beyond which our people should not at present be permitted to settle." On the west side of that line it was stated ". . . that the several nations or tribes of Indians with whom we are connected, and who live under our protection, should not be molested or disturbed in the possession of such parts of our dominions and territories as, not having been ceded to, or purchased by us, are reserved to them, or any of them, as their hunting grounds." Thus with a few legalistic strokes of the pen the "lands beyond the heads or sources of any of the rivers which fall into the Atlantic Ocean on the west or northwest" became, as shown on this map, "Lands Reserved for the Indians." As stated in the document itself, the Proclamation Line was not intended to be a permanent barrier to white expansion, rather its purpose was to quiet the Indians' fears concerning loss of their land. What upset the colonists most, however, was the provision that "all persons, whatever, who have either willfully or inadvertently seated themselves upon any lands within the countries described [are] forthwith to remove themselves from such settlements."

Almost before the ink was dry on the proclamation the British began setting in motion policies designed to work out a carefully negotiated, surveyed, and demarcated boundary to separate those areas open for further colonial expansion from those restricted to the use of the Indians of America's interior. That boundary, known as The Indian Boundary Line, became a major concern of George III's colonial administrators throughout the era that culminated in the Revolutionary War. Some of the most valuable maps of the southern backcountry in this period were based on surveys of the Indian Boundary Line that stretched from the Ohio River to Florida and west to the Mississippi River.

References

De Vorsey, L. *The Indian Boundary in the Southern Colonies, 1763–1775.* Chapel Hill, North Carolina, 1966, pp. 34–40.

Fite, E. D., and A. Freeman. *A Book of Old Maps Delineating American History.* New York, 1969, pp. 218–21.

Reproductions

Plate 60A.

De Vorsey, L. *The Indian Boundary*, p. 37.

Fite, E. D., and A. Freeman. *A Book of Old Maps*, p. 218.

Copies

Congress.

---

337. Wyly 1764 MS

A Map of the Catawba Indians Land surveyed agreeable to an Agreement made with them by His Majesty's Governors of South Carolina, North Carolina, Georgia, & Virginia and Superintendant of Indian Affairs at a Congress lately held at Augusta by His Majesty's Special Command. Surveyed by Order of His Excellency Thomas Boone Esqr Executed and Certifyed by me this 22.^d day of February 1764 Saml Wyly

Size: 25 × 21. *Scale:* None.

Description: This is a map of the 144,000 acres granted to the Catawba Indians on the Catawba or Wateree, not far below the present site of Charlotte, North Carolina.

Reproductions

Hulbert, A. B. *Crown Collection.* [Cleveland, 1914–16], Series III, Plates 31–34.

Original

London. Public Record Office. C.O. 700, Carolina no. 24.

---

338. North and South Carolina Boundary Line 1764 MS

A Plan of the temporary Boundary-Line between the Provinces of North and South Carolina, run agreeable to the Instructions given us by His Excellency Arthur Dobbs, Esq$_3$. Governor of North Carolina, and His Honour William Bull, Esq$_3$. Lieut.^t Governor of South Carolina, and finished; As Witness our Hands this 24.th Septr 1764. Ja. Moore, George Pawley, Saml Wyly, Arthur Mackay, Surv.^rs Laid down by a Scale of one Mile to an Inch.

Size: 96 × 7½. *Scale:* 1" = 1 mile.

Description: This survey (recorded on two sheets) is the second leg of the east-west boundary line, which runs from the old 1737 meadow stake on the main Cheraw–Cape Fear River road and ends where the "Waggon Road from Charles Town" crosses "Waxaw Creek." It represents another step in the long controversy between the two provinces over the boundary line, a dispute which caused much bitter feeling and difficulty for the settlers along the border. In 1764 the Board of Trade and Plantations questioned the validity of the line of 1737; commissioners were authorized by the king; and the line was run from the terminus of the 1737 line to the Charleston-Salisbury road. The first leg of this survey runs 91 miles from the ocean NW to a stake marked "35°." The title given above is found on the second sheet, which records the second leg of the survey and runs 61 miles E-W from the 1737 meadow stake. The commissioners, in spite of the instructions of the Board of Trade and Plantations to check the end of the 1737 line, did not discover that it ran only to 34°49' instead of to 35°. In accepting this erroneous beginning for the line which they ran to the west, they laid the basis for further controversy. See Marvin L. Skaggs, *North Carolina Boundary Disputes Involving Her Southern Line*, Chapel Hill, 1941, pp. 71–74, and C. O. Paullin, *Atlas of the Historical Geography of the United States*, Washington, 1932, p. 82 and Plate 100A.

In the North Carolina State Department of Archives and History is a copy of this map on the same scale and with the same title, with the following certification added: "Endorsed and sealed on 20th November 1806 by Geo. Chalmers, Chief Clerk of the Privy Council and John Pratt Kennell, Land Surveyor of London, that this printed map is a true copy of the original map of 1764." This 1806 copy is probably the map John Steele secured, with other documents, through James Monroe, Minister to England. In the following year, Dr. Caldwell took several observations along the line and found that it varied from 11" 43" 34" south to 9" 34" 7" 20" north of the thirty-fifth parallel. See Skaggs, op. cit., pp. 125–26, and documents there cited.

Reproductions

Hulbert, A. B. *Crown Collection*. [Cleveland, 1914–16], Series III, Nos. 35–42.

Congress (photocopy).

Original

London. Public Record Office. C.O., Carolina. no. 25.

---

339. Tyby 1764 MS

View of Tiby Light House at the Entrance of Savanna River Georgia. Decr 1764

Size: 11¼ × 7. *Scale:* None.

Description: This is a colored sketch rather than a map; it is beautifully done. Here may also be mentioned a similar painting entitled "View of Cockspur Fort at the Entrance of Savannah River in Georgia Decr 1764." Both paintings appear to have been the work of the same nonprofessional artist who may have been on a ship anchored off Tybee Island. The Cockspur Fort view was reprinted in Hulbert's *Crown Collection*, Ser. 1, Vol. 2, no. 13; and in the Georgia Historical Society's *Collections*, VIII (1913), p. 224. Colored facsimiles of both views have been available for sale to the public at the British Library in London.

Reproduction

Congress (colored photocopy).

Original

British Library. King George III's Topographical Collection. cxxii. 77b&c.

---

340. Lobb 1764 MS

[Cape Lookout]

Size: 19 × 14¾. *Scale:* 1" = 1,200 feet.

Description: This pen-and-ink drawing has the following legend in the lower right-hand corner, below a list of references: "A Plan of the Harbour of Cape lookout Surveyed and Sounded by His Majesty's Sloop Viper Captain Lobb in Sepr 1764 this Harbour is the best on the Coast of North Carolina being Landlock'd and sheltered from all Winds, and the going in attended with on [out?] difficulty in the Day or Night having no bar it Flows 7 hours full and change and the tide was observed to rise 5 feet and when the wind blows hard from the Eastward it will rise 7 feet. That Strangers may know this Harbour there is a Flag-staff erected by Captain Lobb about 50 feet high on a Sorel Hill near Davis's House and on Landlock Point a Port Which small Vessells may Moor too, they bear NE b E from each other—-Cape lookout is in the Lattd 34°-40" Variation 2°. 40" East."

On April 24, 1769, Governor Tryon wrote to Lord Hillsborough: "Cape Look Out is now well known to his Majestys sloops on this station. A plan of the Bay was taken by the *Viper* sloop of War in 1764, which I understand was transmitted by Captain Lobb to the Lords of Admiralty." *Colonial Records of North Carolina*, VIII, p. 30.

Reference

Sellers, J. R., and P. M. Van Ee. *Maps and Charts of North America and the West Indies 1750–1789*. No. 1509, p. 325.

Original

Congress (Howe Collection).

---

341. Southern Indian District 1764 MS

A Map of the Southern Indian District 1764 [cartouche, lower right]

Size: 22 × 18. *Scale:* 1" = 46 miles.

Description: This colored manuscript map was forwarded to the Board of Trade by the Superintendent for Indian Affairs in the Southern District, John Stuart. It originally accompanied and illustrated a lengthy report Stuart addressed to the Board of Trade on March 9, 1764, from which it is now separated. See L. De Vorsey, *The Indian Boundary in the Southern Colonies, 1763–1775*, pp. 13–15. The map is not signed and its style and lettering do not resemble Stuart's earlier signed map, No. 328, above. It was probably prepared under his direction by an employee or associate. The map covers the area from 28° to 40° N.L. and west to just beyond the Mississippi River. Most of the details shown relate to Indian affairs such as their territorial claims and the locations of their numerous villages, which are named. In his report Stuart provides the name of each village and the number of

"Gun Men" it could muster. Ink washes on the map help define the territorial limits claimed by the main tribal divisions, the Cherokee, Creek, Chickasaw, and Choctaw.

Reproductions

Plate 59F.

Hulbert, A. B. *Crown Collection*. Cleveland, 1908, Series I, Vol. V, Nos. 41–42.

Original

British Library. Add. MS. 14,036. fol. 8.

---

342. Bellin 1764 A

La Caroline dans l'Amérique Septentrionale Suivant les Cartes Angloises [cartouche, bottom right] | Tome 1. N.º 36. [top right, above neat line]

Size: 14 × 8¾. *Scale:* 1" = ca. 40 miles.

In: Bellin, J. N. *Le Petit Atlas Maritime*. Paris, 1764, I, No. 36.

Description: This is a well-drawn map, showing the coastal settlements in some detail, the north and south roads, and the trading routes to the Indians. It shows the Granville line.

Reproductions

Color Plate 21.

Waynick, C. *North Carolina Roads and Their Builders*. Raleigh, 1952, end papers.

Copies

Congress; John Carter Brown; Clements; Duke; Harvard; New York Public; University of North Carolina; Yale.

---

343. Bellin 1764 B

Port et Ville de Charles-Town dans la Caroline | Tome I. No.º 37. [above neat line]

Size: 8¼ × 6. *Scale:* 1" = 1,279 feet.

In: Bellin, J. N. *Le Petit Atlas Maritime*. Paris, 1764, I, No. 37.

Description: A plan of the town, of Fort Johnson, and the vicinity is given. This map is based on the inset of the Crisp 1711 map, though it has less detail.

Copies

Congress; John Carter Brown; Clements; Harvard; Kendall; New York Public; Yale.

---

344. Bellin 1764 C

Carte de la Nouvelle Georgie [cartouche, center left] | Tome I. N.º 38. [top left, above neat line]

Size: 5⅞ × 8⅜. *Scale:* 1" = ca. 40 miles.

In: Bellin, J. N. *Le Petit Atlas Maritime*. Paris, 1764, I, No. 38.

Description: The map covers the territory from "St. Augustin" to "Entrée d S.t Helene." It shows the roads and towns, though it does not extend far to the interior.

Copies

Congress, separate; John Carter Brown; Clements; Duke; Harvard; New York Public; Yale.

---

345. Arredondo-Martinez 1765 MS

Descripcion Geographica dela parte que los Españoles poseèn actualmente enel continente dela Florida. Del Dominio en que estàn los Yngleses con legitimo Tituloso o en virtud del Tratado de Pazes del año de 1670, y dela Jurisdicionque indevidamente hàn Òccupado despues de dicho Tratado en que se manifiestan las Tierras que vsurpan y se definen los limites que deven prescrivirse para una, y otra Nacion, en conformidad del derecho dela Corona de España [title above adjacent table of explanations, which is to right of map] | Fernando Mrnz [Martinez] fec.t et scrip.ta. 1765 Matriti [at bottom of table of explanations, to right of map]

Size: 15 × 15⅞ [27¾ × 15⅞ with table of explanations appended]. *Scale:* 1" = ca. 65 miles.

Description: This is one of two copies of an earlier manuscript map made in this year; see Arredondo 1742 MS.

References

Phillips, P. L. *Lowery Collection*, pp. 342–43, No. 497.

Reproductions

Chatelain, V. *The Defenses of Spanish Florida*. Washington, 1941, Map 1.

Hulbert, A. B. *Crown Collection*. Series III, Vol. V, Nos. 36–37.

Congress (photocopy).

Original

British Library. Additional MS 17648 A.

346. Arredondo–de La Puente 1765 MS

Descripcion Geographica de la parte que los Españoles poseen actualmente en el continente de la Florida. Del Dominio en que estan los Yngleses con lexitimo Titulo solo en virtud del tratado de Pazes del ano de 1670, y de la Jurisdicion que indevidamente han Occupado despues de dicho tratado en que manifiestan las Tierras que usurpan, y se definen los limites que deven prescrivirse p[ar]a una, y otra Nacion, en conformid[a]d del derecho de la Corana de Esp.ª [to side of map, above appended explanatory sheet] | Es copia á letra de su origin.l que para efecto de sacar esta me ha facilitado el Cor.l D. Melchor Felin ultima Gov.or q.e fue de la Plaza de S. Agustin de Florida á quien lo ha debuello. Havana y Mayo 25 de 1765.==Juan Josef Elixio de la Puente. [on appended explanatory sheet]

Size: 15 × 15⅞; 27¾ × 15⅞ with appended table of explanations. *Scale:* 1" = ca. 65 miles.

Description: This is one of two copies of an earlier manuscript map; see Arredondo 1742 MS. The reproductions noted below are rather poor facsimiles.

References

Phillips, P. L. *Lowery Collection*, pp. 342–43, No. 497.

Reproductions

Hamilton, Peter J. *Colonial Mobile.* Boston, 1910, p. 210.

Hamilton, Peter J. *The Colonization of the South.* Philadelphia, 1904, opp. p. 285.

Ruidiaz y Caravia, Eugenio. *La Florida.* Madrid, 1894, p. xliii.

Original

Madrid. Div. de Hidrografia. 9a. 4a. 82.

347. Anon. 1765 MS

Cantonment of the Forces in North America, 11th Oct. 1765

Size: 24 × 20. *Scale:* 1" = 100 miles.

Description: This is a colored map. Another MS map of the "Cantonment of H. M. forces in North America, according to the disposition now made and to be compleated as soon as practicable; taken from the general destribution dated at New York, 29 March, 1766, by Dan.l Paterson, Ass.t Q.r M.r Gen.l," is in the British Library, Add. MS. 11, 288. These maps show an alertness to potential dangers and care in planning to meet the enemy adequately and efficiently on the part of the British authorities.

The Peace of Paris in 1763 gave to England a vast empire; the war with France gave Canada to England. But the war for imperial dominion might, England knew, break out at any moment. Pontiac's war showed clearly the danger on the western frontier. This chart, showing the cantonment of British forces in North America, October 11, 1765, "shows a plan for the disposition of troops admirably conceived to meet every contingency, the location of regiments, companies, half companies, and detachments . . . Grenville, in 1765, thought that the thirteen colonies should share the costs of this defense. So he proposed to levy a stamp tax": R. H. Gabriel, ed., *The Pageant of America*, VI, 118.

Reproductions

Gabriel, R. H., ed. *The Pageant of America.* New Haven, 1927, VI, 118.

Original

British Library. Add. MS. 11,287.

348. Kitchin 1765

A New Map of North & South Carolina, & Georgia. Drawn from the best Authorities: By T. Kitchin Geog.r | For the London Magazine [above neat line]

Size: 9⅛ × 6¾. *Scale:* 1" = 100 miles.

In: The London Magazine, XXXIV (April 1765), opp. p. 168.

Description: Very little new information is given in this map, although its detail is an example of the extent to which geographical knowledge concerning the western region from the mountains to the Mississippi was gradually spreading.

Copies

Congress, separate; American Geographical Society; Clements; Duke; Georgia Hargrett Library, separate; Harvard, separate; New York Public, separate, two copies; Yale.

349. Timberlake 1765

A Draught of the Cherokee country, On the West Side of the Twenty four Mountains, commonly called Over the Hills; Taken by Henry Timberlake, when he was in that country, in March 1762. Likewise the names of the Principal or Head men of each Town, and what Number of Fighting Men they send to War. . . . [the list follows, bottom left corner]

Size: 9½ × 15½. *Scale:* 1" = 1 mile.
In: Timberlake, Lieut. Henry. *The Memoirs of Lieut. Henry Timberlake.* London, 1765, opp. p. 160.

Description: This map follows the course of the Tennessee River from the Great Smoky Mountains to Fort Loudoun. In an effort to gain and hold the allegiance of the Cherokee Indians, the colonial governors of Virginia and South Carolina had two forts constructed on the Little Tennessee River among the Indian settlements of the Cherokees. The Virginian fort was never occupied; but De Brahm, who was not only an able cartographer but also an engineer, constructed a strong and elaborate fortification at Fort Loudoun, near the junction of the Tellequo River with the Little Tennessee. The Cherokees attacked the fort on March 20, 1760; on a later attempt they forced a retreat and succeeded in destroying or capturing all but one of the garrisons by August 9. By November 19, 1761, the Cherokees sued for peace; Henry Timberlake, a young Virginian ensign who had already seen service under George Washington around Winchester and Pittsburgh, volunteered to go on a mission to cement friendship. Timberlake prepared the memoirs of his experiences while in London in 1765. He died before or shortly after their publication. The English poet Robert Southey used Timberlake's book in the preparation of his epic poem, *Madoc* (1805), in which the fictitious adventures of a twelfth-century Welsh chieftain in the American wilderness are narrated.

A finely drawn profile and detailed plan with a description of the events leading to the construction of Fort Loudoun are included in L. De Vorsey, ed., *De Brahm's Report of the General Survey in the Southern District of North America*, Columbia, S.C., 1971, pp. 100–105. A facsimile of the plan found in the manuscript version of *De Brahm's Report*, now in the Houghton Library at Harvard University, was published in R. H. Gabriel, ed., *The Pageant of America*, II, 23. In 1957 Paul Kelly prepared a valuable analysis of Timberlake's map by comparing it with the modern map of the area it covers. See his *Historic Fort Loudoun*, Vonore, Tennessee, 1961.

Reproductions

Avery, E. M. *A History of the United States.* Cleveland 1904–7, IV, 346.

Gabriel, R. H. *The Pageant of America.* New Haven, VI (1927), 109; II (1929), 40.

Williams, Samuel Cole, ed. *Lieut. Henry Timberlake's Memoirs 1756–1765.* Johnson City, Tenn., 1927, opp. p. 27.

Copies

Congress; Clements; Duke; Georgia Hargrett Library; Harvard; New York Public; Sondley Reference; Yale.

In: Jefferys, T. *A General Topography of North America and the West Indies.* London, 1768, No. 64.

Copies

Congress; Boston Public.

---

350. Hawks ca. 1766 MS

Elevation of the North Front of the Governors House to be built at Newbern, North Carolina

Size: 19 × 22 (approx.). *Scale:* 1" = 14 feet.

Description: The front appearance and the blueprints of the governor's house are given, with different scales. This plat was "transmitted with Gov. Tryons letter of Feb 23 1769," which asked the Crown for approval.

There are two sets of plans for Tryon Palace: this one in the Public Record Office, and a fuller set by the architect John Hawks, which is preserved among the Hawks MSS in the New York Historical Society.

The plans were drawn by Hawks in 1766, and the foundations were laid in 1767. By 1769, the date of Sauthier's plan of Newbern (which see), Governor Tryon wrote that the roof of the edifice was covered in. The palace was completed in 1771. The taxes of £15,000 imposed on the settlers for the cost of the building created much discontent among the colonists in the western part of the province and was one of the contributing causes for the uprising of the Regulators. The Hawks plans in the New York Historical Society have been reproduced in *New Bern: Cradle of North Carolina*, Raleigh: The Garden Club of North Carolina, 1941, pp. 12 and 14, and in F. B. Johnston and T. T. Waterman, *The Early Architecture of North Carolina*, Chapel Hill: University of North Carolina Press, 1941, pp. 82, 86.

Reproduction

Congress (photocopy).

Original

London. Public Record Office. C.O. 700, Carolina no. 10.

351. Pickins 1766 MS

By Virtue of a Warrant under the hand and Seal of the Honourable William Bull Esquire Lieutenant Governor &c.ª of South Carolina, to me directed dated the 21.st day of June 1765. I have pursuant to his Honors Instructions marked out a Boundary Line between the province of South Carolina and the Cherokee Indian Country, in presence of the Headsmen of the Upper Middle and Lower Cherokee Towns whose hands and Seals are hereunder affixed which line is Represented in the above deleneated [*sic*] Plat. Witness my hand at Fort Prince George this Eighth day of May 1766. Edward Wilkinson Commiss.r | John Pickins, Surveyor.

Size: 24 × 11½. *Scale:* 1" = 2 miles.

Description: This is the official survey that demarcated the boundary between South Carolina and the Cherokee Indians agreed to at Fort Prince George on October 19, 1765. As such, it became an important segment in the Indian Boundary Line, which was surveyed by John Pickins in 1766, and it continues to form the boundary between Abbeville and Anderson Counties in South Carolina. See L. De Vorsey, *The Indian Boundary in the Southern Colonies, 1763–1775*, pp. 126–35.

An endorsement on the right and below the survey states: "We the Headmen of the Upper Middle & Lower Cherokee Indian Towns, Certify that we were present at the marking out of a Boundary Line at Devises Corner, And that We do in the name and by the directions of the whole Nation Assent and agree that the said Line shall be a Boundary between us and our brothers the English." This endorsement was agreed to by six of the most influential Cherokee leaders, including Kittagusta, Tiftoe of Keowee, Emy of Estatoe, Usteneka Otassitie or Juds Friend, Ukeneka or the Wolf, and Katchee or half-breed Will. They made their marks and seals before the deputy Indian superintendent, Alexander Cameron.

Reproductions

Hulbert, A. B. *Crown Collection of American Maps.* Cleveland, 1914–16, Series III, Plates 43–44.

Original

London. Public Record Office. C.O. 700, Carolina no. 26.

352. Indian Nations 1766 MS

A Map of the Indian Nations in the Southern Department 1766 [cartouche, lower right]

Size: 22½ × 18. *Scale:* 1" = ca. 48 miles.

Description: This map shows the same area as and is similar to the 1764 "A Map of the Southern Indian District," described above, Map 341. Like that depiction, it appears to have been prepared by the Superintendent for Indian Affairs in the Southern District, John Stuart, in an effort to explicate the complex geography of Indian settlement and territorial control in the Southeast. Although it can't be confirmed, this map may have been the one forwarded to the Board of Trade on July 10, 1766, with Stuart's detailed review of Indian affairs, which is now in the British Public Record Office (C.O. 5-67, fols. 45–56). Like the 1764 map, it names a large number of Indian towns. Rather than ink washes to show Indian territories, however, this map utilizes a series a fine dotted lines. The names of the British colonies that are prominent on the earlier map are absent from this, but the tribal identities have been retained.

Although Stuart expert John Richard Alden attributed this map to either Stuart or De Brahm, and Clements Library map curators Lloyd Brown and Christian Brun assigned it to De Brahm, there is no firm evidence to support their contentions. Until better evidence becomes available it will be advisable to leave open the question of its authorship.

References

Alden, John. *John Stuart and the Southern Colonial Frontier, 1754–1775.* Ann Arbor, Michigan, 1944, pp. 365.

Brun, C., comp. *Guide to the Manuscript Maps in the William L. Clements Library.* Ann Arbor, Michigan, 1959, p. 134.

Reproductions

Plate 61.

Alden, J. R. *John Stuart*, in pocket at end of book.

De Vorsey, L. *The Indian Boundary in the Southern Colonies, 1763–1775.* Chapel Hill, North Carolina, 1966, front and back endpapers.

Original

Clements.

353. Reuter 1766 MS

Wachovia or Dobbs Parish in Rowan County N.C. with some additional survey's 1766 Aug 25. Reuter [cartouche, top left] | Explanations [along right border]

Size: 10⅜ × 14⅝. *Scale:* 1" = 600 rods.

Description: Christian Gottlieb Reuter, the author of this map, was born in Steinbach, Germany, in 1717. He was a Royal Surveyor under Prince Hanau before coming to America in 1756. In 1759 he came to Bethabara from Pennsylvania, living in Bethabara until 1772, and thereafter in Salem until his death in 1777. His work is marked by accuracy and detail.

A number of maps by Reuter are reproduced in Adelaide L. Fries, *Records of the Moravians in North Carolina*, Publications of the Moravians in North Carolina, Publications of the North Carolina Historical Commission, Raleigh, Vol. I (1922) and Vol. II (1925): "Bethabara in Wachovia . . . 1766" (I, opp. p. 132), "Bethabara. 1766" (I, opp. p. 273), "Wachovia or Dobbs-Parish . . . 1766" (I, opp. p. 310), "Bethania in Wachovia . . . 1766" (I, opp. p. 375), "Salem Anno MDCCLXI" (I, opp. p. 482), and "Wachovia or Dobbs Parish . . . 1766" (II, opp. p. 616; the map here listed). All of these maps are now the property of the Moravian Archives. The land office, or administration office, of the Moravian Church has turned over most of its early maps to the Archives. Maps that are still in the office are regarded as under the control of the Archives. Some of the maps have been loaned for exhibition to the Wachovia Historical Society Museum and Hall of History.

A biographical sketch of Reuter is found in A. L. Fries, op. cit., I, 477–83.

Original

Museum and Hall of History, Wachovia Historical Society, Winston-Salem, N.C.

354. Charlevoix-Ridge 1766

A Map of the British Dominions in North America as Settled by the late Treaty of Peace 1763. I. Ridge Scu.

Size: 15 × 10¼. *Scale:* 1" = ca. 135 miles.

In: Charlevoix, Father P. F. X. de. *A Voyage to North America: undertaken by Command of the present King of France.* Dublin, 1766, opp. p. 47.

Inset: [Southern part of the Peninsula of Florida.]

Description: "The Oblique Strokes shew what formerly belonged to the French whose Forts are markd with double lines, the English Forts by a single line." The parts indicated are shaded west to Mississippi. "Earle Granville's Property" is divided from South Carolina and Virginia by dotted lines. It is also called "North Carolina."

In September 1720, Charlevoix arrived in Canada on an expedition to check the missions of Canada; he wrote *Histoire et Description Générale de la Nouvelle France, avec le Journal Historique d'un Voyage fait par l'Ordre du Roi dans l'Amerique Septentrionale*, Paris, 1744. This history, based upon first-hand experience, state papers, and the archives of the Jesuit order, ranks foremost among eighteenth-century general histories of Canada.

See Bellin 1744.

Copies

Congress; John Carter Brown; Clements.

355. Speer 1766

Plan of Entrance into Cape Fear Harbour North Carolina.

Size: 14¼ × 10. *Scale:* 1" = 1 mile.

In: Speer, J. S. *The West-India Pilot.* London, 1766, No. 13, opp. p. 53.

Description: On the plate: "The Frying Pan Shoal runs off due N & S from Cape Fear, & is reduced to a small Scale, & placed E & W. for the conveniency of bringing it into the Plate." This copperplate was also used for the 1771 printing of this chart. See Map 403, below.

Copies

Congress 1766, 1771; New York Public; Clements 1766.

356. Cook–Port Royal 1766

A Draught of Port Royal Harbour in South Carolina, with the Marks for going in. Most humbly Inscribed to the Publick. by their Humble Serv.t James Cook. Approv'd of by M.r Joiner 20 Years a Pilot of that Place. | Eman.l Bowen, Sculp.t | Published by the Author according to Act of Parl.t Dec.r 1766.

Size: 29 × 21¼. *Scale:* 1" = 1 mile.

Description: Of this map Professor L. C. Karpinski writes in his *Early Maps of Carolina*, 1937, p. 38: "It is one of the rarest of maps relating to the Carolinas. Apart from the soundings there are many locations, including the forts on Beaufort Island, Beaufort Town, and Sir John Collingwood's estate. On Hilton Head Island it is stated there are twenty-five families, on St. Helena Island, thirty families and one church. Cook was for a time resident in the Carolinas commanding the Grenville schooner. His directions for sailing into the harbour were copied by Des Barres in his map of Port Royal in the 'Atlantic Neptune.' James Cook was one of the editors of the first part of the North American Pilot published in London by Sayer and Bennett in 1777–1778."

See Gascoigne–Port Royal 1729 MS.

Copies

Clements (Clinton Collection, No 321); Kendall; New York Public.

---

357. Cook–West Florida 1766

A Draught of West Florida, from Cape St. Blaze to the River Iberville, with Part of the River Missisipi [cartouche, top left center] | Eman^{l}. Bowen Sculp^{t}. [below title cartouche] | To John Ellis Esq. F. R. S. King's Agent for the Province of West Florida. This Draught is Humbly Inscribed by his Most obliged & obedient humble Servt. James Cook [cartouche, top left]

Size: 50½ × 20¾. *Scale:* 1" = ca. 8 miles.
Insets: 1. A Plan of Pensacola Harbour, with the Marks for going in. *Size:* 10⅞ × 9⅜. *Scale:* 1" = 1½ miles.
2. A Draught of Spirito Sancto and Coast adjacent. *Size:* 15⅞ × 8¾. *Scale:* 1" = 6 miles.

Copies

Clements (Clinton Collection, No. 342).

---

358. Gauld-Pittman [1767] MS

A Sketch of the Entrance from the Sea to Apalatcy. and Part of the Environs. taken by George Gauld Esqr. Surveyor of the Coast. and Lieutenant Philip Pittman Assist^{t}. Engineer [upper right] | To General Haldimand [lower right]

Size: 17¾ × 34. *Scale:* 1" = 1 mile.
Description: This is a survey of the paths or roads from Apalachee Bay by Fort Saint Mark to "Talahassa or Tonabys Town" and farther to "Mikisuki or Newtown, Indian Village." Robert R. Rea provides an excellent description of the contents of this survey in his revision of John D. Ware's *George Gauld: Surveyor and Cartographer of the Gulf Coast*, Gainesville and Tampa, Florida, 1982, pp. 87–91.

Original

Clements (Gage).

---

359. Duckenfield 1767 MS

A Plan of a Tract of Land belonging to Esqr. Dukenfield, Lying between Salmon Creek & Kashy River . . . Resurveyed 22^{d}. Augt 1767. W: Churton. [top left] | A Copy of the Courses of Esqr. Duckenfield's Lattest Pattents . . . [bottom right]

Size: 13⅞ × 16⅞. *Scale:* 1" = 30 chains.
Description: This MS survey of the lands between Salmon Creek and Kashy River, just north of the mouth of the Roanoke River on Albemarle Sound, North Carolina, is found in a memorial filed June 4, 1789, by the loyalist Sir Nathaniel Duckenfield, in his claim against the British government for compensation for lands forfeited and sold by North Carolina. Duckenfield's 6,842 acres were sold for £31,445 (*State Gazette of North Carolina*, I [January 12, 1786], No. x), and Duckenfield's eventual compensation was £3,000. The property involved was bequeathed to William Duckenfield by his kinsman John Ardern according to a will filed on October 22, 1707. Duckenfield had owned the property as early as 1697 and sold it to Ardern in 1702; see *Colonial Records of North Carolina*, XX, pp. 170–72; I, p. 595.

A survey of the same region in the same year as Duckenfield's is in the possession of Dr. William R. Capehart of Bertie County: "A Tracing of Survey of Certain Lands Lying Between Salmon Creek and Cashie River made by W. Churton the 22^{d}. Augt. 1767." It is 17½ × 11½ inches and drawn at a scale of 1" = 1,980 feet. The Southern History Collection of the University of North Carolina holds a photocopy of this survey.

The interest in Duckenfield's plan lies in its being a survey by William Churton, whose surveys and maps of the Granville District were incorporated in Fry-Jefferson 1751, Collet 1770, and Mouzon 1775; in evidence on the map and the accompanying memorial of continued ownership and use of this region throughout the eighteenth century, where "Batts House" is shown on the Comberford 1657 MS map;

and in the shipyard delineated on the survey on the south bank of Salmon Creek. This is probably the earliest definitely located shipyard in the state; recently members of the Bertie County Historical Association have found in the vicinity the large iron cauldron used in pitching the vessels.

"Duckinfield" is given as the landowner south of Salmon Creek in Collet's 1768 MS copy of Churton's lost map of North Carolina; "Dickenson," probably an engraver's error, is found there on Collet 1770; on Mouzon 1775 no name appears.

References

Crittenden, C. C. "Ships and Shipping in North Carolina, 1763–1789." *North Carolina Historical Review*, VIII (January 1931), 7, note 16.

Reproductions

North Carolina Department of Archives and History.

Original

London, Public Record Office. Audit Office 13. Bundle 118.

---

360. Collet 1767 MS

Plan of Johnston Fort at Cape Fear With the project of one Covert way with places of Arms. Survey'd and designed by Cap[n] Collet in d[bre] 1767 [followed by table of Explanation; to left of plan of fort, within neat line]

Size: 28 × 22¾. *Scale:* 1" = 12 feet.

Description: This is a colored plan of Fort Johnston at Cape Fear, which guarded the approaches to Wilmington. This manuscript map, from the papers of the earl of Shelburne, was made by the then recently appointed young Swiss captain John A. Collet, who redesigned, rebuilt, and was in command of Fort Johnston. See Introductory Essay.

Original

Clements.

---

361. Thornton-Fisher 1767

A New Mapp of Carolina By John Thornton at y[e] Platt in y[e] Minories And by Will: Fisher at y[e] Postorn Gale [*sic*, for Gate] on Tower hill in London

Size: 19¾ × 16¼. *Scale:* 1" = 18 miles.

In: The English Pilot. Dublin, 1767, Fourth Book, No. 17.

Inset: A Large Draught of Ashly and Coopers River.

Size: 7¾ × 5½. *Scale:* 1" = 3 miles.

Description: This map is like the previous maps published in *The English Pilot* with a similar title; the Dublin map is, however, from a different plate, with a difference in size and other slight dissimilarities. See Thornton-Fisher 1698.

John Thornton's Carolina map is included in *The sea-atlas: containing an hydrographical description of most of the sea-coasts of the known parts of the world*, London, ca. 1702–7, in the New York Public Library's collection. Because of its rarity and beauty this atlas was reproduced in microfiche form with an index and background notes by Ashley Baynton-Williams by Micro Color International, Inc., in Midland Park, New Jersey. It is distributed under the title *The Samuel Thornton Sea Atlas*.

Copies

Congress 1767; New York Public.

---

362. Collet 1768 MS

To His most excellent Majesty George III[d] by the Grace of God King of Great Britain, France, and Ireland, Defender of the Faith &c. This Accurate Map of the back Country of North Carolina is humbly dedicated and presented by his Majesty's most humble and most obedient Servant and Subject. I. A. Collet Captain and Commander of Fort Johnston. 1768. [cartouche, bottom right; a cartouche at bottom left, blank]

Size: 92 × 53. *Scale:* 1" = ca. 8 miles.

Description: This large MS draft extends from the coast to the Cherokee boundary line of 1767, slightly west of "Quacker Meadows" on the north side of the Catawba River near the present location of Morganton. This is the map, in three parts made of twelve sheets pasted together, and largely copied by Collet and his assistant from the original work of William Churton, which Collet took with him to England in December 1768.

One of the parts, south of the Granville line and extending from New Bern and Bogue Inlet westward to the Cherokee line, does not fit well with the other parts of the map. Most of this third part (68½ × 26) beomes sparse in detail west of the coastal settlements; this is the only area for which Collet added any number of significant details in his printed map of 1770.

See Collet 1770 and Mouzon 1775. For a fuller discussion

of the background of this map and its importance, see the Introductory Essay.

Reproduction

Congress (photocopy).

Original

British Library. King George III's Topographical Collection. cxxii. 52. 2 Table.

---

363. Sauthier 1768 MS

Plan of the Town of Hillsborough in Orange County North Carolina [Reference list omitted] Survey'd & Drawn in October 1768 by C, J, Sauthier.

Size: 20 × 17. *Scale:* 1" = 372 feet.

Description: Between 1768 and 1771, Sauthier surveyed in the several towns and locations in North Carolina listed in the following maps. Some of these maps are in the British Library, others are in the Clinton Collection of the William L. Clements Library.

See Introductory Essay.

Reproductions

Congress (photocopy).

Johnston, F. B., and T. T. Waterman. *The Early Colonial Architecture of North Carolina.* Raleigh, 1952, p. 234.

Original

British Library. King George III's Topographical Collection. cxxii. 59.

---

364. de la Puente 1768 MS

Nueva descripcion dela Costa Oriental y Septentrional, delas Provincias dela Florida corrida por la primera, Desde la Barra de Sta Elena, acuios limetes, y en Virtud delos Tratados de Pas del Año de 1670 [title on attached sheet of explanatory remarks]

Size: 18½ × 14⅝. *Scale:* 1" = ca. 50 miles (varies).

Description: This map of Florida, extending from St. Elena (in South Carolina) to the Mississippi, has an appended text to the right of the map listing 120 places with corresponding numbers on the map and giving latitude and longitude of places on the coast. Also on the additional page of text is "Dĕdicato Al Exmo S^{or} Fr. d^{n} Antonio Bucarely y Vraua."

Two colored maps of the Florida peninsula that are also found in the Geographical Department of the Ministry of War, Madrid, are signed by Elixio de la Puente and dated 1768, and are reproduced in *Cartografia de Ultramar*, II, Plates 53, 54 (with transcriptions of place-names and legends, pp. 296–302). One of these, Servicio Geografico del Ejército, 8a. 1a. a, 12, is very similar to the map described above; it differs chiefly in the names being written on the map itself and in a long "Breve Descripc^{n}" to the left of the map. A photocopy of this map is also in the Karpinski series of reproductions.

Reproductions

Congress (Karpinski series of reproductions).

Original

Madrid. Servicio Geografico del Ejército. 9a. 2a. a, 14.

---

365. Whitefield 1768

Vernon River [on face of map, extreme right]

Size: 12⅞ × 6¾. *Scale:* 1" =½ mile.

In: Whitefield, George. *A Letter to Governor Wright relative to converting the Georgia Orphan-House into a College.* London, 1768, opp. p. 1.

Description: This map shows the land on Vernon River, a Georgia saltwater river, and the lands extending northward for about six miles. In the same volume are two plates showing the existing and intended buildings on the site of the Orphan's Home; they are found in a volume containing letters from Whitefield to Governor Wright and to and from the archbishop of Canterbury. Whitefield's proposal is to have a college in Georgia to equal those founded by the archbishop of Canterbury in New York and Philadelphia.

See an earlier reference to the Georgia Orphan-House at Bethesda under Seale 1741.

References

Catalogue of the Wymberley Jones De Renne Georgia Library. Wormsloe, Georgia, 1931, I, 189–90.

Copies

Georgia Hargrett Library.

---

366. Jefferys 1768

Florida from the Latest Authorities. By T. Jefferys, Geographer to His Majesty.

Size: 14 × 15⅛. *Scale:* 1" = 68 miles.

In: Jefferys, T. *A General Topography of North America and the West Indies.* London, 1768, No. 65. Same as Jefferys 1763 (Map 334).

Description: The map extends from 23° to 36° N.L. along the Atlantic coast and westward to the Mississippi.

Copies

Congress; Boston Public; Clements.

---

367. Jefferys-Mackay 1768

A survey of the Coast about Cape Lookout in North Carolina, taken the 29th of June 1756. This Draught is most Humbly Presented to His Excellency Arthur Dobbs Esqr. His Majesties Captain General, Governor & Commander in Chief in & over the Province of North Carolina, & Vice Admiral of the Same, By His Excellencys Most Obedient & most Devoted Humble Servant Arthur Mackay.

Size: 10¾ × 13¾. *Scale:* 1" = .8 mile.

In: Jefferys, T. *A General Topography of North America and the West Indies.* London, 1768, No. 58.

Description: This is a well-drawn plan of Cape Lookout and the adjacent coast for about ten miles along the shore on either side of the cape. Concerning the bay formed by the fishhook-shaped end of the cape, a legend reads: "The Spanish Privateers kept a Rendezvous in this Bay the latter end of last War." This note refers to the Spanish marauders from St. Augustine in 1747 (cf. *Colonial Records of North Carolina*, IV, p. xi). Arthur Mckay was deputy to the surveyor general of North Carolina in 1763 (*CRNC*, VI, 1011).

Copies

Congress, separate; Boston Public; Huntington; North Carolina Department of Archives and History.

---

368. Hyrne-Jefferys 1768

A New and Exact Plan of Cape Fear River from the Bar to Brunswick, by Edward Hyrne, 1749. [cartouche, top left] | Printed for Robt. Sayer in Fleet Street and Thos. Jefferys the Corner of St. Martins Lane in the Strand. [bottom left, above neat line]

Size: 12½ × 15¼. *Scale:* 1" = 1⅔ miles.

In: Jefferys, T. *A General Topography of North America and the West Indies.* London, 1768, No. 63.

Description: This is the same map, though with a different imprint, as that published in 1753. See Hyrne 1749 MS and Hyrne-Jefferys 1753. In the Library of Congress (Collection of American Maps, No. 20) is a chart, more detailed, with changes in the contour of the cape and with soundings which are lacking in the chart listed here, having the title: "Riviere du Cap Fear de la Bare a Brunswick Traduit de l'Anglais A Paris chez le Rouge des Grands Augustins, 1778." *Size:* 12⅛ × 18⅞. *Scale:* 1" = 1⅓ miles. This map is probably based on "A Plan of Cape Fear River from the Bar to Brunswick ["Remarks"] London, Printed, for R. Sayer & J. Bennett, Map & Chart sellers No. 53 in Fleet Street. Published as the Act directs 1st July 1776." *Size:* 14 × 20½. *Scale:* 1" = 1¼ miles.

Reproduction

Color Plate 22.

Copies

Congress, Duke.

In: North American Pilot, London, 1778, II, Map 8.

Copies

Congress 1778.

368A. Indian Boundary Line 1768 MS

[Board of Trade map of Indian Boundary Line]

Size: 17 × 21. *Scale:* 1" = ca. 87 miles.

Description: This colored manuscript map of eastern North America was prepared to illustrate the Board of Trade's "A Report to the Crown on the Management of Indian Affairs in America. . . ." Throughout the period following the promulgation of the Royal Proclamation of October 7, 1763, the Board of Trade had been striving to formulate a policy to deal with the Indian land and trade problem in the colonies; this report was the product of that labor. Recognizing that the Proclamation Line had been only an expedient that had satisfied neither Indians nor frontier settlers, the lords of trade and plantations ordered that the Indian Boundary as described in their report—and shown on this map—should be "ratified and confirmed in every part and the colonies required to enact the most effectual laws for preventing all settlement beyond such line." See L. De Vorsey, *The Indian Boundary in the Southern Colonies, 1763–1775*, pp. 62–64.

Reproduction

Cumming, W. P., and H. F. Rankin. *The Fate of a Nation.* London, 1975, p. 12.

Original

London. Public Record Office. MPG 280.

---

369. Savery 1769 MS A

To his Excellency, Iames Wright, Esquire Governor and Commander in chief of the Province of Georgia This Sketch of the Boundary Line, as it is now mark'd, between the Afforesaid Province and the Creek Indian Nation Is most humbly Dedicated by his Excellencys Most obedient Servt. Saml. Savery D: Surveyor A Scale of Twenty one English Miles, Four in one Inch.

In this sketch the Boundary Line is distinguished by a deep red Line, all swamps are shaded and Coloured Black, High Lands fit for Planting are bounded by [one inch is torn off; about two words] and Pine or Barren Lands of any kind by Yellow. ["January" partially torn off] 1769. [bottom left]

Size: 55 × 14. *Scale:* 1" = 4 miles.

Description: This is a colored map showing the Georgia-Indian boundary from Williams Creek, southward to St. Marys River on the Florida Boundary; the "deep red Line" crosses the Great Ogechee and Altamaha Rivers. Only the terrain immediately adjacent to the line is shown. This map is similar to the next item, in the William L. Clements Library Clinton Collection MSS, except that the Clinton map is dated later and is certified by Mackay and McGillivray.

Samuel Savery, described as a "proper geometer," had difficulty in receiving compensation for the arduous survey that produced this map. His memorial requesting compensation details many of the hardships that boundary surveyors endured. Among these he mentioned the loss of two horses "killed with fatigue," as well as "the great Injury of his Health" resulting from three months spent surveying the 260-mile line while "exposed to all the inclemencies" of a Georgia summer. See *Colonial Records of the State of Georgia*, XV, p. 402. L. De Vorsey, *The Indian Boundary in the Southern Colonies, 1763–1775*, pp. 149–58, includes a review of the events surrounding Savery's survey as well as a cartographic analysis of the maps based thereon.

Reproductions

Hulbert, A. B. *Crown Collection of American Maps.* Cleveland, 1914–16, Series III, Nos. 144–47.

Congress (photocopy).

Original

London. Public Record Office. C.O. 700, Georgia no. 14.

---

370. Savery 1769 MS B

To Lachlan Mcgillivray Esqr. Deputy Superintendant. This sketch of the Boundary Line between the Province of Georgia and the Creek Nation is address'd by His Most obedient Servt. Saml. Savery, D.S. A Scale [of] Miles, Four in One Inch, 13 March 1769. [bottom left]

Size: 54¾ × 13½. *Scale:* 1" = 4 miles.

Description: See the preceding map. In the lower right corner of the Clinton Collection map is a certification of the accuracy of the map, dated April 11, 1769, and signed by John Mackay and Lachlan McGillivray.

Reproduction

Congress (photocopy).

Original

Clements (Clinton Collection, No. 330).

370A. Savery-Romans 1769 MS C

A true Copy taken from the Original done by Samuel Savory Depty Survr. for Lands in Georgia Certifed 31st March 1769. Pr. Bernard Romans Depy. to Survr. Genl. for the Southern District of America [lower right]

Size: 60 × 14. *Scale:* 1" = 4 miles.

Description: This copy of Savery's survey of the Georgia–Creek Indian boundary was forwarded to London by John Stuart, the superintendent for Indian affairs in the Southern Department, as an enclosure to his letter of April 14, 1769. It is very similar to the two Savery maps described immediately above.

Reference

Maps and Plans in the Public Record Office, London, 1974, No. 2374, p. 407.

Original

London. Public Record Office. MPG 337.

371. Sauthier-Bath 1769 MS

Plan of the Town and Port of Bath in Beaufort County North Carolina. Survey'd and Drawn in May 1769. By C, J, Sauthier.

Size: 21 × 17. *Scale:* 1" = 276 feet.

Description: Another "Plan of the Town of Bath" was copied from deputy surveyor John Forbes's draft, dated February 28, 1766, by James R. Hoyle. Hoyle's copy, dated August 23, 1807, is found in the John Gray Blount Papers, N.C. Department of Archives and History, No. P.C. 877. It is reproduced in H. Pascal, *History of Colonial Bath*, Raleigh, N.C., 1955.

Reproduction

Congress (photocopy).

Original

British Library. King George III's Topographical Collection. cxxii. 53.

372. Sauthier-Brunswick 1769 MS

Plan of the Town and Port of Brunswick in Brunswick County. North Carolina [Reference list omitted] Survey'd & Drawn in April 1769 By C, J, Sauthier

Size: 21 × 17. *Scale:* 1" = 260 feet.

Reproduction

Congress (photocopy).

Original

British Library. King George III's Topographical Collection. cxxii. 55.

373. Sauthier-Edenton 1769 MS A

A Plan of the Town & Port of Edenton in Chowan County North Carolina [Reference list omitted] Survey'd & Drawn in June 1769. By C, J, Sauthier.

Size: 20 × 17. *Scale:* 1" = 372 feet.

Reproductions

Congress (photocopy).

Johnston, F. B., and T. T. Waterman. *The Early Colonial Architecture of North Carolina.* Raleigh, 1952, p. 48.

Original

British Library. King George III's Topographical Collection. cxxii. 57.

374. Sauthier-Edenton 1769 MS B

A Plan of the Town & Port of Edenton in Chowan County. North Carolina [Reference list omitted] Survey'd and Drawn in 1769 by C, J, Sauthier. [cartouche, top left]

Size: 20⅞ × 15½. *Scale:* 1" = 380 feet.

Reproduction

North Carolina Department of Archives and History (photocopy).

Original

Clements (Clinton Collection, No. 292).

375. Sauthier-Halifax 1769 MS

Plan of the Town of Halifax in Halifax County, North Carolina. [Reference list omitted] Survey'd & Drawn in June 1769. By C, J, Sauthier.

Size: 20 × 17. *Scale:* 1" = 270 feet.

Reproductions

Congress (photocopy).

North Carolina Department of Archives and History (photocopy).

Johnston, F. B., and T. T. Waterman. *The Early Colonial Architecture of North Carolina.* Raleigh, 1952, p. 112.

Original

British Library. King George III's Topographical Collection. cxxii. 58.

376. Sauthier–New Bern 1769 MS A

Plan of the Town of Newbern in Craven County North Carolina. [Reference list omitted] Survey'd & Drawn in May 1769. By C, J, Sauthier.

Size: 21 × 17. *Scale:* 1" = 390 feet.

Description: In the plan is shown the governor's palace, commonly known as Tryon Palace.

See Hawks ca. 1766 MS.

Reproductions

Henderson, A. *North Carolina.* Chicago, 1941, I, 204.

Johnston, F. B., and T. T. Waterman. *The Early Colonial Architecture of North Carolina.* Raleigh, 1952, p. 72.

New Bern: Cradle of North Carolina. Raleigh, The Garden Club of North Carolina, 1941, pp. 8–9.

Congress (photocopy).

Original

British Library. King George III's Topographical Collection. cxxii. 60.

---

377. Sauthier–New Bern 1769 MS B

A Plan of the Town of Newbern in Craven County, North Carolina. [Reference list omitted] Survey'd & Drawn in 1769 by C, J, Sauthier [cartouche, top right]

Size: 20¾ × 15¾. *Scale:* 1" = 390 feet.
Description: See Sauthier–New Bern 1769 MS A.

Reproductions

Congress (photocopy).

North Carolina Department of Archives and History (photocopy).

Original

Clements (Clinton Collection 294).

---

378. Sauthier-Wilmington 1769 MS

Plan of the town of Willmington in New Hanover County North Carolina [Reference list omitted] Surveyed and Drawn in December 1769 by C, J, Sauthier. | 214 [below neat line, to right]

Size: 20½ × 16⅝. *Scale:* 1" = ca. 270 feet.

Description: This is a well-drawn plan of Wilmington and its vicinity. For the first map of Wilmington, see Wilmington ca. 1755 MS.

Reproductions

Johnston, F. B., and T. T. Waterman. *The Early Colonial Architecture of North Carolina.* Raleigh, 1952, p. 142.

Sprunt, James. *Chronicles of the Cape Fear River, 1660–1916.* Raleigh, 1916, between pp. 46–47.

North Carolina Department of Archives and History (photocopy).

Original

British Library. King George III's Topographical Collection. cxxii. 62.

---

379. Jefferys 1769

East Florida, from Surveys made since the last Peace, adapted to Dr. Stork's History of that Country. By Thomas Jefferys, Geographer to the King. [bottom left] | T Jefferys Sculp [below neat line, right]

Size: 13⅝ × 16½. *Scale:* 1" = 38 miles.
In: [Stork, William]. *A Description of East-Florida, with a Journal, kept by John Bartram of Philadelphia.* London, 1769, front.

Description: This map extends from the point of Florida northward to the latitude of "Savanah," and westward to Pensacola. It does not include as large an area as Jefferys's map published six years before in Roberts's *An Account of the First Discovery and Natural History of Florida*, 1763, but shows marked improvement in geographical knowledge. The lower peninsula is presented as a solid body of land, rather than as a kind of archipelago; the size of Tampa Bay is smaller and corrected; the interior geography is improved by the knowledge gained through Bartram's journey up the St. Johns River.

The map is not found in the earlier issues of the work published in 1766; it appears first in "Third Edition" of 1769, and in the "Fourth Edition" of 1774. An earlier state of the 1769 edition map has no scale (as in one Library of Congress copy), but the scale is found on maps subsequently issued.

Copies

Congress 1769, three copies; John Carter Brown 1769; Georgia Hargrett Library 1769; Harvard 1769, 1774; Huntington 1769; New York Public 1769, 1774; Yale 1769; Clements 1769.

---

380. Brunswick County 177– MS

[Brunswick County]

Size: 12½ × 7⅞. *Scale:* None.

Description: A rough pen-and-ink sketch of creeks and roads in Brunswick County, North Carolina. It is possibly of the Revolutionary period.

Original
Clements (Clinton MS. 288).

---

381. North Carolina Roads 177– MS
A Plan of part of the principal Roads in the province of N. Carolina. The Bridges in This Plan are at, Tillis's Mill, Monro's on Drowning Creek, Little River, Coles, Cross Creek, Rock Fish, Newberry's, Ben Willis's, Peter Lord's, Colonel Waddell's, Hammond's Creek, Bertram's Mill, Newfields, Bladen County Line, Levingston's, The principal of these are, Levingston, the County Line, Hammond's Creek, Peter Lord's, Rock Fish, Little River, and Cole's at Drowning Creek. The Ferry's are Wilmington, Eagan's, Cheraw Town, Coulston's, Deep River, and Cain Creek. All the rest of the Rivers and Creeks are Fordable.

Size: 21½ × 29¾. *Scale:* 1" = 10 miles.
Description: The roads in North Carolina from Wilmington west and northwest are given with the number of miles from Wilmington as far as Charlotsburg (224 miles) and "The Blue Ridge" (300 miles) above "The Pilot Mountain." The date of the map is uncertain; it may have been made during the Revolution.

Original
Clements (Clinton Collection, No. 290).

---

382. Eagle's Island 177– MS
Eagle's Island [across map face]

Size: 15 × 12. *Scale:* None.
Description: This is a rough pen-and-ink sketch of Eagle's Island, with the roads and houses in the vicinity, in Brunswick County, North Carolina. It is possibly of the Revolutionary period.

Original
Clements (Clinton MS. 287).

---

383. St. Simon's ca. 1770 MS A
[Plan and profile of the forts and redoubts in St. Simon's and Jekyll Islands.]

Size: 48 × 20. *Scale:* None.
Description: The sketches on this map are unfinished.

Original
British Library. King George III's Topographical Collection. cxxii. 71. b, c, and d.

---

384. St. Simon's ca. 1770 MS B
Plan du Fort proposé dans l Isle de S.t Simon Pour la deffence de l Entree du Havre de Jenkins Sownd

Size: 16 × 12. *Scale:* 1" = 100 toises (640 feet).

Reproduction
Congress (photocopy).
Original
British Library. King George III's Topographical Collection. cxxii. 72.

---

385. Albemarle Sound Region post-1770 MS A
[Sound Region of North Carolina]

Size: 27¾ × 31. *Scale:* 1" = ca. 4⅓ miles.
Description: This map extends from "Great Dismal Swamp" to "Neus R." Only Edenton and Bath are shown, with no roads; but it is an accurate map of the coastline. Neither this nor the next map, both in the Clinton Collection, are dated. They may have been made during the Revolutionary War period.

Reproduction
Congress (photocopy).
Copies
Clements (Clinton Collection, No. 283).

---

386. Albemarle Sound Region post-1770 MS B
[Parts of the modern counties of Currituck, Camden, and Pasquotank in North Carolina]

Size: 14½ × 17. *Scale:* 1" = 2 miles.
Description: A minute and careful survey of the land adjacent to and between Little River, Pasquotank, and North River off Albemarle Sound. The names and locations of the landowners are given.

Reproduction
Congress (photocopy).
Original
Clements (Clinton Collection, No. 293).

---

387. South Carolina–Georgia Coast ca. 1770 MS A
[Coast from St. Augustine to Charleston]

Size: 55 × 20. *Scale:* None.

Description: This is a colored map, with minute detail, of the coastal region. The cartouche is blank. The bays, rivers, islands, and inlets seem to be carefully surveyed, but no soundings are given. There is reason to believe that this map may have been compiled much earlier than the "about 1770" date assigned by the British Library.

Reproductions
Hulbert, A. B. *Crown Collection.* Cleveland, 1908, Series I, Vol. V, Maps 32–33.
Congress (photocopy).
Original
British Library. King George III's Topographical Collection. cxxii. 65.

---

388. South Carolina–Georgia Coast ca. 1770 MS B
[Coastal area from Charles Town to St. Augustine.]

Size: 12 × 15½. *Scale:* None.

Description: This is a colored map of the coast of South Carolina, Georgia, and Florida, from Charles Town southward to St. Augustine.

Reproduction
Congress (photocopy).
Original
British Library. King George III's Topographical Collection. cxxii. 66.

---

389. Gaillard-Cook 1770 MS
A Map of South Carolina from an Actual Survey to the Hon[ora]ble Peter Manigault, Esq. Speaker and the Hon[ora]ble the Commons House of Assembly This Draught is most Humbly Inscribed, by Their Obedient Humble Servants, Tacitus Gaillard Iames Cook Feb.ʸ 1770

Size: 63 × 93. *Scale:* 1" = ca. 2 miles.
Insets: Plan of Camden. *Size:* 6¾ × 11. *Scale:* 1" = ca. 11 chains.
Plan of Georgetown. *Size:* 11 × 6¾. *Scale:* 1" = 700 feet.

Description: A manuscript, in color, of the eastern half only of South Carolina. The western half is missing; the condition of the manuscript is poor. It is very similar to but not quite as detailed as the Lodge-Cook 1771 MS, though differences, such as the omission of "Grayston" (on Parris Island) and "Horse I.," are so small as to indicate a very close relationship.

This is apparently part of the original manuscript presented to the Commons House of Assembly of South Carolina on March 21, 1770. Tacitus Gaillard and James Cook were appointed to survey the province, and a sum of £18,000 currency or about £3,000 sterling was allocated for that purpose: see *Historical Commission of South Carolina, Journal of the Commons House of Assembly*, Vol. 38, p. 276.

The MS was given to the South Carolina Historical Society in July 1920 by the Misses Gibbes. Aware of the map's great value, the Society's board of managers in 1973 decided to undertake the cost of its conservation by the Gaylord Donnelly Company of Chicago. The large map now hangs on one of the few wall spaces in the Historical Society's building large enough to accommodate its size.

See Lodge-Cook 1771 MS and Cook 1773.
See Introductory Essay.

Original
South Carolina Historical Society.

---

390. Sauthier-Beaufort 1770 MS
Plan of the Town & Port of Beaufort in Carteret County North Carolina [Reference list omitted] Survey'd and Drawn in August 1770 by C, J, Sauthier

Size: 21 × 17. *Scale:* 1" = 300 feet.

Reproduction
Congress (photocopy).
Original
British Library. King George III's Topographical Collection. cxxii. 54.

391. Sauthier–Cross Creek 1770 MS
Plan of the Town of Cross Creek in Cumberland County. North Carolina. [Reference list omitted] Survey'd & Drawn in March 1770. By C, J, Sauthier.

Size: 20 × 17. *Scale:* 1" = 186 feet.
Description: This is a survey of Fayetteville, North Carolina.

Reproduction
Congress (photocopy).
Original
British Library. King George III's Topographical Collection. cxxii. 56.

392. Sauthier-Salisbury 1770 MS
Plan of the Town of Salisbury in Rowan County North Carolina [Reference list omitted] Survey'd and Drawn in March 1770 By C, J, Sauthier.

Size: 20 × 17. *Scale:* 1" = 62 fathoms (372 feet).

Reproductions
Plate 62.
Congress (photocopy).
Johnston, F. B., and T. T. Waterman. *The Early Colonial Architecture of North Carolina.* Raleigh, 1952, p. 220.
Original
British Library. King George III's Topographical Collection. cxxii. 61.

393. Fuller 1770
To the Right Honourable John Earl of Egmont, &c. This Plate is most humbly Inscribed by his Lordship's most Obedient Humble Servant Will.m Fuller. [cartouche, center] | Published 26 March 1770 according to Act of Parliament by Thomas Jefferys Geographer to the King, in the Strand. [below neat line of map]

Size: 23¾ × 19⅞. *Scale:* Various: see below.
In: Faden, W. *The North American Atlas.* London, 1777, Vol. B, No. 26.
Insets: This plate is a sheet containing three separate maps, as follows:

1. Plan of Amelia Island in East Florida North Point of Amelia Island lyes in 30:55 North Latitude 80:23 W. Longitude from London Taken from De Brahm's Map of South Carolina & Georgia. *Size:* 6¼ × 19¾. *Scale:* 1" = 1 mile.

2. A Chart of the Entrance into St Mary's River taken by Captn W. Fuller in November 1769. *Size:* 17⅜ × 6. *Scale:* 1" =½ mile. This map has a pictorial inset in the upper right corner with the title: A View of the Entrance into St Mary's River.

3. A Chart of the Mouth of Nassau River with the Bar and the Soundings on it taken at Low Water by Captn W. Fuller. *Size:* 17⅜ × 13¾. *Scale:* 1" = ca. ⅓ mile.

Description: The Henry E. Huntington Library has a later edition of this chart, issued for the use of the French navy and transports during the Revolution. It is approximately the same size, with the title and legends in French: "Fuller dédia dette Carte au Comte D'Egmont, La fit publier par Gefferys à Londres en 1770. Traduite de l'Anglais, à Paris. Chez Le Rouge, rue des Gds Augustins, en 1778." Fuller's manuscript draft for Inset 2, "Chart and View of the entrance. Lines of bearing and leading marks with soundings, place names," of St. Marys River is filed as map MPD 173 in the British Public Record Office. See P. A. Penfold, ed., *Maps and Plans in the Public Record Office 2. America and West Indies*, London, 1974, No. 2391, p. 410.

Copies
Congress 1777; American Geographical Society; Georgia Hargrett Library; Harvard; Huntington; Clements (separate).
In: Atlas of the Battles of the American Revolution. [New York, 1845?], No. 29 (a collection of eighteenth-century maps bound together, with no imprint).
Copies
Congress 1845.

394. Collet 1770
A Compleat Map of North-Carolina from an actual Survey. By Captn Collet, Governor of Fort Johnston. Engraved by I. Bayly. [across top of map, above neat line] | To His Most Excellent Majesty George the IIId King of Great Britain, &c, &c, &c, This Map is most humbly dedicated by His Majesty's most humble obedient & dutiful Subject John Collet [cartouche, bottom right] | Publish'd according to Act of

Parliament, May the 1st 1770, by S. Hooper. No. 25 Ludgate Hill, London. [below neat line, center]

Size: 43½ × 28⅝. *Scale:* 1" = ca. 13¾ miles.

Description: This map includes all of North Carolina westward to "Table [Rock] Mountain" in the Blue Ridge, near the present site of Morganton, North Carolina. It is one of the major maps in the history of North Carolina because it greatly surpasses in accuracy and scope any previous map of the region, it records the great western movement of population up to and across the Piedmont during the middle of the eighteenth century, it is the original, even when it is not the immediate, basis for most subsequent maps of North Carolina from its appearance until the Price-Strother map of 1808, and Bayly, the engraver, executed his work with skill and artistry.

The Granville District is based upon surveys and information gathered by the Granville District surveyor, William Churton. Many of the soundings along the coast probably were also made by Churton, who died in December 1767, while making surveys of the maritime and southern part of the province: see *Colonial Records of North Carolina*, VII, pp. 861–62. West of the Catawba River, the details, which were based on fewer surveys, become increasingly scanty and less accurate. See Collet 1768 MS and Mouzon 1775.

See Introductory Essay.

Reproductions

Plates 62A, 63–66.

Cumming, W. P. *North Carolina in Maps*. Raleigh, 1966, Plate VII.

Waynick, C. *North Carolina Roads and Their Builders*. Raleigh, 1952, opp. p. 246 (large folded sheet).

Copies

Congress, two copies; John Carter Brown; Clements (Gage; Clinton 284); Archibald Craige Collection, Winston-Salem, N.C.; Harvard; Kendall; North Carolina Department of Archives and History; North Carolina State Library; University of North Carolina; Wachovia Historical Society, Winston-Salem, N.C.

394A. Calhoun 1770 MS

Map of Lands Near the Salud River Showing Reservations made for Cherokee half-breeds: With Surveyor's Certificate

Size: 18½ × 15. *Scale:* 1" = 3 miles.

Description: Edward Wilkinson was an influential Indian trader who had served as South Carolina's boundary commissioner when John Pickins surveyed the boundary between that colony and the land of Cherokee Indians in 1766 (see Map 351). After the Cherokees had become deeply indebted to him, Wilkinson persuaded the Indians to surrender a large area of their hunting grounds in return for the cancellation of their £8,000 debt. Understanding the legal barriers that forbade the direct transfer of Indian land, Wilkinson sought to have the Crown accept the cession with the provision that he should receive his £8,000 from the proceeds of the land when it was sold or be allowed its free use for ten years. Neither scheme, however, was deemed acceptable by royal authorities. It was not long before other traders followed Wilkinson's lead in attempting to acquire Indian land west of the Indian Boundary Line.

In addition to the 158,236 acres Wilkinson was hoping to gain control of, surveyor Patrick Calhoun's map shows two large rectangular tracts identified as: "Land reserved by the Cherokees for One of their Boys begotten by a White Man Richard Pearis" and "Land reserved for an Indian Boy begotten by a White Man Alexander Cameron." Calhoun's map is also valuable in that it shows the boundary surveyed between the Cherokees and North Carolina from the Reedy River to Mt. Tryon in 1767. No original map of this survey is known to have survived to the present day.

Here should also be mentioned two undated manuscript maps in the William L. Clements Library's Clinton Collection. On the reverse of the first is found the following: "Mr. Wilkinsons Plat of ye: Cherokee Country." Symbolized by a dotted line, the "Road to Keowee" is shown extending and branching through the Cherokee country, where a number of villages and rivers are named. The North Carolina–South Carolina boundary is shown terminating at "Tryon Mountain," the northern end of the "Indian Boundary" in South Carolina. Both Fort Rutledge and Fort Keowee are identified. See Christian Brun, *Guide to the Manuscript Maps in the William L. Clements Library*, Ann Arbor, Michigan, 1959, No. 603, pp. 143–44. In the same *Guide* Map 617 (p. 147) is listed a manuscript survey showing the area of the modern counties of Oconee, Pickens, Anderson, and Greenville, South Carolina. It is also undated and found in the Clinton Papers. An inscription on the verso, which has been partially lost through trimming, suggests that it too should be associated with Patrick Calhoun's surveys in the Cherokee country.

Reference

De Vorsey, L. *The Indian Boundary in the Southern Colonies, 1763–1775*. Chapel Hill, N.C., 1966, pp. 105, 161–62.

Reproduction

De Vorsey, L. *The Indian Boundary in the Southern Colonies, 1763–1775*. Chapel Hill, N.C., 1966, p. 104 (tracing).

Original

London. Public Record Office. MPG 338.

---

395. Sauthier-Alamance 1771 MS A

A Plan of the Camp of Alamance from the 14th to the 19th of May 1771. Composed of the Provincials of North Carolina, Comanded by His Excellency Governor Tryon. On an Expedition against Rebels who styled themselves Regulators. Surveyed and Drawn by C, J, Sauthier. geo:r

Size: 11 13/16 × 14 1/2. *Scale:* None.

Description: This is a well-executed, detailed plan of the region in the vicinity of the battle of Alamance, showing the disposition and movement of the forces from May 14 to May 19. It covers a wider area than the three plans which follow and which show only the battleground. This plan and the next are found in a letter from Governor Tryon, dated August 2, 1771.

The size is taken from the photocopy listed below.

Reproductions

North Carolina Department of Archives and History (photocopy and a tracing in Hall of History).

Original

London. Public Record Office. C.O. 5/314 f. 187.

---

396. Sauthier-Alamance 1771 MS B

A Plan of the Camp and Battle of Alamance, the 16th of May 1771. Between the Provincials of North Carolina Commanded by His Excellency Governor Tryon. and the Rebels who styled themselves Regulators. Survey'd and Drawn by C, J, Sauthier

Size: 22 × 13. *Scale:* 1" = ca. 2.1 miles.

Description: This is a carefully drawn map, similar to the plan by Sauthier in the Clements Library, but differing in minor details. It accompanies the letter from Governor Tryon dated August 2, 1771.

Reproduction

North Carolina Department of Archives and History (photocopy).

Original

London. Public Record Office. C.O. 5/314 f. 188.

---

397. Sauthier-Alamance 1771 MS C

Plan of the Camp and Battle of Alamance the 16 May 1771. Between the Provincials of N:th Carolina. Commanded By His Excellency Governor Tryon. and Rebels who styled themselves Regulators. Surveyed and Drawn by C, J, Sauthier

Size: 22 1/2 × 13. *Scale:* 1" = ca. 2.1 miles.

Description: A finished, colored, topographical map of the Salisbury-Hillsborough Road where it intersects the Great Alamance River.

Reproduction

Congress (photocopy).

Original

Clements (Clinton Collection, No. 295).

---

398. Sauthier-Alamance 1771 MS D

A Plan of the Camp and Battle of Alamance, the 16th May 1771. Between the Provincials of North Carolina, Commanded By His Excellency Governor Tryon. And Rebels who styled themselves Regulators [Reference list omitted] Survey'd and Drawn by C, J, Sauthier

Size: 22 1/4 × 12 1/4. *Scale:* 1" = ca. 1/3 mile.

Description: This map is found pasted in at the end of Governor Tryon's Order Book entitled "Journal of the Expedition against the Insurgents in the Western Frontiers of North Carolina begun 20th April 1771." It is similar to the Clements Library copy with the same title, except that it has more explanatory legends written on the face of the map, the reference list is in the cartouche at the bottom left instead of in a separate box at the top right, and the legend at the bottom center, "NB From the Camp to the field of Battle is about 5 Miles," differs from the Clements copy legend, "N'B. The Distance from the Camp to the field of Battle is 4 Mile." Neither map has a scale; the map which accompanies Tryon's letter, Sauthier-Alamance 1771 MS B, has a scale but no note. This map, unlike the B and C copies, has under

"Reference" the legend "O. Ditto [Enemy's] Flight," and shows on the map the position of the Regulators' flight.

Reproductions

Powell, William S. *The War of the Regulation and the Battle of Alamance, May 16, 1771*. Raleigh, N.C., 1949, end.

Original

North Carolina State Department of Archives and History.

---

399. Lodge-Cook 1771

A Map of South Carolina. With all the Islands, Marshes, Swamps, Bays, Rivers, Creeks, Inland Navigations. And all the Countys, Districts, Towns, Roads, County, Parish, and Provincial Lines; From an Actual Survey. [cartouche, bottom left] | J. Lodge sulp[t]. N[o]. 45 Shoe Lane. [bottom left, below cartouche]

Size: Six parts, each 17½ × 23½. *Scale:* 1" = 4⅔ miles.
Insets: 1. A Plan of George Town [top left of large map] *Size:* 5 × 6¼. *Scale:* 1" = 1,000 feet.

2. A Plan of Carles Town [left center; with "references"] *Size:* 10½ × 13½. *Scale:* 1" = 580 feet.

3. A Chart of the Bar and Harbour of Charles Town [bottom right] *Size:* 14¾ × 14¾. *Scale:* 1" = 1 mile.

4. A Plan of Beaufort, on Port Royal Island [top right] *Size:* 7¾ × 6⅜. *Scale:* 1" = 900 feet.

Description: This is a beautifully executed and detailed nine-sheet printed map of South Carolina from the entrance of Savannah River to North Carolina at 34°50' N.L., and westward to the Cherokee land. Neither the authorship nor date of the map is given. The date "[1771]" is written beneath the neat line to the left in a later hand. Internal evidence shows that this map is later than the Gaillard-Cook 1770 MS, the original draft submitted to the Commons House of Assembly in February–March 1770, and that it is earlier than the Cook 1773, which shows the territorial changes of 1722. It is so close to both in innumerable details that it must have a common authorship. It is closer to Cook 1773 than to Gaillard-Cook 1770 MS; therefore Tacitus Gaillard's name is omitted from the title tag given above. B. R. Carroll, *Historical Collections of South Carolina*, New York, 1836, Vol. I, frontispiece, has "A Map of North & South Carolina. Accurately copied from the old maps of James Cook. Published in 1771, and of Henry Mouzon in 1775." An article clarifying and distinguishing the three contemporary James Cooks who were associated with a variety of cartographic enterprises and often confused is in *The Map Collector*, No. 34 (March 1986). See Jeannette D. Black with R. A. Skelton, "Too Many Cooks."

John Lodge, who signed the map as engraver, engraved maps for *The Political Magazine* in 1780 and for William Russell, *The History of America*, London, 1800. He also engraved Mouzon's "A Map of the Parish of St. Stephen," ca. 1773, and Adair's "Map of the American Indian Nations," 1775.

See Gaillard-Cook 1770 MS and Cook 1773.

See Introductory Essay.

Reproduction

Congress (photocopy).

Original

British Library. K. 122. 63. 1 Tab. (P. 3266). [King Geo. III's Top. Coll.]

---

400. Boss-Brailsford ca. 1775 MS

A Map of South Carolina from the Savannah Sound to S[t]. Helena's Sound, With the several Plantations, their proper Boundary lines, their Names, and the Names of the Proprietors included And the Grants of Lands belonging to Landgrave William Hodgson. coloured Green and edged with Red with additional Plans of those Plots which have been resurvey'd by Order of Mess[rs]. Boss and Brailsford the Purchasers from Hodgson, who inherited from his Ancestors, as they had their Original Grant from the Palatinate Lord Carteret &c

Size: 26½ × 31. *Scale:* 1" = 2¾ miles.

Inset: [List of Lands, Proprietors, etc., lower left corner]

Description: This map is based on a portion of De Brahm's, "A Map of South Carolina and a Part of Georgia," which was published in 1757 (Map 310). It shows the South Carolina coastal zone from the Savannah River to just beyond the South Edisto River with several plantations indicated by small house symbols. In addition to Hodgson's, many landholdings are outlined in red, and their owner's names are given in a lengthy table on the lower left. In the *Report of the Librarian of Congress* for 1914, p. 79, the following description outlines the map's provenance: "This map presented by Frank Morton Jones was forwarded as a donation to the Library by the descendents of William Hodgson, nephew and heir of 'William Hodgson of the six Clerk's office Middlesex, Esquire, Landgrave and Cassique of the Province of South Carolina,' who married Anne, sister of William Lord Craven and who held title to extensive grants in the Province of Carolina. The map is an original on rice paper, and it gives the roads, churches, houses, and names of their occupants. The map is undated, but other papers indicate that the grants were made about the year 1715, and that the map was probably drawn, in part from former surveys, about 1771. The map has been handed down in the Hodgson family, who did not come into possession of the property or profit by its sale, their agents, Boss & Brailsford, mentioned on the map, failing to make any return for the portions sold. The map should have considerable historical interest."

In earlier editions of this work the Boss-Brailsford map was dated 1771 in accordance with the Library of Congress's conclusion. Thanks to the research on the map carried out by South Carolina's former state archivist, Charles E. Lee, that date is shown to be "1774, or even more likely the year 1775, rather than 1771." Lee's unpublished typescript "'Hodgson's Barony'—Tentative Notes on the Boss-Brailsford Map of Lower South Carolina in the 1770's," is in the Subject File (I, c. 2-1) of the South Carolina Department of Archives and History in Columbia. It is an invaluable source that should be consulted by anyone interested in the Boss-Brailsford map and the matters it deals with. Lee goes so far as to speculate that the raising of Hodgson's old land claims in the mid-1770s might have figured in turning South Carolina against the Crown. He correctly concludes that "it is a subject that needs investigation." Two copies of the map, one a photostat in four sheets from the Library of Congress, and the other a blueprint copy with the notation "copied September 1908," are included as II MC 9-6 & 9-7 in the South Carolina Archives files.

Original
Congress.

---

401. Stuart 1771 MS

A Sketch of the Cherakee Boundaries with the Province of Virginia &c. 1771 [cartouche, top center] | N°. 1. In M^r. Stuart's (N°. er) of 24th Septr. 1771.

Size: 16½ × 14. *Scale:* 1" = 30 miles.

Description: This unsigned MS map, probably by Stuart, accompanies the letter of John Stuart, Superintendent of Indian Affairs in the Southern Department, to the earl of Hillsborough. An extract from the letter, which is dated from Pensacola on September 24, 1771, follows: "I begg Pardon for not having complyed with my Promise in my Letter of 30th July 1769, of furnishing your Lordship with the boundary Lines marked on Accurate Map by some good hand; which I hope you will be pleased to believe not to have proceeded from inattention to your Lordship's Commands; but from the impossibility of performing it with such a Degree of Accuracy as To Convey a just Idea of our Boundaries, upon any of the printed Maps that I have seen, in all of which the Natural Boundaries specified in the different Treaties are either erroneously laid down or entirely left out: and there is no possibility of forming a precise Idea of the Extent of the different Cessions made by the Indians, untill such Natural Boundaries, are accuratly [*sic* in transcript] explored and | [p. 670] properly laid down; as The Lines particularly behind this Province are determined by the Courses and confluences of Brooks and Rivers, with which we are not by any means well acquainted; I have sent a proper Person to ascertain their Latitudes distances and Situations: and I hope to accomplish surveying and Marking the Indian Boundaries before I leave West Florida, which with the Materials that I have collected will enable me to make a good Map of this Country: and as there will be actual Surveys of the Lines behind Virginia North and South Carolinas & Georgia accomplished before my return to Charles Town, I flatter myself with the Expectation of having it in my power to lay before your Lordship a Map of my Department which may be depended upon." [P.R.O., C.O. 5. 72/669–70; Library of Congress Transcript.]

The map has "Boundary Line settled with N: Carolina," "Boundary Line with Virginia as settled in 1768," and in red dots from the old boundaries westward along the Virginia–North Carolina line to within a few miles of Long Island on the "Hulston R.," then northeast in a straight line to the juncture of the "Great Conhaway River" with the Ohio River, "The Line settled in 1770." Thence the red dots run along the Ohio to the Little Conhaway River, with the legend "Boundary with the Western Indians." This map showed a very significant shift to the interior of the Indian Boundary that had been mandated by the British Board of Trade in 1768 and illustrated on Map 368A. It was, however, a boundary destined to be shifted even farther to the west when it was actually demarcated by John Donelson, Alexander Cameron, and Cherokee chief Attakullakulla. See L. De Vorsey, "The Virginia-Cherokee Boundary of 1771," *The East Tennessee Historical Society's Publications*, 33, 1961, pp. 17–31.

The branches of the Tennessee River, with some of the Cherokee towns on the upper reaches of "Tannassee River," are given; but the map is in general sparse in detail. It extends westward to the Mississippi River.

Reference

De Vorsey, L. "The Colonial Southeast on an Accurate General Map." *The Southeastern Geographer*, VI, 1966, pp. 20–32.

Reproductions

Congress (Transcripts, MSS. Division).

De Vorsey, L. *The Indian Boundary in the Southern Colonies, 1763–1775*. Chapel Hill, N.C., 1966, p. 69 (tracing).

Original

London. Public Record Office. MPG 348.

401A. Donelson 1771

The Contents in acres Ceded to the Crown By the Treaty of Lochaber in Oct. 1770 is Estimated to 10,000,000 of acres Lying Between the River Louisa Ohio and the Great Kanaway. Sy Jno Donelson. [near Ohio River between "Sandey Cr." and "Louisa" river]

Size: 46¾ × 56½. *Scale:* 1" = 5 miles.

Description: This ink-on-parchment manuscript shows the area of Virginia between the Blue Ridge of the eastern Appalachians and the Ohio River. While not a professional surveyor or cartographer, Donelson produced a valuable map of this little-known region and the crucial Indian Boundary Line blazed through it by a party of Virginians and Cherokees. As discussed above, John Stuart was responsible for establishing the Indian Boundary Line, marking the limits of territory which legally could be acquired for Euroamerican settlement in the colonial Southeast. Virginia had traditionally maintained a claim to the western lands, and many in the Old Dominion resisted the Crown's attempts to establish a boundary which might negate that claim and become a barrier to frontier expansion. In October 1770, in response to expansionist pressure, negotiations were held with about a thousand Cherokees at Lochaber, the plantation of one of Stuart's deputies (the "Lochaber" mentioned in Donselson's statement cum map title). At Lochaber the Indians agreed to permit a westward shift in the boundary. When Virginian John Donelson and the Cherokee chief known as Attakullakulla led a party to survey the boundary in May of 1771, they agreed to mark a line to the headstream of the Kentucky River and down that stream to the Ohio River far to the west of the termination ratified in the formal treaty signed earlier at Lochaber. On Donelson's map the boundary stream is designated "Louisa," an early name used for the modern Kentucky River.

The interpretations of frontier historian Clarence W. Alvord, published in his *The Mississippi Valley in British Politics*, II, Cleveland, 1917, were reached without an analysis of Donelson's map, and as a result are badly flawed. In those interpretations, John Stuart is cast as a willing agent of Virginia expansionists and land speculators, a view which was not corrected until J. R. Alden published his well-researched monograph, *John Stuart and the Southern Colonial Frontier*, Ann Arbor, 1944. For a discussion of the Donelson Map and the events surrounding the survey it depicts see L. De Vorsey, "The Virginia-Cherokee Boundary of 1771," *The East Tennessee Historical Society's Publications*, 33, 1961, pp. 17–31.

Reference

Penfold, P. A., ed. *Maps and Plans in the Public Record Office*, 2. London, 1974, No. 2403, p. 413.

Reproduction

De Vorsey, L. *The Indian Boundary in the Southern Colonies, 1763–1775*, Chapel Hill, 1966, Figure 9, p. 80 (tracing).

Original

London. Public Record Office. C.O. 700, Virginia no. 19.

---

402. Jekyll Sound 1771

Plan du Port de Gouadaquini now called Jekil Sound in the Province of Georgia in North America. Latitude 31°:13' North. [cartouche, top right]

Size: 14⅝ × 10⅝. *Scale:* 1" = 5¾ miles.

In: Speer, J. S. *The West India Pilot*. London, 1771, No. 22.

Description: "Jekil Island called Pallavona in 1721" is on Jekyll Island. On "Island St. Simon" is Fredericksburgh, near "Stocade Fort" and "Fort de Ladrillo," with a road connecting the settlements on the island.

See Map 247.

Reproductions

Georgia Historical Society, *Collections*, VII, Part III (1913), opp. p. 66.

Copies

Congress 1771; Georgia Hargrett Library; John Carter Brown.

403. Speer 1771

A Plan of the Entrance into Cape Fear Harbour North Carolina [cartouche, rectangle, upper left] | To face page 53. [top right, below neat line]

Size: 14½ × 10. *Scale:* 1" = 1 mile; shoal scale is 1" = 1¼ mile.

In: Speer, J. S. *The West India Pilot.* London, 1771, No. 13, opp. p. 53.

Description: The following note accompanies a plan of Frying Pan Shoal: "The Frying Pan Shoal runs off due N & S from Cape Fear, & is reduced to a small Scale, & placed E & W for the conveniency of bringing it into the Plate."

Copies

Congress; Yale; Clements.

404. Mouzon 1772 [1801] MS

A Map of part of the Counties of Mecklenburg and Tryon Lately added to the Province of South Carolina By H. Mouzon Junr. [across top below neat line] | Surveyed agreeable to His Majesty's Royal Instructions the 4th day of June 1772 Under direction of William Moultrie William Thomson Commissioners by James Cook Ephraim Mitchell Surveyors. [bottom left]

Size: 21½ × 16¾. *Scale:* 1" = 4 miles.

Description: A carefully executed pen-and-ink copy of the original drawing of the North Carolina–South Carolina boundary line survey of 1772. Formerly in the office of the secretary of state of South Carolina, this map has been removed to the archives of the South Carolina Historical Commission.

The map is endorsed as follows: Map of the New Acquisition taken off the counties of Mecklenburgh & Tryon 1772 the uper [*sic*] line is the same as that in C 4 June 1772.

The certification of this copy is also endorsed: Whitehall to wit. I, the undersigned, Chief Clerk of the Right Honourable the Lords of the Most Honourable Privy Council, appointed for the Consideration of all matters relating to Trade and Foreign Plantation, hereby certify to all to whom these Presents shall come, that the within Map is an accurate Copy of the original, which was transmitted in Lord Charles Greville Montagu's Letter to the secretary of State, dated the 27th of July, 1772, and which now remains as of Record at Whitehall aforesaid: In Testimony whereof, I have caused Their Lordships' Seal of Office to be affixed hereto, at Whitehall, this 28th Day of April, 1801

[seal] Geo. Chalmers

For an account of this survey by one of the commissioners, see Charles S. Davis, ed., "The Journal of William Moultrie While a Commissioner on the North and South Carolina Boundary Survey, 1772," *Journal of Southern History*, VIII (November 1942), 549–55. See also A. S. Salley, *The Boundary Line between North Carolina and South Carolina, Bulletins of the Historical Commission of South Carolina No. 10*, Columbia, S.C., 1929, pp. 28–30.

Reference

Penfold, P. A., ed. *Maps and Plans in the Public Record Office 2. America and West Indies.* London, 1974, No. 2747, pp. 474–75.

Originals

London. Public Record Office. MPG 524. South Carolina Archives Department, Columbia, S.C.

405. De Brahm [1773] MS A.1

The Special Survey of the Inlet, Haven, City, and Environs of Charlestown

Size: 9 × 14. *Scale:* 1" =⅔ mile.

Description: This detailed chart of Charleston harbor is the first of a collection of maps prepared by William Gerard De Brahm to illustrate his opus magnum, the "Report of the General Survey in the Southern District of North America." De Brahm (1718–99) was a German military engineer who had emigrated to Georgia in 1751 and quickly distinguished himself as a surveyor-cartographer and architect of fortifications and public works. His widely acclaimed published general map is described above (see Map 310 and Introductory Essay). In 1764 De Brahm received a royal appointment to the newly created office of Surveyor General of the Southern District in North America. His counterpart in the north was Samuel Holland. In the words of his official charge, De Brahm was to oversee and carry out the detailed survey and mapping of "All His Majesty's territories on the Continent of North America, which lye to the south of the Potomac River, and of a line drawn due west from the Head of the main branch of that River as far as His Majesty's Dominions extend." As might be expected, however, De Brahm was instructed to place the highest priority on the surveying and mapping of the new British colony of East

Florida—a virtual terra incognita to George III and his advisers. In keeping with the limited space available in this volume only those maps in De Brahm's "Report" depicting places in South Carolina and Georgia can be listed here. For his maps of East Florida and the Atlantic Ocean consult L. De Vorsey, ed., *De Brahm's Report of the General Survey in the Southern District of North America*, Columbia, South Carolina, 1971. In addition to an extended biographical sketch on De Brahm, this edited book includes all the maps and most of the text found in the British Library's two bound manuscript volumes in the collection of manuscripts which belonged to the library of King George III. That collection was presented by George IV in 1823 and is now known as "King's Manuscripts." In 1772 and 1773 De Brahm presented his "Report" personally to George III while retaining a second somewhat less elegant personal copy. In 1798 De Brahm delivered his personal copy of the "Report," along with an extended manuscript entitled "The Continuation of the Atlantic Pilot," to Phineas Bond, then British Consul General in Philadelphia. In the course of time these manuscripts found their way to London, where they were acquired by Henry Stevens, who sold them to Harvard University in 1848. Now in Harvard's Houghton Library, these manuscripts provided material for two nineteenth-century publication efforts. The Harvard De Brahm chapter describing Georgia was privately printed in 1849 in a limited edition of forty-nine copies by George Wymberley-Jones of Wormsloe, Georgia, under the title *History of the Province of Georgia: With maps of the Original Surveys*. In 1856 Plowden C. J. Weston included De Brahm's South Carolina chapter without maps in his *Documents Connected with the History of South Carolina*, printed for private distribution in London. Because of their provenance and the slight differences in many of the maps, the Harvard and British Library versions of De Brahm's "Report" maps are listed separately here.

For some unexplained reason volume III of the *Catalogue of the Manuscript Maps, Charts, and Plans, and of the Topographical Drawings in the British Museum*, London, 1861, describes De Brahm's "Report" maps in King's MSS 210 and 211 as being "colored," when in fact they are uncolored, ink on paper, drawings: see pages 510–16 in that catalogue for descriptions.

Reproductions

Congress (photocopy).

De Vorsey, L. *De Brahm's Report of the General Survey.* Columbia, S.C., 1971, facing p. 90.

Original

British Library. King's MSS 210, 1.

---

406. De Brahm [1773] MS A.2

Plan and Profile of Fort Johnston

Size: 9 × 14. *Scale:* 1" = 300 feet.

Description: This is a plan of a fort near Charles Town designed to protect against a seaborne attack. A lengthy "Explanation" at the foot of the map explains many of the details shown.

Reproductions

Congress (photocopy).

De Vorsey, L. *De Brahm's Report of the General Survey.* Columbia, S.C., 1971, facing p. 91.

Original

British Library. King's MSS 210, 2.

---

407. De Brahm [1773] MS A.3

[Plan of the fortified canal from Ashley to Cooper River]

Size: 14 × 9. *Scale:* 1" = 900 feet.

Description: This was De Brahm's proposal for a fortified entrenchment to protect Charleston from a land-mounted attack. The city occupied the tip of a peninsula, or neck, formed by the Ashley and Cooper Rivers.

Reproductions

Congress (photocopy).

De Vorsey, L. *De Brahm's Report of the General Survey.* Columbia, S.C., 1971, facing p. 90.

Original

British Library. King's MSS 210, 3.

---

408. De Brahm [1773] MS A.4

Plan of the City and Fortification of CHARLESTOWN

Size: 14½ × 8¾. *Scale:* 1" = 900 feet.

Description: While at a much smaller scale than the large colored plan he prepared in 1755 (Map 290), this is an excellent depiction of Charleston's city plan with the system of walls and bastions De Brahm designed for the city's defense. An "EXPLANATION" at the foot of the map describes and identifies many of the structures shown.

Reproductions

Congress (photocopy).

De Vorsey, L. *De Brahm's Report of the General Survey.* Columbia, S.C., 1971, facing p. 98.

Original

British Library. King's MSS 210, 4.

---

409. De Brahm [1773] MS A.5

Plan of the Environs in the Neck of Tanasee and Talequo Rivers about Fort Loudoun and little Tamothly the westernmost of the Upper Cherokee Towns.

Above is the Representation of the 700 Acres of Land which the Cherokee Nation anno 1755 has ceded to the King, for the use of a Fort, bounding N.E.wardly upon Tanasee River, W. and S.W.wardly upon Telequo River; & S.E.wardly on the Indian Land, distinguished by an artificial Line drawn from Tanasee across the North to Talequo River. Surveyed in the year as above pursuant to an order from His Excellency James Glen Esq.r governor in Chief over His Majesty's Province of South Carolina By William Gerard de Brahm Surveyor General [bottom of map below scale bar]

Size: 9 × 14. *Scale:* 1" = 20 chains.

Description: This is a very carefully drawn survey of the topography around the site of ill-fated Fort Loudoun. De Brahm was commissioned to design and oversee its construction (see Map 308). In addition to depicting accurately the rivers and terrain, De Brahm included the layout of the Cherokee town of "Little Tamothly" with its earthen "Town house," as well as the site of "Taskigee old Town this place the Indians proposed for a Fort."

Reproductions

Plate 66A.

De Vorsey, L. *De Brahm's Report of the General Survey,* facing p. 99.

Original

British Library. King's MSS 210, 5.

---

410. De Brahm [1773] MS A.6

Plan and Profiles of Fort Loudoun upon Tanasee River

Size: 9 × 14. *Scale:* 1" = 150 feet.

Description: An artistically drawn plan view of the fort with profiles to the right of the sheet and below the main plan; the explanations of alphabetical symbols A to K are at the bottom of the sheet.

Reproductions

Plate 66B.

Congress (photocopy).

De Vorsey, L. *De Brahm's Report of the General Survey,* facing p. 105.

Original

British Library. King's MSS 210, 6.

---

411. De Brahm [1773] MS A.7

The Saltzburger's Settlement in Georgia

Size: 9 × 14. *Scale:* 1" = 80 chains.

Description: This is a detailed map of the landholdings and names of the German-speaking settlers around the town of New Ebenezer on the Savannah River.

Reproduction

De Vorsey, L. *De Brahm's Report of the General Survey,* facing p. 142.

Original

British Library. King's MSS 210, 7.

---

412. De Brahm [1773] MS A.8

Plan of the Town Ebenezer and its Fort

Size: 9 × 14. *Scale:* 1" = 200 feet.

Description: This map shows the plan of the town that Oglethorpe permitted Georgia's German-speaking "Salzburger's" to establish on the Savannah River after they complained of the remoteness and sterility of their first settlement site near the headwaters of Ebenezer Creek. Its pattern of straight streets and open squares flanked by trust lots is reminiscent of that of Savannah's but with the north-south streets separating tythings omitted. "William de Brahm Esq.r his Land" extends prominently on both sides of the "Savannah Stream," above the town. See Urlsperger-Seutter 1747 (Map 265).

Reproductions

Congress (photocopy).

De Vorsey, L. *DeBrahm's Report of the General Survey*, facing p. 143.
Jones, Charles C. *Dead Towns of Georgia*. Savannah, Morning News Steam Printing House, 1878.
Jones, Charles C. *History of the Province of Georgia, 1849*. Wormsloe Quartos No. 2.

Original
British Library. King's MSS 210, 8.

---

413. De Brahm [1773] MS A.9
The Bethanian Settlement in Georgia

Size: 9 × 14. *Scale:* 1" = 80 chains.

Description: This map is a detailed survey of the tracts of land upstream of Ebenezer, along the Savannah River and its branches. De Brahm's 300 acres opposite Ebenezer are shown.

Reproductions
Congress (photocopy).
De Vorsey, L. *DeBrahm's Report of the General Survey*, facing p. 142.

Original
British Library. King's MSS 210, 9.

---

414. De Brahm [1773] MS A.10
Plan of the City of Savannah and Fortifications

Size: 9 × 14. *Scale:* 1" = 400 feet.

Description: Below the plan is a profile view of the city, ". . . upon a N.15.E: Line, shewing the Streets, Houses, Bay, Wharfs and Fortification."

Reproductions
Congress (photocopy).
De Vorsey, L. *DeBrahm's Report of the General Survey*, facing p. 155.

Original
British Library. King's MSS 210, 10.

---

415. De Brahm [1773] MS A.11
Chart of the Savannah Sound

Size: 9 × 14. *Scale:* 1" = 40 chains.

Description: This chartlet provides an important historical glimpse of the extremely dynamic conditions prevailing at the navigation entrance to the Savannah River adjacent to Tybee and Cockspur Islands. At the foot of his map, De Brahm drew a "Profile of the Lands North & South of Savannah Sound as it appears due West." It is interesting to note that wide areas he indicated on his map as "shoal, dry at low Water," are now submerged at all tidal states.

Reproductions
Hulbert, A. B. *Crown Collection*, Series II, Vol. I, No. 34.
Congress (photocopy).
De Vorsey, L. *DeBrahm's Report of the General Survey*, facing p. 158.

Original
British Library. King's MSS 210, 11.

---

416. De Brahm [1773] MS A.12
Plan and Profile of Fort George, on Coxpur Island.

Size: 9 × 14. *Scale:* 1" = 50 feet.

Description: In addition to the plan view of the fort and its immediate surroundings, De Brahm drew an artistic "Profile Upon a North and South Line," of this small defensive blockhouse. It should be compared with the anonymous painting described with Map 339. An "EXPLANATION" below the map and profile indicates that Fort George was built with a "wooden Tower Bastion" surrounded by a "redoubt built of Earth and faced with Cabage Trees."

Reproductions
Congress (photocopy).
De Vorsey, L. *DeBrahm's Report of the General Survey*, facing p. 159.

Original
British Library. King's MSS 210, 12.

---

417. De Brahm [1773] MS A.13
The Environs of Fort Barrington

Size: 9 × 14. *Scale:* 1" = 8 chains.

Description: Fort Barrington was located on the Altamaha River about twelve miles upstream from Darien, Georgia. It was built as a response to a formal request Darien's residents made in 1751 for a defensive garrison to guard their land route to Savannah. This map shows the fort's site on a bluff just upstream of the ford where the crossing of trails identified as the "Path from Augustin" [St. Augustine] and "Path to Savannah" was located. The fort was named in honor of Josiah Barrington, a military colleague of Georgia's founder, James Edward Oglethorpe.

Reference

"Georgia Forts: Fort Barrington," *Georgia Magazine*, March 1971, pp. 14–20.

Reproductions

Hulbert, A. B. *Crown Collection*, Series II, Vol. I, No. 35.

Congress (photocopy).

De Vorsey, L. *DeBrahm's Report of the General Survey*, facing p. 158.

Original

British Library. King's MSS 210, 13.

---

418. De Brahm [1773] MS A.14

View and Plan of Fort Barrington consisting of a wooden Tower Bastionee, and four wooden Caponieres, all built of Renching Timber

Size: 9 × 14. *Scale:* 1" = 8 chains.

Description: This is a carefully drawn profile and plan view of an elaborate timber fortification. In an "EXPLANATION" at the bottom De Brahm provided information on its various parts.

Reproductions

Congress (photocopy).

De Vorsey, L. *DeBrahm's Report of the General Survey*, facing p. 159.

Original

British Library. King's MSS 210, 14.

---

419. De Brahm [1773] MS B.1

A special Survey of the Inlet, Harbour, City, and Environs of Charles-Town.

Size: 14¼ × 9¼. *Scale:* 1" = ca. 2 miles.

In: De Brahm, William Gerard, "History of the three Provinces South Carolina Georgia and East Florida," [1773], f.26.

Description: This De Brahm map and the several that follow are now in Harvard University's Houghton Library, where they form parts of another manuscript version of his "Report of the General Survey in the Northern Department of North America." See L. De Vorsey, ed., *De Brahm's Report of the General Survey in the Northern Department of North America*, Columbia, S.C., 1971, pp. 273–74. Except for a covering letter attached to that manuscript and addressed "to the High Commissioners of the Treasury" and a few other inscriptions, the writing appears to be scrivener's copperplate and is very similar to the script found in the British Library's version. A comment found in De Brahm's letter of transmittal supports the view that Harvard's version of the "Report" and its maps formed De Brahm's personal or file copy he retained until 1798, when he delivered it with other materials to Phineas Bond, British Consul to the United States. There are a number of differences to be noted between the maps in the British Library and Harvard versions of De Brahm's "Report." Found only at Harvard is De Brahm's later manuscript, "The Continuation of the Atlantic Pilot," with its important map showing the surface circulation system of the North Atlantic Ocean. See L. De Vorsey, "William De Brahm's 'Continuation of the Atlantic Pilot,' An Empirically Supported Eighteenth-Century Model of North Atlantic Surface Circulation," M. Sears and D. Merriman, eds., *Oceanography: The Past*, New York, Heidelberg, and Berlin, 1980, pp. 718–33.

For additional discussion of De Brahm and his cartography, consult the Introductory Essay, Map 310, and Map 405.

Original

Harvard.

---

420. De Brahm [1773] MS B.2

Plan and Profile of Fort Johnston

Size: 14¼ × 9¼. *Scale:* 1" = 300 feet.

In: De Brahm, William Gerard, "History of the three Provinces . . . ," [1773], f.28.

Inset: (Profile and Explanation at bottom of folio.)

Description: See De Brahm [1773] MS A.1 and De Brahm [1773] MS B.1.

Reproductions

Adams, J. T., ed. *Album of American History*. New York, 1944, I, 200.

Original

Harvard.

421. De Brahm [1773] MS B.3

Plan of the fortified Canal from Ashley to Cooper River

Size: 14¼ × 9¼. *Scale:* 1" = 900 feet.
In: De Brahm, William Gerard, "History of the three Provinces . . . ," [1773], f.30.
Description: See De Brahm [1773] MS A.1 and De Brahm [1773] MS A.2.

Original
Harvard.

422. De Brahm [1773] MS B.4

Plan of the City and Fortification of Charlestown

Size: 14¼ × 9¼. *Scale:* 1" = 900 feet.
In: De Brahm, William Gerard, "History of the three Provinces . . . ," [1773], f.40.
Description: An "Explanation" lists the chief houses in the city at that time.
See De Brahm [1773] MS A.1 and De Brahm [1773] MS B.1.

Reproductions
Adams, J. T., ed. *Album of American History.* New York, 1944, I, 200.
Original
Harvard.

423. De Brahm [1773] MS B.5

Plan of the Environs in the Neck of Tanassee & Talequo Rivers, about Fort Loudoun and little Tamothly, the westernmost of the upper Cherokee Towns.

Size: 14¼ × 9¼. *Scale:* 1" = 20 chains.
In: De Brahm, William Gerard, "History of the three Provinces . . . ," [1773], f.42.
Description: A note by De Brahm at the bottom states that this represents the 700 acres of land ceded to the king by the Cherokee nation in 1755, and surveyed that same year by De Brahm, pursuant to an order from Governor James Glen of South Carolina.
See De Brahm [1773] MS A.1 and De Brahm [1773] MS B.1.

Original
Harvard.

424. De Brahm [1773] MS B.6

Plan and Profiles of Fort Loudoun upon Tanassee River Projected and Constructed by William Gerard de Brahm.

Size: 14¼ × 9¼. *Scale:* 1" = 150 feet.
In: De Brahm, William Gerard, "History of the three Provinces . . . ," [1773], f.49.
Description: Profiles at bottom and at right side.
See De Brahm [1773] MS A.1 and De Brahm [1773] MS B.1.

Reproductions
Adams, J. T., ed. *Album of American History.* New York, 1944–49, I, 349.
Gabriel, R. H. *The Pageant of America*, II, 23.
Williams S. C. "De Brahm's Account (1756)" in his *Early Travels in the Tennessee Country, 1540–1800.* Johnson City, Tenn., 1928, opp. p. 187.
Original
Harvard.

425. De Brahm [1773] MS B.7

The Salzburger's Settlement in Georgia.

Size: 14¼ × 9¼. *Scale:* 1" = 80 chains.
In: De Brahm, William Gerard, "History of the three Provinces . . . ," [1773], f.109.
Description: A careful survey of land belonging to the Salzburgers on and near Ebenezer Island on the Savannah River.
See De Brahm [1773] MS A.1 and De Brahm [1773] MS B.1.

Original
Harvard.

426. De Brahm [1773] MS B.8

Plan of the Town Ebenezer and its Fort

Size: 14¼ × 9¼. *Scale:* 1" = 400 feet.
In: De Brahm, William Gerard, "History of the three Provinces . . . ," [1773], f.111.
Description: See De Brahm [1773] MS A.1 and De Brahm [1773] MS B.1.

Reproductions
Adams, J. T., ed. *Album of American History.* New York, 1944, I, 222.

De Brahm, William Gerard. *History of the Province of Georgia* [ed. George Wymberley Jones]. Wormsloe, 1849, opp. p. 24.
Georgia Historical Society. *Collections*, IV (1878), opp. p. 11.
Jones, C. C. *The History of Georgia*. Boston, 1833, I, opp. p. 212.

Original
Harvard.

---

427. De Brahm [1773] MS B.9
The Bethanian Settlement in Georgia.

Size: 14¼ × 9¼. *Scale:* 1" = 80 chains.
In: De Brahm, William Gerard, "History of the three Provinces . . . ," [1773], f.113.
Description: See De Brahm [1773] A.1 and De Brahm [1773] B.1.

Original
Harvard.

---

428. De Brahm [1773] MS B.10
Plan of the City Savannah and Fortification

Size: 14¼ × 9¼. *Scale:* 1" = 400 feet.
In: De Brahm, William Gerard, "History of the three Provinces . . . ," [1773], f.126.
Inset: Profile "upon a N. 15 E. line shewing Streets, Houses, Bay, Wharfs, and Fortification" [at bottom of sheet]
Description: See De Brahm [1773] MS A.1 and De Brahm [1773] MS B.1.

Reproduction
De Brahm, William Gerard [ed. George Wimberley Jones]. op. cit., opp. p. 36.

Original
Harvard.

---

429. De Brahm [1773] MS B.11
Chart of the Savannah Sound

Size: 14¼ × 9¼. *Scale:* 1" = 37½ chains.
In: De Brahm, William Gerard, "History of the three Provinces . . . ," [1773], f. 132.
Inset: Profile of the Lands North and South of Savannah Sound as it appears due West
Description: See De Brahm [1773] MS A.1 and De Brahm [1773] MS B.1.

Reproductions
De Brahm, William Gerard [ed. George Wimberley Jones]. op. cit., opp. p. 44.

Original
Harvard.

---

430. De Brahm [1773] MS B.12
Plan and Profile of Fort George on Coxpur Island

Size: 14¼ × 9¼. *Scale:* 1" = 50 feet.
In: De Brahm, William Gerard, "History of the three Provinces . . . ," [1773], f.134.
Inset: Profile upon a North and South Line.
Description: See De Brahm [1773] MS A.1 and De Brahm [1773] MS B.1.

Reproductions
De Brahm, William Gerard [ed. George Wimberley Jones]. op. cit., after p. 44.

Original
Harvard.

---

431. De Brahm [1773] MS B.13
The Environs of Fort Barrington

Size: 14¼ × 9¼. *Scale:* 1" = 8 chains.
In: De Brahm, William Gerard, "History of the three Provinces . . . ," [1773], f.136.
Description: See De Brahm [1773] MS A.1 and De Brahm [1773] MS B.1.

Reproductions
De Brahm, William Gerard [ed. George Wimberley Jones]. op. cit, after p. 44.

Original
Harvard.

432. De Brahm [1773] MS B.14
Plan and View of Fort Barrington, consisting of a wooden Tower Bastionee, and four wooden Caponieres, all built of Renching Timber.

Size: 14¼ × 9¼. *Scale:* 1" = 8 chains.
In: De Brahm, William Gerard, "History of the three Provinces . . . ," [1773], f.138.
Inset: A South view of the Fort and the ground plans are given.
Description: See De Brahm [1773] MS A.1 and De Brahm [1773] MS B.1.

Reproductions
Adams, J. T., ed. *Album of American History.* New York, 1944, I, 224.
De Brahm, William Gerard [ed. George Wimberley Jones]. op. cit., after p. 44.
Original
Harvard.

433. N.C.-S.C. Boundary 1772 MS A
A Plan of the Province Line from the Cherokee Line to Salisbury Road Between North and South Carolina Certified by Us this June 4th 1772.

Thomas Rutherfurd, Thomas Polk } Surveyors for North Carolina
James Cook, Ephraim Mitchell } Surveyors for South Carolina
John Rutherfurd, William Dry } Commissioners for North Carolina
William Moultrie, William Thomson } Commissioners for South Carolina

Size: 38 × 14½. *Scale:* 1" = 2½ miles.
Description: This map is one of the original copies of the survey, extending from below the "Catawba Nation" tract northwest and then westward to a corner nine and a half miles south of Tryon mountain. The survey begins at Lat. 34°48' on the Salisbury road. The Catawba River and its branches in the Catawba Nation tract are given, and the river is followed to Lat. 35°8', whence a straight westerly line is surveyed for 65 miles.

Included in Governor Martin's letter of November 5, 1773, this map appears to be the fair copy made from the original draft, Map 434.

In the Map Division of the Library of Congress is a manuscript form of this map, from the title evidently post-Revolutionary in date: "A Plan of that part of the Boundary between the States of North and South Carolina Lying between that part of said boundary marked in 1764 and the Old Cherokee boundary-line. Fixed and marked by Commissioners appointed by each respective Colony, now States, in June AD 1772." *Size:* 28¼ × 20½. *Scale:* 1" = 4 miles.

Reproduction
North Carolina Department of Archives and History (photocopy).
Reference
Penfold, P. A., ed. *Maps and Plans in the Public Record Office*, No. 2746, p. 474.
Original
London. Public Record Office. MPG 362.

434. N.C.-S.C. Boundary 1772 MS B
An Exact map of the Boundary Line between the Provinces of North & South Carolina Agreeable to the Royal Instruction Certified by us June 4th 1772 |

Thomas Rutherfurd, Thomas Polk } Surveyors for North Carolina
James Cook, Ephraim Mitchell } Surveyors for South Carolina
John Rutherfurd, William Dry } Commissioners for North Carolina
William Moultrie, William Thomson } Commissioners for South Carolina

Size: 36½ × 12½. *Scale:* 1" = 2½ miles.
Description: In his communication of April 20, 1773, Governor Martin apologized for the quality of this map, terming it to be "an Original Draft made upon the Spot."

Reference
Penfold, P. A., ed. *Maps and Plans in the Public Record Office*, No. 2744, p. 474.
Reproduction
North Carolina Department of Archives and History (photocopy).

Original

London. Public Record Office. MPG 282.

---

435. N.C.-S.C. Boundary 1772 MS C

A Plan of the Province Line from the Cherokee Line to Salisbury Road Between North and South Carolina Certified by Us this 4th June 1772.

Thomas Rutherfurd, Thomas Polk } Surveyors for North Carolina

James Cook, Ephraim Mitchell } Surveyors for South Carolina

John Rutherfurd, William Dry } Commissioners for North Carolina

William Moultrie, William Thomson } Commissioners for South Carolina

Size: 29 × 14½. *Scale:* 1" = 3½ miles.

Description: This map is based on Map 434 but is on a smaller scale. More carefully drawn than that map, it includes the latitudes observed by the survey party and gives the acreage of the Catawba Indians' tract.

Reference

Penfold, P. A., ed. *Maps and Plans in the Public Record Office*, No. 2745, p. 474.

Reproduction

North Carolina Department of Archives and History (photocopy).

Original

London. Public Record Office. MPG 283.

---

436. Clements 1772 MS

South & North Carolina. An Exact Map of the Boundary line Between the Provinces of South & North Carolina Agreeable to the Royal Instructions Certified by us this fourth day of June 1772.

South Carolina Secretarys Office. A true Copy taken from the Original Map and Examined per Thos Skottowe Secry.

William Moultrie, William Thomson } Commissioners for South Carolina

John Rutherfurd, William Dry } Commissioners for North Carolina

James Cook, Ephraim Mitchell } Surveyors for South Carolina

Thomas Rutherfurd, Thomas Polk } Surveyors for North Carolina

Size: 42 × 16¼. *Scale:* 1" = 2½ miles.

Description: A finished, pen-and-ink survey of the area from 34°48' W. to 35°8' W., including the valleys of the Packolet, Broad, and Catawba Rivers.

Reference

Adams, R. G. *British Headquarters Maps and Sketches*. Ann Arbor, Michigan, 1928, p. 95.

Reproduction

Congress (photocopy).

Original

Clements (Clinton Collection, No. 297).

---

437. Stuart-Georgia 1772 MS

"The Lands which the Cherokees have assign'd for payment of their debts," [1772]

Size: 37 × 40. *Scale:* 1" = ca. 5½ miles.

Description: This map was made under the direction of John Stuart and sent by him to the earl of Hillsborough, then secretary of state for the Colonies, accompanying Stuart's letter dated "Charles Town 13th June 1772." It shows "Ft Agusta," "Ft Moore," the situation of the lands suggested for cession by the Cherokee Indians in Georgia. A careful copy, hand-drawn from the original, is in the Library of Congress.

References

Library of Congress, Noteworthy Maps, No. 2. Accessions. 1926–27, p. 3, item 10.

Penfold, P. A., ed. *Maps and Plans in the Public Record Office*, No. 2375, p. 407.

Reproduction

Congress (photocopy).

Original

London. Public Record Office. MR 18.

438. Stuart-Purcell [1775] MS A

A MAP of the Southern Indian District of NORTH AMERICA Compiled under The Direction of John Stuart Esqr His MAJESTY's Superintendent of Indian Affairs. By Joseph Purcell [cartouche, lower left]

Size: 69 × 76 15/16. *Scale:* 1" = 16½ miles.

Description: This colored manuscript map was incorrectly dated 1773 in earlier editions of this work. It and Map 439 were compiled by Joseph Purcell in an effort to carry out John Stuart's longstanding promise to provide George III's advisers with a reliable general map of the Southeast. The copy now in the Public Record Office (Map 439), was forwarded to London by Stuart on February 10, 1776. Stuart, in his role as "His Majesty's Superintendent of Indian Affairs for the Southern District of North America," had responsibility for negotiating the Indian Boundary Line that had been mandated to separate the southern colonies from the lands of their Indian neighbors. See L. De Vorsey, *The Indian Boundary in the Southern Colonies, 1763–1775*, Chapel Hill, N.C., 1966.

For other maps associated with the Southern Indian Boundary Line, see Maps 336A, 341, 351, 352, 368A, 369, 370, 394A, 401, 437, 440, 440A, 440B, and 446.

Early in 1769 the secretary of state for the Southern Department, Lord Hillsborough, had expressed his desire that Stuart employ a skillful cartographer "to lay down upon some accurate general map of America . . . for His Majesty's information the several lines agreed upon and marked out, for the want of which it is difficult to distinguish with precision in what manner the several lines unite and the courses they follow." Needless to say, Stuart did not find this an easy order to obey given the absence of any printed maps that even began to approach the sort of topographic detail and accuracy required to portray correctly the many segments of the Southern Indian Boundary Line, located, as they were, deep in the unsurveyed interior. In 1773 Stuart forwarded to London a copy of Mitchell's map (see Map 293) with the Indian Boundary Line inked on it. Although far from satisfactory in Stuart's judgment, it met with approval in London, where even an imperfect general map was deemed better than the disconnected and confusing collection of original surveyor's sketches and Indian treaties that were at hand. Regrettably Stuart's annotated copy of the Mitchell map has not been identified. Stuart and his capable surveyor-cartographer Joseph Purcell did not, however, rest content but pushed ahead in their effort to compile "an accurate general map" of the Southeast based on the best and most current surveys and maps at their disposal.

In addition to showing much of the region's hydrography and topography with a completeness and accuracy not improved upon for several decades, the 1775 Stuart-Purcell map contains a wealth of first-hand information on the Southeastern Indians. Some 45 Creek, 57 Choctaw, 11 Chickasaw and 42 Cherokee towns are indicated and named in tables of "References" associated with each tribal grouping. Also shown are a number of green inked lines by which Stuart appears to have been attempting to define the limits of the "Lands Claimed" by each of these major Southeastern tribal groupings. He was probably overly optimistic in indicating that these tribal land divisions had been accepted by the Indians. In the last analysis, the most significant boundaries shown were the purpose for which the map was compiled. These were the several colonial segments making up the Southern Indian Boundary Line. As explained in a block entitled "General References," located in the upper left corner of the Ayer Collection version and the upper center of the Public Record Office copy of the 1775 Stuart-Purcell map, lines "Colour'd Blue or Red are the Boundaries between the Lands Ceded to His Majesty by the Different Indian Nations and the Lands belonging to the said Indians." The distinction defined by these two colors is further explained: "Thoes [*sic*] Blue are the Boundaries Laid out and Marked, and Thoes Red are the Boundaries agreed upon by Treaty, as yet not Laid out nor Marked." In addition to the several Indian boundaries, Purcell employed yellow ink to indicate "the Boundaries of the Different Provinces as laid out and by Charter," and brown ink to illustrate "the Bounds of the Lands Intended for the New Government on the Ohio," the abortive Vandalia colony scheme in the Virginia backcountry. Brown ink is also used to outline a large tract of land in present-day Alabama that had been "Ceded to His Majesty by the Chactaws but claimed by the Creeks." The region's churches, courthouses, forts, trails, bridges, and ferries are also indicated on the map.

See Stuart-Gage 1773 MS.

See Introductory Essay.

References

De Vorsey, L. "The Colonial Southeast on an Accurate General Map," *The Southeastern Geographer* VI, 1966, pp. 20–32.

Smith, Clara A. *List of Manuscript Maps in the Edward E.*

Ayer Collection, Newberry Library. Chicago, 1927, pp. 62–63, No. 228.

Reproductions

Plates 68–71.

De Vorsey, L. "The Colonial Southeast on an Accurate General Map," *The Southeastern Geographer* VI, Plates 1–4, pp. 26–29.

Swanton, J. R. *Early History of the Creek Indians*. Smithsonian Institution, Bureau of American Ethnology, Bulletin 73. Washington, 1922, Plate 7 in pocket (redrawing of Indian Towns area only).

Congress (photocopy).

Original

Newberry (Ayer Collection).

438A. Stuart-Purcell [1775] MS B

A Map of the Southern Indian District of NORTH AMERICA Compiled under the Direction of John Stuart Esq.r His MAJESTY's Superintendant of Indian Affairs. and by him Humbly Inscribed to the EARL of DARTMOUTH His MAJESTY's Principal Secretary of State for the Colonies etc. [cartouche, upper left] | This MAP of the Southern Indian District of NORTH AMERICA wase [*sic*] Compiled under the Direction of the Hon.ble John Stuart Esq.r His MAJESTY's Superintendant of Indian Affairs in Said District. By Joseph Purcell [upper right]

Size: 75 × 79. *Scale:* 1" = 16½ miles.

Description: This large colored manuscript map is very similar to Map 438. It was sent to London by John Stuart, the superintendent for Indian affairs, in February 1776 and reflects his knowledge of Southeastern Indian settlement geography and land claims as they had evolved through 1775. Details concerning the provenance of Map 438, now in the Newberry Library's Edward E. Ayer Collection, are lacking but it seems safe to conclude that it was retained in America as Stuart's reference or file copy when this inscribed copy was forwarded to London. Most of the descriptive details given for Map 438 above apply to this copy of Purcell's large and detailed compilation and should be consulted.

In the course of his research for the book *The Indian Boundary in the Southern Colonies, 1763–1775*, L. De Vorsey identified two separately catalogued map fragments in the British Public Record Office as portions of an additional and heretofore unknown copy of the 1775 Stuart-Purcell Southern Indian District map. These fragments were joined and are now catalogued as map MR 919 in that archive. See P. A. Penfold, ed., *Maps and Plans in the Public Record Office*, No. 2036, p. 349.

When John Stuart died in 1779, the position of Superintendent of Indian Affairs in the east was filled by Thomas Brown. Purcell continued his cartographic efforts under Brown's administration and produced another compilation of the Southeast in 1781. Like the 1775 maps it is an impressively large and detailed manuscript measuring 73½ × 79 inches. Now in the British Public Record Office, where it is catalogued as C.O. 700, North American Colonies General no. 15, it reflects Purcell's improved knowledge of the region as well as his evolving cartographic technique and style. See P. A. Penfold, ed., *Maps and Plans in the Public Record Office*, 2, No. 2041, p. 350. This 1781 Brown-Purcell Map, or a copy of it, came to the attention of the U.S. Army Corps of Topographical Engineers (1838–63), and a somewhat simplified tracing was prepared from it. Coastal depth soundings and remarks concerning Indian groups to the west of the Mississippi indicate that this map, now in the National Archives (R.G. 77, Map—U.S. 113), contains information not present on the Public Record Office copy of the 1781 Brown-Purcell map.

References

Alden, John R. *John Stuart and the Southern Colonial Frontier*. New York, 1966.

De Vorsey, L. "The Colonial Southeast on an Accurate General Map," *The Southeastern Geographer* VI, 1966, pp. 20–32.

Guide to Cartographic Records in the National Archives. Washington, 1971, pp. 72–76.

Penfold, P. A., ed. *Maps and Plans in the Public Record Office* 2. London, 1974, Nos. 2035, 2036, and 2041, pp. 348–50.

Original

London. Public Record Office. C.O. 700, North American Colonies General no. 12.

---

439. Mouzon ca. 1773

A Map of the Parish of S.t Stephen, in Craven County; Exhibiting a View of the several Places Practicable for making a Navigable Canal, between Santee and Cooper Rivers, from an Actual Survey by Henry Mouzon Jun.r [cartouche, top right] | Engrav'd by John Lodge No 45 Shoe Lane London. [below cartouche]

Size: 33¼ × 25½. *Scale:* 1" = 1 mile.

Description: This is a detailed topographical map of St. Stephen's Parish, with three or four practicable routes for a canal between the Cooper and Santee Rivers. Houses with the names of their owners are shown. In the bottom left corner is an engraving of an indigo plantation, with the steps in the preparation and making of indigo shown. Above this engraved section is a drawing of an indigo plant and two enlarged figures of indigo leaves and a bunch of the seeds.

Though Mouzon's plan for the Santee-Cooper River Canal was best, Major Senf's route surveyed after the Revolution was adopted for the canal built in 1792–1800. For references, see D. D. Wallace, *The History of South Carolina*, New York, 1934, II, 399.

In the *South Carolina Gazette*, January 7, 1773, p. 1, col. 2, appears the following advertisement of this map

> Proposals—for Engraving by Subscription—a Map of the Parish of St. Stephen in Craven County, Exhibiting a View of the Several Places practicable for making a Navigable Canal between Santee and Cooper Rivers; from an actual Survey.
>
> *Conditions* The dimensions of the Map will be Twenty-seven Inches by Thirty-four Inches including a considerable Part of the Parishes of St. John and St. James, and Part of the Parishes of St. Thomas and St. Denis; to be nearly done on a Scale of Eighty Chains, or one Mile to an Inch: In which with proper Distinctions, will be laid down all public and private Roads, customary Paths, Creeks, Lakes, Islands, Rivers and Inland Swamps, Bays etc.—The Compartment to be embellished with a curious Representation of manufacturing Indico, now the principal Article of Produce in the parish.
>
> The Price to Subscribers will be Five Pounds Currency each to be paid on the Delivery of the Map, which will be some time in July next.
>
> Subscriptions are taken by Messrs. Bonneau and Slann, Merchants; Mr. William Doughty, Merchant; Mr. James Oliphant, Jeweller; Mr. Nicholas Langford, Bookseller; Mr. Charles Crouch, Printer; and at the Printing Office on the Bay, near the Exchange; and by several Gentlemen in the different Parishes in the Country.
>
> Henry Mouzon, jun.
>
> The Author having met with Success in the Undertaking, acquaints his Subscribers, that he has considerably enlarged his Design, both as to the Extent and the Execution of the Map, which is now sent to England to be engraved by one of the best Artists there—and flatters himself that the Corrections and Elegance of this Engraving (which he will spare no Expence to have executed in the best Manner) will make amends for a few Months of unavoidable Delay in the Publication.

Copies

Charleston Library Society; Duke.

---

440. Stuart-Gage 1773 MS

A MAP OF WEST FLORIDA part of E^t^ FLORIDA. GEORGIA part of S^o^ CAROLINA i[n]cluding [torn] & Chactaw Chickasaw & Creek Nations with [missing] [r?]oad the[?] Pensacola through ye: Creek Nation to Augusitus & CharlesTown. Compiled under the directi[on] of ye: So[missing] John Stuart Esq: His Majesty's [partly obliterated] Superintendant of Indian affairs in [torn: the Southern Department?] of Nth: America & by him humbly [D?] r[esented?] to His Excellency ye: Honble: Thomas Gage Esqr: General & Commandr in Chief of all His Majesty's: Forces in Nth America &ca. &ca. &ca.

Size: 102 × 65½. *Scale:* 1" = ca. 8 miles.

Description: This enormous two-panel manuscript map shows the coastal Southeast from approximately the latitude of Charleston (lost in present condition of map) to approximately 27°30', on the Florida coast. A second interior panel shows the Gulf Coast from "Deadman's Bay," in the east, to slightly west of the "Savin [Sabine] River." Much of the interior space of the map is taken up by an elaborate cartouche with landscape scene [upper right], a set of descriptive "Remarks" by Bernard Romans and David Taitt [upper center], "A Table of the distances in Computed Miles from Pensacola through the Upper & lower Creek Nations to Charlestown also the distances of one town to another," and a block containing "Names of Chicasaw Villages," and "Names of Choctaw Villages." An interesting and unique feature of this map, correctly ascribed by Lloyd Brown, formerly of the Clements Library, to the pen of Joseph Purcell, is the dual scale of latitude employed to separate the eastern and western panels. A careful examination reveals that the latitude graticule of the eastern map is approximately 25 minutes north of that on the western panel. This skewing is obscured by the fact that the "Alatmaha" River appears to flow uninterruptedly from one panel to the other. Along the Indian Boundary Line, the rivers and major trails to the Indian villages, detail is impressive while most other spaces are left blank.

Joseph Purcell, Stuart's chief cartograher and compiler of this map, arrived in East Florida with the large contingent of Mediterranean people transported there through the effort of Dr. Andrew Turnbull to establish New Smyrna. Purcell joined Stuart's staff after having been employed as a "Draughtsman, Mathematician and Navigator" by William Gerard De Brahm, the Surveyor General of the Southern District in North America. On October 23, 1773, De Brahm wrote complainingly to Lord Dartmouth to report that his "geometer," Joseph Purcell, had secretly taken copies of his surveys and entered Stuart's service as Bernard Romans had done earlier. De Brahm went on to claim that Stuart's appointment to the office of Indian Superintendent was "entirely the effect" of his recommendation. See Historical Manuscript Commission, *The Dartmouth MSS* II (Fourteenth Report), London, 1895, p. 178. Based in Charleston, South Carolina, after the Revolutionary War, Purcell worked as a land surveyor and prepared the first map of the Southeast to be published by Jedidiah Morse in his immensely successful *The American Geography.* For other Purcell maps, see Maps 438 and 438A.

See Introductory Essay.

References

Brun, Christian, comp. *Guide to the Manuscript Maps in the William L. Clements Library.* Ann Arbor, Michigan, 1959, No. 663, pp. 160–61.

De Vorsey, L. "The Colonial Southeast on an Accurate General Map," *The Southeastern Geographer* VI, 1966, pp. 20–32.

Phillips, P. L. *Notes on the Life and Works of Bernard Romans* [A facsimile reproduction of the 1924 edition with an introduction and index by John D. Ware]. Gainesville, Florida, 1975, pp. 40–44.

Original

Clements (Gage).

440A. Andrew Way 1773 MS

Pursuant to Instructions from His Excellency Sir James Wright Governor and Commander in Chief of his Majesty's Said Province Directed to Thomas Carter and William Jones Esq.rs dated the 30th day of October 1773 with the assistance & directions of above said Thomas Carter and William Jones Esq.rs Commissioners I have Traversed and Marked the Boundary Line, Between the Province of Georgia And Nation of Creek Indians, (in Presence of their Deputys the Pumkin and Chehaw Kings and Telechee) from Alatamaha to Ogechee Rivers, the Markes whereof both Natural and Artificial are Represented in the above Delineated Plan.

Certified this 30th Day of Nov. 1773 By And.w Way Dept.y Surv. [lower left]

Size: 26 × 14½. *Scale:* 1" = 4 miles.

Description: This pen-and-ink map shows the old and new Indian Boundary Lines between the Ogeechee and Altamaha Rivers in Georgia. It was forwarded to London in a letter from the colony's governor, James Wright, to Lord Dartmouth, dated April 26, 1774. Wright had taken advantage of Indian willingness to surrender portions of their hunting lands for relief of indebtedness and secured for Georgia a large two-parcel cession in 1773. The northern portion of that cession was entirely on the Piedmont, and it immediately became the scene of active settlement. The portion surveyed and depicted in this map by Way, however, was on the Georgia coastal plain where soils were not perceived as nearly so productive or desirable. As a consequence it did not gain the place in history that is occupied by the larger and rapidly settled northern portion of Georgia's "New Purchase," or "Ceded Lands," as Wright's 1773 land acquisitions came to be known. On the face of his map Way noted, "This Platt contains 674,000 acres," which he described in various places as "Low Pine Land," "High Hilly Pine Land," "High Pine Land Intermixed with Brown Pebble Stones," none of which would have excited much interest on the part of would-be farmers. Way incorrectly shows the former or "Old" Indian Boundary Line as being formed by a straight line connecting the Ogeechee and Canoochee Rivers and another straight line from the Canoochee to the Altamaha River. For an accurate depiction of that boundary see Map 369, Savery 1769 MS A; and Map 370A, Savery-Romans 1769 MS C. For a discussion of Georgia's Indian boundaries see L. De Vorsey, "Indian Boundaries in Colonial Georgia," *The Georgia Historical Quarterly* LVI (Spring 1970), pp. 63–78.

Reference

Penfold, P. A., ed. *Maps and Plans in the Public Record Office,* 2. London, 1974, No. 2376, p. 407.

Reproduction

De Vorsey, L. *The Indian Boundary in the Southern Colonies, 1763–1775*. Chapel Hill, 1966, Figure 23, p. 177.

Original

London. Public Record Office. MPG 357(2).

440B. Philip Yonge 1773 MS

A Map of the Lands Ceded to His Majesty by the Creek and Cherokee Indians at a Congress held in Augusta the 1st June 1773 By His Excellency Sir James Wright Bart. Captain General Governor and Commander in Cheif [*sic*] of the Province of Georgia The Honorable John Stuart Esq[r] Agent and Superintendent of Indian Affairs and the said Indians—Containing 1616298 Acres. Copy delineated by Philip Yonge Depy. Srv. [cartouche lower right] | Survey performed in the year 1773 By Edward Barnard LeRoy Hammond Philip Yonge Joseph Purcell and William Barnard [following "A Description" text upper right]

Size: 29½ × 41½. *Scale:* 1" = 2½ miles.

Description: This artistically drawn manuscript map shows a large Indian cession in northeastern Georgia originally designated Wilkes County in the state's first constitution in 1777. Today it includes all or portions of Hart, Elbert, Franklin, Madison, Clarke, Oglethorpe, Greene, Taliaferro, Warren, Lincoln, and Wilkes Counties. Known in Georgia history as the "New Purchase," or "Ceded Lands," it was the largest of two tracts acquired from the Cherokee and Creek Indians by Georgia's royal governor, Sir James Wright, in return for the Crown's assuming the large debts the Indians owed to a consortium of merchant-traders. Located on the Piedmont and larger in area than the state of Delaware, the "Ceded Lands" became an eagerly sought out settlement region as thousands of pioneer farmers moved into the southern Piedmont during the Revolutionary War era. See *The Colonial Records of the State of Georgia*, XII, pp. 371–76. Because the Crown planned to recoupe the expense of paying off the Indian debts by selling the land in small tracts directly to settlers, more than usual attention was paid to the quality of the soils found in the Ceded Lands by surveyor-cartographer Philip Yonge and William Bartram, the well-known Quaker naturalist who accompanied the boundary survey party. See Francis Harper, ed., *The Travels of William Bartram: Naturalist's Edition*, New Haven, Connecticut, 1958, pp. 22–30, 342–45. One soil scientist has identified this map by Yonge as a soil map and found that soil information it contains could be correlated with his own field observations made in 1992. See T. G. Macfie, "William Bartram's Observations on Soils, 1773," *Soil Survey Horizons* 33 (no. 4), 1992, pp. 96–102. Yonge used capital letters placed on the map to indicate the various types of soil described in the text block "A Description of the Natural Produce and Soil of this Map with References." It is interesting to note that Yonge's father was also an accomplished surveyor-cartographer (see Map 280, Yonge 1751 MS).

For preliminary cartographic depictions of the region surveyed and mapped by Philip Yonge in 1773 see Map 323, Georgia Indian Lands ca. 1771 MS, and Map 437, Stuart-Georgia [1772].

Reproduction

Plate 66C.

References

De Vorsey, L. *The Indian Boundary in the Southern Colonies, 1763–1775*. Chapel Hill, 1966, pp. 161–72 (tracing of map, Figure 22, p. 176).

Penfold, P. A., ed. *Maps and Plans in the Public Record Office*, 2, No. 2377, pp. 407–8.

Original

London. Public Record Office. MPG 2.

---

441. Romans 1773 MS

A Map of part of West Florida done under direction of the Honourable John Stuart Esq[e] & by him humbly inscribed to his Excellency Thomas Gage Esquire General and Commander in Chief of all his Majesty's Forces in North–America Survey'd & drawn by Bernard Romans. between The Month of June 1772 & January 1773

Size: 29¾ × 21¼. *Scale:* 1" = 8 miles.

Description: This carefully drawn coastal map extends from the eastern bank of the Mississippi River to Pensacola. It lacks a latitude-longitude graticule, but Romans gave the latitude of the "Principall [*sic*] Pass" or navigation entrance to the Mississippi's birdfoot delta as "in Lat[d] 29°11'. He also paid particular attention to the area termed the "Island of Orleans," which included the site of New Orleans. In 1763 the Peace of Paris had excluded that territorial fragment from the cession to Britain of French and Spanish lands lying east of the Mississippi. The forces under Gage's command in the area spent a great deal of effort in trying to find a waterway that would allow them to bypass Spanish-held New Orleans by entering the Mississippi via Lake Pontchartrain and the Iberville River.

References

Brun, Christian, comp. *Guide to the Manuscript Maps in the William L. Clements Library*. Ann Arbor, Michigan, 1959, No. 662, p. 160.

Phillips, P. L. *Notes on the Life and Works of Bernard Romans* [facsimile reproduction of the 1924 edition, with an introduction and index by John D. Ware]. Gainesville, Florida, 1975, pp. xlix–lx.

Reproduction

Brun, *Guide to the Manuscript Maps in the William L. Clements Library*, facing p. 160.

Original

Clements (Gage).

441A. Lyford 1773 MS

A Chart of Tibee Inlet In Georgia [lower right with drawing of tree and corn plants] | A: B: July 30th. 1776 [under drawing]

Size: 14¾ × 18½. *Scale:* 1" = 1 mile.

Description: This colored manuscript chartlet is marked "p. 339 Dartmouth Vol.I," in the upper right-hand corner. The 1776 date given on the face of the map may indicate when it was catalogued in England since pilot William Lyford's statement on the reverse is dated "13th Decr 1773. Identifying himself as "Branch Pilot for the Barr and River of Savannah in Georgia," Lyford provides detailed navigation instructions for bringing a ship over the Savannah's bar and to safe anchorage off the point of Cockspur Island. Lyford, like his contemporaries De Brahm and Yonge, showed broad offshore areas marked "Shoals" and "Shoal Water" that are now submerged at all tidal states. The "Lazzaretto" or quarantine station and "Light House" are shown on Tybee Island, while a small symbol marks the site of Fort George on Cockspur Island. It is interesting to note that the name "Tybee" is correctly spelled on the map and in Lyford's remarks but incorrectly spelled in the map's title. It may be that "A: B:" and not William Lyford provided the decorative title for this chart.

There is another copy of this map in the British Public Record Office. It was included in correspondence from Georgia's royal governor Sir James Wright dated December 20, 1773, just a week after Lyford signed his descriptive remarks on the reverse. The Public Record Office map is filed as M.P.G. 357.

References

Brun, Christian, comp. *Guide to the Manuscript Maps in the William L. Clements Library*. Ann Arbor, 1959, No. 634, pp. 151–52.

De Vorsey, L. *The Georgia–South Carolina Boundary, A Problem in Historical Geography*. Athens, 1982, p. 75.

Penfold, P. A., ed. *Maps and Plans in the Public Record Office*, 2. London, 1974, No. 2393, p. 410.

Original

Clements.

---

442. Sharpe 1773 [1847] MS

[A Map of Fourth Creek Congregation, By William Sharpe, Esq. 1773]

Size: 22½ × 25⅝. *Scale:* 1" = 1 mile.

Description: The title and description of this map is taken from the lithographed copy of the original pen-and-ink map which was made for Professor E. F. Rockwell of Davidson College in 1847. As late as 1867 the original map was in the possession of Alexander Nisbett of Iredell County (see Rockwell's article listed below, p. 84).

The center of the map is about two miles northwest of the then nonexistent town of Statesville, N.C. From this center the map is laid off in a series of eleven concentric circles, each having an increase in radius of one inch and indicating an increase in distance of one mile. Thus the map embraces a tract twenty-two miles in diameter. This land was included in the congregation of the Fourth Creek Church, now the Presbyterian church of Statesville.

The map contains the names and locations of all the heads of families belonging to the congregation at that time. This settlement was begun about 1750–51 by Scotch-Irish emigrants from Pennsylvania who had previously come from Ireland. Soon after they settled, a colony of Highland Scotch came, establishing themselves about eight miles west of what is now Statesville, in "New Scotland." This map, therefore, gives the names of the pioneers in what was the westernmost settlement in that part of North Carolina at that time. One hundred and ninety-six families are given, including one hundred and eleven different names. Rocky

Creek, Catawba River, South Yadkin, and numerous other creeks and their tributaries are given.

William Sharpe, the author of the map and whose name is given in the vicinity of Snow Creek, was born in Cecil County, Maryland, in 1742, and migrated to Mecklenburg County when he came of age. In 1768 he married a daughter of David Reese, one of the signers of the Declaration of Independence. Sharpe himself, known as "Lawyer William," was one of the leading men of the region; he aided in the establishment of the "Clio's Nursery," the famous academy to the south of Statesville, was appointed one of the deputies from Rowan County to the Provincial Congress in New Bern, and in 1779 was the representative at Philadelphia.

The apparent occasion for making the map was a proposal to abolish the old meeting house and to establish two new churches, one in the extreme northeast quarter of the congregation, near Rocky Creek, and another at "Beattie's Old Field" in the northwest corner. In a document entitled "A Remonstrance to the North Carolina Presbytery which is to sit in April, 1773," published by Professor Rockwell (see reference below, p. 84), the division of the congregation is opposed because it would leave the south part of the congregation without a meeting house within reasonable traveling distance. The map itself is not connected with this Remonstrance; but the survey for the map was presumably made between the fall of 1772 (when the new churches, according to the Remonstrance, were proposed) and the meeting of the Presbytery in the spring of 1773.

The map gives one of the fullest and most detailed surveys of any part of North Carolina up to that time. Copies of the small lithographed edition are now very rare. Professor Rockwell's article, listed below, gives a full account of the map, lists the names on it, and preserves many details of the early history of the region.

A nineteenth-century manuscript copy of Sharpe's map is preserved in the North Carolina Department of Archives and History. It is drawn by T. C. Harris, is reduced in scale by half, and gives no names of settlers.

The late Rev. Dr. C. E. Raynal of the First Presbyterian Church of Statesville, N.C., assisted W. P. Cumming in an attempt to find the original Sharpe manuscript. Although Alexander Nisbett's map, from which the Rockwell lithograph was made, was not found, William Sharpe apparently made several holograph copies of the map which are still in the possession of Statesville families. The late John M. Sharpe, a direct descendant of William Sharpe, owned one of the original copies. Miss Mattie R. Hall of Statesville has a copy in four sheets, possibly not an original, which was reproduced with slight changes by the newspaper *Statesville Record & Landmark* in its bicentennial celebration edition of the founding of Statesville on September 11, 1953. Mr. Francis J. Marschner of the Agricultural Research Service, U.S. Department of Agriculture, Washington, D.C., reproduced a facsimile of the map in his monograph *Land Use and Its Patterns in the United States*; Mr. Marschner notes that it is an excellent example of the method of indiscriminate settlement practiced in the Carolina Piedmont, which stands out in strong contrast to the later settlements in the Public Domain States.

References

Marschner, Francis J. *Land Use and Its Patterns in the United States* [Handbook No. 153, U.S. Department of Agriculture]. Washington, D.C., April, 1959, pp. 11–13.

Rockwell, Professor E. F. "An Ancient Map of the Central Part of Iredell County, N.C." *The Historical Magazine, and Notes and Queries Concerning the Antiquities History and Biography of America*, XII [Vol. II, Second Series] (August 1867), pp. 84–90.

Reproductions

(Rockwell's lithographed edition) Congress; Davidson College, Davidson, N.C.; North Carolina Department of Archives and History; Historical Foundation, Montreat, N.C.

Statesville Record & Landmark. Statesville, N.C., September 11, 1953 (bicentennial edition).

Marschner, Francis J. *Land Use and Its Patterns*, p. 12.

Original

Unknown (see Description).

---

443. Cook 1773

A Map of the Province of South Carolina with all the Rivers, Creeks, Bays, Inletts, Islands, Inland Navigation, Soundings, Time of High Water on the Sea Coast, Roads, Marshes, Ferrys, Bridges, Swamps, Parishes Churches, Towns, Townships, County Parish District and Provincial Lines. Humbly inscribed to the Hon^{bl}. Lawlins Lowndes Esqr. Speaker & the rest of the Members of the Honble the Commons House of Assembly of the Province by their most Obed^{t}. & faithful Servt Jam^{s} Cook [cartouche, lower left] |

Publish'd according to Act of Parliament July 7th. 1773. and Sold by H. Parker in Cornhill. [below neat line, center] | Thos. Bowen, Sculpt. 1773. [below neat line, left]

Size: 31⅞ × 30¾. *Scale:* 1" = 9¼ miles.
Insets: 1. A Plan of Beaufort on Port Royal Island. *Size:* 5 × 5⅜. *Scale:* 1" = 1,450 feet.
2. A Plan of Camden. *Size:* 5¾ × 4½. *Scale:* 1" = 26 chains.
3. A Plan of Georgetown. *Size:* 4¼ × 4¼. *Scale:* 1" = 1,400 feet.
4. A Draught of Port Royal Harbour in South Carolina with the marks for going in. *Size:* 10½ × 7¼. *Scale:* 1" = 3 miles.
5. A Plan of Charles Town. *Size:* 5 × 5¾. *Scale:* 1" = 1,400 feet.
6. A Chart of the Bar and Harbour of Charles Town. *Size:* 12 × 12 (triangular). *Scale:* 1" = 1¼ miles.
Description: This is the most detailed and accurate printed map of South Carolina yet to appear, especially for the interior. It includes the whole state as far west as the Cherokee land and shows the Indian Boundary. Gaillard-Cook 1770 MS, Lodge-Cook 1771 MS, and Cook 1773 appear to be successive stages in the development of essentially the same map. Cook 1773 has the new acquisition given by the 1773 boundary agreement with North Carolina and even extends north of it to include surveys up to the south fork of the Catawba River. But along the coast Lodge-Cook 1771 MS has more careful draftmanship and more detail than either the Gaillard-Cook 1770 MS or Cook 1773. It alone has "Horse I." to the west of Parris Island. "Graystons" on Parris Island is on the 1771 map; "Grastons" is on the 1773 map, but it is not on the Gaillard-Cook 1770 MS. For the area shown on the 1771 and 1773 maps, the topographical details, location of houses, and spelling of names are so nearly identical that only close scrutiny provides noticeable differences. North of 34°50' in Cook 1773 the engraving is from a different plate on a separate sheet which has been pasted on the large bottom sheet. Two roads running north on the lower sheet are not continued on the top sheet.

Both Mouzon's 1775 map and Stuart's 1780 edition of De Brahm's map take much detail from Cook's map. Mouzon's map extends farther west, however, and Stuart's map has many surveys, rectangles, and lines not in Cook; both maps show additional and later information. Though the insets of Charles Town on Cook's and Mouzon's map are different in shape, they have the same legends and soundings and are the result of the same survey.

The Harvard copy, one of the Clements Library copies (Clinton Collection, No. 298), and a copy in the Bibliothèque Nationale, Paris (137.5.1.D), are examples of an early state of Cook's map, without the imprint, after the date 1773.

See De Brahm 1757, Gaillard-Cook 1770 MS, Lodge-Cook 1771 MS, and Mouzon 1775.

See Introductory Essay.

Reproduction
Plate 67.
Copies
Congress; Clements, two copies; Harvard.

---

444. Bellin 1773

Carte de la Caroline et Georgie Pour servir a l'Hist. des Etablissements Europeens. Tiree des Auteurs Anglois par M. B. Ing. de la Marine. A. v. Krevelt, Sculpsit, Amsterdam, 1773.

Size: 11 × 7⅜. *Scale:* 1" = ca. 21½ common French leagues.
In: Robert de Vaugondy, D. *Atlas Portatif.* Amsterdam, 1773.
Description: This map, extending from 30° to 37° N.L., gives many coastal names. It is a copy of the Bellin 1757 map.

Copies
Harvard 1773.
In: Raynal, Guillaume T. F. *Atlas Portatif.* Amsterdam, 1773, No. 40.
Copies
Congress 1773.

---

445. Smith 1774 MS

Sketch of Clinch River by Mr. Daniel Smith July 18th 1774—see his Letter of that date [on reverse of map with several sets of figures and floral doodles]

Size: 8⅝ × 11⅝. *Scale:* 1" = 10 miles.
Description: This map was incorrectly described in earlier editions of this work. It is a pen-and-ink manuscript drawing, in good condition, and now part of the Draper Collec-

tion of the State Historical Society of Wisconsin, where it is catalogued as Draper MSS 4XX62. The upper reaches of the Clinch and North Fork of the Holston with their many named tributary "forks" are shown in the area between "Laurel Ridge," on the north, and "Walker's Mountain," to the south. To the northwest the area is captioned "Parts uninhabited and not fully discovered." A straight line near the western edge of the map is captioned "Col. Donelson's Indian Line being as I've been informed, N 45° W." See Map 401 for a discussion of John Donelson's survey and map of the Virginia-Cherokee Boundary in May 1771.

Reproduction

Congress (photocopy).

Original

State Historical Society of Wisconsin (Draper MSS).

---

446. Stuart-Lewis 1774 MS

A New Map of West Florida, Georgia, & South Carolina; with part of Louisiana. The whole laid down from different Actual Surveys, and other best Authorities, under the direction, and by Order of John Stuart Esq.r: superintendant of Indian Affairs, in the Southern District of North America.

This map reduced and copied by Samuel Lewis, February the 4,th 1774. Draftsman to the Plantation Office [cartouche, top right center. Lewis signature and date very finely written between scroll flourishes]

Size: 6'9" × 4'. *Scale:* 1" = ca. 11 miles.

Description: This map was made from the large Stuart-Gage map (Map 440, now in the William L. Clements Library) while General Gage was in London. The workmanship and finish are exceptionally fine; the map may have been made especially for presentation to the king.

References

Brun, Christian, comp. *Guide to the Manuscript Maps in the William L. Clements Library*. Ann Arbor, 1959, No. 663, pp. 160–61.

Catalogue of the Manuscript Maps, Charts, and Plans, . . . in the British Museum, III. London, 1861, p. 508.

Original

British Library. King George III's Topographical Collection. cxxii. 89.

---

447. Tardieu-Valet ca. 1775

Carte de la Caroline Meridionale et Septentrionale et de la Virginie [cartouche, bottom right] | P. F. Tardieu, Sculpsit [below neat line, left] | P. J. Valet scripsit [below neat line, right] | No. 135 [above neat line, right]

Size: 16⅞ × 12⅝. *Scale:* 1" = 46 miles.

Description: This map extends from 32° to 40° N.L. (Port Royal in South Carolina to Philadelphia). It is identical to Map 291 above. The boundary line between North and South Carolina is that run in 1772. The villages of the Cherokees are shown west of South Carolina, possibly after the Mouzon map of 1775. Tardieu and Valet made maps for the Mentelle-Chanlaire atlas, ca. 1797; see C. E. Le Gear, *A List of Geographical Atlases*, Washington, 1958, No. 6011, p. 290.

Copies

Harvard; Clements, 1806.

---

448. Adair 1775

A Map of the American Indian Nations, adjoining to the Missisippi, West & East Florida, Georgia, S. & N. Carolina. Virginia, &c [bottom, right] | Jn.o Lodge Sculp. [below neat line, right]

Size: 9½ × 12¾. *Scale:* 1" = 100 miles.

In: Adair, James. *The History of the American Indians.* London, 1775, front, or opp. p. 1.

Description: The map gives the location of the Indian tribes in the territory designated by the title. Between Virginia and North Carolina is "Granvil," the territory ceded by the king to one of the heirs of the lords proprietors. The lateral boundaries of the other Southeastern colonies are shown by fine dotted lines that run from the Atlantic coast to the Mississippi River. Notably, there is no attempt to show the Indian Boundary, which did in fact bound those colonies in the west when Adair's book was published. This map illustrates a Southeast with the Indians safely tucked away in the interior wilderness, exactly the condition Adair's readers would have approved of. John Stuart and Crown administrators, on the other hand, knew the Indian land question was far more complex and potentially dangerous. Maps prepared by Stuart and Purcell and others associated with the office of the Superintendent for Indian Affairs should be consulted for a more accurate view of Indian-Colonial conditions in the pre-Revolutionary Southeast. See Maps 323, 331, 336A, 337, 341, 352, 369, 370, 370A, 401, 437, 438, 438A, 440, 440A, 440B, and 446.

The John Carter Brown Library copy of the German translation of this work *Geschichte der Amerikanischen Indianer*, Breslau, 1782, has no map.

Reproductions

Bartholomew, J. G. *A Literary & Historical Atlas of America*. (Everyman Library), New York, n.d., p. 126.

Gabriel, R. H., ed. *The Pageant of America*. New Haven, 1929, II, 25.

Williams, S. C., ed. *Adair's History of the American Indians*. Johnson City, Tennessee, 1930, front.

Copies

Congress; American Geographical Society; John Carter Brown; Clements; Georgia Hargrett Library; Harvard; Kendall Collection; New York Public, two copies; Sondley Reference; University of North Carolina; Yale.

---

449. Fry-Jefferson 1775

A Map of the most Inhabited part of Virginia containing the whole Province of Maryland with Part of Pensilvania, New Jersey and North Carolina. Drawn by Joshua Fry and Peter Jefferson in 1775. [cartouche, lower right corner] | To the Right Honourable, George Dunk Earl of Halifax First Lord Commissioner; and to the Rest of the Right Honourable and Honourable Commissioners for Trade and Plantations, This Map is most humbly Inscribed to their Lordship's By their Lordship's Most Obedient & most devoted Humble serv^t. Thos. Jefferys. [bottom right, below picture in cartouche] | Printed for Robt. Sayer at N°. 53 in Fleet Street, & Thos. Jefferys at the Corner of S^t. Martins Lane, Charing Cross, London. [bottom right, above neat line]

In: Jefferys, T. *The American Atlas*. London, 1775, No. 20–21.

Description: The 1775 impression of the Fry-Jefferson map differs from the 1761? issue only in the change of date from 1751 to 1775.

In 1794 Laurie and Whittle, the successors to Sayer and Bennett, published *The North American Pilot*, with a changed imprint for the Fry and Jefferson map: "Printed for Robt. Sayer at No. 53 in Fleet Street, London" [the remaining part of the imprint as given above for the 1775 edition is deleted]. Copies of the 1794 issue are in several libraries, including that of the University of Virginia. Mr. Coolie Verner furnished the information concerning the 1794 imprint from his copy.

See Fry-Jefferson 1751 [1753] and Introductory Essay.

References

De Vorsey L. "Introductory Notes." *North America at the Time of the Revolution: A Collection of Eighteenth Century Maps*, II. Lympne Castle, Kent, England, 1974, p. 1.

Sanchez-Saavedra, E. M. *A Description of the Country: Virginia's Cartographers and Their Maps, 1607–1881*. Richmond, 1975, pp. 25–34.

Verner, C. "The Fry and Jefferson Map," *Imago Mundi* 21 (1967), pp. 70–94.

Reproductions

Color Plate 23.

Bloom, Sol, Director General, United States Constitution Sesquicentennial Committee. Washington, D.C., 1936, separate sheet.

North America at the Time of the Revolution, Sheets FJ-1 through FJ-4 (full-scale facsimile).

A Description of the Country, Map No. 3, parts 1 through 4 (colored facsimile in portfolio).

The American Revolution 1775–1783: An Atlas of 18th Century Maps, Theatres of Operations. Washington, 1972, Map 12 (facsimile of eastern seaboard area with cartouche, in portfolio).

Copies

Congress 1775, 1776, 1778, 1782, two separates; American Geographical Society, separate, and upper half of another copy; John Carter Brown (*United States of America Atlas, American Atlas*, two separates); Clements; Duke; Huntington, three separates; New York Public 1776, 1778, separate; South Carolina Historical Society; Yale 1776, three copies.

In: Faden, W. *North American Atlas*. London, Nos. 27–28.

Copies

Congress, two copies; Clements; New York Public, two copies; Yale, two copies.

---

450. Mouzon 1775

An Accurate Map of North and South Carolina, with their Indian Frontiers, Shewing in a distinct Manner all the Mountains, Rivers, Swamps, Marshes, Bays, Creeks, Harbours, Sandbanks and Soundings on the Coasts; with The Roads and Indian Paths; as well as the Boundary or Provincial Lines. The Several Townships and other divisions of the Land In Both the Provinces; The whole from Actual Surveys. By Henry Mouzon and Others. London. Printed for Robt Sayer and J. Bennett, Map and Print-sellers, No 53 in Fleet Street. Publish'd as the Act directs May 30th 1775. [cartouche, top left] | Sparrow Sc. [below cartouche] | Publish'd as the Act directs May 30th 1775. by R. Sayer and J. Bennett [below neat line, left center] | Publish'd as the Act directs, May 30th 1775. by R. Sayer and J. Bennett [below neat line, right center]

Size: 56½ × 39¾; four sheets, each 28¼ × 19⅞.
Scale: 1" = 8⅔ miles.
In: Jefferys, T. *The American Atlas*. London, 1775, Nos. 22–23.
Insets: 1. "The Harbour of Port Royal." *Size:* 7⅛ × 10⅝. *Scale:* 1" = 3 miles.
2. "The Bar and Harbour of Charlestown." *Size:* 10 × 10⅝. *Scale:* 1" = 1⅐ miles.

Description: This map, which extends from the coastal area of the two Carolinas westward to the Appalachian mountains in the Cherokee country, was the chief type map for the region during the forty or fifty years following its publication. It was used by both British and American forces during the Revolutionary War.

The following advertisement by Mouzon, which explains the purpose and nature of the map, appeared in *The South Carolina and American General Advertiser*, May 6, 1774–June 3, 1774:

> Proposals for Publishing by Subscription an Improved Map of the Colony of South Carolina, corrected from Actual Surveys by Henry Mouzon, jun. and Ephraim Mitchell.
>
> Conditions—The Map will be sent to England in a few weeks, will be engraved in a very elegant Manner by one of the best Artists on a scale of Seven Miles to an Inch, being Thirty Inches in Depth and Thirty four Miles [?]—The Sea Coast, Islands, Rivers—Towns, Court Houses, Churches—Forts—Subdivisions of the Colony into Counties, Districts—Parishes—Catawba Tracts etc. etc.
>
> The Price to Subscribers Five Pounds each on Delivery about the fifth day of January next.—
>
> A few Pocket Maps on a scale of Twenty Miles to an Inch will be struck off on Parchment for the use of the Gentlemen who ride the Circuits, and others who travel. Price to Subscribers, One Dollar each.
>
> As there are already two Maps of this Colony extant, it may be necessary to give some reasons for undertaking a Third, and as this is chiefly owing to the Inaccuracies observed in the others, we need only particularize the several Alterations, Corrections and Additions in the present one, that the Publick may judge whether they are of Consequence sufficient to induce such a Publication.
>
> In doing this, we hope we shall not offend the Contributors of those Maps, as the Errours we are under the Necessity of pointing out are not imputed to want ot Abilities, but to the Nature of the Subject, which will not admit of Perfection but by gradual Advances.
>
> I. The several District Lines are laid down from actual Surveys, which in the first Map are all, except one, omitted, and in the second, drawn from the Directions of the Circuit Court Act, the Consequence of which is, that many Houses and other remarkable Places represented by them to be in one District, are in Reality found to be many miles within another, to which we may add, that in the last Map, one half of the District of Ninety-Six is included in that of Camden. We have, in the course of our Survey carefully noted every River, Creek, Swamp, Bay, etc. etc. in our Way, and reduced them to their proper Places, and made many Additions not to be found in the other Map.
>
> II. The Townships of Williamsburgh, Orangeburgh, Fredericksburgh, Kingston, Hillsborough, and Belfast, (the two last misnamed New Bordeaux) being erroneously laid down, are considerably corrected, and those of Purrysburgh, Queensborough, Boonesborough and the Welch Tract, omitted in the other Maps are here added. The Bounds of the Parishes of Prince Frederick, St. Mark and St. David, not defined in the first Map, and erroneously laid down in the second, are laid down from actual Surveys; and the Bounds of St. Matthew and St. James Goose Creek, likewise undetermined in both, are here inserted.
>
> III. Exclusive of the Advantages derived from laying out the several District Lines the Publishers have been able in the Course of their Practice to make many other Corrections and Additions from actual Surveys, particularly of the whole Parishes of St. Stephen and St. James

Santee; a considerable Part of the Parishes St. John Berkley, St. Thomas, Christ Church, St. Matthew, St. Mark, St. David, Prince Frederick, St. Paul and St. John in Colleton County, the New Acquisition, also Santee River for the Distance of one hundred Miles, including the Creeks and Islands near its Mouth and the Harbour of North Edisto from its Entrance to the Confluence of North and South Edisto Rivers.

IV. The Meridians are drawn as respecting the Number of Miles contained in a Degree of Longitude in these Latitudes showing the Difference of Time between the City of London and Charlestown, or between any two Places in the Colony, a Circumstance neglected in the other Maps.

V. In the nominal Part the Orthography is corrected, and many Places appearing without Names, that Defect is here supplied.

VI. The several Houses are omitted for this Reason, that as the Proprietors of them are continually changing, the Insertion of them would serve no other End in the Course of a few Years, than to perplex and confuse; besides, as it would be impossible to insert them all, we might incur the Charge of Partiality. We hope this will be considered as an Improvement rather than Defect, as every Gentleman may determine the precise Situation of his own House, by observing the River, Creek or Swamp on which it is situated or its Distance from any Church, Chapel, Tavern, Road, Bridge or Ferry.

Lastly we may mention the Size as another Advantage as in every Map or Plan it is necessary for the Eye to Command the Whole at one View, in order to form a just Idea of the Relation and Connection between the several Parts while at the same Time it should be so large as to avoid the Confusion occasioned by being too much crowded.

Several lines in The Charleston Library Society copy of the newspaper from which this is taken are badly torn. This transcription was made by Mrs. R. W. Hutson of Charleston, S.C.

A French edition of Mouzon's map was published in George L. Le Rouge, *Atlas Amériquain Septentrionale*, Paris, 1778, No. 22, with the title in English and also in French: "Caroline Septentrionale et Méridionale en 4 Feuilles, Traduite de l'Anglais a Paris, Chez Le Rouge, Ingénieur—Géographe du Roi rue des grands Augustins, 1777." Below the scale in the cartouche is: "Printed at Paris for M. Le Rouge Ingineer Géographer Augustin Street 1777. With Priviledge." *Size:* 56¼ × 38¾. In the lower right is an inset: "Attaques du Fort Sulivan par James Lieutenant Colonel d'Artillerie," with a further legend that the fort was attacked by the English on June 28, 1776.

A manuscript copy of Mouzon's map was offered for sale in 1940 by The Old Print Bookshop, New York (*Old Maps of America*, No. 98). On the verso of one of the four sheets (each 22¾ × 16¼; the entire map 45½ × 32½) is the following endorsement: "Original Drawings of 4 Sheet North & South Carolina made by Mr. De La Rochette in which are the Harbours of Charleston & Port Royal—part of the stock of The late Mr. Sayer—Will'd to Laurie & Whittle." Robt. Laurie and Whittle (1794–1812) were the successors of R. Sayer and J. Bennett (1770–87). The Old Print Bookshop states that these drawings are the originals from which the Mouzon 1775 engraved plates were made. In 1945 W. P. Cumming examined the original MS sheets with Mr. Rush of Chapel Hill, N.C., and found a number of differences, besides the obvious one of size, between them and the printed Mouzon. A road in the upper center of the printed map is only half completed in the MS; several small sailing ships together with some dotted lines marked, "A Good Channel," leading into some of the coastal harbors, appear only on the printed map. The cartouche is left blank. Since the engraver usually follows with the utmost exactness the drawing made for him, it is at least questionable whether this is an example of that rare object, a preserved engraver's copy.

The original plates of Mouzon's map were used by Laurie and Whittle near the end of the century. The map itself, the title, and the engraver's signature are unchanged; under the scale in the cartouche, the Sayer and Bennett imprint has been changed to: "Published by Laurie & Whittle, 53, Fleet Street. 12th May 1794." Copies of this issue are in the Library of Congress, the South Carolina Historical Society, and the Sondley Reference Library.

Reference

Cumming, W. P. *North Carolina in Maps*. Raleigh, 1966, pp. 21–22.

Reproductions

Color Plate 24.

Cartografia de Ultramar, II, Plate 46 (two plates; 1777 French edition; transcriptions of place-names, pp. 264–80).

Cumming, W. P. *North Carolina in Maps*. Raleigh, 1966, Plate VIII (reduced-size facsimile).

Greenwood, W. B., comp. *The American Revolution 1775–1783: An Atlas of 18th Century Maps and Charts.* Washington, 1972, Map Nos. 14 and 15 (facsimiles of eastern map sheets in portfolio).

Waynick, C. *North Carolina Roads and Their Builders.* Raleigh, 1952, opp. p. 168 (large folded insert).

Sol Bloom, Director General, United States Constitution Sesquicentennial Committee, Washington, D.C., 1938 (issued as separate sheet).

Copies

Congress 1775, 1776, 1778, 1782, separate; American Geographical Society 1776, 1778, two separates; John Carter Brown; Charleston Library Society; Charleston Museum; Clements (Clinton Collection, No. 285, two separates); Duke; Harvard 1775, 1776, separate; H. E. Huntington, three separates; New York Public 1776, 1778, separate; North Carolina Department of Archives and History, upper half; North Carolina State Library, lower half; South Carolina Historical Society; University of North Carolina, three separates; University of South Carolina; Yale 1776, two copies, separate.

In: Faden, W. *North American Atlas.* London, Nos. 29–30.

Copies

Congress, two copies; Clements, upper half a separate; New York Public, three copies; Yale.

COLOR PLATE 1. Sebastian Münster. 1540 [1575].
Tauola dell' isole nuoue, le quali son nominate occidentale, & indiane per diuersi rispetti.

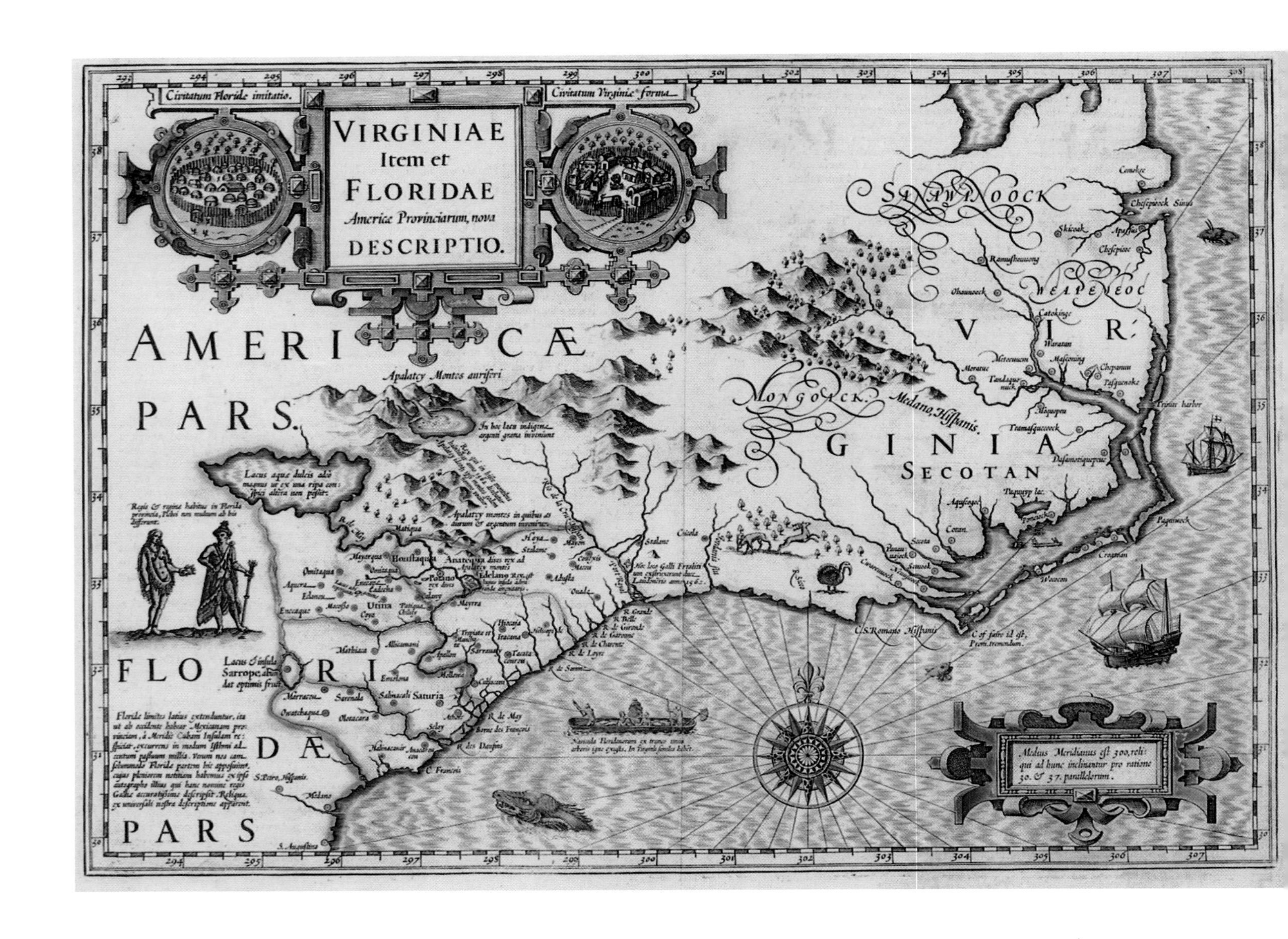

COLOR PLATE 2. Gerard Mercator and Jodocus Hondius. 1606.
Virginiae Item et Floridae Americae Provinciarum nova Descriptio. [Map 26].

COLOR PLATE 3. Gerard Mercator and Jodocus Hondius. 1632. Nova Virginiæ Tabula.

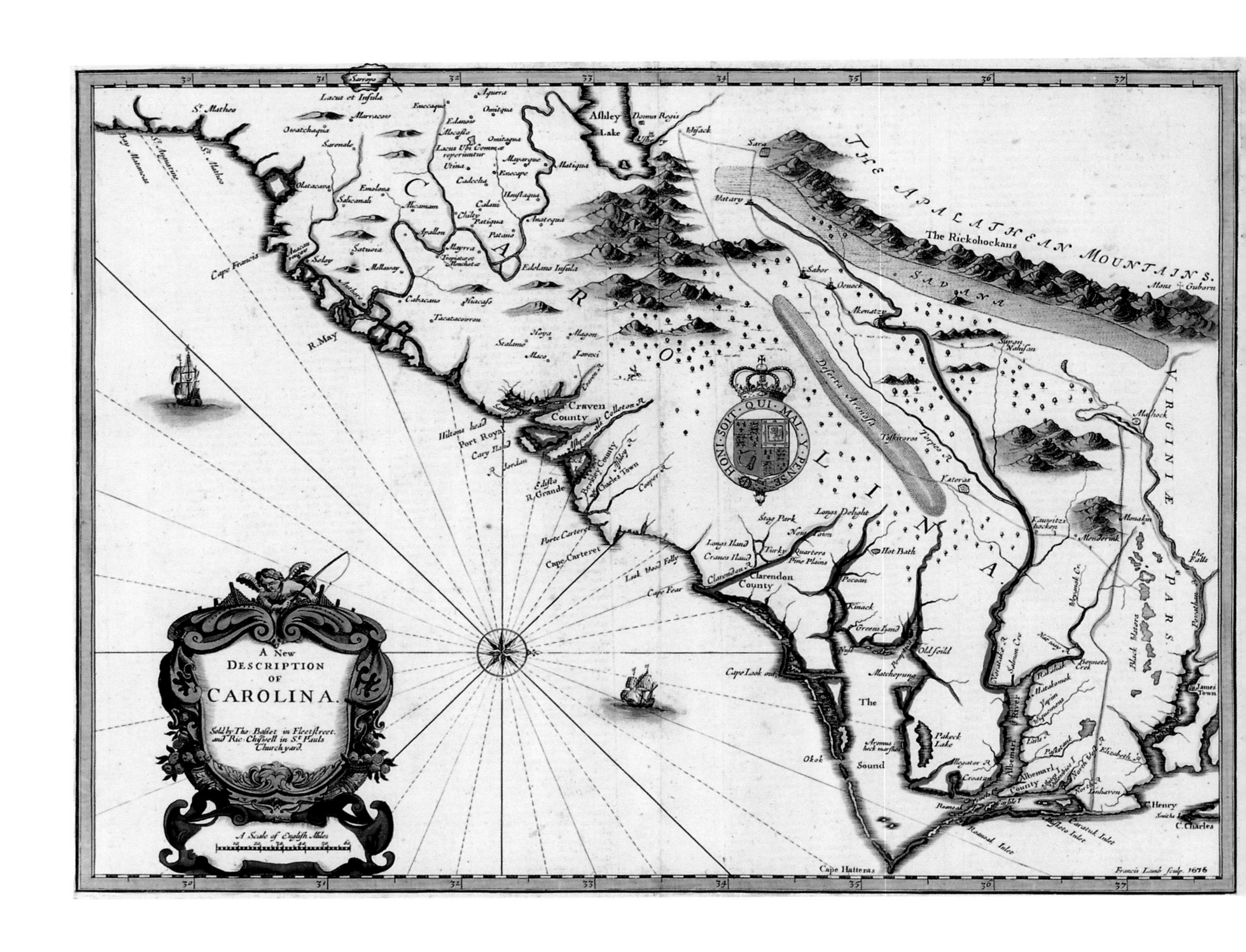

COLOR PLATE 4. John Speed. A New Description of Carolina. 1676. [Map 77].

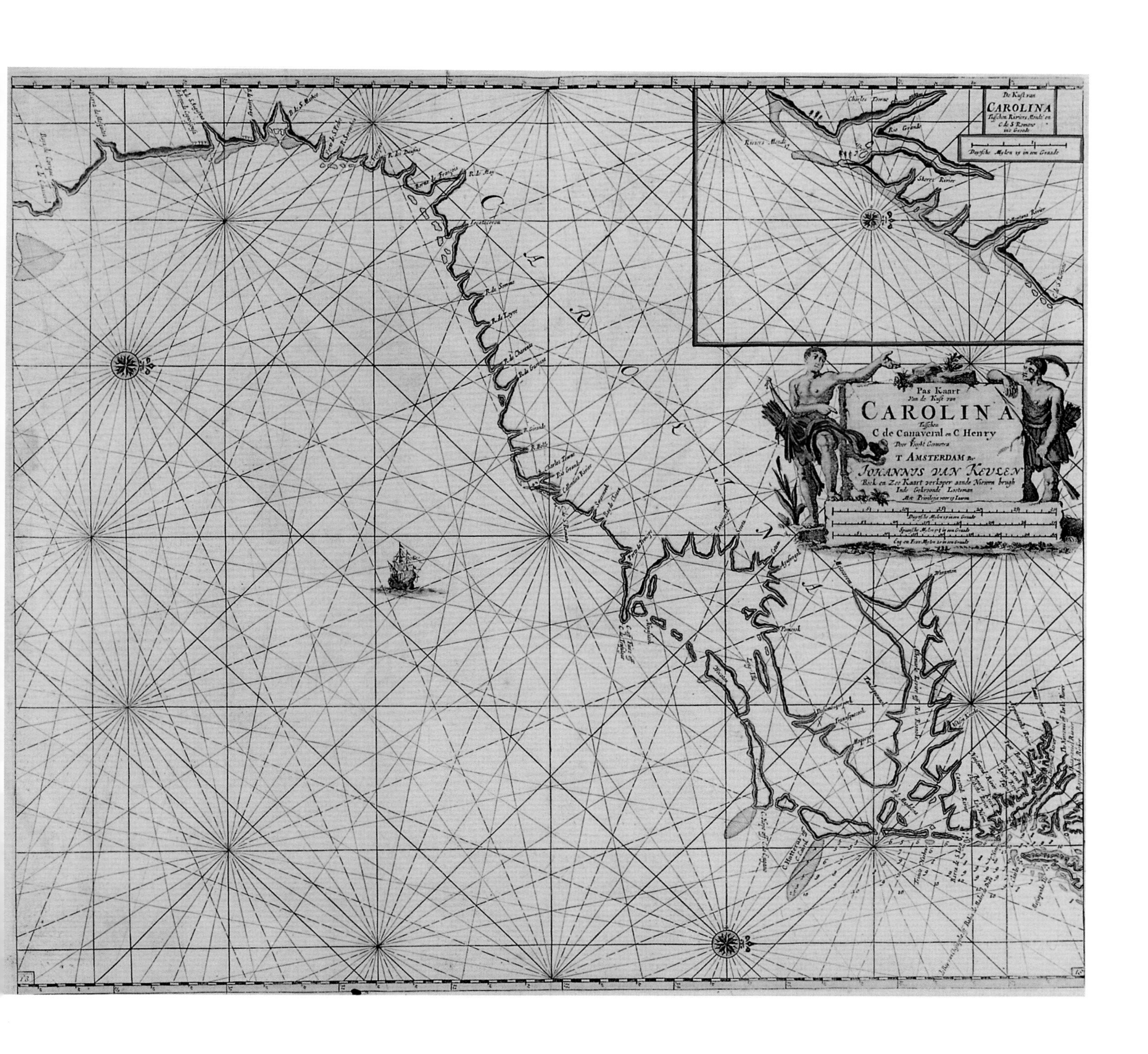

COLOR PLATE 5. Johannis van Keulen. Pas Kaart Van de Kust van Carolina. 1682. [Map 91].

COLOR PLATE 6. John Thornton, Robert Morden, and Philip Lea. A New Map of Carolina [A]. Ca. 1685. [Map 104].

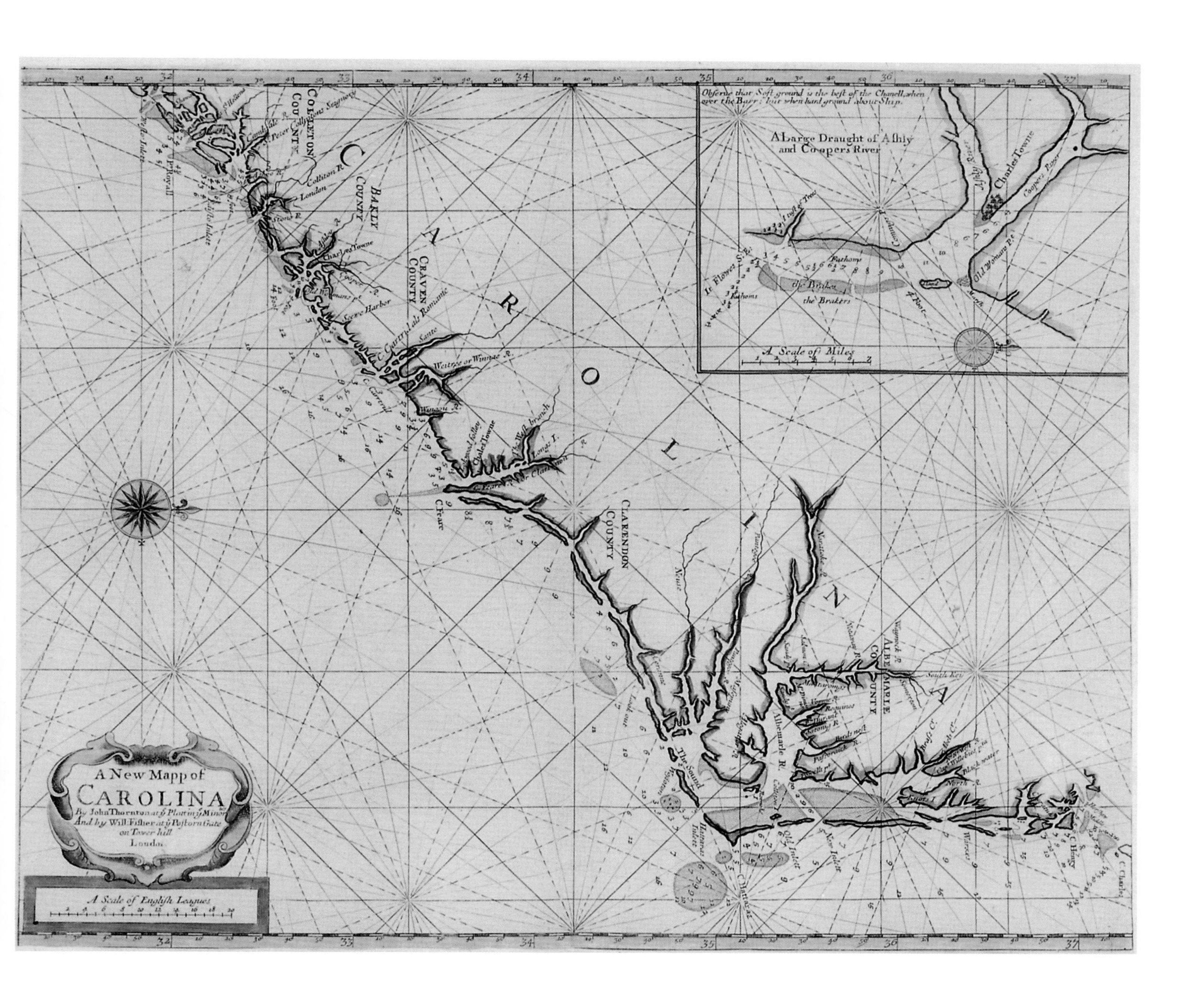

COLOR PLATE 7. John Thornton and Will: Fisher. A New Mapp of Carolina. 1698. [Map 123].

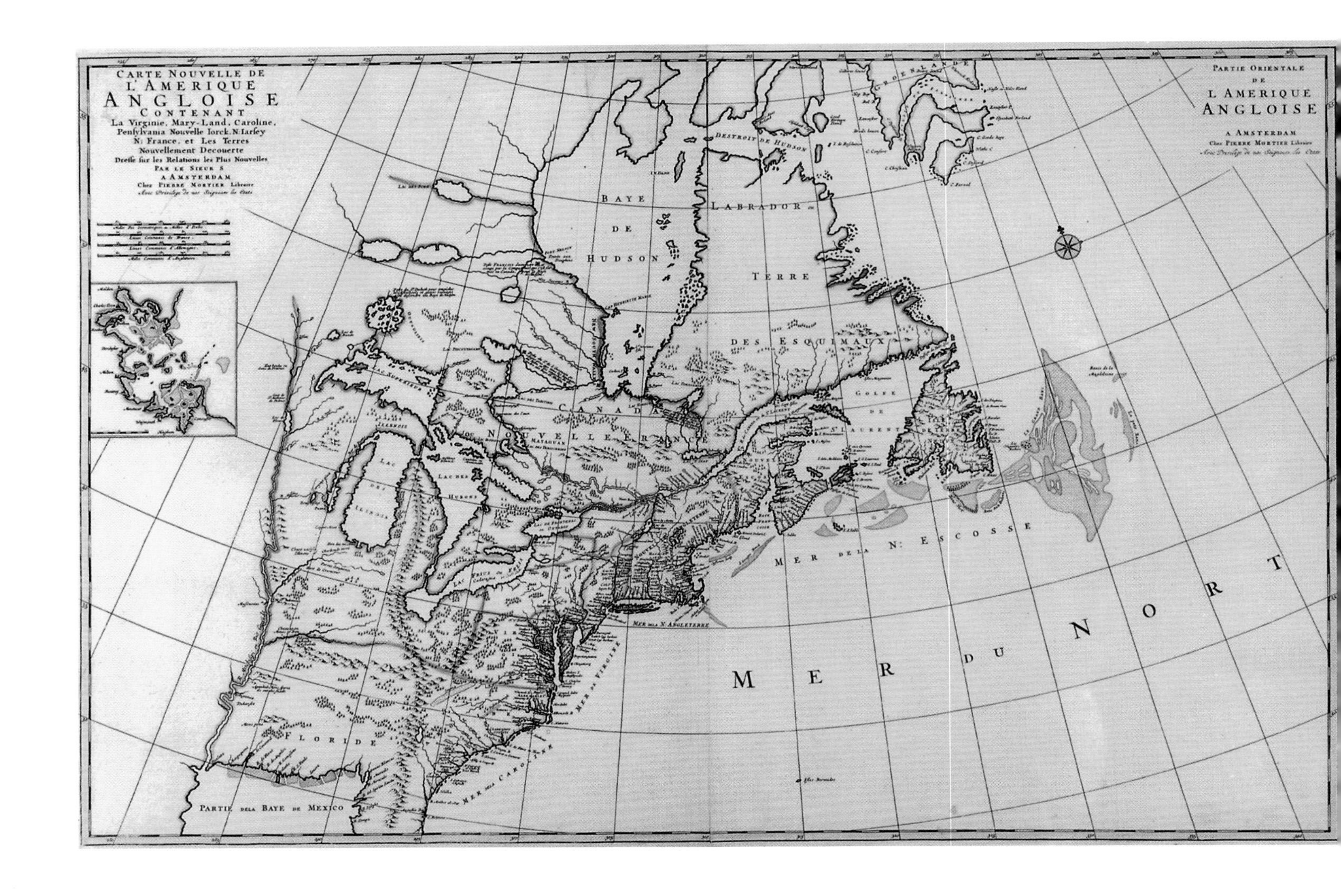

COLOR PLATE 8. Nicolas Sanson d'Abbeville. Carte Nouvelle de l'Amerique Angloise. 1700. [Map 129].

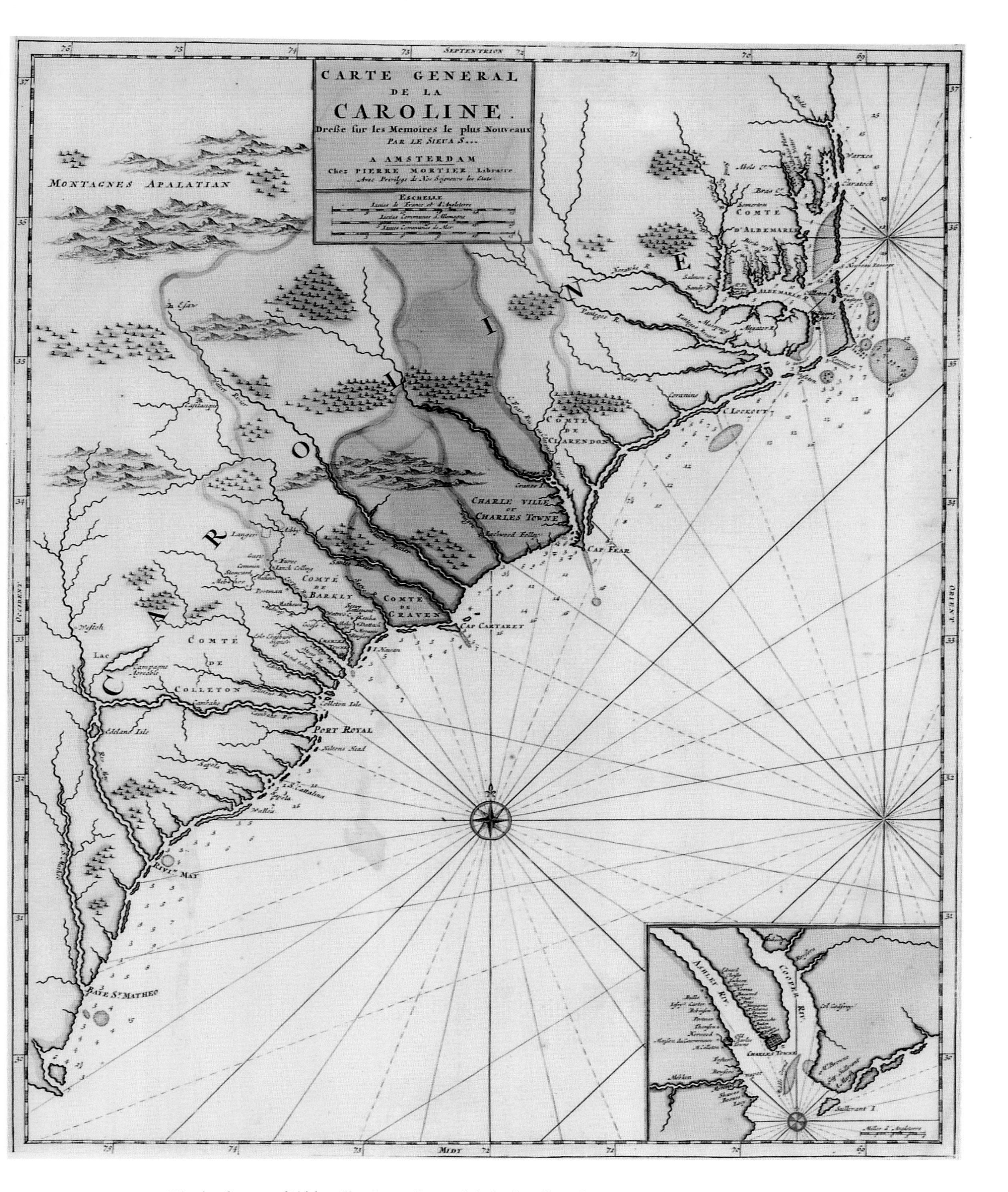

COLOR PLATE 9. Nicolas Sanson d'Abbeville. Carte General de la Caroline. Ca. 1696 A. [Map 120].

COLOR PLATE 10. Nicolas Sanson d'Abbeville. Carte Particuliere de la Caroline. Ca. 1696 B. [Map 121].

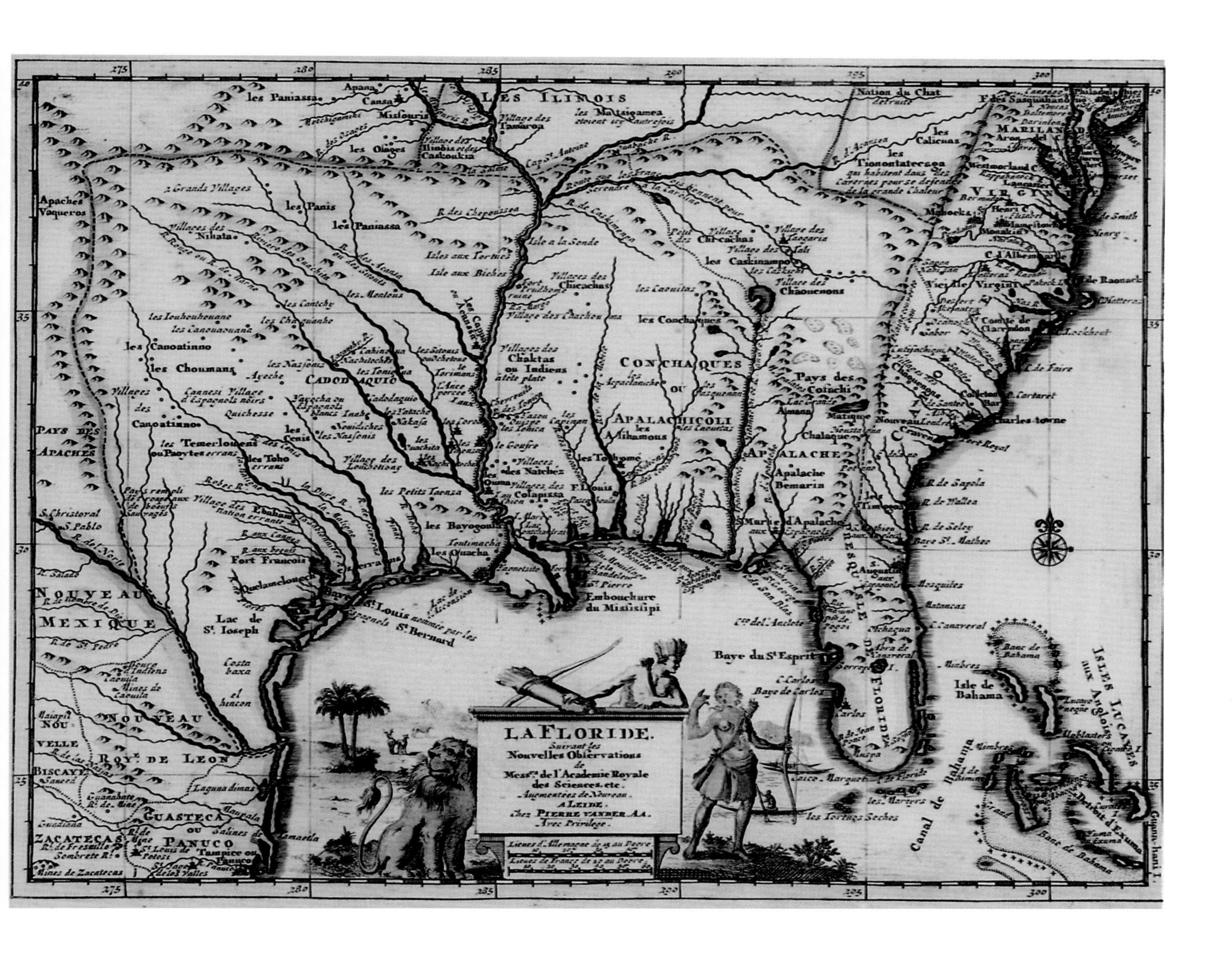

COLOR PLATE 11. Pierre vander Aa. La Floride. 1713. [Map 155].

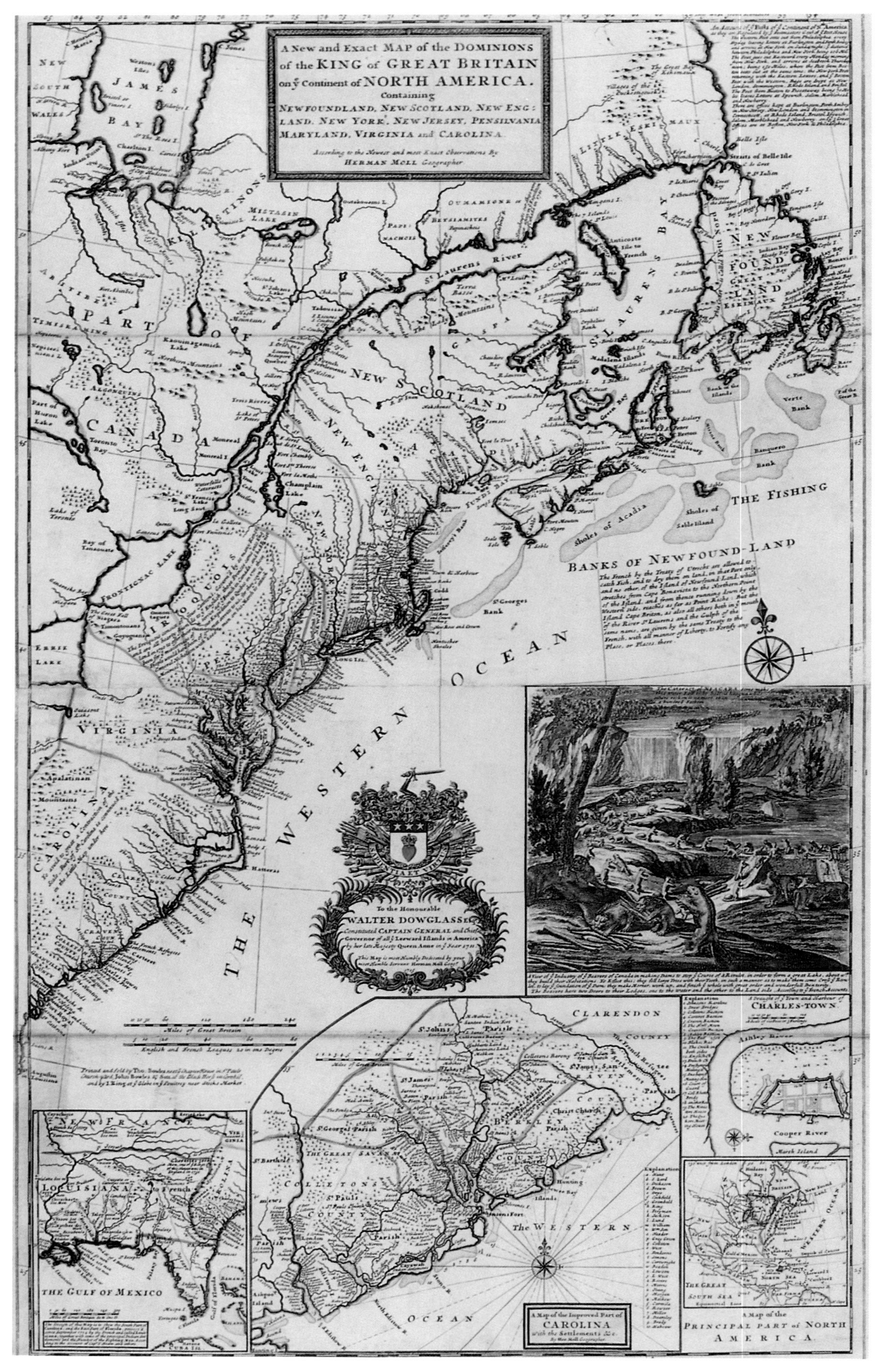

COLOR PLATE 12. Herman Moll. A New and Exact Map of the Dominions of the King of Great Britain on y^e Continent of North America. 1715. [Map 158].

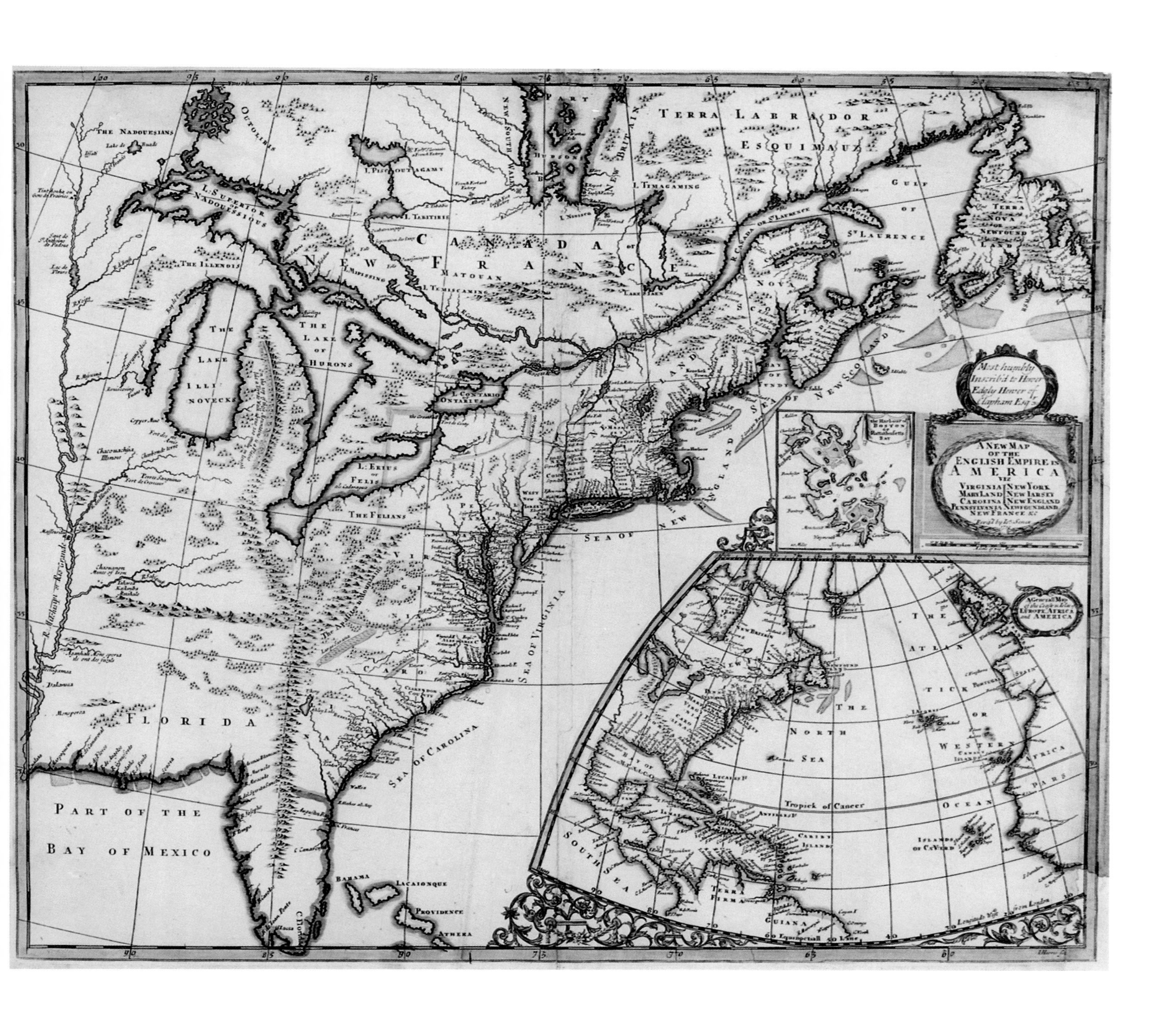

COLOR PLATE 13. John Senex. A New Map of the English Empire in America. 1719. [Map 172].

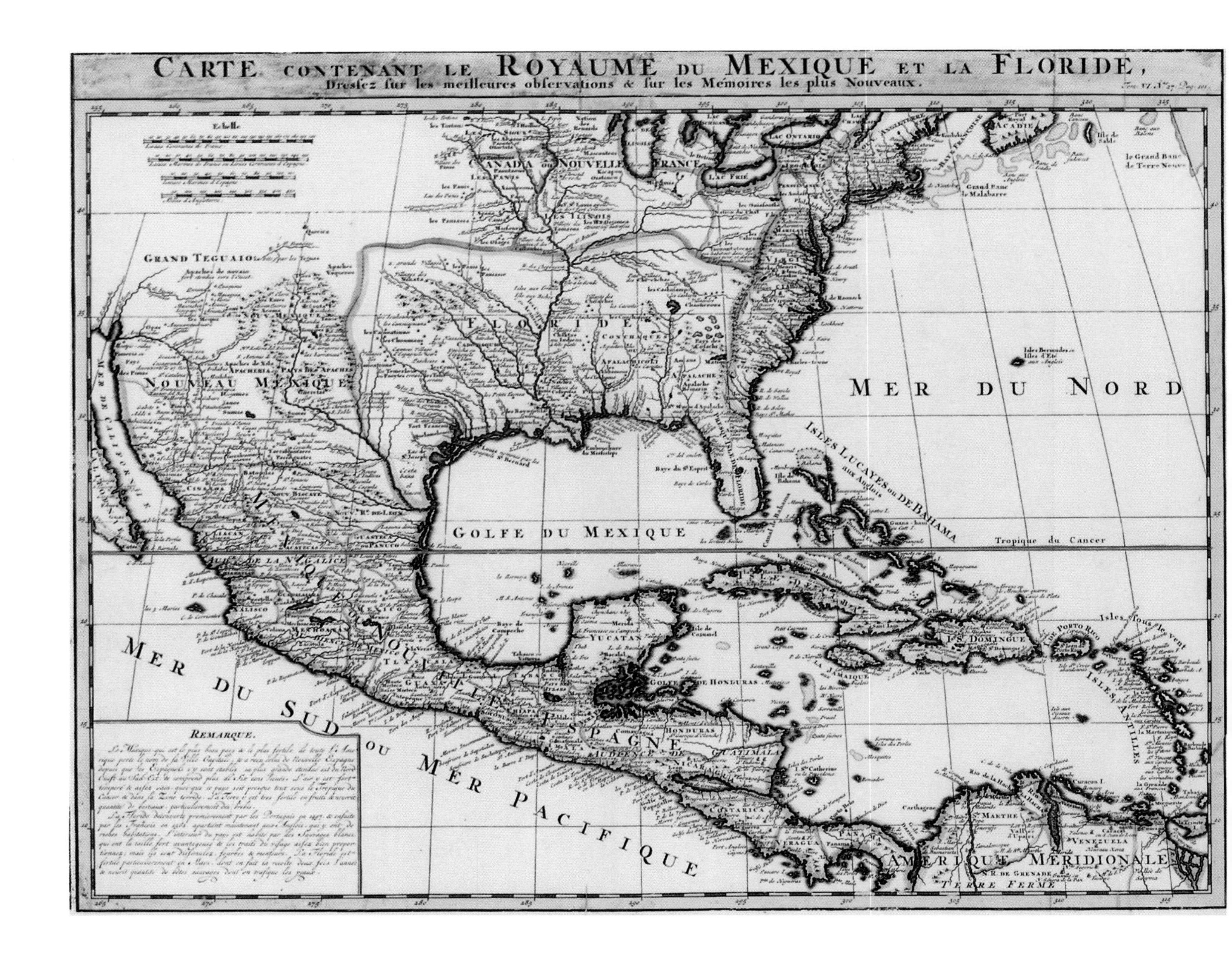

COLOR PLATE 14. Henri A. Chatelain. Carte contenant le Royaume du Mexique et la Floride. 1719. [Map 173].

COLOR PLATE 15. Guillaume De l'Isle. Tabula Geographica Mexicæ et Floridæ &c. 1722. [Map 191].

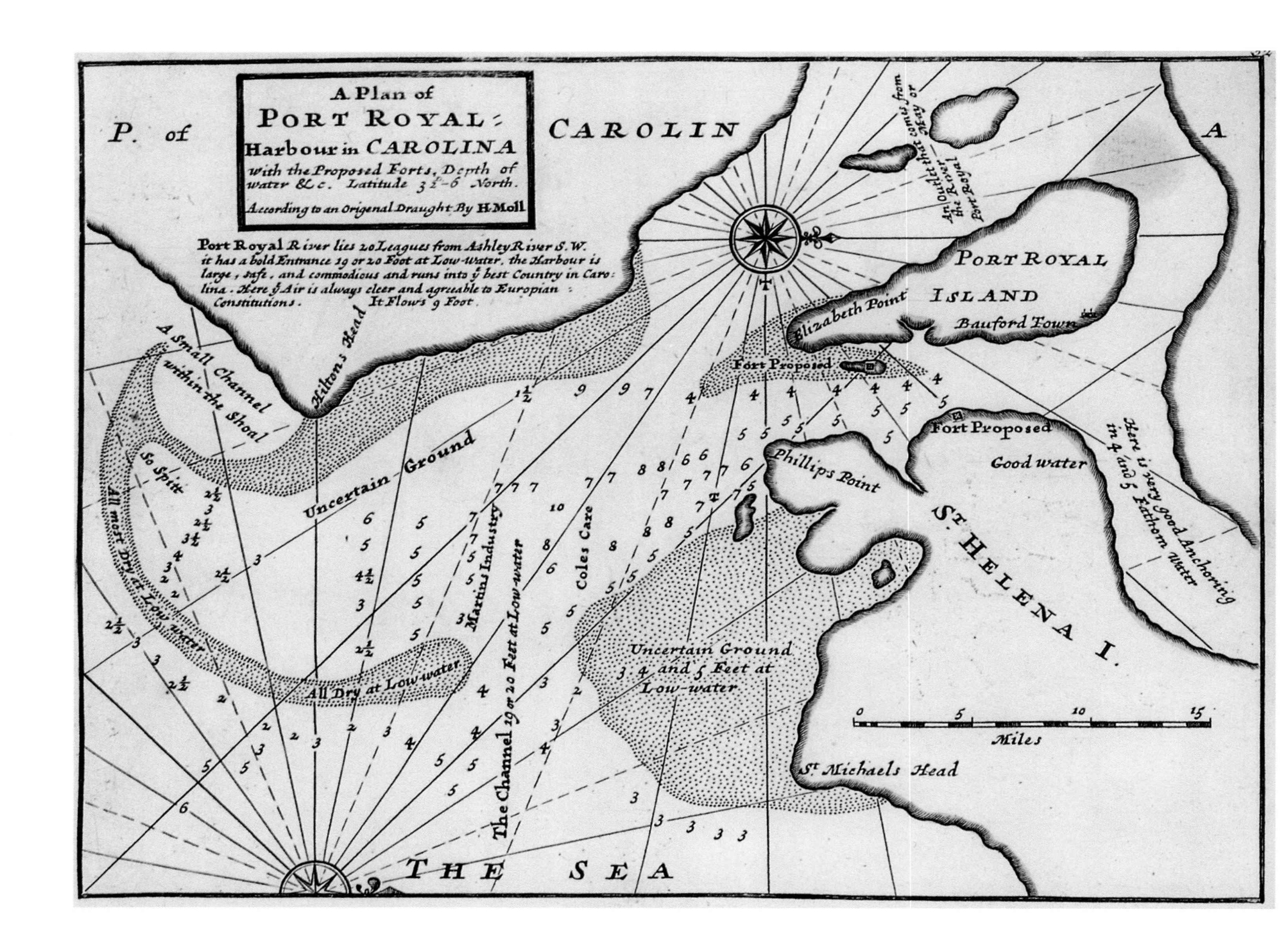

COLOR PLATE 16. Herman Moll. A Plan of Port Royal=Harbour in Carolina. 1728 A. [Map 200].

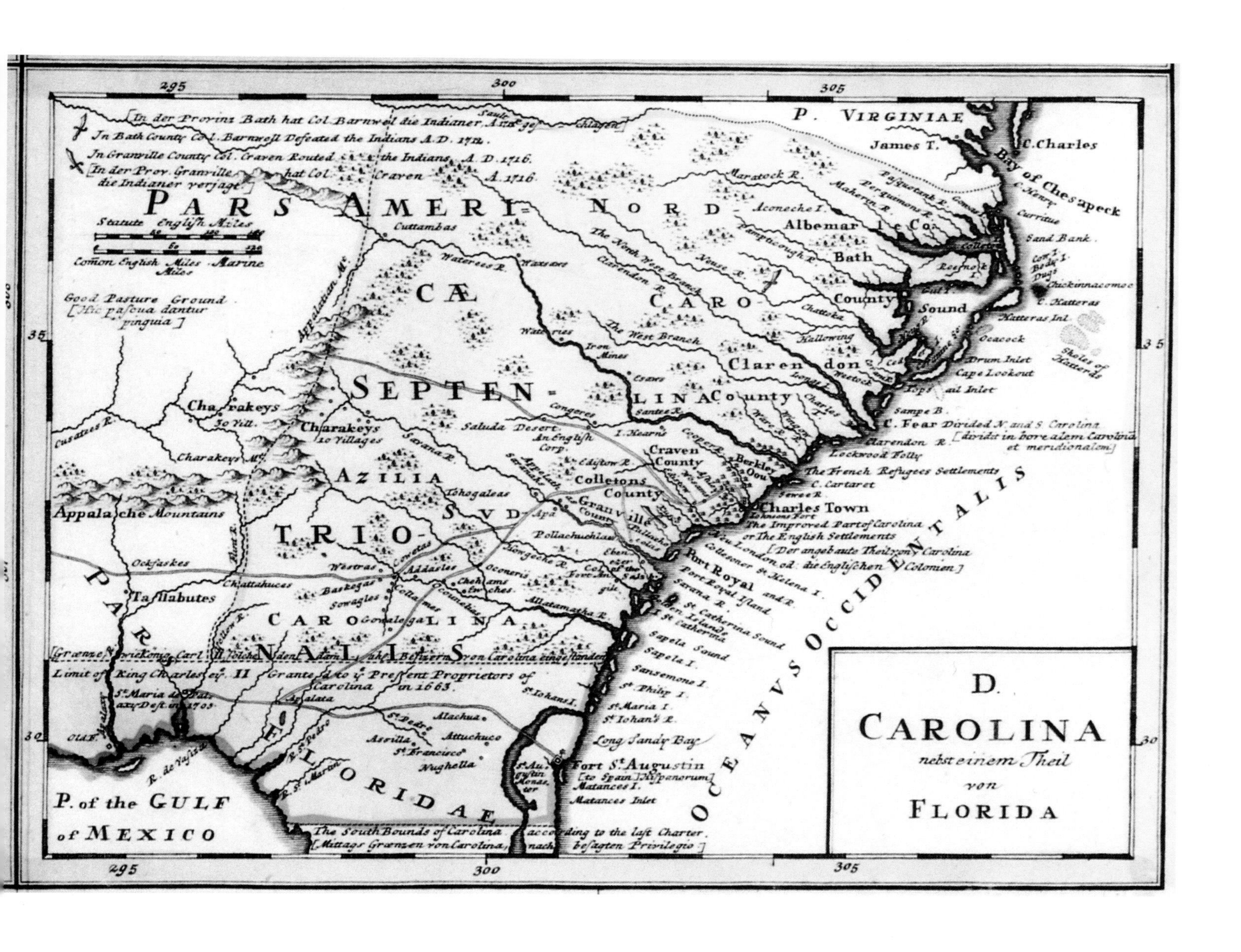

COLOR PLATE 17. Johann Baptist Homann D. Carolina nebst einem Theil von Florida. 1737. [Map 233].

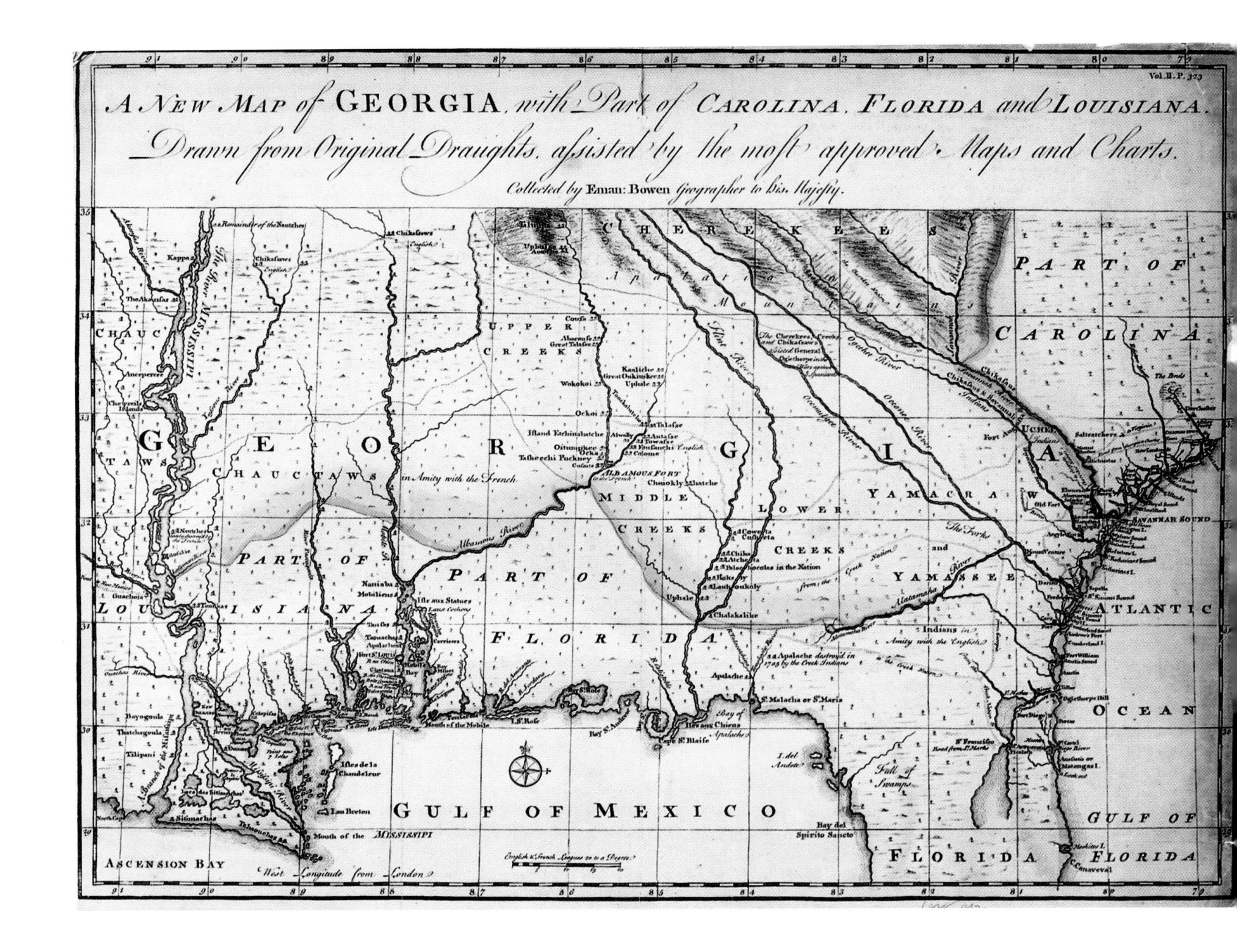

COLOR PLATE 18. Emanuel Bowen. A New Map of Georgia, with Part of Carolina, Florida and Louisiana. 1748. [Map 267].

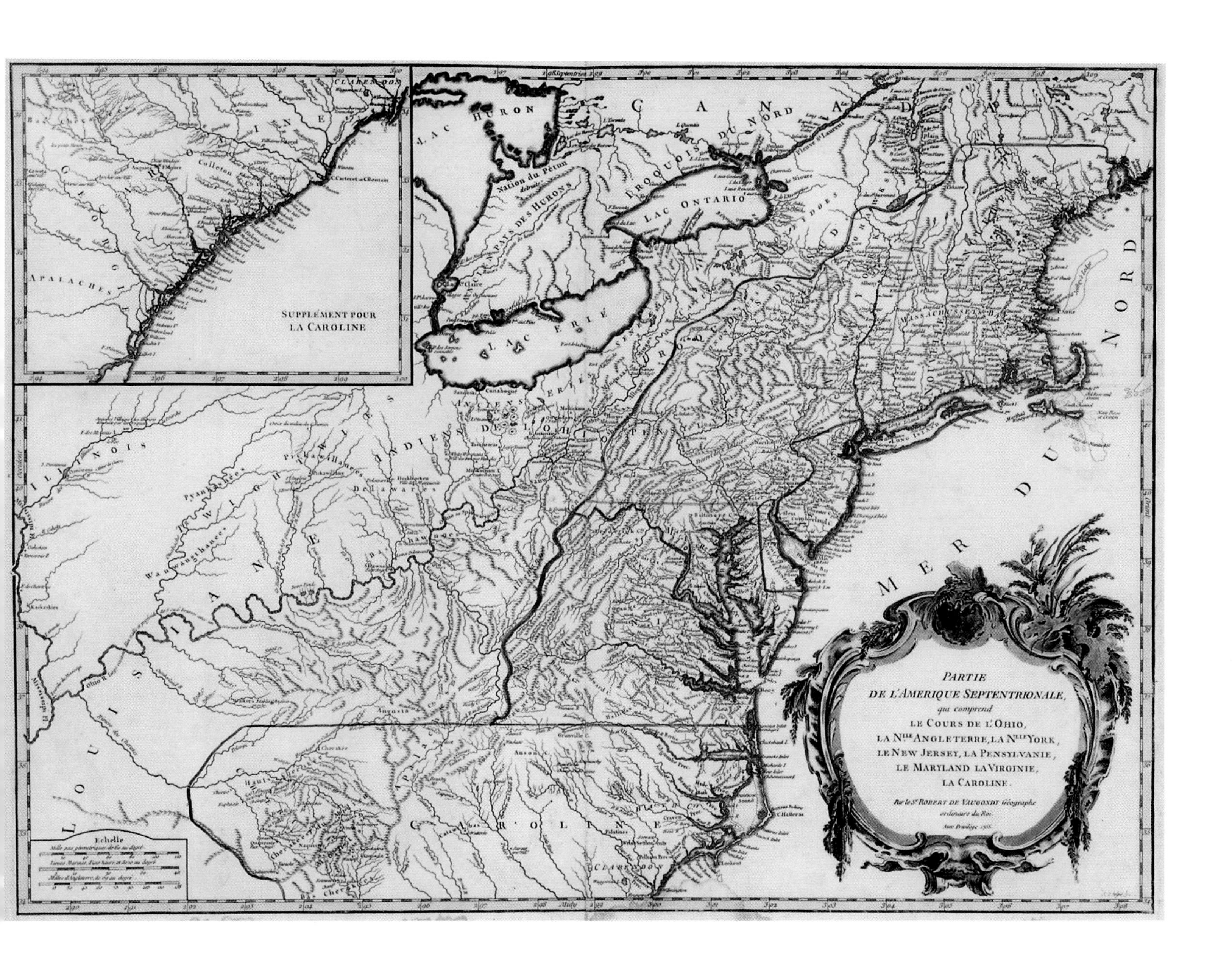

COLOR PLATE 19. Robert de Vaugondy. Partie de l'Amerique Septentrionale. 1755. [Map 295].

COLOR PLATE 20. Rigobert Bonne. Carte de la Lousiane et de la Floride. Ca. 1760. [Map 326].

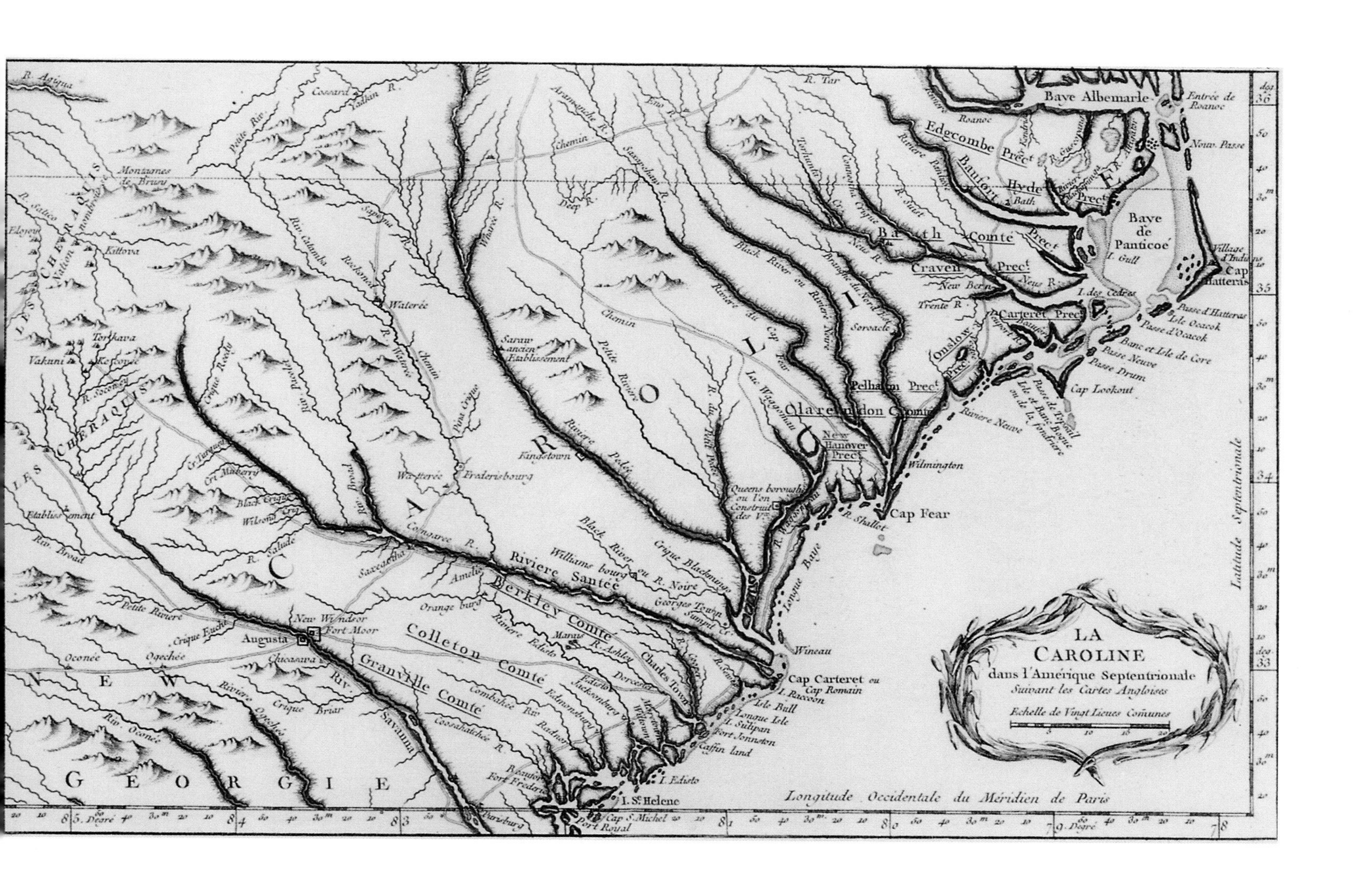

COLOR PLATE 21. Jacques Nicolas Bellin. La Caroline dans l'Amérique Septentrionale. 1764 A. [Map 342].

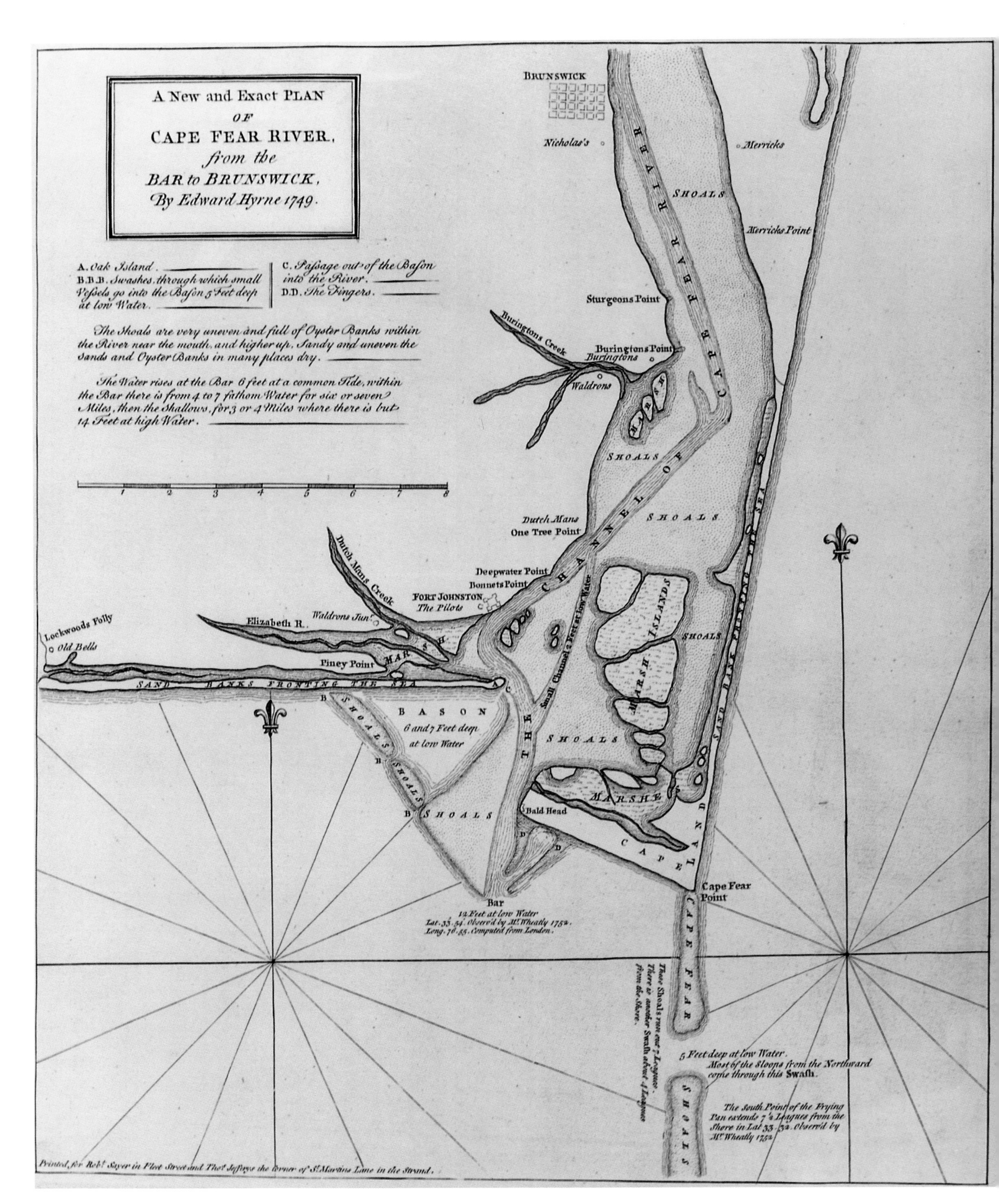

COLOR PLATE 22. Edward Hyrne. A New and Exact Plan of Cape Fear River from the Bar to Brunswick. 1768. [Map 368].

COLOR PLATE 23. Joshua Fry and Peter Jefferson. A Map of the most Inhabited part of Virginia containing the whole Province of Maryland with Part of Pensilvania, New Jersey and North Carolina. 1775. [Map 449].

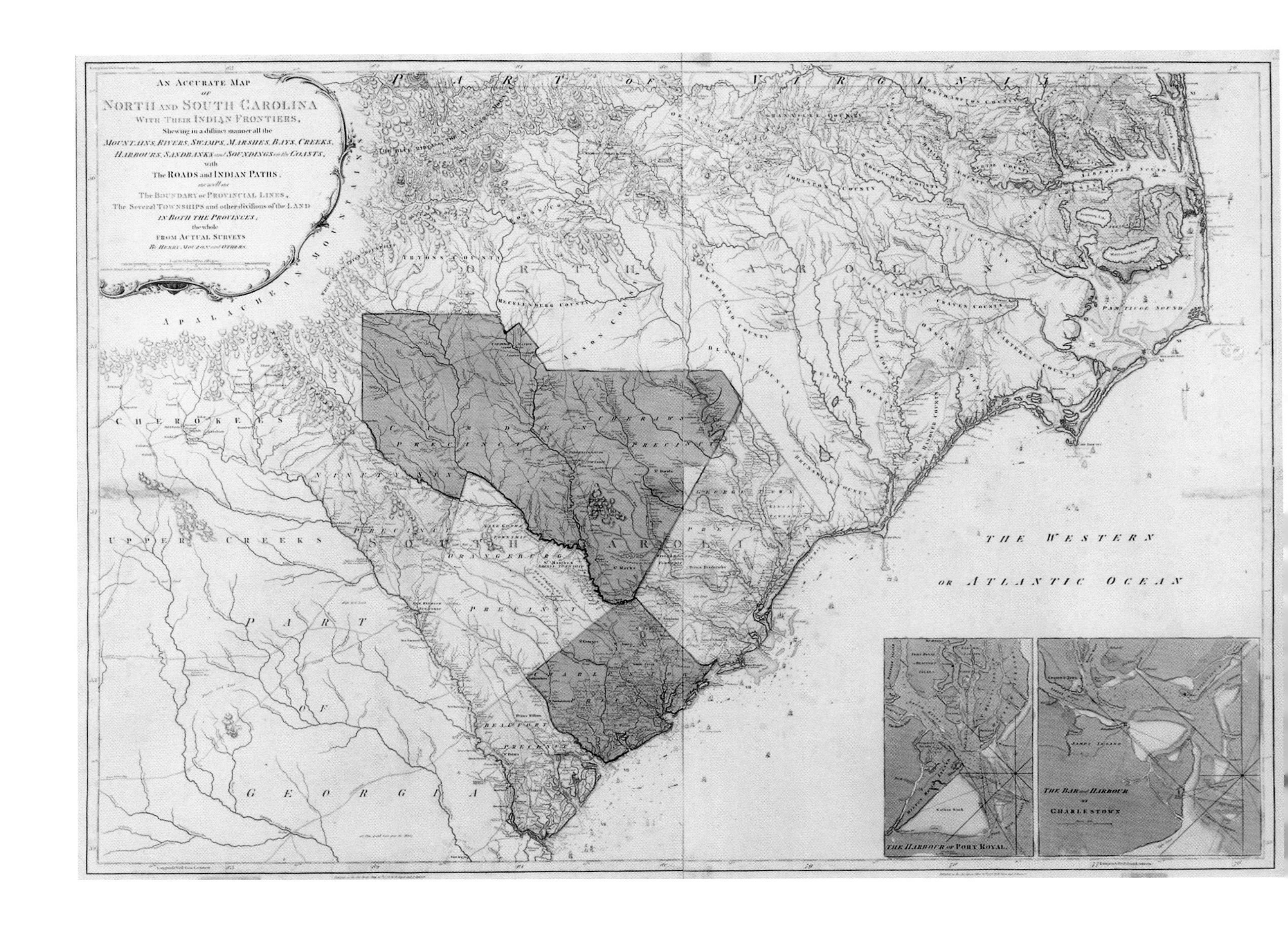

COLOR PLATE 24. Henry Mouzon. An Accurate Map of North and South Carolina, with their Indian Frontiers. 1775. [Map 450]

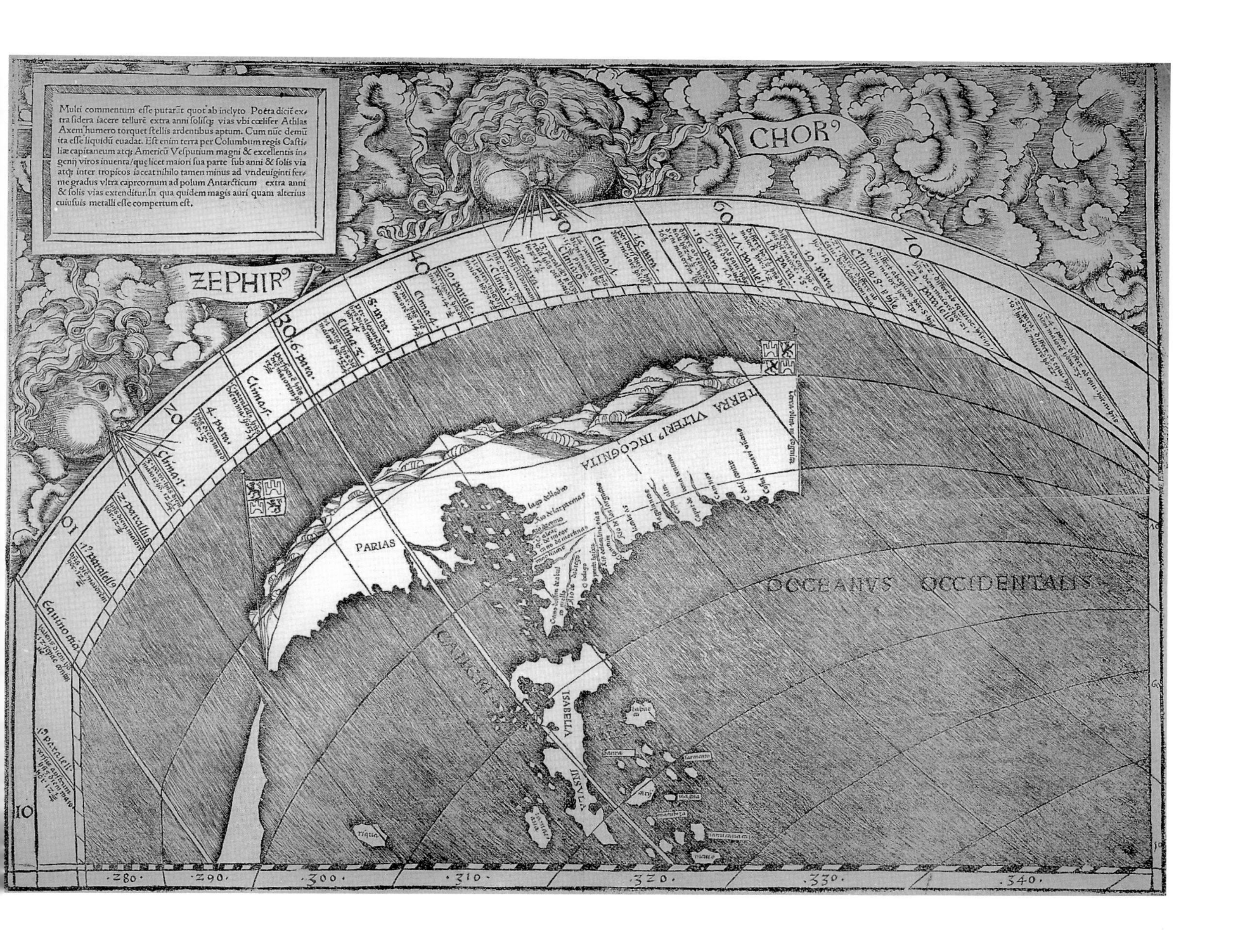

PLATE I. Martin Waldseemüller (Hylacomylus). Universalis Cosmographia [detail]. 1507. [Map 01].

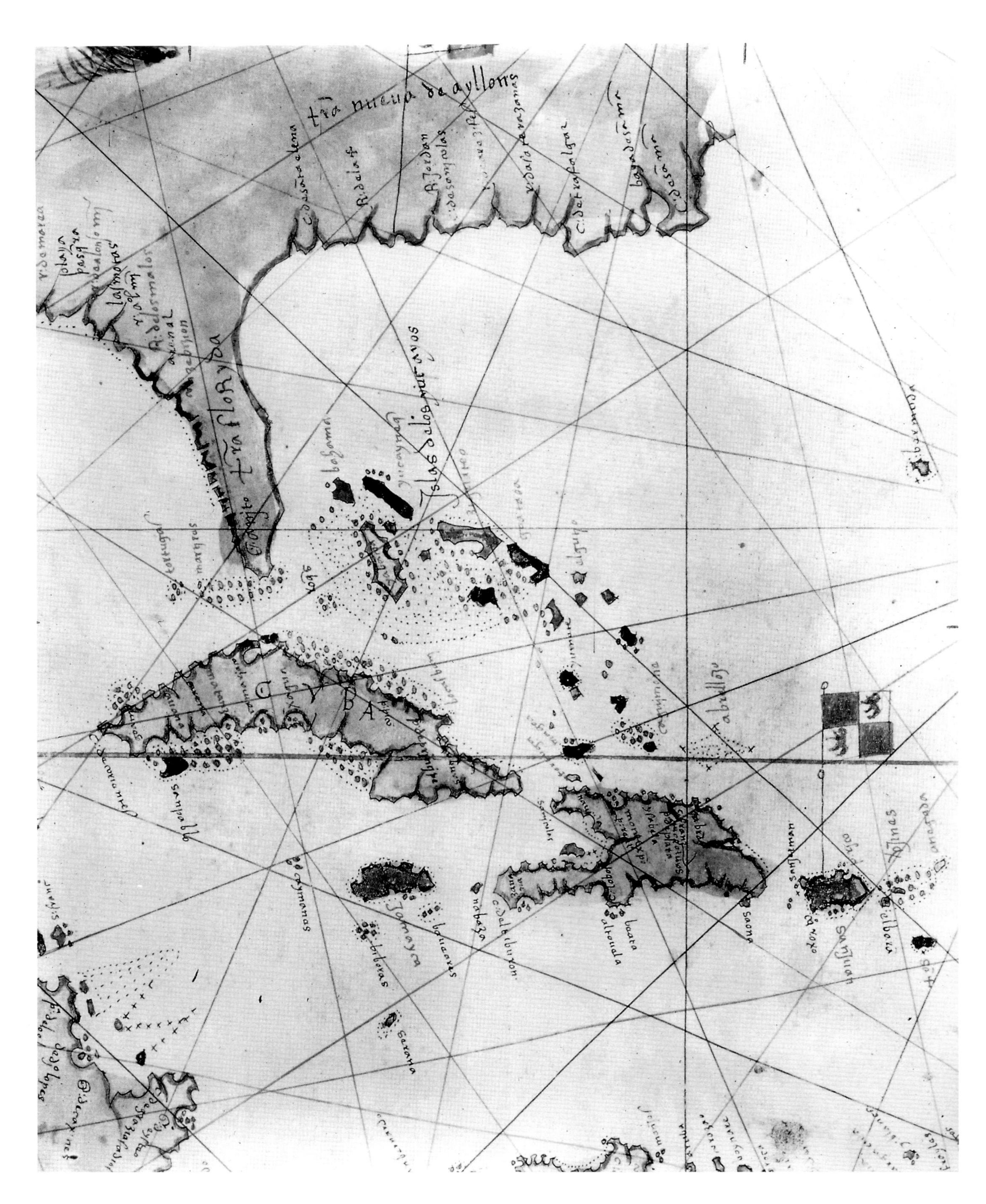

PLATE 2. Juan Vespucci. World Map [detail]. 1526 MS. [Map 02].

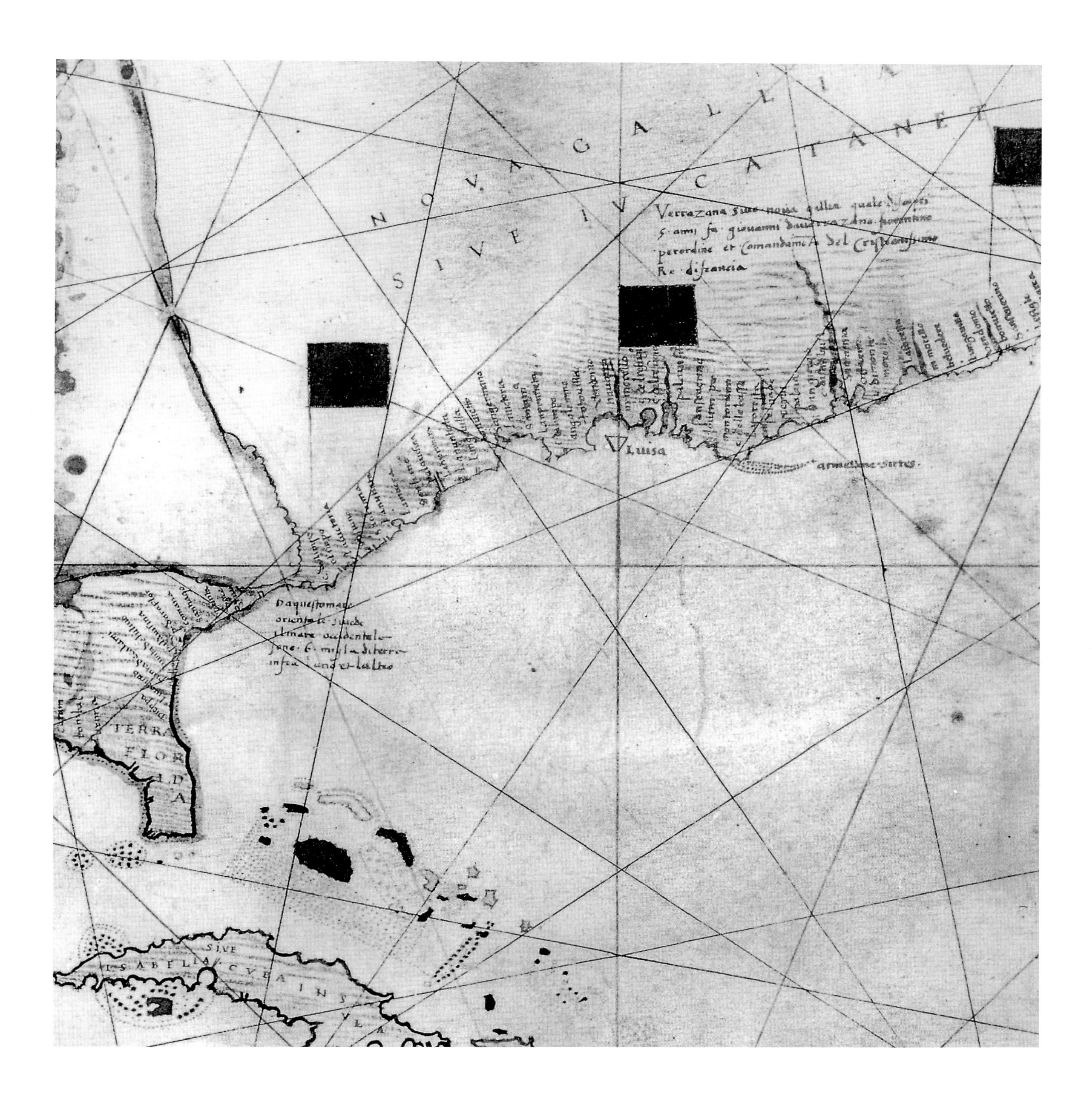

PLATE 3. Gerolamo da Verrazano. World Map [detail]. 1529 MS. [Map 03].

PLATE 4. Diego Ribero. World Map [detail]. 1529 MS. [Map 04].

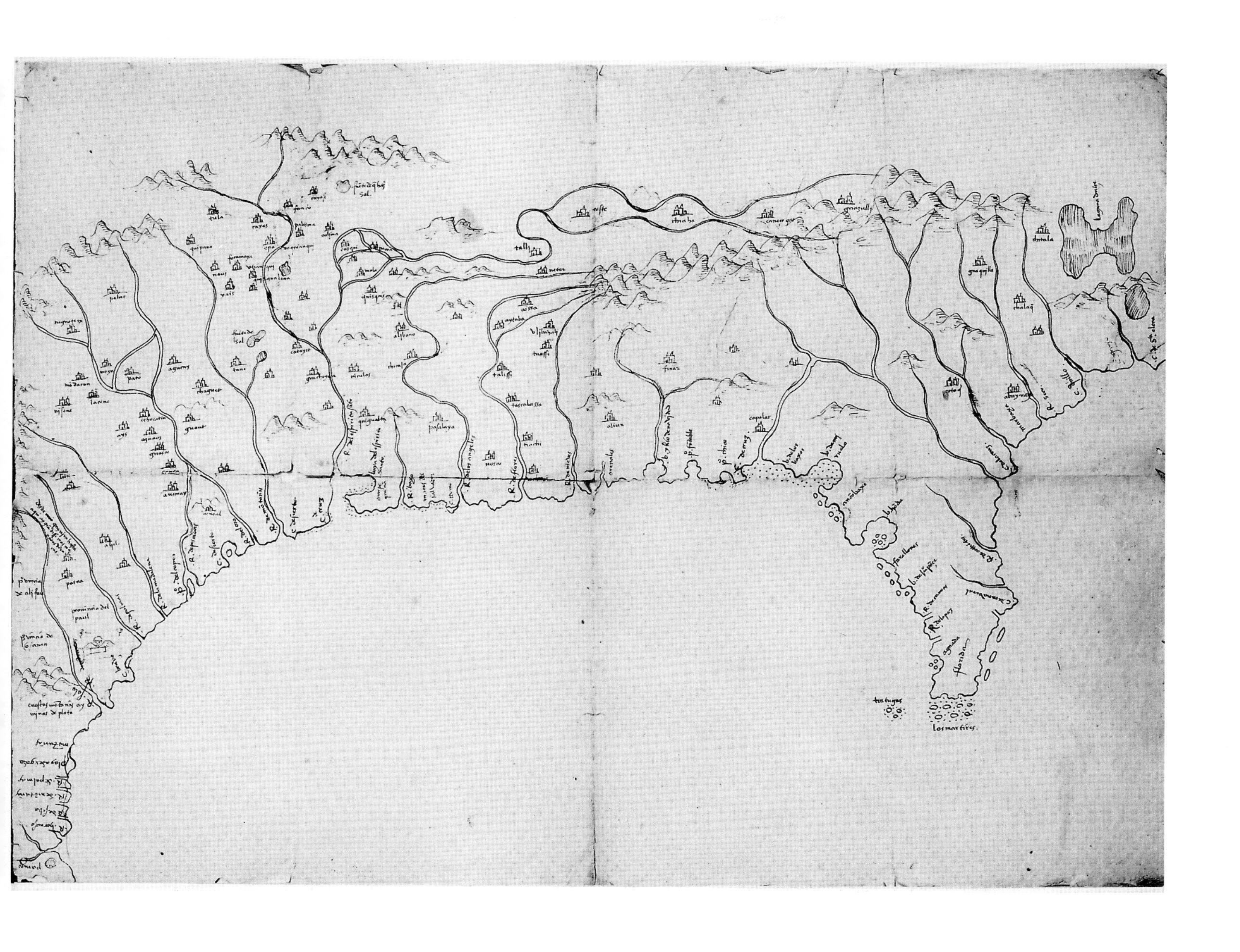

PLATE 5. De Soto. Mapa del Golfo y costa de la Nueva España. Ca. 1544 MS. [Map 1].

PLATE 6. Diego Gutiérrez. Americae . . . Nova Et Exactissima Descriptio [detail]. 1562. [Map 2].

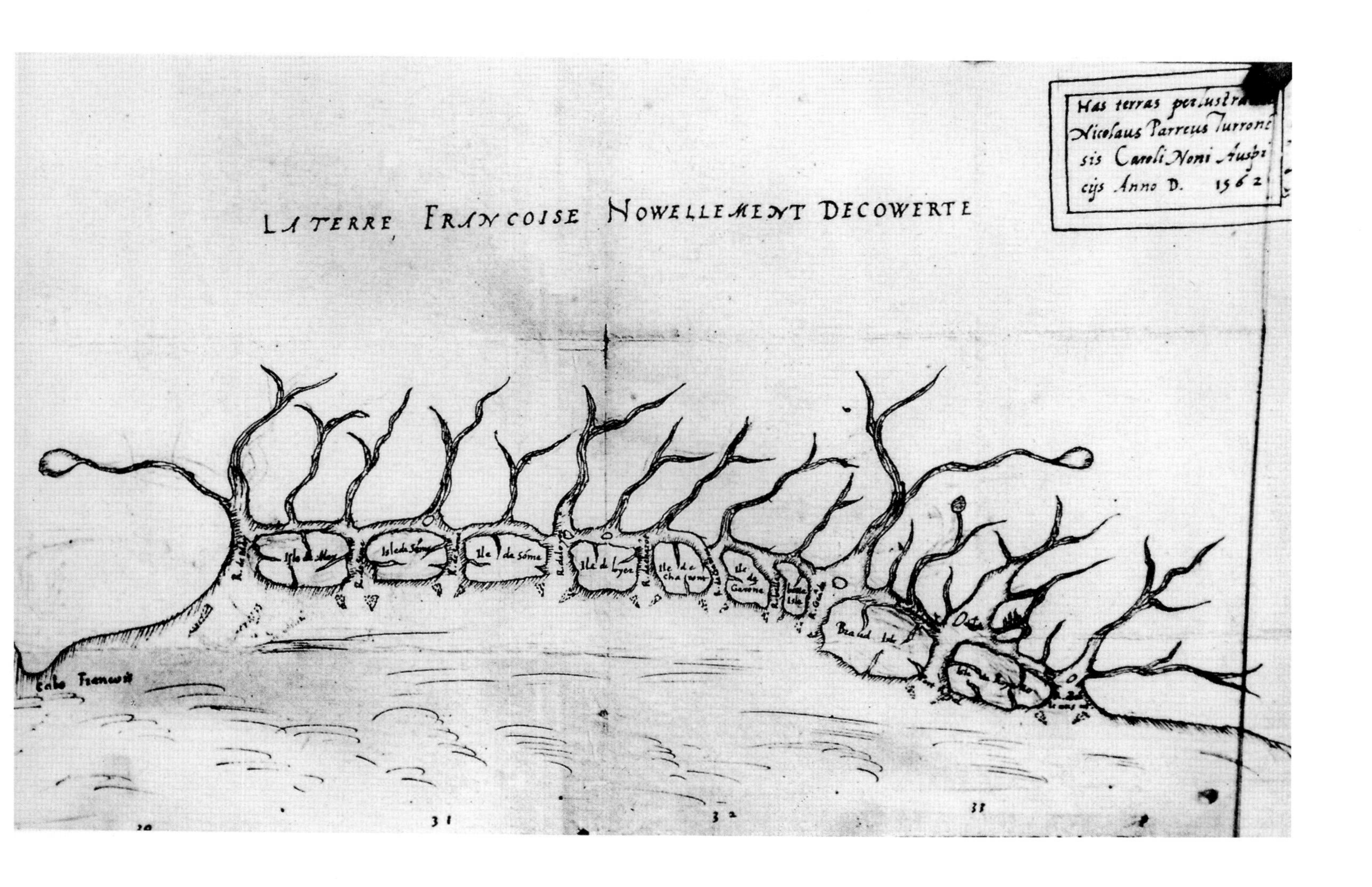

PLATE 6A. Parreus. French Florida. 1562 [1563] MS. [Map 2A].

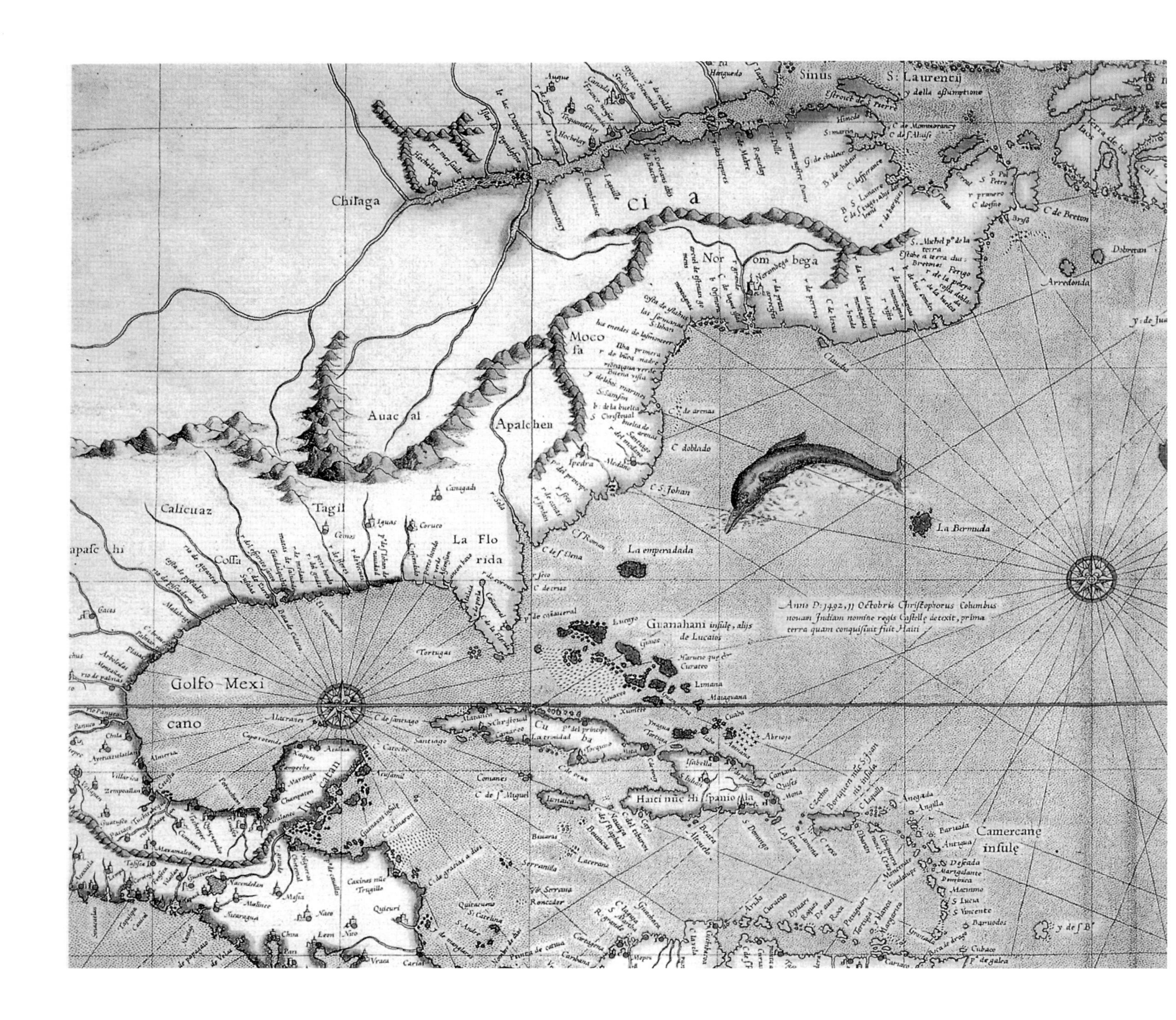

PLATE 7. Gerard Mercator. Nova Et Aucta Orbis Terrae Descriptio . . . [detail]. 1569. [Map 3A].

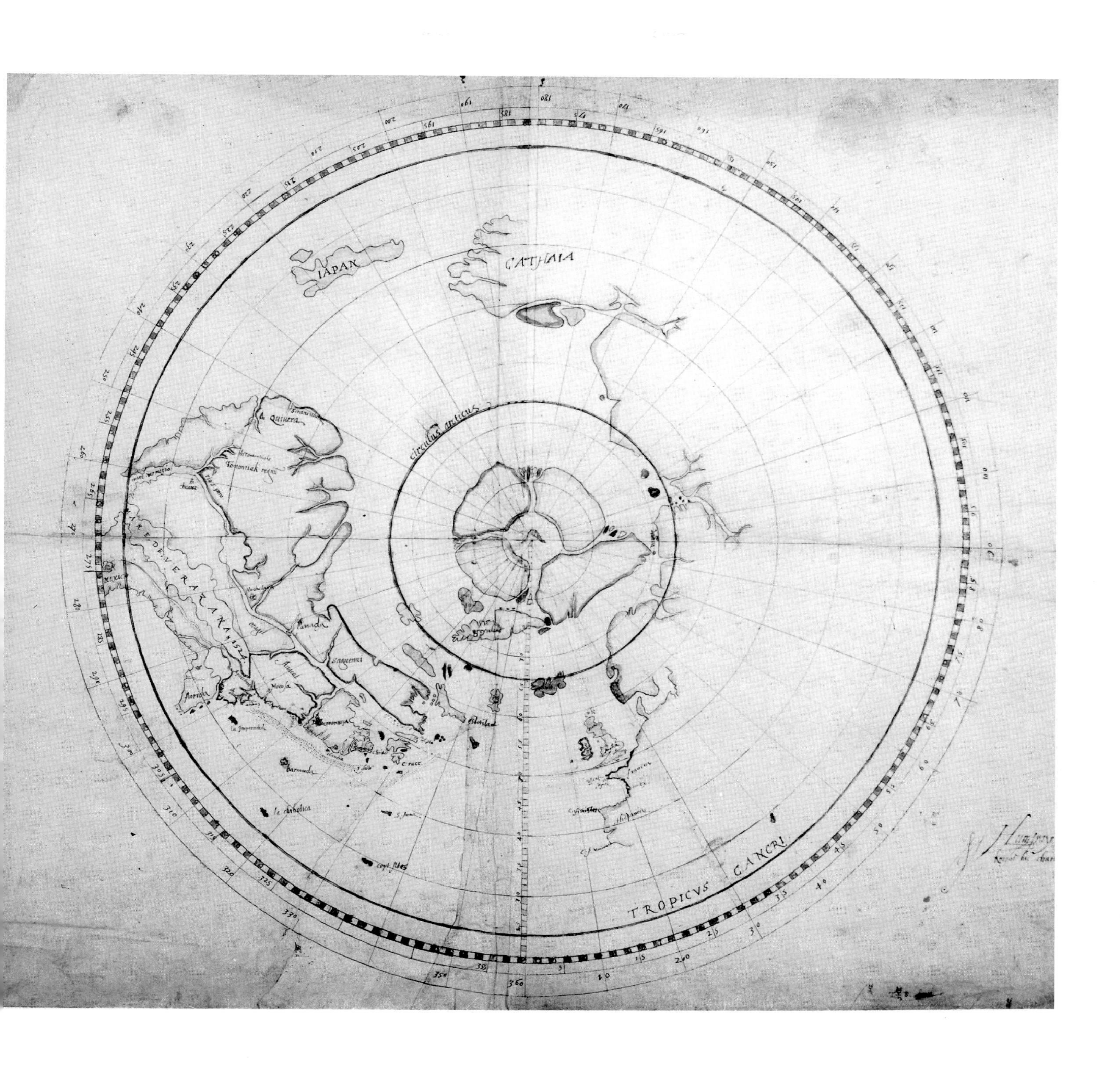

PLATE 8. John Dee. [North Pole and part of the Northern Hemisphere]. Ca. 1582 MS. [Map 4B].

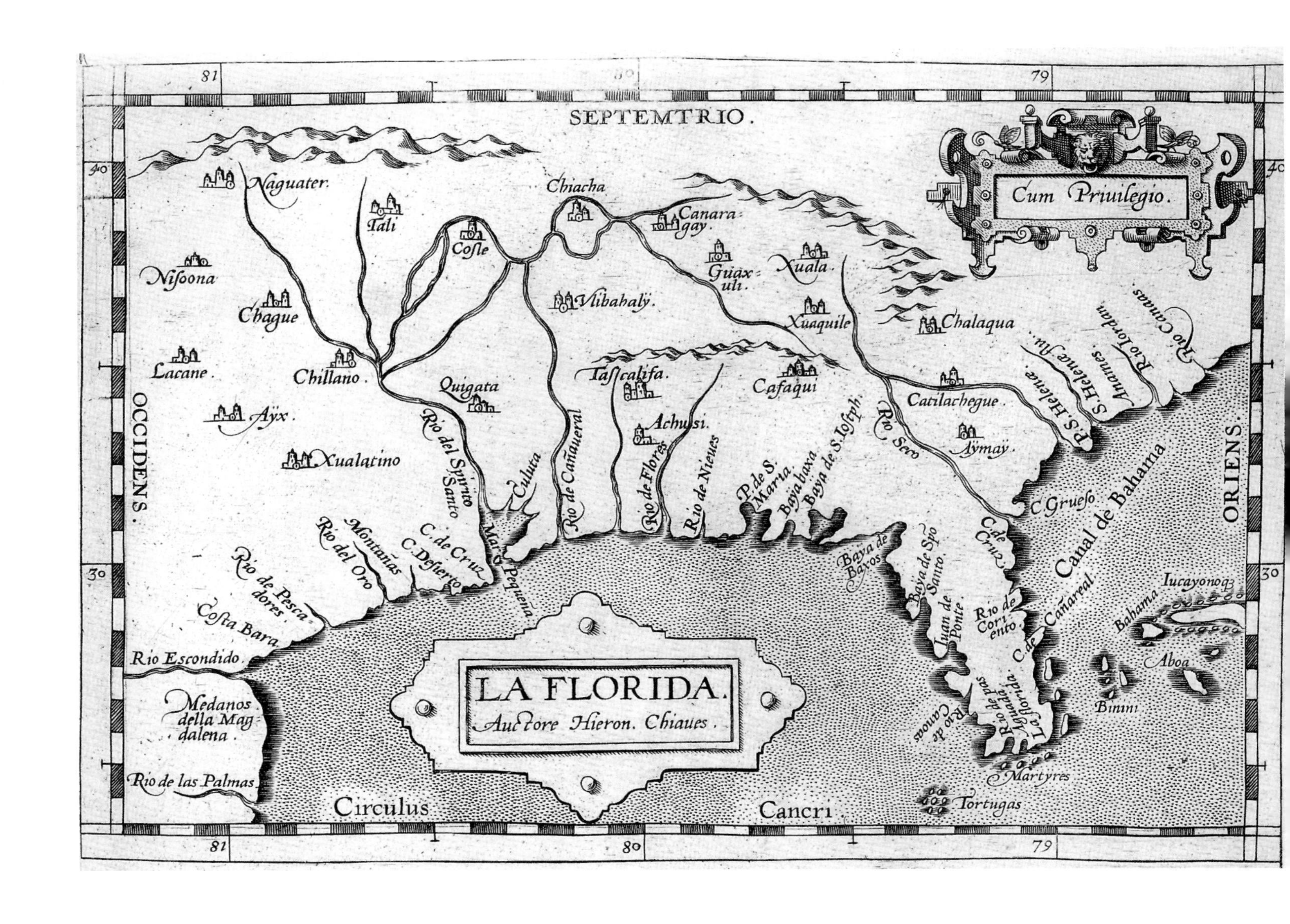

PLATE 9. Ortelius-Chiaves. La Florida. 1584. [Map 5].

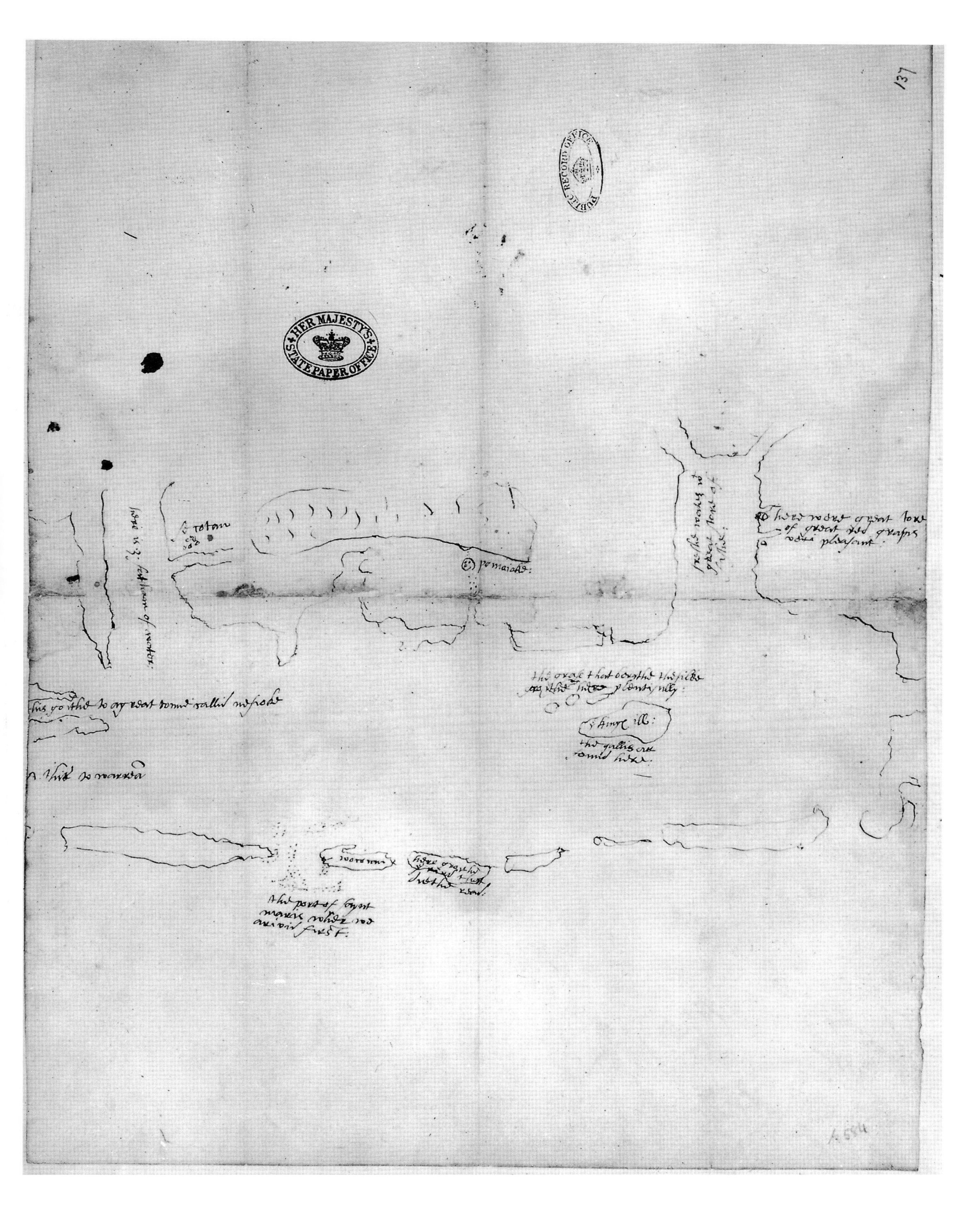

PLATE 10. Anonymous. A discription of the land of virginia. Ca. 1585 MS. [Map 6].

PLATE 11. White. La Virgenia Pars. 1585 MS A. [Map 7].

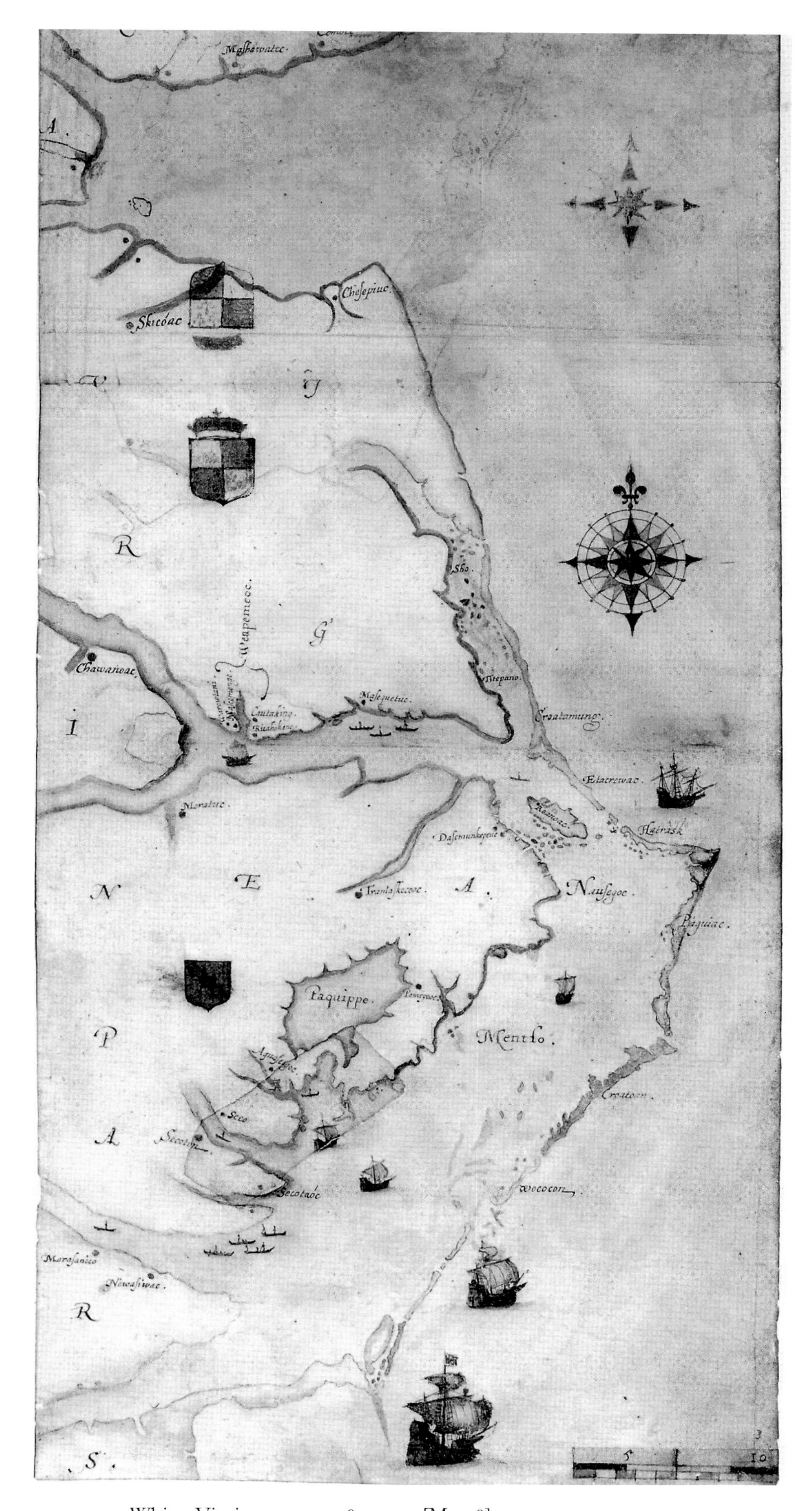

PLATE 12. White. Virginea pars. 1585 MS B. [Map 8].

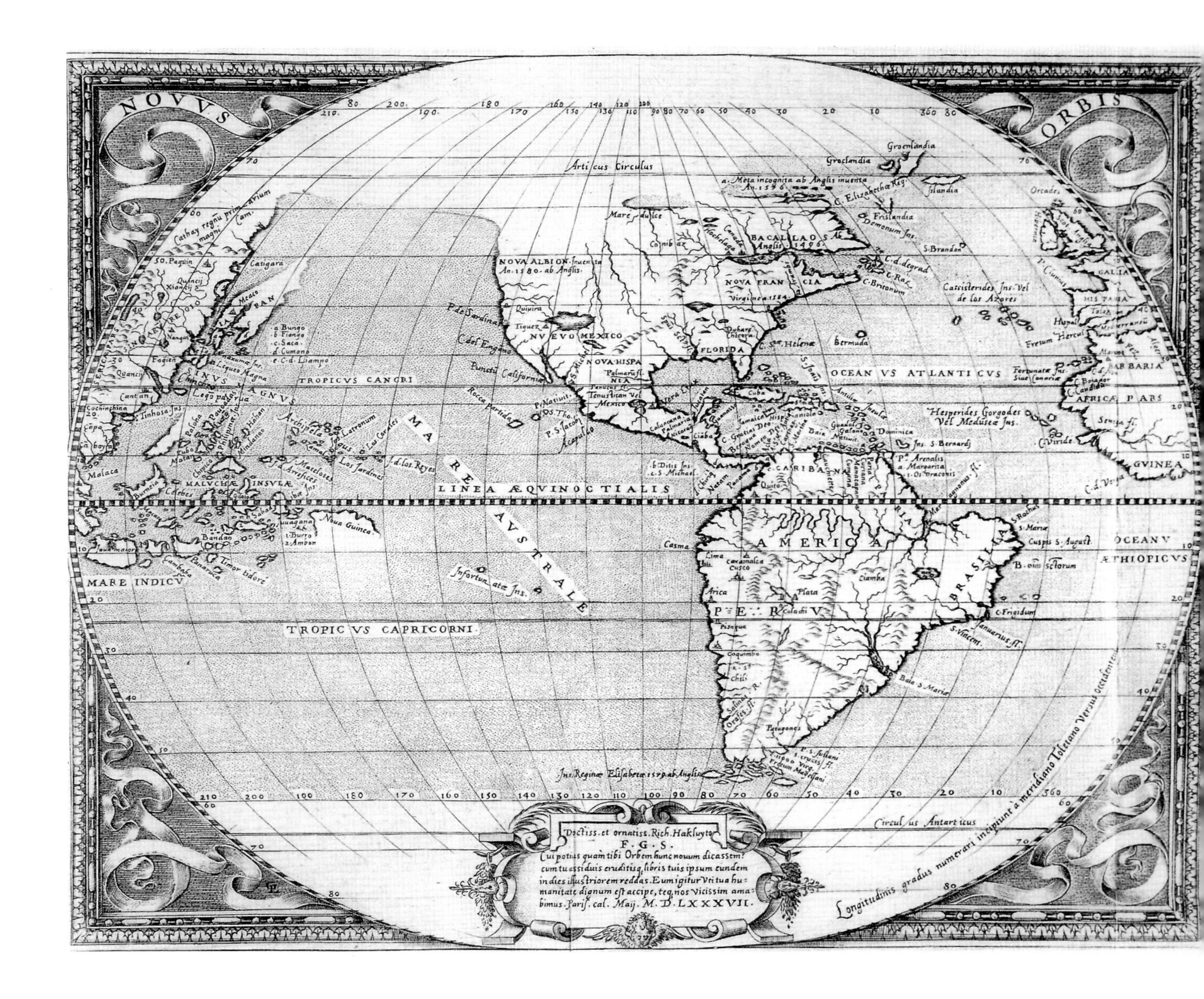

PLATE 13. Filips Galle. Novvs Orbis. 1587. [Map. 9].

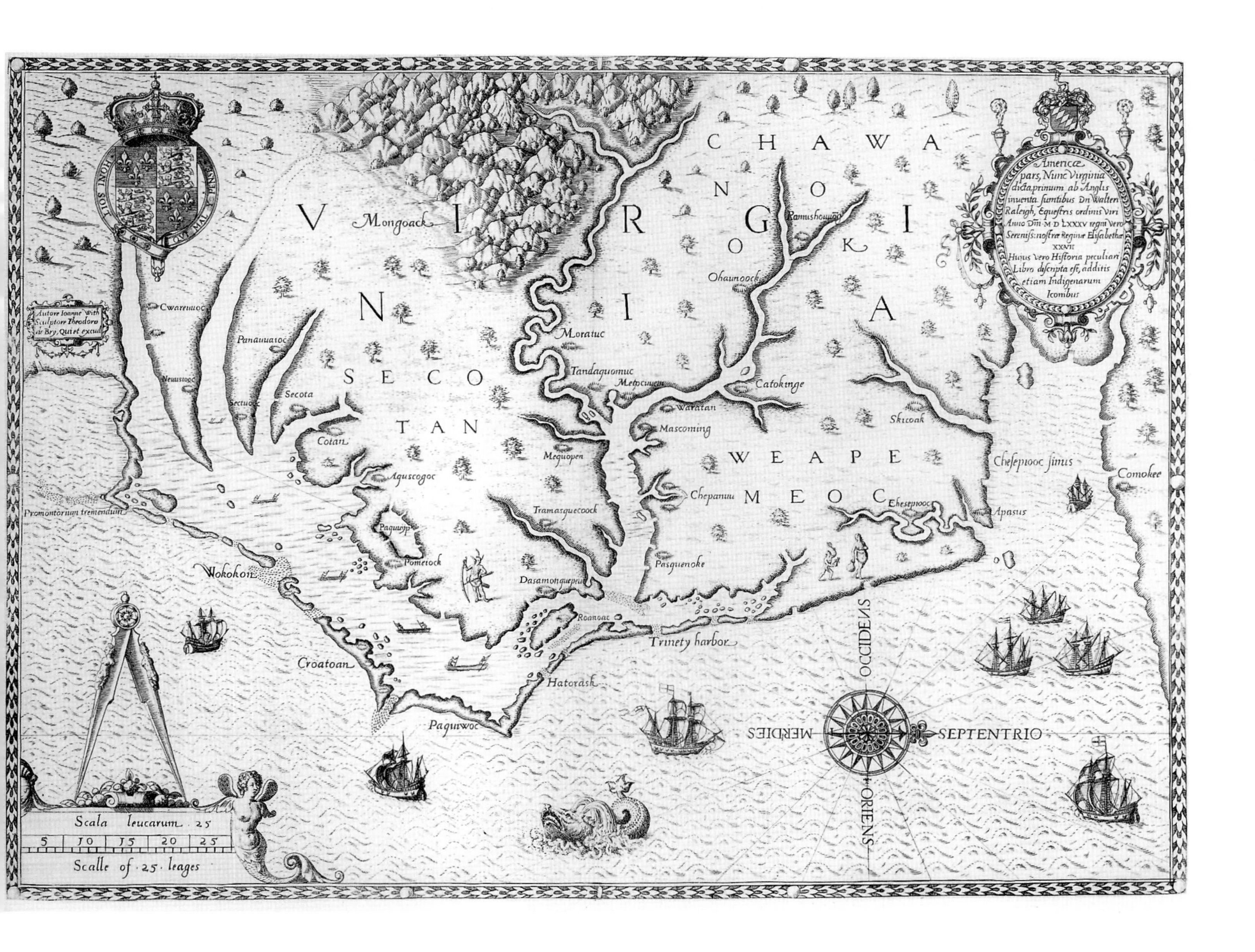

PLATE 14. White–De Bry. Americae pars, Nunc Virginia dicta. 1590 A. [Map 12].

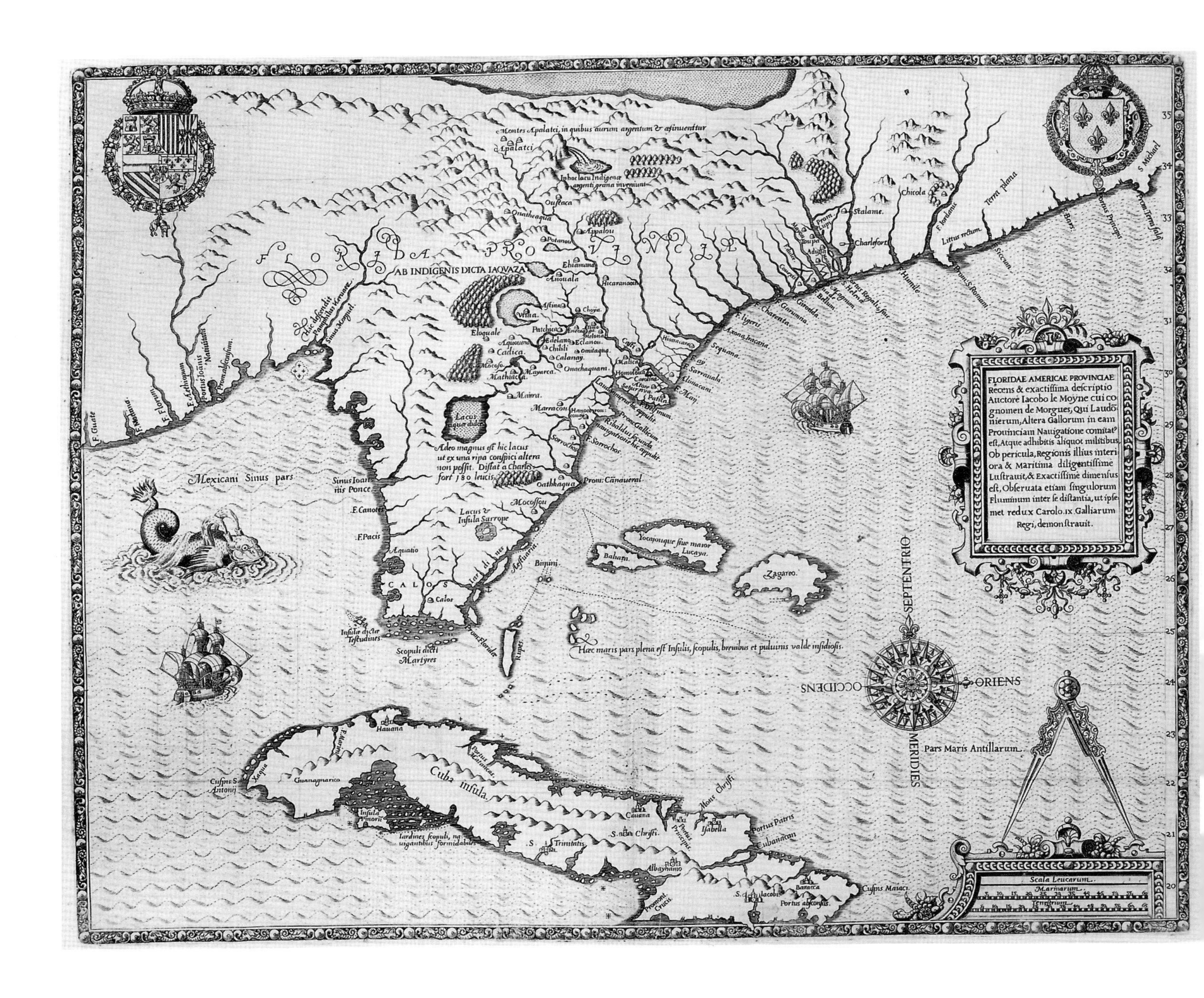

PLATE 15. Le Moyne. Floridae Americae Provinciae . . . descriptio. 1591. [Map 14].

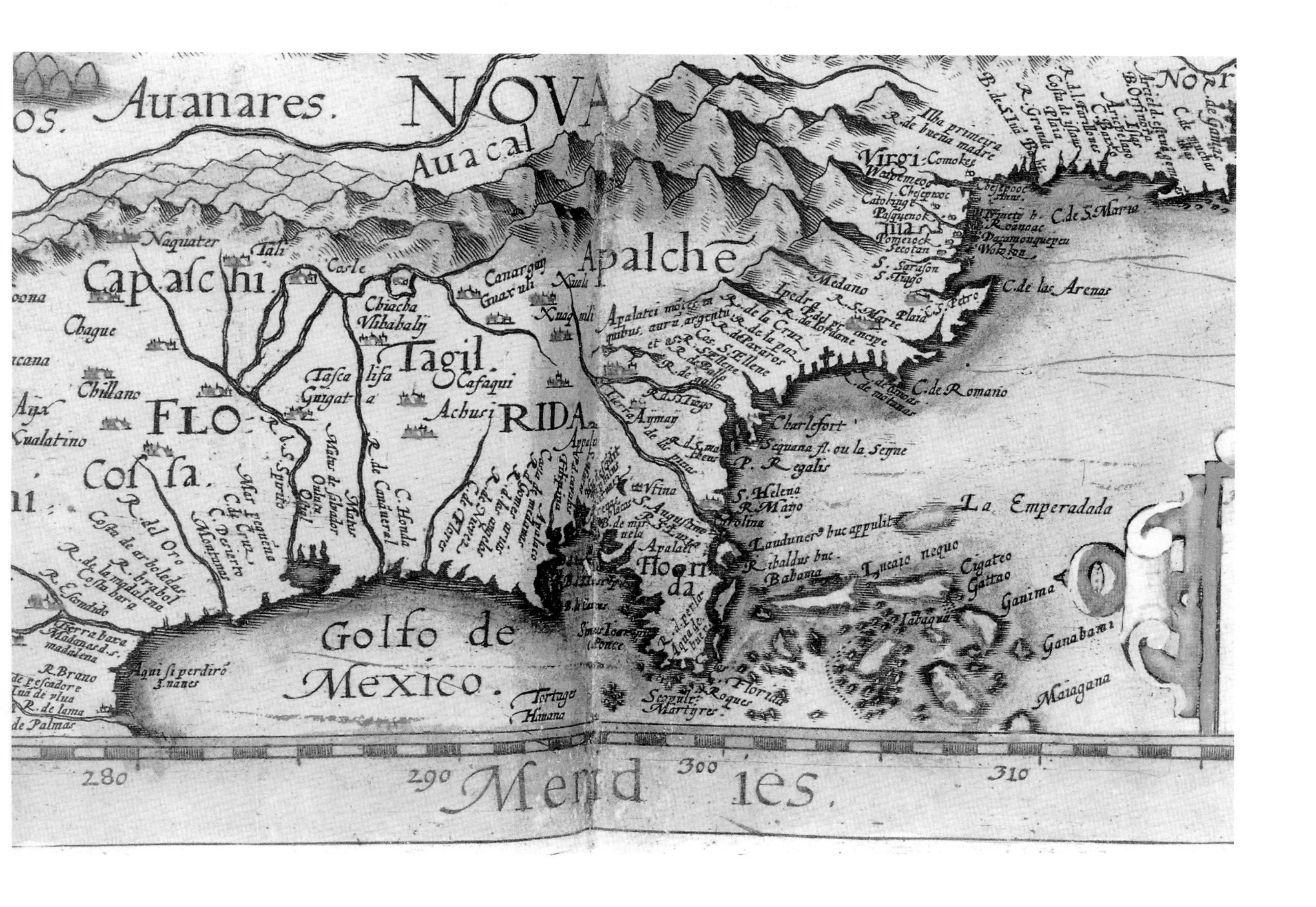

PLATE 16. Cornelis de Jode. Americæ Pars Borealis [detail]. 1593. [Map 16].

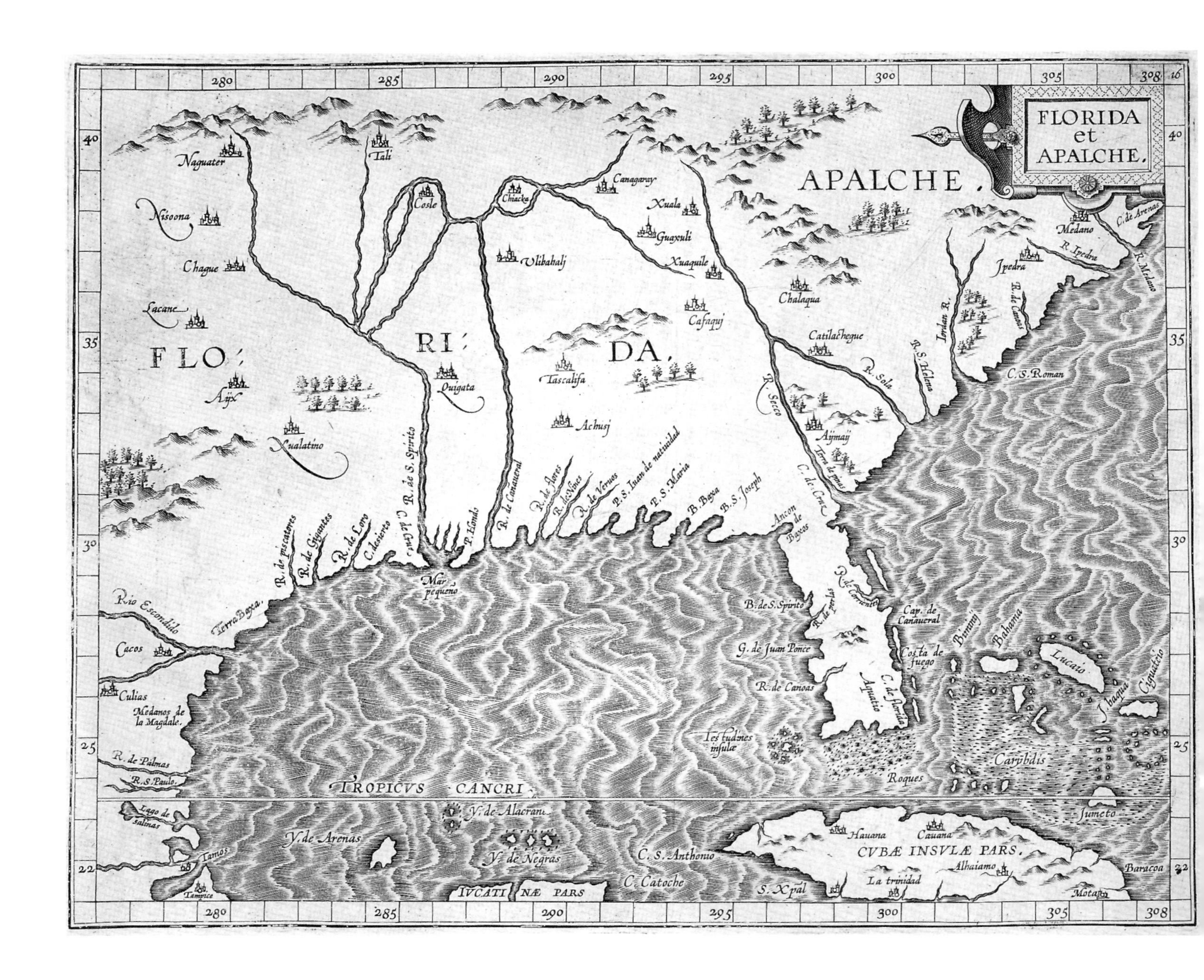

PLATE 17. Corneille Wytfliet. Florida et Apalche. 1597. [Map 18].

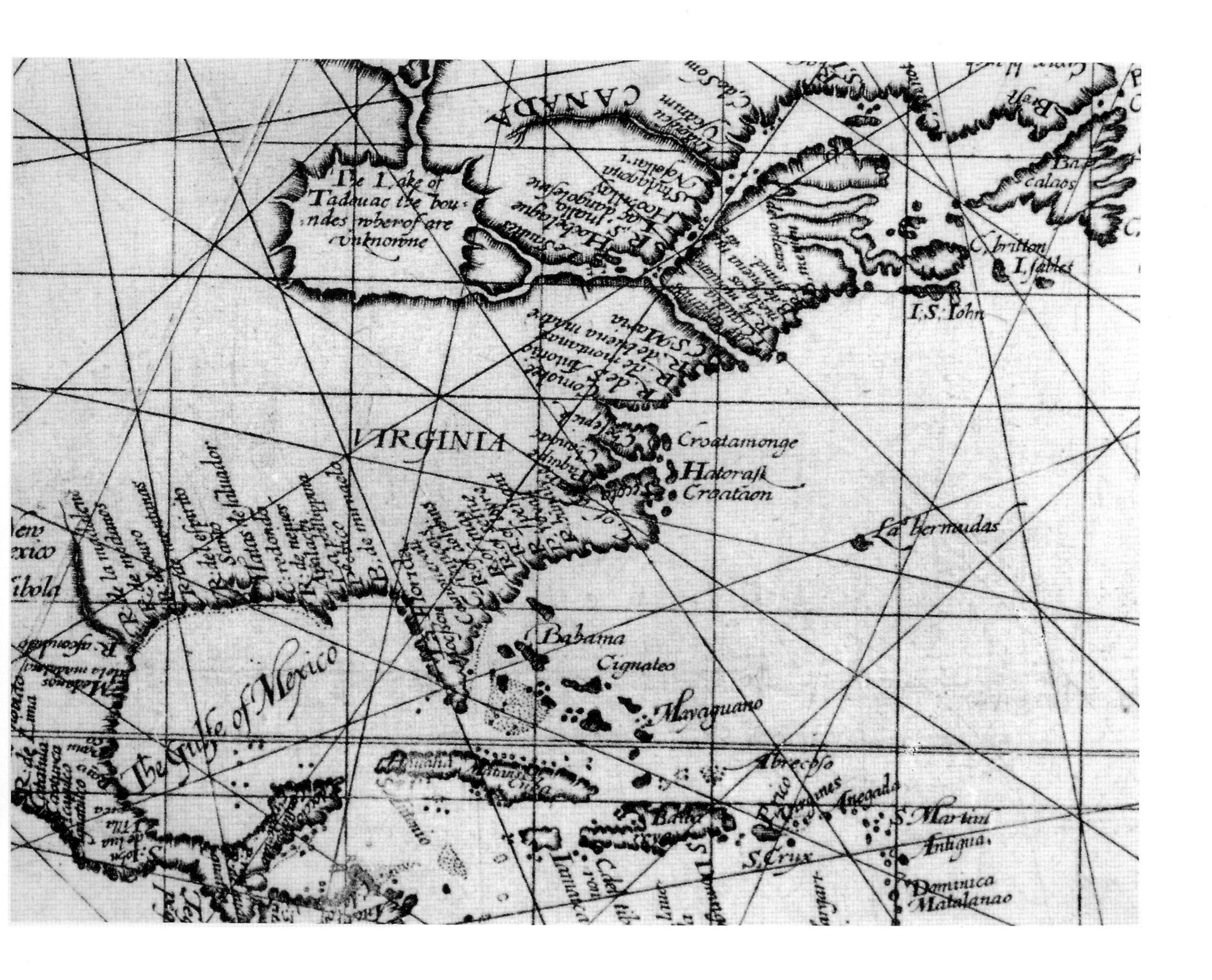

PLATE 18. Wright. [A Chart of the World on Mercator's Projection (detail)]. 1599. [Map 21A].

PLATE 19. Tatton. Noua et rece Terraum et regnorum Californiæ. 1616 [1600]. [Map 25].

PLATE 20. Mercator-Hondius. Virginiae Item et Floridae. 1606. [Map 26].

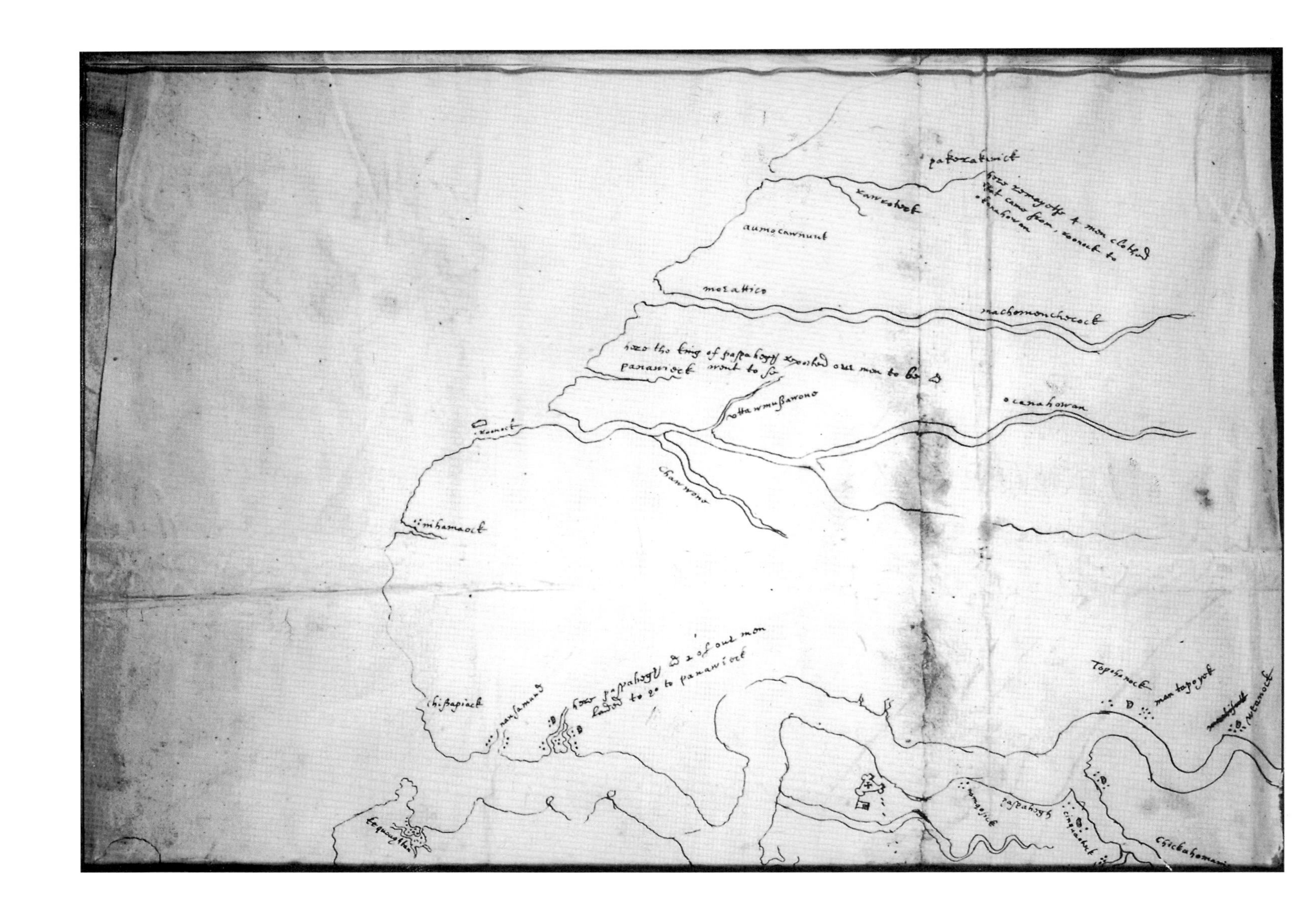

PLATE 21. Zuñiga. [Chart of Virginia (southern part)]. 1608 MS. [Map 28].

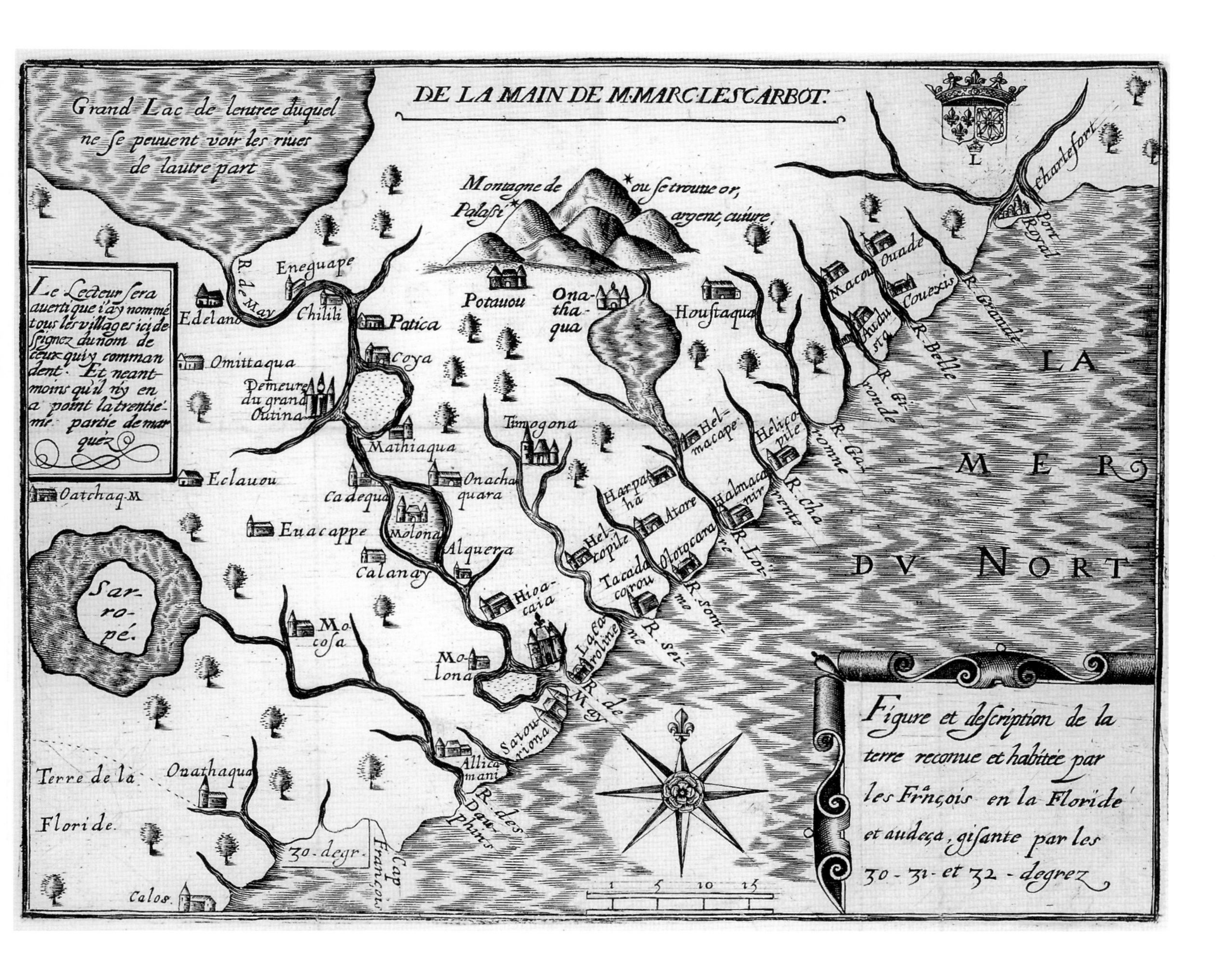

PLATE 22. Marc Lescarbot. Figure et description de la . . . Floride. 1612. [Map 30].

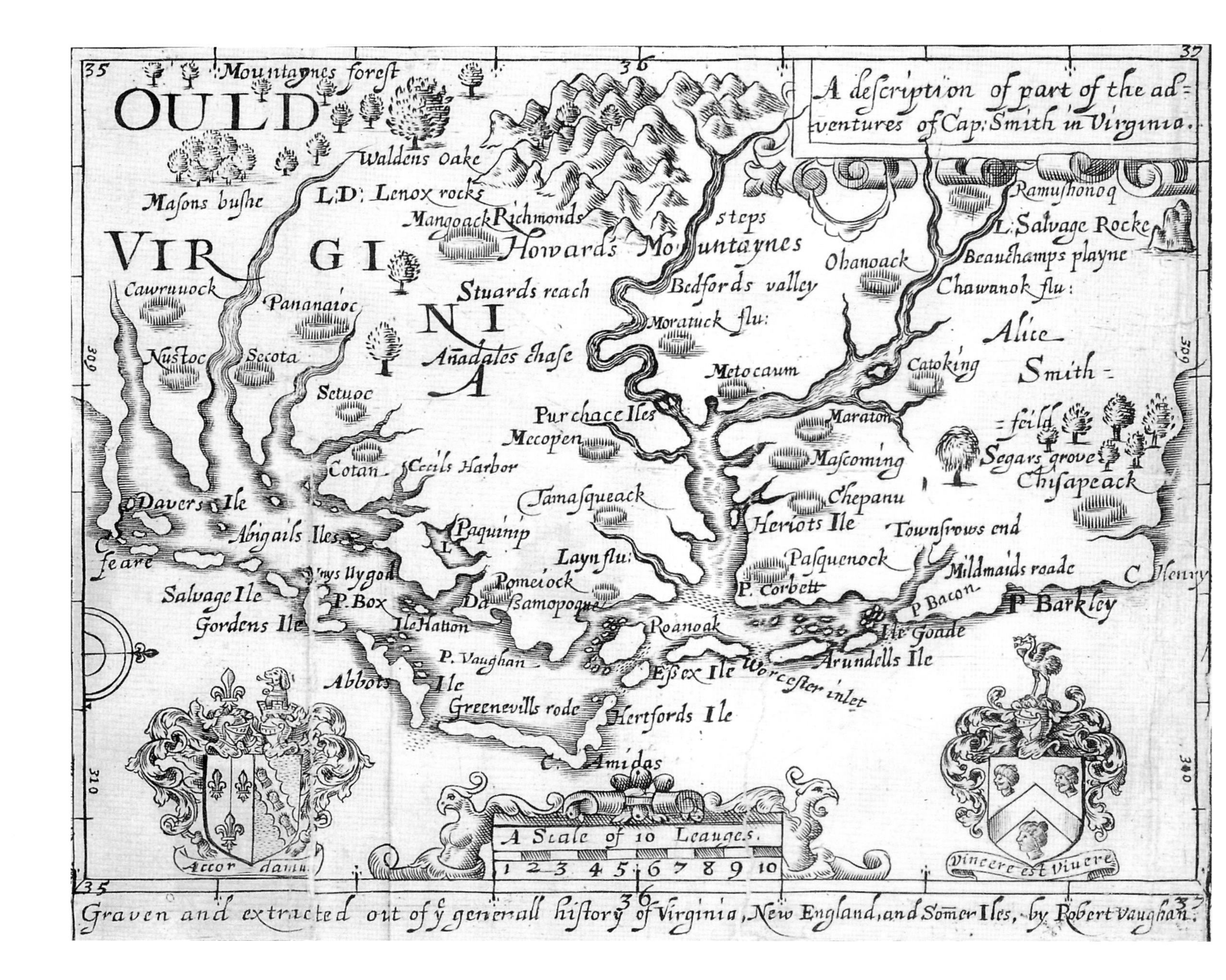

PLATE 23. John Smith. Ould Virginia. 1624. [Map 32].

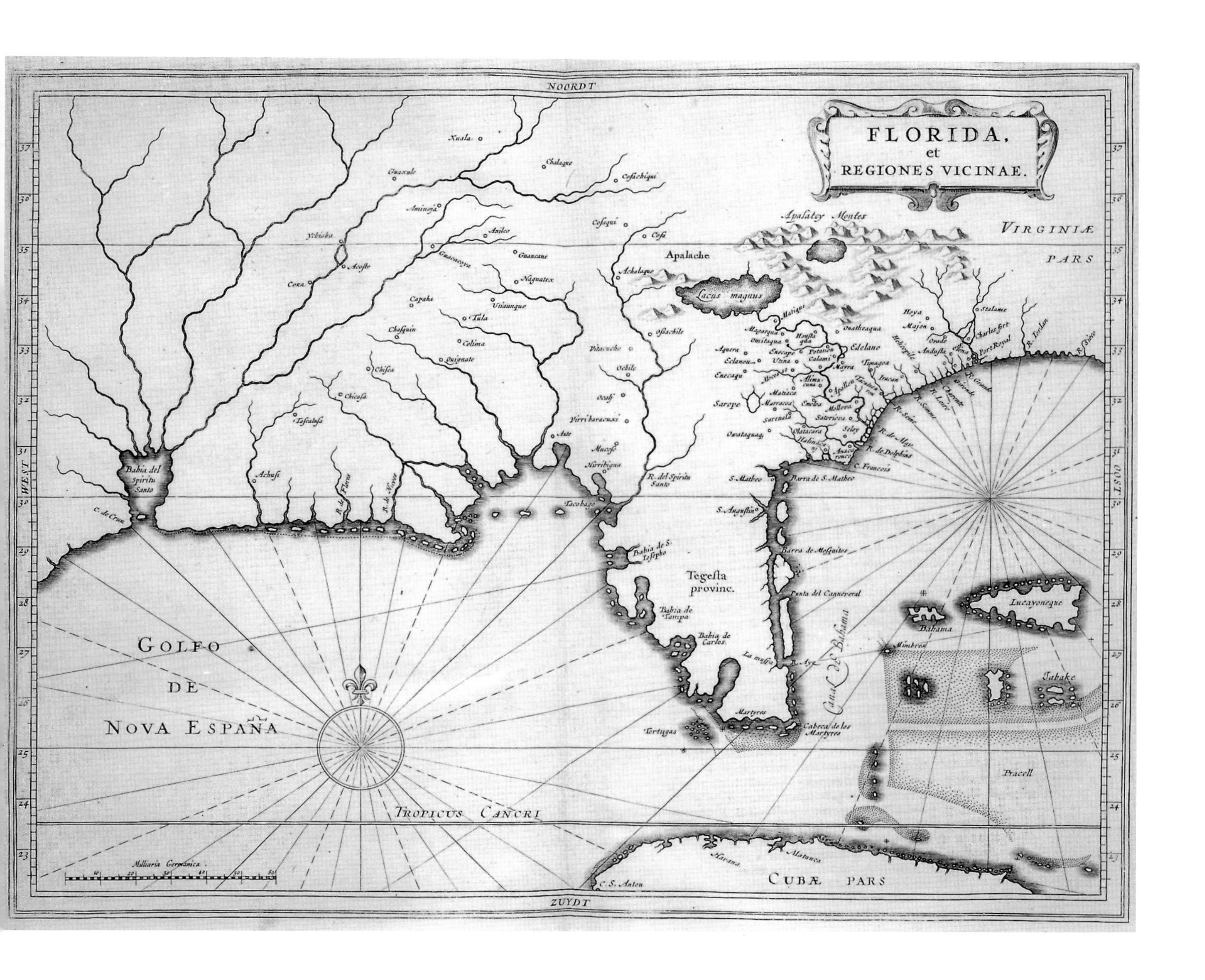

PLATE 24. Joannes de Laët. Florida et Regiones Vicinae. 1630 A. [Map 34].

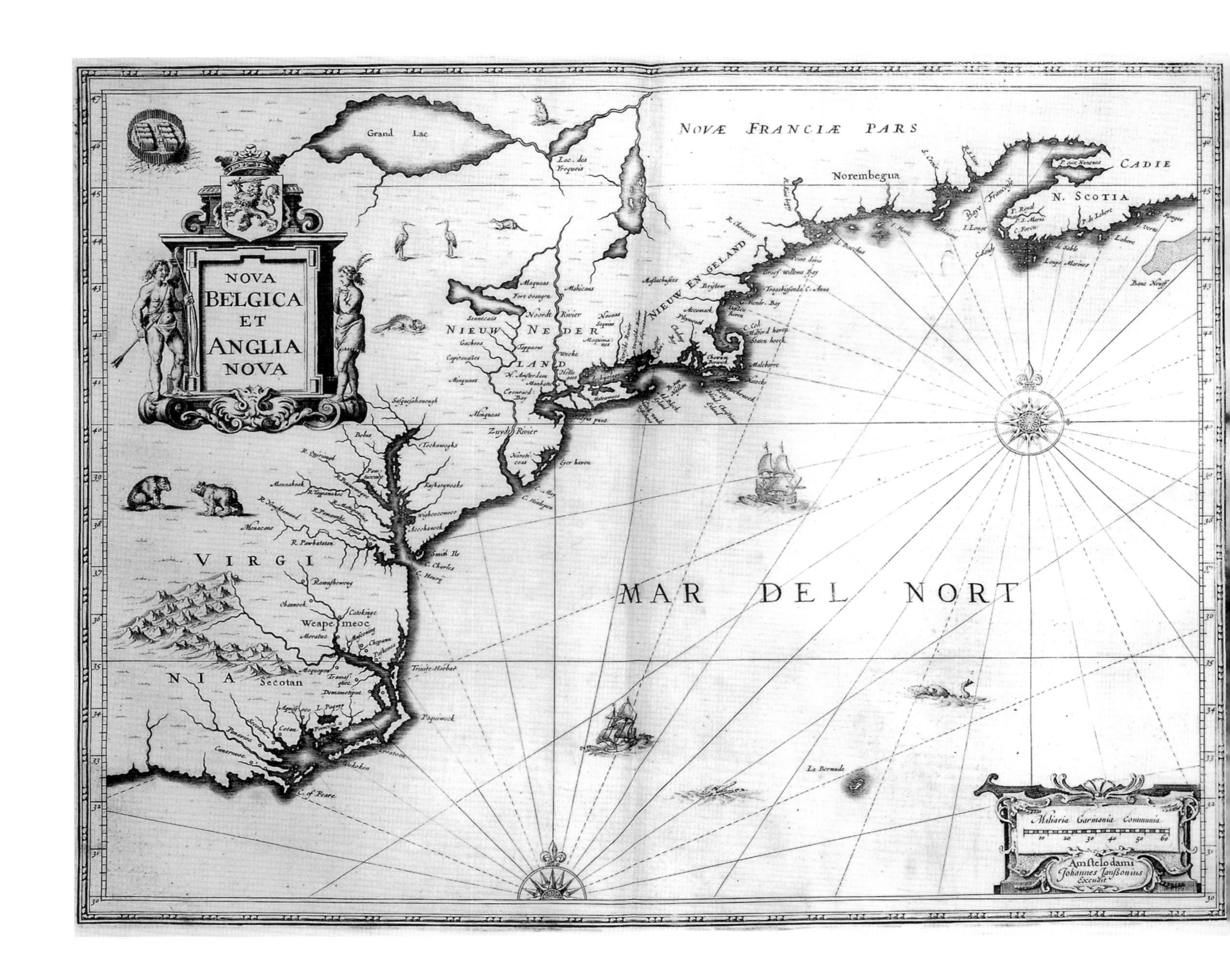

PLATE 25. Jan Jansson. Nova Belgica et Anglia Nova. 1647. [Map 43].

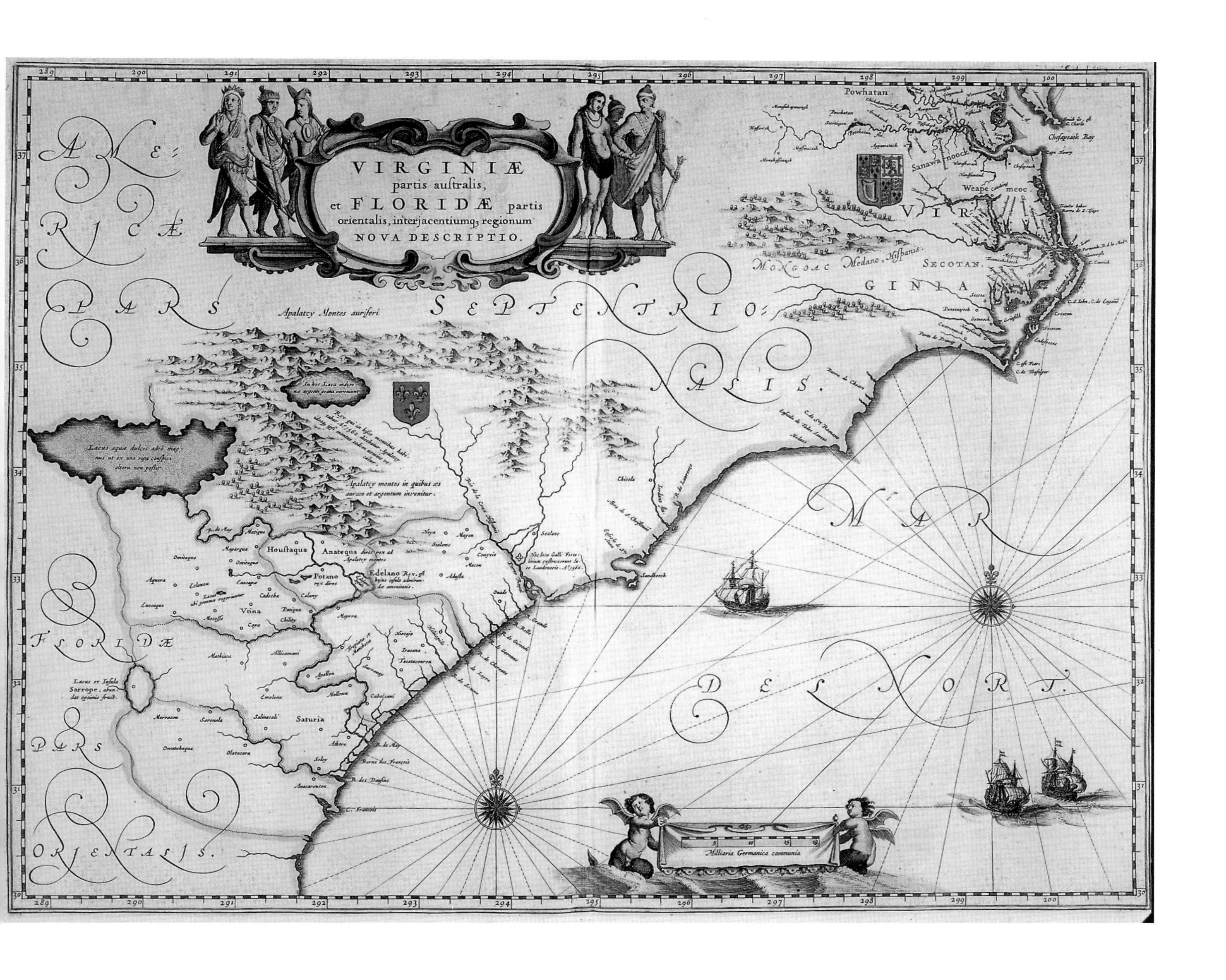

PLATE 26. Willem Janszoon Blaeu. Virginiæ partis australis, et Floridæ. 1640. [Map 41].

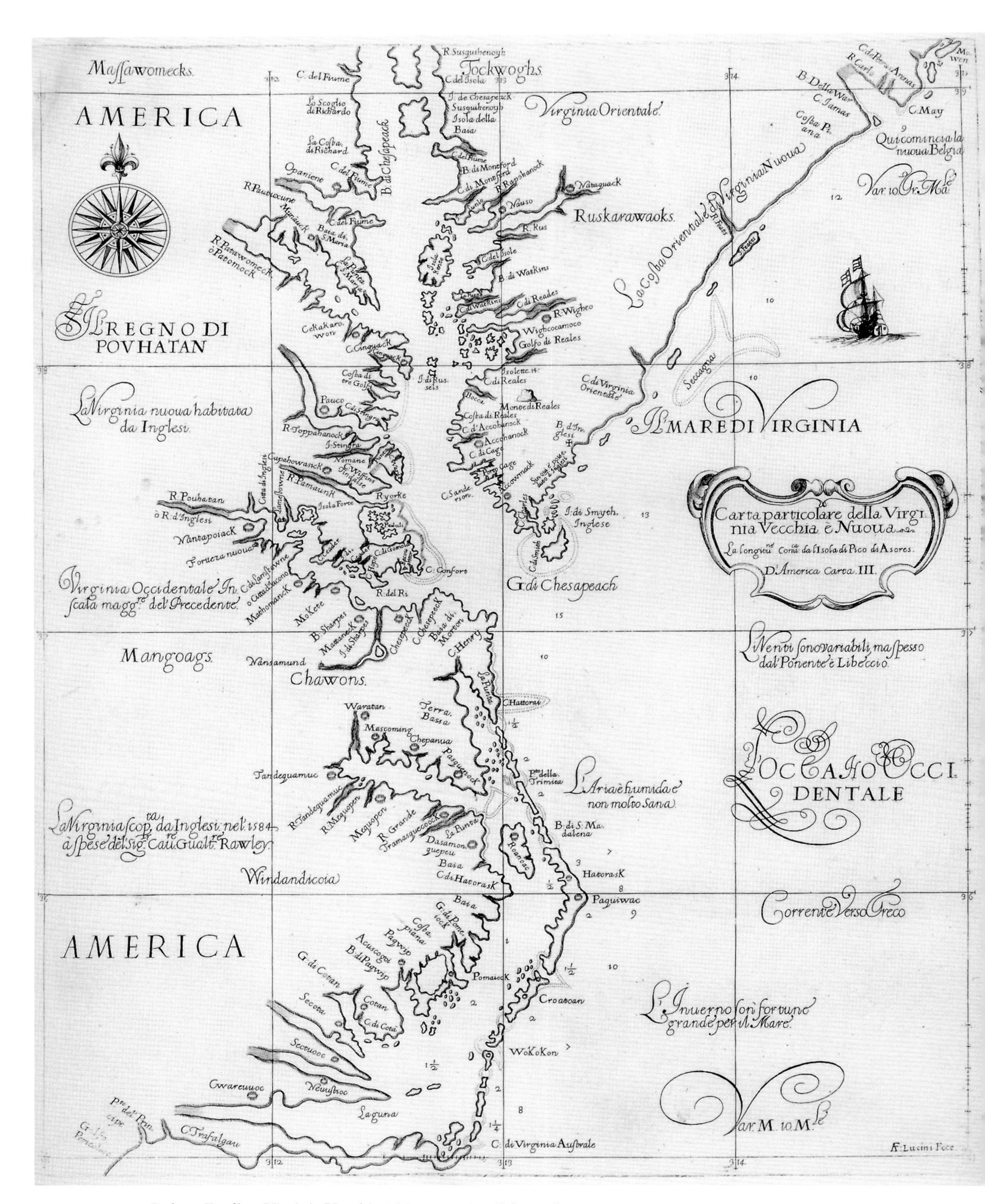

PLATE 27. Robert Dudley. Virginia Vecchia è Nuoua. 1647. [Map 44].

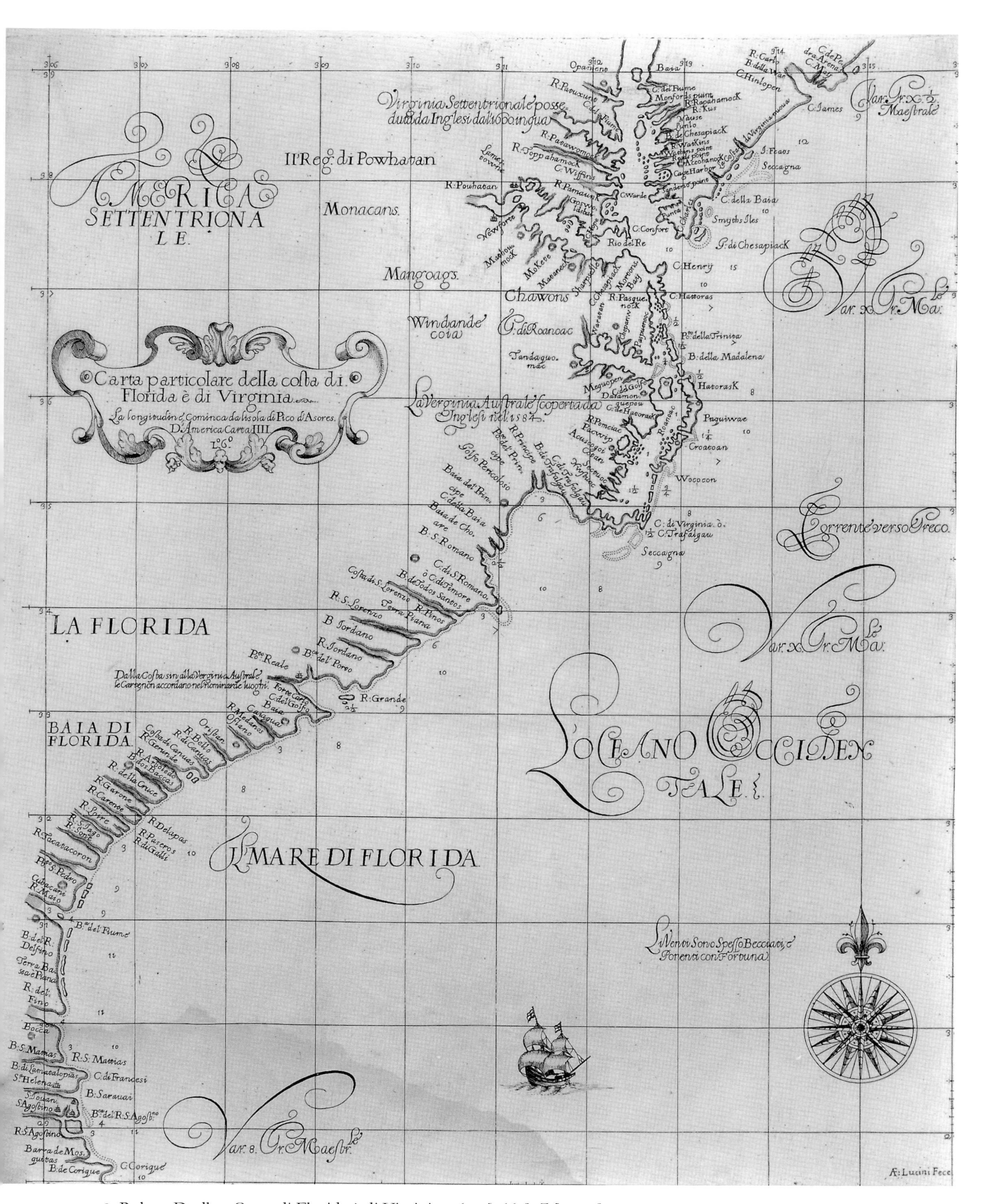

PLATE 28. Robert Dudley. Costa di Florida è di Virginia. 1647 [1661]. [Map 45].

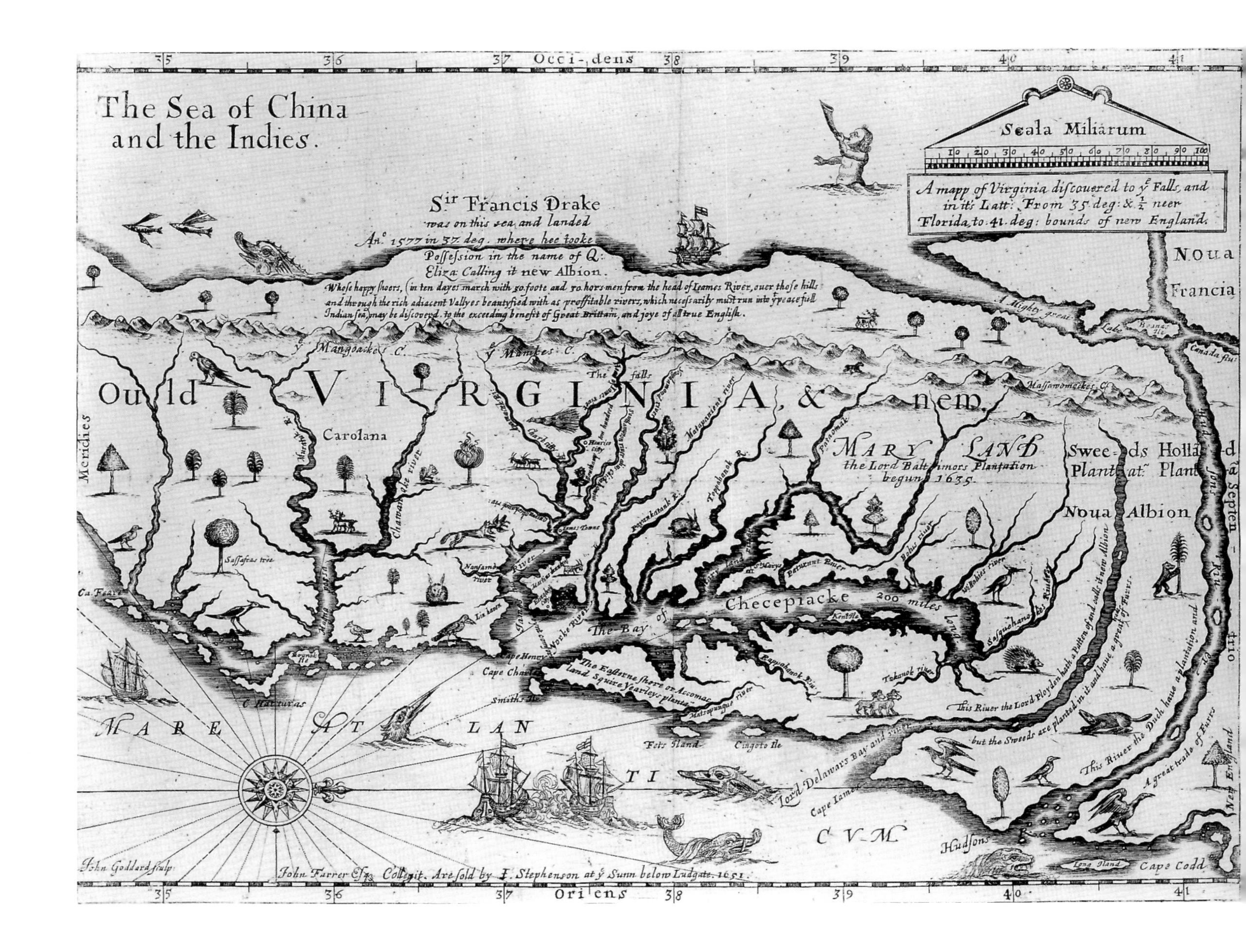

PLATE 29. John Farrer. A mapp of Virginia. 1651. [Map 47].

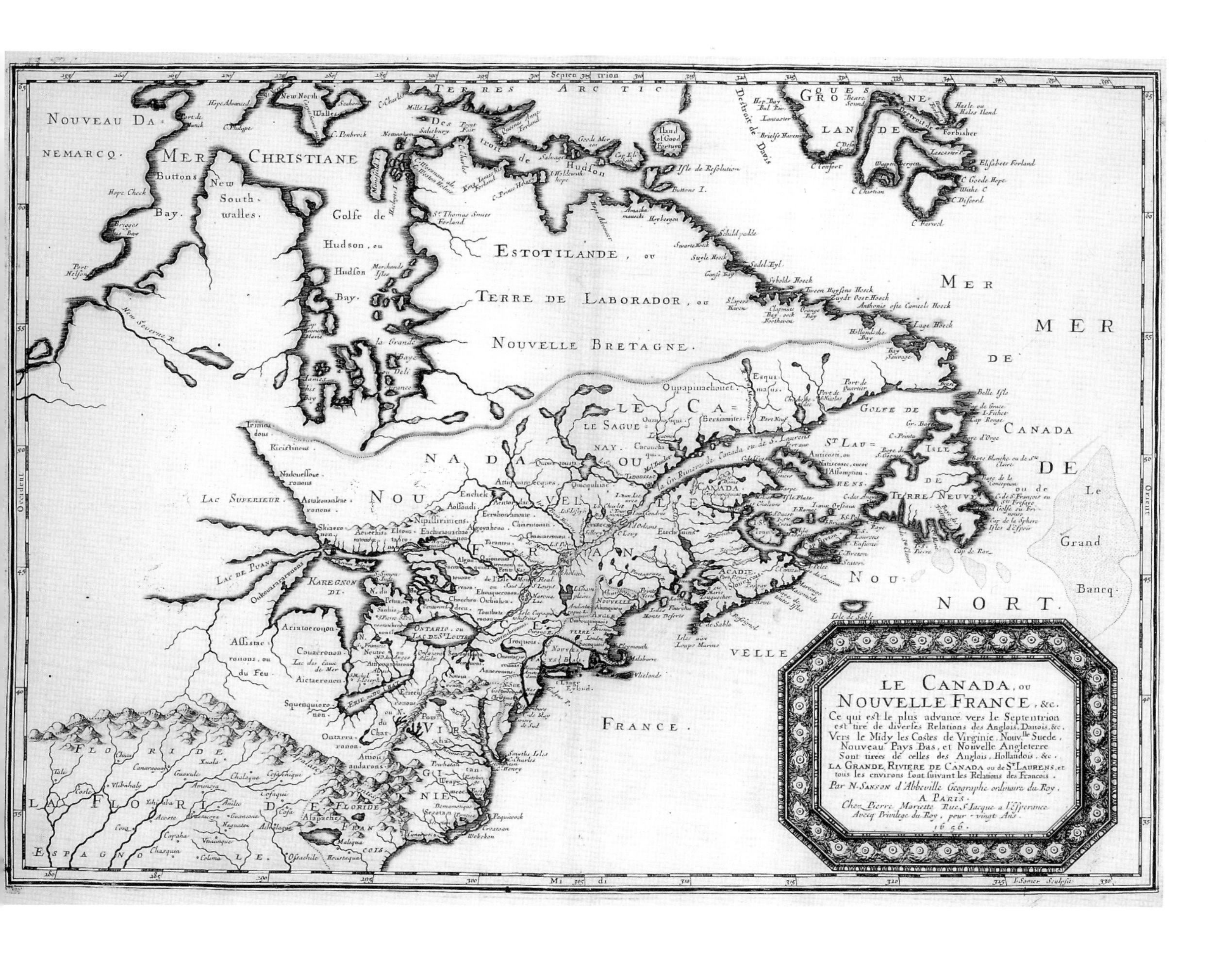

PLATE 30. Nicolas Sanson. Canada. 1656. [Map 48].

PLATE 31. Nicolas Sanson. Nouveau Mexique, et La Floride [detail]. 1656. [Map. 49].

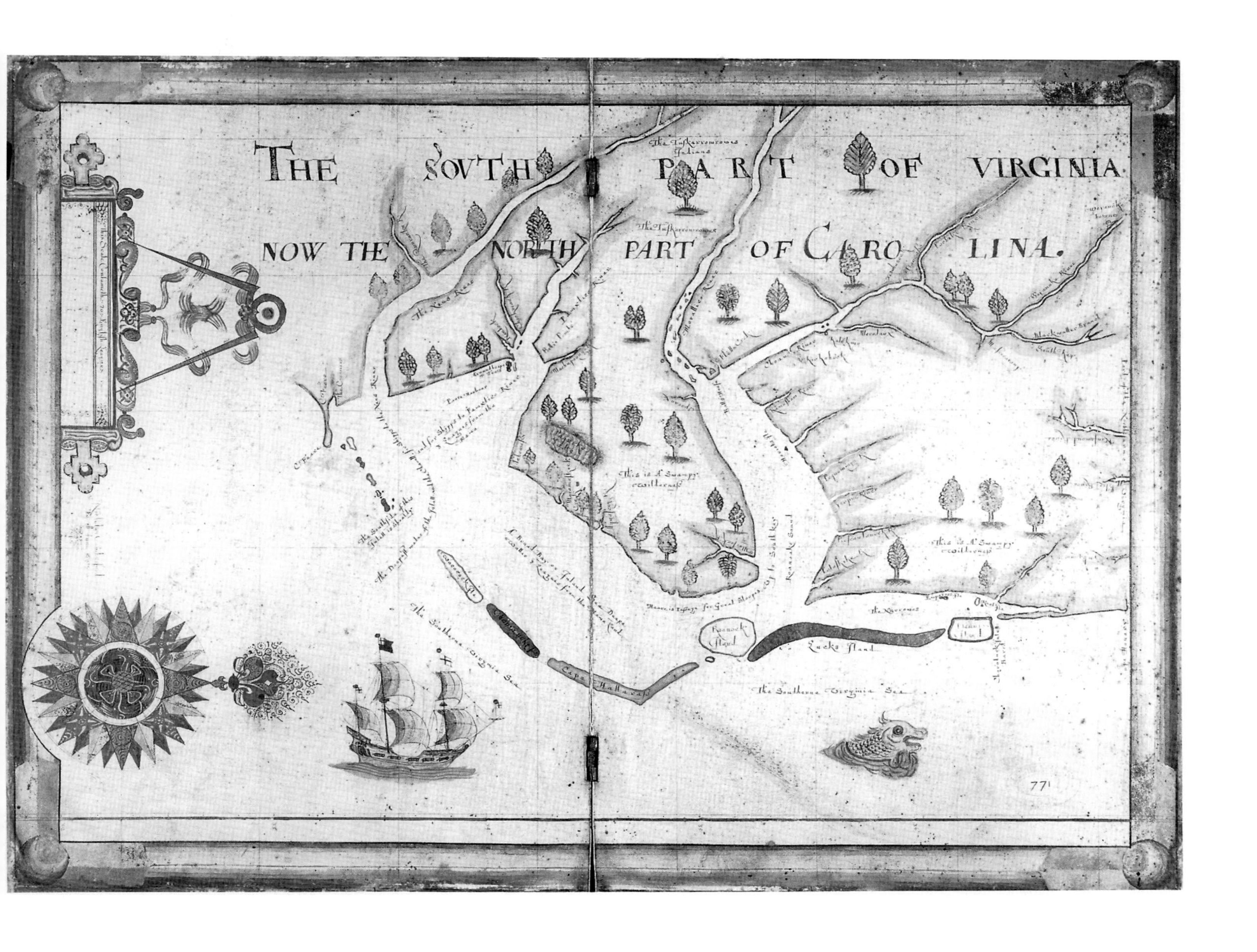

PLATE 32. Nicholas Comberford. The Sovth Part of Virginia. 1657 MS A. [Map 50].

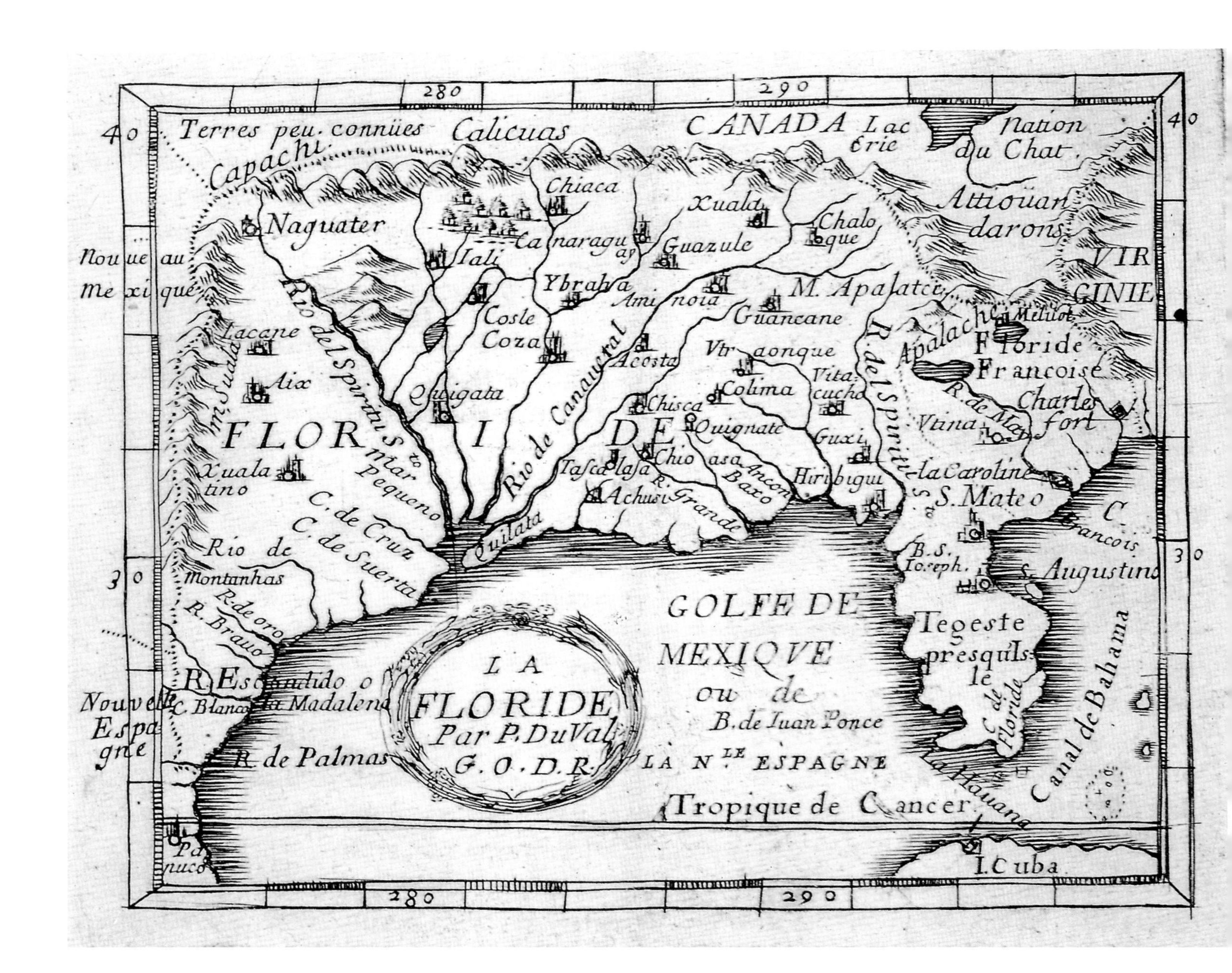

PLATE 33. Du Val. La Floride. 1663. [Map 57].

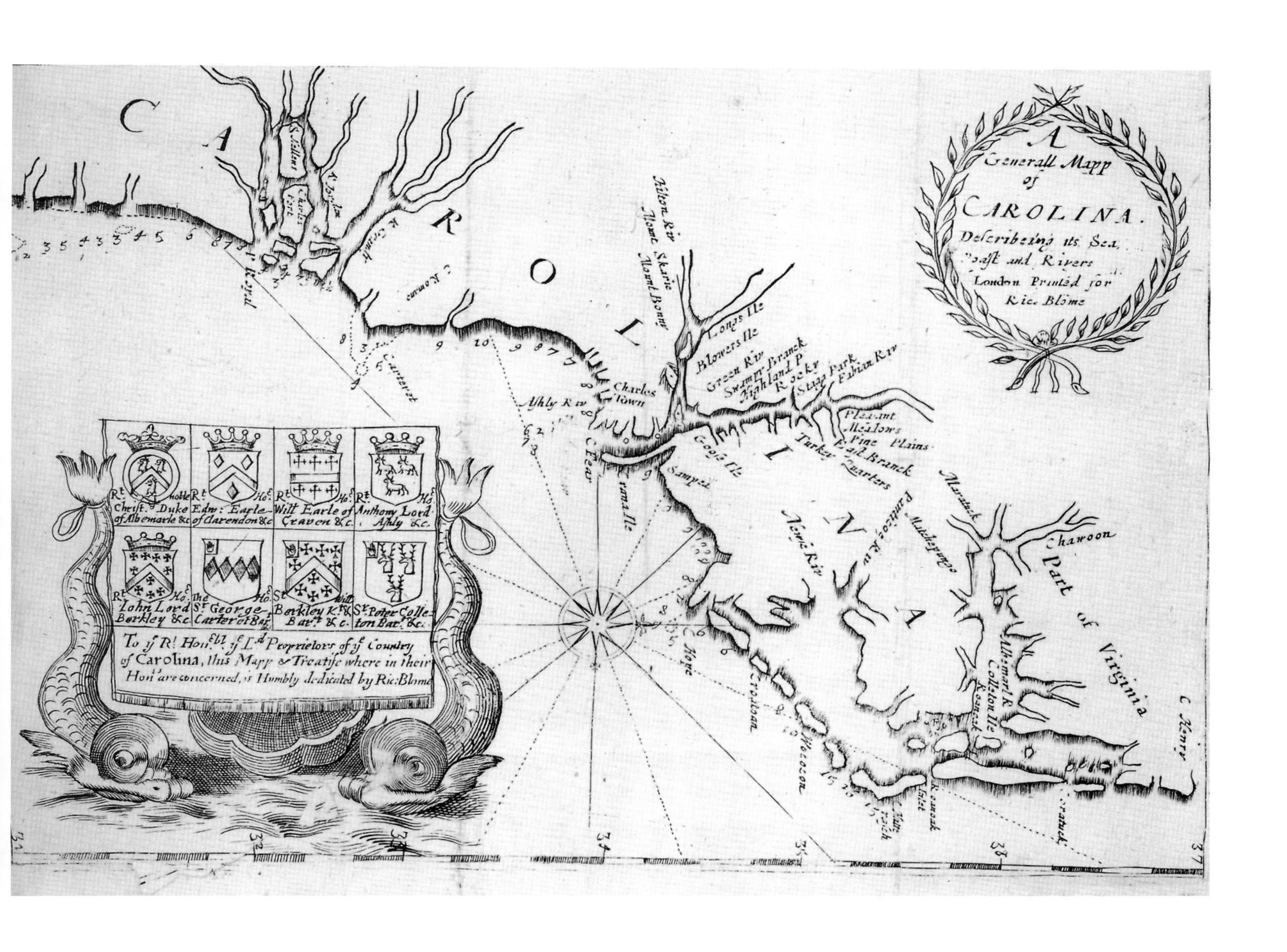

PLATE 34. Richard Blome. A Generall Mapp of Carolina. 1672. [Map 69].

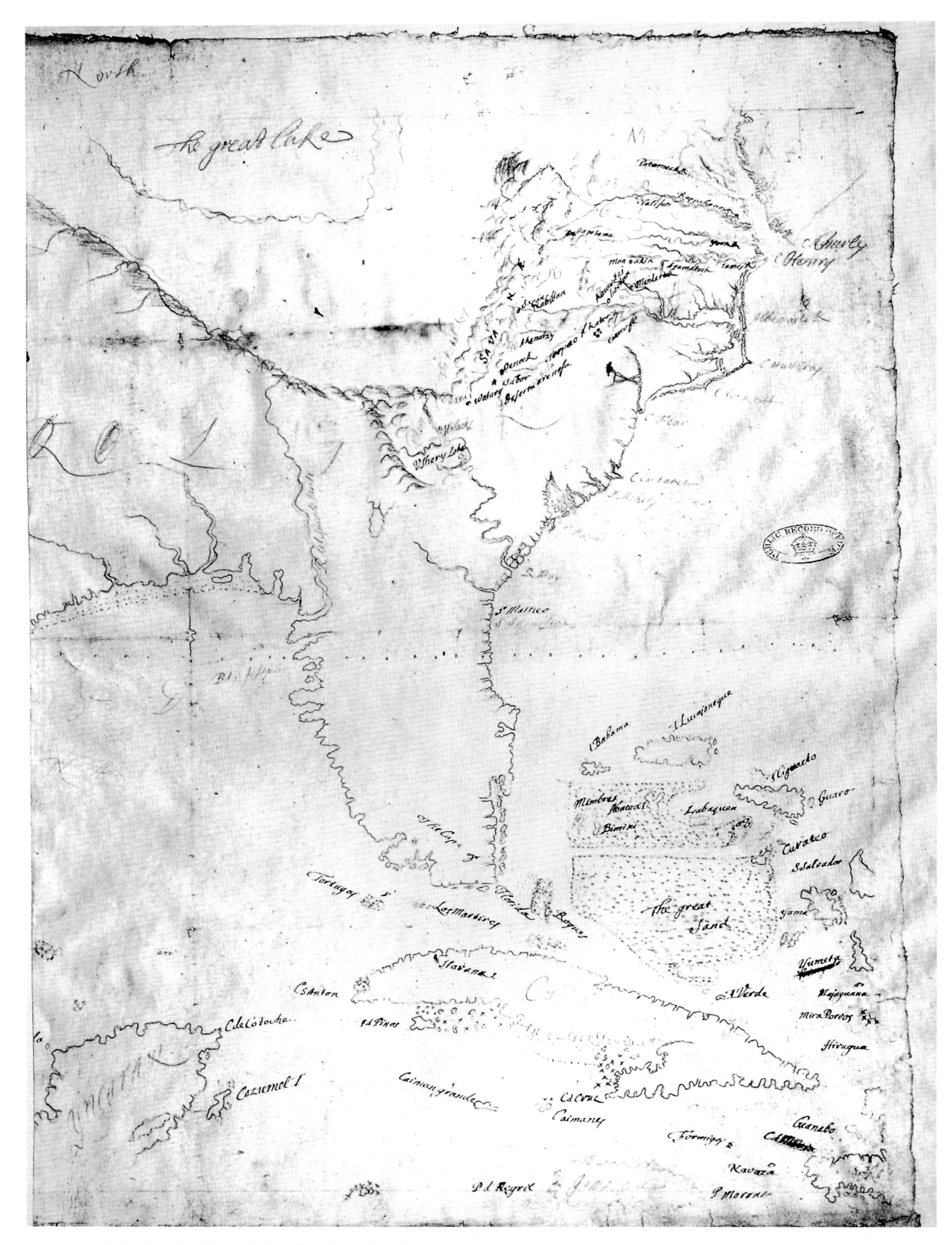

PLATE 35. John Locke. Map of Carolina [detail]. 1671 MS. [Map 65].

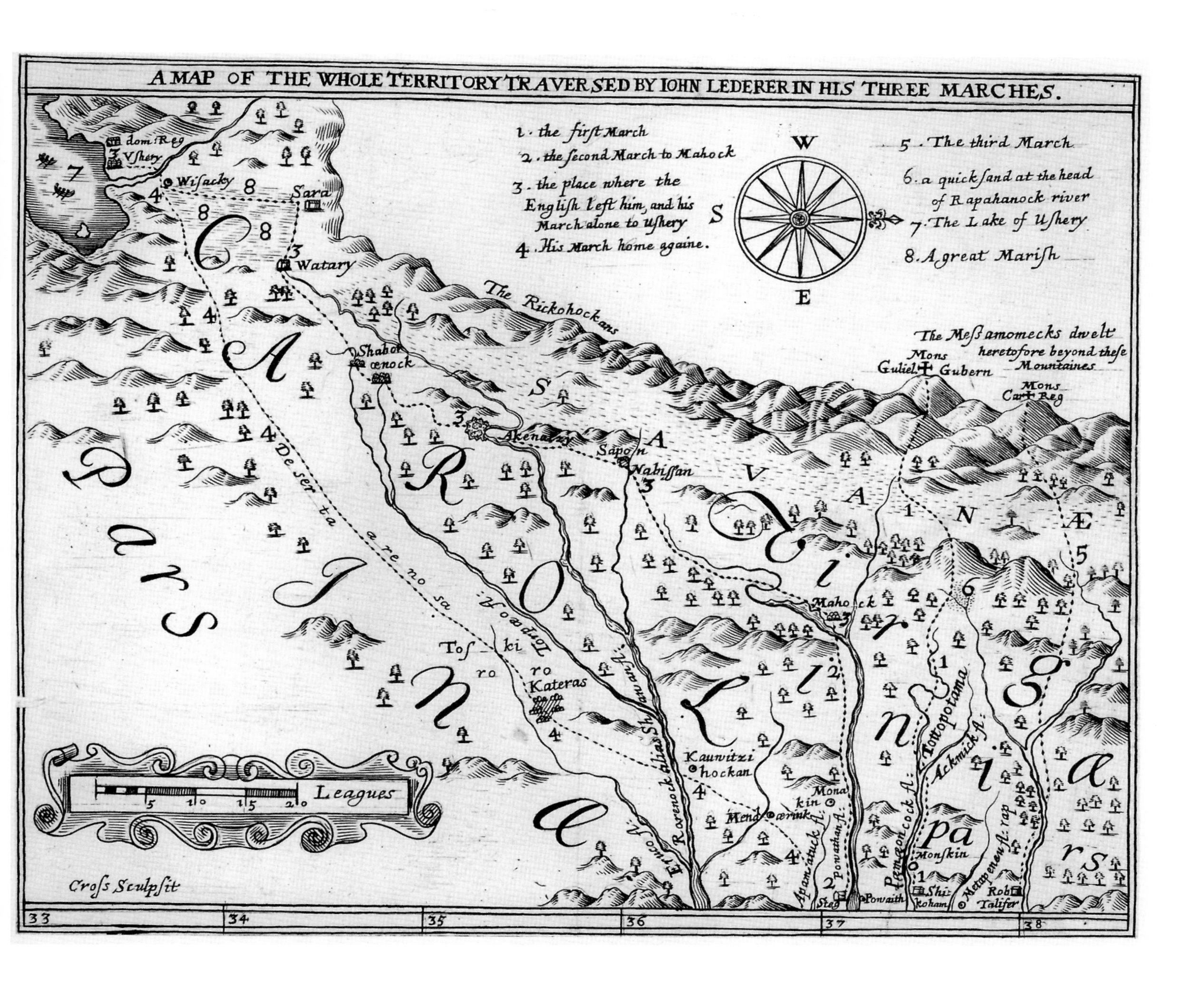

PLATE 36. John Lederer. A Map of the Whole Territory Traversed. 1672. [Map 68].

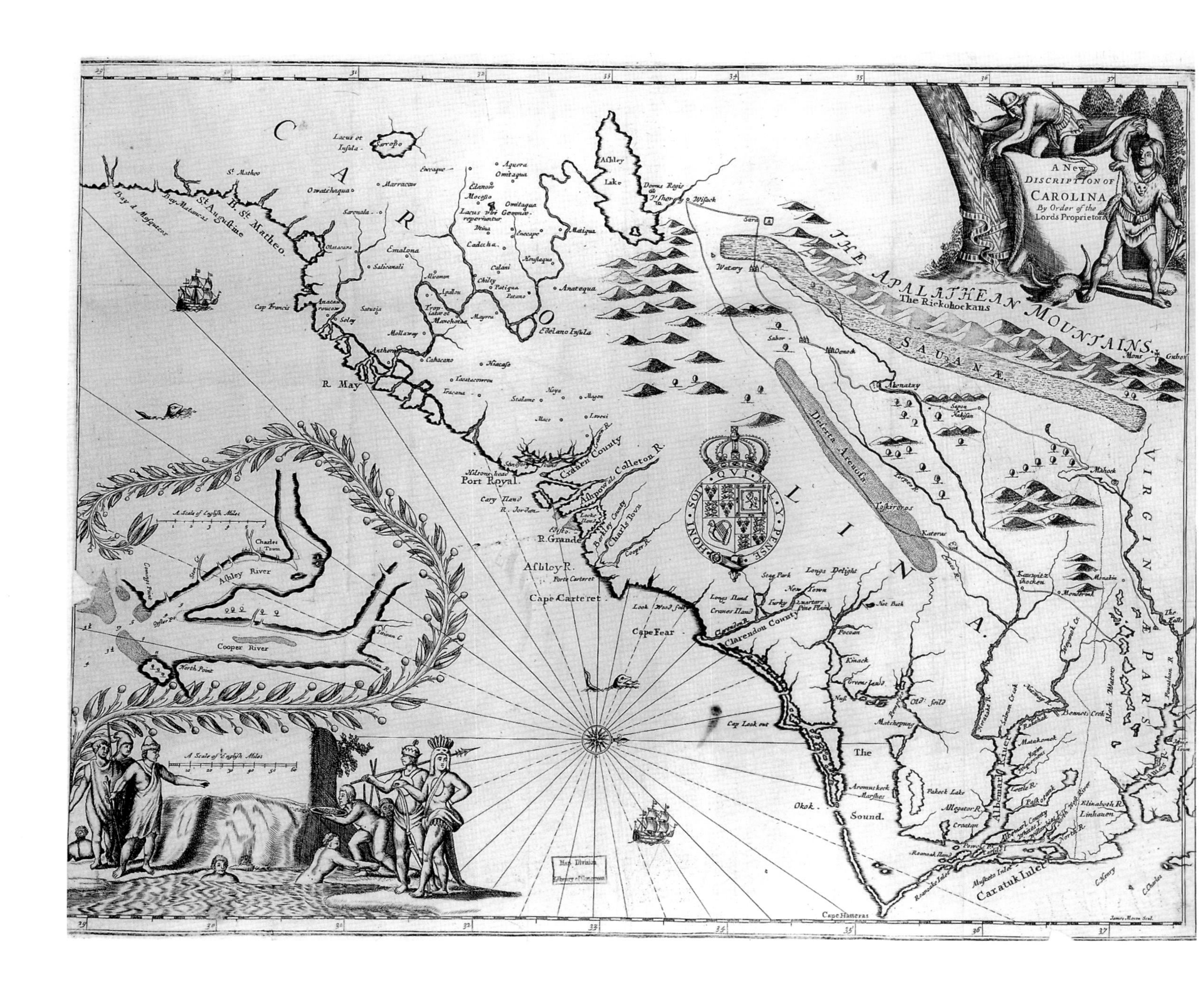

PLATE 37. John Ogilby–James Moxon. A New Discription of Carolina. Ca. 1672. [Map 70].

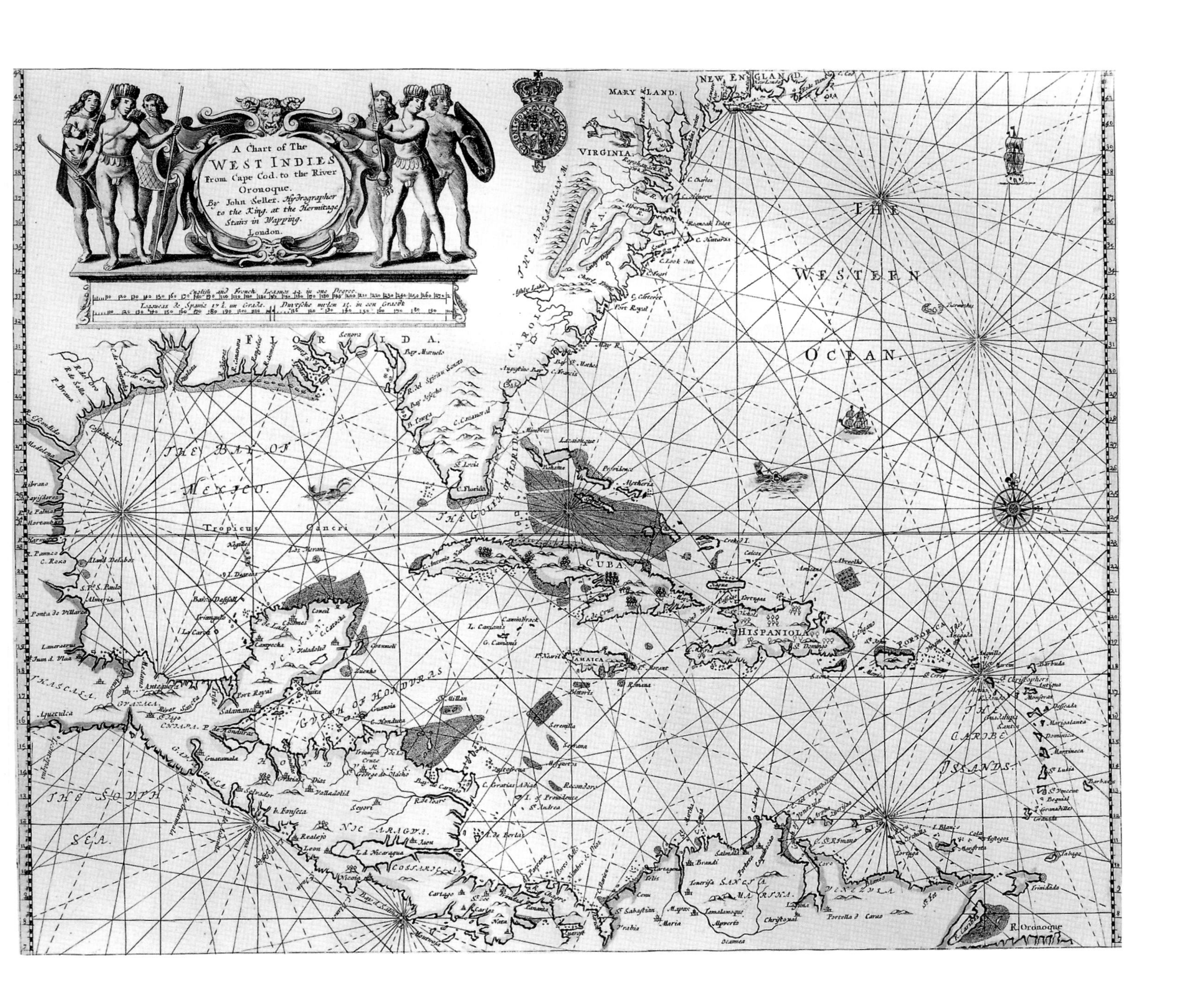

PLATE 38. John Seller. A Chart of the West Indies. Ca. 1675. [Map 75].

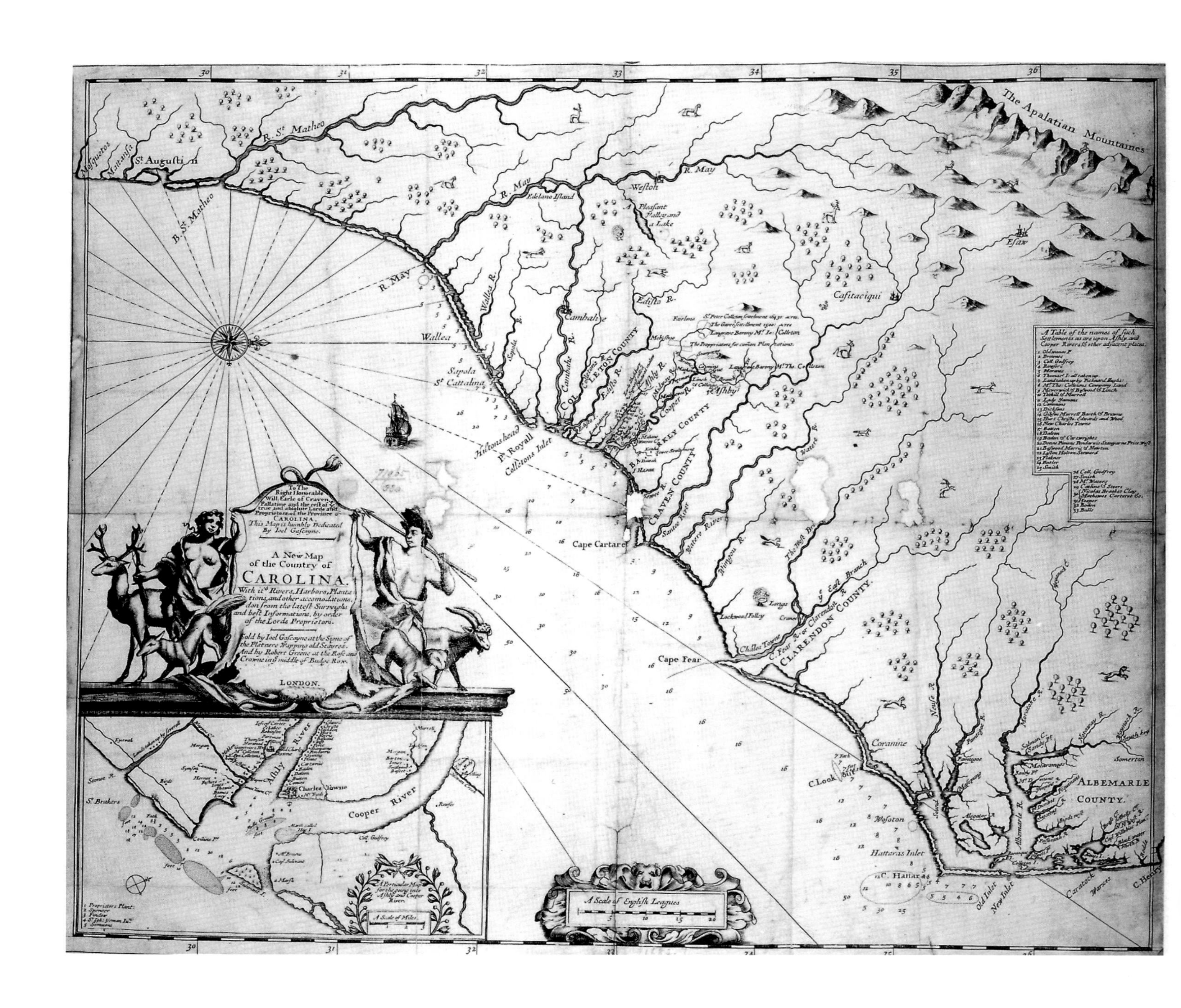

PLATE 39. Joel Gascoyne. A New Map of the Country of Carolina. 1682. [Map 92].

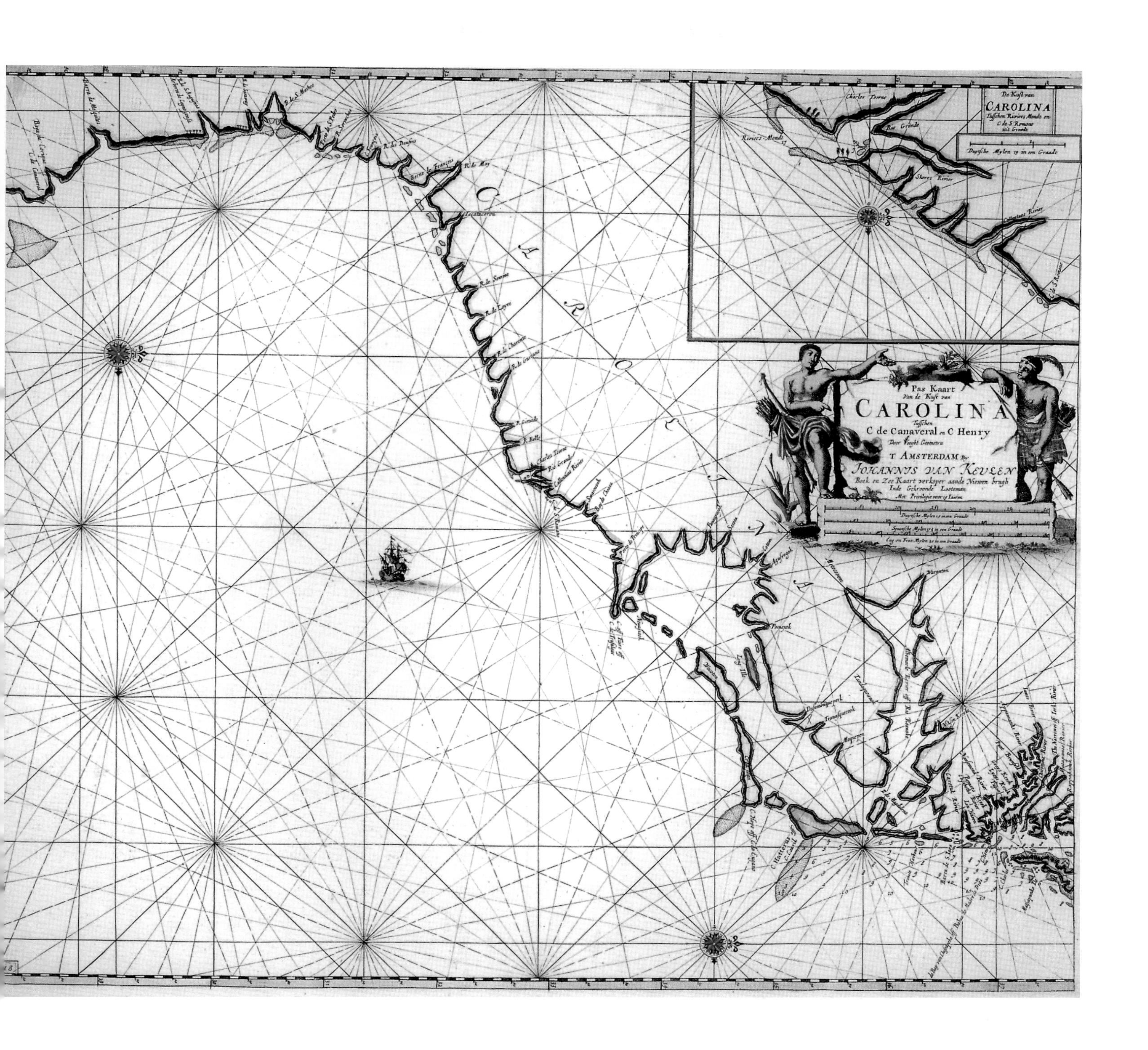

PLATE 40. Johannis van Keulen. Pas Kaart Van de Kust van Carolina. 1682? [Map 91].

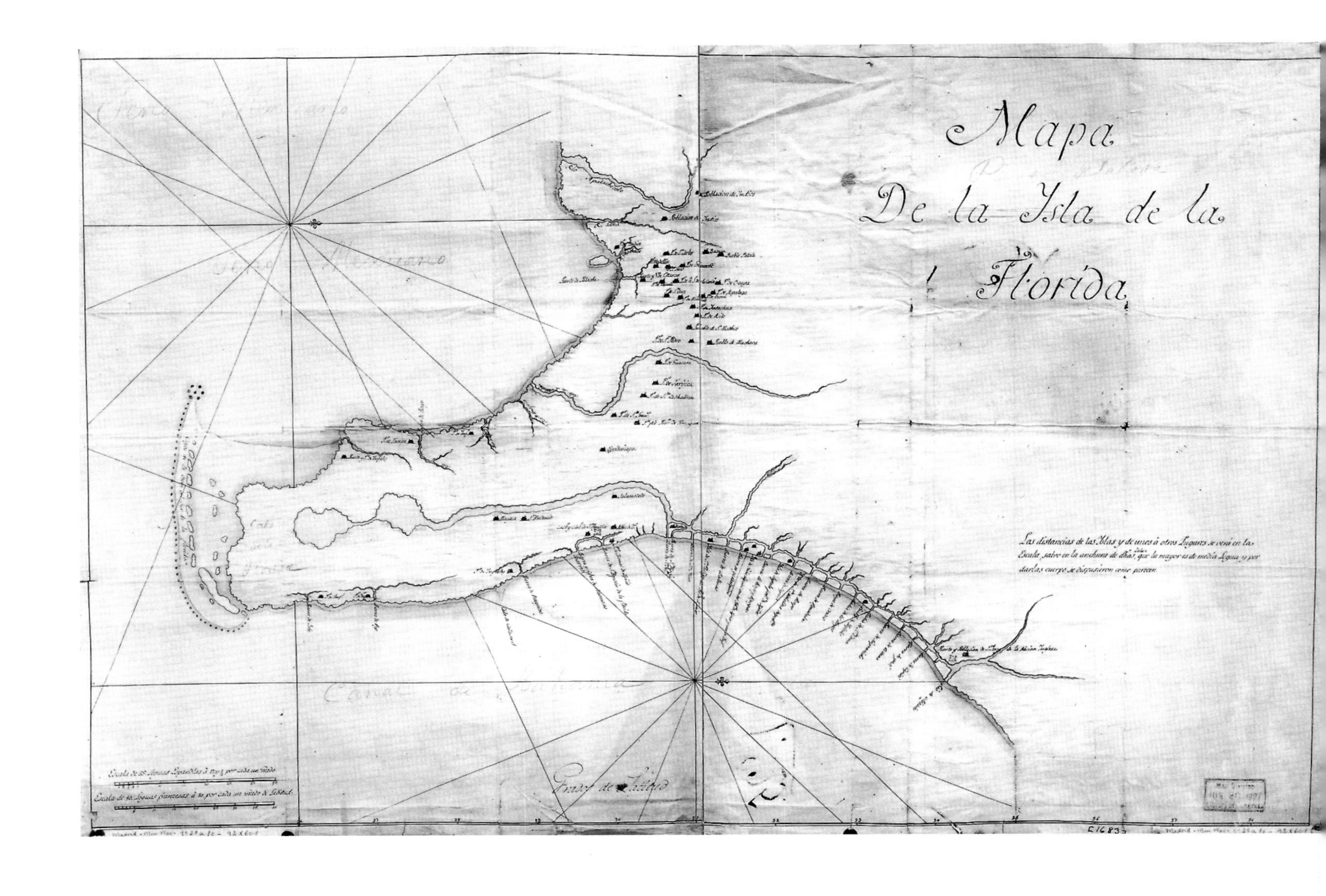

PLATE 40A. Anonymous, Spanish. Mapa De la Ysla de la Florida. 1683 MS. [Map 94].

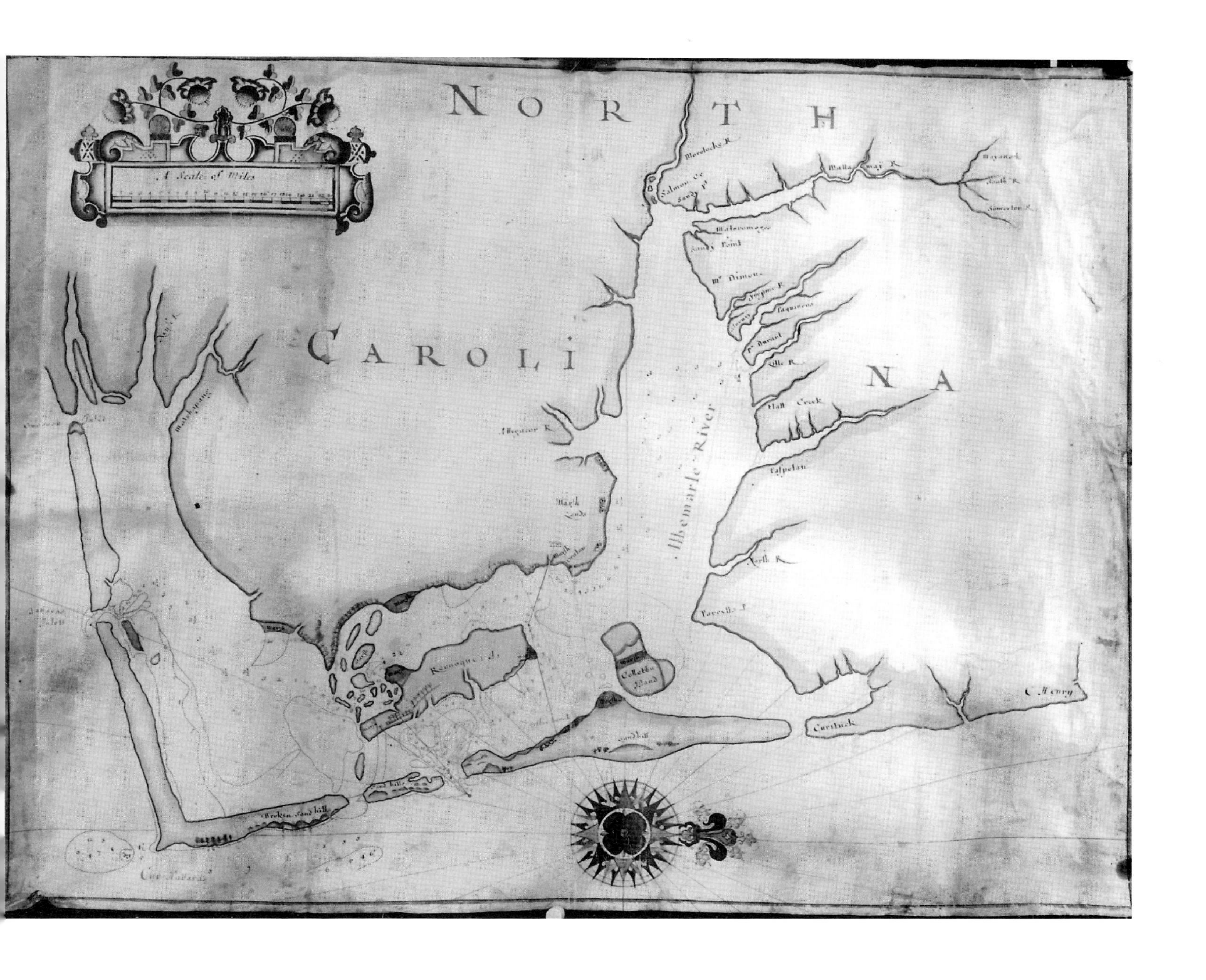

PLATE 41. Anonymous, Thames School. North Carolina. Ca. 1684 MS B. [Map 99].

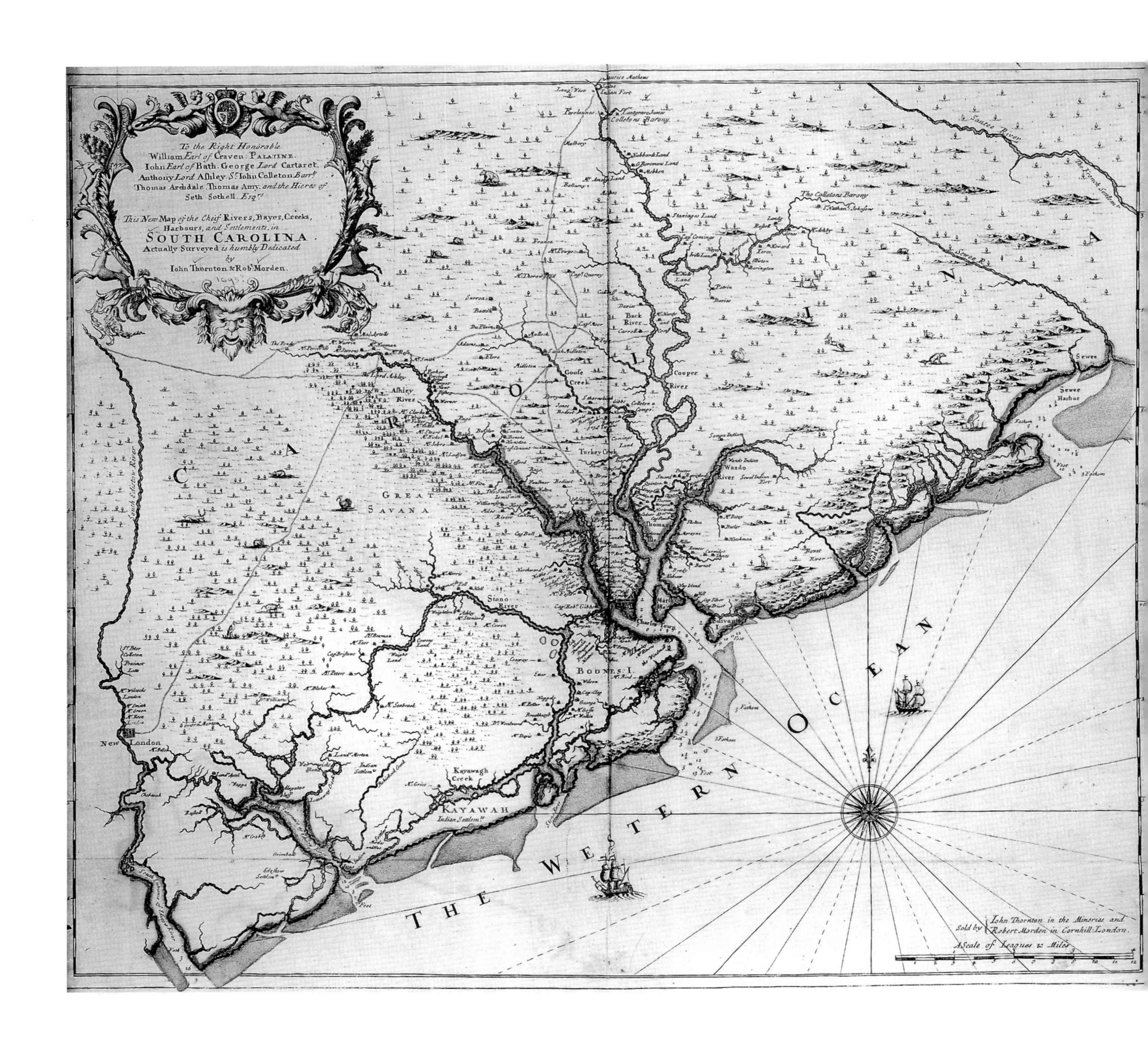

PLATE 42. John Thornton and Robert Morden. South Carolina. Ca. 1695. [Map 118].

PLATE 43. Guillaume Delisle. Carte du Mexique et de la Floride. 1703. [Map 137].

PLATE 43A. Lamhatty. [Map of His Odyssey from the Gulf Coast to Virginia]. 1708 MS. [Map 146].

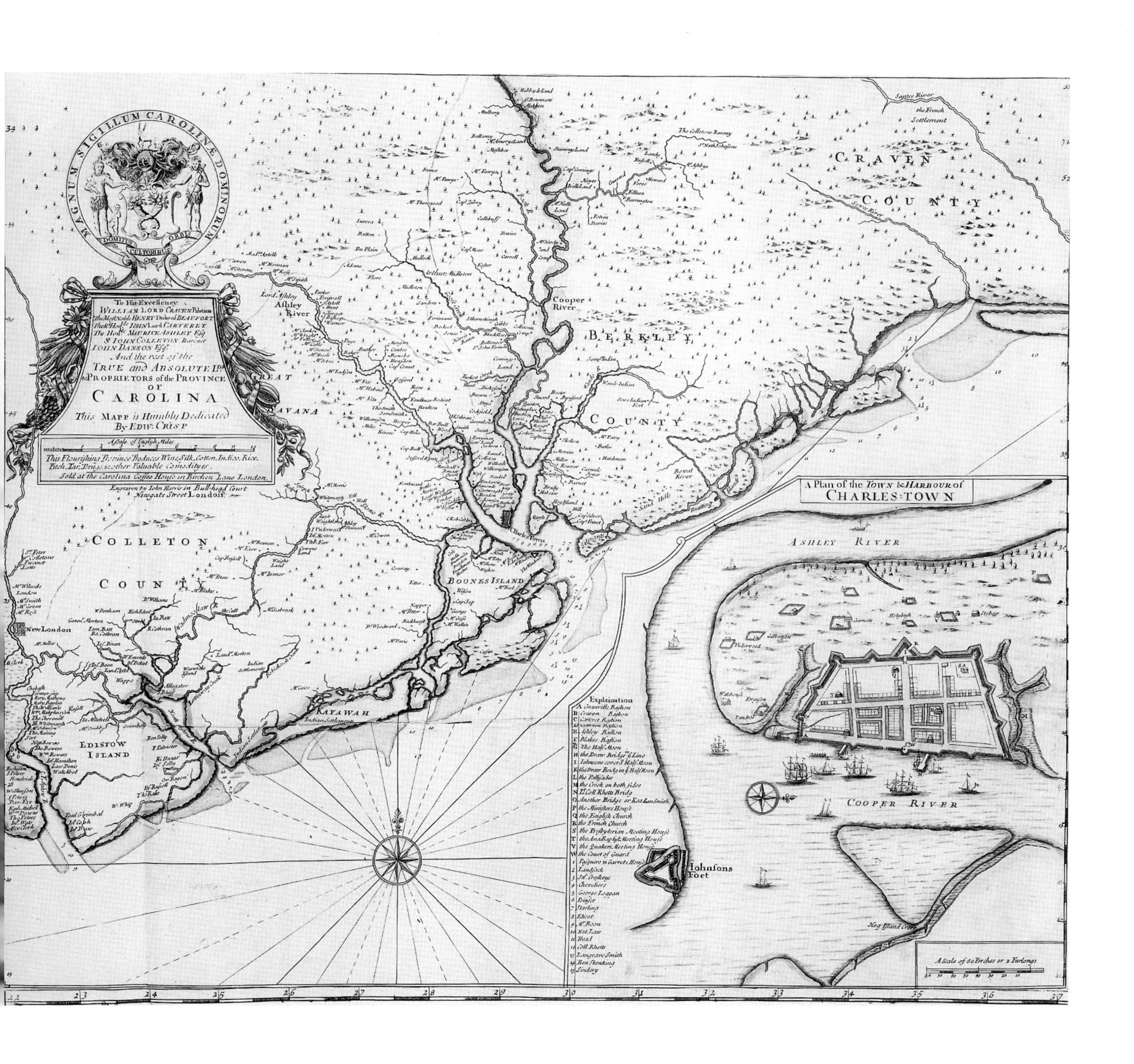

PLATE 44. Edward Crisp. A Compleat Description of the Province of Carolina [detail, with Charles-Town inset]. [1711]. [Map 151].

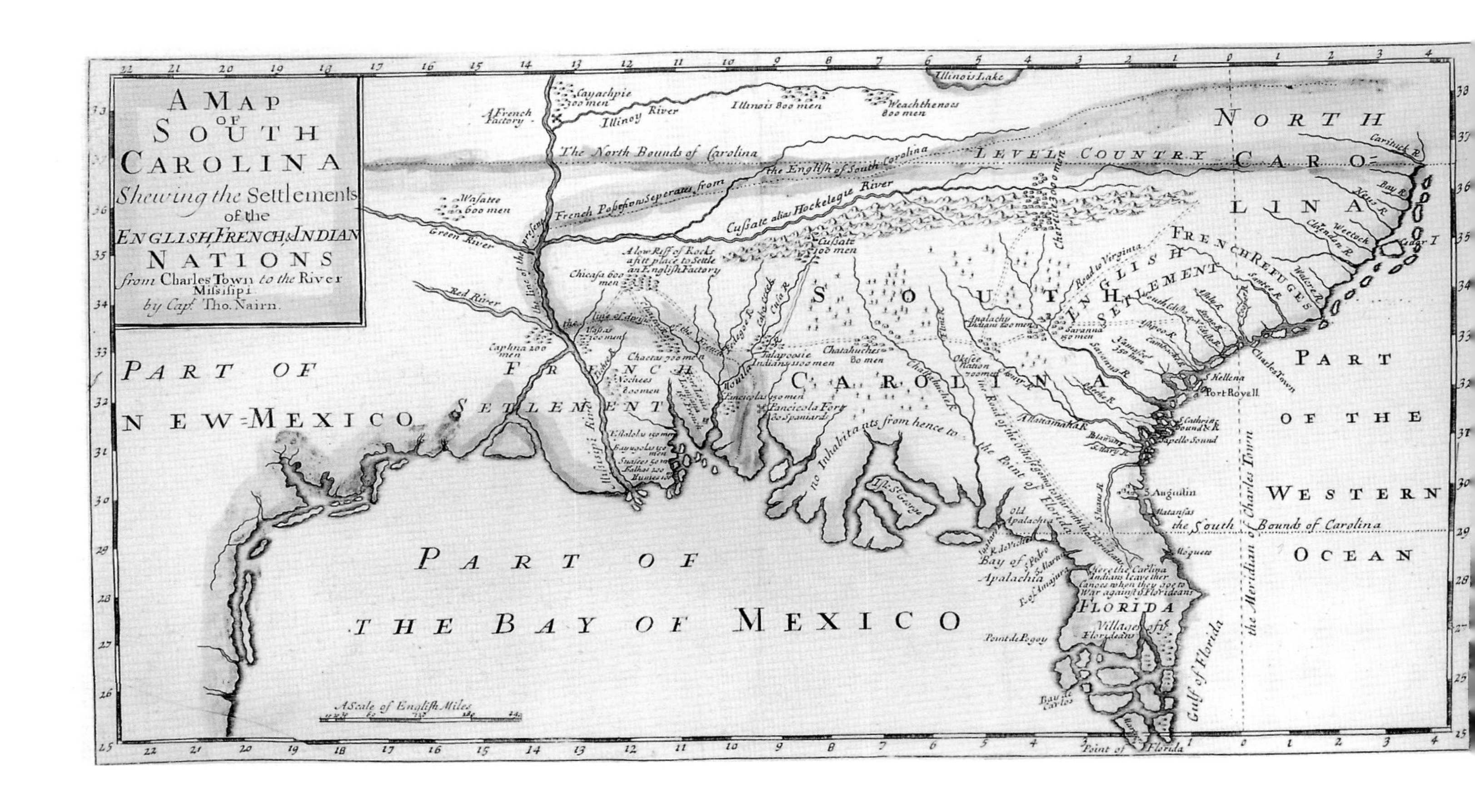

PLATE 45. Edward Crisp. A Compleat Description of the Province of Carolina. [Thomas Nairn map inset]. [1711]. [Map. 151].

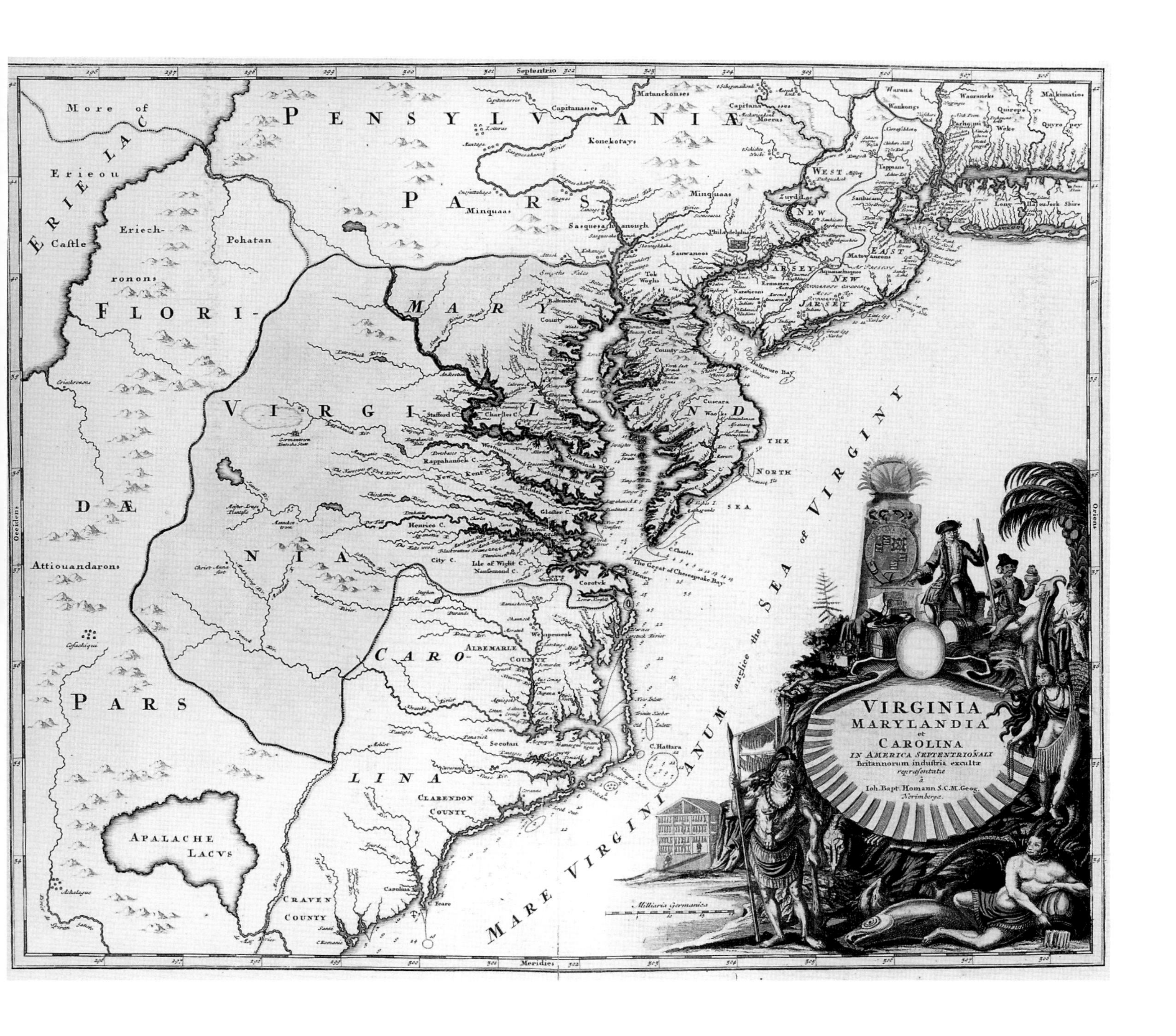

PLATE 46. Johann B. Homann. Virginia Marylandia et Carolina. 1714. [Map 156].

PLATE 46A. Anonymous. [Indian Villages]. Ca 1715 MS. [Map 157].

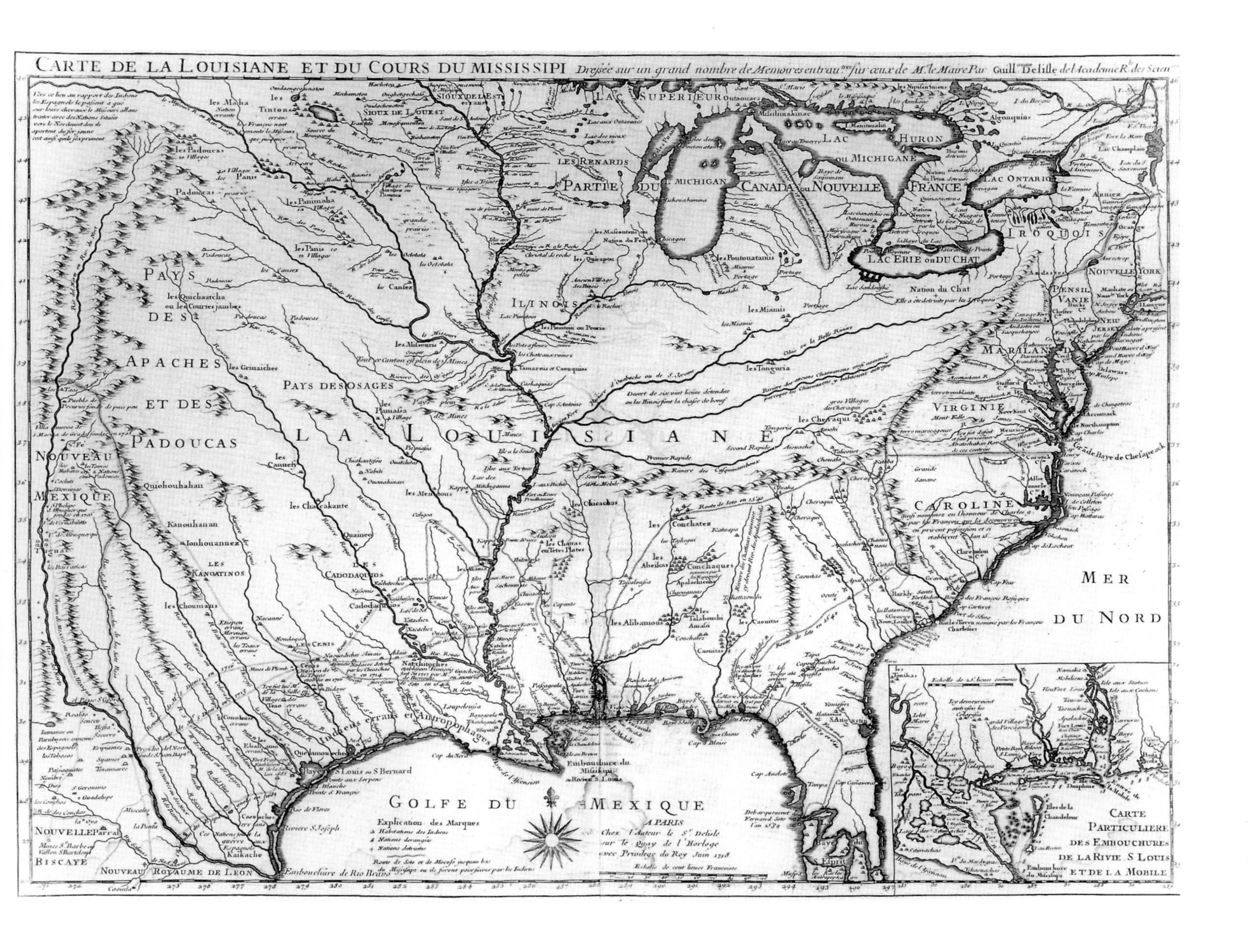

PLATE 47. Guillaume Delisle. Carte de la Louisiane. 1718. [Map 170].

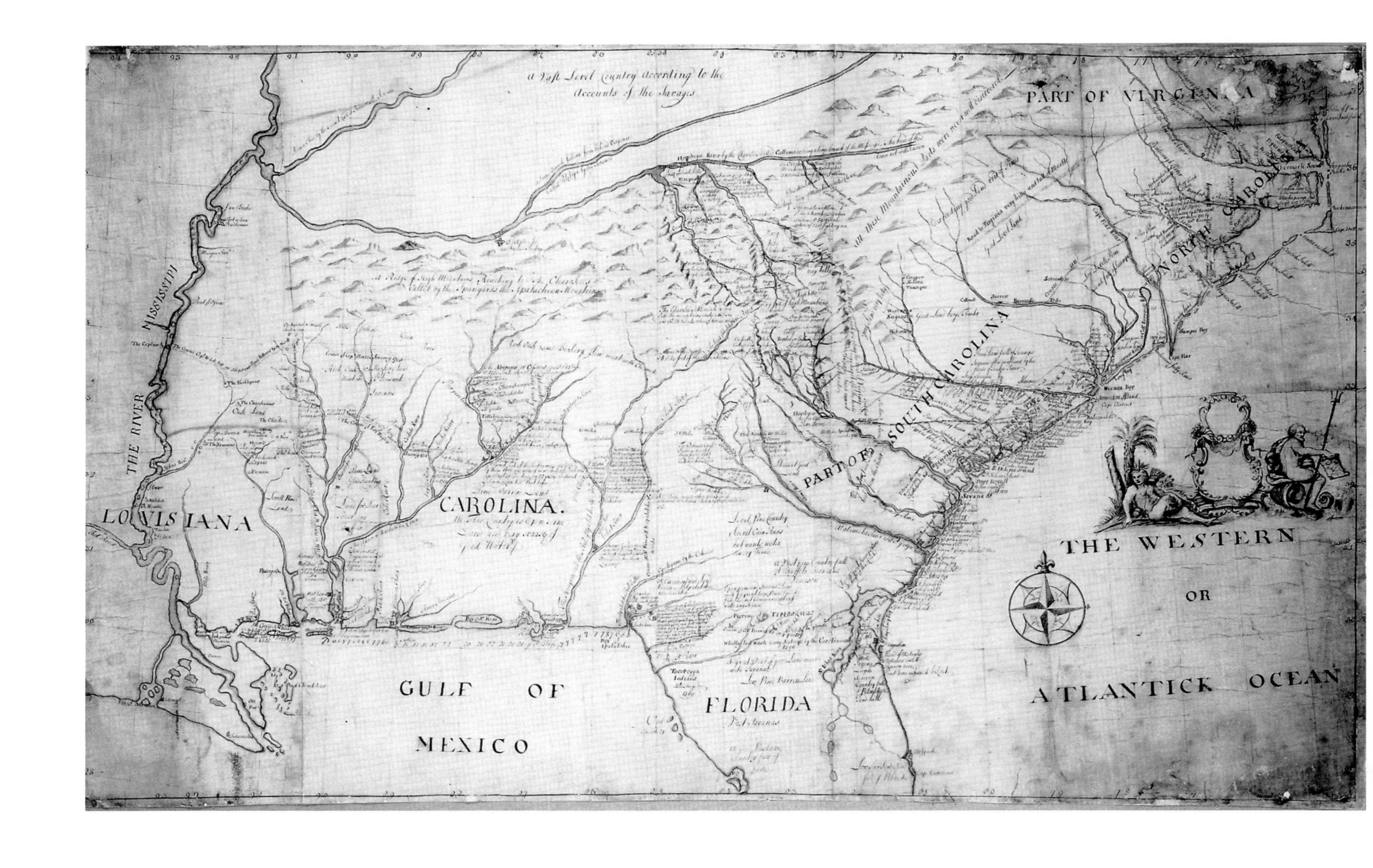

PLATE 48. Barnwell-Hammerton. [Southeastern North America]. Ca. 1721 MS. [Map 184A].

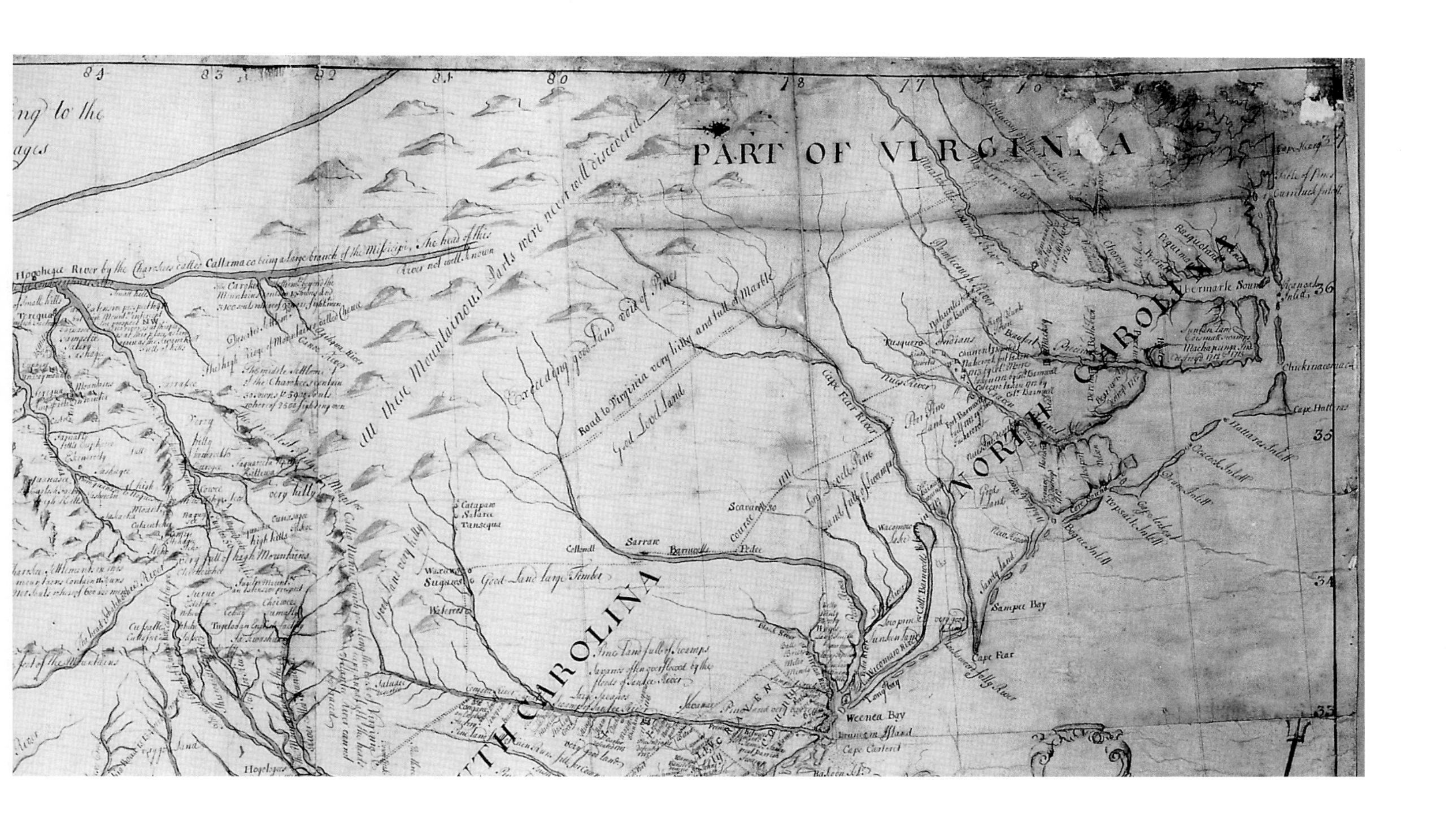

PLATE 48A. Barnwell-Hammerton. [Southeastern North America (northeast quadrant)]. Ca. 1721 MS. [Map 184A].

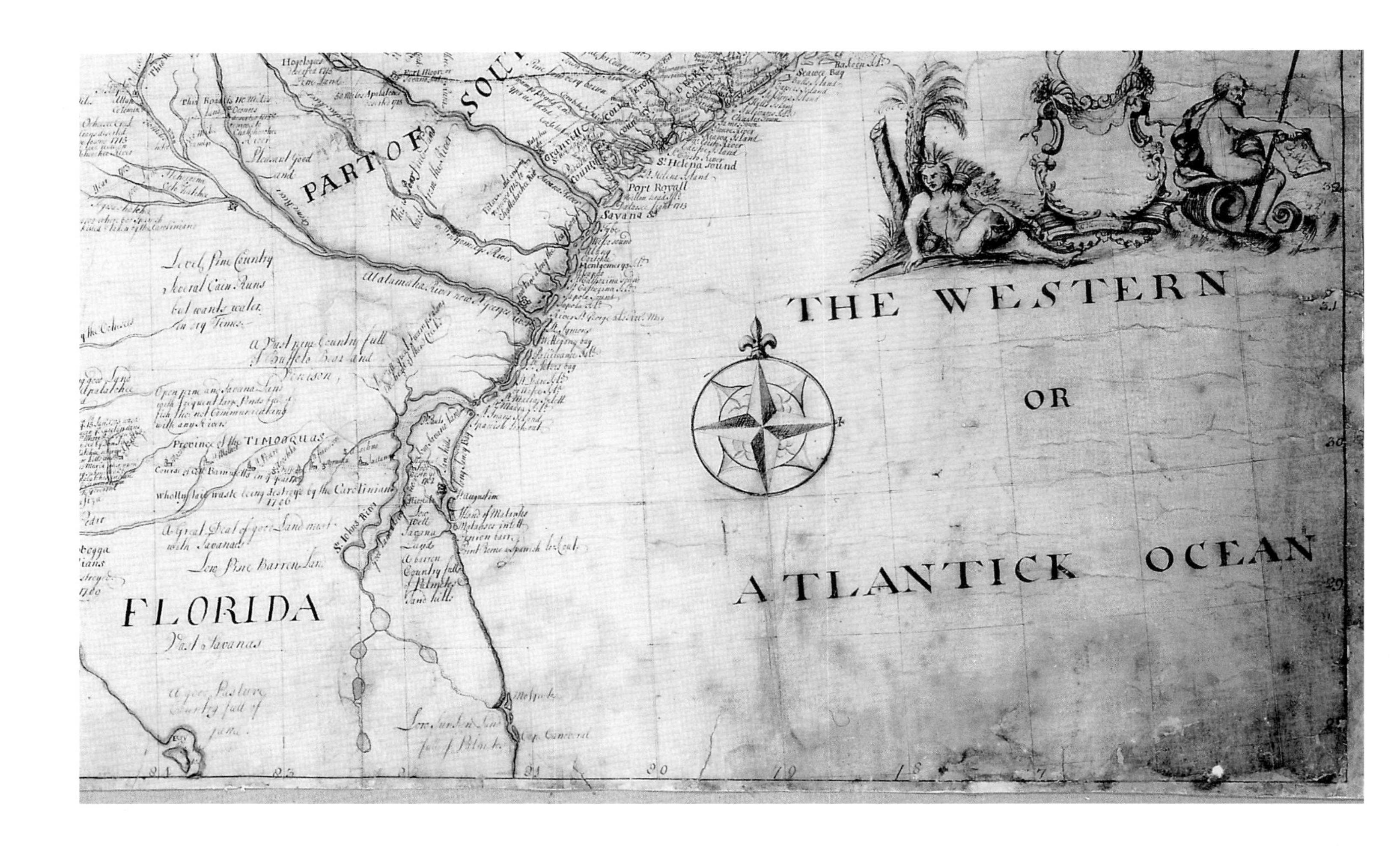

PLATE 48B. Barnwell-Hammerton. [Southeastern North America (southeast quadrant)]. Ca. 1721 MS. [Map 184A].

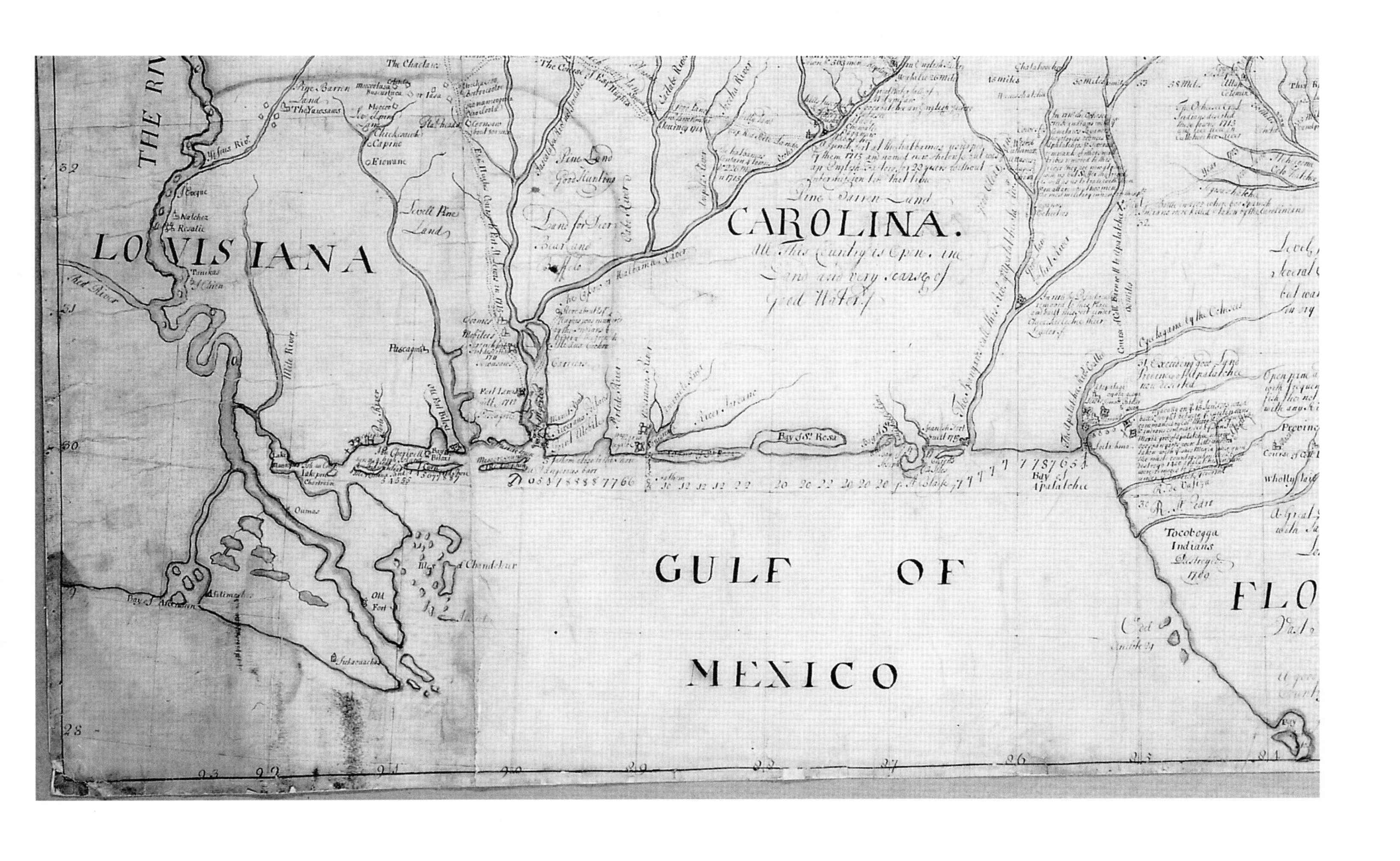

PLATE 48C. Barnwell-Hammerton. [Southeastern North America (southwest quadrant)]. Ca. 1721 MS. [Map 184A].

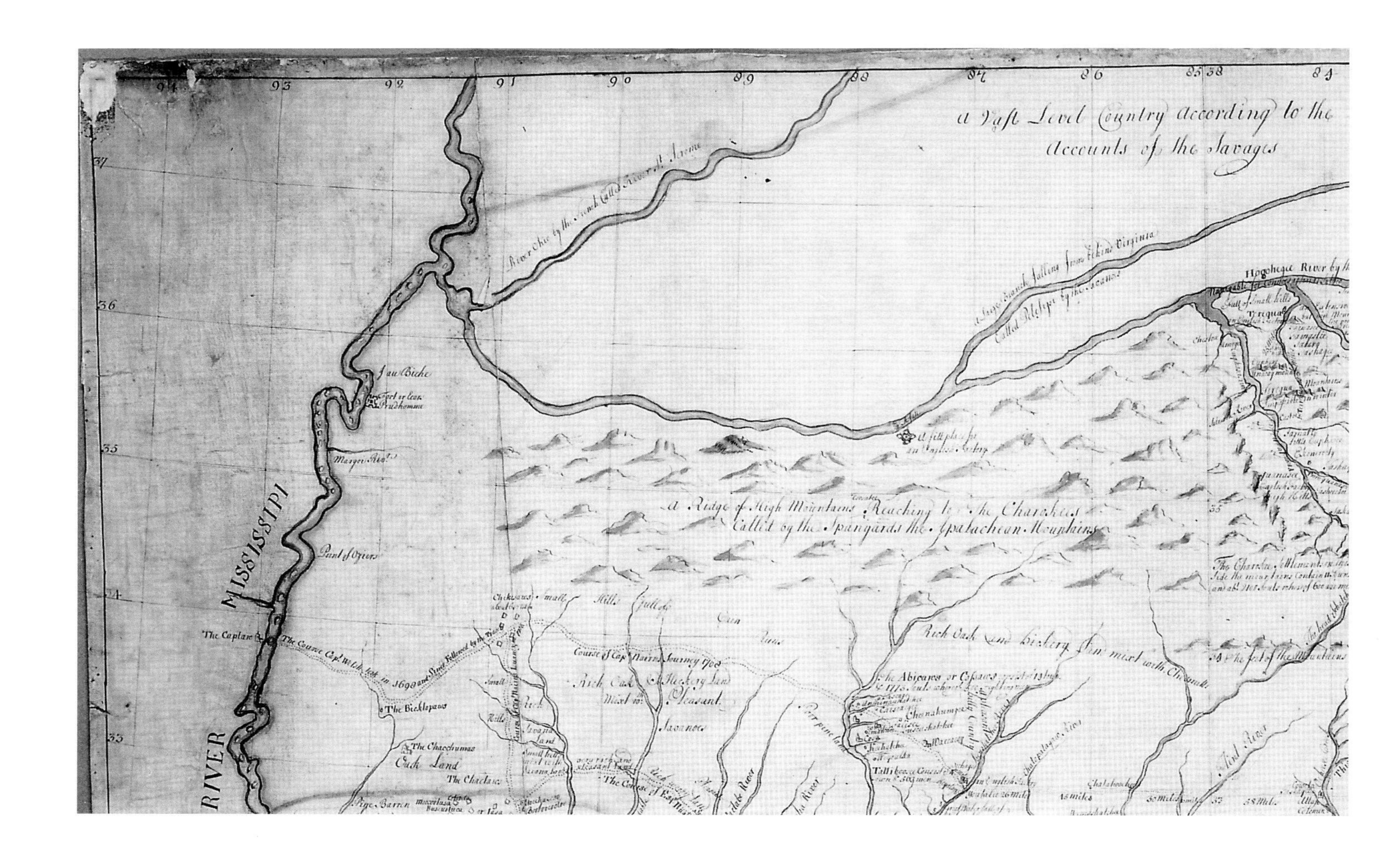

PLATE 48D. Barnwell-Hammerton. [Southeastern North America (northwest quadrant)]. Ca. 1721 MS. [Map 184A].

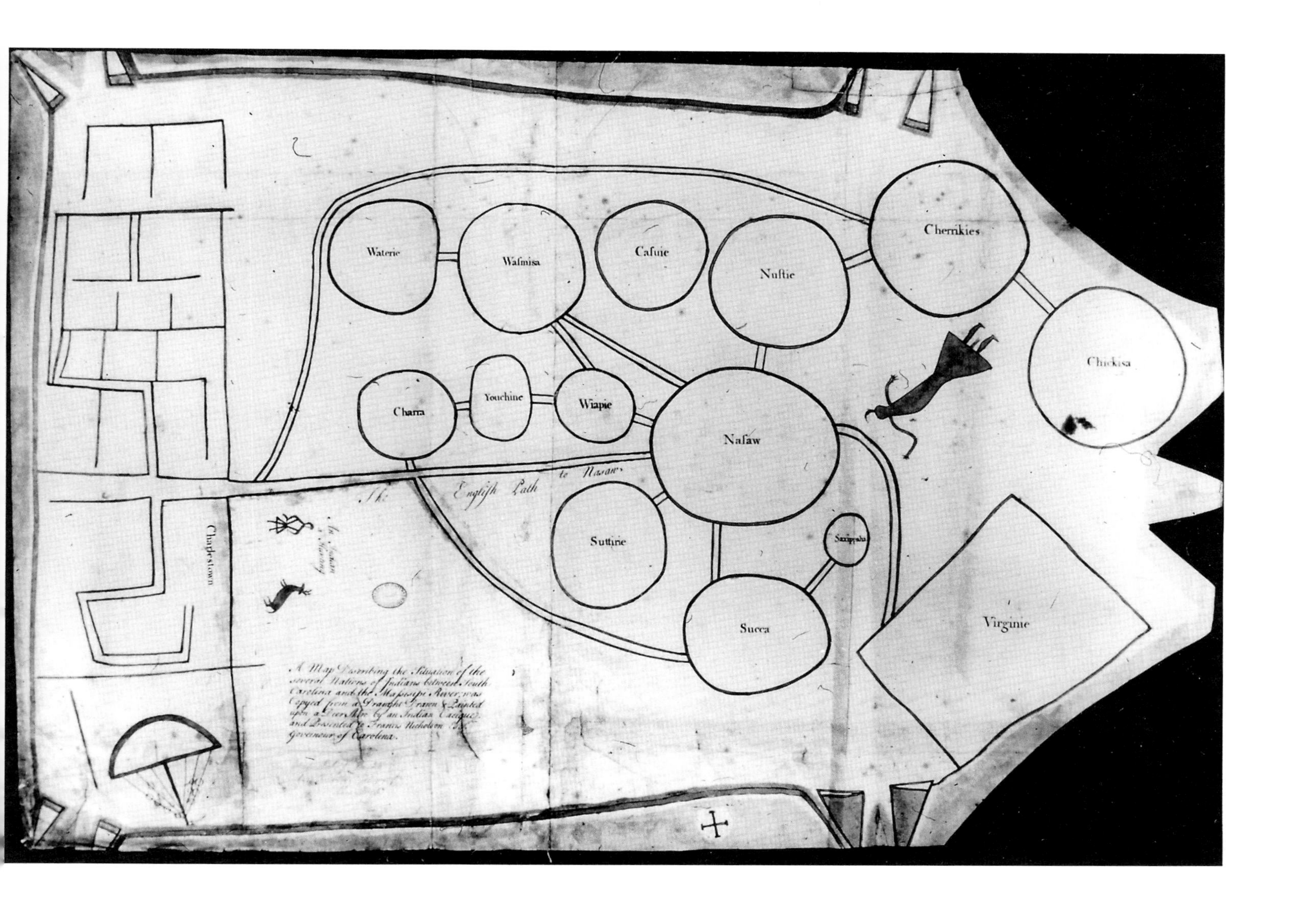

PLATE 48E. An Indian Cacique. A Map Describing the Situation of the Several Nations of Indians. Ca. 1724 MS B. [Map 193].

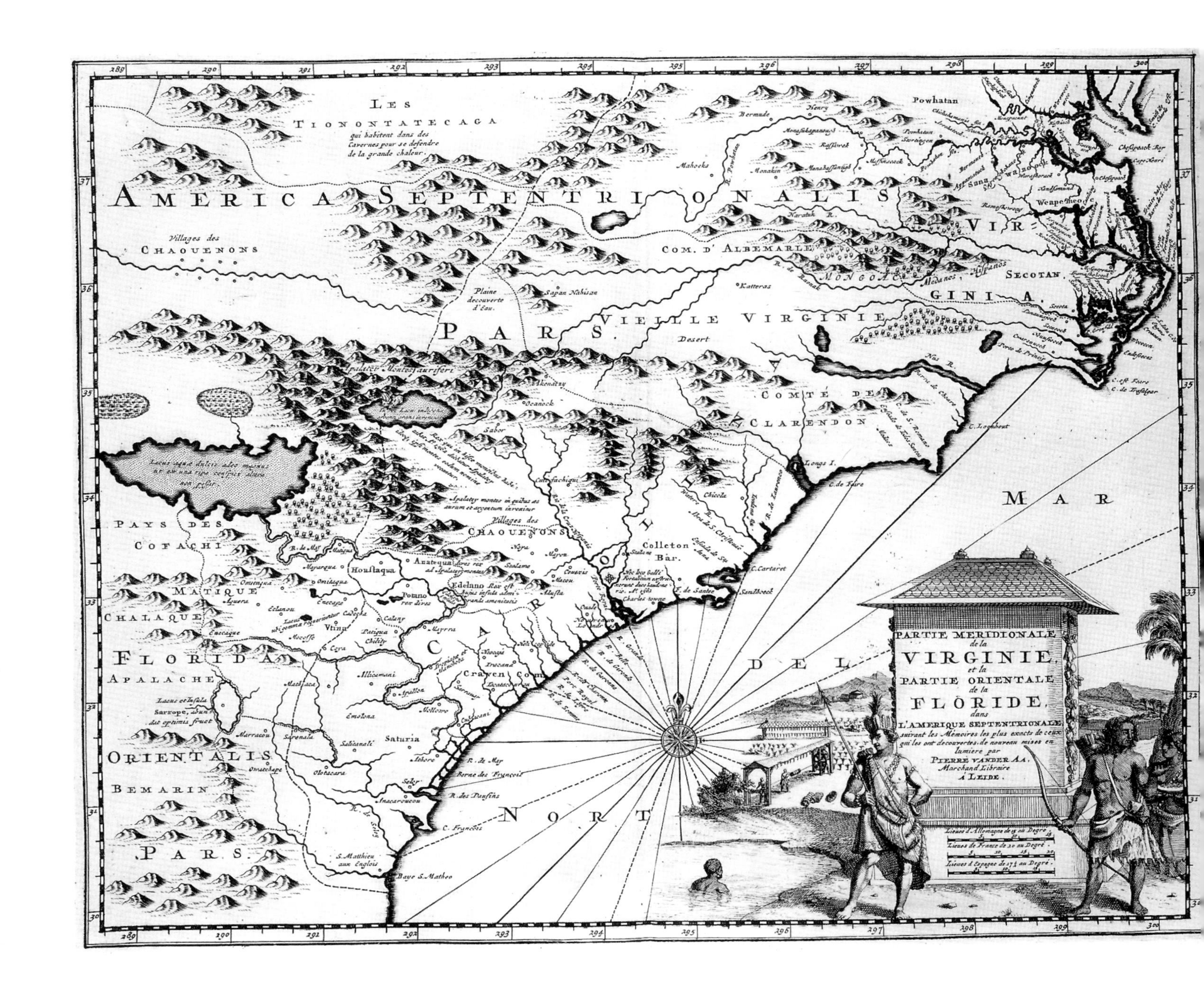

PLATE 49. Pieter vander Aa. Partie Meridionale de la Virginie, et la Partie Orientale de la Floride. 1729. [Map 205].

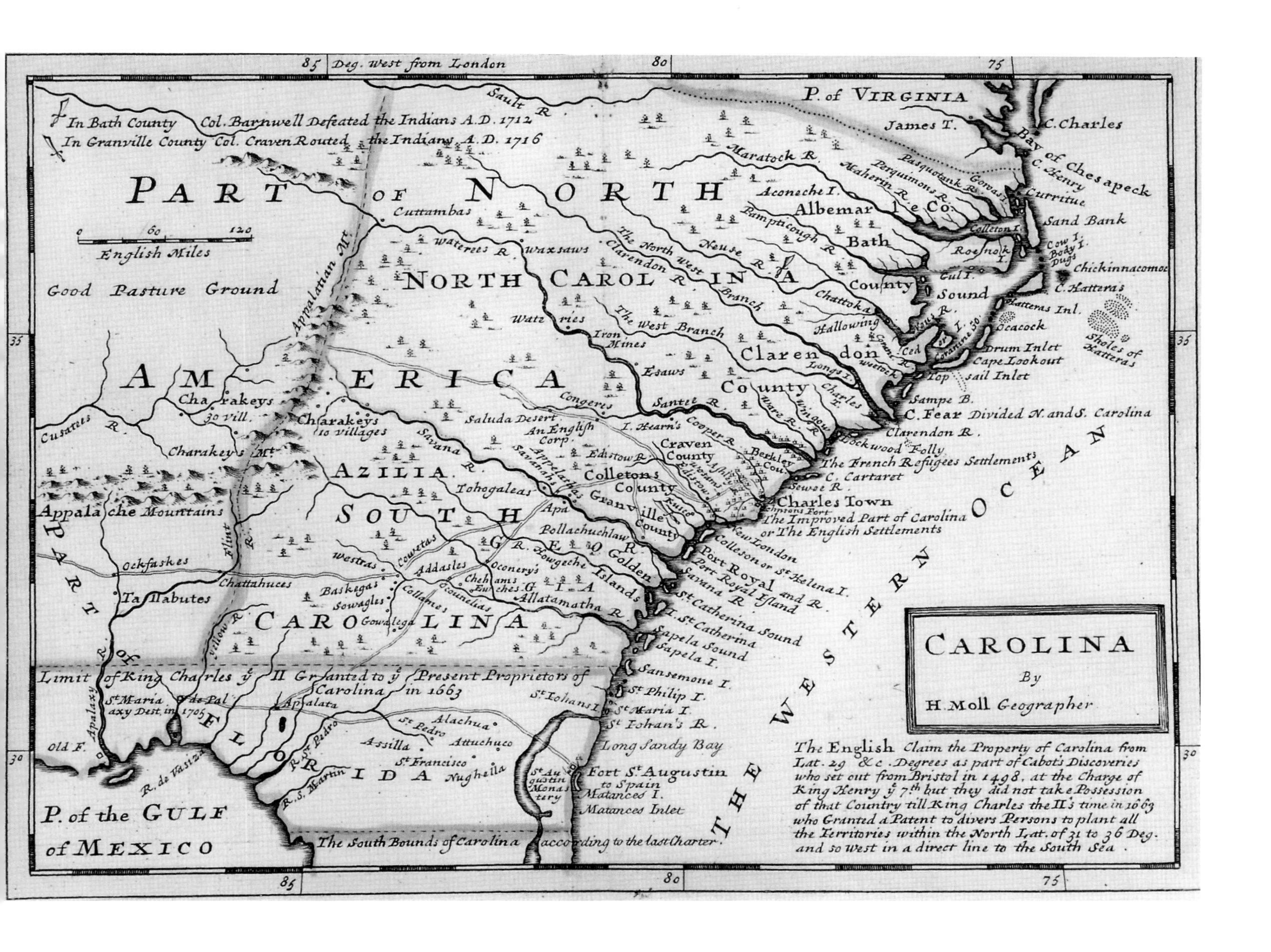

PLATE 50. Herman Moll. Carolina. 1729. [Map 206].

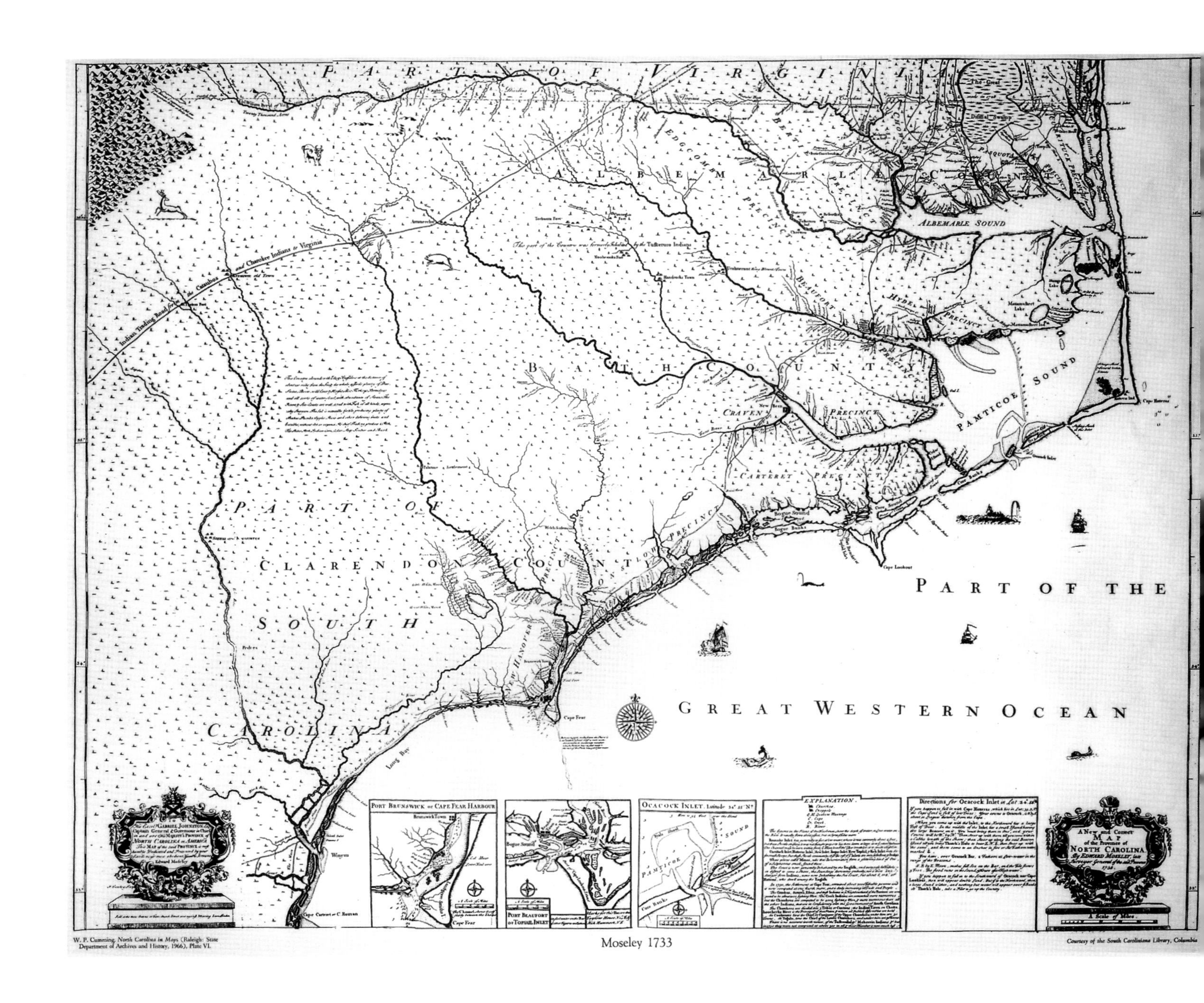

PLATE 50A. Edward Moseley. A New and Correct Map of the Province of North Carolina. 1733. [Map 218].

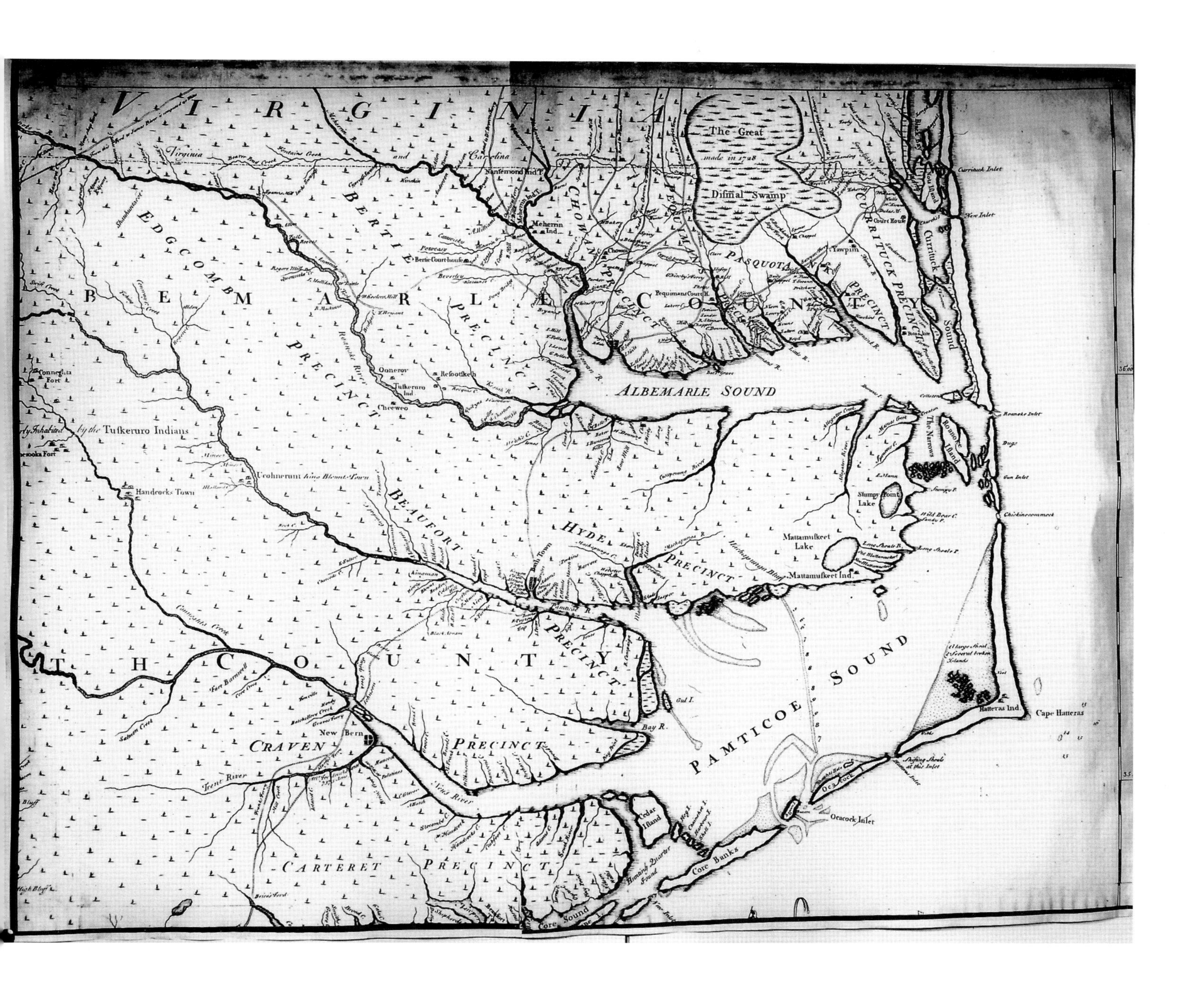

PLATE 51. Edward Moseley. A New and Correct Map of the Province of North Carolina [northeast quadrant]. 1733. [Map 218].

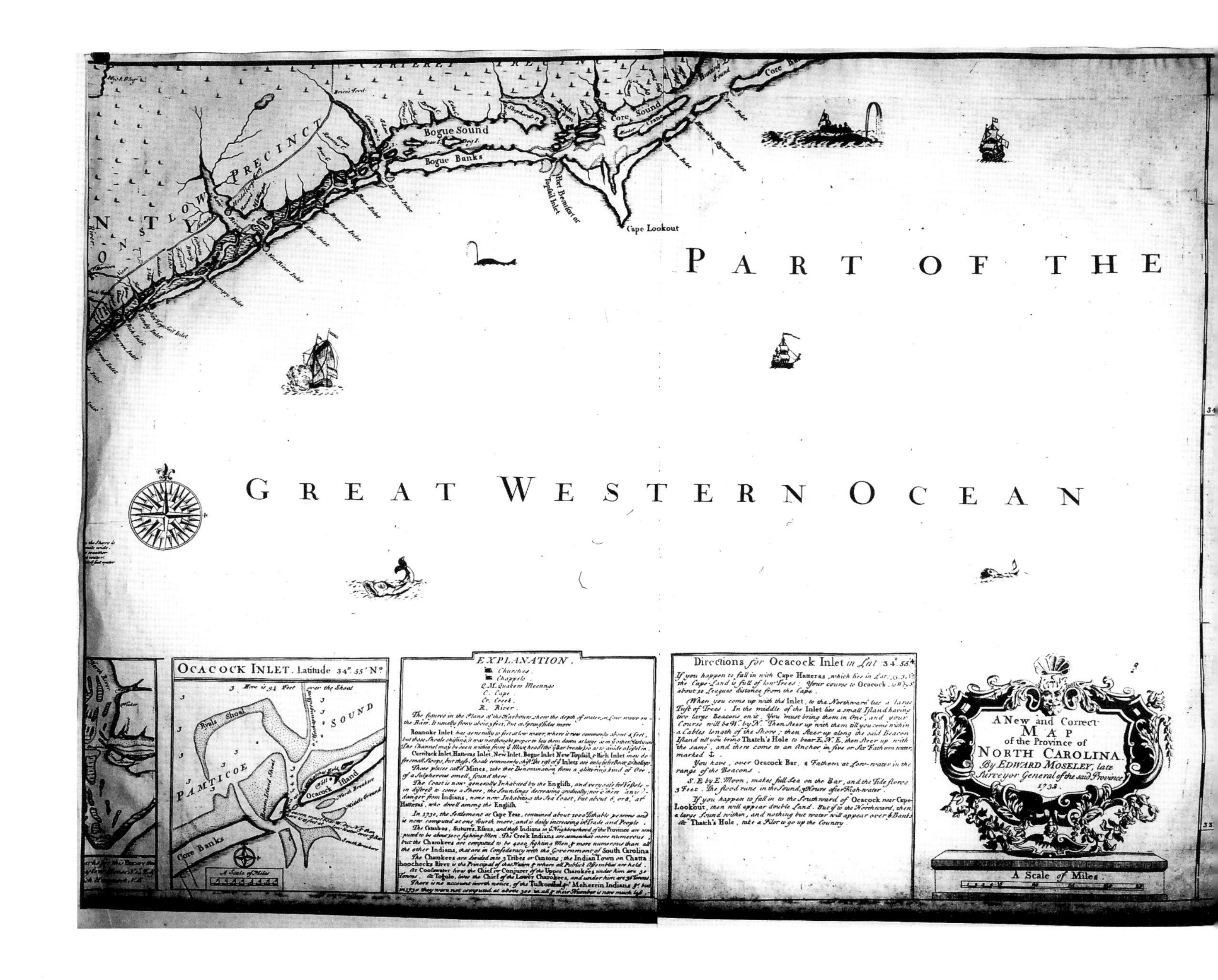

PLATE 52. Edward Moseley. A New and Correct Map of the Province of North Carolina [southeast quadrant]. 1733. [Map 218].

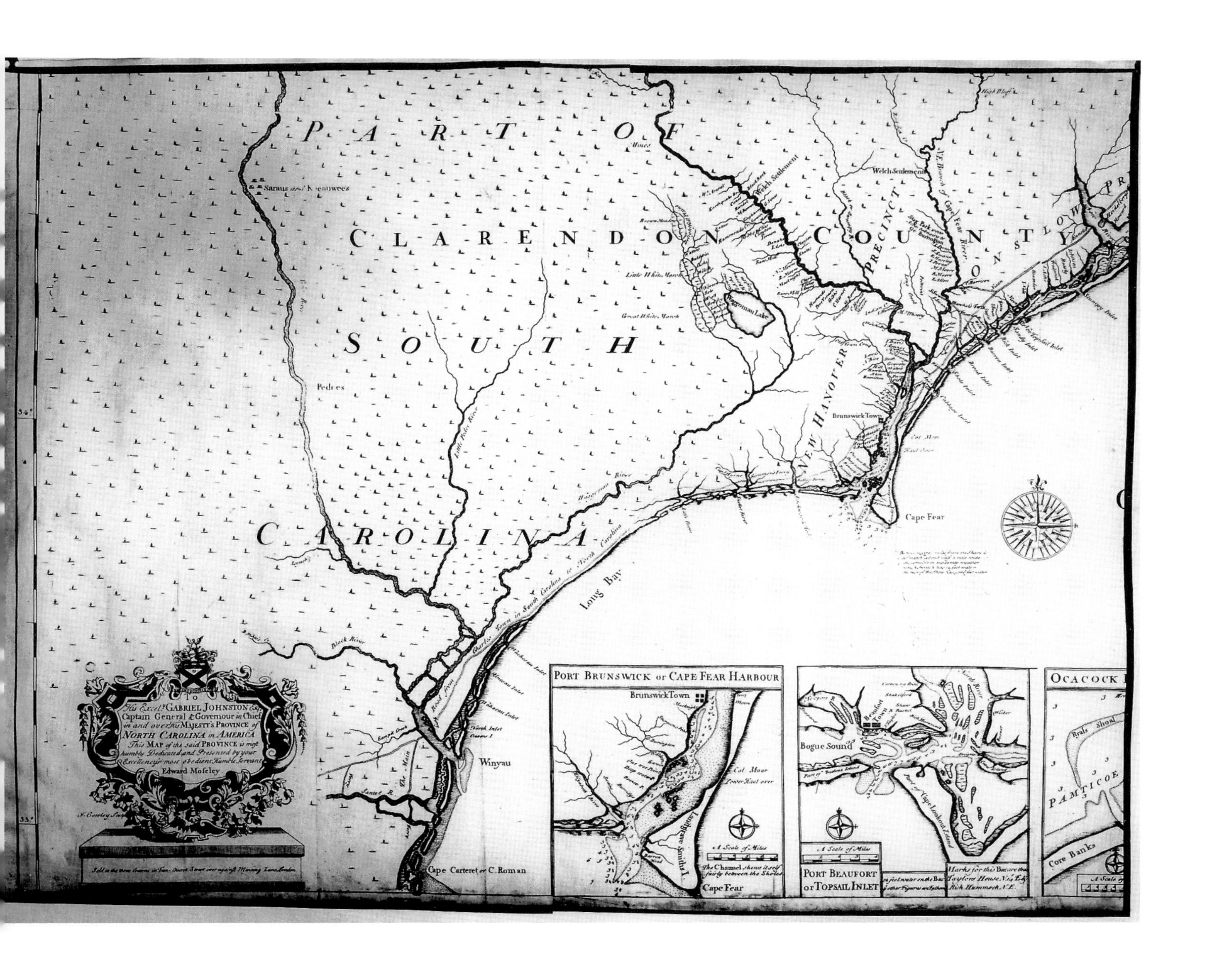

PLATE 53. Edward Moseley. A New and Correct Map of the Province of North Carolina [southwest quadrant]. 1733. [Map 218].

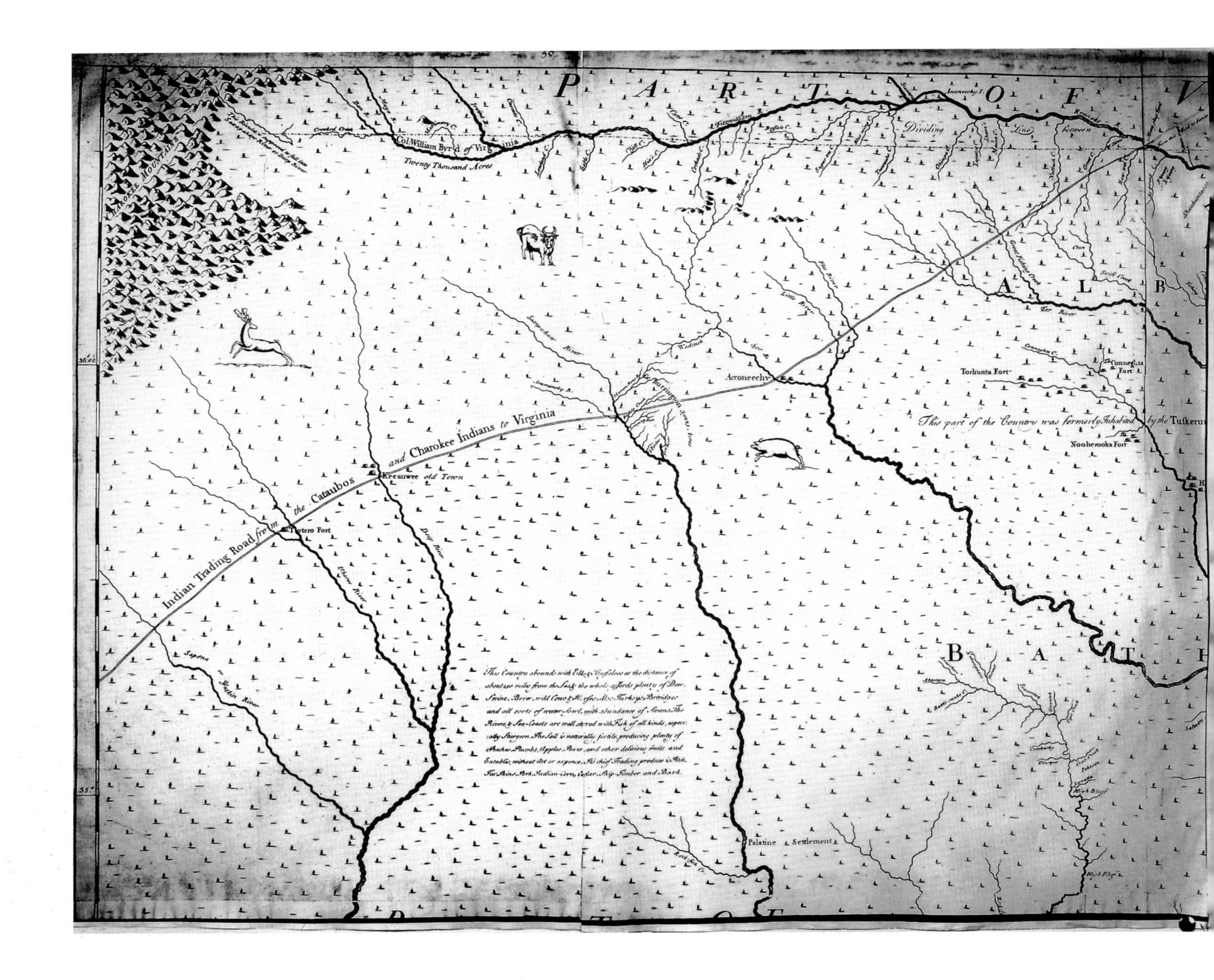

PLATE 54. Edward Moseley. A New and Correct Map of the Province of North Carolina [northwest quadrant]. 1733. [Map 218].

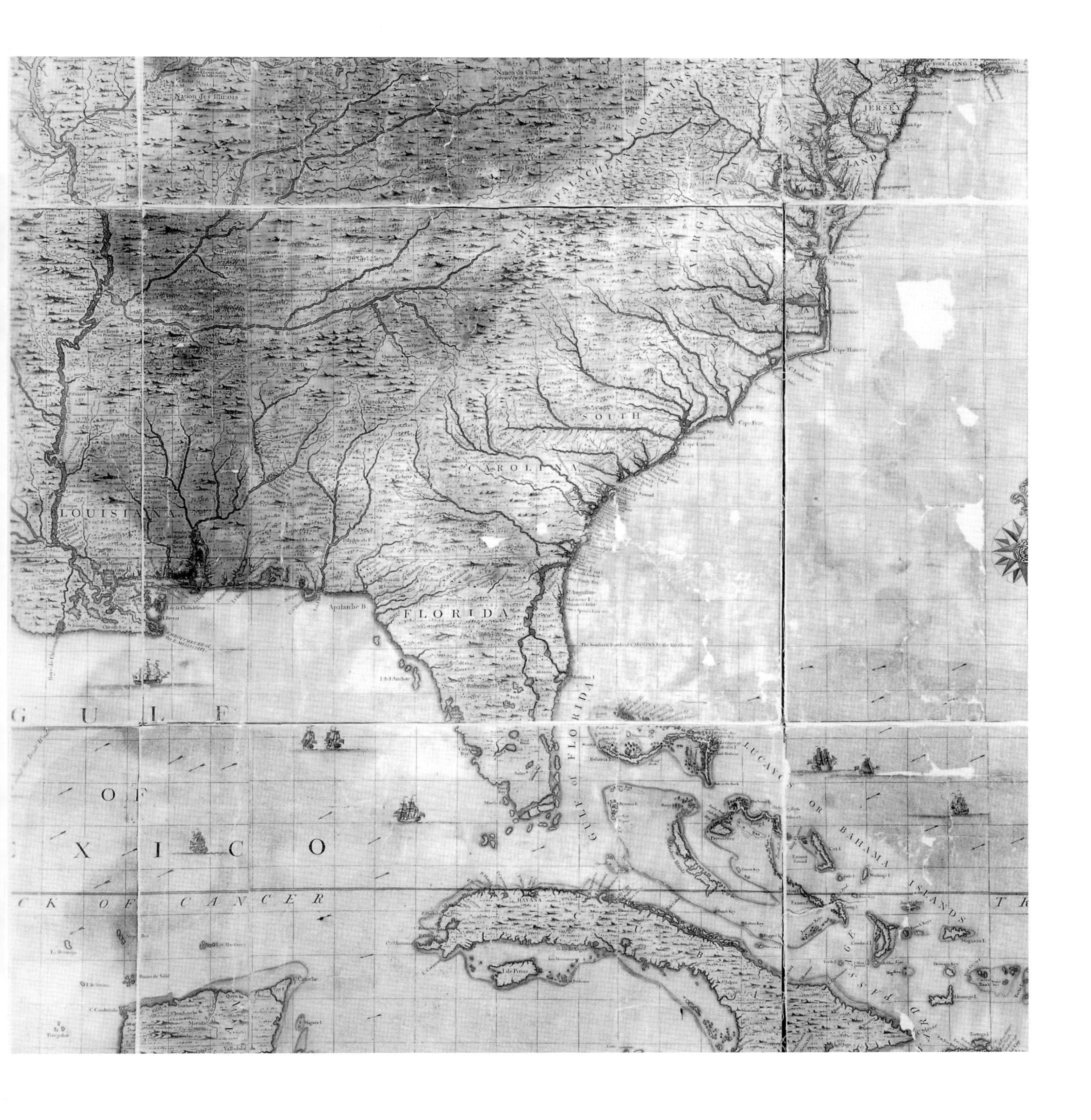

PLATE 55. Henry Popple. A Map of the British Empire in America [detail]. 1733. [Map 217].

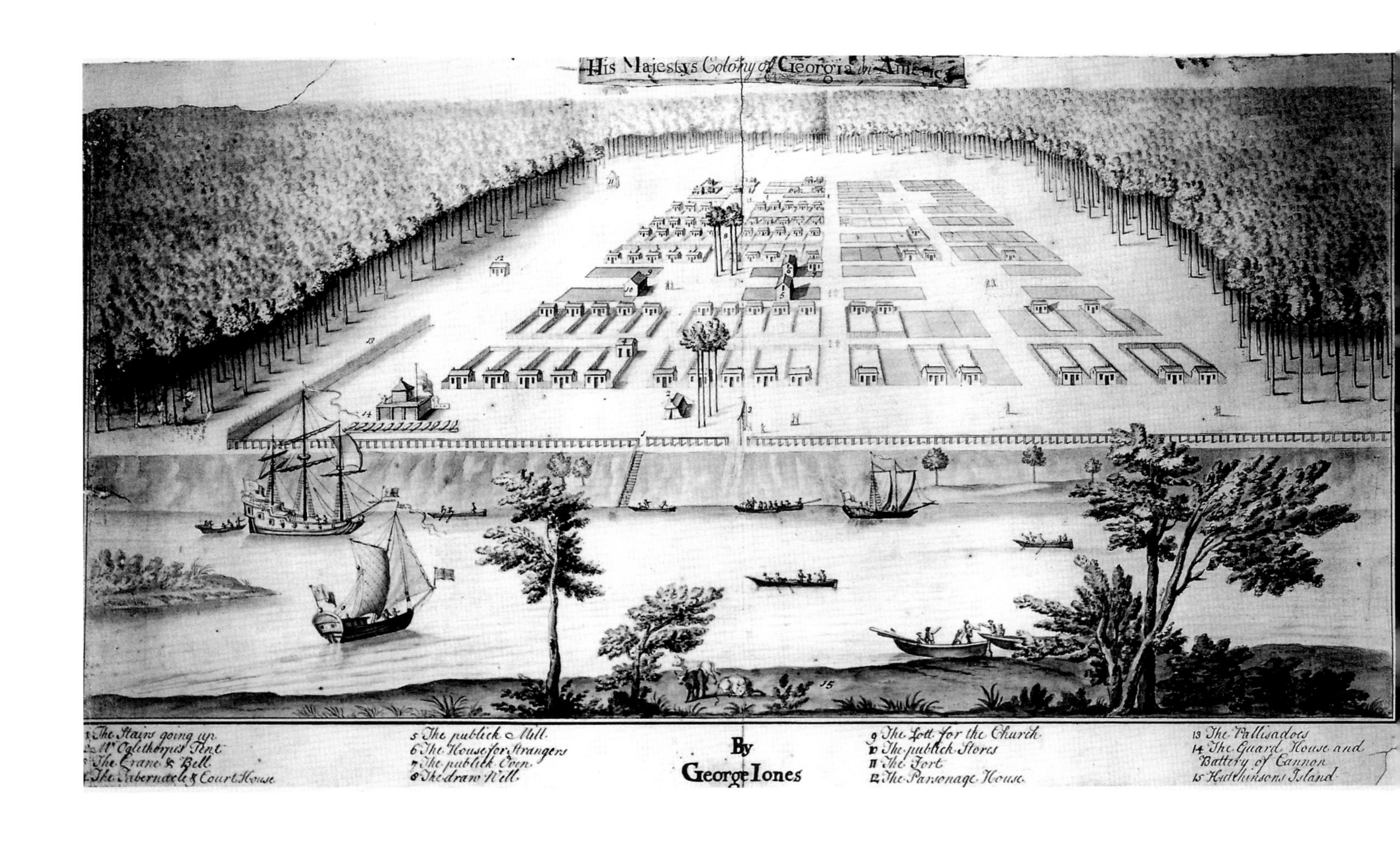

PLATE 55A. George Jones. His Majestys Colony of Georgia in America. 1734 MS. [Map 219A].

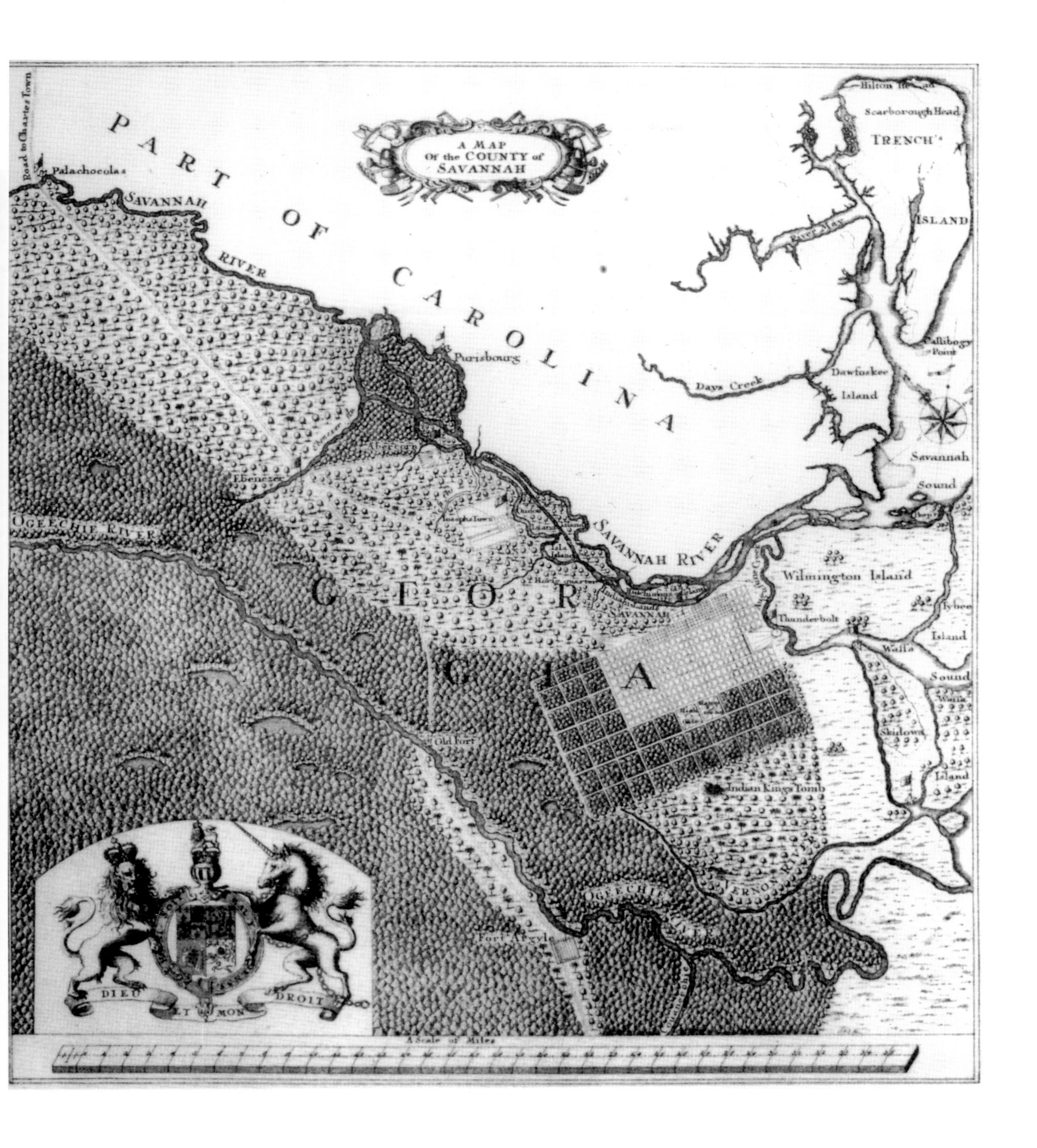

PLATE 55B. Anonymous. A Map of the County of Savannah. [1735]. [Map 246].

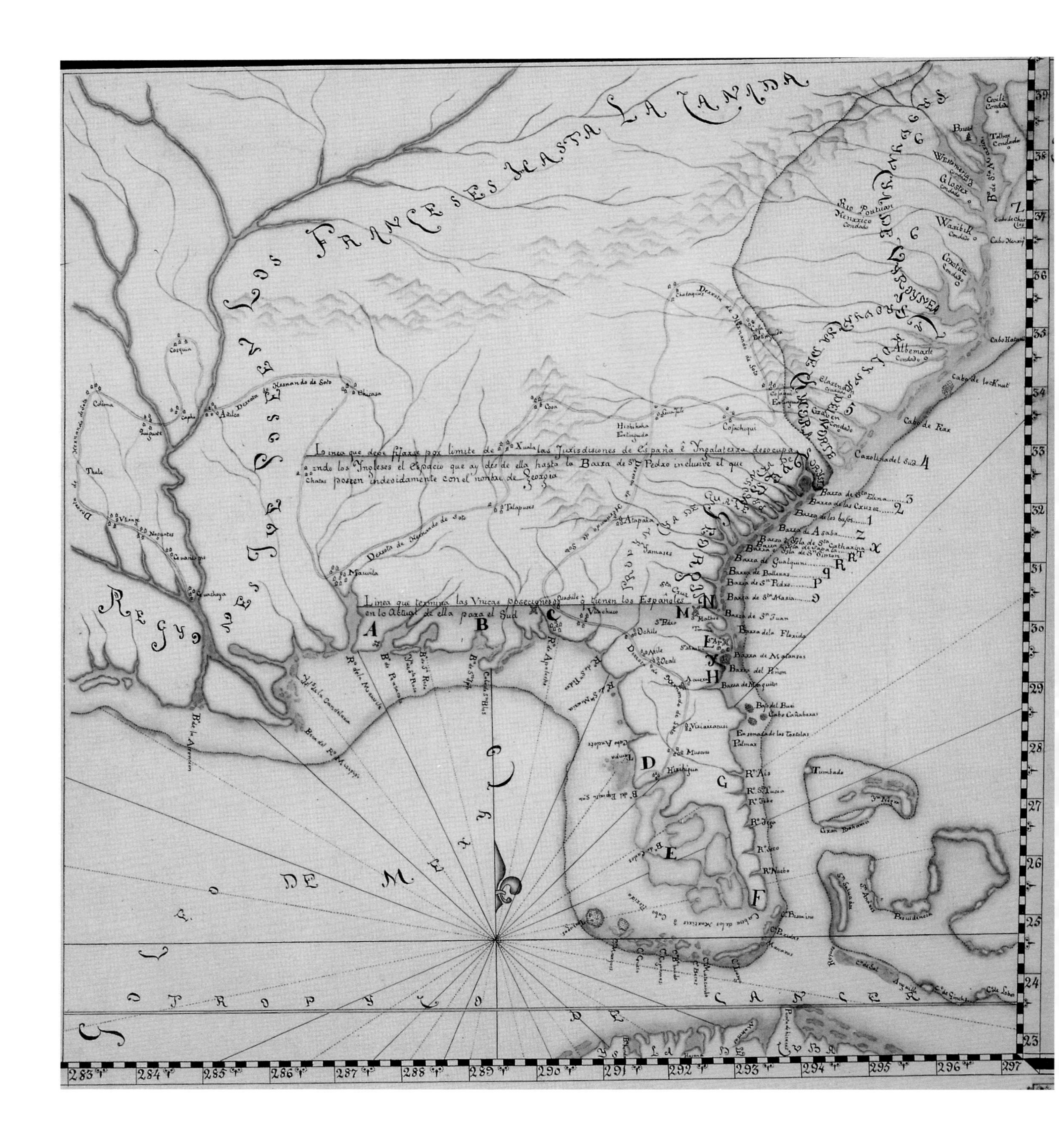

PLATE 56. Antonio de Arredondo. Descriptio Geographica . . . de la Florida. 1742 MS. [Map 253].

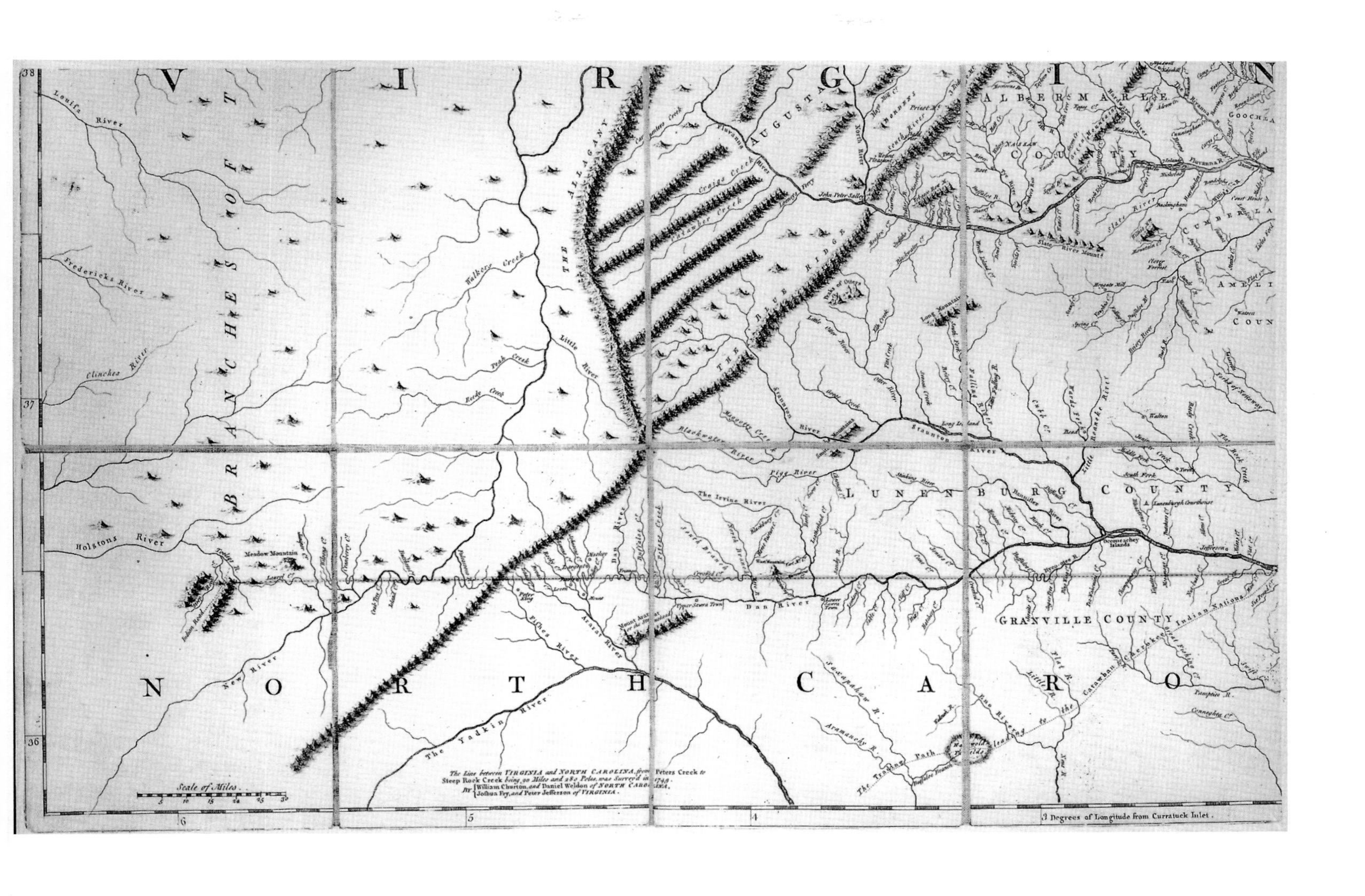

PLATE 57. Joshua Fry–Peter Jefferson. A Map of the Inhabited Part of Virginia [detail]. 1751 [1753]. [Map 281].

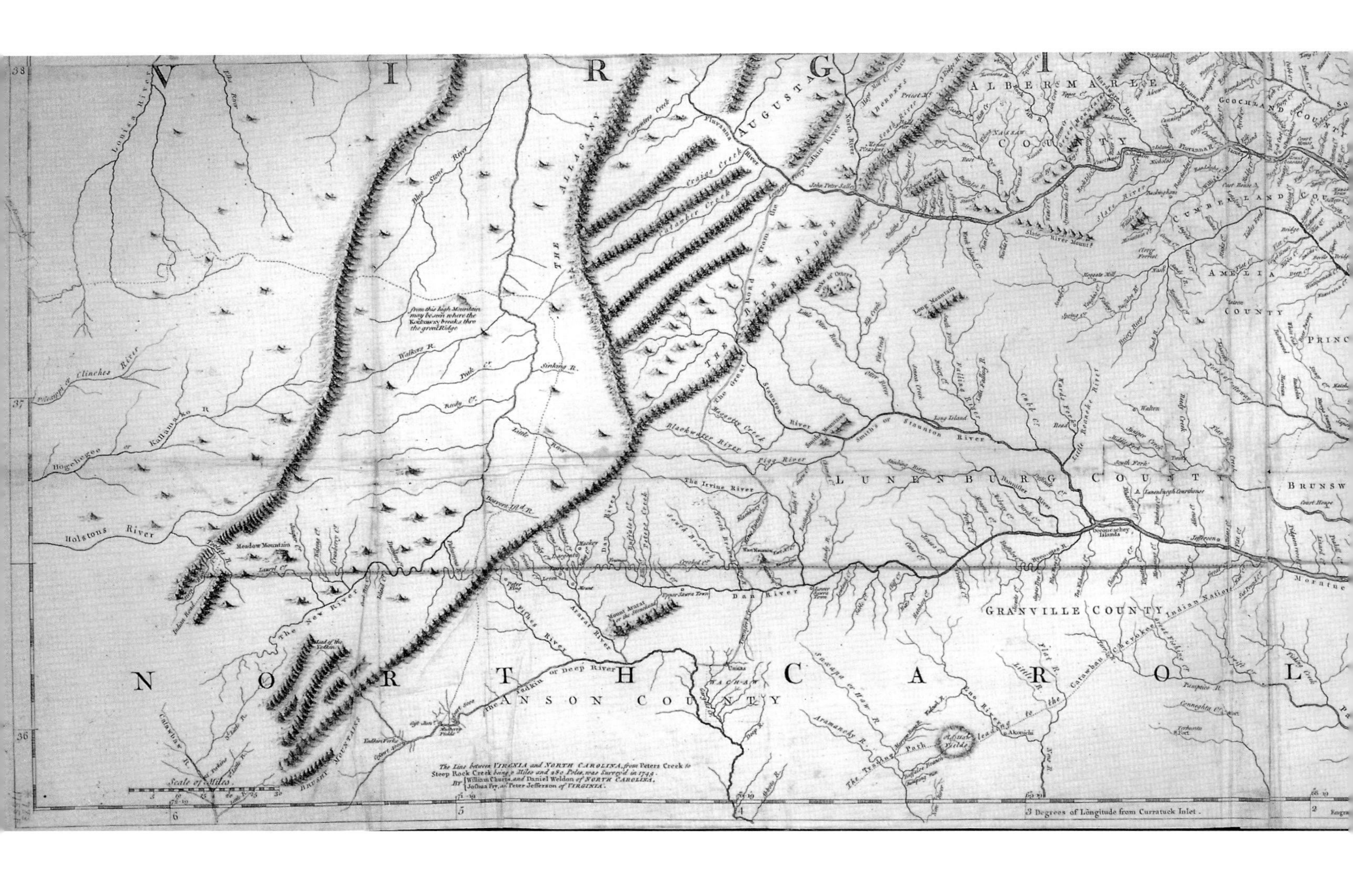

PLATE 58. Joshua Fry–Peter Jefferson. A Map of the Inhabited Part of Virginia, [detail]. 1751 [1755]. [Map 281].

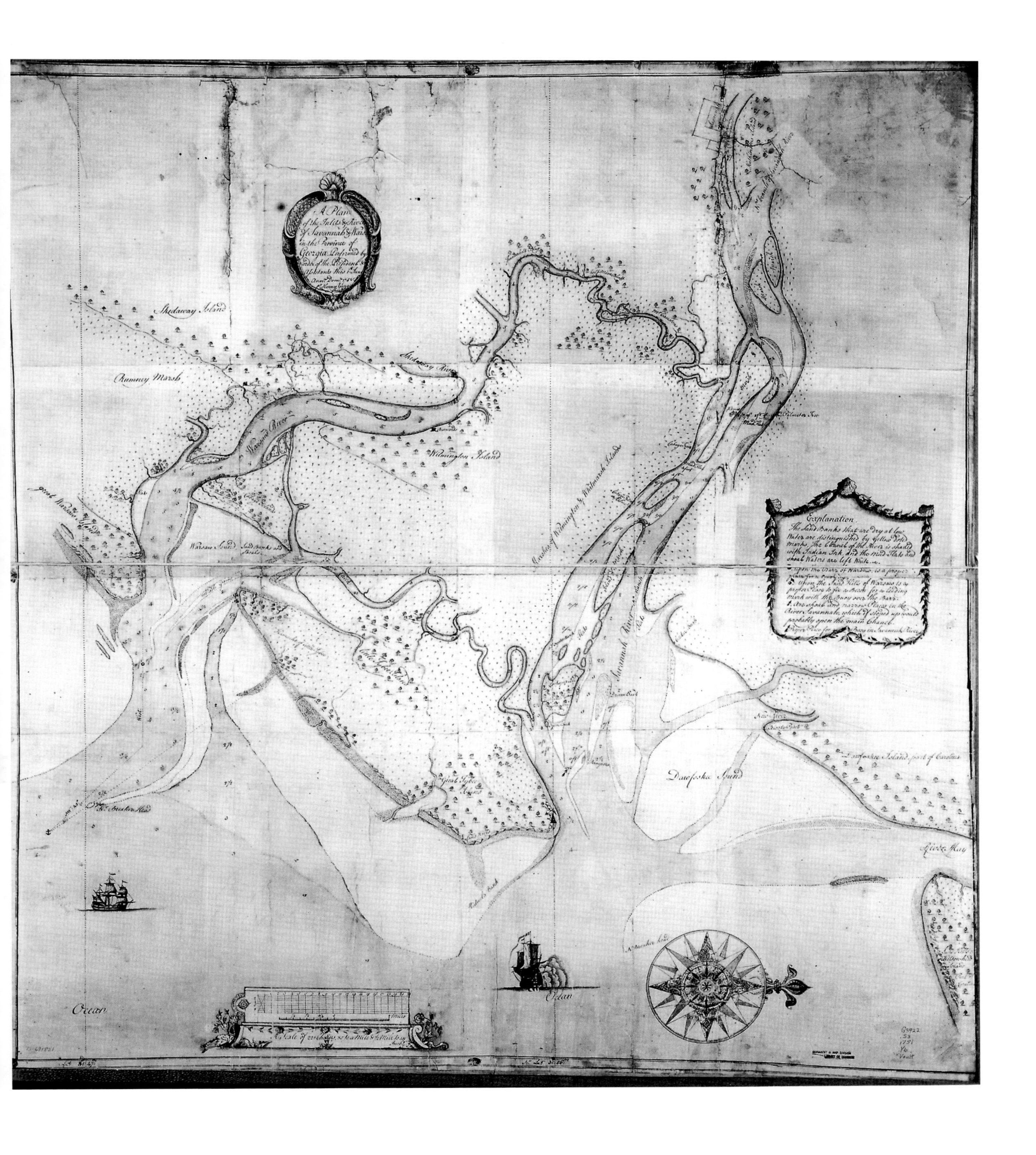

PLATE 58A. Henry Yonge. A Plan of the Inlets, & Rivers of Savannah & Warsaw in the Province of Georgia. 1751 MS. [Map 280].

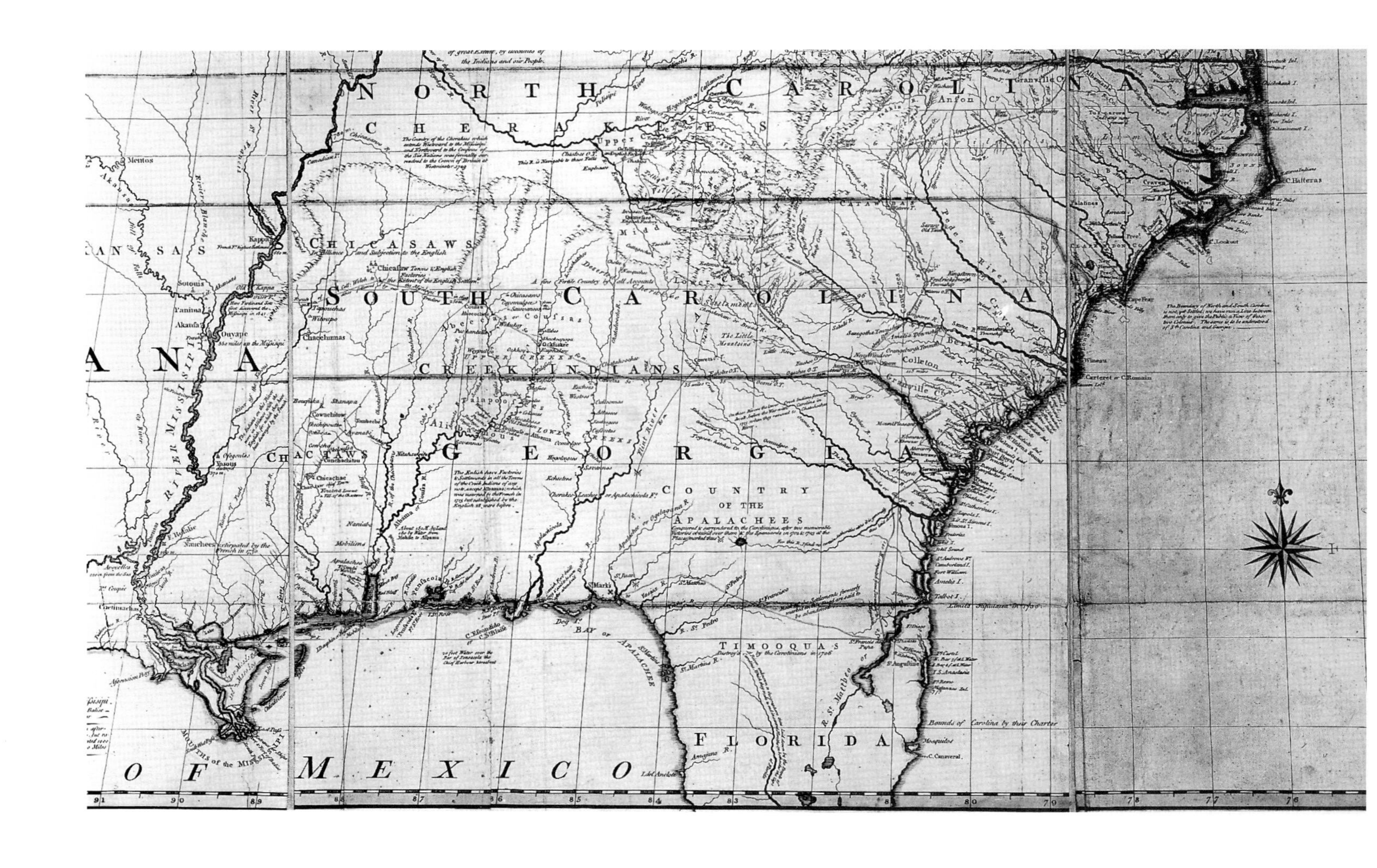

PLATE 59. John Mitchell. A Map of the British and French Dominions in North America [detail]. 1755. [Map 293].

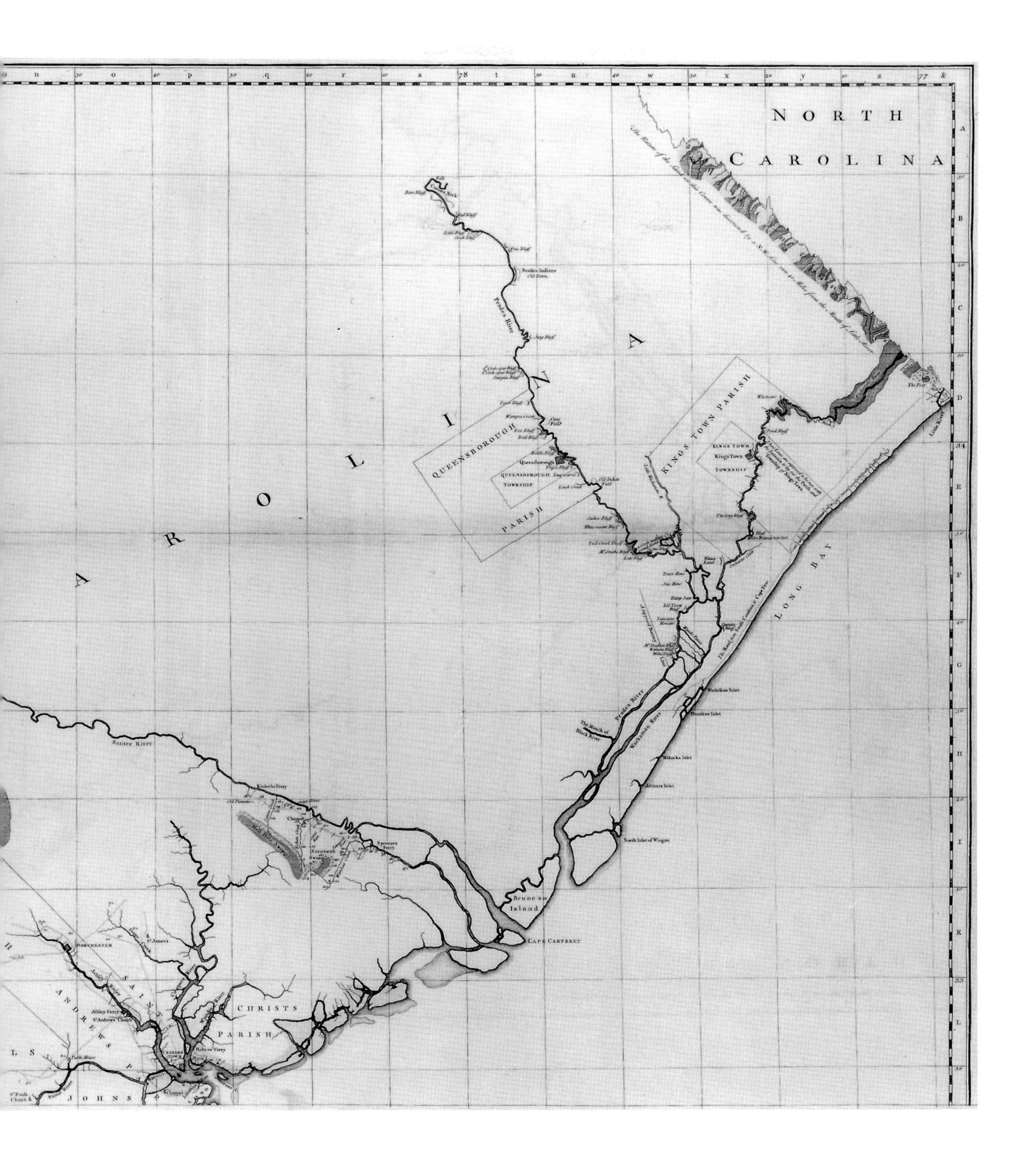

PLATE 59A. William De Brahm. A Map of South Carolina and a Part of Georgia [northeast quadrant]. 1757. [Map 310].

PLATE 59B. William De Brahm. A Map of South Carolina and a Part of Georgia [southeast quadrant]. 1757. [Map 310].

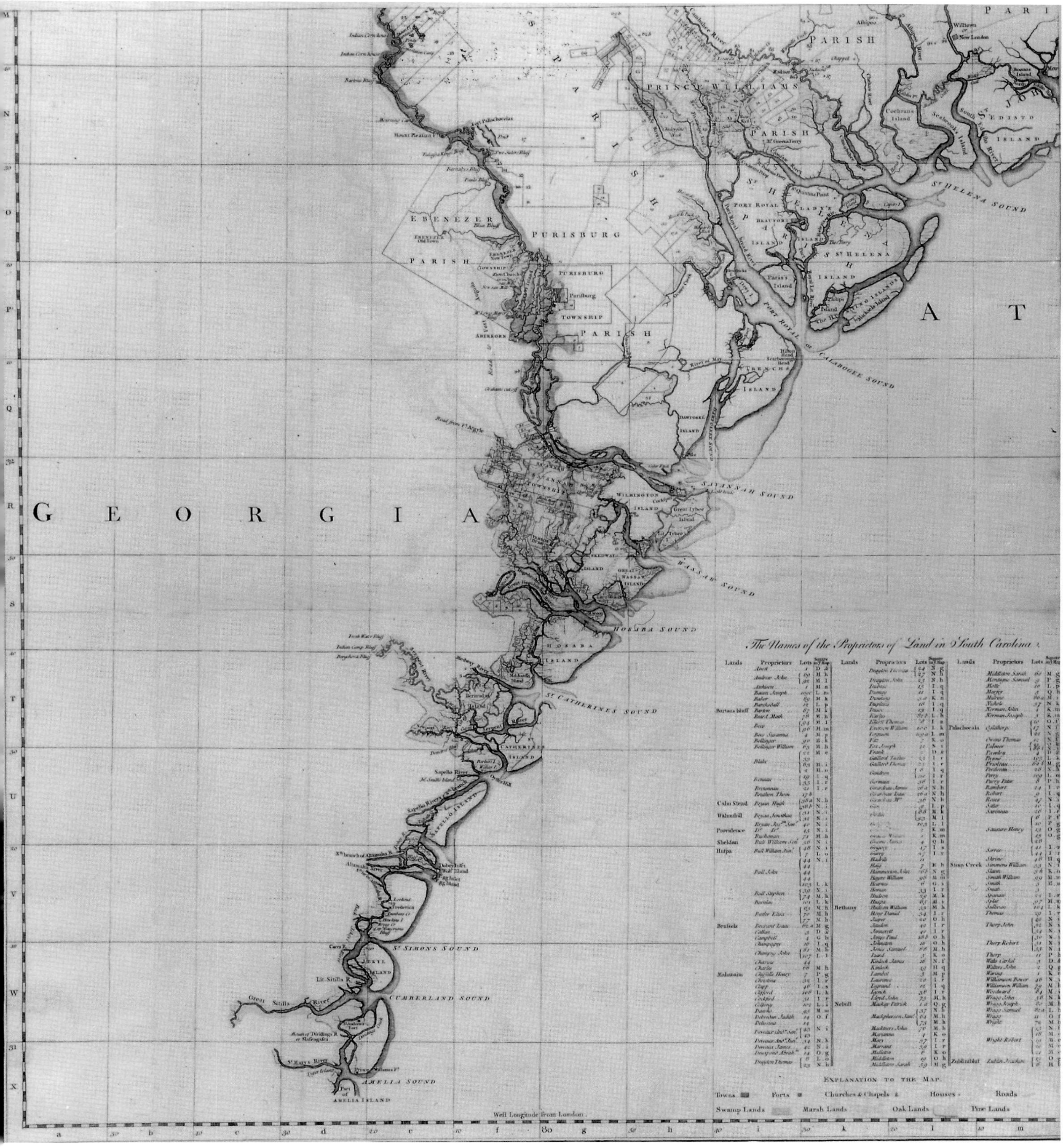

PLATE 59C. William De Brahm. A Map of South Carolina and a Part of Georgia [southwest quadrant]. 1757. [Map 310].

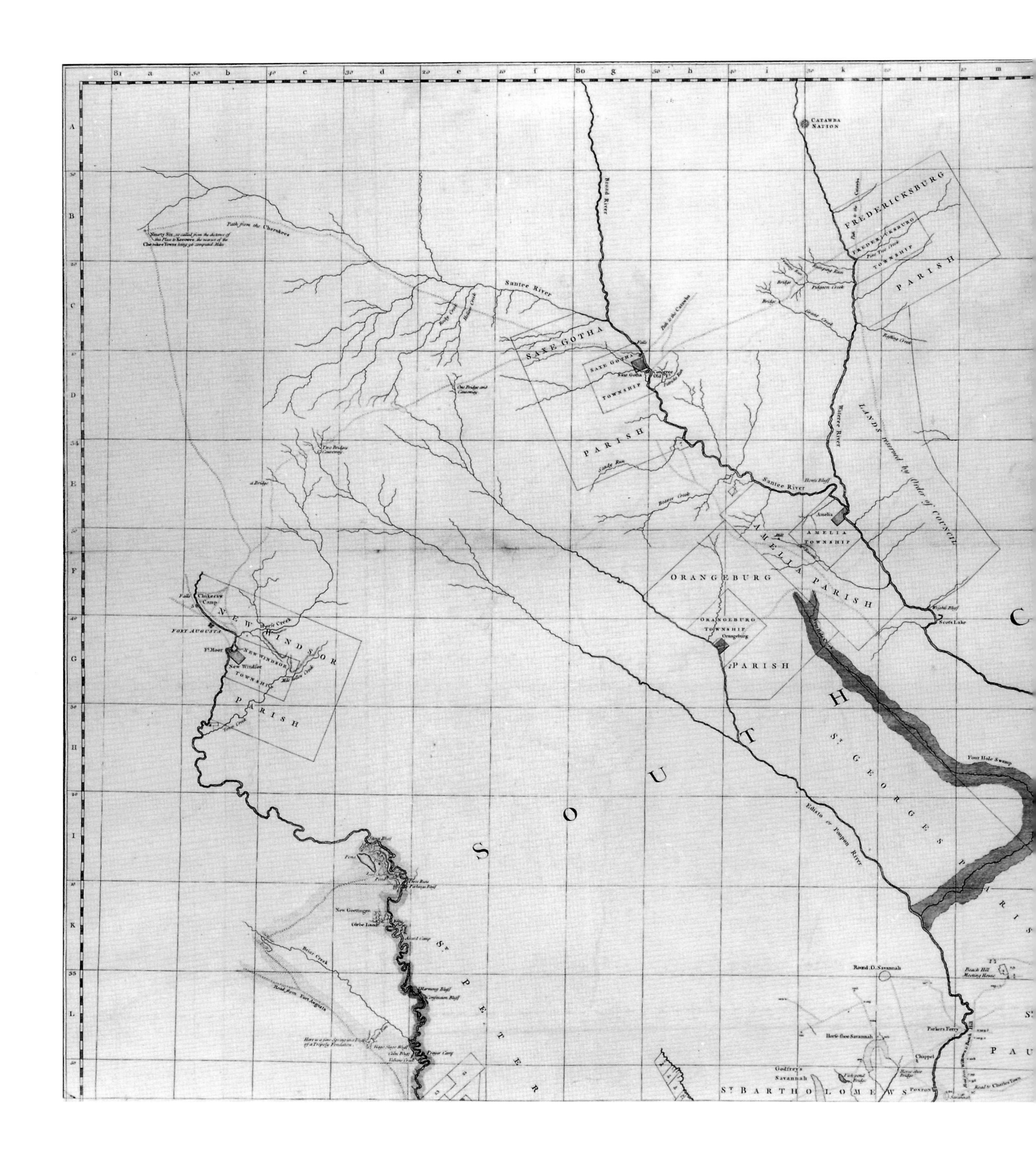

PLATE 59D. William De Brahm. A Map of South Carolina and a Part of Georgia [northwest quadrant]. 1757. [Map 310].

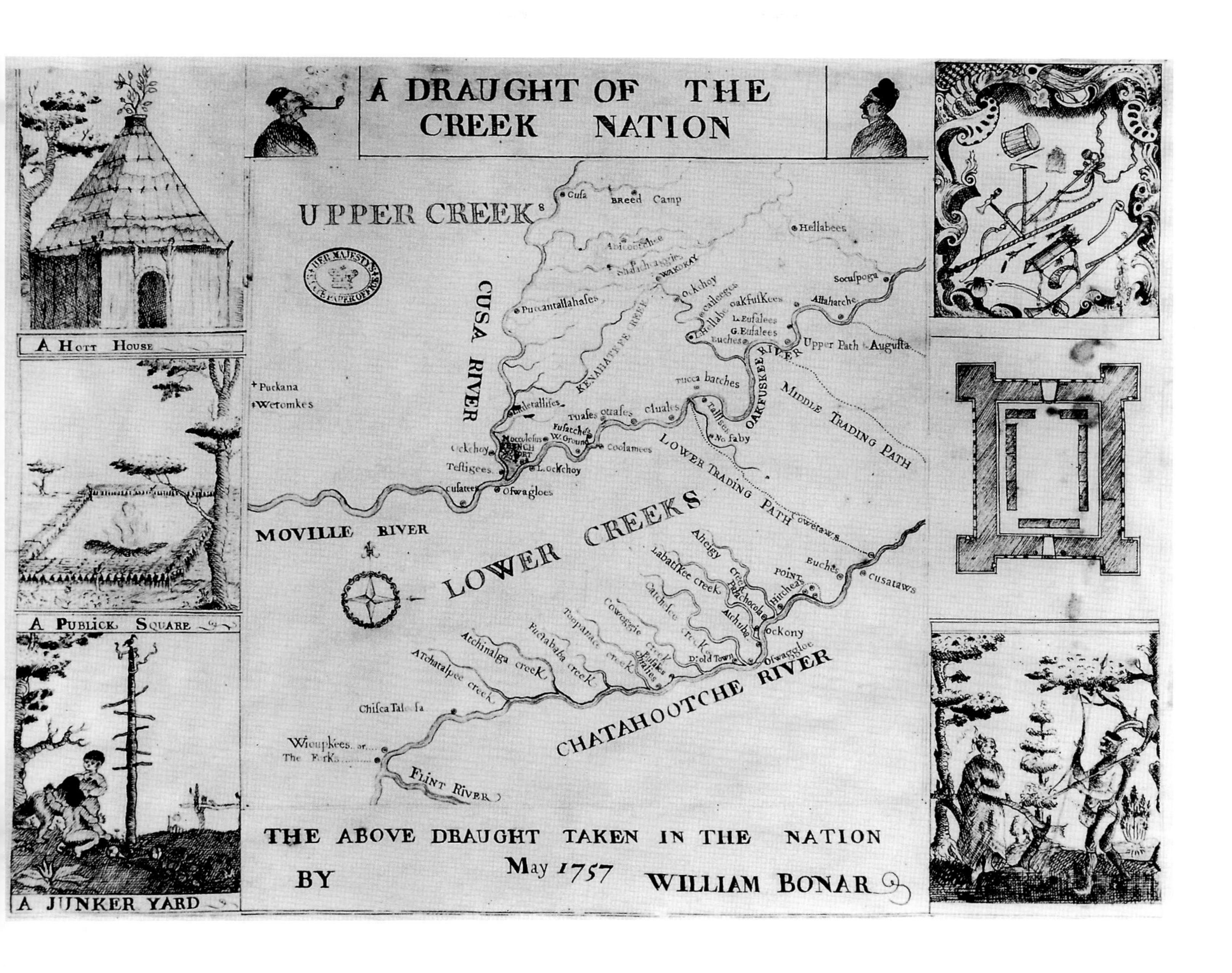

PLATE 59E. William Bonar. A Draught of the Creek Nation. 1757 MS A. [Map 306].

PLATE 59F. John Stuart. A Map of the Southern Indian District. 1764 MS. [Map 341].

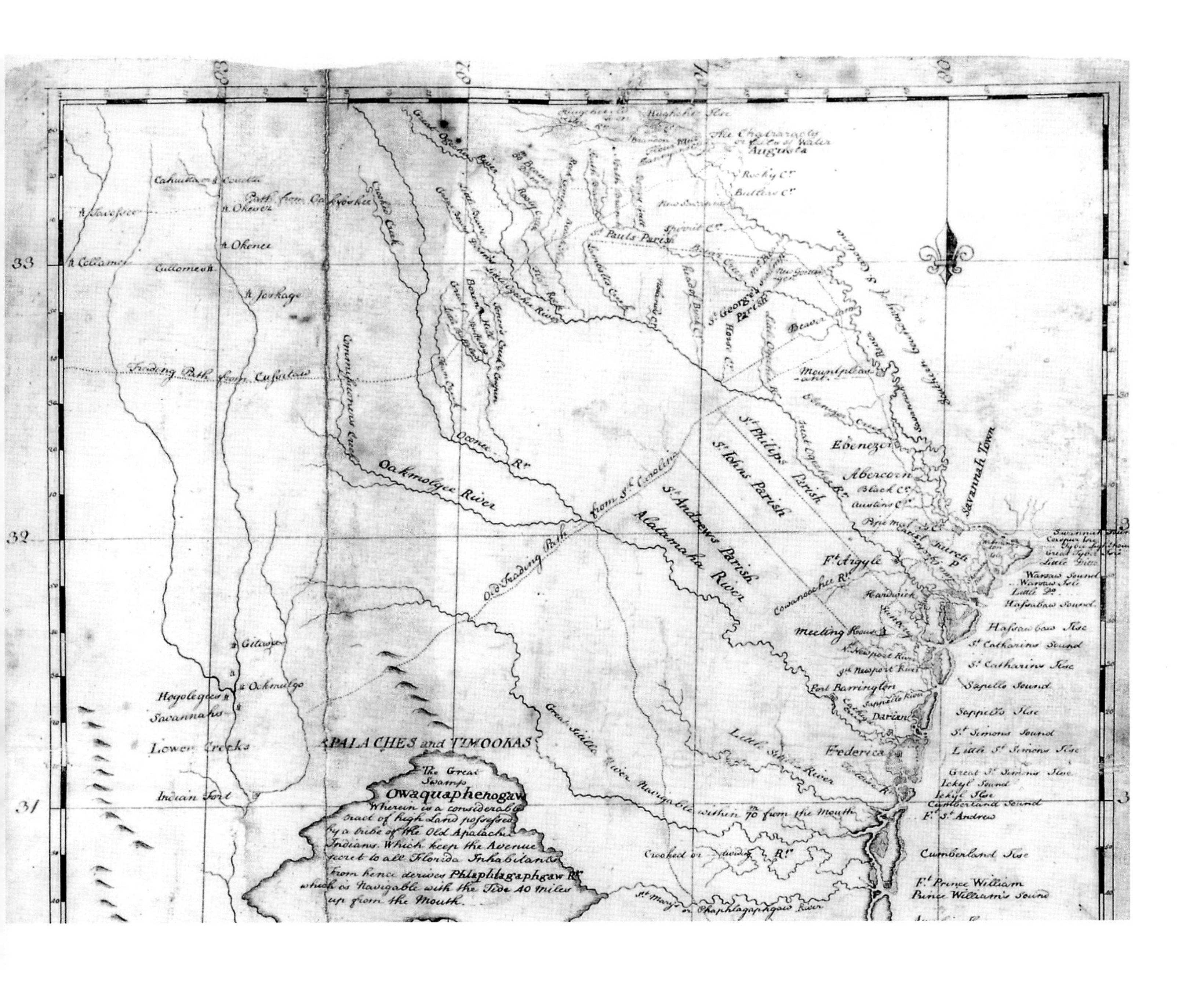

PLATE 60. Thomas Wright. A Map of Georgia and Florida [northern portion]. 1763 MS. [Map 333].

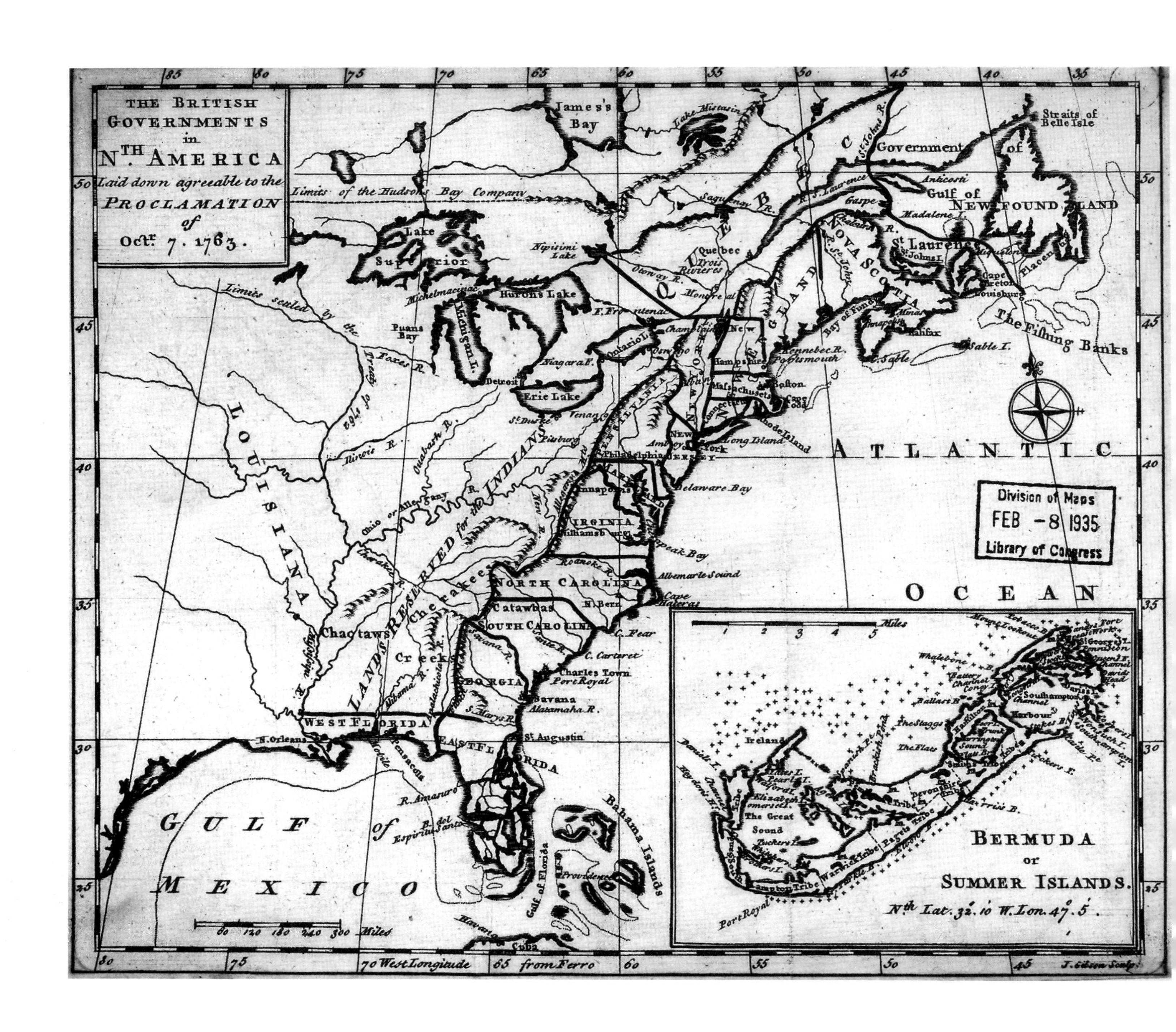

PLATE 60A. Anonymous. The British Governments in N^th. America laid down agreeable to the Proclamation of Octr. 7. 1763. 1763. [Map 336A].

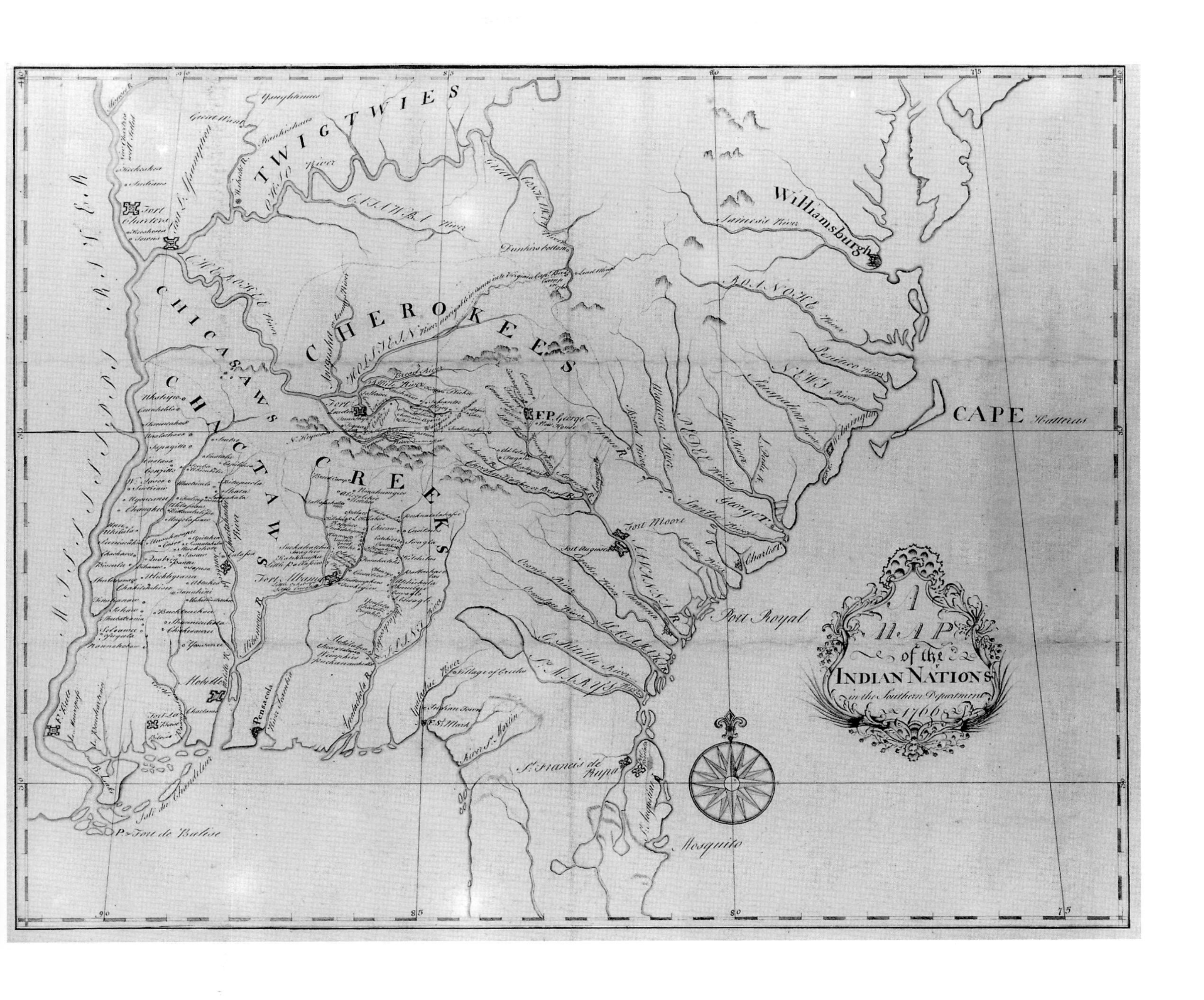

PLATE 61. John Stuart. A Map of the Indian Nations in the Southern Department 1766. 1766 MS. [Map 352].

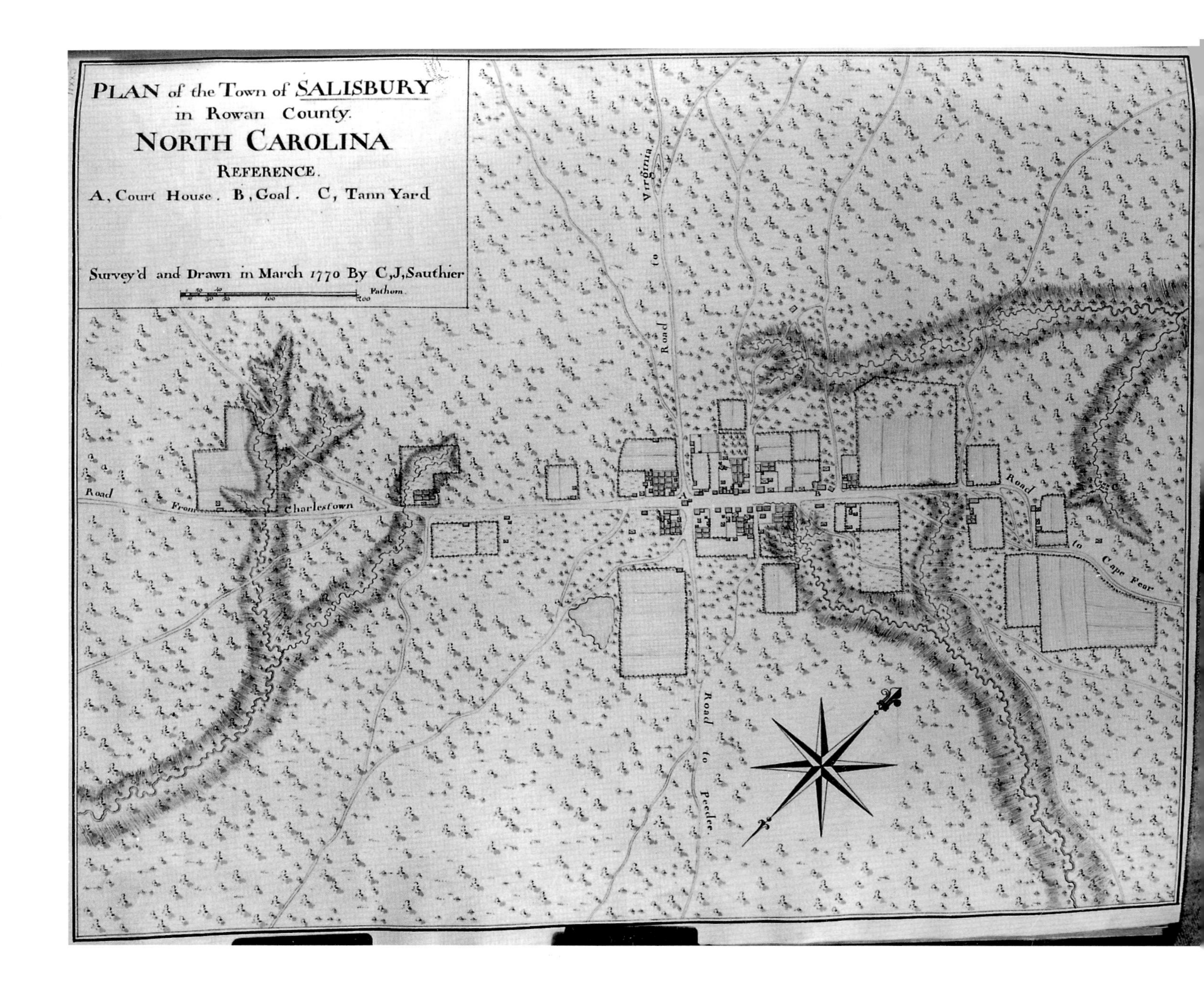

PLATE 62. Claude Joseph Sauthier. Plan of the Town of Salisbury. 1770 MS. [Map 392].

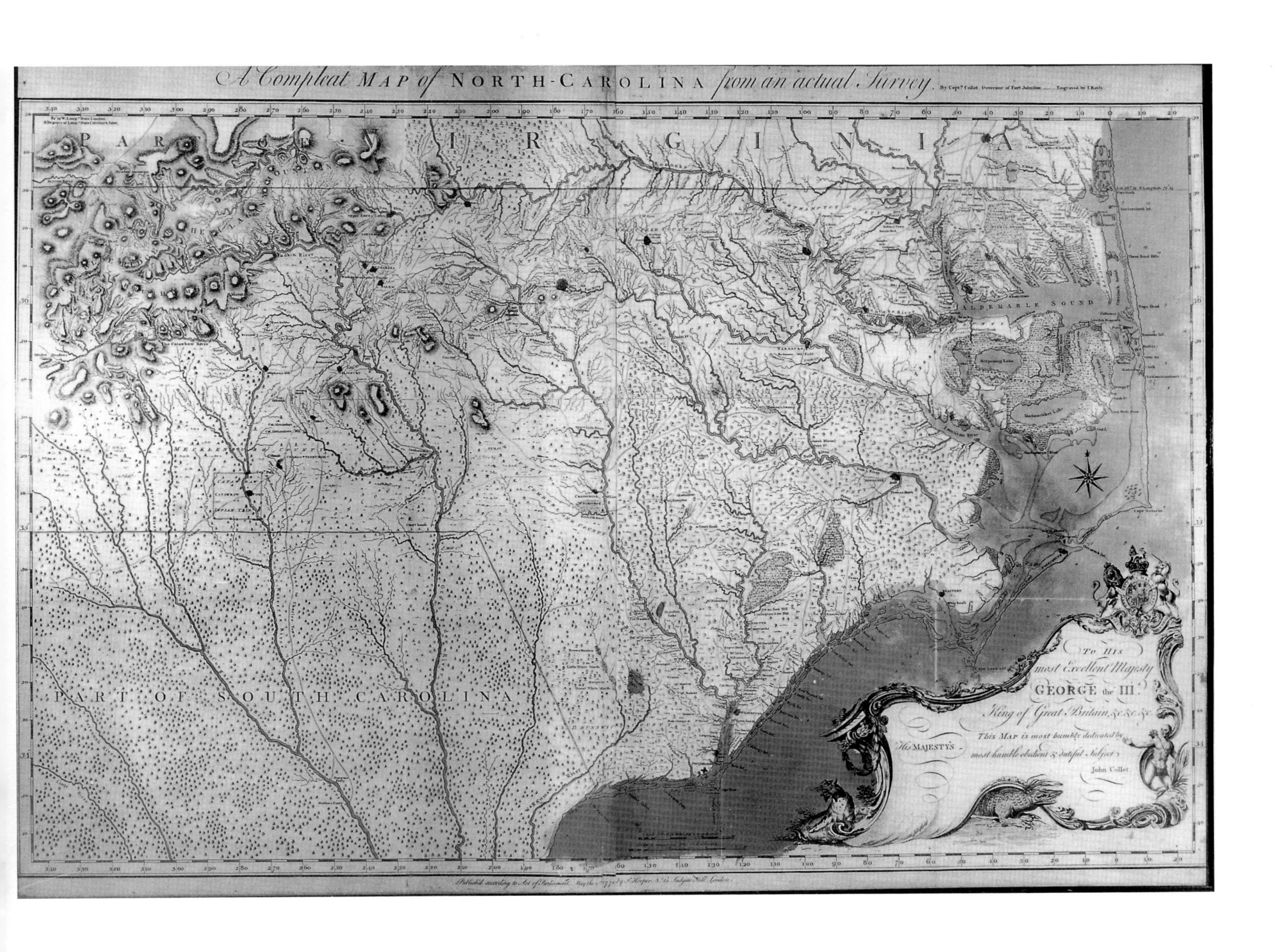

PLATE 62A. John A. Collet. A Compleat Map of North-Carolina. 1770. [Map 394].

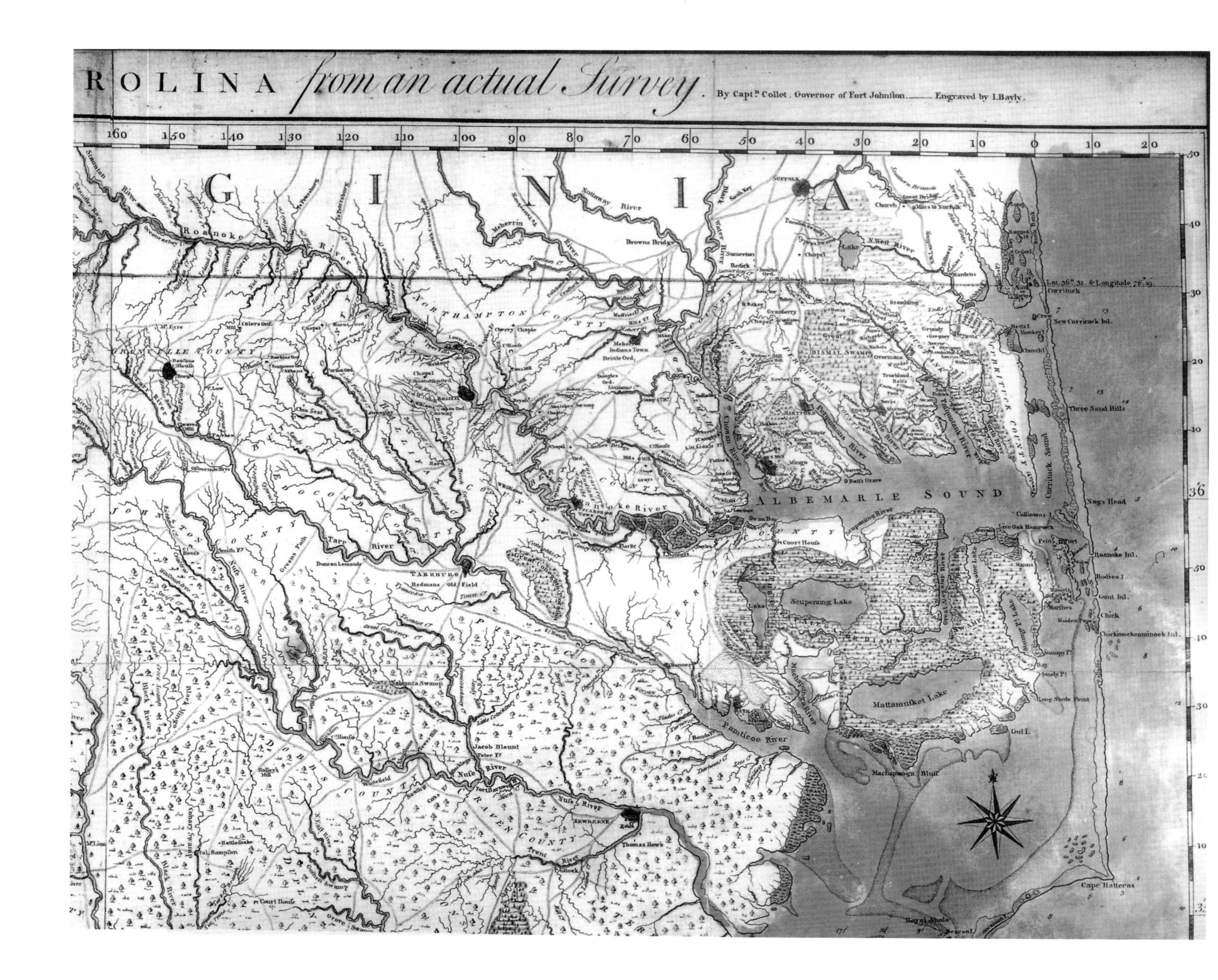

PLATE 63. John A. Collet. A Compleat Map of North-Carolina [northeast quadrant]. [Map 394].

PLATE 64. John A. Collet. A Compleat Map of North-Carolina [southeast quadrant]. [Map 394].

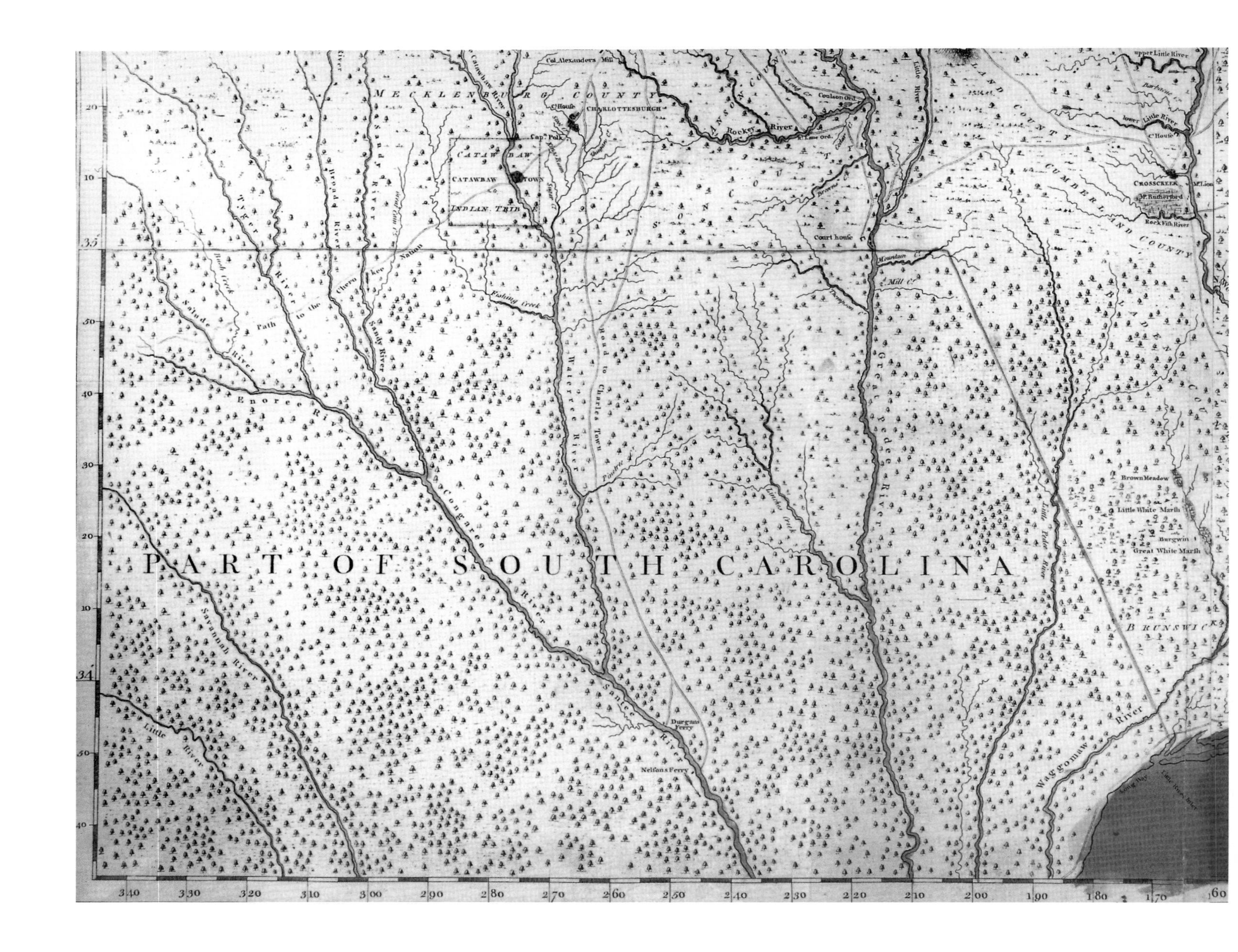

PLATE 65. John A. Collet. A Compleat Map of North-Carolina [southwest quadrant]. [Map 394].

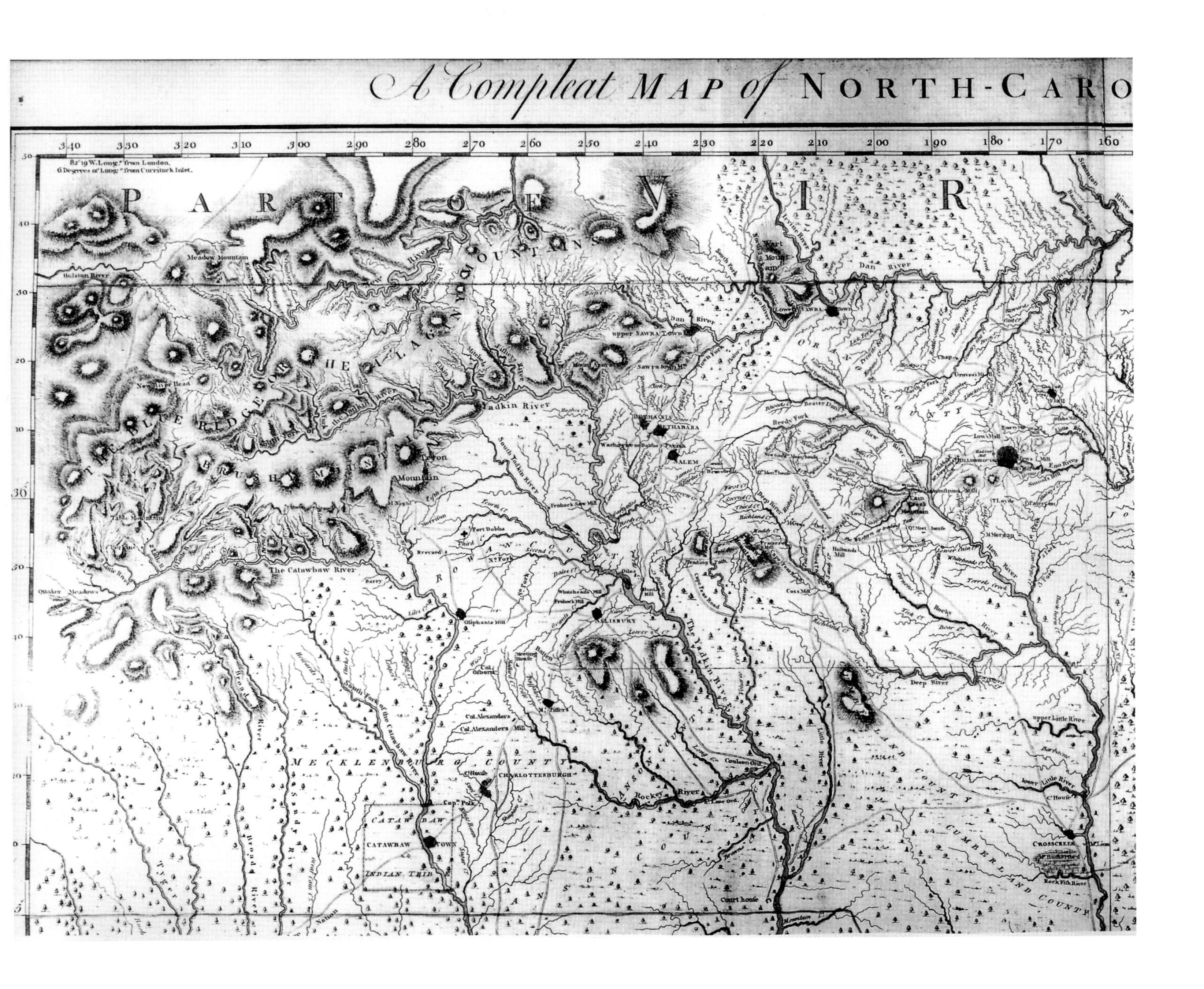

PLATE 66. John A. Collet. A Compleat Map of North-Carolina [northwest quadrant]. [Map 394].

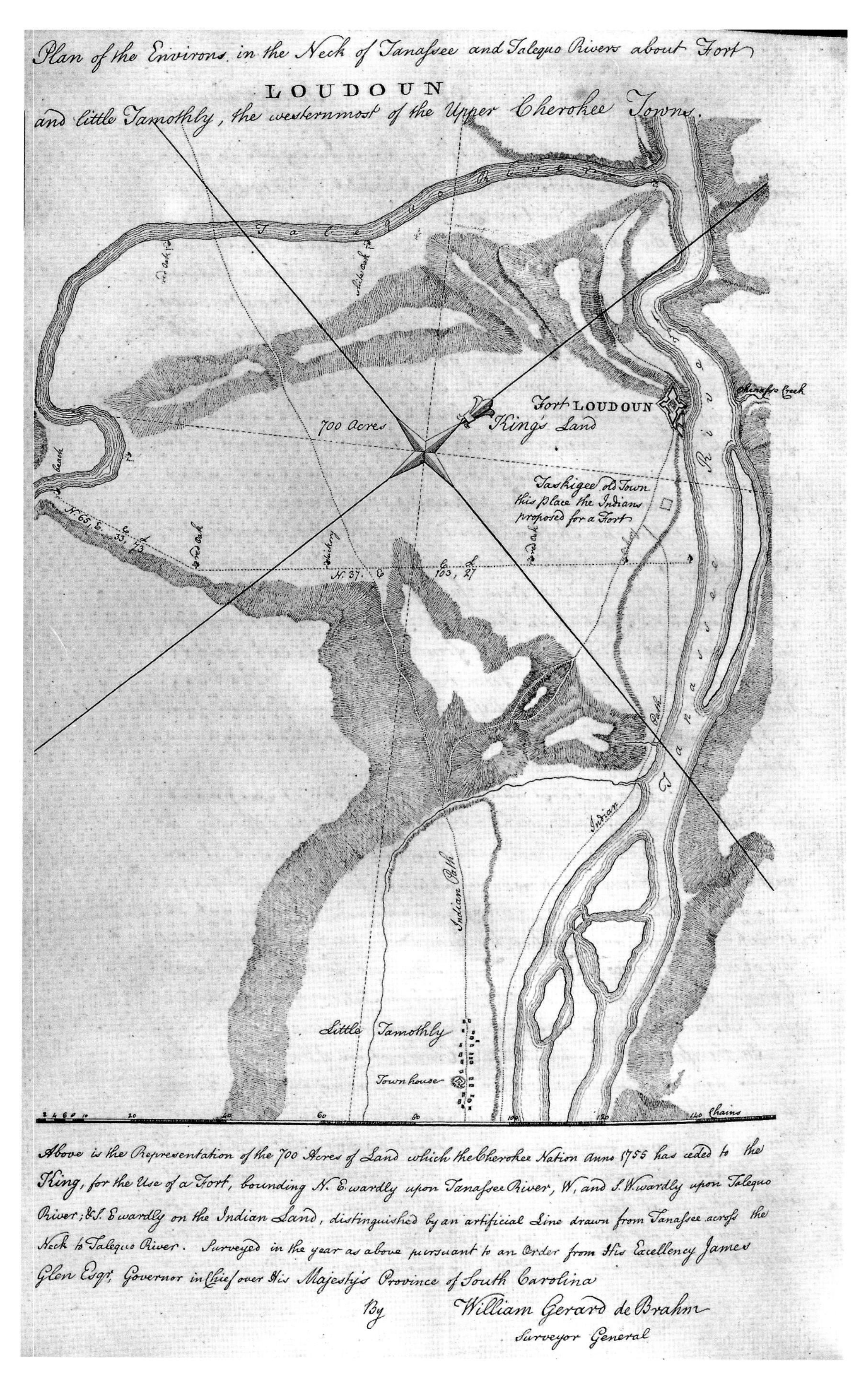

PLATE 66A. William De Brahm. Plan of the Environs in the Neck of Tanasee and Talequo Rivers about Fort Loudoun. . . . [1773] MS A.5. [Map 409].

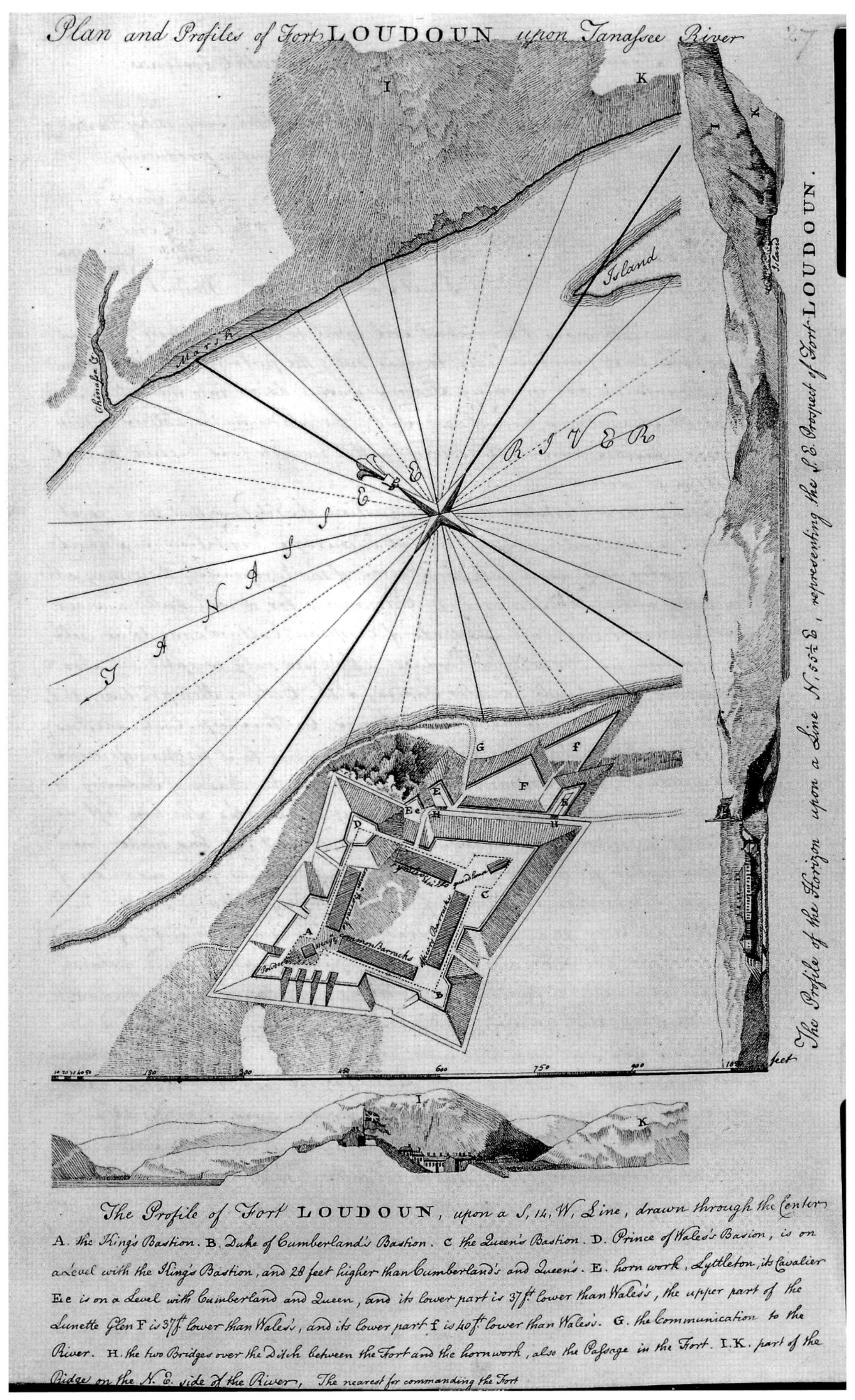

PLATE 66B. William De Brahm. Plan and Profiles of Fort Loudoun upon Tanasee River. [1773] MS A.6. [Map 410].

PLATE 66C. Philip Yonge. A Map of the Lands Ceded to His Majesty by the Creek and Cherokee Indians. . . . 1773 MS. [Map 440B].

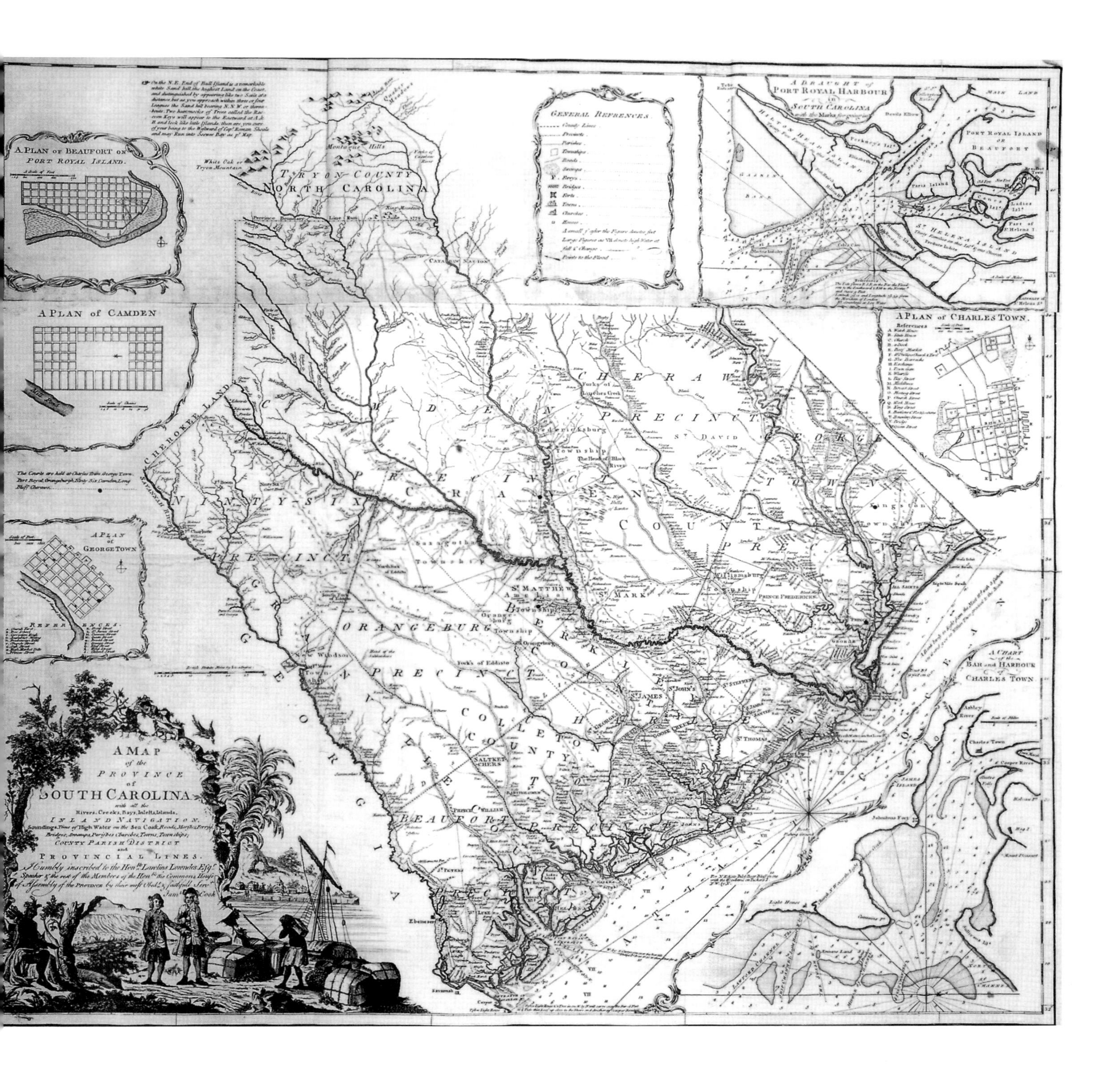

PLATE 67. James Cook. A Map of the Province of South Carolina. 1773. [Map 443].

PLATE 68. Stuart-Purcell. A Map of the Southern Indian District of North America . . . [northeast quadrant]. [1775] MS A. [Map 438].

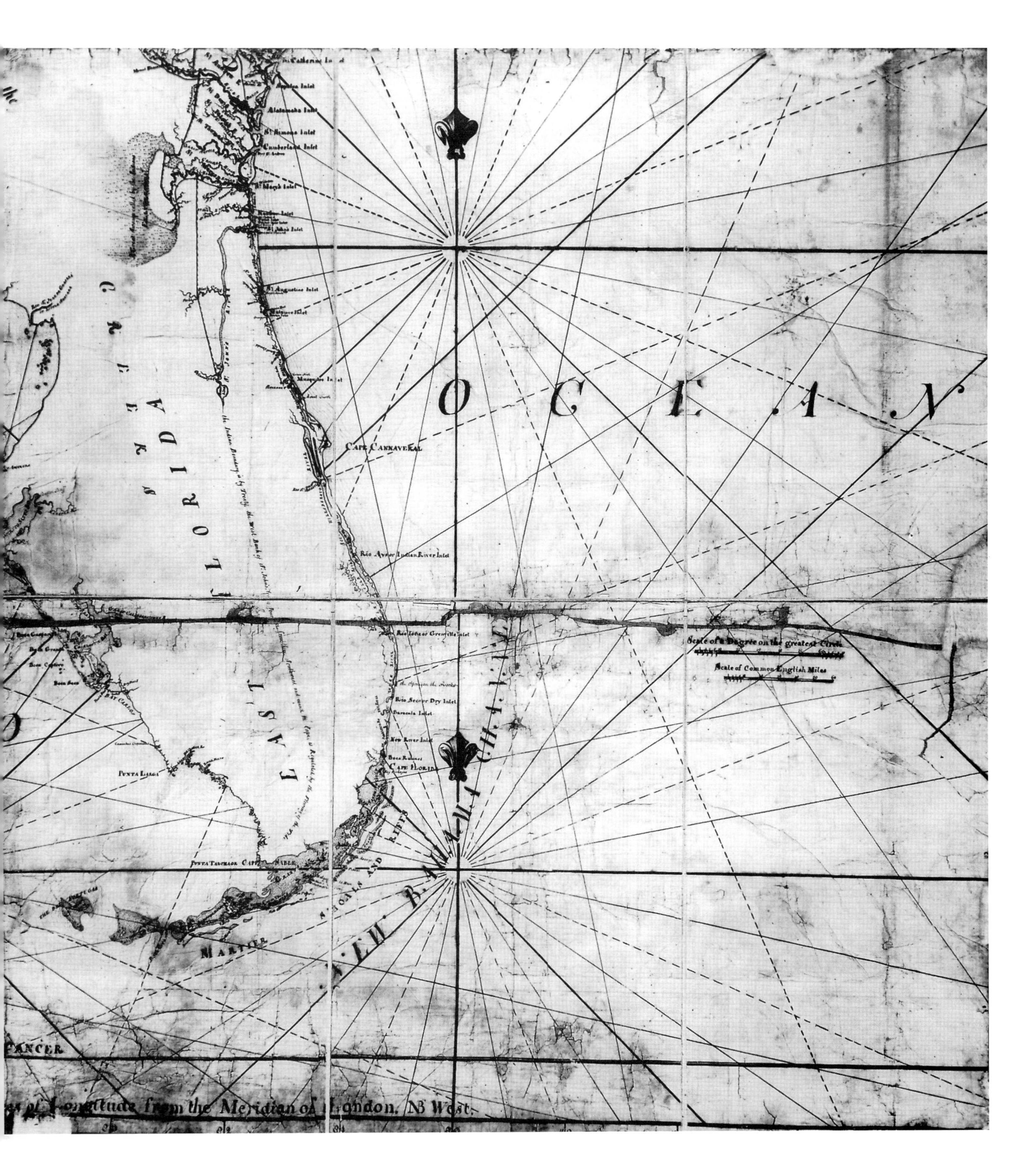

PLATE 69. Stuart-Purcell. A Map of the Southern Indian District of North America . . . [southeast quadrant]. [1775] MS A. [Map 438].

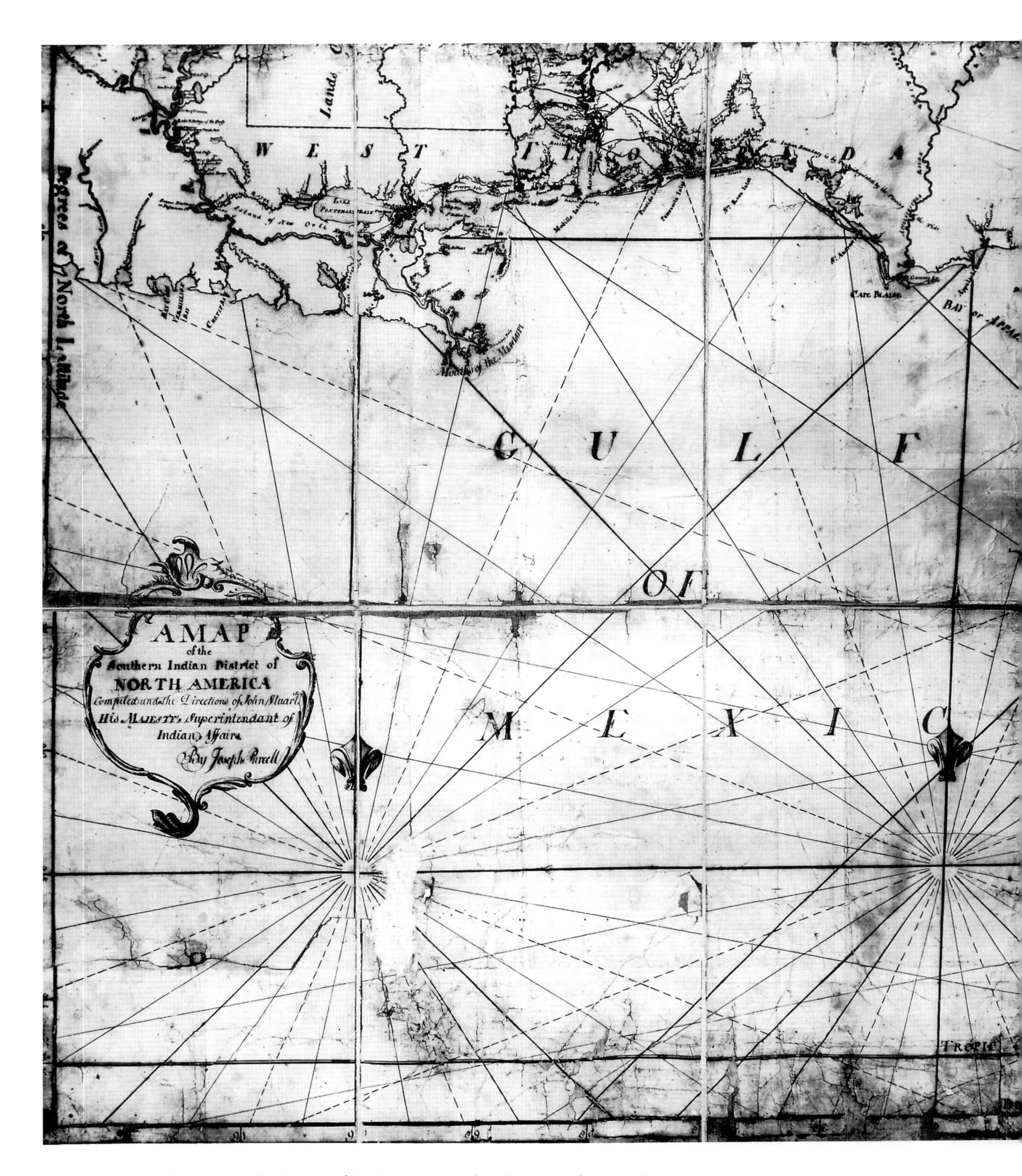

PLATE 70. Stuart-Purcell. A Map of the Southern Indian District of North America . . . [southwest quadrant]. [1775] MS A. [Map 438].

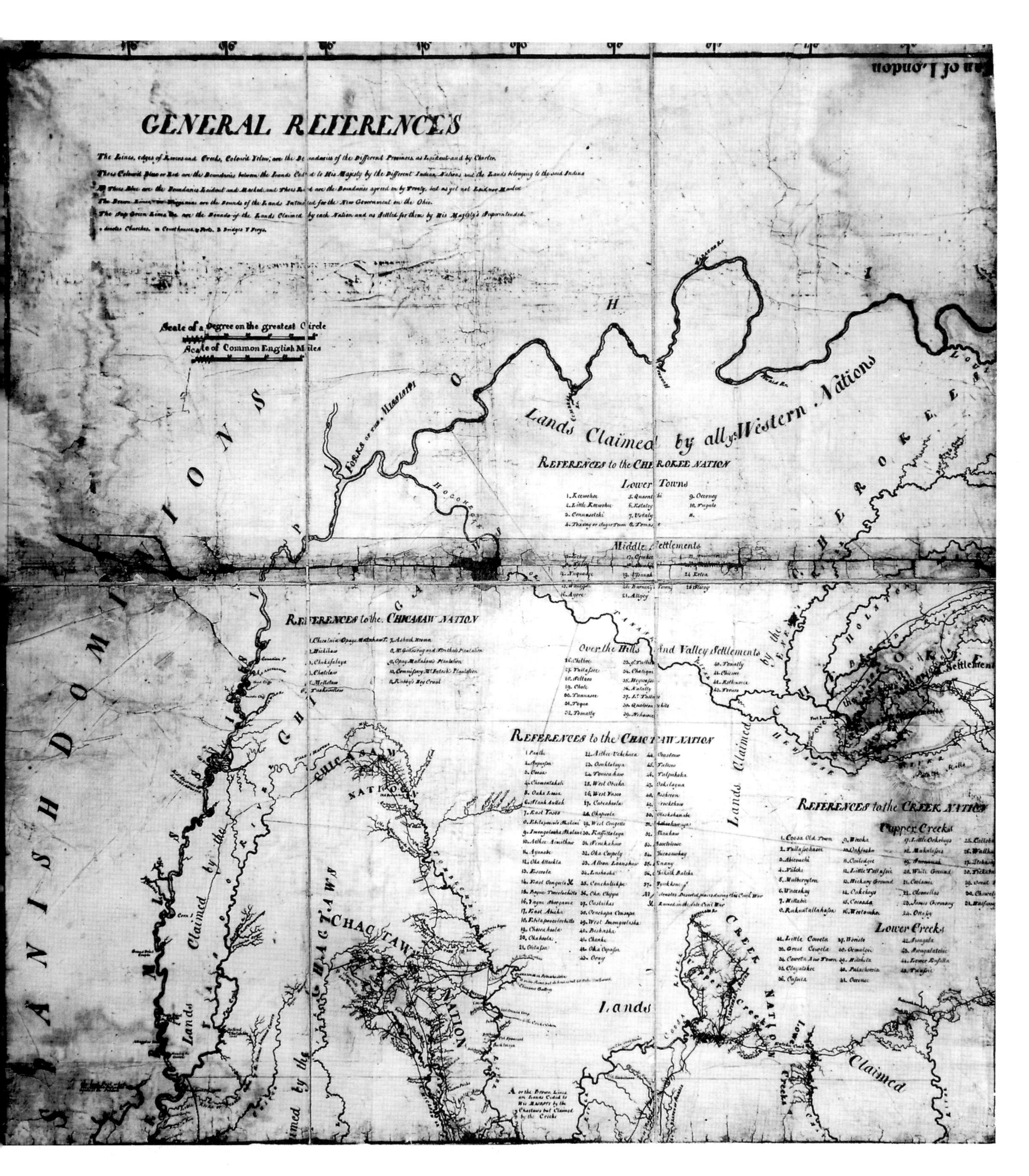

PLATE 71. Stuart-Purcell. A Map of the Southern Indian District of North America . . . [northwest quadrant]. [1775] MS A. [Map 438].

Chronological Title List of Maps

MAP TITLE	TITLE TAG	MAP NUMBER
Universalis Cosmographia Secundum Ptholomaei Traditionem et Americi Vespucii Alioruque Lustrationes	Waldseemüller 1507	01
[detail from untitled world map by Juan Vespucci]	Vespucci 1526 MS	02
[detail from untitled world map by Gerolamo da Verrazano]	da Verrazano 1529 MS	03
[detail from world map prepared in Seville]	Ribero 1529 MS	04
[endorsed, Mapa del Golfo [*sic*] y costa de la Nueva España]	De Soto ca. 1544 MS	1
Americae Sive qvartae Orbis Partis	Gutiérrez 1562	2
La Terre Francoise Nowellement Decowerte	Parreus French Florida 1562 [1563] MS	2A
La Carreline	Norment-Bruneau 1565	3
Nova Et Aucta Orbis Terrae Descriptio . . .	Gerard Mercator 1569	3A
[Second Spanish Fort at Santa Elena]	San Marcos ca. 1578 MS	4
The Cownterfet of Mr Fernando Simon his Sea carte . . .	Fernandez-Dee 1580 MS	4A
[Northern Hemisphere, from pole to Tropic of Cancer]	John Dee ca. 1582 MS	4B
La Florida. Auctore Hieron. Chiaves	Ortelius-Chiaves 1584	5
A discription of the land of virginia	Virginia ca. 1585 MS	6
La Virgenia Pars	White 1585 MS A	7
Virginea Pars	White 1585 MS B	8
Novvs Orbis	Galle 1587	9
The Famouse West Indian voyadge	Boazio 1588	10
Americae et Proximarvm Regionvm	Hogenberg 1589	11
Americae pars, Nunc Virginia dicta	White–De Bry 1590 A	12
Anglorum in Virginiam aduentus	White–De Bry 1590 B	13
Floridae Americae Provinciae Recens & exactissima descriptio	Le Moyne 1591	14
El fuerte de san ta Elena	Mestas 1593 MS	15
Americæ Pars Borealis, Florida, Baccalaos Canada	Jode 1593	16
Planta de la costa de la florida y . . . la Guna Maymi	Maymi ca. 1595 MS	17
Florida et Apalche	Wytfliet-Florida 1597	18
Norvmbega et Virginia	Wytfliet-Norvmbega 1597	19
Florida et Apalche	Acosta-Florida 1598	20
Norvmbega et Virginia	Acosta-Norvmbega 1598	21
[a true hydrographical description]	Wright 1599	21A
Norvmbega et Virginia	Matal-Norvmbega 1600	22
Florida et Apalche	Matal-Florida 1600	23
Novi Orbis Pars Borealis	Quad 1600	24
Noua et rece Terraum	Tatton 1600	25
Virginae Item et Floridae Americae Provinciarum nova Descriptio	Mercator-Hondius 1606	26
Virginia et Florida	Mercator-Hondius 1607	27
The draught by Robert Tindall of Virginia, anno 1608	Tindall Virginia 1608 MS	27A
[Chart of Virginia]	Zuñiga 1608 MS	28
[East Coast of North America]	Velasco 1611 MS	29
Figure et description de la terre reconue et habitée par les Fr[a]nçois	Lescarbot 1612	30

MAP TITLE	TITLE TAG	MAP NUMBER
Virginia et Nova Francia	Bertius 1616	31
Ould Virginia	Smith 1624	32
Virginia et Florida	Purchas (Hondius) 1625	33
Florida, et Regiones Vicinae	Laët 1630 A	34
Nova Anglia, Novvm Belgivm et Virginia	Laët 1630 B	35
Virginiae Item Floridae Americae Provinciarum nova Descriptio	Mercator-Hondius 1630	36
Virginia et Florida	Mercator-Hondius-Saltonstall 1635	37
[Map of the East Coast]	Dudley ca. 1636 MS	38
Nova Anglia Novvm Belgivm et Virginia	Jansson 1636	39
[Carolina Coast]	Vingboons ca. 1639 MS	40
Virginiæ partis australis, et Floridæ partis orientalis	Blaeu 1640	41
Virginiæ partis australis, et Floridæ partis orientalis	Jansson 1641	42
Nova Belgica et Anglia Nova	Jansson 1647	43
Carta particolare della Virginia Vecchia è Nuoua	Dudley-Virginia 1647	44
Carta particolare della costa di Florida è di Virginia	Dudley-Florida 1647	45
Ould Virginia 1584, now Carolana 1650	Farrer 1650 MS	46
A mapp of Virginia discouered to ye Falls	Farrer 1651	47
Le Canada, ou Nouvelle France, &c. . . .	Sanson-Canada 1656	48
Le Nouveau Mexique, et La Floride	Sanson-Floride 1656	49
The Sovth Part of Virginia Now the North Part of Carolina	Comberford 1657 MS A	50
The Sovth Part of Virginia	Comberford 1657 MS B	51
Le Canada, ou Nouvelle France, &c.	Sanson 1657 A	52
La Floride, Par N. Sanson d'Abbeville	Sanson 1657 B	53
La Virginie	Du Val 1659	54
Floride	Du Val ca. 1659	55
Carte de La Virginie	Du Val–Virginie 1663	56
La Floride	Du Val–Floride 1663	57
Discouery made by William Hilton	Shapley 1662 MS	58
Americae Septentrionalis Pars	Moxon 1664	59
La Floride Francoise Dressee sur La Relation des Voiages	Du Val 1665	61
Carolina Described	Horne 1666	60
[Albemarle and Pamlico Sounds]	Anon. ca. 1670 MS	62
Culpepers Draught of Ashley Copia vera	Culpeper 1671 MS A	63
Culpepers Draught of the L^{ds} P^{rs} Plantacon	Culpeper 1671 MS B	64
Map of Carolina [16]71	Locke 1671 MS	65
[Ashley and Cooper Rivers]	Ashley-Cooper 1671 MS	66
Virginiae partis australis, et Floridae partis orientalis	Montanus 1671	67
A Map of the Whole Territory Traversed by Iohn Lederer	Lederer 1672	68
A Generall Mapp of Carolina	Blome 1672	69
A New Discription of Carolina	Ogilby-Moxon ca. 1672	70
Plott of ye Lords Prop[rietors'] plon[tation] 44½ Acres lond	Culpeper 1672/3 MS	71
Virginia and Maryland	Herrman 1673	72
A New Mapp of the north part of America	Thornton [1673]	89
A New Map of the English Plantations in America	Morden-Berry 1673–77?	73
Caerte vande Cust van Florida tot de Verginis	Roggeveen ca. 1675	74
A Chart of the West Indies from Cape Cod to the River Oronoque	Seller ca. 1675	75

MAP TITLE	TITLE TAG	MAP NUMBER
Carolina	Speed 1675	76
A New Description of Carolina	Speed 1676	77
The Mapp of Carolina	Blathwayt ca. 1679 (1664–83) MS A	78
[Carolina]	Blathwayt ca. 1679 (1662–83) MS B	79
Carolina	Blathwayt ca. 1679? (1677) MS C	80
Carolina	Lancaster 1679 MS D	81
A Map of ye English Empire in ye Continent of America	Daniel ca. 1679	82
Canada, sive Nova Francia &c.	Sanson 1679 A	83
Florida	Sanson 1679 B	84
Virginia	Du Val 1679 A	85
Florida	Du Val 1679 B	86
Florida	Seller 1679	87
Carolina Virginia Mary Land & New Iarsey	Morden 1680	88
Florida	Moore 1681	90
Pas Kaart Van de Kust van Carolina Tusschen C de Canaveral	Keulen 1682?	91
Carolinass	Mathews 1682 MS	91A
A New Map of the Country of Carolina	Gascoyne 1682	92
Carolina Newly Discribed	Seller 1682	93
Mapa De la Ysla de la Florida	Spanish 1683 MS	94
Floride De L'Amerique	Mallet 1683	95
Le Canada, ou Nouvelle France, &c.	Sanson-Canada 1683	96
La Floride Par N. Sanson d.'Abbeville	Sanson-Florida 1683	97
Carolina	Thames School ca. 1684 MS A	98
North Carolina	Thames School ca. 1684 MS B	99
Carolina	Hack 1684 MS	100
A Plat of the Province of Carolina	Mathews ca. 1685 MS	101
[Florida and West Indies]	Spanish ca. 1685 MS	102
A Map of ye English Empire in ye Continent of America	Morden ca. 1685	103
A New Map of Carolina	Thornton-Morden-Lea ca. 1685	104
The English Empire in America	Crouch 1685	105
Floride/das Landt Florida	Mallet 1686	106
A New Map of Carolina	Morden 1687	107
Nouvelle Carte de la Caroline	Blome 1688	108
A Map of Florida and ye Great Lakes of Canada	Morden 1688	109
A New Map of Carolina	Morden 1688	110
A New Mapp of Carolina	Thornton-Fisher 1689	110A
Viage que el año 1690	de Leon 1690 MS	110B
Virginia	Müller-Virginia 1692	111
Florida	Müller-Florida 1692	112
A platt of ye Towne	Archdale ca. 1695 MS A	113
The Draughts of the Forts	Archdale ca. 1695 MS B	114
[Charles Town harbor and neighboring coastline]	Archdale ca. 1695 MS C	115
The Mapp of Ashley & Cooper rivers: &:	Archdale ca. 1695 MS D	116
A rough draught of North Carolina Coast to Santee	Archdale ca. 1695 MS E	117
To the Right Honorable William Earl of Craven; Palatine, Iohn Earl of Bath . . . South Carolina	Thornton-Morden ca. 1695	118

MAP TITLE	TITLE TAG	MAP NUMBER
A New Map of the English Empire in America	Morden-Brown ca. 1695	119
Carte Generale de la Caroline	Sanson ca. 1696 A	120
Carte Particuliere de la Caroline	Sanson ca. 1696 B	121
Das Englische America	Crouch 1697	122
A Map of the English Possessions in North America	English Possessions 1699 MS	124
[Coast from Port Royall to Charles Town]	S.C. Coast ca. 1700 MS	125
[Carolina]	French ca. 1700 MS	126
A Chart of Carolina	Carolina ca. 1700	127
Ashley & Cooper River	Lea 1700?	128
Carte Nouvelle de L'Amerique Angloise	Sanson 1700	129
A New Map of the most Considerable Plantations of the English In America	Wells 1700	130
Carte des environs du Missisipi	Delisle ca. 1701 MS	131
The English Empire in America	Moll 1701 A	132
Mexico, or New Spain	Moll 1701 B	133
Carte du Canada et du Missisipi	Delisle 1702 MS	134
Virginia | V. c[ärtlein]	Müller 1702 A	135
[Florida] | VI. c[ärtlein]	Müller 1702 B	136
Carte du Mexique et de la Floride	Delisle 1703	137
A Plan of Charles Town from a survey of Edwd Crisp	Crisp 1704	138
Florida zoo als het van de Spaanschen	Sanson 1705	139
Caerte vande Cust Carolina Tusschen B. de S. Matheo	Loots 1706	140
De Vaste Kust Van Chicora	Aa 1706 A	141
't Amerikaans Gewest van Florida	Aa 1706 B	142
Zee en Land Togten der Franszen Gedaan	Aa 1706 C	143
A Large Draft of South Carolina from Cape Roman to Port Royall	Thornton 1706	144
A Large Draught of Port Royall Harbour in Carolina	Thornton–Port Royall 1706	145
Mr. Robert Beverley's Acco^{t}. of Lamhatty	Lamhatty 1708 MS	146
[Plan of the town of Low Wickham]	Low 1708 MS	147
Carolina By Hermann Moll Geographer	Moll-Oldmixon 1708	148
Virginia Nord Carolina Sud Carolina	Kocherthal 1709	149
To His Excellency William Lord Palatine; This Map is Humbly Dedicated by Io.^{n} Lawson	Lawson 1709	150
A Compleat Description of the Province of Carolina	Crisp [1711]	151
Die Provinz Nord und Sud Carolina	Ochs 1711	152
Die vornehmste Êigenthums Herren und Besitzer von Carolina	Lawson-Visher 1712	153
[Colonel Moore's attack on Fort Nohucke]	Fort Nohucke ca. 1713 MS	154
La Floride, Suivant les Nouvelles Observations	Aa 1713	155
Virginia Marylandia et Carolina	Homann 1714	156
[Map of North and South Carolina and Florida[	Indian Villages ca. 1715 MS	157
A New and Exact Map of the Dominions	Moll 1715	158
[Carolina]	Indian Wars ca. 1716 MS	159
Nord Carolina	Graffenried 1716 MS A	160
Baronie de Bernberi	Graffenried 1716 MS B	161
Plan der Schwijtzereschen Coloney In Carolina	Graffenried 1716 MS C	162
Carte Nouvelle de la Louisiane et païs circonvoisons	Le Maire 1716 MS A	163

MAP TITLE	TITLE TAG	MAP NUMBER
Carte nouvelle de la Louisiane et Paÿs Circonvoisons	Le Maire 1716 MS B	164
Carte Generale de la Louisiane	Vermale 1717 MS	165
A Plan representing the Form of Setling the Districts, or County Divisions in the Margravate of Azilia	Montgomery 1717	166
This Plane . . . represents the Island of Roan-Oak	Maule-Roanoke 1718 MS	167
A Map of the English Plantations in America	Lawson 1718	168
Partie Méridionale de la Rivière de Missisipi	de Fer 1718	169
Carte de la Louisiane et du Cours du Mississipi	Delisle 1718	170
Floride | das Landt Florida	Mallet 1719	171
Most humbly Inscrib'd to Hewer Edgly Hewer . . . A New Map of the English Empire	Senex 1719	172
Nouvelle Carte de la Caroline	Chatelain 1719	173
A Map of the Country adjacent to the River Misisipi	Hughes ca. 1720 MS	174
Carte de la Louisiane et du Cours du Mississipi	Delisle ca. 1720	175
A New Map of Louisiana and the River Mississipi	Delisle-Louisiana 1720	176
A Map or Plan of the Mouth of Alatamahaw River	Altamaha River ca. 1721 MS	177
The Northern B[ra]nch of Alatama River	Barnwell 1721 MS	178
A Chart of S^t. Simon's Harbour	St. Simon's 1721 MS	179
Mapp of Beaufort in South Carolina	Beaufort 1721 MS	180
South Carolina: The Ichnography or Plann of the Fortifications of Charlestown	Herbert 1721 MS	181
A Map of Louisiana and of the River Mississipi	Senex 1721	182
Carolina door Herman Mol	Moll-Kyser 1721	183
[Southeastern North America]	Barnwell ca. 1721 MS	184
[Southeastern North America]	Barnwell-Hammerton 1721 MS	184A
A Plan of King George's Fort at Allatamaha, South Carolina	King George's Fort 1722 MS	185
[Plan of Fort King George and part of the Altamaha River]	Stollard 1722 MS	186
[A. Fort King George. B. Part of St. Simon's Island, Georgia]	Fort King George– St. Simon's Island ca. 1722 MS	187
The Ishnography or Plan of Fort King George	Fort King George ca. 1722 MS A	188
[A plan of Fort King George]	Fort King George ca. 1722 MS B	189
Plan of Beaufort Town	Beaufort ca. 1722	189A
Map of Carolana and of the River Meschacebe	Coxe 1722	190
Tabula Geographica Mexicæ et Floridæ	Delisle 1722	191
A Map Describing the Situation of the several Nations of Indians between South Carolina and the Massissipi	Southeastern Indians ca. 1724 MS A	192
A Map Describing the Situation of the several Nations of Indians between South Carolina and the Massissipi	Southeastern Indians ca. 1724 MS B	193
This map describing the scituation of the Several Nations of Indians to the N W. of South Carolina	Southeastern Indians ca. 1724 MS C	194
A Platt of Charles Town	Charles Town ca. 1725 MS	195
A Plan of Fort King George as it's now Fortifyed, 1726	Fort King George 1726 MS	196
A Map of English and French Possessions on the Continent of North America	Popple 1727 MS	196A
We the underwritten Commissioners . . . the Boundary betwixt the governments of Virginia & North Carolina	Virginia–North Carolina Line 1728 MS A	197

MAP TITLE	TITLE TAG	MAP NUMBER
A Map of the Boundary between Virginia and Carolina	Virginia-Carolina Line 1728 MS B	198
[1. Copy of 1728 MS A above. 2. Copy of 1728 MS B above]	Virginia-Carolina Line 1728 MS C	199
A Plan of Port Royal = Harbour in Carolina	Moll 1728 A	200
Florida Calle'd by y^e French Louisiana	Moll 1728 B	201
The English Empire in America	Crouch 1728	202
This Plan represents Roanoke Island	Maule-Moseley 1729 MS	203
A Plan of Port-Royal, South Carolina	Gascoigne-Swaine ca. 1729 MS	204
Partie Meridionale de la Virginie, et la Partie Orientale de la Floride	Aa 1729	205
Carolina	Moll 1729	206
[Cherokee Nation and the Traders' Path from Charles Town via Congaree]	Hunter 1730 MS	207
Carte de la Louisiane et du Cours du Mississipi	Cóvens-Mortier-Delisle ca. 1730	208
A Map of the Province of Carolina, Divided into its Parishes	Moll 1730	209
[Southeastern North America]	Georgia 1732	211
Carte de la Caroline du nord et du sud	Carolina-Georgia post-1732 MS	212
Beaufort S^o Carolina	Beaufort ca. 1733 MS	213
My Plat of 20,000 acres in N^o Carolina	Byrd 1733 MS	214
Carte De partie de la Louisianne	Crenay 1733 MS	215
A Large and aisect Drafe [exact Draft] of the Sea Coast of N^o Carolina	Wimble 1733 MS	215A
A Map of the British Empire	Popple 1733 A	216
America Septentrionalis	Popple 1733 B	217
To His Excel^y Gabriel Johnston . . . North Carolina	Moseley 1733	218
A Map of Carolina, Florida and the Bahama Islands	Catesby [1734]	210
A View of Savanah [*sic*] as it stood the 29th of March, 1734	Oglethorpe-Jones-Gordon 1734	219
His Majestys Colony of Georgia in America	George Jones 1734 MS	219A
Carte de la Louisiane et du Cours du Mississipi	Bernard-Delisle 1734	220
A Plan of the Township of Charlottenbourg on the South side of the River Alatamaha in Georgia	Charlottenbourg ca. 1735 MS	221
[Tract of Land in Albemarle County]	Albemarle ca. 1735 MS	222
A Map of the Division Line between the Province of North & South Carolina	Gray 1735 MS	223
La Louisiane, Suivant les Nouvelles Observations	Cóvens-Mortier ca. 1735	224
A Map of the County of Savannah	Georgia [1735]	246
North Carolina. Mr. McCulloh's Draught	McCulloh 1736 MS	225
North Carolina	Jenner 1736 MS	226
A Plan of Frederika	Frederika 1736 MS	227
The Fort at Frederica in Georgia	Frederica Fort 1736 MS	228
Plano de la entrada de Gvaliqvini Rio de S^n Simon	Arredondo 1737 MS A	229
Plano de la Entrada de Gvaliqvini Rio de San Simon	Arredondo 1737 MS B	230
Carte Particulière de L'Amerique Septentrionale	Cóvens-Mortier-Popple ca. 1737	231
Eden in Virginia	Eden 1737	232
Dominia Anglorum in America Septentrionali	Homann Heirs 1737	233
New map of Georgia	Bernard 1737	234
Carte de La Louisiane et du Cours du Mississipi	Delisle-Bernard 1737	235
A Map of North Carolina	Brickell 1737	236
A True Copy of a Draught of the Harbour	Gascoigne [1737] MS	236A

MAP TITLE	TITLE TAG	MAP NUMBER
Nations Amies Et Enemies Des Tchikachas	Ouma–De Batz 1737 MS	236B
Plan Et Scituation Des Villages Tchikachas	Pakana–De Batz 1737 MS	236C
This Chart was transmitted by Col.º Bull	Bull 1738 MS	237
Plano y Perfil . . . de S.n Francisco de Pupo	Ruiz 1738 MS	238
Virginiae Pars Carolinae Pars	Virginia-Carolina Line ca. 1738	239
A New Map of Virginia	Keith 1738	240
To His Grace Thomas Hollis Pelham . . . This Chart of his Majesties Province of North Carolina	Wimble 1738	241
[A plan of St. Simon's and Jekyll Islands]	Thomas ca. 1738 MS	318
[Plans and profiles of the forts and redoubts in St. Simon's and Jekyll Islands]	Thomas ca. 1738 MS	318A
2e Plan dune Batterie & d'une Redoute	Thomas ca. 1738 MS	319
Plan dun petit Fort pour L'Isle de S.t Andre	Thomas ca. 1738 MS	322
New Map of the Country of Carolina	Verelst 1739 MS	242
This Chart was transmitted by Col.º Bull	Bull-Verelst 1739 MS	243
The Ichnography of Charles-Town at High Water	Roberts-Toms 1739	244
[A Map of St. Simon's and Jekyll Islands on the coast of Georgia]	Auspouger ca. 1739 MS	317
Plan de la Ville et du Port de Charles Town	Desbruslins 1740	245
A Map of the Islands of St. Simon and Jekyll	Thomas 1740 MS	246A
Plan du Port de Gouadaquini	Jekyll Sound 1741	247
Georgia	Seale 1741	248
The Town and Harbour of Charles Town	Popple-Cóvens-Mortier ca. 1741	249
A View of the Town of Savanah	Savannah 1741	250
Ciudad de Carolina	Charles Town 1742 MS	251
Mapa de la Costa de la Florida	Ayala ca. 1742 MS	252
Plano y Descripcion Del Puerto	Zathalin 1742	252A
Descriptio Geographica . . . de la Florida	Arredondo 1742 MS	253
Amerique Septentrionale Suivant la Carte de Pople	Le Rouge–Popple 1742	254
The Original of this Map was drawn by Col. Barnevelt	Barnevelt ca. 1744 MS	255
A New Mapp of His Maiestys Flourishing Province of South Carolina	Hunter-Herbert 1744 MS	256
Map of Lord Carterets 1/8 part of Carolina	Carteret-Toms 1744 MS	257
[Map of the eighth part of Carolina]	Carteret-Toms 1744	258
Carte Des Costes De La Floride Françoise Suivant les premieres découvertes	Bellin 1744	259
Carolina Von Herman Moll	Moll 1744	260
This Plan sheweth part of the Southern Boundary of the Lands Granted . . . to . . . Lord Carteret	Carteret 1746 MS	261
The Town and Harbour of Charles Town in South Carolina	Bowen ca. 1747	262
A New & Accurate Map of the Province of North & South Carolina, Georgia, &c.	Bowen 1747	263
Georgia	Lotter 1747	264
Plan Von Neu Ebenezer	Urlsperger-Seutter 1747	265
Map N.º I . . . Hudson's River . . . James Turner	Turner 1747	266
A New Map of Georgia	Bowen 1748	267
A Plan of the Line between Virginia and North Carolina	Virginia–North Carolina Boundary 1749 MS	268

MAP TITLE	TITLE TAG	MAP NUMBER
This is a Plan of the Line between Virginia and North Carolina	Fry-Jefferson ca. 1749 MS	269
A New and Exact Plan of Cape Fear River from the Bar to Brunswick	Hyrne 1749 MS	270
La Floride divisee en Floride et Caroline	Robert de Vaugondy 1749	271
A Map of the British American Plantations	Bowen 1749	272
[Sketch Map of the Rivers Santee, Wateree, Saludee, &c., with the road to the Cuttauboes]	Carolina Rivers ca. 1750 MS	273
Plano de Charles Town Capital de la Carolina	Juan de la Cruz ca. 1750	274
[Town of Savannah]	Savannah ca. 1779 MSS	275
Plan of the Harbour of Charles Town	Condor ca. 1750	276
Carta Geographica Della Florida	Delisle 1750	277
[Indian Country, Western Carolina]	Haig 1751 MS	278
[Tybee Island and Cockspur Island]	Tybee Island ca. 1751 MS	279
A Plan of the Inlets, & Rivers of Savannah & Warsaw in the Province of Georgia	Yonge 1751 MS	280
A Map of the Inhabited part of Virginia	Fry-Jefferson 1751 [1753]	281
Map of the Virginia and North Carolina dividing line	Fontaine 1752 MS	282
A Map of Savannah River	De Brahm 1752 MS	282A
[Edisto River] A Plan of Twenty different Tracts of Land	Warner 17[5]3 MS	283
A New and Exact Plan of Cape Fear River	Hyrne-Jefferys 1753	284
Gegend der Provinz Bemarin im Königsreich Apalacha	Schröter 1753	285
A Plan of Fort Johnston with some Alterations	Dobbs–Fort Johnston 1754 MS	286
An Exact=Plan of GEORGE=TOWN	Yonge 1754 MS	286A
Plan of Wilmington Scituate on the East side of the North East Branch of Cape Fear River	Wilmington ca. 1755 MS	287
The Profile of the whole Citadelle of Frederica	De Brahm 1755 MS A	288
A Map of the inhabited Part of Georgia	De Brahm 1755 MS B	289
Plan of a Project to Fortifie Charlestown	De Brahm 1755 MS C	290
Carolinae Floridae nec non Insularum Bahamensivm	Catesby-Seligmann 1755	292
A Map of the British and French Dominions in North America	Mitchell 1755	293
A Map of Virginia, North and South Carolina, Georgia, Maryland with part of New Jersey, &c.	Baldwin 1755	294
Partie de l'Amerique Septentrionale	Robert de Vaugondy 1755	295
Canada Louisiane et Terres Angloises	Anville 1755	296
Charles-Town, Capitale de la Caroline	Le Rouge 1755	297
[Indian Tribes of the Tombigbee, Alabama, region]	Upper Creek Indians ca. 1756 MS	298
An Actual Survey of the Lands	Claud Thomson [1786?] MS	300
Plano I. Descripcion de la Costa . . . de Cañaveral, . . . de Florida, Georgia, y Carolinas	Linares 1756 MS	301
Virginia, North Carolina, and Parts adjacent	Seale–North Carolina ca. 1756	302
South Carolina, Georgia, and Parts adjacent	Seale–South Carolina ca. 1756	303
A Map of Georgia . . . the Places . . . to be Fortified	Reynolds 1756 MS	304
Karte von den Küsten des Französischen Florida	Bellin 1756	305
A Draught of the Creek Nation	Bonar 1757 MS A	306
A Draught of the Upper Creek Nation	Bonar 1757 MS B	307
The Plan and Profile of ffort Loudoun	De Brahm ca. 1757 MS	308
Copy of Mr De Brahms Plan for fortifying Charles Town	De Brahm–Charles Town 1757 MS	309

MAP TITLE	TITLE TAG	MAP NUMBER
A Map of South Carolina and a Part of Georgia	De Brahm 1757	310
Carte de la Caroline et Georgie	Bellin 1757	311
Fort Littleton at Port Royal	Hess ca. 1758 MS	312
Karte von Carolina und Georgien	Bellin 1758	313
Carolina and Georgia	Gibson 1758	314
La Florida	Lopez 1758	315
Carte de la Louisiane, Mary-land, Virginie, Caroline, Georgia	Sepp 1758	316
Plan Dun Fort convenable	St. Simon's ca. 1760 MS	320
[Fort and redoubt for St. Simon's Island]	St. Simon's ca. 1760 MS	321
Tractus I der Wachau in Nord Carolina	Wachau ca. 1760 MS	324
A Plan of the Islands of Sappola	Yonge–De Brahm 1760 MS	325
Carte de la Louisiane et de la Floride	Bonne ca. 1760	326
A New Map of the Cherokee Nation	Kitchin 1760	327
A New Map of the Cherokee Country	Stuart ca. 1761 MS	328
Sketch of the Cherokee Country and march of the Troops	Cherokee Country 1761 MS	328A
The Number of Farm's Improv'd by British Planters the Colony of Virginia	Bullitt [1762?]	328B
A New & correct map of the Provinces of North & South Carolina, Georgia, & Florida	Gibson 1762	329
Above is the Map of Coxspur Island in the Enbuschure [*sic*] of Savannah River	De Brahm 1762 MS	329A
A New Map of Carolina and Likewise a plan of Y^{e} Cherokee Nation	Powder Horn 1762 MS	329B
[Carolina and Granville, Spanish, and French claims]	Anon. ca. 1763 MS	330
A Map of the Sea Coast of Georgia	Yonge–De Brahm's Georgia 1763 MS A	331
A Map of the Sea Coast of Georgia	Yonge–De Brahm's Georgia 1763 MS B	332
A map of Georgia and Florida	Wright 1763 MS	333
Florida from the Latest Authorities	Jefferys 1763	334
Carte Rappresentante la Penisola della Florida	Scacciati-Pazzi 1763	335
A Map of the New Governments, of East & West Florida	Gibson 1763	336
The British Governments in N.th America	Proclamation Boundaries 1763	336A
A Map of the Catawba Indians Land	Wyly 1764 MS	337
A Plan of the temporary Boundary-Line between the Provinces of North and South Carolina	North & South Carolina Boundary Line 1764 MS	338
View of Tiby Light House	Tyby 1764 MS	339
[Cape Lookout]	Lobb 1764 MS	340
A Map of the Southern Indian District 1764	Southern Indian District 1764 MS	341
La Caroline dans l'Amérique Septentrionale	Bellin 1764 A	342
Port et Ville de Charles-Town dans la Caroline	Bellin 1764 B	343
Carte de la Nouvelle Georgie	Bellin 1764 C	344
Descripcion Geographica . . . dela Florida	Arredondo-Martinez 1765 MS	345
Description Geographica . . . de la Florida	Arredondo–de La Puente 1765 MS	346
Cantonment of the Forces in North America	Anon. 1765 MS	347
A New Map of North & South Carolina, & Georgia	Kitchin 1765	348
A Draught of the Cherokee country	Timberlake 1765	349

MAP TITLE	TITLE TAG	MAP NUMBER
Elevation of the North Front of the Governors House to be built at Newbern, North Carolina	Hawks ca. 1766 MS	350
By Virtue of a Warrant . . . a Boundary Line between the province of South Carolina and the Cherokee Indian Country	Pickins 1766 MS	351
A Map of the Indian Nations	Indian Nations 1766 MS	352
Wachovia or Dobbs Parish	Reuter 1766 MS	353
A Map of the British Dominions in North America	Charlevoix-Ridge 1766	354
Plan of Entrance into Cape Fear Harbour	Speer 1766	355
A Draught of Port Royal Harbour in South Carolina	Cook–Port Royal 1766	356
A Draught of West Florida, from Cape St. Blaze to the River Iberville	Cook–West Florida 1766	357
A Sketch of the Entrance from the Sea to Apalatcy	Gauld-Pittman [1767] MS	358
A Plan of a Tract of Land [in Bertie County, N.C.]	Duckenfield 1767 MS	359
Plan of Johnston Fort at Cape Fear	Collet 1767 MS	360
A New Mapp of Carolina	Thornton-Fisher 1767	361
To His most excellent Majesty . . . This Accurate Map of the back Country of North Carolina	Collet 1768 MS	362
Plan of the Town of Hillsborough	Sauthier 1768 MS	363
Nueva descripcion dela Costa . . . dela Florida	de la Puente 1768 MS	364
Vernon River	Whitefield 1768	365
Florida from the Latest Authorities	Jefferys 1768	366
A survey of the Coast about Cape Lookout	Jefferys-Mackay 1768	367
A New and Exact Plan of Cape Fear River from the Bar to Brunswick	Hyrne-Jefferys 1768	368
[Board of Trade map of Indian Boundary Line]	Indian Boundary Line 1768 MS	368A
To his Excellency, Iames Wright, . . . the Boundary Line between [Georgia] and the Creek Indian Nation	Savery 1769 MS A	369
To Lachlan Mcgillivray . . . the Boundary Line between the Province of Georgia and the Creek Nation	Savery 1769 MS B	370
A true Copy taken from the Original . . . for Lands in Georgia	Savery-Romans 1769 MS C	370A
Plan of the Town and Port of Bath	Sauthier-Bath 1769 MS	371
Plan of the Town and Port of Brunswick	Sauthier-Brunswick 1769 MS	372
A Plan of the Town & Port of Edenton	Sauthier-Edenton 1769 MS A	373
A Plan of the Town & Port of Edenton	Sauthier-Edenton 1769 MS B	374
Plan of the Town of Halifax	Sauthier-Halifax 1769 MS	375
Plan of the Town of Newbern	Sauthier–New Bern 1769 MS A	376
A Plan of the Town of Newbern	Sauthier–New Bern 1769 MS B	377
Plan of the town of Willmington	Sauthier-Wilmington 1769 MS	378
East Florida	Jefferys 1769	379
[Brunswick County]	Brunswick County 177– MS	380
A Plan of part of the principal Roads in . . . N. Carolina	North Carolina Roads 177– MS	381
Eagle's Island	Eagle's Island 177– MS	382
[Plan and profile of the forts and redoubts in St. Simon's and Jekyll Islands]	St. Simon's ca. 1770 MS A	383
Plan du Fort proposé dans l'Isle de St Simon	St. Simon's ca. 1770 MS B	384
[Sound Region of North Carolina]	Albemarle Sound Region post-1770 MS A	385

MAP TITLE	TITLE TAG	MAP NUMBER
[Currituck, etc.]	Albemarle Sound Region post-1770 MS B	386
[Coast from St. Augustine to Charleston]	South Carolina–Georgia Coast ca. 1770 MS A	387
[Coastal area from Charles Town to St. Augustine]	South Carolina–Georgia Coast ca. 1770 MS B	388
A Map of South Carolina from an Actual Survey	Gaillard-Cook 1770 MS	389
Plan of the Town & Port of Beaufort in Carteret County	Sauthier-Beaufort 1770 MS	390
Plan of the Town of Cross Creek	Sauthier–Cross Creek 1770 MS	391
Plan of the Town of Salisbury	Sauthier-Salisbury 1770 MS	392
To the Right Honourable John Earl of Egmont, &c.	Fuller 1770	393
A Compleat Map of North-Carolina	Collet 1770	394
Map of Lands Near the Salud River	Calhoun 1770 MS	394A
A Plan of the Camp of Alamance	Sauthier-Alamance 1771 MS A	395
Plan of the Camp and Battle of Alamance	Sauthier-Alamance 1771 MS B	396
Plan of the Camp and Battle of Alamance	Sauthier-Alamance 1771 MS C	397
A Plan of the Camp and Battle of Alamance	Sauthier-Alamance 1771 MS D	398
A Map of South Carolina	Lodge-Cook 1771	399
A Map of South Carolina from the Savannah Sound to St. Helena's Sound	Boss-Brailsford ca. 1771 MS	400
A Sketch of the Cherakee Boundaries	Stuart 1771 MS	401
The Contents in acres Ceded to the Crown By the Treaty of Lochaber	Donelson 1771	401A
Plan du Port de Gouadaquini	Jekyll Sound 1771	402
A Plan of the Entrance into Cape Fear Harbour	Speer 1771	403
Unpurchased Land the Whole Extraordinary good	Georgia Indian Lands ca. 1771 MS	323
A Map of part of the Counties of Mecklenburg and Tryon	Mouzon 1772 [1801] MS	404
A Plan of the Province Line from the Cherokee Line to Salisbury Road	N.C.-S.C. Boundary 1772 MS A	433
An Exact map of the Boundary Line between the Provinces of North & South Carolina	N.C.-S.C. Boundary 1772 MS B	434
A Plan of the Province Line from the Cherokee Line to Salisbury Road	N.C.-S.C. Boundary 1772 MS C	435
South & North Carolina. An Exact Map of the Boundary line	Clements 1772 MS	436
"The Lands which the Cherokees have assign'd for payment of their debts"	Stuart-Georgia 1772 MS	437
The Special Survey of the Inlet, Haven, City, and Environs of Charlestown	De Brahm [1773] MS A.1	405
Plan and Profile of Fort Johnston	De Brahm [1773] MS A.2	406
[Plan of the fortified canal from Ashley to Cooper River]	De Brahm [1773] MS A.3	407
Plan of the City and Fortification of Charlestown	De Brahm [1773] MS A.4	408
Plan of the Environs in the Neck of Tanasee and Talequo Rivers about Fort Loudoun	De Brahm [1773] MS A.5	409
Plan and Profiles of Fort Loudoun upon Tanasee River	De Brahm [1773] MS A.6	410
The Saltzburger's Settlement in Georgia	De Brahm [1773] MS A.7	411
Plan of the Town Ebenezer and its Fort	De Brahm [1773] MS A.8	412
The Bethanian Settlement in Georgia	De Brahm [1773] MS A.9	413
Plan of the City of Savannah and Fortifications	De Brahm [1773] MS A.10	414

MAP TITLE	TITLE TAG	MAP NUMBER
Chart of the Savannah Sound	De Brahm [1773] MS A.11	415
Plan and Profile of Fort George, on Coxpur Island	De Brahm [1773] MS A.12	416
The Environs of Fort Barrington	De Brahm [1773] MS A.13	417
View and Plan of Fort Barrington	De Brahm [1773] MS A.14	418
A special Survey of the Inlet, Harbour, City, and Environs of CharlesTown	De Brahm [1773] MS B.1	419
Plan and Profile of Fort Johnston	De Brahm [1773] MS B.2	420
Plan of the fortified Canal from Ashley to Cooper River	De Brahm [1773] MS B.3	421
Plan of the City and Fortification of Charlestown	De Brahm [1773] MS B.4	422
Plan of the Environs in the Neck of Tanassee & Talequo Rivers, about Fort Loudoun	De Brahm [1773] MS B.5	423
Plan and Profiles of Fort Loudoun	De Brahm [1773] MS B.6	424
The Salzburger's Settlement in Georgia	De Brahm [1773] MS B.7	425
Plan of the Town Ebenezer and its Fort	De Brahm [1773] MS B.8	426
The Bethanian Settlement in Georgia	De Brahm [1773] MS B.9	427
Plan of the City Savannah and Fortification	De Brahm [1773] MS B.10	428
Chart of the Savannah Sound	De Brahm [1773] MS B.11	429
Plan and Profile of Fort George on Coxpur Island	De Brahm [1773] MS B.12	430
The Environs of Fort Barrington	De Brahm [1773] MS B.13	431
Plan and View of Fort Barrington	De Brahm [1773] MS B.14	432
A Map of the Parish of St. Stephen, in Craven County	Mouzon ca. 1773	439
A Map of West Florida part of Et. Florida. Georgia part of So. Carolina	Stuart-Gage 1773 MS	440
Boundary Line, Between Province of Georgia And Nation of Creek Indians	Andrew Way 1773 MS	440A
A Map of the Lands Ceded by the Creek and Cherokee Indians	Philip Yonge 1773 MS	440B
A Map of part of West Florida	Romans 1773 MS	441
A Chart of Tibee Inlet In Georgia	Lyford 1773 MS	441A
[A Map of Fourth Creek Congregation]	Sharpe 1773 [1847] MS	442
A Map of the Province of South Carolina	Cook 1773	443
Carte de la Caroline et Georgie	Bellin 1773	444
Sketch of Clinch River	Smith 1774 MS	445
A New Map of West Florida, Georgia, & South Carolina	Stuart-Lewis 1774 MS	446
Carte de la Caroline Meridionale et Septentrionale et de la Virginie	Tardieu-Valet ca. 1775	447
A Map of the American Indian Nations	Adair 1775	448
A Map of the most Inhabited part of Virginia containing . . . Part of . . . North Carolina	Fry-Jefferson 1775	449
An Accurate Map of North and South Carolina, with their Indian Frontiers	Mouzon 1775	450
A Map of the Southern Indian District	Stuart-Purcell [1775] MS A	438
A Map of the Southern Indian District	Stuart-Purcell [1775] MS B	438A
Sketch of the Country between the River Alatamaha and Musqueta Inlet	Georgia-Florida Coast ca. 1777 MS	299
[Town of Savannah]	Savannah ca. 1779 MSS	275
An Actual Survey of the Lands	Claud Thomson [1786?] MS	300
Carte de la Caroline	Anon. ca. 1797	291

Alphabetical Short-Title List of Maps

Numerals refer to numbers in the List of Maps.
The initial articles *a*, *an*, and *the* are disregarded in the alphabetical order of titles.

Index to the Introductory Essays

Illustration Credits

Illustrations have been provided courtesy of the following institutions, which have granted permission for their reproduction in this volume.

Bibliothèque Nationale, Paris
  Plate 7. Mercator 1569
British Library, Map Library, London
  Plate 62. Sauthier 1770
  Figures 3, 4. Smith/Hole 1606
British Library, Manuscripts Department, London
  Plate 41. Thames School 1684 (ADD 5415.G.6)
  Plate 59F. Indian District 1764 (ADD 14036.D f8v-9)
  Plate 66A. De Brahm 1773 (Kings 210 f23v)
  Plate 66B. De Brahm 1773 (Kings 210 f27)
British Museum, Department of Prints and Drawings, London
  Plate 11. White 1585, MS A
  Plate 12. White 1585, MS B
John Carter Brown Library, Brown University, Providence, R.I.
  Plate 14. White–DeBry 1590
  Plate 15. LeMoyne 1591
  Plate 33A. Horne 1666
  Plate 36. Lederer 1672
  Plate 38. Seller 1675
  Plate 42. Thornton-Morden 1695
  Plates 44, 45. Crisp 1711
William Cumming Collection, Davidson College Library, Davidson, N.C.
  Plate 47. Delisle 1718
  Plate 50. Moll 1729
  Plates 59A–D. De Brahm 1757
  Color Plate 1. Münster 1540
  Color Plate 2. Mercator-Hondius 1606
  Color Plate 3. Mercator-Hondius 1632
  Color Plate 4. Speed 1676
  Color Plate 5. Keulen 1682?
  Color Plate 6. Thornton-Morden-Lea 1685
  Color Plate 7. Thornton-Fisher 1698
  Color Plate 8. Sanson 1700
  Color Plate 9. Sanson 1696
  Color Plate 10. Sanson 1696
  Color Plate 11. Aa 1713
  Color Plate 12. Moll 1715
  Color Plate 13. Senex 1719
  Color Plate 14. Chatelain 1719
  Color Plate 15. Delisle 1722
  Color Plate 16. Moll 1728
  Color Plate 17. Homann 1737
  Color Plate 18. Bowen 1748
  Color Plate 19. Vaugondy 1755
  Color Plate 20. Bonne 1760
  Color Plate 21. Bellin 1764
  Color Plate 22. Hyrne-Jefferys 1768
  Color Plate 23. Fry-Jefferson 1775
  Color Plate 24. Mouzon 1775
Duke University, Special Collections Library, Durham, N.C.
  Figure 5. Bernard Romans 1775
Rare Book Department, Free Library of Philadelphia
  Plate 8. Dee 1582
Hargrett Rare Book and Manuscript Library, University of Georgia, Athens
  Plate 55A. Jones 1734
  Plate 55B. Georgia 1735
Harvard Map Collection, Harvard College Library, Cambridge, Mass.
  Plate 67. Cook 1773
Library of the Hispanic Society of America, New York City
  Plate 2. Vespucci 1526
Huntington Library, San Marino, Calif.
  Plate 29. Farrer 1651
John Work Garrett Library, Johns Hopkins University, Baltimore
  Plate 13. Galle 1587
  Plate 18. Wright 1599
Library of Congress, Washington, D.C.
  Plate 5. Desoto 1544
  Plate 6. Gutiérriez 1562
  Plate 27. Dudley 1647
  Plate 40. Spanish 1683
  Plate 55. Popple 1733
  Plate 56. Arredondo 1742
  Plate 58A. Yonge 1761
  Plate 60A. Proclamation Boundaries 1763
  Figure 1. Martyr 1511
William L. Clements Library, University of Michigan, Ann Arbor
  Plate 61. Indian Nations 1766
Museo Naval, Navarette Colección, Madrid
  Plate 6A. Parreus 1562
Edward E. Ayer Collection, Newberry Library, Chicago
  Plates 68–71. Stuart-Purcell 1775
Rare Books and Manuscripts Division, New York Public Library, Astor, Lenox and Tilden Foundations, New York City
  Plate 32. Comberford 1657
North Carolina Collection, University of North Carolina, Chapel Hill
  Plates 50A, 51–54. Moseley 1733
  Plates 62A, 63–66. Collet 1770

Private collection
Plate 19. Tatton 1600
Public Record Office, Surrey, United Kingdom
Plate 10. Virginia 1585
Plate 35. Locke 1671
Plate 46A. Indian Villages 1715
Plate 48E. Southeastern Indians 1724
Plate 59E. Bonar 1757
Plate 60. Wright 1763
Plate 66C. Yonge 1773
Archivo General de Simancas, Simancas, Valladolid
Plate 21. Zuñiga 1608
Biblioteca Apostolica Vaticana, Rome
Plate 3. Verrazano 1529 (MS. Borgiano I)
Plate 4. Ribero 1529 (MS. Borgiano III)
Virginia Historical Society, Richmond
Plate 43A. Lamhatty 1708 (Lee Family Papers, 163, MSS1L51f677)
Tracy W. McGregor Library of American History, Special Collections Department, University of Virginia Library, Charlottesville
Plate 57. Fry-Jefferson 1751
Yale Center for British Art, New Haven, Conn., Gift of the Acorn Foundation, Inc., Alexander O. Vietor, '36 President, in honor of Paul Mellon
Plates 48, 48A–D. Barnwell 1721

Beinecke Rare Book and Manuscript Library, Yale University, New Haven, Conn.
Plate 1. Waldseemüller 1507
Plate 9. Ortelius 1584
Plate 16. Jode 1593
Plate 17. Wytfliet 1597
Plate 20. Mercator-Hondius 1606
Plate 22. Lescarbot 1612
Plate 23. Smith 1624
Plate 24. Laët 1630
Plate 25. Jansson 1647
Plate 26. Blaeu 1640
Plate 28. Dudley 1647
Plate 30. Sanson 1656
Plate 31. Sanson 1656
Plate 33. DuVal 1663
Plate 34. Blome 1672
Plate 37. Ogilby 1672
Plate 39. Gascoyne 1682
Plate 40. Keulen 1682
Plate 43. Delisle 1703
Plate 46. Homann 1714
Plate 49. Aa 1729
Plate 58. Fry-Jefferson 1755
Plate 59. Mitchell 1755